Problematic Wildlife II

Francesco Maria Angelici • Lorenzo Rossi
Editors

Problematic Wildlife II

New Conservation and Management Challenges in the Human-Wildlife Interactions

Springer

Editors
Francesco Maria Angelici
FIZV
Rome, Italy

Lorenzo Rossi
Museo dell'Ecologia di Cesena
Cesena, FC, Italy

ISBN 978-3-030-42334-6 ISBN 978-3-030-42335-3 (eBook)
https://doi.org/10.1007/978-3-030-42335-3

This Springer imprint is published by the registered company Springer Nature Switzerland AG
The registered company address is: Gewerbestrasse 11, 6330 Cham, Switzerland

Preface

I think having land and not ruining it is the most beautiful art that anybody could ever want.

–Andy Warhol

The Earth has a skin and that skin has diseases; one of these diseases is called "man."

–Friedrich W. Nietzsche

It is with mixed emotions that we present to the readers *Problematic Wildlife, Vol. II*. If on one hand we are proud of the work done and the quality of the chapters it contains, on the other hand, we are worried about the biodiversity increasingly threatened by human activities.

Indeed, all the qualified referees who judged the contents of this volume in the editorial proposal phase agreed about the great relevance the topics covered will have in the next future. This makes us understand that there is much to do regarding conflicts arising from the interaction between humans and fauna.

Originally, our plan was not a series of volumes, but the opportunity to analyze some issues not covered in the first book arose in 2016, when we organized the "III Congresso Nazionale Fauna Problematica" (*III Problematic Fauna National Congress*) in Cesena, Italy. The large participation of renowned experts and the many case studies presented provided us with new ideas to explore in this book.

Compared to the first volume, in addition to birds, mammals, and reptiles, we have also discussed amphibians.

The general structure is the same as the previous book: the chapters are divided into thematic parts with a brief introductory chapter.

We had also included chapters that seemed very important to us such as the problem of deforestation of tropical areas to create rice paddies, intensive oil palm plantations, or climate changes in the polar areas and their repercussions on the fauna. In some cases, it was not possible to involve the experts we contacted to develop the topics, while other planned manuscripts have been rejected by our referee system, because they were not considered suitable for the quality standard of the book. These are ordinary things that must always be taken into account.

Because of that, we greatly thank all the authors and all the specialists who have read, commented, and improved the manuscripts.

We initially thought we would finish the work earlier. Several papers were ready in 2018, but unfortunately, some long and complex referee processes have involved the reviewers, the authors, and ourselves for much longer. So, even if the book comes out a little later than expected, we hope the authors can forgive us. We all have worked very hard, with passion and professionalism.

This book is dedicated to some zoologists who have recently passed away: Colin Groves, world-leading systematic mammologist and primatologist, whose latest work we are honored to publish here; Bernardino Ragni (he had prepared an incomplete manuscript that we could not be able to work out in consistent with the author objectives), Felidae international expert; and Augusto Vigna Taglianti, one of the Italian pioneers in the study of "wildlife biology."

The chapters (including five rejected manuscripts) have been reviewed and improved by the critical referee processes of the following people: Franco Andreone, Francesco M. Angelici, Eva V. Baermann, Natale E. Baldaccini, Anke Benten, Simon A. Black, Andrew Bowkett, Francesca Cini, Longino Contoli Amante, Chris R. Dickman, Emmanuel Do Linh San, Filippo Favilli, Stefano Filacorda, Arash Ghoddousi, Spartaco Gippoliti, Fred Kraus, Aritra Kshettry, Harvey B. Lillywhite, Sandro Lovari, Thomas Madsen, Terry L. Maple, Garry Marvin, L. David Mech, Alberto Meriggi, Terry A. Messmer, Emiliano Mori, Zjef Pereboom, Robert N. Reed, David L. Roberts, Lorenzo Rossi, Danilo Russo, Josefina C. Santana, Valter Trocchi, Sebastian Vetter, Dietmar Zinner, Marco A.L. Zuffi, and four anonymous referees.

The English text was improved by Madeleine Cléa Montanari.

We hope that this second book will be as successful as the first one: we really care that everyone's efforts are somehow gratified.

But the most important thing is that our work will be useful to students, scientists, professionals, operators in the field of ecology, environmentalists, hunters, and all the people directly involved in the fields of natural sciences and conservation.

Rome, Italy
Cesena, Italy
22 September 2019

Francesco M. Angelici
Lorenzo Rossi

Contents

Contributors

Franco Andreone Museo Regionale di Scienze Naturali, Torino, Italy

Francesco Maria Angelici Fondazione Italiana per la Zoologia dei Vertebrati (FIZV), Rome, Italy

Enrico Benussi Studio Immagine Natura, Trieste, Italy

Manisha Bhardwaj Department of Ecology, Grimsö Wildlife Research Station, Swedish University of Agricultural Sciences, Uppsala, Sweden

Simon A. Black Durrell Institute of Conservation & Ecology, University of Kent, Canterbury, UK

Clint Boal Department of Natural Resource Management, Texas Tech University, Lubbock, TX, USA

David Britton San Marcos Aquatic Resources Center, U.S. Fish & Wildlife Service, San Marcos, TX, USA

C. Guillermo Bueno Institute of Ecology and Earth Sciences, Department of Botany, University of Tartu, Tartu, Estonia

Angelica Crottini CIBIO, Research Centre in Biodiversity and Genetic Resources, InBIO, Universidade do Porto, Porto, Portugal

Indraneil Das Institute of Biodiversity and Environmental Conservation, Universiti Malaysia Sarawak, Kota Samarahan, Sarawak, Malaysia

Olivia Dondina Dipartimento di Scienze dell'Ambiente e della Terra, Università degli Studi di Milano-Bicocca, Milan, Italy

Ernesta Martina Esposito Universita' degli studi di Teramo, facoltà di Medicina Veterinaria, Piano d'Accio, Tassarnas Djurklinik, Göteborg, Sweden

Pedro Fernández-Llario Innovación en Gestión de Ungulados SL, Cáceres, Spain

Francesco Ferretti Research Unit of Behavioural Ecology, Ethology and Wildlife Management, Department of Life Sciences, University of Siena, Siena, Italy

Carlos Fonseca CESAM and Departamento de Biologia, Universidade de Aveiro, Aveiro, Portugal

Maurizio Fraissinet Wildlife Research Unit, Department of Agriculture, Federico II University of Napoli, Portici (Napoli), Italy

Association of Southern Italian Ornithological Studies – ASOIM, San Giorgio a Cremano (Napoli), Italy

Karen Freeman Madagascar Fauna and Flora Group, Toamasina, Madagascar

Ricardo García-González Pyrenean Institute of Ecology (CSIC), Jaca, Spain

Alberto Giménez-Anaya Area of Ecology, Department of Agrarian and Environmental Sciences, Technical School, University of Zaragoza, Huesca, Spain

Spartaco Gippoliti Società Italiana per la Storia della Fauna "Giuseppe Altobello", Rome, Italy

Colin P. Groves (Deceased) School of Archaeology & Anthropology, Australian National University, Canberra, ACT, Australia

Scott E. Henke Caesar Kleberg Wildlife Research Institute, Texas A&M University-Kingsville, Kingsville, TX, USA

Juan Herrero Area of Ecology, Department of Agrarian and Environmental Sciences, Technical School, University of Zaragoza, Huesca, Spain

Geoff Hosey University of Bolton, Bolton, UK

Samantha S. Kahl Department of Natural Resource Management, Texas Tech University, Lubbock, TX, USA

Uzma Khan Research Unit of Behavioural Ecology, Ethology and Wildlife Management, Department of Life Sciences, University of Siena, Siena, Italy

WWF Pakistan, Lahore, Pakistan

Mark Lacy Department of Internal Medicine and Pediatrics, Texas Tech University Health Sciences Center, Lubbock, TX, USA

Fulvio Licata CIBIO, Research Centre in Biodiversity and GeneticResources, InBIO, Universidade do Porto, Porto, Portugal

Sandro Lovari Research Unit of Behavioural Ecology, Ethology and Wildlife Management, Department of Life Sciences, University of Siena, Siena, Italy

Maremma Natural History Museum, Grosseto, Italy

Pia Lucidi Facoltà di Bioscienze e Tecnologie Agro-Alimentari e Ambientali, Università degli Studi di Teramo, Polo Silvio Spaventa, Teramo, Italy

Vicky Melfi Hartpury University, Gloucester, UK

Lubomír Melichar Zoo Liberec, Liberec, Czechia

Alberto Meriggi Dipartimento di Scienze della Terra e dell'Ambiente, Università degli Studi di Pavia, Pavia, Italy

John C. Murphy Science & Education, Field Museum, Chicago, IL, USA

Carlos Nores INDUROT, University of Oviedo, Oviedo, Spain

Davide Palumbo Museo dell'Ecologia di Cesena, Cesena, Italy

Franco Perco Montelupone (MC), Italy

Dan Perry Arava Institute for Environmental Studies, Kibbutz Ketura, D.N. Hevel Eilot, Israel

Gad Perry Department of Natural Resource Management, Texas Tech University, Lubbock, TX, USA

Sahondra Rabesihanaka Ministère de l'Environnement et du Développement Durable, Antananarivo, Madagascar

James T. Reardon Department of Conservation, Fiordland District Office, Te Anau, New Zealand

Jan Robovský Faculty of Science, University of South Bohemia, České Budějovice, Czechia

Eric Robsomanitrandrasana Ministère de l'Environnement et du Développement Durable, Antananarivo, Madagascar

Carme Rosell Animal Biology Department, University of Barcelona, Barcelona, Spain

Lorenzo Rossi Museo dell'Ecologia di Cesena, Cesena, FC, Italy

Melanie A. Sarge College of Media and Communication, Texas Tech University, Lubbock, TX, USA

The Media School, Indiana University, Bloomington, IN, USA

Carmelo Maria Scuzzarella Università di Parma, Facoltà di Scienze Politiche e delle relazioni internazionali, Parma, Italy

Andreas Seiler Department of Ecology, Grimsö Wildlife Research Station, Swedish University of Agricultural Sciences, Uppsala, Sweden

Safdar Ali Shah Khyber Pakhtunkhwa Wildlife Department, Peshawar, Pakistan

Suzanne M. Shepherd Department of Emergency Medicine, Hospital of the University of Pennsylvania, University of Pennsylvania Healthcare System, Environmental Education & Animal Behavior Intern (EEABI), The Philadelphia Zoo, Philadelphia, PA, USA

Elisa Torretta Dipartimento di Scienze della Terra e dell'Ambiente, Università degli Studi di Pavia, Pavia, Italy

Robin Verble Department of Natural Resource Management, Texas Tech University, Lubbock, TX, USA

Mark Wallace Department of Natural Resource Management, Texas Tech University, Lubbock, TX, USA

Missouri S&T, Ozark Research Field Station, Lubbock, TX, USA

Samantha J. Ward School of Animal Rural and Environmental Sciences, Nottingham Trent University, Nottingham, UK

Part I
Introduction to ‘Problematic Wildlife II’: Problematic Species Are Increasing, in a World that Is Constantly Changing

Chapter 1
The Need and Relevance of the Book: Problematic Wildlife and the Modern World

Francesco Maria Angelici and Lorenzo Rossi

In Wildness is the preservation of the world
(Henry David Thoreau)

Oh nature, nature, why do you withhold what first you promise?
Why do you so deceive these sons of yours?
(Giacomo Leopardi)

We must not force nature but persuade it
(Epicurus)

1.1 What Is the Inspiration Behind this New Book?

This book is the natural continuation of a previous volume titled *Problematic Wildlife* (Angelici 2015). *Problematic Wildlife* was well-received by the public and critics (Ramanan and Khapugin 2017). The public acclaim for the book has been attributed to its very structure, which addressed diverse themes and needs, incorporating research syntheses and case studies, to develop a broad, innovative, and comprehensive definition of the phase "problematic wildlife."

The primary stimulus for a second volume on the same theme was the Third International Congress on conflicts and interactions between man and wildlife ("III Congresso Nazionale Fauna Problematica") held in Cesena, Italy, in 2016 (Angelici and Rossi 2016). Some of the themes and topics which emerged on the Congress that were not addressed in *Problematic Wildlife* included sustainable hunting, invasive species, urban wildlife, and the dynamic relationship between science and animal activists.

F. M. Angelici (✉)
Fondazione Italiana per la Zoologia dei Vertebrati (FIZV), Rome, Italy

L. Rossi
Museo dell'Ecologia di Cesena, Cesena, FC, Italy
e-mail: info@associazioneorango.com

F. M. Angelici, L. Rossi (eds.), *Problematic Wildlife II*,
https://doi.org/10.1007/978-3-030-42335-3_1

We further explore these topics in the new volume because of their relevance to emerging issues related to the ecology of the planet in the broadest sense and the inevitable overlap between ecosystems, habitats, wildlife conservation, and human activities (Rosell and Llimona 2012; Clement and Standish 2018; Tucker et al. 2018).

It is becoming increasingly clear that the interactions between humans and wildlife will become common and more challenging. As such, increased human interventions will be needed to better manage human-wildlife conflicts to achieve some level of coexistence with wildlife (Frank and Glikman 2019). However, even increased human intervention does not always ensure a possibility of success (Angelici et al. 2015).

As we reviewed the term "problematic wildlife" as defined in the introductory chapter of the first volume (see Angelici 2015), we realized that the definition can also be applied to a multitude of diverse, dynamic, and complex interactions presented in the second volume. And at the same time, we can see new case studies and management experiences merging in this "composite of discipline."

1.2 Problematic Wildlife: From Direct Danger to Humans to Negative Impact on Human Activities

In some cases, wildlife can pose a serious danger to human health and safety (e.g., Conover et al. 1995; Dickman and Hazzah 2016). This is due to the uncontrolled increase of the human population, which in many areas is overlapping with the range of potentially dangerous species (e.g., Landy 2017), or because the habitats of these species are eroded or irreparably altered, such as when natural prey disappear (e.g., Linnell et al. 2002).

Some tipologies of risks to humans posed by large predators (Harris and Herrero 2007; Caldicott et al. 2005) or by venomous species (Alirol et al. 2010) were presented in the first volume (see McLennan and Hockings 2016; Linnell and Alleau 2016). In the present volume, this theme is further examined with new cases (Perry et al. 2020c; Murphy 2020; Sheperd 2020).

The negative human interactions with wildlife can also be linked to anthropogenic activities and include increased predation on livestock and/or crop damage (Mishra et al. 2016; Atickem et al. 2010). Moreover, the uneasy coexistence between humans and some predators (i.e., the gray wolf *Canis lupus*) in North America or Europe has created new economic impacts and social realities which also raise new ethical issues regarding the role or right of humans to the management of wildlife (Mech et al. 1996; Messmer et al. 2001; Nie 2002; Treves et al. 2003; Imbert et al. 2016). All these issues are addressed in this volume (see Sheperd 2020; Khan et al. 2020; Meriggi et al. 2020).

1.3 Problematic Wildlife: Urban Wildlife Conflicts Are an Emerging Problem

Wildlife and their management in urban landscapes is a theme that has been emerging for a long time (Leedy et al. 1978; Goode and Goode 1989; Messmer 2000; Adams 2005). Today many species of large homoeothermic vertebrates have adapted to live in urban or suburban environments (Gloor et al. 2001; Graser et al. 2012), sometimes by changing their habits, as diet or activity patterns (see Ditchkoff et al. 2006; Lowry et al. 2013; Gunther et al. 2018). Often these scenarios created new problems (Hong et al. 2014), to include road traffic accidents (Collinson et al. 2014; Sáenz-de-Santa-María and Tellería 2015; Found and Boyce 2011; Gunther et al. 2018) and the increased risk of disease transmission (de Mattos et al. 2013; Vos et al. 2012) that will need to be addressed.

In recent years, the distribution of wild boar (*Sus scrofa*, feral swine) populations has expanded worldwide (see Massei et al. 2015), and they now inhabit peri-urban and urban areas (Stillfried et al. 2017). Feral swine have also colonized large metropolitan areas in Europe (Castillo-Contreras et al. 2018). Additionally many mammals of medium and large size, from red foxes (*Vulpes vulpes*) to coyotes (*Canis latrans*), have adapted well to urban life (Lawrence and Krausman 2011; Scott et al. 2014; Baker and Timm 2017). The unique conflicts the wildlife creates in urban environment will require innovative solution. Perry et al. (2020a) provide an extensive review of the issues in the book.

Also the populations of many species of birds have been increasing in urban areas (Shochat et al. 2010) and have impacts on urban ecosystems that are often difficult to assess (Blackwell et al. 2013; Manville II 2016; Menchetti et al. 2016). For example, the problem of urban seagulls is increasingly worldwide (Belant 1997), as reviewed in the chapter by Benussi and Fraissinet (2020).

1.4 Problematic Wildlife: Hunting and Ecotourism – Possible Mechanisms for Conservation and Coexistence?

The role of hunting in wildlife conservation is a theme that has often caused rifts between hunters and scientists (Johannesen 2005; Hames 2007; Lindsey et al. 2007; Treves 2009; Organ et al. 2012; Paulson 2012; Delibes-Mateos et al. 2014; Benítez-López et al. 2017). We know how different human visions and values of ecosystems and the environment, together with ethical and practical contents, can lead to diametrically opposed positions, but in some cases, these two worlds may not be completely incompatible and can coexist to pursue the same goals, even synergistically (Redpath et al. 2017), while maintaining their perspectives.

As an example, there is a lack of consensus in the role of trophy hunting in Africa (Lindsey et al. 2013; Packer et al. 2011). Trophy hunting under specific and strictly controlled cases allows the removal of selected animals and generates resources for the conservation of all species (Di Minin et al. 2016; Lindsey et al. 2006), but this industry must be managed in a strictly ethicall way to avoid repercussions on ecosystems (Messmer et al. 1998; Lindsey 2008; Lindsey et al. 2009).

We know that many hunted species also are a source of conflict for agriculture producers and pastoralists (Messmer et al. 1998, Frank et al. 2015). Many of these prized game species are artificially managed via introduction, reintroductions, or restocking operations. This form of management may have consequent impacts and repercussions on native wildlife populations and habitats (Champagnon et al. 2012; Goedbloed et al. 2013). Emblematic cases that exemplify this situation include the European hare (*Lepus europaeus*) (Canu et al. 2018) and wild boar (Giménez-Anaya et al. 2020).

Finally, even ecotourism, an emerging aspect of human-wildlife interactions, under certain conditions, can make a contribution to wildlife conservation (Krüger 2005). In fact, ecotourism is in continuous development and represents a sustainable activity that will allow to make the wildlife and its importance for ecosystems more widely known and to raise funds for the development of depressed areas (Hunt et al. 2015) and for the protected areas (Brandt and Buckley 2018).

1.5 Problematic Wildlife: Species Extinction

Establishing with certainty the extinction of an animal species and the underlying causes is a complex and often subjective task (Collen et al. 2010). In fact, many taxa were rediscovered tens and even hundreds of years after declared extinct (Scheffers et al. 2011).

Furthermore, in several cases, on the basis of discoveries of documents and samples, or using statistical-mathematical models, the year of extinction of a species has been debated (Black et al. 2013; Turvey et al. 2017).

Being certain of the extinction of a species is an important aspect of conservation, as species considered prematurely extinct cannot be effectively protected (Collar 1998). Black (2020) illustrates the challenges in the conservation of animal species that are difficult to observe.

A special category is represented by the big cats, which despite their size can be very cryptic and difficult to observe (Brassine and Parker 2015). Examples of this are the rediscovery of the lion *Panthera leo*, in Gabon (Barnett et al. 2018), and the survival of the Caspian tiger *Panthera tigris virgata* (once considered extinct in the 1970s), until the 1990s (Emre 2004).

The volume contains a review by Rossi et al. (2020) on this topic, with several case studies of big cats of the genus *Panthera* whose true extinction date remains difficult to determine.

1.6 Problematic Wildlife: Zoos, Conservation, and Animal Rights

The role and importance of the zoos in species conserving biological diversity are still subject for debate (Mazur and Clark 2001; Minteer and Collins 2013; Scanes 2017); ex situ conservation engages zoos through their recovery and reproduction centers. But the importance of captive breeding for the protection of animal species continues to raise questions (Ebenhard 1995; Bowkett 2009). In captivity it was possible to save not only *taxa* reduced to few individuals (Collar et al. 2012) but also species extirpated in nature (Maddison et al. 2012). Zoos are also involved in in situ conservation through recovery programs for endangered species in collaboration with government authorities and local communities (Tribe and Booth 2003).

Themes, from the animal welfare to management challenges, animal rights, and conflicts between scientists and animal activists, are examined in the volume by Perry et al. (2020b) and Robovský et al. (2020).

Another novel aspect, relative to zoos and changing public opinion, are roles zoos play in the conservation of "charismatic species" such as elephants, big apes, or cetaceans. On one hand, charismatic species may be considered problematic for zoos because they attract more animal welfare-related concern from animal activist groups. But on the other hand, their popularity helps zoos to achieve their mission, increasing funding available for field conservation (Hosey et al. 2020).

The role human dimensions play with respect to human coexistence with certain emblematic species (i.e., the large carnivores) is a contemporary debated theme (Dickman et al. 2013; Lewis et al. 2017). A field of the so-called "human dimensions" is to analyze the perception that people, especially those living in areas where the large carnivores exist, have in regard to predators, their ecological role, and their own "right to exist" (Madden 2004). In terms of conservation, the public perception of the taxonomic status of a species plays an important role because it directly affects the targeting of resources and the priority of interventions (Master 1991; Messmer et al. 1999, 2001). However, some studies have not found significant effects of taxonomy on conservation (Morrison et al. 2009); there are cases in which populations considered as endemic species were instead introduced species that should not be considered as conservation priorities (Messmer et al. 1999; Helgen and Wilson 2003). Finally, it should be noted that the only way that a taxon can be legally protected is through its formal recognition as species or subspecies and subsequent assessment of extinction risk according to internationally accepted procedures (ICZN 1999). It follows that the taxonomic revisions of populations then ascribed to new species or subspecies (Hrbek et al. 2014) can be very important for conservation. These topics, including two case studies, are discussed in this book by Gippoliti and Groves (2020).

1.7 Problematic Wildlife: Humans and Herpetofauna

In the second volume, specific cases and research and management experiences are further examined concerning homoeothermic (i.e., mammals and birds) and poikilothermic vertebrates mainly reptiles (Perry et al. 2020c). It is rather intuitive to understand how several reptile species can represent a danger to man and his activities (Alves et al. 2012).

We know that venomous snakes represent a threat to human life (Chippaux 1998). It is estimated that at least 20,000 people die each year from snake bites, but the number, due to data deficits from many countries where venomous snakes are widespread, may reach almost 100,000 (McNamee 2001; Williams et al. 2019). The total number of snake bites is more than 15 times the reported fatalities (Chippaux 1998).

In rare cases, the snakes may be dangerous even if not venomous. Despite the fact that few constricting species, mostly pythons, *Python sebae* and *Malayopython reticulatus*, and the green anaconda, *Eunectes murinus*, are implicated in humans' death, the fear that these large reptiles cause has a global dimension (Murphy and Henderson 1997). A very exhaustive and in-depth review on the conflicts between man and big snakes is reviewed in the book (Murphy 2020).

Annually, there are also human mortalities attributed to attacks by crocodiles (mainly *Crocodylus niloticus* and *Crocodylus porosus* but also lesser and occasional other species), alligators (*Alligator mississippiensis*), or more rarely *Melanosuchus niger* and other species; however these are a few in number (Conover et al. 1995; Caldicott et al. 2005).

Others problems can be created by reptiles, and even amphibians, if introduced in areas where they are not originally present (Kahl et al. 2020; Licata et al. 2020). Invasive herpetofauna can cause ecological upheavals, on the native fauna (Kraus 2015), and to find a solution is necessary to invest in new resources even without the certain of a positive outcome (Kraus 2009; Licata et al. 2020).

Acknowledgments We have much gratitude for the revisions and suggestions of Emmanuel Do Linh San and Terry Messmer who have greatly improved the chapter. In addition, Terry Messmer also impeccably reviewed and improved the English grammar and style.

References

Adams LW (2005) Urban wildlife ecology and conservation: a brief history of the discipline. Urban Ecosyst 8:139–156. https://doi.org/10.1007/s11252-005-4377-7

Alirol E, Sharma SK, Bawaskar HS et al (2010) Snake bite in South Asia: a review. PLoS Negl Trop Dis doi 4:e603. https://doi.org/10.1371/journal.pntd.0000603

Alves RRN, Vieira KS, Santana GG, Vieira WLS, Almeida WO, Souto WMS, Pereira Montenegro PFG (2012) A review on human attitudes towards reptiles in Brazil. Environ Monit Assess 184:6877–6901. https://doi.org/10.1007/s10661-011-2465-0

Angelici FM (2015) Problematic wildlife: a cross-disciplinary approach. Springer International, Cham, 603 p

Angelici FM, Rossi L (eds) (2016) Atti del III Congresso Nazionale Fauna Problematica. Cesena, Italy p 240

Angelici FM, Mahama A, Rossi L (2015) The lion in Ghana: its historical and current status. Anim Biodivers Conserv 2:151–162

Atickem A, Williams S, Bekele A, Thirgood S (2010) Livestock predation in the Bale Mountains. Ethiopia Afr J Ecol 48:1076–1082. https://doi.org/10.1111/j.1365-2028.2010.01214.x

Baker RO, Timm RM (2017) Coyote attacks on humans, 1970–2015: implications for reducing the risks. Human–Wildl Inter 11. https://doi.org/10.26077/jy37-s271

Barnett R, Sinding MHS, Vieira FG, Mendoza MLZ, Bonnet M, Araldi A et al (2018) No longer locally extinct? Tracing the origins of a lion (*Panthera leo*) living in Gabon. Conserv Genet 19:611–618

Belant JL (1997) Gulls in urban environments: landscape-level management to reduce conflict. Landsc Urban Plan 38:245–258. https://doi.org/10.1016/S0169-2046(97)00037-6

Benítez-López A, Alkemade R, Schipper AM, Ingram DJ, Verweij PA, Eikelboom JAJ, Huijbregts MAJ (2017) The impact of hunting on tropical mammal and bird populations. Science 356:180–183

Benussi E, Fraissinet M (2020) The colonization of the Western Yellow-legged Gull (*Larus michahellis*) in an Italian city: evolution and management of the phenomenon. In: Angelici FM, Rossi L (eds) Problematic wildlife II. New conservation and management challenges in the human-wildlife interactions. Springer International, Cham, pp 191–212

Black SA (2020) Assessing presence, decline and extinction for the conservation of difficult-to-observe species. In: Angelici FM, Rossi L (eds) Problematic wildlife II. New conservation and management challenges in the human-wildlife interactions. Springer International, Cham, pp 359–392

Black SA, Fellous A, Yamaguchi N, Roberts DL (2013) Examining the extinction of the Barbary lion and its implications for felid conservation. PLoS One 8:e60174. https://doi.org/10.1371/journal.pone.006017

Blackwell BF, Seamans TW, Schmidt PM, Devault TL, Belant JL, Whittingham MJ, Martin JA, Fernández-Juricic E (2013) A framework for managing airport grasslands and birds amidst conflicting priorities. Ibis 155:189–193

Bowkett AE (2009) Recent captive-breeding proposals and the return of the ark concept to global species conservation. Conserv Biol 23:773–776. https://doi.org/10.1111/j.1523-1739.2008.01157.x

Brandt JS, Buckley RC (2018) A global systematic review of empirical evidence of ecotourism impacts on forests in biodiversity hotspots. Curr Opin Environ Sustain 32:112–118

Brassine E, Parker D (2015) Trapping elusive cats: using intensive camera trapping to estimate the density of a rare African felid. PLoS One 10:e0142508. https://doi.org/10.1371/journal.pone.0142508

Caldicott DGE, Croser D, Manolis C, Webb G, Britton A (2005) Crocodile attack in Australia: an analysis of its incidence and review of the pathology and management of crocodilian attacks in general. Wilderness Environ Med 16(143):159

Canu A, Apollonio M, Scandura M (2018) Unmasking the invader: genetic identity of invasive wild boar from three minor islands off Sardinia (Italy). Mamm Biol 93:29–37. https://doi.org/10.1016/j.mambio.2018.07.008

Castillo-Contreras R, Carvalho J, Serrano E, Mentaberre G, Fernández-Aguilar X, Colom A et al (2018) Urban wild boars prefer fragmented areas with food resources near natural corridors. Sci Total Environ 615:282–288

Champagnon J, Elmberg J, Guillemain M, Gauthier-Clerc M, Lebreton JD (2012) Conspecifics can be aliens too: a review of effects of restocking practices in vertebrates. J Nat Conserv 20:231–241. https://doi.org/10.1016/j.jnc.2012.02.002

Chippaux JP (1998) Snake-bites: appraisal of the global situation. Bull World Health Organ 76:515–524

Clement S, Standish RJ (2018) Novel ecosystems: governance and conservation in the age of the Anthropocene. J Environ Manage 208:36–45

Collar NJ (1998) Extinction by assumption; or, the Romeo error on Cebu. Oryx 32:239–244
Collar NJ, Gardner L, Jeggo DF et al (2012) Conservation breeding and the most threatened birds in Asia. BirdingASIA 18:50–57
Collen B, Purvis A, Mace GM (2010) When is a species really extinct? Testing extinction inference from a sighting record to inform conservation assessment. Divers Distrib 16:755–764. https://doi.org/10.1111/j.1472-4642.2010.00689.x
Collinson WJ, Parker DM, Bernard RTF et al (2014) Wildlife road traffic accidents: a standardized protocol for counting flattened fauna. Ecol Evol 4:3060–3071. https://doi.org/10.1002/ece3.1097
Conover M, Pitt W, Kessler K, DuBow T, Sanborn W (1995) Review of human injuries, illnesses, and economic losses caused by wildlife in the United States. Wildl Soc Bull (1973–2006) 23:407–414
de Mattos CA, Favi M, Yung V et al (2013) Bat rabies in urban centers in Chile. J Wildl Dis 36:231–240. https://doi.org/10.7589/0090-3558-36.2.231
Delibes-Mateos M, Ferreira C, Rouco C et al (2014) Conservationists, hunters and farmers: the European rabbit *Oryctolagus cuniculus* management conflict in the Iberian Peninsula. Mamm Rev 44:190–203
Di Minin E, Leader-Williams N, Bradshaw CJA (2016) Banning trophy hunting will exacerbate biodiversity loss. Trends Ecol Evol 31:99–102
Dickman AJ, Hazzah L (2016) Money, myths and man-eaters: complexities of human-wildlife conflict. In: Angelici FM (ed) Problematic wildlife: a cross-disciplinary approach. Springer International, Cham, pp 339–356
Dickman AJ, Marchini S, Manfredo M (2013) The human dimension in addressing conflict with large carnivores (pp 110–126). In: Macdonald DW, Willis KJ (eds) Key topics in conservation biology 2. Wiley, Hoboken, 510 p
Ditchkoff SS, Saalfeld ST, Gibson GJ (2006) Animal behavior in urban ecosystems: modifications due to human-induced stress. Urban Ecosyst 9:5–12. https://doi.org/10.1007/s11252-006-3262-3
Ebenhard T (1995) Conservation breeding as a tool for saving animal species from extinction. Trends Ecol Evol 10:438–443
Emre O (2004) Status, conservation and management of large carnivores in Turkey. T-PVS/Inf 8:1–28. Strasbourg, Council of Europe
Found R, Boyce MS (2011) Predicting deer-vehicle collisions in an urban area. J Environ Manag 92:2486–2493. https://doi.org/10.1016/j.jenvman.2011.05.010
Frank B, Glikman J (2019) Human–wildlife conflicts and the need to include coexistence. In: Frank B, Glikman J, Marchini S (eds) Human–wildlife interactions: turning conflict into coexistence (conservation biology). Cambridge University Press, Cambridge, pp 1–19. https://doi.org/10.1017/9781108235730.004
Giménez-Anaya A, Bueno G, Fernández-Llario P, Fonseca P, García-González R, Herrero J, Nores C, Rosell C (2020) What do we know about wild boar in Iberia? In: Angelici FM, Rossi L (eds) Problematic wildlife II. New conservation and management challenges in the human-wildlife interactions. Springer International, Cham, pp 251–271
Gippoliti S, Groves C (2020) Cryptic problematic species and troublesome taxonomists: a tale of the Apennine bear and the Nile white rhinoceros. In: Angelici FM, Rossi L (eds) Problematic wildlife II. New conservation and management challenges in the human-wildlife interactions. Springer International, Cham, pp 509–527
Gloor S, Bontadina F, Hegglin D et al (2001) The rise of urban fox populations in Switzerland. Mamm Biol 66:155–164
Goedbloed DJ, , van Hooft P, Megens HJ, Langenbeck K, Lutz W, Crooijmans RPMA, van Wieren SE, Ydenberg RC, Prins HHT (2013) Reintroductions and genetic introgression from domestic pigs have shaped the genetic population structure of Northwest European wild boar. BMC Genet 14:43. https://doi.org/10.1186/1471-2156-14-43

Goode ADA, Goode DA (1989) Urban nature conservation in Britain. J Appl Ecol 26:859–873. https://doi.org/10.2307/2403697
Graser WH, Gehrt SD, Hungerford LL, Anchor C (2012) Variation in demographic patterns and population structure of raccoons across an urban landscape. J Wildl Manag 76:976–986. https://doi.org/10.1002/jwmg.344
Gunther KA, Wilmot KR, Cain SL, Wyman TC, Reinertson EG, Bramblett AM (2018) Managing human-habituated bears to enhance survival, habitat effectiveness, and public viewing. Human–Wildl Inter 12. https://doi.org/10.26077/83cn-mh23
Hames R (2007) Game conservation or efficient hunting? In: Penn DJ, Mysterud I (eds) Evolutionary perspectives on environmental problems. Aldine Transaction, Piscataway, pp 53–66
Harris GA, Herrero S (2007) Bear attacks: their causes and avoidance. J Range Manag 40:96. https://doi.org/10.2307/3899374
Helgen KM, Wilson DE (2003) Taxonomic status and conservation relevance of the raccoons (Procyon spp.) of the West Indies. J Zool 259:69–76. https://doi.org/10.1017/S0952836902002972
Hong SH, Shim WJ, Han GM et al (2014) Levels and profiles of persistent organic pollutants in resident and migratory birds from an urbanized coastal region of South Korea. Sci Total Environ 1:1463–1470. https://doi.org/10.1016/j.scitotenv.2013.07.089
Hosey G, Melfi V, Ward SJ (2020) Problematic animals in the zoo: the issue of charismatic megafauna. In: Angelici FM, Rossi L (eds) Problematic wildlife II. New conservation and management challenges in the human-wildlife interactions. Springer International, Cham, pp 485–508
Hrbek T, Da Silva VMF, Dutra N, Gravena W, Martin AR, Farias IP (2014) A new species of river dolphin from Brazil or: how little do we know our biodiversity. PLoS One 9:e83623
Hunt CA, Durham WH, Driscoll L, Honey M (2015) Can ecotourism deliver real economic, social, and environmental benefits? A study of the Osa Peninsula. Costa Rica J Sustain Tour 23:339–357. https://doi.org/10.1080/09669582.2014.965176
ICZN (1999) International code of zoological nomenclature, 4th edn. International Trust for Zoological Nomenclature, London
Imbert C, Caniglia R, Fabbri E et al (2016) Why do wolves eat livestock?: factors influencing wolf diet in northern Italy. Biol Conserv 195:156–168. https://doi.org/10.1016/j.biocon.2016.01.003
Johannesen AB (2005) Wildlife conservation policies and incentives to hunt: an empirical analysis of illegal hunting in western Serengeti. Tanzania Environ Dev Econ 10:271–292. https://doi.org/10.1017/S1355770X0500207X
Kahl SS, Henke SE, Britton D et al (2020) Risk Assessment Model for Brown Treesnake Introduction into the Continental United States. In: Angelici FM, Rossi L (eds) Problematic wildlife II. New conservation and management challenges in the human-wildlife interactions. Springer International, Cham, pp 603–616
Khan U, Ferretti F, Ali Shah S, Lovari S (2020) In: Angelici FM, Rossi L (eds) Problematic wildlife II. New conservation and management challenges in the human-wildlife interactions. Springer International, Cham, pp 93–110
Kraus F (2009) Alien reptiles and amphibians: a scientific compendium and analysis. Springer Science + Business Media, Berlin, 563 p
Kraus F (2015) Impacts from invasive reptiles and amphibians. Annu Rev Ecol Evol Syst 46:75–97. https://doi.org/10.1146/annurev-ecolsys-112414-054450
Krüger O (2005) The role of ecotourism in conservation: panacea or Pandora's box? Biodivers Conserv 14:579–600. https://doi.org/10.1007/s10531-004-3917-4
Landy F (2017) Urban leopards are good cartographers. Human-nonhuman and spatial conflicts at Sanjay Gandhi National Park, Mumbai. In: Rademacher A, Sivaramakrishnan K (eds) Places of nature in ecologies of urbanism. https://doi.org/10.5790/hongkong/9789888390595.001.0001
Lawrence SE, Krausman PR (2011) Reactions of the public to urban coyotes (*Canis latrans*). Southwest Nat 56:404–409. https://doi.org/10.1894/n03-mlk-06.1
Leedy DL, Maestro RM, Franklin TM (1978) Planning for wildlife in cities and suburbs. Fish and Wildlife Service, U.S. Department of the Interior, Ellicott City. 75 pp

Lewis PM, Burns GL, Jones D (2017) Response and responsibility: humans as apex predators and ethical actors in a changing societal environment. Food Webs 12:49–55. https://doi.org/10.1016/j.fooweb.2016.09.001

Licata F, Andreone F, Freeman K, Rabesihanaka S, Robsomanitrandrasana E, Reardon J, Crottini A (2020) Giant Snake-human relationships. In: Angelici FM, Rossi L (eds) Problematic wildlife II. New conservation and management challenges in the human-wildlife interactions. Springer International, Cham, pp 617–638

Lindsey PA (2008) Trophy hunting in Sub-Saharan Africa: economic scale and conservation significance. FAO (Rome) Best Practices in Sustainable Hunting, pp 41–47

Lindsey PA, Alexander R, Frank LG et al (2006) Potential of trophy hunting to create incentives for wildlife conservation in Africa where alternative wildlife-based land uses may not be viable. Anim Conserv 9:282–291. https://doi.org/10.1111/j.1469-1795.2006.00034.x

Lindsey PA, Frank LG, Alexander R et al (2007) Trophy hunting and conservation in Africa: problems and one potential solution. Conserv Biol 21:880–883. https://doi.org/10.1111/j.1523-1739.2006.00594.x

Lindsey PA, Romañach SS, Davies-Mostert HT (2009) The importance of conservancies for enhancing the value of game ranch land for large mammal conservation in southern Africa. J Zool 277:99–105. https://doi.org/10.1111/j.1469-7998.2008.00529.x

Lindsey PA, Balme GA, Funston P et al (2013) The trophy hunting of African Lions: scale, current management practices and factors undermining sustainability. PLoS One 8:e73808. https://doi.org/10.1371/journal.pone.0073808

Linnell JDC, Alleau J (2016) Predators that kill humans: myth, reality, context and the politics of wolf attacks on people. In: Angelici FM (ed) Problematic wildlife: a cross-disciplinary approach. Springer International, Cham

Linnell JDC, Andersen R, Andersone Ž et al (2002) The fear of wolves: a review of wolf attacks on humans. A large carnivore initiative for Europe. Trondheim, 65 pp

Lowry H, Lill A, Wong BBM (2013) Behavioural responses of wildlife to urban environments. Biol Rev 88:537–549. https://doi.org/10.1111/brv.12012

Madden F (2004) Creating coexistence between humans and wildlife: global perspectives on local efforts to address human–wildlife conflict. Hum Dimens Wildl 9:247–257. https://doi.org/10.1080/10871200490505675

Maddison NR, Zhigang J, Boyd M (2012) The Père David's deer: clinging on by a cloven hoof. WAZA Magazine 13:29–32

Manville AM II (2016) Impacts to birds and bats due to collisions and electrocutions from some tall structures in the United States: wires, towers, turbines, and solar arrays-state of the art in addressing the problems. In: Angelici FM (ed) Problematic wildlife: a cross-disciplinary approach. Springer International, Cham, pp 415–442

Massei G, Kindberg J, Licoppe A et al (2015) Wild boar populations up, numbers of hunters down? A review of trends and implications for Europe. Pest Manag Sci 71:492–500. https://doi.org/10.1002/ps.3965

Master LL (1991) Assessing threats and setting priorities for conservation. Conserv Biol 5:559–563

Mazur NA, Clark TW (2001) Zoos and conservation: policy making and organizational challenges. Yale School Forest Environ Stud Bull Ser 96(105):185–201. https://elischolar.library.yale.edu/yale_fes_bulletin/96

McLennan MR, Hockings KJ (2016) The aggressive apes? Causes and contexts of great ape attacks on local persons. In: Angelici FM (ed) Problematic wildlife: a cross-disciplinary approach. Springer International, Cham, pp 373–394

McNamee D (2001) Tackling venomous snake bites worldwide. Lancet 357:1680. https://doi.org/10.1016/S0140-6736(00)04868-6

Mech DL, Fritts SH, Nelson ME (1996) Wolf management in the 21st century: from public input to sterilization. J Wildl Res 1:195–198

Menchetti M, Mori E, Angelici FM (2016) Effects of the recent world invasion by ring-necked parakeets *Psittacula krameri*. In: Angelici FM (ed) Problematic wildlife: a cross-disciplinary approach. Springer International, Cham, pp 253–266

Meriggi A, Torretta E, Dondina O (2020) Recent changes in wolf habitat occupancy and feeding habits in Italy: implications for conservation and reducing conflict with humans. In: Angelici FM, Rossi L (eds) Problematic wildlife II. New conservation and management challenges in the human-wildlife interactions. Springer International, Cham, pp 111–138

Messmer TA (2000) The emergence of human–wildlife conflict management: turning challenges into opportunities. Int Biodeterior Biodegradation 45:97–102

Messmer TA, Dixon CE, Shields W, Barras SC, Schroeder SA (1998) Cooperative wildlife management units: achieving hunter, landowner, and wildlife management agency objectives. Wildl Soc Bull 26:325–332

Messmer TA, Brunson M, Reiter D, Hewitt D (1999) United States public attitudes regarding predators and their management to enhance avian recruitment. Wildl Soc Bull (1973–2006), 27: 75–85

Messmer TA, Reiter D, West BC (2001) Enhancing wildlife sciences' linkage to public policy: lessons from the predator-control pendulum. Wildl Soc Bull 29:1253–1259

Minteer BA, Collins JP (2013) Ecological ethics in captivity: balancing values and responsibilities in zoo and aquarium research under rapid global change. ILAR J 54:41–51. https://doi.org/10.1093/ilar/ilt009

Mishra C, Redpath SR, Suryawanshi KR (2016) Livestock predation by snow leopards: conflicts and the search for solutions. In: McCharty T, Mallon D (eds) Snow leopards: biodiversity of the world: conservation from genes to landscapes. Academic Press, Oxford, pp 59–67

Morrison WR, Lohr JL, Duchen P et al (2009) The impact of taxonomic change on conservation: does it kill, can it save, or is it just irrelevant? Biol Conserv 142:3201–3206. https://doi.org/10.1016/j.biocon.2009.07.019

Murphy JC (2020) Giant Snake-human relationships. In: Angelici FM, Rossi L (eds) Problematic wildlife II. New conservation and management challenges in the human-wildlife interactions. Springer International, Cham, pp 581–602

Murphy JC, Henderson RW (1997) Tales of giant snakes: a historical natural history of anacondas and pythons. Krieger, Malabar, 221 p

Nie MA (2002) Wolf recovery and management as value-based political conflict. Ethics Place Environ 5:65–71. https://doi.org/10.1080/13668790220146465

Organ, JF, Geist V, Mahoney SP, Williams S, Krausman PR, Batcheller GR, Decker TA, Carmichael R, Nanjappa P, Regan R, Medellin RA, Cantu R, McCabe RE, Craven S, Vecellio GM, Decker DJ (2012) The North American model of wildlife conservation. The Wildlife Society Technical Review 12-04. The Wildlife Society, Bethesda, Maryland, USA, ix + 47 p

Packer C, Brink H, Kissui BM et al (2011) Effects of trophy hunting on lion and leopard populations in Tanzania. Source Conserv Biol 25:142–153

Paulson N (2012) The place of hunters in global conservation advocacy. Conserv Soc 10:53–62. https://doi.org/10.4103/0972-4923.92195

Perry G, Boal C, Verble R, Wallace M (2020a) "Good" and "bad" urban wildlife. In: Angelici FM, Rossi L (eds) Problematic wildlife II. New conservation and management challenges in the human-wildlife interactions. Springer International, Cham, pp 141–170

Perry G, Sarge MA, Perry D (2020b) Alternative facts and alternative views: scientists, managers, and animal rights activists. In: Angelici FM, Rossi L (eds) Problematic wildlife II. New conservation and management challenges in the human-wildlife interactions. Springer International, Cham, pp 421–450

Perry G, Lacy M, Das I (2020c) Snakes, snakebites, and humans. In: Angelici FM, Rossi L (eds) Problematic wildlife II. New conservation and management challenges in the human-wildlife interactions. Springer International, Cham, pp 561–580

Ramanan SS, Khapugin AA (2017) Book review: problematic wildlife: a cross-disciplinary approach. Nat Conserv Res. https://doi.org/10.24189/ncr.2017.009

Redpath SM, Linnell JDC, Festa-Bianchet M et al (2017) Don't forget to look down – collaborative approaches to predator conservation. Biol Rev 92:2157–2163. https://doi.org/10.1111/brv.12326

Robovský J, Melichar L, Gippoliti S (2020) Zoos and conservation in the Anthropocene: opportunities and problematic issues. In: Angelici FM, Rossi L (eds) Problematic wildlife II. New conservation and management challenges in the human-wildlife interactions. Springer International, Cham, pp 451–484

Rosell C, Llimona F (2012) Human-wildlife interactions. Anim Biodivers Conserv 35:219–220

Rossi L, Scuzzarella CM, Angelici FM (2020) Extinct (or perhaps extinct) big cats. Their uncertain and unexpected stories. In: Angelici FM, Rossi L (eds) Problematic wildlife II. New conservation and management challenges in the human-wildlife interactions. Springer International, Cham, pp 393–417

Sáenz-de-Santa-María A, Tellería JL (2015) Wildlife-vehicle collisions in Spain. Eur J Wildl Res 61:399–406. https://doi.org/10.1007/s10344-015-0907-7

Scanes CG (2017) Animals in entertainment. In: Scanes CG, Toukhsati SR (eds) Animals and human society. Academic Press, London, pp 225–255

Scheffers BR, Yong DL, Harris JBC et al (2011) The world's rediscovered species: Back from the brink? PLoS One 6:e22531. https://doi.org/10.1371/journal.pone.0022531

Scott DM, Berg MJ, Tolhurst BA et al (2014) Changes in the distribution of red foxes (Vulpes vulpes) in urban areas in Great Britain: findings and limitations of a media-driven nationwide survey. PLoS One 9:e99059. https://doi.org/10.1371/journal.pone.0099059

Sheperd SM (2020) Large felid man eaters. In: Angelici FM, Rossi L (eds) Problematic wildlife II. New conservation and management challenges in the human-wildlife interactions. Springer International, Cham, pp 17–91

Shochat E, Lerman S, Fernández-Juricic E (2010) Birds in urban ecosystems: population dynamics, community structure, biodiversity, and conservation. Urban Ecosyst Ecol. https://doi.org/10.2134/agronmonogr55.c4

Stillfried M, Fickel J, Börner K et al (2017) Do cities represent sources, sinks or isolated islands for urban wild boar population structure? J Appl Ecol 54:272–281. https://doi.org/10.1111/1365-2664.12756

Treves A (2009) Hunting for large carnivore conservation. J Appl Ecol 46:1350–1356

Treves A, Grossberg R, Naughton-Treves L (2003) Paying for tolerance: rural citizens attitudes toward wolf depredation and compensation. Conserv Biol 17:1500–1511. https://doi.org/10.1111/j.1523-1739.2003.00060.x

Tribe A, Booth R (2003) Assessing the role of zoos in wildlife conservation. Hum Dimens Wildl 8:65–74. https://doi.org/10.1080/10871200390180163

Tucker MA, Böhning-Gaese K, Fagan WF et al (2018) Moving in the Anthropocene: global reductions in terrestrial mammalian movements. Science 369:466–469. https://doi.org/10.1126/science.aam9712

Turvey ST, Barnes I, Marr M, Brace S (2017) Imperial trophy or island relict? A new extinction paradigm for père david's deer: a Chinese conservation icon. R Soc Open Sci 4:171096. https://doi.org/10.1098/rsos.171096

Vos A, Ortmann S, Kretzschmar AS et al (2012) The raccoon (Procyon lotor) as potential rabies reservoir species in Germany: a risk assessment. Berl Munch Tierarztl Wochenschr 125:228–235

Williams DJ, Faiz MA, Abela-Ridder B, Ainsworth S, Bulfone TC, Nickerson AD, Habib AG, Junghanss T, Wen Fan H, Turner M, Harrison RA, Warrell DA (2019) Strategy for a globally coordinated response to a priority neglected tropical disease: snakebite envenoming. PLoS Negl Trop Dis 13(2):e0007059. https://doi.org/10.1371/journal.pntd.0007059

Part II
From Direct Danger to Humans to Negative Impact on Human Activities

The conflicts between large carnivores and humans are an increasingly widespread phenomenon (Fascione et al. 2004) that include the acceptance of the presence of potentially or actually dangerous species (the so-called human dimension of the problem; see Treves et al. 2006) and the ecological preservation of habitats where large carnivores coexist with humans and their activities (e.g., Oriol-Cotterill et al. 2015; Woodroffe et al. 2005).

This part consists of three chapters. The first is a review of the topic by Sheperd (2020) that concerns usual habitats, ecology, and predatory behaviors of the big cats and the human-felid conflicts considering also approaches to conserve these endangered predators.

It is followed by Khan et al. (2020) work: a study about the presence of the common leopard (*Panthera pardus*) in a densely inhabited region of Pakistan. The authors examine its impact on livestock and solutions proposed to mitigate this conflict. It deals also on social impact (attacks on people), conservation (leopard is Critically Endangered in Pakistan), and ecology (landscape is in continuous transformation due to anthropization).

The third chapter (Meriggi et al. 2020) is a case study concerning the Apennine wolf (*Canis lupus italicus*) in Italy. This predator has increased in number over the last decades because of some favorable circumstances, re-colonizing areas of the Italian Peninsula where the species had been absent for at least a century and generating many interactions with humans and their activities.

References

Fascione N, Delach A, Smith ME (eds) (2004) People and predators: from conflict to coexistence. Island Press, Washington, DC

Khan U, Ferretti F, Ali Shah S et al. (2020) A large carnivore among people and livestock: the common leopard. In: Angelici FM, Rossi L (eds) Problematic wildlife II. New conservation and management challenges in the human-wildlife interactions. Springer, Cham, pp 93–110

Meriggi A, Torretta E, Dondina O (2020) Evolution of wolf habitat occupancy and feeding habits in Italy: implications for species conservation and conflict resolution with humans. In: Angelici FM, Rossi L (eds) Problematic wildlife II. New conservation and management challenges in the human-wildlife interactions. Springer, Cham, pp 111–138

Oriol-Cotterill A, Valeix M, Frank LG et al. (2015) Landscapes of coexistence for terrestrial carnivores: the ecological consequences of being downgraded from ultimate to penultimate predator by humans. Oikos 124:1263–1273

Sheperd SM (2020) Large felid man eaters: can humans successfully manage endangered felid species in a shrinking wild environment to preserve these species and prevent the development of man eaters or problem animals that are accidental livestock or man killers. In: Angelici FM, Rossi L (eds) Problematic wildlife II. New conservation and management challenges in the human-wildlife interactions. Springer, Cham, pp 17–91

Treves A, Wallace RB, Naughton-Treves L, et al. (2006) Co-managing human–wildlife conflicts: a review. Hum Dimens Wildl 11:383–396

Woodroffe R, Thirgood S, Rabinowitz A (2005) People and wildlife, conflict or co-existence? Cambridge University Press, Cambridge

Chapter 2
Large Felid Predators and "Man-Eaters": Can We Successfully Balance Conservation of Endangered Apex Predators with the Safety and Needs of Rapidly Expanding Human Populations?

Suzanne M. Shepherd

"Tiger is a large-hearted gentleman with boundless courage and that when he is exterminated-as exterminated he will be unless public opinion rallies to his support-India will be the poorer, having lost the finest of her fauna." – Jim Corbett, Man-eaters of Kumaon. 1944.

2.1 Introduction

As predators and scavengers, the large felid carnivores have had a long co-evolutionary relationship with hominins, documented and studied in archeological sites in Europe, Asia, Africa, and the Americas (Schaller and Lowther 1969; Brantingham 1998; Lewis 1997; Arribas and Palmqvist 1999; van Valkenburgh 2001; Mercader et al. 2002; Smilie 2002; Stiner 2012; Grayson and Meltzer 2003; Pickering et al. 2004; Blasco et al. 2010; Saladié et al. 2014; Camarós et al. 2015: Daujeard et al. 2016; Lewis 2017). Until relatively recently however, humans did not play a dominant role in determining the survival of large felid carnivores. These cats are among the most endangered, and the most challenging, species to conserve on this increasingly human-dominated planet. The extensive natural home ranges of these obligate carnivores, and continued human and livestock expansion into their territorial wildlands, increasingly cause these big cats to compete directly with humans for space and food (Treves and Karanth 2003; DeFries et al. 2004;

S. M. Shepherd (✉)
Department of Emergency Medicine, Hospital of the University of Pennsylvania, University of Pennsylvania Healthcare System, Environmental Education & Animal Behavior Intern (EEABI), The Philadelphia Zoo, Philadelphia, PA, USA
e-mail: suzanne613@comcast.net

F. M. Angelici, L. Rossi (eds.), *Problematic Wildlife II*,
https://doi.org/10.1007/978-3-030-42335-3_2

Inskip and Zimmerman 2009; Conover 2008; Carter and Linnell 2016; Khan et al. 2018). Resultant decreases in the populations of these large carnivores, and their wild prey, not only threaten their existence as species but also deregulate the structure and function of entire ecosystems (Kellert et al. 1996; Polisar 2000; Terborgh et al. 2002; DeFries et al. 2004; Grange and Duncan 2006; Conover 2008; Mondol et al. 2009; Seidensticker 2010; Athreya et al. 2013; Karanth et al. 2013; Carter and Linnell 2016; Khan et al. 2018). As such, their conservation serves a larger role than just the preservation of these iconic cats. It has become a global conservation priority. Balancing measures to accommodate growing human populations, often exceedingly poor agrarian peoples living in multiuse lands bordering small protected areas, and those to conserve these big cats are complex and difficult problems (Karanth et al. 1999; Maskey et al. 2001; Treves and Karanth 2003; Patterson 2006; Karanth 2002; Graham et al. 2005; Kolowski and Holekamp 2006; Lagendijk and Gusset 2008; Lindsey et al. 2013a; Inskip and Zimmerman 2009; Karanth et al. 2012; Schwartz 2017; Kshettry et al. 2017; Krafte Holland et al. 2018). Basic questions that remain to be clarified are multiple. What are the essential requirements to maintain sustainable apex carnivore populations? These requirements include the determination of appropriate natural prey and predator numbers to maintain species viability. The optimal size and connectivity of protected areas to allow for maintenance of both individual carnivore populations and judicious intermixing of these populations to preserve maximal genetic diversity must also be determined (Simberloff and Cox 1987; Athreya et al. 2007; Karanth et al. 2013; Shrestha et al. 2014; Wolf and Ripple 2018). As most of these big cats are specialized to hunt ungulates, livestock in areas abutting their territories provide more readily available and easily hunted targets than shrinking populations of native prey species. How can these prey populations be balanced to preserve their viability as species, maintain ecosystem balance, and also meet both human and carnivore needs (Polisar 2000; Mech et al. 2000; Kolowski and Holekamp 2006; Graham et al. 2005; Odden et al. 2008; Schwartz 2017; Kulbhushansingh et al. 2017; Khan et al. 2018)?

Diseases introduced from human livestock and companion canines, such as rabies, distemper, and tuberculosis, further negatively impact the health of both carnivores and their natural prey (Tessaro 1986; Bengis et al. 2002; Dybas 2009; Sweeney and Miller 2010; Viljoen et al. 2015; Viana et al. 2015). Studies in the Serengeti Park system have found that rabies is maintained in the dog population in this area (Lembo et al. 2008; Hampson et al. 2015). In 1994, co-infection with babesiosis from the ingestion of infected cape buffalo and a canine distemper virus (CDV) epidemic killed 30% of lions and affected other carnivores, in the Serengeti National Park (Roelke-Parker et al. 1996; Dybas 2009). In 2001, a similar percentage of lions were also killed in the nearby Ngorongoro Crater (Dybas 2009). One question raised was whether local dogs also served as the maintenance population for the CDV virus. In follow-up, a study looked at the utility of mass vaccination of dogs around the park system. Unfortunately, the study demonstrated that over time *Morbillivirus* infection peaks in lions had become asynchronous with those in canines, and while vaccination significantly decreased dog outbreaks, it did not prevent lion infections. This suggested that other wildlife species had become involved in maintenance of the CDV infection in the park system. The authors suggested that further investigation of transmission patterns would be necessary to inform further disease control strategies (Viana et al. 2015).

As these highly intelligent predators become accustomed to the increasing presence of humans and their livestock, they no longer fear them and these become potential prey (Maskey et al. 2001; Löe and Röskaft 2004; Packer et al. 2005; Kolowski and Holekamp 2006; Conover 2008; Valeix et al. 2012; Athreya et al. 2013; Lindsey et al. 2013a, b, c; Acharya et al. 2016). Humans have little biological weaponry to compete with these very large, evolutionarily honed apex predators. As such, attacks occur, and the outcomes are usually adverse for the human(s). Due to a number of factors, these big cats may turn to either accidental or incidental human predation or to true man-eating behavior (Anderson 1954; Capstick 1998; Neiburger and Patterson 2000; Peterhans et al. 2001; Maskey et al. 2001; Patterson et al. 2003; Löe and Röskaft 2004; Yeakel et al. 2009; Chanchani et al. 2015; Penteriani et al. 2016; Acharya et al. 2016; DeSantis and Patterson 2017). Conservation and control efforts for "problem animals" must be balanced to engage and appropriately understand and address the concerns of all stakeholders in the process. Control efforts to date have largely involved a three-pronged approach: eradication, translocation, and preservation with nonlethal deterrence (Chellam and Johnsingh 1993; Nowell and Jackson 1996; Anderson and Ozolins 2000; Polisar 2000; Fox 2001; Andelt 2001; Treves et al. 2002; Mishra et al. 2003; Ogada et al. 2003; Naughton-Treves et al. 2003; Montag 2003; Shivik et al. 2003; Bradley et al. 2005; Graham et al. 2005; Woodroffe et al. 2006; Athreya et al. 2007; Dickman 2010; Weilenmann et al. 2010; Massei et al. 2010; Lichtenfeld et al. 2014; Athreya et al. 2011; Fontúrbal and Simonetti 2012; Karanth et al. 2012; Bhatia et al. 2013; Ordiz et al. 2013; Banarjee et al. 2013; McManus et al. 2014; Weise et al. 2015; Athreya et al. 2015; Miller et al. 2016a; Hale and Koprowski 2018). Creative, multipronged, scientifically valid, globally involved, and well-funded measures to conserve these iconic cats also must incorporate attention to the needs of local humans and their livestock (Jalais 2009; Seidensticker 2010; Suryawanshi et al. 2017; Redpath et al. 2013; Chapron et al. 2014; Shepherd et al. 2014; Harihar et al. 2015; Carter and Linnell 2016; Miller et al. 2016a; Krafte Holland et al. 2018; Wolf and Ripple 2018; Sanderson and Walston 2019). Without government, media, and local population education, involvement, and buy-in, these efforts will ultimately fail (Bhatia et al. 2013; Shepherd et al. 2014; Tyrrell and Western 2017; Krafte Holland et al. 2018; Harihar et al. 2018; Sanderson and Walston 2019).

2.2 Evolution and Environmental Niche of the Large Felid Carnivores

The large Felidae, suborder Feliformes, contains the subfamilies Pantherinae (tiger (*Panthera tigris*), leopard (*Panthera pardus*), snow leopard (*Panthera uncia*), clouded leopard (*Neofelis nebulosa*), jaguar (*Panthera onca*), and lion (*Panthera leo*)) and Felinae (cheetah (*Acinonyx jubatus*) and puma (*Felis concolor*)) (also known as cougar, catamount, or mountain lion). These large Felidae are obligate carnivores that function naturally as apex predators in their biologic communities. They are, aside from lions, generally solitary, often nocturnal, and secretive. These carnivores, via predation, serve to regulate both prey and lower-level predator populations and thus directly and irreplaceably impact the structure and function of their entire ecosystem (Schaller 1972; Terborgh et al. 2002; Estes et al. 2011; Thinley et al. 2018).

2.3 Evolution

The first cat, *Proailurus*, was felt to have emerged during the Oligocene, approximately 25 million years ago in Asia (Hunt 1987). Its successors, 11 *Pseudaelurus* species, ancestral to the cats in the current subfamilies and to the now extinct subfamily Machairodontinae (saber-toothed cats), lived in the Miocene era approximately 20 to eight million years ago (Werdelin et al. 2010). Nuclear and mitochondrial DNA analyses have shown that the ancient cats evolved into eight main lineages that diverged over multiple migrations from continent to continent via the Bering land bridge and the Isthmus of Panama (Johnson et al. 2006).

Most of the large cats of the genus *Panthera*, despite their different skull and body shapes (Sakamoto and Ruta 2012), were suggested through gene mapping studies to have evolved approximately 4.6 million years ago (O'Brien and Johnson 2007; Figuero et al. 2017). Fossil discovery on the Tibetan Plateau of *Panthera blytheae*, distinctly of the snow leopard line, also suggests that the pantherine big cats split off early into their own group from a common Asian ancestor in the Miocene era, dating to 4.1 to 5.95 Ma (Tseng et al. 2013; Holland 2013). The earliest *Panthera* fossils from Laetoli belong to two different species, one lion sized and one leopard sized (Werdelin et al. 2010). Of interest, gene mapping studies also have suggested that the big cats have engaged in at least some degree of crossbreeding over the course of their evolutions (Figuero et al. 2017). Previously, the oldest pantherine fossils had been found in Africa, seemingly contradicting molecular phylogeny studies (Yu and Zhang 2005) that pointed to Asia as the region of origin of the big cats.

Of the *Panthera* species, the ancestor of the clouded leopard (*Neofelis* spp.) was apparently the first to diverge, approximately 6.37 million years ago (Johnson et al. 2006; Werdelin et al. 2010). *Neofelis nebulosa* (the clouded leopard) and *Neofelis diardi* (the Sunda clouded leopard) do not roar (Klemuk et al. 2011; Kitchener et al. 2016). These large felid predators live in Asia, with habitats extending from Sumatra and Borneo (*Neofelis diardi*), through mainland Southeast Asia, into the Himalayan foothills (*Neofelis nebulosa*). Their habitat in Southeast Asia is undergoing one of the world’s fastest deforestation rates (Victor 2017; Sullivan 2018). This habitat loss, direct exploitation for the exotic pet and skin trades, and tissue harvesting for the food and traditional medicine markets have contributed to a significant decline in their numbers in the wild (IUCN Red List. Cites Appendix 1; Platt 2010). Genetic hair sample analysis of the two subspecies found that they diverged approximately 1.4 million years ago, after utilizing a now submerged land bridge between the Asian mainland and Sumatra and Borneo (Buckley-Beason et al. 2006; Kitchener et al. 2006; Wilting et al. 2010). The clouded leopard is felt to be an evolutionary link between the big cat and smaller cat species (Hemmer 1968). In addition to a number of body adaptations for its largely arboreal life, such as the longest tail of the big cats, it has the largest canines of the current big cats in proportion to its body. The canines of the significantly smaller clouded leopard, 5 centimeters in length, are of the same size as those of the approximately ten times larger tiger (Guggisberg 1975).

P. tigris and *P. uncia* branched off next (approximately 3.9 Ma). The tiger developed into a unique species toward the end of the Pliocene era (approximately 3.2 Ma) (Davis et al. 2010). The earliest (estimated 2.55–2.16 Ma) complete skull of a pantherine (*P. zdanskyi*), closely related to the tiger, was found in Gansu Province, China. This suggests that the distinctive tiger skull morphology was established early in its evolution, while later changes included increase in body size and a reduction in relative size of the dentition. This was felt to be coupled with the increase in size during the tiger’s evolution of its primary prey species (Mazak et al. 2011). Since the early twentieth century, tiger populations are estimated to have lost more than 90% of their original historical range. Their habitat had extended in forested areas from eastern Turkey to the coast of the Sea of Japan and from the Indian subcontinent across Asia to the Sunda Islands (Kitchener and Yamaguchi 2010; Sunquist 2010). Three of the original nine subspecies of the current tiger (*Panthera tigris*) have become extinct during the twentieth century. The Javan tiger (*Panthera tigris sondaica*), the Caspian tiger (*Panthera tigris virgata*), and the Balinese tiger (*Panthera tigris balica*) have disappeared (Sunquist 2010; IUCN.org, accessed 12/31/2018).

Snow leopards (*Panthera uncia*) do not roar (Klemuk et al. 2011, Kitchener et al. 2016). Their tails are thick and as long as their bodies to facilitate balance (Kitchener et al. 2016) (see Fig. 2.1). Their skins are paler and their coats are thicker, including fur in between their paw pads, in adaptation to the colder, more mountainous, higher altitudes in Asia that they inhabit (Kitchener et al. 2016). A recent genetic analysis of global snow leopard populations suggests that there are three subspecies. This study posited that the snow leopard experienced a genetic bottleneck approximately 8000 years ago during the Holocene, coinciding with an episode of global warming. The authors suggested that conservation efforts might focus most effectively on programs specific to the ecology of each subspecies. They also recommended that

Fig. 2.1 Female snow leopard Maya. Note the relatively small round ears, long thick tail, and thick paler fur in this cat who naturally lives in colder, higher altitudes. (Photograph taken by author at Philadelphia Zoo. January 11, 2019)

natural connections should be maintained between the areas containing these populations, rather than creating artificial corridors (Janecka et al. 2017).

The divergence of the ancestors of the tiger and snow leopard was followed by branching off of the ancestors of *P. onca*, *P. pardus*, and *P. leo*, approximately 4.3–3.8 Ma. The ancestor of the jaguar (*P. onca*) diverged from the ancestor of lions and leopards approximately 3.6–2.5 Ma. The ancestors of lions (*P. leo*) and leopards (*P. pardus*) split approximately 2 Ma (Davis et al. 2010).

The jaguar (*Panthera onca*) is believed to have originated in Eurasia and moved across the Bering land bridge during the Early Pleistocene (Kurtén and Anderson 1980). Until the early twentieth century, the jaguar roamed from Northern Argentina northwest into the area that now encompasses Nebraska, Northern California, and Washington in the United States (Kurtén 1973; Mccain and Childs 2008; Christiansen 2008b). The earliest fossil remains to date, from the Middle Pleistocene, were found in California (Hemmer et al. 2010). The exact evolutionary relationship of the current jaguar to the larger European jaguar (*P. gombaszoegensis*) remains unclear (Hemmer et al. 2001; Hemmer and Kahlke 2005; Werdelin et al. 2010; Mol et al. 2011). While jaguars recently have been spotted in Mexico and Arizona, it remains to be seen if jaguars will be able to regain and hold their former more northern territories (Grant 2016).

Ancestors of the lion *(Panthera leo)* were larger. The earliest lion-like cat (*Panthera leo fossilis*), dating to the Late Pliocene (5–1.8 million years ago), was discovered in Tanzania (Barry 1987). The first true lion fossil was discovered at Olduvai Bed I in Tanzania and dated to 1.87–1.7 million years ago (Petter 1973). With the change from forest to savannah-like grasslands in the Late Pliocene and Early Pleistocene, and the concomitant rise of ungulates and ground-dwelling primates, lion populations apparently flourished (Cerling et al. 1998; Turner 1999). Lions lived in a significantly larger range, and probably developed group living behaviors, before migrating approximately 21,000 years ago from Eastern and Southern Africa to Northern Africa and parts of Europe and Asia (Yamaguchi et al. 2006; Riggio et al. 2013; Barnett et al. 2014). Mitochondrial DNA studies, on both living and extinct lions, demonstrated that lions expanded from North Africa first into India and then later into the Middle East, Asia, North America, and South America as far as Peru (Turner and Antón 1997). The American and northern Eurasian lions became extinct at the end of the Pleistocene (Turner and Antón 1997). In the 1960s, lions became extinct in most of North Africa, except the southern Sudan (Black et al. 2013; Riggio et al. 2013). Today, only fragmented populations remain in sub-Saharan Africa and a critically endangered Asian population in the Gir Forest National Park in the Gujarat State in Western India (Turner and Antón 1997; Yamaguchi et al. 2006; Burger et al. 2004; Barnett et al. 2006). Genetic studies have demonstrated that current lion populations in Central and West Africa are more closely related to Asian lions. As such, the authors suggest that these populations not be combined with Southern African lions in conservation programs in order to maintain maximal species diversity (Barnett et al. 2014). The earliest male lions were suggested to have been maneless. Maned lions are presumed to have arisen approximately 320,000–190,000 years ago and expanded and replaced earlier maneless males.Gir lions today have a shorter and sparser mane than Southern African lions (West and Packer 2002; Yamaguchi et al. 2006). Darker manes predict males with higher testosterone levels and a better chance of survival and protecting vulnerable cubs, but this is at the cost of increased heat stress (West and Packer 2002). The Eurasian cave lion, *Panthera leo spelaea*, genetically distinct but in appearance similar to the modern lion, lived approximately 300,000 to 13,000 years ago in areas of Europe and Asia. The cave lion may have been the largest conical-toothed big cat to have existed (Barnett et al. 2009). The French Chauvet cave paintings, dating back 17,000 years, clearly illustrate these maneless cave lions. They also provide the first depiction of a leopard (Clottes 2003) (see Fig. 2.2). Lions have been clearly associated with humans and livestock and pack animal predation since ancient times (see Fig. 2.3). Lion hunting is well described in ancient historical records (see Fig. 2.4). In 440 BC (*The Histories*), Herodotus described lions

Fig. 2.2 Chauvet cave paintings of maneless European cave lions (*Panthera spelaea*) Gif-sur-Yvette, France. The paintings are believed to date to the very beginning of the Upper Paleolithic era, approximately 30,000 years ago. The original is in the Anthropos Museum, City of Brno, South Moravia, Czech Republic. This is an open-sourced photographic reproduction of a two-dimensional public domain work of art, obtained from WikiCommons on the World Wide Web

attacking camels in Xerxes' caravans during the march toward Therma (Rawlinson 1992; Peterhans et al. 2001; Rollinger 2012).

The modern leopard is believed to have evolved in Africa 0.5–0.8 million years ago. Extremely adaptable, the leopard has the widest range and habitats of any of the big cats (Jacobson et al. 2016). At the Pliocene/Pleistocene Swartkrans site in South Africa, *Australopithecus robustus* was clearly shown by fossil remains to be a prey item for leopards (Brain 1970, 1981). Of the original 27 subspecies described, 9 leopard subspecies, based on mitochondrial analysis, currently are found from sub-Saharan Africa to temperate and tropical Asia, including the Southern Russia/Northern Chinese border (Amur leopard) and Java (Javan leopard) (Uphyrkina et al. 2001). Unfortunately, these habitats are increasingly threatened. Leopards have disappeared from almost 75% of their original range, including Japan, Hong Kong, Singapore, Kuwait, Syria, Libya, Tunisia, and Europe (Diedrich 2013; Ghezzo and Rook 2015; Izawa et al. 2015; Stein et al. 2016; Jacobson et al. 2016; Williams et al. 2017). Leopards are one of the big cat species whose ancestors did not traverse the Bering Strait to the Americas (Nowell and Jackson 1996; Uphyrkina et al. 2001).

Fig. 2.3 Lion Panel. The processional way to Babylon's Ishtar Gate. Circa 575 BC. British Museum, London. This was originally one of the Seven Wonders of the Ancient World, until replaced by the Lighthouse of Alexandria. (Photograph taken by Author in the British Museum, London. September 11, 2003)

Fig. 2.4 Portion of a sculpted bas-relief depicting Ashurbanipal, the Assyrian King, hunting lions. From the North Palace of Nineveh, Iraq, C. 645–635 BC. In ancient Assyria, lion hunting was considered a royal sport. (Picture taken by the author in the British Museum, London. September 11, 2003)

In the 1990s, mitochondrial and nuclear DNA evidence demonstrated that the cheetah, which had been placed alone in the genus *Acinonyx*, was actually most closely related to cougars and jaguarundis in the genus *Puma* (Janczewski et al. 1995; Johnson and O'Brien 1997; Mattern and McLennan 2000; Kitchener et al. 2016). These DNA studies were further supported by morphological studies of the cranium in these species (Salles 1992; Li et al. 2016; Segura 2013). It is felt that these acinonychins diverged from other cats approximately 6.7 Ma ago. Subsequently, the ancestral cheetah diverged from the puma line approximately 4.9 million years ago (Johnson et al. 2006; Culver et al. 2000; Li et al. 2016). It is believed that approximately 2 to possibly as many as 11 species of cheetah have existed (Spassov 2011). Results of a phylogeographic study on cheetah subspecies indicate that African and Asian cheetah populations, which are genetically distinct, diverged approximately 32,000–67,000 years ago (Charruau et al. 2011). It remains unclear whether cheetah ancestors diverged from the puma lineage in the Americas and then migrated back across the land bridges into Asia and Africa or whether they diverged in Asia (Culver et al. 2000; Barnett et al. 2005; Yu and Zhang 2005). Cheetahs remain at increased risk of extinction, due to the lack of genetic diversity in the modern cheetah populations. This was demonstrated to be caused by genetic bottlenecking due to two mass extinction events approximately 100,000 and then 12,000 years ago (Dobrynin et al. 2015; Barnett et al. 2005). Inbreeding may have left the cheetah population less likely to survive if attacked by a lethal infection and also less likely to successfully breed due to an accumulation of deleterious recessive mutations, like sperm malformation (Menotti-Raymond and O'Brien 1993; Merola 1994; Castro-Prieto et al. 2010; Dobrynin et al. 2015; Terrell et al. 2015). Of interest, cheetahs may have found a way to improve genetic variability in their offspring: female cheetahs mate with multiple males from different geographic areas, allowing individual litters of cubs to have multiple unrelated fathers (Gottelli et al. 2007).

Felis concolor, the puma, is in a separate genus of the smaller cats. Like the leopard, they have adapted to live in a large number of habitats in the Americas, ranging from forest and tropical jungle to desert regions and from relatively isolated mountainous areas to peri-urban flatlands and swamp (Florida panther). Their range extends from the southern Andes in South America to the Canadian Yukon (Sweanor et al. 2005; Conroy et al. 2006; Ernest et al. 2014; Knopff et al. 2014). The oldest known member of the *Puma* genus, *P. pardoides*, was found in the Pliocene-Pleistocene boundary areas in Western Europe (Cherin et al. 2013), which suggests that this genus first appeared in Eurasia and then crossed the land bridge into the Americas approximately 4 million years ago.

2.4 Developmental Advantages in the Large Felid Carnivores

Felidae have evolved as highly complex organisms to most effectively ambush, hunt prey via short pursuit, or, in the case of lions, cooperatively hunt in their respective habitats (Schaller 1972; Taylor and Rowntree 1973; Packer et al. 1990; Estes 1991;

Packer and Pusey 1997; Stander 1992; Hayward and Kerley 2005; Marker and Dickman 2005; Hayward et al. 2006; Christiansen and Wroe 2007; Owen-Smith and Mills 2008; Mosser and Packer 2009; Werdelin et al. 2010; Williams et al. 2014; Moore and Bieweber 2015; Sakai et al. 2016; Bjordal 2016; Wilson et al. 2018). These predators have also developed a number of crucial specializations that allow them to adapt to their environment and improve their success as apex predators and propagators of their species (Berlow et al. 2009; Chamoli and Wroe 2011; Schmitz 2017). Most possess large, muscular, lithe bodies, enabling them to dominate other predators; to tackle large, dangerous, and agile prey; and to carry heavy carcasses to hide and protect them for later consumption (Carbone et al. 1999; van Valkenburgh and Wayne 2010). The cheetah, snow leopard, and clouded leopard have long tails to provide balance during rapid chase and climbing (see Fig. 2.1) (Chadwick 2008; Patel et al. 2016). The big cats have developed large soft paws, retractable hook-shaped claws, and flexible and strong forelimbs. These allow them to approach stealthily, climb agilely, and better catch and hold moving prey (Ortolani 1999). In addition, several of these predators have developed additional trophic adaptations to their special environments and hunting styles. Cheetahs have nonretractable claws and tough pads to assist in traction when running at high speed (Wilson et al. 2013a, b). Clouded leopards have unfused radii and ulnae, allowing them to have greater range of motion when climbing or hunting, including the ability to hang upside down from branches and easily climb down trees headfirst (Guggisberg 1975). Relatively large eyes are forward facing to provide binocular vision, allowing good depth perception. Their eyes also contain a relatively large proportion of retinal rods and a tapetum lucidum to aid in night vision (Hall et al. 2012). All of the big cats possess excellent senses of hearing and smell (Padodara and Jacob 2014). They have evolved a number of cranial and dental adaptations: long, sharp, and strong conical incisors, well-developed carnassials, large cranio-mandibular muscles, and the ability of the jaws to move up and down rather than side to side. These modifications allow them to bite quickly, securely, and strongly and to fully penetrate, tear, and shear tough skin and flesh (see Fig. 2.5) (van Valkenburgh and Ruff 1987; Cohle et al. 1990; Christiansen and Wroe 2007; Christiansen 2008a, b; Sicuro 2011; Mazak et al. 2011). While lions and tigers exert greater absolute bite strength (>1000 PSI), jaguars possess the strongest bite force relative to size of the large felid carnivores (Therrien 2005). This makes sense when one considers the fact that heavily shelled turtles and scaled caiman form a regular part of their diet (Hartstone-Rose et al. 2012; Everton et al. 2016). Big cat coats are often rosette-patterned to camouflage approach and concealment. Rosettes simulate the shifting shadows in shaded vegetation and break up their body outline (see Fig. 2.6) (Ortolani 1999; Allen et al. 2010). Tiger's stripes are elongated forms of the camouflaging rosettes, providing excellent concealment in the light-dappled heavily forested areas that serve as their native habitat (Godfrey et al. 1987). Lion and cougar cubs are spotted as protective camouflage, and spots may persist on the legs and abdomen of females into adulthood. Lions are the only big cat to show obvious sexual dimorphism between males and females: other cats differ largely in size and weight (West and Packer 2002; Yamaguchi et al. 2006). As previously noted, the manes of lions may have evolved

Fig. 2.5 Jaguar Zia easily and rapidly using her carnassials and heavy jaw muscles to shear a large piece of meat at the Philadelphia Zoo. (Picture taken by the author. January 18, 2019)

due to the more social nature of this species (West and Packer 2002; Trivedi 2002; Yamaguchi et al. 2006). The pantherine big cats are also defined by having the ability to roar instead of purr, with the exceptions of snow and clouded leopards. Cheetahs and pumas also do not roar (Volodina 2000). This ability to roar is due to structural adaptations in the morphology and function of the hyoid apparatus, pharynx, and vocal folds (Hast 1989; Weissengruber et al. 2002; Titze et al. 2010; Klemuk et al. 2011).

2.5 Hunting Methods

All of the big cats are obligate carnivores. Based on evolutionary advantages, each species of big cat kills prey in a fashion typical of their size, armamentarium of weapons, hunting skills, and environment (Hildebrand 1961; Eaton 1970;

Fig. 2.6 Female Amur leopard Kira at Philadelphia Zoo demonstrating how coat patterns break up the animal's outline in dappled shade. The reader may also note the relatively longer legs and slightly lighter coloration in this predator adapted to living in the snowy climate in the Amur area between Northern China and Southern Russia. (Photograph taken by author at the Philadelphia Zoo, September 21, 2018)

Scheel 1993; Karanth and Sunquist 1995; Miquelle et al. 1996; Karanth and Sunquist 2000; Funston et al. 2001; Hopcraft et al. 2005; Hayward et al. 2006; Ramesh et al. 2009; Hudson et al. 2011; Farhadinia et al. 2012; Davidson et al. 2012;Wilson et al. 2013a, b; Rostro-Garcia et al. 2015; Hilborn et al. 2018; Shrestha et al. 2018). A study on pumas and cheetahs found that the cat's hunting strategies have evolved to optimally balance energy expended during hunting with the sustenance received from their prey (Williams et al. 2014). Foraging for prey is a "two player high stakes game of stealth and fear" (Brown et al. 1999), with the predator often giving up and losing (van Orsdol 1984; Laundré 2010; Scantlebury et al. 2014; Williams et al. 2014). Death secondary to significant trauma, including severe vascular, airway, and neuromuscular injuries, is well described after large cat attacks (Shepherd et al. 2014). Injuries from large cat attacks tend to fall into specific patterns based on the circumstances and terrain of the attack, the ability of the prey to attempt to fight back, and the individual attacking animal (Loeffler 1997). Tigers, lions, and jaguars are estimated to exert between 690 and 1000 PSI (pounds per square inch) of bite strength, with larger males on the higher end of the spectrum (Christiansen and Wroe 2007). This permits these large felid carnivores to securely bite and then twist and tear prey. This bite strength also produces skull and spine fractures and penetration of the intracranial vault or spinal cord during bites to the head and neck (Kohout et al. 1989; Prasad et al. 1991; Kadesky et al. 1998; Bahram et al. 2004; Anderson et al. 2008). Proprioceptive receptors in the jaw and teeth allow the cat to align the teeth between the victim's cervical vertebrae, severing the spinal cord as the neck is hyperextended (Morgan 1999; Bahram et al. 2004; Chum and Ng 2011). As such, animals, and humans, attacked in an open area by large cats sustain severe injuries to the head, neck, spine, torso, and extremities. Tigers, pumas, and jaguars commonly ambush prey from behind. They approach as closely as they can and then rush and pounce, biting the victim's neck and occiput. Larger animals are held by the throat and suffocated, while smaller animals are killed by a bite to the back of the neck (Schaller 1967; Schaller and Vaseconselos 1978; Kohout et al. 1989; Cohle et al. 1990; Rollins and Spencer 1995; Beier et al. 1995; Wiens and Harrison 1996; Kadesky et al. 1998; Bank and Franklin 1998; Morgan 1999; Bahram et al. 2004; Anderson et al. 2008; Meachen-Samuels and Van Valkenburgh 2009). Jaguars also utilize some more unique prey-dependent killing methods. They bite directly through the skull of prey such as capybara, piercing the brain (Schaller and Vaseconselos 1978). Jaguars crack open turtle shells, kill caimans, and may bite the heads off of sea turtles (Emmons 1987). Leopards may either attack from behind or bite the victim's throat depending on the circumstances of the attack (Loeffler 1997). Stalk and ambush predators are generally less selective in the individuals chosen for pursuit. Their prey, chosen in part by size, are usually those individuals on the edge of the herd or traveling alone or in small groups (Stiner 1990). Pursuit predators scout and stalk to a closer distance so as not to exhaust their metabolic resources too quickly. They then give chase either solo (cheetah) or as a coordinated group (lion) (Taylor and Rowntree 1973; Stander 1992). Cheetah will trip their prey during the chase: agility, not just speed, gives cheetahs an advantage (Scantlebury et al. 2014). Lions and cheetah typically bite their prey's throat or neck nape. This crushes the larynx and strangles the prey. It also can lacerate or sever the carotid arteries and jugular veins, producing rapid exsanguination. Other lionesses, in a well-coordinated attack, will mount and claw the prey's rump and flanks to further impede escape (Schaller 1968; Eaton 1970; Schaller 1972; Elliott et al. 1977; Scheel and Packer 1991; Stander 1992). These pursuit predators tend to target young, old, or otherwise weakened individuals of an appropriate size for them to handle (Husseman et al. 2003).

2.6 Humans as Large Feline Prey

Unfortunately, human attacks worldwide are underreported and poorly characterized due to the lack of requirements for mandatory reporting and lack of standardized reporting protocols in the areas in which they typically occur (Löe and Röskaft 2004). Similar injuries to those in prey species have been reported in humans who have been attacked by the big cats. Injuries include abrasions, lacerations, tissue avulsions, limb disarticulations, crush injuries, and penetration of the skull, spine, and body cavities (Shepherd et al. 2014; Pawar et al. 2018). These injuries often involve the head and neck (Shepherd et al. 2014). Most large predators can, and will, scavenge human carcasses or pursue humans as prey (Löe and Röskaft 2004). Children often fall victim to attack as they are smaller and less able to defend themselves and often are similar in size to common prey (Wiens and Harrison 1996). Lion attacks on humans are frequent compared with most other carnivores; however tigers remain the most frequently documented killers of humans worldwide (Löe 2002; Löe and Röskaft 2004; Nyhus et al. 2010). It is estimated that only 3000–4000 wild tigers remain in Asia in a range restricted to just 7% of where they once roamed. Thus, these cats are increasingly forced to share their habitat with expanding local human populations (Mondol et al. 2009). In the Bangladesh Sundarbans alone, a total of 1396 deaths were recorded over a 63-year period (Barlow et al. 2013). The most comprehensive study of tiger attack-related deaths estimates that at least 373,000 people died from these attacks between 1800 and 2009 (Nyhus et al. 2010). Of interest, and for unclear reasons, reported tiger-related human deaths in certain parts of Asia are extremely rare, including much of the Indian subcontinent, Thailand, Myanmar, Malaysia, and Sumatra (Nyhus et al. 2010). While tigers definitely attack humans as prey, they appear to more commonly attack people defensively to protect themselves or their cubs. This has been especially described when they have been previously injured by hunters, snares, individuals foraging in their forested areas, or villagers protecting their livestock (Gurung et al. 2008; Goodrich 2010).

Due to the large force involved, most humans who experience penetration of the cranio-cervical compartment by a big cat attack do not survive (Christiansen and Wroe 2007; Anderson et al. 2008; Emami et al. 2012; Pathak et al. 2013). Tiger attacks commonly result in immediate loss of limb or death. In one report, a single tiger paw blow fractured a skull (Prasad et al. 1991). In another report, a 39-year-old male attacked by a tiger sustained vascular and pharyngeal injuries, as well as fractures of the C1 and C2 vertebral bodies with an accompanying spinal cord laceration (Kohout et al. 1989). Another tiger attack victim sustained a pharyngeal injury and cervical spine fracture (Wiens and Harrison 1996). A very interesting case involved an individual who developed a post-traumatic syrinx and tethered spinal cord more than 25 years after being bitten in the neck by a Bengal tiger (Papadoupoulos and Tubridy 1999). An 11-year-old child sustained a comminuted, open C1 anterior arch fracture, with associated paralysis, after bites to the back of the neck and occiput. The child then developed purulent meningitis within 24 hours of the tiger bite (Burdge et al. 1985). Burdge et al. described deep forearm and lower leg lacerations, with the delayed presentation of a septic open shoulder dislocation, in a Maasai tribesman attacked by a lion (Burdge et al. 1985). In another series of ten lion attack victims, a variety of serious injuries occurred. These included five individuals who sustained bilateral penetration of the neck by the teeth, a jugular injury, a compound fracture of a vertebral body, and pharyngeal, laryngeal, and tracheal injuries (Loeffler 1997). Leopard attacks on humans appear to vary by geographic area. Attacks are most commonly reported in India and Nepal (Inskip and Zimmerman 2009; Athreya 2012; Maskey et al. 2001). In his experience with leopard attacks on humans, Loeffler noted that they occurred more frequently than lion attacks and were likely to result in more superficial injuries (Loeffler 1997). Other authors report more massive tissue damage from leopard bites, including fatal bites to the head and neck (Cohle et al. 1990; Bahram et al. 2004; Nabi et al. 2009; Hejna 2010; Pawar et al. 2018). Loeffler also described two individuals who sustained arm and hand bite wounds from a cheetah (Loeffler 1997). Serious injuries and death are also reported in mountain lion attacks on children and adults (Conrad 1992; Rollins and Spencer 1995; Kadesky et al. 1998; Iserson and Francis 2015). A serious scalp degloving injury requiring microsurgical reconstruction and deep hand lacerations were described in a male hiker attacked by a mountain lion (Hazani et al. 2008). Serious injuries and death from jaguar attacks on both adults and children have been reported, including skull fracture with direct brain injury and spinal transection (Neto et al. 2011; Iserson and Francis 2015).

In those humans who survive an attack by a big cat, multi-bacterial infections from the oral flora complicate recovery in 5–30% of victims (Bahram et al. 2004). Before the advent of antibiotics, and in areas where antibiotics were not readily available, approximately 75% died from these infections (Mcgeachie 1958; Weber et al. 1984; Burdge et al. 1985; Woolfrey 1985; Isolato and Edgar 2000; Capitini and Herero 2002). Surviving humans may also develop tetanus and rabies if the carnivore was infected (Andrade dos Santos and Tokarnia 1960; McKee 2003; Shepherd et al. 2014; Iserson and Francis 2015).

2.7 Human Large Felid Conflict

The survival of large felid carnivores faces a number of challenges. Human-carnivore conflict has developed for multiple reasons. Their extensive home ranges and required protein-rich diet cause these cats often to directly compete with humans for food and space (Treves and Karanth 2003; DeFries et al. 2004; Inskip and Zimmerman 2009; Conover 2008; Carter and Linnell 2016; Khan et al. 2018). Human-carnivore conflict has been increasing in many areas as human populations continue to grow and expand into previously "wild" areas (Treves and Karanth 2003; DeFries et al. 2004; Inskip and Zimmerman 2009; Conover 2008; Carter and Linnell 2016; Khan et al. 2018). Introduced human livestock also compete directly with wild herbivores for food and habitat (Gordon 2009; Kartzinel et al. 2015). As wild prey populations and their natural hunting areas decrease, carnivores increasingly are driven to develop new hunting areas and hunt domesticated animals, thus becoming "problem animals" (Treves et al. 2002; Naughton-Treves et al. 2003; Treves and Karanth 2003; Kulbhushansingh et al. 2017; Athreya et al. 2015). Furthermore, as many of the large felid carnivores worldwide are specialized to hunt ungulates, more readily available and easily hunted domestic livestock often become prey (Linnell et al. 1999; Polisar 2000; Mech et al. 2000; Kolowski and Holekamp 2006; Graham et al. 2005; Odden et al. 2008; Suryawanshi et al. 2017; Schwartz 2017; Kulbhushansingh et al. 2017; Khan et al. 2018). Livestock loss in turn may motivate local humans to "retaliatorily kill" the predator or kill it to prevent further predation (Sunde et al. 1998; Linnell et al. 1999; Karanth and Madhusudan 2002; Naughton-Treves et al. 2003; Ikanda and Packer 2008; Raza et al. 2012; McManus et al. 2014; Miller et al. 2016a, b; Athreya et al. 2015; van Eeden et al. 2017). Increased perception of risk from predators negatively influences human tolerance of these cats and the likelihood of sponsoring their conservation (Treves and Karanth 2003; Ripple et al. 2014; Knopf et al. 2016; Hathaway et al. 2017). In addition to the previously noted diseases introduced by dogs, human livestock also negatively impact the health and population of both wild predators and prey due to diseases that they carry and transmit (e.g., tuberculosis, babesiosis via ticks, and rinderpest) (Tessaro 1986; Bengis et al. 2002; Dybas 2009; Sweeney and Miller 2010; Viljoen et al. 2015; Viana et al. 2015). This is particularly concerning when one considers the low genetic diversity present in many of these Felidae species. In this case, a serious or deadly introduced disease could rapidly kill an entire susceptible species or reduce their population to unsurvivable numbers (O'Brien and Johnson 2005).

Attempts to maintain carnivore wildlands and recover endangered carnivore and wild prey populations face many obstacles. Wildlife conservation funded and promoted recovery programs for endangered carnivores often ignore, or appear to minimize, the concerns and economics of local populations that face direct risk from these apex predators (Montag 2003; Jalais 2009; Dickman 2010; Karanth et al. 2012; Banarjee et al. 2013; Redpath et al. 2013; Ducarme et al. 2013; Shrestha et al. 2014; Harihar et al. 2015; Schwartz 2017). Many of the large carnivores, like tigers, have not been able to adapt well to urban landscapes and the fragmentation of their large required hunting areas (Tian et al. 2011; Bateman and Fleming 2012; Joshi et al. 2013). Ecologically motivated land use practices, such as forest regrowth programs and creation of wildlife corridors, also create sociopolitical problems in largely agrarian communities (Simberloff and Cox 1987; Bangs et al. 1998; Joshi et al. 2013). Opponents of carnivore recovery, and those that benefit economically from carnivores (international trade in cubs for the exotic pet market, trophy hunting and purveyors of "canned hunting farms," sales of exotic animal skins, as well as sales of organs and tissue for traditional medicine use), often directly oppose conservation activities by national and international governmental organizations as well as those of private conservation and animal welfare societies. These anti-recovery groups and market factors have significantly contributed to the drastic reduction of large cat populations in the modern era (Bangs et al. 1998; Landa et al. 1999; Fox 2001; Treves and Karanth 2003; Hoogesteijn and Hoogesteijn 2008;

Datta et al. 2008; Tian et al. 2011; Davidson et al. 2011; Joshi et al. 2013; Bjordal 2016; Ripple et al. 2016; Bale 2017; Angula et al. 2018; Batavia et al. 2019).

Large felid carnivore-related injuries and deaths continue to adversely affect human populations in largely agrarian and economically disadvantaged areas of the world. This negatively impacts local efforts at their conservation (Maskey et al. 2001; Packer et al. 2005; Barlow et al. 2013; Bhatia et al. 2013; Acharya et al. 2016; Bjordal 2016). In other less densely populated areas, such as North America, fatalities due to wild large felid carnivore attacks, although still relatively rare, are increasing. When an attack occurs, it is frequently highly and sensationally publicized compared with those from other wildlife (Penteriani et al. 2016; Bombieri et al. 2018). Large carnivore populations, such as those of the puma, are estimated to be increasing, but this is not the sole reason for increased reporting of attacks. Humans increasingly engage in outdoor activities around increasingly expanding cities and towns and in national wildlands. During these outdoor activities, some individuals act in manners that place them at higher risk for an adverse encounter. These activities include leaving children unattended, walking alone in wooded areas in the evening or early morning, or entering areas of prior attacks. It the case of puma attacks, it is estimated that these human behaviors contribute to approximately one half of attacks (Beier 1991; Mattson et al. 2011). This underlines the need for further objective data collection on these events to better inform the public about risk avoidance. These data will also counteract the impact of sensationalized single accounts, which adversely affect important conservation efforts and governmental management plans (McCombs and Shaw 1972; Beier 1991; Kellert et al. 1996; Conover 2008; Larue et al. 2012; Knopff et al. 2014; Chapron et al. 2014; Wolfe et al. 2014; Penteriani et al. 2016; Crown and Doubleday 2017).

2.8 Man-Eaters and Human Predation

"Man-eater" is a lay term applied to animals that have developed a pattern/habit of actively hunting human prey and incorporating human flesh into their regular diet (Corbett 1944, 1949) (see Fig. 2.7). One of the earliest depictions of a lion eating battlefield victims from wars between the Egyptians and Libyans is approximately 5000 years old (Aldred 1980; Peterhans et al. 2001). The term man-eater was probably first applied to the tiger in India in the 1800s (Campbell 1845). Jim Corbett, the legendary hunter of man-eating tigers and leopards in India, who later turned to conservation efforts, said:

> It is a tiger that has been compelled through stress of circumstances beyond its control, to adopt a diet that it alien to it. The stress is, in nine cases out of ten, wounds, and in the tenth case, old age. Human beings are not natural prey of tigers, and it is only when tigers have been incapacitated through wounds or old age that, in order to survive, they are compelled to take to a diet of human flesh. (Corbett 1944)

Fig. 2.7 The Gunsore man-eater after it was shot on April 21, 1901, by British Officer W.A. Conduitt in the village of Somnapur, Seoni District, India. The leopard was shot lying on top of the child that was the last victim (Conduitt WA, 1903). (Figure obtained from open-sourced materials available on the World Wide Web from the public domain of India. Created originally on April 22, 1901)

A single human attack when the animal is hungry or needs to provide food for their young does not equal man-eating. An incidental ingestion of a human that the animal has killed in self-defense also does not define a man-eater. Scavenging of corpses does not define a man-eater (Corbett 1949). Retaliatory leopard attacks on humans who have appropriated the leopard's kill (kleptoparasitism) also do not equal man-eating (Treves and Naughton-Treves 1999). However, these behaviors may allow the animal to lose fear of humans and habituate the animal to eating human flesh, possibly contributing to later man-eating activity (Corbett 1949; Linnell et al. 1999; Athreya 2012). Leopards, especially in the setting of habitat and wild prey loss, are attracted to human settlements by livestock and domestic dogs and cats. Dogs appear to be particularly attractive to leopards (*Panthera pardus*) (Bhattacharjee 2006; Schiess-Meier et al. 2007; Butler et al. 2014; Athreya et al. 2015; Braczkowski et al. 2018). The downside is that once in/near human habitation, their presence may also result in attacks on humans, including herdsmen, who attempt to protect their animals or are incidentally in their path (Guggisberg 1975;

McDougal 1987; Treves and Naughton-Treves 1999; Peterhans et al. 2001). While this risk is clear, a new study points out that there may be a potential public health upside of increased leopard populations in large areas of human habitation. In a study from Mumbai, their predation on stray dogs potentially decreases human exposure to rabies and injuries from dog bites (Braczkowski et al. 2018). However, this author would suggest that at least one downside is subsequent infection of these leopards with rabies, which may make them more likely to attack humans, cause significant trauma, and transmit the disease.

While a large cat that has killed a human previously in a predatory manner may attack humans again, only a small percentage of large cats become true man-eaters (Brain 1981; Löe and Röskaft 2004). Paleontological studies demonstrate that primates have been/are a natural food source for leopards, tigers, lions, and jaguars, and archeological findings suggest that developing man was no exception (Schaller and Lowther 1969; Fay et al. 1995; Lee-Thorp et al. 2000; Srivastava et al. 1996; Smilie 2002). Leopards are perhaps the oldest feline predator of man, with their bite marks found in the fossilized bones of early hominids (Brain 1970; Gommery et al. 2007). Tigers, lions, and leopards have been found to be the culprits in most reported cases of man-eating behavior among the large felid carnivores (Corbett 1944). During a 5-year period in India in the 1920s alone, 7000 human deaths from tigers were reported (McDougal 1987; Nyhus et al. 2010). In Tanzania in the 1930s alone, three generations of one pride of lions killed 1500 individuals (Rushby 1965).

Understanding the circumstances that lead large carnivores to attack humans is a crucial step in attempting to avoid human carnivore conflict and promote conservation of these unique creatures (Frank 1998). As previously noted, wild and captive felid attack data are frequently incomplete and suffer from significant flaws in collection; as such, available numbers vary widely and conclusions may be incorrect (Löe and Röskaft 2004; Shepherd et al. 2014). In the case of the Tsavo man-eaters, a number of theories have been suggested to explain their activities. One lion was found to have a severely infected canine tooth root, making it more difficult for him to hunt his traditional larger, more agile prey. Relatively hairless, soft, weaker human prey would have been much easier to catch and eat (Peterhans et al. 2001; Tucker 2009; DeSantis and Patterson 2017). This lion was found by hair analysis studies to have eaten more people, although both were also shown to have eaten other prey before their deaths (Neiburger and Patterson 2000; Peterhans et al. 2001; Patterson et al. 2003; Yeakel et al. 2009; DeSantis and Patterson 2017). Dental studies on the lions also did not show telltale evidence of bone eating, which would be seen if they had been starving. Instead, their teeth wear suggested a softer diet (DeSantis and Patterson 2017). The Tsavo brothers were also large, adult but not old, and maneless (see Fig. 2.8). In fact, not all man-eaters are old or infirm. Two studies of dentition in lions suggest that dental injuries and disease are common, although rarely accompanied by severe pathology. As such, while in rare instances severe dental injury and disease may help to explain the development of man-eating, this is probably not the only factor in most cases (Patterson et al. 2003; DeSantis and Patterson 2017). A review of "problem" lion skulls demonstrated that the majority were subadults (3–4 years old) and individuals in their prime (5–7 years old),

Fig. 2.8 The Tsavo man-eaters. Responsible for a number of deaths during construction of the Kenya-Uganda Railway from March to December, 1898. Shot by Lieutenant-Colonel John Henry Patterson. Reconstructed at the Chicago Field Museum. (Figure obtained from open-sourced materials available on the World Wide Web from the Wikimedia Foundation)

rather than older individuals. Males also predominated. As Kerbis-Peterhans and Gnoske also note in this study, subadult males are forced to leave the support structure of their birth prides at this age (Schaller 1972; Peterhans et al. 2001). Male predominance was noted in "problem" Indian lions (Saberwal et al. 1994) and leopards (Turnball-Kemp 1967).

Many human factors have also been implicated in the development of man-eating behavior among these apex predators. In the case of the Tsavo lions, an 1898 outbreak of rinderpest, a viral disease brought in with imported cattle, devastated not only local cattle but also buffalo, warthogs, and other large native herbivores that the local lions relied upon for food (Mettam 1937; Peterhans et al. 2001: Patterson 2004). At that time, depletion of keystone local elephants due to poaching allowed the overgrowth of dense thorny thickets, decreasing the availability of good grazing for herbivores and providing the big cats cover during hunting (Thorbahn 1979; Peterhans et al. 2001; Funston et al. 2001). Big cats also scavenge as a way to obtain enough protein to survive. In scavenging, they may develop a taste for humans and become accustomed to their smell (Corbett 1949). Slave caravans bound for Zanzibar used the Tsavo River crossing and abandoned slaves that were too ill to travel further or had expired during transit. Ivory caravans also left porters who were severely ill or dead from injury or illness along the route. Of note, the Uganda Railway was built along one such caravan route in Tsavo (Patterson 1907; Peterhans et al. 2001). Local Kikuyu, Somali, and Maasai tribes also abandoned their near-dead and dead in the bush (Saitoti 1980; Peterhans et al. 2001). Additionally, substantial number of Hindu and Moslem workers died during the large-scale railroad building in Tsavo and often received incomplete cremation and shallow burial (Patterson 1907; Peterhans et al. 2001). Similarly, the Rudraprayag man-eating leopard (1918–1926) began its activities after a severe outbreak of influenza. During this outbreak, so many individuals died that bodies were left unburied in the countryside. The Rudraprayag leopard initially had scavenged dead bodies (Corbett 1949; Linnell et al. 1999; Athreya 2012). Similarly, the Panar leopard's man-eating (1905–1907) followed a severe cholera outbreak, where again many bodies were left uncremated and shallowly buried (Corbett 1954). In the case of the man-eating Njombe lion pride in Tanzania (1932–1947), an outbreak of rinderpest threatened livestock in that area. The local government decided that the best way to prevent spread of the epidemic was to cull all animals capable of carrying the disease. They then culled a large portion of the available wild prey. This was followed by three generations of this local lion pride turning to man-eating (Packer 2009). One thousand five hundred deaths were attributed to Njombe pride (Rushby 1965; Packer 2009). Lionesses in the pride taught their cubs that scavenged corpses or live humans were an available meal. The cubs then followed this behavior, incorporating humans into their regular diet (Packer 2009). Man killing and man-eating have also been suggested by other researchers to be learned behaviors (Rushby 1965; Funston et al. 2001; Peterhans et al. 2001; Taylor 1959). Man-eating has also been described in human civil unrest and conflict zones, including World Wars I and II and the Vietnam War, where carnivores have had prior access to corpses (Jackson 1985; McDougal 1987; Peterhans et al. 2001). Provisioning wild carnivores with food has also been shown to correlate with man-eating. In the Gir Forest, provisioning the lions to attempt to avoid livestock predation was halted after it was followed by man-eating outbreaks (Saberwal et al. 1994). Similarly, provisioning of leopards was discontinued in several African national parks after an increase in attacks on tourists was noted in these areas (Peterhans et al. 2001) (Table 2.1).

No comprehensive global database of fatal big cat attacks exists, and many countries do not keep precise records (Shepherd et al. 2014). Tigers have been recorded in the largest number of Felid man-eating episodes and human deaths (Nowak and Paradiso 1983; Boomgaard 2001). This may represent reporting bias, as the British in India, and the Indian government going forward, have kept better records on man-eating big cats than many other countries (Corbett 1944, 1949; McDougal 1987; Boomgaard 2001; Rangarajan 2001; Daniel 2009; Nyhus et al. 2010). In many parts of Africa, local tribes kept no written records of individuals lost to lion attacks or leopard attacks (Saitoti 1980). Often, African provincial and national governments did not concern themselves with the loss of local tribal people to attack by lions and leopards (Peterhans et al. 2001). Certainly, reports of man-eaters, and animal attacks on local populations and tourists, would suggest that lions, and less

Table 2.1 Famous man-eaters. Historical information available from open-sourced materials available on the World Wide Web *(see alleged file)*

Name	Dates active	Location	# Victims	Potential contributing events	Illness/injury animal	Notes	What happened to animal	Reference
Asia								
Tigers								
Tigers of Chowgarh	1925–1930	3900 km^2 eastern Kumaon, Uttarakhand, India, killed near village Kala agar	64		Tigress's claws and canine broken, teeth worn down	2 tigers – Old mother and subadult son	Both shot by Jim Corbett	Corbett (1944)
Tigress of Champawat	Late 1800s–1907s	Western Nepal, Kumaon, Uttarakhand, India, killed near village Champawat	436	Hunter shot her and broke her teeth	Tigress's upper and lower right canines broken	Eventually began killing in broad daylight; unit Nepalese Army were sent to capture or kill tiger and failed	Shot by Jim Corbett	Corbett (1944); Huckelbridge (2019)
Tigress of Jowlagiri	1950s	Jowlagiri, Tamil Nadu, India, to Gundalam, Afghanistan, to Mysore, Karnataka, India	15	Poacher killed tigress's mate			Shot by Kenneth Anderson	Anderson (1954)

(continued)

Table 2.1 (continued)

Name	Dates active	Location	# Victims	Potential contributing events	Illness/ injury animal	Notes	What happened to animal	Reference
Tiger of Segur	1950s	District Malabar Wayanad, below Blue Mountains, Tamil Nadu, to Gudalur, Nilgiris district, Tamil Nadu, to Segur River	5		Subadult male, blind in one eye from shotgun slug		Shot by Kenneth Anderson	Anderson (1954)
Tiger of Mundachipallam	1950s	Around village Pennagaram, near Hogenakkal falls in Tamil Nadu, India	7		Healthy adult male	Killed over the body of last victim	Shot by Kenneth Anderson	Anderson (1954)
Man-eater of Calcutta	1903	Not specified	200			Only information on photo	Captured and placed in Calcutta zoo, where it died	Harper stereograph collection, https://ark.digitalcommonwealth.org/ark:/5959/sq87dj9
Man-eater of Bhimashankar	1940s – 2 yrs.	Bhimashankar Forest near Pune	100+			Only 2 bodies ever found	Shot by Ishmael, local hunter	Africa Hunting.com. Accessed January 25, 2007
Tara of Dudhwa national park	1978	Dudhwa National Park, Palia Kalan, Uttar Pradesh, India	24	Brought to area from Twycross zoo by conservationist Singh	Did not have hunting skills and associated humans with food			Africa Hunting.com. Accessed January 25, 2007

Name	Dates active	Location	# Victims	Potential contributing events	Illness/ injury animal	Notes	What happened to animal	Reference
Man-eater of Changa Nala	1966 to 1969	Changa Nala, India close to Nepal border	8 to 12		Male adult	Known as the "big terror"	Shot by Dr. David Coleman, Pasadena, California	Africa Hunting.com. Accessed January 25, 2007
Thak man-eater	Sept.–Nov. 1938	Between villages Thak, Chuka, Kot Kindri, and Sem, Uttarakhand, India	4		Female	Shot twice, wound in shoulder became septic, limiting her ability to hunt. She also hunted Corbett and his team	Last man-eater shot by Jim Corbett	Corbett (1944)
The Striped Terror of Chamala Valley	1937	Chamala Valley, Andhra Pradesh, southeastern India	7	Healthy	Female		Shot by Kenneth Anderson	Anderson (1954)
The man-eater of Yemmedoddi	1948	Town Birur, Hogar Khan foothills of the Baba Budan's mountain range, Karnataka, India	12+	Small	Male	Began by cattle raiding, wounded in jaw by villager he turned to man-eating	Shot by Kenneth Anderson	Anderson (1954)

(continued)

Table 2.1 (continued)

Name	Dates active	Location	# Victims	Potential contributing events	Illness/injury animal	Notes	What happened to animal	Reference
The Kosi man-eaters	2010–2011	Villages Ramnagar, Nullah, and Sundarkhal, buffer zone Corbett National Park, Uttarakhand, northern India	6 to 7		2 healthy adult males	Unclear what precipitated attacks	One shot by park ranger and one taken away and released	Eldredge (2014)
TI, Avni	2016–2018	Yavatmal district, eastern Maharashtra Vidarbha region	13		Female with 2 cubs	Case went to supreme court, animal rights activists appealed death sentence and wanted tiger placed in zoo	Shot by government ranger	Pandey (2018)
Leopards								
	Not stated by Anderson	Gumlapur village, 650 km^2 area in Korutla Mandal Karimnagar District, northern Telangana, Deccan plateau, Central India	42+	Human bodies left unburied after cholera epidemic	2 porcupine quills lodge between toes right forefoot	"The spotted devil of Gummalapur"	Killed by Kenneth Anderson	Anderson (1954)

Name	Dates active	Location	# Victims	Potential contributing events	Illness/ injury animal	Notes	What happened to animal	Reference
The leopard of Gummalapur	Not stated by Anderson	Jolarpettai railway town, Vellore District, Tamil Nadu, India	3		Old male with worn down teeth and blunt claws	Attacked victims in daylight	Killed by Kenneth Anderson	Anderson (1957)
The leopard of the Yelagiri Hills	1918–1925	Benji Village, road between Kedarnath and Badrinath Hindu shrines, Chamoli and Rudraprayag districts, Uttarakhand, Himalayas, India	125+	? Human bodies left unburied after 1918 flu epidemic	Began young, few scars	Male, dragged victims from homes, failed attempts to kill leopard via poisoning, traps, Gurkha soldiers	Killed by Jim Corbett on may 2, 1926	Corbett (1949)
The leopard of Rudraprayag	Early 1900s to 1910	Panar region of Almora District, Kumaon, northern India	400+	? Human bodies left unburied after cholera epidemic	Male, injured by hunter and unable to hunt usual prey	Avoided traps, took individuals from homes in front of family	Killed by Jim Corbett as he was charged	Corbett (1954)
The Panar leopard	1857–1860	Seoni District	200		Male, axe wound to head, missing toe	Usually killed with bite to throat, reputation after death of being a "were-leopard"	Killed by local shikari, Kurria	Sterndale (1877); Newman (2012)

(continued)

Table 2.1 (continued)

Name	Dates active	Location	# Victims	Potential contributing events	Illness/ injury animal	Notes	What happened to animal	Reference
The Kahani man-eater	1995–1997	Dugadda, Uttarakhand	3+				Shot by local hunter	Raheja (2015)
Poojari, the man-eater	Early 1970s	Kotdwar near Duggada	30+				Captured and sent to Lucknow zoo	Raheja (2015)
Poojari man-eater	1900–1901	Sround village Kahani, Seoni District, district of Madhya Pradesh, Central India	20+				Shot by British officer W.A. Conduitt	Conduit (1903)
The Gunsore man-eater	Early twentieth century	Central provinces of British India, covered parts Madhya Pradesh, Chhattisgarh and Maharashtra states, killings 20–30 miles apart	150+		Healthy male, mother likely man-eater	AKA "devilish cunning panther"	Killed with a gas pipe propelled projectile	Bethell (1933)

Name	Dates active	Location	# Victims	Potential contributing events	Illness/ injury animal	Notes	What happened to animal	Reference
Leopard of the central provinces	1899–1903	Mulher Valley, Dhang and Nashik districts, Maharashtra, Deccan plateau, western peninsular India	30+	Indian famine of 1899–1900	Adult male	Osmaston postulated that leopard began eating humans after killing and eating dying famine victim	Killed by LS Osmaston of Imperial forestry service	Osmaston (1904)
Leopard of the Mulher Valley	1924	Town Punanai, near Batticaloa in eastern Sri Lanka	12+		Large male, number of knife wounds from victims	Attacked several victims at time. Stuffed leopard is in National Museum of Sri Lanka in Colombo	Killed by hunter roper Shelton agar	Jayawardene (2014); Ondaatje (1992)
Leopard of Punanai	1950s	Jalahalli region, northern Bangalore, India	11 mauled, 3 killed	Caught in nets during rabbit hunt	Previously wounded by policeman	Not true man-eater. Leopard caught in nets during rabbit hunt, mauled individuals attempting to escape	Beaten by villagers, succumbed to wounds	Anderson (1954)
The killer of Jalahalli								
Africa								

(continued)

Table 2.1 (continued)

Name	Dates active	Location	# Victims	Potential contributing events	Illness/ injury animal	Notes	What happened to animal	Reference
Lions								
	1898	Tsavo River, Kenya	35 (originally 140 reported)	See text	Maneless brothers, one with significant dental disease	Bodies on display at Chicago Field Museum	Shot by John Henry Patterson	Patterson (1907); Capstick (1998)
Man-eaters of Tsavo		Kasawa, Zambia	43					Tucker (2009)
Namvelieza, the cunning one	1991	Near Msoro Mission, Luangwa River Valley, Eastern Zambia	6		Maneless lion, 10-feet long	Body on display at Chicago Field Museum	Shot by California hunter Wayne Hosek in blind	Tucker (2009)
The Mfuwe lion	1909	Chiengi, northern Rhodesia	90	Hunted with 2 other lions	Pale colored, missing half of his tail	Eluded traps and excellent marksmen	Shot in a gun trap	Tucker (2009)
Chiengi Charlie	2002–2004	8 villages around Rufiji, Tanzania	50+	Part of pride, did not hunt alone, mother ate humans	Large abscess on molar, died at 3.5 yrs. of age		Shot by game scouts, April 2004	Tucker (2009)
Osama	1929	Msoro Mission, Luangwa River Valley, Eastern Zambia	Large number			Eluded numerous traps	Disappeared	Tucker (2009)

Name	Dates active	Location	# Victims	Potential contributing events	Illness/ injury animal	Notes	What happened to animal	Reference
Msoro Monty	1932–1947	Southern Tanzania	1500+	Government had eliminated prey 2/2 Rinderpest outbreak	3 generations in large pride lions	Killed in afternoons, traveled 15–20 miles at night between villages	15 shot by game warden George Rushby, the rest left area	Rushby (1965); Tucker (2009)
Man-eaters of Njombe	1970s	Kruger National Park, South Africa at border of Mozambique	Multiple	Refugees from Mozambique famine and persecution	Multiple lions of estimated 2000 lions in park	Killed at night attempting to traverse park		Frump (2006)
Man-eaters of Eden								
Leopards								
	1899	Golis Mountains of British Somaliland	100+		Male	Lay in wait where large rocks overlooked dark, rough jungle path	Unable to be tracked in rough, rocky terrain	Swayne (1899)
Leopard of the Golis range								

so leopards, took and continue to take significant human life (Patterson 1907; Rushby 1965; Peterhans et al. 2001; Packer et al. 2005; Frump 2006; Packer et al. 2011a, b).

2.8.1 *Tigers*

The most comprehensive study of deaths due to tiger attacks estimates that at least 373,000 humans died from tiger attacks between 1800 and 2009, the majority of which occurred in South and Southeast Asia (Nyhus et al. 2010). Approximately 1000 human victims were reported in each year in India during the early 1900s. One especially active tigress, The Man-Eater of Champawat was credited with 430 deaths (Huckelbridge 2019) (see Fig. 2.9). Tigers still kill a significant number of

Fig. 2.9 The Man-Eater of Champawat. This female Bengal tiger killed 436 documented victims in Nepal and India. British Hunter Jim Corbett killed her in 1907 outside of the town of Champawat, Uttarakhand, India, just hours after she killed her final 16-year-old female victim. A postmortem of the tigress demonstrated that her right upper and lower canines were badly broken, preventing her from hunting her natural prey. A hunter had shot her in the mouth. These injuries were described to have left her in constant pain. (Figure obtained from open-sourced materials available on the World Wide Web from the public domain of India)

individuals each year. The largest number recorded, 129 individuals from 1969 to 1971, was by Bengal tigers in the Sundarbans mangrove forests extending from India to Pakistan. Unlike reports of man-eating lions and leopards, tigers are rarely reported to enter human habitations to obtain their prey (Seidensticker and Lumpkin 1990). The majority of victims are reportedly in the tiger's hunting territory during daylight hours when attacked. On average, tigers naturally make one large kill every week (Seidensticker and Lumpkin 1990). A number of theories have been advanced to explain why so many attacks occur in the Sundarbans. Habitat loss, and the relative density of tigers and humans often engaging in outdoor activities in close proximity, may lead to attacks and then man-eating (Barlow et al. 2013; Inskip et al. 2013, 2014). Tigers were also suggested to become habituated to human smell and taste by eating corpses washed into this area after major storms or major cholera outbreaks. This was then suggested to promote their attacks on live humans. Tigers in this area are forced to drink brackish water, which cause the tigers to sustain liver and renal injuries. This was suggested to cause them to be in constant discomfort and therefore more aggressive toward humans (Hendrichs 1975; Mallick 2007). Unfortunately, tigers are also some of the most endangered big cats in the wild (Seidensticker 2010; Joshi et al. 2013; Karanth 2016; Sanderson et al. 2010). Tiger populations number approximately 3000–4000 in the wild globally, with 2226 on last census in India, and they occupy fragmented areas comprising approximately 5% to 7% of their prior range (Seidensticker 2010; Joshi et al. 2013; Karanth 2016; Sanderson et al. 2010). Loss of prey species to human encroachment and hunting, deforestation and loss of habitat, exotic pet trade and pelt poaching, and, most importantly, the extensive use of tiger parts for the Asian traditional medical market have plummeted global tiger populations (Joshi et al. 2013; Karanth 2016). Due to this, governments and wildlife conservation organizations continue to attempt to prevent human loss but also conserve these iconic animals. Models have been set up in key tiger population sites to protect and attempt to allow recovery of tiger populations (Karanth et al. 1999; Karanth and Madhusudan 2002; Karanth 2002, 2016; Sanderson et al. 2010; Chanchani et al. 2015; Barlow et al. 2013). A number of different methods have been tried to decrease tiger attacks. Fresh water ponds were provided in the Sundarbans; however no change in attacks was noted. The government wildlife department issued mannequins to place in work areas, which did not show much effect. As tigers were known to attack from the rear, the government also issued rear-facing rubber figural facemasks for census takers and local individuals working in the Sundarbans to wear. The thinking was that this would cause the tigers to think that the potential victims were observing them and that they would then break off an attack. These showed some success initially and were adapted by the local population; however tigers began to attack individuals again. Presumably, they had become habituated to the masks (Chakraborty 2001; Montgomery 2001; Rishi 2018; Cirino and Safina 2016). Government census workers working in the area were also issued helmets and stiff spiked metal pads to wear over the back of the neck to prevent tiger bites from penetrating the skull and spine. Finally, the Indian government has subsidized the periodic release of domesticated animals into the mangrove swamps to provide the tigers with food (Jalais 2009).

2.8.2 Lions

Lions have been reported to attack an average of 550–750 people a year, 100 per year in Tanzania alone (Tucker 2009). Five hundred sixty-three deaths, and at least 308 injuries, were reported between 1990 and 2004 (Packer et al. 2005). This has been attributed to the significant increase in Tanzania's human population since 1988. These attacks have continued despite a fairly steady decline in lion populations (Frank 2006). Man-eating behavior in lions has been the subject of a number of studies (Peterhans et al. 2001; Frump 2006; Yeakel et al. 2009; Kushnir et al. 2010; DeSantis and Patterson 2017). Unlike tigers, man-eating lions have been documented to enter villages both at night and during the day and to force their way into thatched homes to snatch victims. Attacks may be most common at night in the days immediately following the full moon (Packer et al. 2011b). Lions become man-eaters under circumstances similar to those that drive tigers, including age and infirmity (Dickinson 2015; Baldus 2004; DeSantis and Patterson 2017), habitat encroachment, and loss of regular prey (Quigley and Herraro 2003; Löe and Röskaft 2004; Patterson 2004, 2006; Packer et al. 2005). Learned behavior from parents, and in the case of lions other pride members, and scavenging of human carcasses also promote this behavior. Specific lion populations have been found to specialize in killing certain prey populations, a behavioral tradition passed down from one generation to the next (Guggisberg 1975; Mloszewski 1983; Peterhans et al. 2001; Malkin 2003; Holden 2003). In rural Tanzania near Selous Game Reserve and in Lindi Province near the border with Mozambique, this behavior increased significantly from 1990 to 2005, with approximately 563 villagers attacked and many eaten (Kushnir et al. 2010). Many man-eating lions remain nameless. While lists of famous man-eaters include mostly males, females are actually responsible for more deaths (Tucker 2009). One wonders if this might, at least in part, be a function of the larger role that lionesses play in prey hunting in general.

2.8.3 Leopards

While leopards generally avoid humans, they are more adaptable to habitat disturbance and intrusions by humans than are lions and tigers (Athreya 2012; Bhattacharjee and Parthasarathy 2013; Athreya et al. 2013, 2015; Odden et al. 2014; Shehzad et al. 2015). The leopard population in India increased from 6830 in 1993 to 9850 in 2001, according to wildlife service reports (Singh 2005). An India-wide count of leopards in 2015 estimated the total leopard population to be 12,000–14,000 individuals (Bhattacharya 2015; Singh 2019). Unlike those of tigers, leopard ranges in India also remain largely contiguous (Bhattacharya 2015; Athreya et al. 2015). Unfortunately, their natural ranges are increasingly encroached upon, causing them to progressively access areas of human habitation for food (Quamman 2004; Athreya 2012; Dollar 2016). Optimally, a leopard needs approximately 25–40 square kilometers (10–15 $mile^2$)), with enough prey, water, and shelter, to exist (Odden and Wegge 2005). In areas where tiger populations are high or increasing, leopards are driven to live in areas closer to human habitation (Odden et al. 2010). In the Gir National Park, leopards and lions are sympatric (Singh et al. 1999). In the Himalayas, leopard and snow leopard populations overlap, but leopards usually are found at lower altitudes than snow leopards (Lovari et al. 2013). In more southern Asia, leopards are sympatric with the more arboreal clouded leopards (Borah et al. 2013).

Leopards continue to cause human fatalities yearly, especially in urban and rural areas that abut forest conservation reserves and other leopard habitats in Pakistan, India, and Nepal. Attacks are less commonly reported in Africa. This, as previously noted, probably represents underreporting bias (Maskey et al. 2001; Löe and Röskaft 2004; Athreya et al. 2004; Karanth et al. 2013; Kumar et al. 2017). A number of factors make leopards excellent human predators, although they less commonly do so than lions and tigers (Quigley and Herraro 2003; Inskip and Zimmerman 2009). They are smaller, faster, and more agile. They are also stronger and easily climb trees, often carrying heavy prey. Leopards are easily camouflaged in their natural habitats (Corbett 1944; Quamman 2004). Some, especially those that have hunted them, say they also are more cunning than the larger *Panthera* (Corbett 1944, 1949; Anderson 1954; Quamman 2004). Leopards have been reported to stalk and attack humans at night, dispatching them with a single bite to the back of the head and neck (Corbett 1949; Cohle et al. 1990; Maskey et al. 2001; Bahram et al. 2004). They have broken into homes and dragged victims from their beds (Quamman 2004; Löe and Röskaft 2004). Humans may also be injured attempting to protect domesticated cattle and dogs, which leopards often hunt when their usual prey are depleted (Linnell et al. 1999; Quamman 2004; Shehzad et al. 2015). Hunters have also become the hunted (Anderson 1954; Daniel 2009). As Peter Capstick pointed out in his book *Maneaters*:

> A man eating leopard is the most difficult and dangerous of all cats to hunt because of its' unnerving ability to reverse situations in its favor. (Capstick 1998)

It has also been suggested that leopards develop a decided taste for human flesh after corpse scavenging. Jim Corbett noted in his discussion of the man-eating leopards of Kumaon:

> A leopard, in an area in which his natural food is scarce, finding these bodies very soon acquires a taste for human flesh, and when the disease dies down and normal conditions are established, he very naturally, on finding his food supply cut off, takes to killing human beings. (Corbett 1944; Brain 1981).

Corbett also pointed out that the Rudraprayag man-eating leopard broke into a pen with 40 goats and only ate the sleeping 14-year-old boy that was guarding them (Corbett 1949) (see Fig. 2.10). In Uganda, when starving villagers expropriated leopard kills (kleptoparasitism), attacks on humans increased (Treves and Naughton-Treves 1999; Treves et al. 2002). Historically, several man-eating leopards, including the Leopard of Rudraprayag, the Panar Leopard, the Leopard of the Central Provinces, and the Leopard of the Golis Range in Somaliland, had more than 100

Fig. 2.10 British hunter Jim Corbett posed with the Rudraprayag Leopard on May 2, 1926. (Figure obtained from open-sourced materials available on the World Wide Web from the public domain of India)

human deaths attributed to them (Swayne 1899; Bethell 1933; Corbett 1944, 1949; Corbett 1954). In many parts of India in recent years, leopard attacks, including those with human fatalities, have outnumbered those from all other large carnivores combined. This may in part be a function of the decline in tiger populations, while leopard populations have increased (Kimothi 2011; Athreya et al. 2004; Maskey et al. 2001). In 2011, Raina reported in the *New York Times*:

> Several rural villages in India have suffered leopard attacks in recent months. In July alone, nearly 16 people were mauled in four different attacks across the country. The most serious was in the eastern state of West Bengal, where a leopard wandered into a village and injured 11 people. (Raina 2011)

In October of that same year, Pagnamenta reported for the Times of London:

> Leopards are terrorizing people in the Mumbai suburbs as the city's overflowing slums press further into India's biggest urban nature reserve [the 100 KM2 Sanjay Gandhi National Park]. In the northern district of Borivali, children are being kept inside after dusk as residents say leopards are feasting on stray dogs, chickens and goats, as well as at local rubbish pits. (Pagnamenta 2011)

Local wildlife experts stated that several factors appeared to contribute to this problem. Due to a near doubling of the population in the prior 20 years, at least 200,000 human beings had steadily moved into the nature reserve (Inskip and Zimmerman 2009). Thirty-five leopards were believed to live in the park at baseline. Additionally, captured "problem" leopards from other regions had been translocated into the park. It was felt that the translocated leopards, attempting to adjust to the new area, were severely stressed and unable to compete for hunting areas with the already established leopards. They then attacked feral dogs and humans (Inskip and Zimmerman 2009; Athreya et al. 2011). Traditionally, translocation and more commonly lethal control of problem animals have been the primary methods of addressing human-big cat conflict in most countries (Treves and Karanth 2003). India has more strict guidelines. Under Schedule I of the Indian Wildlife Protection Act of 1972, only confirmed man-eaters that are likely to continue to prey on humans may be killed. Man-eating is established when the animal has been confirmed to habitually stalk human beings and avoid its natural prey. Once this is established, under section 11 of the same Act, only the Chief Wildlife Warden of the state has the authority to issue in writing the reasons that the animal must be killed. Execution may only occur after all other options, including potential capture and placement in a recognized zoo, have been exhausted. At this point, an individual in the Game Department with appropriate skills, or a hired expert, is issued a permit to kill the animal (Updated Guidelines for Human-Leopard Conflict Management, the Indian Wildlife Protection Act of 1972, 2011; Athreya 2012).

Leopards have also become the hunted. Due to losses that are financially substantial for people that often subsist at the margin, farmers kill leopards to protect their livestock (Bhattacharjee 2006; Kissui 2008; Inskip and Zimmerman 2009; Shehzad et al. 2015; Athreya et al. 2015; Swanepoel et al. 2015; Muriuki et al. 2017; Kshettry et al. 2017; Naha et al. 2018). Leopards are easily baited and poisoned due to their habits of frequently following the same pathways, caching their kills in specific spots, and recurrently killing livestock and domesticated animal in the same locations (Butler et al. 2014; Athreya et al. 2015; Dobb 2018). Bush and tree clearing for agriculture and firewood also reduces the cover that leopards require to hunt and raise their cubs (Athreya et al. 2004; Marker and Dickman 2005; Minnie et al. 2015). Leopards, as well as lions and tigers, also may be caught in snares set for other animals (de Bruin 2017; Antram 2018). Unfortunately, leopards also continue to be trophy hunted, poached for the exotic pet trade, killed for their skins to supply pelts for the fashion industry, and killed for body parts to supply the Asian "medicinal" market (Nowell and Jackson 1996; Banks et al. 2006; Aryal 2009; Shankar 2008; Wassener 2012; Krofel et al. 2015; Swanepoel et al. 2015; Stein et al. 2016; Cruise 2016; Kumar et al. 2017; Aggarwal 2018).

2.8.4 *Other Large Cats*

Several of the other big cats have been implicated in accidental and predatory human attacks, but not true man-eating activity. Jaguars rarely attack humans today. When they do, the outcome is frequently fatal (Neto et al. 2011; Iserson and Francis 2015).

Jaguars have been described knocking humans from boats or leaping into the water after humans. They are excellent swimmers and can carry large prey, such as a caiman, while swimming (Neto et al. 2011). Similar to attacks by other large predators, loss and fragmentation of habitats, decreasing wild prey population, increasing human presence in wildlands during mating season and cubbing season, and intentional feeding to provide tourists a sighting are all felt to contribute to attacks (Rabinowitz 1986; Michalski et al. 2006; Neto et al. 2011; Jedrzejewski et al. 2011; Iserson and Francis 2015).

Cheetahs also rarely attack humans. They have been tamed by a number of civilizations as pets and hunting animals (O'Brien and Wildt 1986; Caro 1994). Of note, they do not coexist well with humans in the wild. Humans appear to disturb their hunting and consumption of prey. This is a serious problem for a species that are also easily deprived of prey by lions, tigers, leopards, wild dogs, and hyenas (O'Brien and Wildt 1986; Caro 1987; Laurenson and Caro 1994). Human attacks have been attributed to the animal being cornered, protecting cubs, or to animals that have become rabid (Hunter and Hinde 2005).

Pumas, despite loss of wild habitat to encroaching human habitation and agricultural expansion, rarely attack and kill humans. Attacks are however increasing in frequency (McKee 2003; Valeix et al. 2012; Knopff et al. 2014; Darimont et al. 2015; Smith et al. 2017). In pumas, prey recognition is a learned behavior from the mother. Pumas usually have not been found to perceive humans as prey (Fitzhugh 1989; Sweanor et al. 2005; Mattson et al. 2011). They are, however, associated with significant livestock loss in the Americas (de las Mercedes et al. 2017). Livestock loss often triggers retaliatory hunting. Of note, retaliatory hunting was shown to have a paradoxical effect on both livestock predation and human-puma conflict. This was attributed to loss of older pumas that had learned to avoid humans. It was suggested that these pumas were replaced by younger animals that had not yet learned human avoidance (Beier 1991; Torres et al. 1996; Peebles et al. 2013). Between 1890 and 1989, in Canada and the United States, only 53 human attacks by pumas were recorded, 10 of which resulted in human deaths. These attacks were attributed to approximately 15 animals. All of these pumas were killed, and 80% were found to be underweight or ill (Beier 1991; Mattson et al. 2011; Moussaieff Masson 2014). Most commonly, attacks have been attributed to protection of young, habituation to humans, defense of territory, the presence of attended or unattended children, or starvation due to habitat encroachment and natural prey loss (Kizer 1989; Conrad 1992; Kadesky et al. 1998; Sweanor et al. 2005). Some data would suggest that attacks are more commonly reported in late spring and summer, when juvenile pumas are forced out on their own to search for new territory (Rollins and Spencer 1995; Kertson et al. 2013). Contrary to these data however, Mattson et al. found that older pumas were more likely to kill humans than younger animals (Mattson et al. 2011). This apparent contradiction remains to be clarified.

To date, there have been no authenticated cases of a human losing a life to a snow leopard attack. Only two attacks have been reported (Heptner and Sluskii 1992; Inskip and Zimmerman 2009). One of these snow leopards was found to be toothless, emaciated, and elderly. The other was rabid. Snow leopards have been found, in a GPS collaring study of 16 individuals, to require territories significantly larger than those currently designated as protected areas in Central Asia. This is due to the low density of prey populations in the cold, mountainous regions in which they live (Mccarthy et al. 2010). Wild prey have also become dietary sources of protein to the local human populations and a target of sport hunting (Theile 2003; Mishra et al. 2016). Unfortunately, as humans have encroached on their habitat and wild prey populations have decreased, domestic animals, especially goats, now represent a small but important portion of the snow leopard diet (Bagchi and Mishra 2006; Shehzad et al. 2012; Johansson et al. 2016). One or two goats represent a significant loss to the subsistence-level agro-pastoralists that live in these areas (Ikeda 2004; Jackson et al. 2010). Even when they are cornered by herders in livestock pens, snow leopards do not attack and are easily driven away or killed (Sunquist and Sunquist 2002). This is not due to size, as snow leopards are able to kill large prey, such as argali sheep, markhor goats, deer, yaks, camels, and young horses (Shehzad et al. 2012; Johansson et al. 2015). Retaliatory killings, and the loss of snow leopards to poaching, remain significant conservation concerns, although losses may have decreased after the Chinese government issued the Protection of Wildlife Law in 1998 (Li and Lu 2013).

Clouded leopards hunt both terrestrial and arboreal prey. They are under increased survival pressure by ongoing depletion of their prey (Sunquist and Sunquist 2002; Wolf and Ripple 2016). Clouded leopards are reclusive and found most commonly in primary tropical evergreen forests, but they have been seen to venture onto logging roads to change locations and hunt (Gordon and Stewart 2007). They are primarily nocturnal but also have been found to hunt during daylight hours (Davis 1990). Clouded leopards consume a variety of animals and, where their natural prey has been depleted and human habitation encroaches on their territories, have been observed to feed on domestic animals (Chiang and Allen 2017). Humans are the main predators of the clouded leopard due to the exotic pet trade, the bush meat trade, their fur, and body parts for the traditional medicine market. They tend to avoid human, in part because hunters use dogs to track and corner them (Wilting et al. 2006; D'Cruze and Macdonald 2015; Nijman and Shepherd 2015). They have also lost significant habitat to logging and palm oil harvesting. Many remaining forested areas are too small to ensure their persistence (Wilting et al. 2006, 2010). This author was unable to find any reports of documented attacks on humans in the wild attributed to clouded leopards.

2.8.5 Current Management Strategies

Governments have focused on three main strategies to manage wild carnivores. The choice of strategy depends on the society's perception of carnivores, the society's perception of the sanctity of life, the importance to the society of the concept of endangered species, the value of carnivores in ecosystem management, and economic cost-benefit goals of preserving these carnivores (Treves and Karanth 2003;

Löe and Röskaft 2004; Inskip and Zimmerman 2009; Treves et al. 2006, 2016; Athreya et al. 2013; Chapron et al. 2014; van Eeden et al. 2017). Carnivore species have been eradicated, regulated by specific harvest, translocated, or preserved in place. Many governments have chosen to utilize a combination of these management plans over time, largely dependent on the existing biological and geopolitical landscapes (Torres et al. 1996; Breitenmoser 1998; Treves and Naughton-Treves 1999; Sunde et al. 1998; Anderson and Ozolins 2000; Rangarajan 2001; Karanth and Madhusudan 2002; Treves and Karanth 2003).

2.8.6 Eradication

In the past, governments largely eliminated problem carnivores via hunting by trained government agents or bounties paid to private hunters (Treves and Naughton-Treves 1999; Rangarajan 2001). In India alone, between 1860 and 1875, 4708 leopards and tigers were exterminated (Boomgaard 2001). At this time, campaigns to eradicate carnivore populations have largely been abandoned unless the importation of exotic carnivores threatens native fauna (Treves and Karanth 2003). Even "problem" carnivores that kill livestock, and sometimes even humans, may be non-lethally deterred, or translocated or relocated, instead of being eradicated. This is in part due to significant declines in large carnivore populations (Sunde et al. 1998; Karanth et al. 1999; Rangarajan 2001). The important ecological health role that carnivores play in regulating the population of herbivores, via the removal of weak, aged, and less biologically viable prey, has been increasingly recognized by most jurisdictions around the world (Karanth and Sunquist 1995; Treves and Karanth 2003). A review of lethal control suggested that 11 to 71% of carnivores killed to prevent conflict showed no actual evidence of having been involved in recent predation activities (Treves et al. 2016). In the case of the Rufiji man-eater, at least eight other lions were killed before the actual man-eater was shot (Baldus 2006). Studies also show that conflicts recur in the same location even after removal of a few targeted individuals (Karanth and Madhusudan 2002; Treves and Karanth 2003; Treves et al. 2016). Nonlethal methods have also been suggested to be more cost-effective in one study from South Africa (McManus et al. 2014). Also, the culture of a region that experiences big cat conflict also plays a role in the use of specific management tools (Treves et al. 2006; Dickman et al. 2013). It had been suggested that highly selective humane removal of repeat offenders might facilitate public approval of protection for the rest. This argument has met with mixed reviews by the public and conservation organizations and has placed local and state wildlife management officials in very uncomfortable positions (Fox 2001; Treves and Karanth 2003; Athreya 2012; Karanth 2013). More well-performed comprehensive studies regarding the effectiveness of this management plan are needed. Certainly, it would be preferable to humanely execute one individual rather than remove additional innocent animals by cruder cruel methods such as poisoning. It has also been suggested that elimination of problem carnivores, and preservation of those that avoid humans and their livestock, might allow this preferred behavior to be genetically passed or learned by their offspring (Jorgensen et al. 1978; Treves and Karanth 2003).

While trophy hunting, or regulated harvest or consumptive tourism, currently does not target "problem" animals, it might be thought about as another form of lethal removal or eradication of animals. Basically, it involves the payment of a fee by an individual for a hunting experience, usually guided by an expert, to kill one or more individuals of a particular species and to obtain a trophy, such as the head or entire animal, to be taken home. It is regulated by government wildlife agencies of the country in which the animal is hunted. Importation of trophies also may be regulated by the hunter's home country (Campbell 2013; Trophy Hunting by the Numbers 2016). One hundred seven different nations (104 importing and 106 exporting) participated in the trophy hunting trade between 2004 and 2014. An estimated 1.7 million trophies were traded between these nations during that period (Flocken 2016). Of these trophies, African lions showed the highest statistically significant increase in animals hunted. African leopards were also included in the six most traded (Flocken 2016). The Republic of South Africa has one of the largest hunting industries, generating just over US$341 million annually (Saayman et al. 2018). Smaller industries exist in Namibia, Zimbabwe, Botswana, and smaller still in Zambia and Mozambique. Tanzania has the largest industry in East Africa (Flocken 2016). Trophy hunting remains at the center of significant debate. Trophy hunting has been promoted to provide economic incentives to conserve large carnivores and their natural ecosystems (Baker 1997; Hurt and Ravn 2000; Lindsey et al. 2006; Packer et al. 2011a, b; Nelson et al. 2013; Lindsey et al. 2015; Di Minin et al. 2016). Trophy hunting generates more income per client than does ecotourism; however the total amount generated by this is significantly less than the billions generated by ecotourism (Lindsey 2008; Campbell 2013; Mossaz et al. 2015). Trophy hunting can also occur in remote areas. Monies generated also help to maintain hunting areas as wildlands. Less infrastructure is required for hunting than for ecotourism, with less potential effect on the local ecosystem (Lindsey et al. 2006; Buckley 2009). In some countries, e.g., Tanzania, Zimbabwe, and Zambia, lease agreements require assistance from trophy hunting companies in anti-poaching activities (Lindsey 2008). Trophy hunting has also been promoted to create local jobs and increase local income (van der Merwe et al. 2014; Naidoo et al. 2016).

Arguments against trophy hunting as a conservation method include questioning the morality of this practice, especially when animals are excessively harvested, are baited out of protected wildlife sanctuaries, and, if not immediately killed, are allowed to die in often prolonged, severe pain. In "put and take," disoriented animals may be dropped into unfamiliar areas just before the hunter is prepared to shoot them (Damm 2008). Animals may also be raised on "farms." These animals are raised to not fear humans and to be shot in small enclosures with no chance of escape or ability to protect themselves (Croes et al. 2011; Groom et al. 2014; Rosenblatt et al. 2014; van der Merwe et al. 2014; Saayman et al. 2018). Opponents of trophy hunting also point to ongoing decline in targeted predator species, especially lions. Lion populations are significantly decreasing across Africa, in part because of hunting (Packer et al. 2011a, b, 2013; Riggio et al. 2013; Lindsey et al.

2013a, b, c; Creel et al. 2016). Furthermore, in addition to population numbers, the health, behaviors, and social structure of targeted species have been found to be negatively influenced by trophy hunting (Milner et al. 2007; Loveridge et al. 2007; Davidson et al. 2011; Lindsey et al. 2013a; b, c; Peebles et al. 2013; Teichman et al. 2016; Cruise 2016; Ripple et al. 2016; Batavia et al. 2019). Trophy hunters often request male lions with the biggest, darkest mane or the largest, healthiest appearing male tiger or leopard. These are the very ones that biologically would be selected to father the most viable next generation (Briggs 2017). There is also significant concern that the monies generated from these activities largely do not go toward conservation and nor do they reach the local peoples that are at largest risk from predators and might truly financially benefit from these funds (Childs 2000; Frost and Bond 2008; Campbell 2013; Nelson et al. 2013). One group has suggested that a certification system be designed, which would rate hunting operators on their commitment to conservation, how much they benefit and involve local peoples, and the degree to which they comply with ethical standards (Lindsey et al. 2007). In lions, Creel et al. suggested that increased hunting fees due to decreasing populations and a strategy combining periods of population recovery, an age limit greater than 7 years for targeted animals, and a maximum quota of 0.5 lions shot/1000 km^2 would yield a risk of extermination of less than 10% (Creel et al. 2016).

2.8.7 *Preservation*

Preservation of big cat populations carries significantly higher investments of time and resources than does translocation or eradication of problem animals (Stander et al. 1997; Karanth and Madhusudan 2002; Shivik et al. 2003; Mishra et al. 2003; Montag 2003; DeFries et al. 2004). A number of strategies, incorporating ecological and sociocultural approaches, have been suggested to address the complexities, conflicts, and risks inherent in the preservation of these populations (Dickman 2010; Waylen et al. 2010; Redpath et al. 2013; Larson et al. 2016; Pooley et al. 2016). The stakes are high, with possible outcomes of significant economic loss due to livestock predation for local human populations and death for either human or predator populations (Ogada et al. 2003; Woodroffe et al. 2006; Balme et al. 2009; Kushnir et al. 2014). Unfortunately, at this time, a significant discrepancy exists between suggested conflict mitigation recommendations and the number of these which have actually been systematically evaluated (Madden and McQuinn 2014; Eklund et al. 2017; Krafte Holland et al. 2018). Zoning, including the development of specifically dedicated parklands encompassing the natural habitats of large carnivores and voluntary monetarily compensated resettlement of any resident humans, has been utilized by a number of countries (Chellam and Johnsingh 1993; Karanth and Madhusudan 2002; Karanth 2002; West et al. 2006; Shrestha et al. 2014). One key piece to zoning is the creation of adequate safe corridors to allow interbreeding among cat populations in these fragmented areas. This is an attempt to maximize species vigor in declining predator populations (Simberloff and Cox 1987;

Fig. 2.11 Google Earth Map of a wildlife crossing in Scotch Plains, New Jersey. These crossings have been erected in a number of countries around the world to allow connections between habitats to combat fragmentation. They also protect animals from being hit by rapidly moving vehicles and human car passengers from becoming injured when they hit a large animal. Google Earth Image downloaded by author on January 15, 2019

Rosenberg et al. 1995; Bauer et al. 2010; Joshi et al. 2013; Chanchani et al. 2015; Olsoy et al. 2016; Thapa et al. 2018). Corridors also allow animals to move into new areas where natural resources may be better than in their home area. Seasonal migration is also safer utilizing protected corridors (Rosenberg et al. 1995). The world's first interconnected jaguar sanctuary was created in Belize (Guynup 2011). A network of protected areas for tigers have been created in Myanmar (Lynam et al. 2006). Very small corridors include underpasses or overpasses, to allow animals' safe passage across busy highways (Ng et al. 2004) (see Fig. 2.11). Key pieces to zoning are to adequately understand and address the concerns of the local human populations who are being relocated and the impact on those that will remain in closely surrounding areas (Mech et al. 2000; Quamman 2004; Dickman 2010; Waylen et al. 2010; Dickman et al. 2013; Bruskotter et al. 2015; Baylis et al. 2016; Naha et al. 2018). These individuals, families, and communities should be relocated to areas that provide similar, or improved, living conditions to those of their prior homes. They should be provided adequate monetary reparation to assist in developing their new lives and as compensation for that which they have given up (Quamman 2004). If not, there will be no buy-in for predator conservation. Also key is that humans continuing to live around the new protected area should be appropriately educated about its structure and how they may appropriately interact with it (Bajracharya et al. 2006; Pooley et al. 2016). Reparations for livestock and human life lost to predation in and around protected areas must also be addressed (Treves et al. 2002; Naughton-Treves et al. 2003; Montag 2003; Dickman 2010; Jackson et al. 2010; Karanth et al. 2012; Karanth et al. 2013; Banarjee et al. 2013; Bhattacharjee and Parthasarathy 2013; Mishra et al. 2016; Muriuki et al. 2017).

Adequate reparations must be provided in a fair and timely manner if human populations are expected to coexist, without retaliation, with these large predators. Ideally, representatives of these populations would be actively included in planning the new zoning activities (Quamman 2004; West et al. 2006; Athreya 2012; Harihar et al. 2015; Macura et al. 2016).

In addition to the development and maintenance of sanctuaries and parklands for existing large felid carnivore populations, due to significant population declines, these keystone species may be reintroduced into, rewilded into, or reestablished in areas that they previously inhabited (Hale and Koprowski 2018). The first, and perhaps one of the most famous, example(s) of this is the 1200 km^2 Sariska Tiger Reserve in Rajasthan, India. This reserve was originally established in 1955, but all of the tigers were poached in 2004. Three tigers from Ranthambore Tiger Reserve were relocated in 2008 and 2009, with six more cats relocated over the next 4 years. By 2014, there were 14 tigers in the reserve (Doubleday 2018). Reintroduced tigers were seen as "foreigners" by the local population. In discussions with the local population, these new tigers were not felt to understand the boundaries that had been established over time between the prior tiger population and local human inhabitants (Doubleday 2018). This concept of learned adaptation and coexistence in carnivore populations has also been explored in the Gir Forest lion population. These lions were found to demonstrate adaptability to coexistence with humans (Rangarajan 2013). Mutual adaptations by both predator and human populations appear to be key to the establishment of long-term coexistence and need to be incorporated when planning, and evaluating, future reintroductions (Odden et al. 2014; Carter and Linnell 2016).

2.8.8 Translocation

Translocation, or the movement of "problem" carnivores to another location, has been proposed as another more humane alternative to their extermination (Boast et al. 2015). Translocation is proposed to allow these animals to live in environments less likely to provide carnivore-human conflict. These animals would still be allowed to contribute to gene pool diversity (Weise et al. 2015). Unfortunately, translocation has shown limited success, in part due to poor choice of relocation site, homing behavior, continued livestock predation, and increased aggression and lethal attacks on people. Translocation is also expensive (Stander et al. 1997; Bradley et al. 2005; Athreya et al. 2007; Massei et al. 2010; Weilenmann et al. 2010; Athreya et al. 2011; Fontúrbal and Simonetti 2012; Weise et al. 2014; Athreya et al. 2015). One small more recent experience in Namibia showed some success, with four of six leopards translocated to a new 117,613 km^2 area surviving and successfully establishing new territories (Weise et al. 2015). Both surviving females also produced cubs. When compared with 12 resident leopards in the new area, translocated leopards showed no significant difference in survival, range behavior, or likelihood of conflict. Livestock predation at their capture site decreased for 16 months, alleviating conflict with local farmers/herders (Weise et al. 2015). A study of translocated cheetahs showed different results, with decreased cheetah survival (80%), little change in predation of livestock, and significant financial costs from the process (Boast et al. 2015). A more recent review, which noted significant data gaps and need for further research, found that compensation schemes and livestock management strategies appeared to be more effective than either translocation or eradication or community interventions (Krafte Holland et al. 2018).

2.8.9 Management Strategies Going Forward

To date, success ratios suggest that livestock management and compensation schemas have shown the most promise (Krafte Holland et al. 2018). Going forward, a number of strategies have been suggested to address key issues in big cat preservation and avoidance of human-big cat conflict. In order to be effective, these strategies must meet a number of goals. Strategies should be cost-effective and have appropriately contextual funding for the community targeted (Hussain 2000; Balmford and Whitten 2003; Dickman et al. 2011; Redpath et al. 2013; Pooley et al. 2016). They should be socio-politically acceptable to all important stakeholders (West et al. 2006; Carter et al. 2012; Pooley et al. 2016). Strategies should be scientifically driven and based on verified information regarding the targeted large cat populations and their ecosystems and the culture(s), needs, and growth of adjacent human population(s) (Madden 2004; Roberge and Angelstam 2004; Bhatia et al. 2013; Redpath et al. 2015; Newsome et al. 2017; Krafte Holland et al. 2018). Ideally, schema should be evidence based, and studies should systematically evaluate each strategy in an unbiased manner. Unfortunately, this is often not the case in current studies (Krafte Holland et al. 2018). Strategies should be flexible and innovative and address multiple aspects of the problem (Redpath et al. 2013). They should include policies that limit further human encroachment on wildlife reserves, limit further protected reserve fragmentation as many of these large carnivores are wide ranging, and provide corridors to safely connect core wilderness areas (Woodroffe 2001; Frank 2006; Watson et al. 2015; Nyhus 2016). At the same time, they must validate and balance these goals with the needs of adjacent humans, limit the intersection of human and carnivore activities, and successfully mitigate conflict (Walston et al. 2010; Lindsey et al. 2015; Carter and Linnell 2016; Waldron et al. 2017; Macdonald et al. 2017; Shrivastav and Singh 2017; Krafte Holland et al. 2018; Launay and Scanlon 2018). Modern scientific tools, such as molecular genetics, multispecies ethnography, and spatial mapping, will further help to best inform survival strategies for these big cats (O'Brien and Johnson 2005; Pooley et al. 2016). Prevention and control of current and emerging disease spillover from livestock and human companion animals to shrinking wild predator populations will be crucial as they are continually driven closer together (Dybas 2009; Shrivastav and Singh 2017). They should provide education on the importance of conservation and evaluate the efficacy of these educational interventions (Krafte Holland et al. 2018). Finally, strategies should address poaching and other human predation of these big cats and conversely address those cats that have become man-killers and man-eaters. Multidisciplinary approaches that evaluate conflict from all aspects will be the most likely to succeed (Redpath et al. 2015; Ghosal and Kjosavik 2015; Pooley et al. 2016; Cronin 2018). As the late Alan Rabinowitz of Panthera, an organization dedicated to the conservation of the world's wild cats, stated:

> Unless we figure out a way that these big cats live within the human landscape, live with people or at least get to move through these human landscapes to the next protected area, then we're going to lose all our big cats eventually because all we'll end up having are disjunct populations, bio-zoos of a sort. (Rabinowitz 1986)

Rabinowitz's suggestion is fully supported by data from researchers in India studying multi-predator-human landscapes (Athreya et al. 2013, 2016).

2.8.10 Multifactorial Approach

Consensus is emerging that the use of multiple cost-effective nonlethal defenses simultaneously, and periodic modification of these defenses to avoid habituation, may be the most successful strategy (Winterbach et al. 2013; Krafte Holland et al. 2018). Cooperative research across different areas of specialization, and focused on different carnivores, may also provide common working principles that might more effectively guide predator conservation and control (Miller et al. 2016a, b). A significant amount of research has focused on the physical aspects of big cat-human conflict. One nonlethal defense that has shown promise in Africa and India as part of a multifactorial approach is the use of physical structures to protect humans and livestock (Chapron et al. 2014; Lichtenfeld et al. 2014). Stall grazing rather than open-field grazing has been suggested to be effective (Carter and Linnell 2016). Night corrals, such as the use of sustainable combined chain link/thick living thorn bush fencing to protect domesticated animals and villages from predation, have also been found to be effective (Lichtenfeld et al. 2014; Karanth et al. 2012; Hazzah et al. 2014). However, lions and leopards do attack livestock in enclosures at night (Ogada et al. 2003; Patterson et al. 2004). Enclosures without thick thorn, and with potential footholds, have not proven effective against excellent climbers, like leopards (Kolowski and Holekamp 2006). Structures must also be well built (Frank and Eklund 2017). Electric fencing did not show efficacy in areas where wolves were studied and is expensive to utilize in extremely poor agrarian communities (Reinhardt et al. 2012). The downside of enclosures is that they may lead to surplus killing of livestock that cannot escape a predator that has gained entrance (Ogada et al. 2003; Patterson et al. 2004). Light and sound devices have also shown promise in some locations (Carter and Linnell 2016). In other locations, these interventions have not shown as much promise, for reasons including cost, perception of lack of local involvement, or infringement on local freedoms (Barua et al. 2013).

Other methods of optimization of livestock management have also been studied. Ideal choices of grazing areas and types of vegetation are suggested to be different in one study depending on the big cat in question (Minnie et al. 2015). When tigers are an issue, stock should be grazed away from more isolated, forested areas. In the case of leopards and snow leopards, the study suggested avoiding early morning and late evening grazing and strengthening overnight enclosures (Minnie et al. 2015; Johansson et al. 2015; Miller et al. 2016a, b). Seasonal herding changes were found to influence the rate of lion predation in common use lands adjacent to a protected area. In the wet season when herds were farthest afield and wild prey populations were reduced, lion predation was significantly higher (Kuiper et al. 2015). Other studies have suggested that the size of livestock is important, as specific cats prefer different sized prey. Lions prefer cattle, while cheetahs prefer sheep and goats (Hayward et al. 2006; Balme et al. 2009; Pitman et al. 2013; Ghouddousi et al. 2016). Wildlife camera trapping has also been suggested as a method of proactively monitoring the activities of local predator populations, including livestock predation and the frequency of their routine proximity to human habitation (Burton et al. 2015). Another study that examined tiger-human conflict suggested that local populations should be involved in radio-collaring and monitoring the activities of potentially dangerous tigers (Gurung et al. 2008). Ongoing, real-time spatial mapping of livestock and human attack hotspots may further inform big cat management strategies to avoid predator-human conflict (Miller 2015).

2.8.11 *Increasing Wild Prey Populations*

The availability of adequate numbers of appropriately sized prey is one significant factor that determines the suitability of an area for large carnivore conservation (Fuller and Sievert 2001; Hayward and Kerley 2005; Hayward et al. 2006). Increasing wild prey populations might seem to be an easy answer for large carnivore conservation, and this has shown positive results when lions, tigers, and snow leopards have been studied (Karanth and Sith 1999; Valeix et al. 2012; Wolf and Ripple 2016). Unfortunately, human consumption of wild prey animals as a cheaper source of protein and lack of funding for conservation of prey animals continue to limit this management strategy in the poorer countries where many of the big cats are found (Balmford and Whitten 2003; Waldron et al. 2013, 2017). This strategy also was not found to be viable for all big cats. Increased populations of wild prey will attract, and increase, populations of large predators. These in turn will also prey on livestock, although usually less frequently than wild quarry (Mech et al. 2000; Stahl and Vandel 2001; Andelt 2001; Ogada et al. 2003; Treves and Karanth 2003; Graham et al. 2005; Kolowski and Holekamp 2006; Kulbhushansingh et al. 2017). Snow leopard populations exhibited this dynamic. Modeled livestock predation, due to the rise in snow leopard numbers, also increased. As such, the researchers noted that this method of snow leopard conservation would need increased local assistance for livestock protection and economic offsets for increased livestock loss (Stahl and Vandel 2001; Odden et al. 2008; Suryawanshi et al. 2017; Kulbhushansingh et al. 2017).

2.8.12 Changes in Animal Husbandry

In addition to changes in livestock containment and choices of grazing locations, alterations in animal husbandry and guarding practices may also decrease the likelihood of predation (Balme et al. 2009; Eklund et al. 2017; Amit and Jacobson 2017). Risk has been shown to increase when high densities of livestock exist, sick or pregnant animals are allowed to roam far from humans and buildings, and herds roam near natural cover in lynx, jaguar, and puma territories (Woodroffe et al. 2006; Zarco-Gonzalez et al. 2013). Results were mixed when lions were studied (Van Bommel et al. 2007). Losses also increased when animals were allowed to roam near protected areas in tiger and leopard studies (Karanth et al. 2013; Tyrrell and Western 2017). Increased loss was also found when jaguars and pumas were studied in Brazil (Palmeira et al. 2008). Use of livestock animals that are more able to defend themselves, e.g., buffalo rather than cattle, has also been suggested in Brazil and Africa (Linnell et al. 1999; Hoogesteijn and Hoogesteijn 2008; Tortato et al. 2015). Use of wild animals as livestock may also take advantage of the fact that these herbivores have been shown to avoid risky habitats as an anti-predator strategy (Laundré 2010; Thaker et al. 2011). Use of older herders rather than children and adding additional numbers of human herders also were suggested to decrease depredation (Ogada et al. 2003). The use of human guards, a relatively expensive proposition, was only studied quantitatively one time in a small population (Lindsey et al. 2013a, b, c). Use of guard dogs with shepherds also may deter predators due to their increased cost of time to hunt and ambush prey in this setting (Andelt 2001; Mukherjee and Heithaus 2013; Marker and Boast 2015). One study also looked at the role of livestock guarding dogs as predator deterrents for leopards and cheetahs. Fewer livestock were lost. One dog killed a cheetah. Two of the dogs killed smaller nontarget carnivores and 15 killed prey species. Thus, this study challenged the categorization of livestock guarding dogs as nonlethal deterrents. The authors also suggested that corrective training must be implemented for dogs that killed nontarget species (Potgieter et al. 2016). Many studies only focused on a single species of carnivore in one geographic area, thus limiting their potential application to the control of other predators. For example, the use of guard dogs has been successfully studied in cheetah and puma. However, this might be expected to worsen the problem in the case of leopards: dogs may actually attract leopard predation (Andelt 2001). Miller et al. reviewed 66 peer-reviewed publications that quantitatively measure wildlife predation. They found that deterrents and husbandry changes demonstrated the greatest potentials (depredation reduced by 42–100% in the case of husbandry) but also the widest variability in effectiveness, in all strategies studied (Miller et al. 2016a, b). Decrease in depredations is essential in promoting stakeholder support for conservation efforts (Barua et al. 2013).

2.8.13 Understanding and Managing Human Factors

Human injury and death, and livestock loss, from big cat attacks clearly result in feelings of hostility toward these carnivores and undermine public support for their conservation (Inskip and Zimmerman 2009; Barua et al. 2013). For example, in certain native African societies, such as the Maasai, while lions are regarded as part of their heritage, cattle are their currency, used for food and milk, debt settlement, dowries, and as a sign of wealth (van der Meer et al. 2015). The loss of even a few cattle to poor rural pastoralists may significantly undermine the family's ability to survive (Muriuki et al. 2017; Schwartz 2017). Many cultures also have different views of carnivores, and these need to be understood to appropriately incorporate them when devising a conservation plan that will actually succeed (Loveridge et al. 2010; Kansky and Knight 2014; Pooley et al. 2016). Many Tibetan Buddhist monasteries protect snow leopards and their habitats due to their religious and cultural values (Li et al. 2014). In Amboseli, Kenya, the Lion Guardians initiative allows young Maasai men to gain social status through the skills and income that they achieve tracking and protecting lions rather than killing them (Hazzah et al. 2014). Outreach education, active local involvement in tourism and conservation activities, re-education and removal of monetary incentives to kill big cats, and other means of translating carnivore value to the local populations may increase local buy-in for, and understanding of, carnivore conservation (Kassam 2009; Lindsey et al. 2013a, b, c; Fitzherbert et al. 2014; Mossaz et al. 2015; Steinberg 2016). The losses that local people face of life, livelihood, and peace of mind, due to large predator attacks, cannot be overemphasized (Marker 2008). These must be acknowledged and incorporated into discussions on big cat conservation, and responses to attacks and man-eating behavior, to allow both animals and humans to benefit going forward. Otherwise, these efforts will ultimately fail (Quamman 2004; West et al. 2006; Lagendijk and Gusset 2008; Winterbach et al. 2013; Redpath et al. 2015; Pooley et al. 2016). Disputes over protected area must be acknowledged and addressed (Redpath et al. 2013).

Compensation programs may provide monetary reparations for losses of livestock, medical expenses when people are attacked, or compensation to a family when a life is lost (Wagner et al. 1997; Nyhus et al. 2006; Handwerk 2013). Provision of monetary compensation and active assistance for losses of well-being, food sources, and livelihood may help to motivate positive behavioral changes in these often desperately poor local human populations (Mishra et al. 2003; Naughton-Treves et al. 2003; Bruyere et al. 2009; Dickman 2010; Dickman et al. 2011; Barua et al. 2013). Monetary compensation should be adequate and timely, as perceived undervaluation of human life by inadequate or tardy compensation will further undermine conservation efforts at the local level (Montag 2003; Baldus 2004; Loveridge et al. 2010; Barua et al. 2013). In addition to sustainability issues with high amounts of payouts, and inadequacy of loss compensation, compensation programs also suffer from difficulty in claims verification, false claims, and government corruption (Madhusudan 2003; Ravanelle and Nyhus 2017; Karanth et al.

2018). Timely access, provision and coverage of medical expenses when people are attacked may also help (Goodrich 2010; Shepherd et al. 2014). While local communities must have a stake, funding from the international conservation community may prove critical to the preservation of these predators in the poorer communities and countries directly dealing with predation (Hussain 2000; Nyhus et al. 2006; Ravanelle and Nyhus 2017). Compensation should be part of a wider conservation plan to most effectively address local predator-human conflict (Hussain 2000; Handwerk 2013). Of note however, one study looked at drivers of human engagement and attitudes toward human-carnivore relations. After analyzing current studies, the authors found that most studies measured variables with a low likelihood of actually explaining the human drivers of these conflicts. They found that intangible costs were the most important controllers of attitudes rather than intangible or tangible benefits. Sociodemographic variables including education and wealth did not explain attitudes. Exposure, prior experience with predators, and stakeholding more accurately did so. The authors concluded that further studies to better understand these intangible costs driving human-predator conflict were needed to better inform management strategies going forward (Kansky and Knight 2014). Ongoing research in this area may also help to explain why apparently scientifically sound and effective mitigation strategies are often not utilized or stopped (Pooley et al. 2016; Ravanelle and Nyhus 2017).

Voluntary resettlement, often in the wake of human casualties, with net benefits for participants has been utilized for over 40 years in India and has resulted in a substantial recovery of carnivore populations and reduction in human-carnivore conflict (Karanth et al. 1999; Karanth 2002; Karanth and Madhusudan 2002; Patterson 2006; Lagendijk and Gusset 2008; Dickman 2010; Banarjee et al. 2013). Tradition, livelihood, and free local resources motivated one studied population to resist relocation (Karanth and Ranganathan 2018). In another study, culture, educational level, and the level of amenities and potential for livelihood provided at a new location determined potential acceptance (Ramesh et al. 2019). Zoning, including specifically dedicated parklands and voluntary transparent resettlement, needs additional study to fully understand best utilization practices in different areas of the globe, including workable measures to ameliorate the ultimate toll on resettled human populations (Quamman 2004; Ramesh et al. 2019).

2.9 Conclusions

Problem large felid carnivore attacks on humans continue as a not infrequent and potentially ameliorable cause of human morbidity and mortality. Ongoing, scientifically valid research into methods to ameliorate these activities must continue (Krafte Holland et al. 2018). Preservation of remaining wild large felid carnivore populations has become a global conservation priority, as many have plummeted over the century. These cats serve a keystone function for their ecosystems, as well as being iconic representatives of our natural world (Winterbach et al. 2013). Balancing measures to ensure sustainable populations of these irreplaceable cats must be weighed against the needs of growing populations of poor and often relatively disenfranchised people, and their livestock, that live in the multiuse lands bordering protected areas (Madden 2004; Lagendijk and Gusset 2008; Karanth and Chellam 2009; Marker and Boast 2015; Pooley et al. 2016; Carter and Linnell 2016). Without national and local government commitment, and local population education and buy-in, to conservation efforts, they will fail. These processes must be based on a solid understanding of the social, economic, and political drivers of potential human-carnivore conflicts (Lynam et al. 2006; Dickman et al. 2013; Madden and McQuinn 2014; Harihar et al. 2015; Ravanelle and Nyhus 2017; Naha et al. 2018). Without local, governmental, and international commitment to end poaching losses, and the use of animal tissue for unproven "medical" uses, conservation efforts will also fail (Lindsey et al. 2015). Scientifically sound international databases need to be developed to better understand global issues in problem wildlife-human encounters. These will develop more widely translatable measures to promote attack avoidance and safer, sustainable predator-human co-existence (Shepherd et al. 2014). At the same time, specific local issues that may derail big cat conservation and human needs must also be fully understood to fully inform critical decision-making (Löe and Röskaft 2004; Miller et al. 2016a, b). More recent studies also suggest the importance of including media in wildlife management, as framing the message properly can play a large role in managing human wildlife conflict (McCombs and Shaw 1972; Bhatia et al. 2013; Crown and Doubleday 2017; Hathaway et al. 2017; Bombieri et al. 2018).

An area that remains to be adequately addressed is that of medical care for human victims that survive their initial attack. The provision of adequate, cost-effective management of attack victims that could be initiated at the local level should be encouraged. Local first aid providers might be trained to safely rescue and stabilize individuals. Stabilized individuals might then be appropriately transferred to larger regional medical centers that could provide more advanced care. Perhaps some conservation monies, and monies generated from hunting fees, could support the development and management of such care. In addition to saving lives and livelihoods for those severely injured, this might go a long way toward engendering good will among local populations entering partnership in big cat conservation efforts (Shepherd et al. 2014; Iserson and Francis 2015).

Acknowledgments and Declarations The author has received no support for this work. The author would like to acknowledge the considerable effort expended by the many field biologists; archeologists; paleontologists; anthropologists; conservation societies; local, state, and national governments; local peoples; and other interested parties, who have contributed and continue to develop the current knowledge base in this area and work on ways to promote sustainable and successful conservation of these iconic large felid carnivores.

References

Acharya KP, Paudel PK, Neupane PR et al (2016) Human-wildlife conflicts in Nepal: patterns of human fatalities and injuries caused by large mammals. PLoS One 11:e0161717. https://doi.org/10.1371/journal/pone.0161717

Aggarwal M (2018) With tigers getting the spotlight, is poaching of leopards increasing in the blind spot? https://india.mongabay.com/2018/12/with-tigers-getting-the-spotlight-is-poaching-of-leopards-increasing-in-the-blind-spot/ Accessed 5 Jan 2019

Aldred C (1980) Egyptian art. Thames and Hudson, Ltd., London

Allen WL, Cuthill IC, Scott-Samuel NE et al (2010) Why the leopard got its spots: relating pattern development to ecology in felids. Proc Royal Soc B 278(1710):1373–1380

Amit R, Jacobson SK (2017) Stakeholder barriers and benefits associated with improved livestock husbandry to prevent jaguar and puma depredation. Hum Dimens Wildl 21:427–444

Andelt WE (2001) Effectiveness of livestock guarding animals for reducing predation of livestock. Endangered Species Update 18:182–185

Anderson K (1954) Nine man-eaters and one rogue. E. P. Dutton, New York, pp 36–51

Anderson K (1957) The man-eating panther of the Yelagiri Hills. In: The Black Panther of Sivanipalli and Other Tales from the Indian Jungle. Allen & Unwin Ltd, Crows Nest, New South Wales, Australia

Anderson Z, Ozolins J (2000) First results of public involvement in wolf research in Latvia. Folia Therologica Estonica 5:7–14

Anderson M, Utter P, Szakowski J et al (2008) Cervical spine injury: tiger attack. Orthopedics 31:12–15

Andrade dos Santos J, Tokarnia CH (1960) Diseases of wild animals held in captivity. Arch Inst Biol Anim 3:1–24

Angula HN, Stuart-Hill G, Ward D et al (2018) Local perceptions of trophy hunting on communal lands in Namibia. Biol Conserv 218:26–31

Antram A (2018). https://www.rainforesttrust.org/rainforest-trust-supports-effort-to-rescue-snared-african-leopard/. Accessed 29 Oct 2018

Arribas A, Palmqvist P (1999) On the ecological connection between sabre-tooths and hominids: faunal dispersal events in the Lower Pleistocene and a review of the evidence for the first human arrival in Europe. J Archaeol Sci 26:571–585

Aryal RS (2009) CITES: implementation in Nepal and India, policy and practice, 2nd edn. Bhrikuti Academic Publications, Kathmandu

Athreya V (2012) Conflict resolution and leopard conservation in a human dominated landscape. Dissertation, Manipal University

Athreya VR, Thakur SS, Chaudhuri S, et al (2004) A study of the man-leopard conflict in the Junnar Forest Division, Pune District, Maharashtra, Office of the Chief Wildlife Warden, Maharashtra State Forest Department, and the Wildlife Protection Society of India. New Delhi, India

Athreya VR, Thakur SS, Chaudhuri S et al (2007) Leopards in human-dominated areas: a spillover from sustained translocations into nearby forests? J Bombay Nat Hist Soc 104:45–50

Athreya V, Odden M, Linnell JDC (2011) Translocation as a tool for mitigating conflict with leopards in human-dominated landscapes of India. Conserv Biol 25:133–141

Athreya V, Odden M, Linnell JDC et al (2013) Big cats in our backyards: persistence of large carnivores in a human dominated landscape in India. PLoS One 8(3):e57872. https://doi.org/10.1371/journal.pone.0057872

Athreya V, Srivathsa A, Puri M et al (2015) Spotted in the news: using media reports to examine leopard distribution, depredation, and management practices outside protected areas in southern India. PLoS One. https://doi.org/10.1371/journal.pone.0142647

Bagchi S, Mishra C (2006) Living with large carnivores: predation on livestock by the snow leopard *(Uncia uncia)*. J Zool. https://doi.org/10.1111/j.1469-7998.2005.00030.x

Bahram R, Burke JE, Lanzi GL (2004) Head and neck injury from a leopard attack: case report and review of the literature. J Oral Maxillofac Surg 62:247–249

Bajracharya SB, Furley PA, Newton AC (2006) Impacts of community-based conservation on local communities in the Annapurna Conservation Area, Nepal. Biodivers Conserv 15:2765–2786
Baker JE (1997) Trophy hunting as a sustainable use of wildlife resources in Southern and Eastern Africa. J Sustain Tourism 5:306–321
Baldus RD (2004) Lion conservation in Tanzania leads to serious human lion conflicts. With a case study of a man-eating lion killing 35 people. Tanzania Wildlife Discussion Paper No. 41. Deutsche Gesellschaft für Technische Zusammenarbeit (GTZ), Dar es Salaam
Baldus RD (2006) A man-eating lion *(Panthera leo)* from Tanzania with a toothache. Eur J Wildlife Res 52:59–62
Bale R (2017) On the trail of jaguar poachers. National Geographic, pp 92–95
Balme GA, Slotow R, Hunter LTB (2009) Impact of conservation interventions on the dynamics and persistence of a persecuted leopard (Panthera pardus) population. Biol Conserv 142:2681–2690
Balmford A, Whitten A (2003) Who should pay for tropical conservation, and how could the costs be met? Oryx 37:238–250
Banarjee K, Jala YV, Chauhan KS, Dave CV (2013) Living with lions: the economics of coexistence in the Gir Forest, India. PLoS One 8:e49457. https://doi.org/10.1371/journal.pone.0049457
Bangs EE, Fritts SH, Fontaine JA et al (1998) Status of gray wolf restoration in Montana, Idaho and Wyoming. Wildlife Soc Bull 26:785–793
Bank MS, Franklin WL (1998) Puma *(Puma concolor patagonica)* feeding observations and attacks on guanacos *(Lama guanicoe)*. Mammalia 62:599–605
Banks D, Lawson S, Wright B (2006) Skinning the cat: crime and politics of the big cat skin trade. Environmental Investigation Agency, Wildlife Protection Society of India
Barlow ACD, Udin Ahmad I, Smith JID (2013) Profiling tigers *(Panthera tigris)* to formulate management responses to human-killing in the Bangladesh Sundarbans. Wildlife Biol Pract 9:30–39
Barnett R, Barnes I, Phillips MJ et al (2005) Genomic legacy of the African cheetah, *Acinonyx jubatus*. Genome Biol 15:589–590
Barnett R, Yamaguchi N, Barnes I et al (2006) The origin, current diversity and future conservation of the modern lion *(Panthera leo)*. Proc Biol Sci 27391598:2119–2125
Barnett R, Shapiro B, Barnes IAN et al (2009) Phylogeography of lions *(Panthera leo spp.)* reveals three distinct taxa and a late Pleistocene reduction in genetic diversity. Mol Ecol 18:1668–1677
Barnett R, Yamaguchi N, Shapiro B et al (2014) Revealing the maternal demographic history of *Panthera leo* using ancient DNA and a spatially explicit genealogical analysis. BMC Evol Biol 14:70–74
Barry JC (1987) Large carnivores (Canidae, Hyaenidae, Felidae) from Laetoli. In: Leakey MD, Harris JM (eds) Laetoli: a Pliocene site in northern Tanzania. Clarendon Press, Oxford, pp 235–258
Barua M, Baghwat SA, Sushrut J (2013) The hidden dimensions of human-wildlife conflict: health impacts, opportunity and transaction costs. Biol Conserv 157:309–316
Batavia C, Nelson MP, Darimont CT et al (2019) The elephant (head) in the room: a critical look at trophy hunting. Conserv Lett 12:e12565. https://doi.org/10.1111/conf.12565
Bateman PW, Fleming PA (2012) Big city life: carnivores in urban environments. J Zool 287:1–23
Bauer H, de Longh H, Sogbohossou E (2010) Assessment and mitigation of human-lion conflict in West and Central Africa. Mammalia 74:363–367
Baylis K, Honey-Rosés J, Börner J, Corbera E et al (2016) Mainstreaming impact evaluation in nature conservation. Conserv Lett 9:58–64
Beier P (1991) Cougar attacks on humans in the United States and Canada. Wildl Soc Bull 19:403–412
Beier P, Choate D, Barrett RH (1995) Movement patterns of mountain lions during different behaviors. J Mammal 76:1056–1070
Bengis RG, Kock RA, Fischer J (2002) Infectious animal diseases: the wildlife/livestock interface. Rev Sci Tech 21:53

Berlow EL, Dunne JA, Martinez ND et al (2009) Simple prediction of interaction strengths in complex food webs. Proc Natl Acad Sci U S A 106:187–191

Bethell LA (ed) (1933) A devilish cunning panther. "Blackwood" tales from the outposts. William Blackwood and Sons, Edinburgh

Bhatia S, Athreya V, Grenyer R et al (2013) Understanding the role of representations of human-leopard conflict in Mumbai through media-content analysis. Conserv Biol 27:588–594

Bhattacharjee S (2006) Monitoring leopard *(Panthera pardus)* and its prey in different human-leopard conflict zones (high, medium and low), Pauri Garhwal, Uttaranchal. Forest Research Institute. Wildlife Institute of India, Dehradun, India

Bhattacharjee A, Parthasarathy N (2013) Coexisting with large carnivores: a case study from Western Duars, India. Hum Dimens Wildl 18:20–31

Bhattacharya A (2015) Finally, India gets a count of it is leopard numbers:12,000–14,000. https://timesofindia.indiatimes.com/home/environment/flora-fauna/Finally-India-gets-a-count-of-its-leopard-numbers-12000-14000/articleshow/48850420.cms. Accessed 2 Feb 2017

Bjordal MD (2016) Why big cats are at high risk of extinction due to their exceptional predatory abilities: what conservation strategies are needed? U Saskatchewan. Undergrad Res J. https://doi.org/10.32396/usurj.v2i2.122

Black SA, Fellous A, Yamaguchi N et al (2013) Examining the extinction of the Barbary lion and its implications for felid conservation. PLoS One 8:e60174. https://doi.org/10.1371/journal.pone.0060174

Blasco R, Rosell J, Arsuaga JL et al (2010) The hunted hunter: the capture of a lion *(Panthera leo fossilis)* at the Gran Dolina site, Sierra de Atapuerca, Spain. J Archeol Sci 37:2015–2060

Boast LK, Good K, Klein R (2015) Translocation of problem predators: is it an effective way to mitigate conflict between farmers and cheetahs Acinonyx jubatus in Botswana? Onyx 50:1–8

Bombieri G, Nanni V, del Mar Delgado M et al (2018) Content analysis of media reports on predator attacks on humans: toward an understanding of human risk perception and predator acceptance. Bio Sci 68:577–584

Boomgaard P (2001) Frontiers of fear: tigers and people in the Malay World, 1600–1950. Yale University Press, New Haven

Borah J, Sharma T, Das D et al (2013) Abundance and density estimates for common leopard *Panthera pardus* and clouded leopard Neofelis nebulosa in Manas National Park, Assam, India. Oryx 48:149–155

Braczkowski A, O'Bryan CJ, Stringer MJ et al (2018) Leopards provide public health benefits in Mumbai. India Front Ecol. https://doi.org/10.1002/fee.1776

Bradley EH, Pletscher DH, Bangs EF et al (2005) Evaluating wolf translocation as a nonlethal method to reduce livestock conflicts in the northwestern United States. Conserv Biol 9:1498–1508

Brain CK (1970) New finds at Swartkrans australopithecine site. Nature 225:1112–1119

Brain CK (1981) The hunters or the hunted: an introduction to African Cave Taphonomy. University of Chicago Press, Chicago

Brantingham PJ (1998) Hominid-carnivore coevolution and invasion of the predatory guild. J Anthropol Archaeol 17(4):327–353

Breitenmoser U (1998) Large predators in the Alps: the fall and rise of man's competitors. Biol Conserv 83:279–289

Briggs H (2017) Trophy hunting removes "good genes". BBC News. November 29, 2017. Accessed 30 Nov 2017

Bruskotter JT, Singh A, Fulton DC et al (2015) Assessing tolerance for wildlife: clarifying relations between concepts and measures. Hum Dimens Wildl 20(3):255–270

Bruyere BL, Beh AW, Lelengula G (2009) Differences in perceptions of communication, tourism benefits, and management issues in a protected area of rural Kenya. Environ Manag 43:49–59

Buckley R (2009) Evaluating the net effects of ecotourism on the environment: a framework, first assessment and future research. J Sustain Tour 17:643–672

Buckley-Beason VA, Johnson WE, Nash WG et al (2006) Molecular evidence for species-level distinctions in clouded leopards. Curr Biol 16(23):2371–2376
Burdge DR, Scheifele D, Speert DP (1985) Serious Pasteurella multocida infections from lion and tiger bites. JAMA 253:3296–3297
Burger J, Rosendahl W, Loreille O et al (2004) Molecular phylogeny of the extinct cave lion Panthera leo spelaea. Mol Phylogenet Evol 30:841–849
Burton AC, Neilson E, Moreira D et al (2015) Wildlife camera trapping: a review and recommendations for linking surveys to ecological processes. J Appl Ecol 52:675–685
Butler JR, Linnell JDC, Morrant DS et al (2014) Dog eat dog, cat eat dog: social-ecological dimensions of dog predation by wild carnivores. In: Bompper ME (ed) Free-ranging dogs and wildlife conservation. Oxford University Press, pp 117–143
Camarós E, Cueto M, Lorenzo C et al (2015) Large carnivore attacks on hominins during the Pleistocene: a forensic approach with a Neanderthal example. Archaeol Anthropol Sci.. published online. https://doi.org/10.1007/s12520-105-0248-1
Campbell W (1845) The old forest ranger; or wild sports of India on the Neilgherry Hills. Jeremiah How, London
Campbell R (2013) The $200 million question. How much does trophy hunting really contribute to African communities? A report for the African Lion Coalition. Economists at Large. Melbourne, Australia
Capitini C, Herero I (2002) Wound infection with *N. weaveri* and a novel subspecies of *P. multocida* in a child who sustained a tiger bite. Clin Infect Dis 34:e74–e76
Capstick P (1998) Maneaters. Safari Press, Huntington Beach
Carbone C, Mace G, Roberts S, Macdonald D (1999) Energetic constraints on the diet of terrestrial carnivores. Nature 402:286–288
Caro TM (1987) Cheetah mother's vigilance: looking out for prey or predators. Behav Ecol Sociobiol 20(5):351–361
Caro TM (1994) Cheetahs of the Serengeti Plains: group living in an asocial species. University of Chicago Press, Chicago
Carter NH, Linnell JDC (2016) Co-adaptation is key to co-existing with large carnivores. Trends Ecol Evol 31:575–578
Carter NH, Riley SJ, Liu J (2012) Utility of a psychological framework for carnivore conservation. Oryx 46:525–535
Castro-Prieto A, Wachter B, Sommer S (2010) Cheetah paradigm revisited: MHC diversity in the world's largest free-ranging population. Mol Biol Evol 28:1455–1468
Cerling TE, Ehleringer JR, Harris JM (1998) Carbon dioxide starvation, the development of C4 ecosystems, and mammalian evolution. Phil Trans R Soc Lond B 353:159–171
Chadwick DH (2008) Out of the shadows. National Geographic. Accessed 23 Jul 2018
Chakraborty M (2001) Census men wear masks to fool tigers. The Times of India. https://timesofindia.indiatimes.com/india/Census-men-wear-masks-to-fool-tigers/articleshow/1540929308.cms. Accessed 13 Dec 2001
Chamoli U, Wroe S (2011) Allometry in the distribution of material properties and geometry of the felid skull: why larger species may need to change and how they may achieve it. J Theoret Biol 283(1):217–226. https://doi.org/10.1016/j.jtbi.2011.o5.020
Chanchani P, Noon BR, Bailey LL, Warrier RA (2015) Conserving tigers in working landscapes. Conserv Biol 30(3). https://doi.org/10.1111/cobi.12633
Chapron G, Kaczensky P, Linnell JDC et al (2014) Recovery of large carnivores in Europe's modern human dominated landscapes. Science 346:1517–1519
Charruau P, Fernandes C, Orozco-Terwengel P et al (2011) Phylogeography, genetic structure and population divergence time of cheetahs in Africa and Asia: evidence for long-term geographic isolates. Mol Ecol 20:706–724
Chellam R, Johnsingh AJT (1993) Management of Asiatic lions in the Gir Forest, India. Symp Zool Soc London 65:409–424

Cherin M, Iurino DA, Sardella R (2013) Earliest occurrence of *Puma pardoides* (Owen, 1846) (Carnivora, Felidae) at the Plio/Pleistocene transition in western Europe: new evidence from the middle Villafranchian assemblage of Montopoli, Italy. Comptes Rendus Palevol 12:165–169
Chiang P-J, Allen MA (2017) A review of our current knowledge of clouded leopards *(Neofelis nebulosa)*. Intl J Avian Wildl Biol 2:00032. https://doi.org/10.15406/ijawb.2017.02.00032
Childs B (2000) Making wildlife pay: converting wildlife's comparative advantage into real incentives for having wildlife in African savannas, case studies from Zimbabwe and Zambia. In: Prins HET, Grootenhuis JG, Dolan TT (eds) Wildlife conservation by sustainable use. Kluwer Academic Publishers, Boston, pp 335–387
Christiansen P (2008a) Evolution of skull and mandible shape in cats (Carnivora: Felidae). PLoS One 3(7):e2807. https://doi.org/10.1371/journal.pone.0002807
Christiansen P (2008b) Phylogeny of the great cats (Felidae: Pantherinae) and the influence of fossil taxa and missing characters. Cladistics 24:97–992
Christiansen P, Wroe S (2007) Bite forces and evolutionary adaptations to feeding ecology in carnivores. Ecology 88:347–358
Chum M, Ng WP (2011) Traumatic tiger attack. J Neurosurg Pediatr 8:530–534
Cirino E, Safina C (2016) Keeping peace with predators can cut livestock deaths. National Geographic. https://blog.nationalgeographic.org/2016/10/12/keeping-peace-with-predators-can-cut-livestock-deaths/
Clottes J (2003) Chauvet cave: the art of earliest times. University of Utah Press, Utah
Cohle SD, Harlan CW, Harlan G (1990) Fatal big cat attacks. Am J Foren Med Path 11:208–212
Conduit WA (1903) A man-eating panther. J Bombay Nat Hist Soc 14:595–597
Conover MR (2008) Why are so many people attacked by predators? Hum Wildl Confl 2:139–140
Conrad L (1992) Cougar attack: case report of a fatality. J Wild Med 3:387–396
Conroy MJ, Beier P, Quigley H et al (2006) Improving the use of science in conservation: lessons from the Florida panther. J Wildl Mgmt 70:1–7
Corbett J (1944) Man-eaters of Kumaon. Oxford University Press, London, pp 8–13
Corbett J (1949) The man-eating leopard of Rudraprayag. Oxford University Press, London
Corbett J (1954) The Temple Tiger and more man-eaters of Kumaon. Oxford University Press, London
Creel S, M'soka J, Droge E et al (2016) Assessing the sustainability of African lion trophy hunting, with recommendations for policy. Ecol Appl 26. https://doi.org/10.1002/eap.1377
Croes BM, Funston PJ, Rasmussen G et al (2011) The impact of trophy hunting on lions *(Panthera leo)* and other large carnivores in the Bénoué complex, northern Cameroon. Biol Conserv 144:3064–3072
Cronin DT (2018) Implementing SMART to conserve big cats globally. https://blog.nationalgeographic.org/2018/02/28/implementing-smart-to-conserve-big-cats-globally/. Accessed 12 Nov 2018
Crown CA, Doubleday KF (2017) "Man-eaters" in the media: representation of human-leopard interactions in India across local, national, and international media. Conserv Soc 15:304–312
Cruise A (2016) The effects of trophy hunting on five of Africa's iconic wild animal populations in six countries-Analysis. Accessed 19 Sep 2017
Culver M, Johnson WE, Pecon-Slattery J, O'Brien SJ (2000) Genomic ancestry of the American puma. J Hered 91:186–197
D'Cruze N, Macdonald CW (2015) Clouded in mystery: the global trade in clouded leopards. Biodivers Conserv 24:3505–3526
Damm G (2008) Recreational trophy hunting: what do we know and what should we do? Best practices in sustainable hunting: a guide to best practices from around the world. International Council for Game and Wildlife Conservation. CIC Technical Series Publication No. 1. Budakeszi, Hungary, pp 5–11
Daniel JC (2009) The leopard in India: a natural history. Nataraj Books, Springfield, Virginia, Dehradun

Darimont CT, Fox CH, Bryan HM et al (2015) The unique ecology of human predators. Science 349:858–860
Datta A, Anand MO, Naniwadekar R (2008) Empty forests: large carnivore and prey abundance in Namdapha National Park, north-east India. Biol Conserv 141:1429–1435
Daujeard C, Geraads D, Galloti R et al (2016) Pleistocene Hominins as a resource for carnivores: a c.500,000 year old human femur bearing tooth-marks in North Africa (Thomas quarry I, Morocco). PLOS One 11(4):e0152284. https://doi.org/10.1371/journal.pone.0152284
Davidson Z, Valeix M, Loveridge AJ et al (2011) Socio-spatial behaviour of an African lion population following perturbation by sport hunting. Biol Conserv 144:114–121
Davidson Z, Valeix M, Loveridge AJ et al (2012) Environmental determinants of habitat and kill site selection in a large carnivore: scale matters. J Mammal 93:677–685
Davis RG (1990) Sighting of a clouded leopard *(Neofelis nebulosa)* in a troop of pigtail macaques *(Macaca nemestrina)* in Khao Yai National Park, Thailand. Siam Soc Nat Hist Bull 38:95–96
Davis BW, Li G, Murphy WJ (2010) Supermatrix and species tree methods resolve phylogenetic relationships within the big cats, Panthera (Carnivora: Felidae). Mol Phylogenet Evol 56:64–76
de Bruin Z (2017) Leopard cruelly caught in snare. http://www.krugerpark.co.za/krugerpark-times-5-6-leopard-cruelly-caught-in-snare-24932html.html. Accessed 1 Jan 2019
de las Mercedes Guerisoli M, Vidal EL, Franchini M et al (2017) Characterization of puma-livestock conflicts in rangelands of Central Argentina. Royal Soc Open Sci 4(12). https://doi.org/10.1098/rsos.170852
DeFries RS, Foley JA, Asner GP (2004) Land use choices: balancing human needs and ecosystem function. Front Ecol Environ 2:249–257
DeSantis LRG, Patterson BD (2017) A dietary behavior of man-eating lions as revealed by dental microwear textures. Sci Rep 7:904. https://doi.org/10.1038/s41598-017-00948-5
Di Minin D, Leader-Williams N, Bradshaw CJA (2016) Banning trophy hunting will exacerbate biodiversity loss. Trends Ecol Evol 31:99–102
Dickinson D. (2015) Toothache made lion eat humans'. http://news.bbc.co.uk/2/hi/africa/3756180.stm. Accessed 1 Nov 2015
Dickman AJ (2010) Complexities of conflict: the importance of considering social factors for effectively resolving human-wildlife conflict. Anim Conserv 13:458–466
Dickman AJ, Macdonald EA, Macdonald DW (2011) A review of financial instruments to pay for predator conservation and encourage human-carnivore co-existence. Proc Nat Acad Sci 108:13937–13944
Dickman A, Marchini S, Manfredo M (2013) The human dimension in addressing conflict with wild carnivores. Key Topics Conserv Biol 2:110–126
Diedrich CG (2013) Late Pleistocene leopards across Europe-northernmost European German population, highest elevated records in the Swiss Alps, complete skeletons in the Boznia Herzegowina Dinarids and comparison to the Ice Age cave art. Quart Sci Rev 76:167–193
Dobb E (2018) Why poison is a growing threat to Africa's wildlife. National Geographic. https://www.nationalgeographic.com/magazine/2018/08/poisoning-africa-kenya-maasai-pesticides-lions-poachers-conservationists/. Accessed 23 Aug 2018
Dobrynin P, Liu S, Tamazian G et al (2015) Genomic legacy of the African cheetah, *Acinonyx jubatus*. Genome Biol 16:1–20
Dollar L (2016) Leopards of India's Silicon City. https://blog.nationalgeographic.org/2016/01/04/leopards-of-indias-silicon-city/. Accessed 3 Mar 2017
Doubleday K (2018) Human-Tiger (re)negotiations: a case study from Sariska Tiger Reserve, India. Soc Animals 26:148–170
Ducarme F, Luque GM, Courchamp F (2013) What are "charismatic species" for conservation biologists. BioSci Master Rev 10:1–8
Dybas CL (2009) Infectious diseases subdue Serengeti lions. Bioscience 59:8–13
Eaton RL (1970) Hunting behavior of the cheetah. J Wildl Mgmt 34:56–67

Eklund A, Lopez-Bao JV, Tourani M et al (2017) Limited evidence on the effectiveness of interventions to reduce livestock predation by large carnivores. Sci Rep 7:2097. https://doi.org/10.1038/s42598-017-02323-w.pmid:28522834

Eldredge N (2014) The Kosi Maneaters. http://www.nileseldredge.com/the-kosi-maneaters.html. Accessed 2 Jan 2019

Elliott JPI, Cowan MT, Holling CS (1977) Prey capture by the African lion. Can J Zool 55:1811–1828

Emami P, Kaiser TM, Regelsberger J et al (2012) Case report: surviving a tiger attack. Neurosurg Rev 35:621–624

Emmons LH (1987) Comparative feeding ecology of felids in a neotropical rainforest. Beh Ecol Sociobiol 20:271–283

Ernest HB, Wayne RK, Dalbeck L et al (2014) Individual behaviors dominate the dynamics of an urban Mountain Lion population isolated by roads. Curr Biol 24:1989–1994

Estes R (1991) The behavior guide to African mammals, including hoofed mammals, carnivores, primates. University of California Press, Los Angeles

Estes JA, Terborgh J, Brashares JS, Power ME, Berger J et al (2011) Trophic downgrading of planet earth. Sci 333:301–306

Everton BP, Miranda J, De Menezes FS et al (2016) Reptiles as principal prey? Adaptations for durophagy and prey selection by jaguar *(Panthera onca)*. J Nat Hist 50:31–32

Farhadinia MS, Hosseini-Zavarei F, Nezami B et al (2012) Feeding ecology of the Asiatic cheetah *Acinonyx jubatus venaticus* in low prey habitats in northeastern Iran: implications for effective conservation. J Arid Environ 87:206–211

Fay JM, Carroll R, Kerbis-Peterhans JC et al (1995) Leopard attack on and consumption of gorillas in the Central African Republic. J Hum Evol 29:93–99

Figuero HV, Li G, Trindade FJ et al (2017) Genome-wide signatures of complex introgression and adaptive evolution in the big cats. Sci Adv 3(7):e1700299. https://doi.org/10.1126/sciadv.1700299

Fitzherbert E, Caro T, Johnson PJ et al (2014) From avengers to hunters: leveraging collective action for the conservation of endangered lions. Biol Conserv 174:84–92

Fitzhugh EL (1989) Managing with potential for lion attacks against humans. In: Proceedings of the Third Mountain Lion Workshop. December 5–8, 1988

Flocken J (2016) Killing for trophies: report analyses trophy hunting around the world. https://d1jyxxz9imt9yb.cloudfront.net/resource/36/attachment/regular/Killing_For_Trophies.pdf. Accessed 2 Nov 2017

Fontúrbal FE, Simonetti JA (2012) Translocations and human-carnivore conflicts problem solving or problem creating? Wildl Biol 17:217–224

Fox CH (2001) Taxpayers say "no" to killing predators. Animal Issues 32:1–2

Frank LG (1998) Living with lions: carnivore conservation and livestock in Laikipia District, Kenya. United States AID contract number 623-0247-C-00-3002-00. Mpala Research Centre, Nanyuki, Kenya

Frank LG (2006) Lions, conflict and conservation in East Africa. Felid Conservation and Biology Conference. January 2006. http://www.felidae.org/JOBURG/LIONS,%20CONFLICT%20AND%20CONSERVATION.doc

Frank J, Eklund A (2017) Poor construction, not time, takes its toll on subsidized fences designed to deter large carnivores. PLoS One 12(4):e0175211. https://doi.org/10.1371/journal.pone.0175211

Frost PGH, Bond I (2008) The CAMPFIRE programme in Zimbabwe: payments for wildlife services. Ecol Econ 65:776–787

Frump RR (2006) The man-eaters of eden: life and death in Kruger National Park. The Lyons Press, Lanham

Fuller TK, Sievert PR (2001) Carnivore demography an the consequences of changes in prey availability. In: Gittleman JL, Funk SM, Macdonald DW, Wayne RK (eds) Carnivore conservation. Cambridge University Press, Cambridge, pp 163–178

Funston PJ, Mills MGL, Biggs HC (2001) Factors affecting hunting success of male and female lions in the Kruger National Park. J Zool 253:419–431
Ghezzo E, Rook L (2015) The remarkable Panthera pardus (Felidae, Mammalia) record from Equi (Massa, Italy): taphonomy, morphology, and paleoecology. Quart Sci Rev 110:131–151
Ghosal S, Kjosavik DJ (2015) Living with leopards: negotiating morality and modernity in Western India. Soc Nat Res 28:1092–1107
Ghouddousi A, Soofi M, Hamidi AK, Lumetsberger T et al (2016) Assessing the role of livestock in big cat prey choice using spatiotemporal availability patterns. PLoS One 11(4):e0153439
Godfrey D, Lythgoe JM, Rumball DA (1987) Zebra stripes and tiger stripes: the spatial frequency distribution of the pattern compared to that of the background is significant in display and crypsis. Biol J Linnean Soc 32:427–433
Gommery G, Pickford M, Senut B (2007) A case of carnivore-inflicted damage to fossil femur from Swartkrans, compared to that on a hominid femur representing *Orrorin tugenensis*, BAR 1003′00 (Kenya). Ann Transvaal Museum 44:215–218
Goodrich JM (2010) Human-tiger conflict: a review and call for comprehensive plans. Integr Zool 5(4). https://doi.org/10.1111/j.1749-4877.2010.00218.x
Gordon IJ (2009) What is the future for wild, large herbivores in human-modified agricultural landscapes? Wildl Biol 15:109
Gordon CH, Stewart AE (2007) The use of logging roads by clouded leopards. Cat News 47:12–13
Gottelli D, Wang J, Bashir S, Durant SM (2007) Genetic analysis reveals promiscuity among female cheetahs. Proc Royal Soc B 274:1993–2001
Graham K, Beckerman AP, Thirgood S (2005) Human predator-prey conflicts: ecological correlates, prey losses and patterns of management. Biol Conserv 122:159–171
Grange S, Duncan B (2006) Bottom-up and top-down processes in African ungulate communities: resources and predation acting on the relative abundance of zebra and grazing bovids. Ecography 29:899–907
Grant R (2016) The return of the great American jaguar. https://www.smithsonianmag.com/science-nature/return-great-american-jaguar-180960443/
Grayson DK, Meltzer DJ (2003) A requiem for North American overkill. J Archaeol Sci 30:585–593
Groom RJ, Funston PJ, Mandisodza R (2014) Surveys of lions Panthera leo in protected areas in Zimbabwe yield disturbing results: what is driving the population collapse? Oryx 48:385–393
Guggisberg CAW (1975) Clouded leopard *Neofelis nebulosa*. In: Wild cats of the world. Taplinger Publishing, New York, pp 125–130. isbn:0-8008-8324-1
Gurung B, Smith J, Mcdougal C et al (2008) Factors associated with human-killing tigers in Chitwan National Park, Nepal. Biol Conserv 141:3069–3078
Guynup S (2011) The Jaguar freeway. https://www.smithsonianmag.com/science-nature/the-jaguar-freeway-73586097/. Accessed 18 Sep 2017
Hale SL, Koprowski JL (2018) Ecosystem-level effects of keystone species reintroduction: a literature review. Restor Ecol 26:1–7
Hall MI, Kamilar JM, Kirk EC (2012) Eye shape and the nocturnal bottleneck of mammals. Proc Biol Sci 279:4962–4968
Hampson K, Coudeville L, Lembo T et al (2015) Estimating the global burden of endemic canine rabies. PLoS Neglected Trop Dis 9(4):30003709
Handwerk B (2013) Predator insurance: when livestock becomes prey, conservationists pay. https://www.nationalgeographic.com/news/2013/12/131210-wildlife-conservation-compensation-predators-science/. Accessed 10 Dec 2017
Harihar A, Verissimo D, MacMillan DC (2015) Beyond compensation: integrating local communities; livelihood choices in large carnivore conservation. Glob Environ Chang 33:122–130
Harihar A, Ghosh-Harihar M, MacMillan DC (2018) Losing time for the tiger Panthera tigris: delayed action puts a globally threatened species at risk of total extinction. Oryx 52:78–88
Hartstone-Rose A, Perry JM, Morrow CJ (2012) Bite force estimation and the fiber architecture of felid masticatory muscles. Anat Rec 295:1336–1351
Hast MH (1989) The larynx of roaring and non-roaring cats. J Anat 163:117–121

Hathaway RS, Bryant A-EM, Draheim MM et al (2017) From fear to understanding. Changes in media representation of leopard incidence after media awareness workshops in Mumbai, India. J Urban Ecol 3:1–6

Hayward MW, Kerley GIH (2005) Prey preferences of the lion (Panthera leo). J Zool 267:309–322

Hayward MW, Hofmeyr M, O'Brien SJ, Kerley GIH (2006) Prey preferences of the cheetah *(Acinonyx jubatus)* (Felidae: Carnivora): morphological limitations or the need to capture rapidly consumable prey before kleptoparasites arrive. J Zool 270:615–627

Hazani R, Buntic RF, Brooks D (2008) Microsurgical scalp reconstruction after a mountain lion attack. Ann Plastic Surg 61:265–268

Hazzah L, Dolrenry S, Naughton L, Edwards CTT, Mwebi O et al (2014) Efficacy of two lion conservation programs in Maasailand, Kenya. Conserv Biol 28:851–860

Hejna P (2010) A fatal leopard attack. J Forensic Sci 55(3):832–834

Hemmer H (1968) Researching the phylogenetic history of the Pantherinae II: studies into the ethology of the clouded leopard Neofelis nebulosa and snow leopard *Uncia uncia*. Veröffentlichungen der Zoologischen Staatssammlung München 12:155–247

Hemmer H, Kahlke RD (2005) Nachweis des Jaguars (Panthera onca gombaszoegensis aus dem späten Unter-oder frühen Mittelpleistozän der Niederlande. Deinsea. Annual of the Natural History Museum, Rotterdam, pp 47–57

Hemmer H, Kahlke RD, Vekua AK (2001) The Jaguar-Panthera onca gombaszoegensis (Kretzoi, 1938) (Carnivora:Felidae) in the late Lower Pleistocene of Akhalkalaki (South Georgia; Transcaucasia) and its evolutionary and ecological significance. Geobias 34:475–486

Hemmer H, Kahlke RD, Vekua AK (2010) *Panthera onca georgica* spp. novi. From the early Pleistocene of Dmanisi (Republic of Georgia) and the phylogeography of jaguars (Mammalia, Carnivora, Felidae). Neues Jahrbuch für Geologie und Paläontologie-Abhandlungen 257:115–127

Hendrichs H (1975) The status of the tiger Panthera tigris in the Sundarbans mangrove forest (Bay of Bengal). Saugertierliche Mitteilungen 23:161–199

Heptner VG, Sluskii AA (1992) Mammals of the Soviet Union, vol 2, Part 2. Amerind Publishing, New Delhi, p 308

Hilborn A, Pettorelli N, Caro T et al (2018) Cheetahs modify their prey handling behavior depending on risks from top predators. Behav Ecol Sociobiol 72(4). https://doi.org/10.1007/s00265-018-2481-y

Hildebrand M (1961) Further studies on locomotion of the cheetah. J Mammal 42:84–96

Holden C (2003) The man-eating habit. Science 299:657a. https://doi.org/10.1126/science.299.5607.657a

Holland JS (2013) Fossil find clears up big cat origins. https://www.nationalgeographic.com/news/2013/11/131112-big-cats-origin-tibet-animals-science/

Hoogesteijn R, Hoogesteijn A (2008) Conflicts between cattle ranching and large predators in Venezuela: could use of water buffalo facilitate felid conservation? Oryx 42:132–138

Hopcraft GC, Sinclair ARE, Packer C (2005) Planning for success: serengeti lions seek prey accessibility rather than abundance. J Animal Ecol 74:559–566

Huckelbridge D (2019) No beast so fierce: the terrifying true story of the Champawat Tiger, the deadliest man-eater in history. William Morrow, New York City

Hudson PE, Corr SA, Payne-Davis RC et al (2011) Functional anatomy of the cheetah *(Acinonyx jubatus)* hindlimb. J Anat 218:363–374

Hunt RM (1987) Evolution of the aeluroid Carnivora: diversity of the earliest Aeluroids from Eurasia. Am Mus Novi 2886:1–74

Hunter L, Hinde G (2005) Cats of Africa: behaviour, ecology, and conservation. Struik Publishers, Cape Town, pp 1–172

Hurt R, Ravn P (2000) Hunting and its benefits: an overview of hunting in Africa with special reference to Tanzania. In: Prins HHT, Grootenhuis JG, Dolan TT (eds) Wildlife conservation by sustainable use, Conservation biology series, vol Vol 12. Springer, Dordrecht

Hussain S (2000) Protecting the snow leopard and enhancing farmer's livelihoods: a pilot insurance scheme in Baltistan. Mountain Res Dev 20:226–231

Husseman JS, Murray DL, Power G et al (2003) Assessing differential prey selection patterns between two sympatric large carnivores. Oikos 101:591–601

Ikanda D, Packer C (2008) Ritual vs retaliatory killing of African lions in the Ngorongoro Conservation Area, Tanzania. Endangered Species Res 6:67–74

Ikeda N (2004) Economic impacts of livestock depredation by snow leopard *Uncia uncia* in the Kanchenjunga Conservation Area, Nepal Himalaya. Environ Conserv 31:322–330

Inskip C, Zimmerman A (2009) Human-felid conflict: a review of patterns and priorities worldwide. Oryx 43:18–34

Inskip C, Ridout M, Fahad Z et al (2013) Human-tiger conflict in context: risks to lives and livelihoods in the Bangladesh Sundarbans. Hum Ecol 41:169–186

Inskip C, Fahad Z, Tully R et al (2014) Understanding carnivore killing behaviour: exploring the motivation for tiger killing in the Sundarbans, Bangladesh. Biol Conserv 180:42–50

Iserson KV, Francis AM (2015) Jaguar attack on a child: case report and literature review. West J Emerg Med 16:303–309

Isolato P, Edgar D (2000) Polymicrobial tenosynovitis with Pasteurella multocida and other gram negative bacilli after a tiger bite. J Clin Pathol 51:871–874

Jackson P (1985) Man-eaters. Int Wildl 15:4–11

Jackson RM, Mishra C, McCarthy TM et al (2010) Snow leopards: conflict and conservation. Biol Conserv Wild Felids 1:417–430

Jacobson AP, Gerngross P, Lemeris JR Jr et al (2016) Leopard *(Panthera pardus)* status, distribution, and the research efforts across its range. Peer J 4:e1974. https://doi.org/10.7717/peerj.1974. PMC 4861552

Jalais A (2009) Forest of tigers: people, politics and environment in the Sundarbans. Routledge, Taylor and Francis Group, New York

Janczewski DN, Modi WS, Stephens JC (1995) Molecular evolution of mitochondrial 12S RNA and cytochrome B sequences in the pantherine lineage of Felidae. Mol Biol Evol 12:690–707

Janecka JE, Zhang Y, Li D et al (2017) Range-wide snow leopard phylogeography supports three subspecies. J Hered 108:597–607

Jayawardene J (2014) The man-eater of Punanai. http://archives.sundayobserver.lk/2014/03/02/spe10.asp. Accessed 18 Sep 2017

Jedrzejewski W, Abarca M, Viloria A et al (2011) Jaguar conservation in Venezuela against the backdrop of current knowledge on its biology and evolution. Interciencia-Caracas 2011:36954–36966

Johansson Ö, McCarthy T, Samelius G et al (2015) Snow leopard predation in a livestock dominated landscape in Mongolia. Biol Conserv 184:251–258

Johansson Ö, Rauset GR, Samelius G et al (2016) Sharing is essential for snow leopard conservation. Biol Conserv 203:1–7

Johnson WE, O'Brien SJ (1997) Phylogenetic reconstruction of the Felidae using 16S rRNA and NADH-5 mitochondrial genes. J Mol Evol 44:S98–S116

Johnson WE, Eizirik E, Pecon-Slattery J et al (2006) The late Miocene radiation of modern Felidae: a genetic assessment. Science 311:73–77

Jorgensen CJ, Conley RH, Hamilton RJ et al (1978) Management of black bear depredation problems. Proc Eastern Workshop Black Bear Manag Res 4:297–321

Joshi A, Vaidyanathan S, Mondol S et al (2013) Connectivity of tiger *(Panthera tigris)* populations in the human-influenced forest mosaic of Central India. PLoS One 8(11):e77980. https://doi.org/10.1371/journal.pone.0077980

Kadesky K, Manarey C, Blair GK et al (1998) Cougar attacks on children: injury patterns and treatment. Pediatr Surg 33:863–865

Kansky R, Knight AT (2014) Key factors driving attitudes towards large mammals in conflict with humans. Biol Conserv 179:93–105

Karanth KU (2002) Nagarahole: limits and opportunities in wildlife conservation. In: Terborgh J, van Schalik CP, Rao M, Davenport LC (eds) Making parks work: identifying key factors in implementing parks in the tropics. Island Press, Covelo, pp 189–202
Karanth U (2013) Man-eaters- when caring less may actually help. http://www.conservationindia.org/resources/man-eaters-when-caring-less-may-actually-help. Accessed 15 Sep 2017
Karanth KU (2016) Tracking tigers. Sci Am 315:54–59
Karanth KU, Chellam R (2009) Carnivore conservation at the crossroads. Oryx 43:1–2
Karanth KU, Madhusudan MD (2002) Mitigating human-wildlife conflicts in southern Asia. In: Terborgh J, van Schalik CP, Rao M, Davenport LC (eds) Making parks work: identifying key factors in implementing parks in the tropics. Island Press, Covelo, pp 250–264
Karanth KK, Ranganathan P (2018) Assessing human-wildlife interactions in a forest settlement in Sathyamangalam and Mudumalai Tiger Reserves. Trop Conserv Sci 11:1–14
Karanth KU, Sith BM (1999) Prey depletion as a critical determinant of tiger population viability. In: Seidensticker J, Christie S, Jackson P (eds) Ridding the Tiger. Tiger conservation in human-dominated landscapes. Cambridge University Press, Cambridge, pp 100–122
Karanth KU, Sunquist ME (1995) Prey selection by Tiger, Leopard and Dhole in tropical forests. J Animal Ecol 64:439–450
Karanth KU, Sunquist ME (2000) Behavioural correlates of predation by tiger *(Panthera tigris)*, leopard *(Panthera pardus)* and dhole *(Cuon alpinus)* in Nagarahole, India. J Zool 250:255–265
Karanth KU, Sunquist ME, Chinnappa KM (1999) Long term monitoring of tigers: lessons from Nagarahole. In: Seidensticker J, Christie S, Jackson P (eds) Riding the Tiger: tiger conservation in human-dominated landscapes. Cambridge University Press, Cambridge, pp 114–122
Karanth KK, Gopalaswamy AM, DeFries R et al (2012) Assessing patterns of human-wildlife conflicts and compensation around a central Indian protected area. PLoS One 7(12):e50433. https://doi.org/10.1371/journal.pone.0050433
Karanth KK, Naughton-Torres L, DeFries RS et al (2013) Living with wildlife and mitigating conflicts around three Indian protected areas. Environ Manag 52:1320–1332
Karanth KK, Gupta S, Vanamamalai A (2018) Compensation payments, procedures and policies towards human-wildlife conflict management: insights from India. Biol Conserv 227:383–389
Kartzinel TR, Chen PA, Coverdale TC et al (2015) DNA metabarcoding illuminates dietary niche partitioning by African large herbivores. PNAS 112(26):8019–8024
Kellert SR, Black M, Rush CR et al (1996) Human culture and large carnivore conservation in North America. Conserv Biol 10:977–990
Kertson BN, Spencer RD, Grue CE (2013) Demographic influences on cougar residential use and interactions with people in western Washington. J Mammal 94:269–281
Khan U, Lovari S, Shah SA et al (2018) Predator, prey and humans in a mountain area: loss of biological diversity leads to trouble. Biodivers Conserv 27:2795–2813
Kimothi P (2011) Losers on both sides as man-animal war rages. https://web.archive.org/web/20110302143321/; http://www.dailypioneer.com/315623/Losers-on-both-sides-as-man-animal-war-rages.html. Accessed 15 Sep 2017
Kissui BM (2008) Livestock predation by lions, leopards, spotted hyenas, and their vulnerability to retaliatory killing in the Maasai Steppe, Tanzania. Anim Conserv 11:422–432
Kitchener AC, Yamaguchi N (2010) What is a tiger? Biogeography, morphology, and taxonomy. In: Tilson R, Nyhus PJ (eds) Tigers of the world: the science, politics and conservation of Panthera tigris, 2nd edn. Academic Press, London, pp 53–84
Kitchener AC, Beaumont MA, Richardson D (2006) Geographical variation in the clouded leopard, *Neofelis nebulosa*, reveals two species. Curr Biol 16:2377–2383
Kitchener AC, Driscoll CA, Nobuyuki Y (2016) What is a snow leopard? Taxonomy, morphology, and phylogeny. In: McCarthy T, Mallon D (eds) Snow leopards. Biodiversity of the world: conservation from genes to landscapes. Elsevier Science, Oxford, pp 3–11
Kizer KW (1989) *Pasteurella multocida* infection from a cougar bite. West J Med 150:87–90
Klemuk SA, Riede A, Walsh EJ et al (2011) Adapted to roar: functional morphology of tiger and lion vocal folds. PLoS One 6(11):e27029

Knopf AA, Knopf KH, St Clair CC (2016) Tolerance for cougars diminish by high perception of risk. Ecol Soc 21:33

Knopff KH, Webb NF, Boyce MS (2014) Cougar population status and range expansion in Alberta during 1991–2010. Wildl Soc Bull 38:116–121

Kohout MP, Percy J, Sears W, Yeo JD (1989) Tiger mauling: fatal spinal injury. Aus NZ Surg 59:505–506

Kolowski J, Holekamp KEK (2006) Spatial, temporal, and physical characteristics of livestock depredations by large carnivores along a Kenyan reserve border. Biol Conserv 128:529–541

Krafte Holland K, Larson LR, Powell RB (2018) Characterizing conflict between humans and big cats Panthera spp: a systematic review of research trends and management opportunities. PLoS One 13(9):e0203877. https://doi.org/10.1371/journal.pone.0203877

Krofel M, Treves A, Ripple WJ et al (2015) Hunted carnivores at outsized risk. Science 350:518–519

Kshettry A, Vaidyanathan S, Athreya V (2017) Leopard in a tea-cup: a study of leopard habitat use and human-leopard interactions in North-Eastern India. PLoS One 12(5):e0177013. https://doi.org/10.1371/journal.pone.0177013

Kuiper TR, Loveridge AJ, Parker DM et al (2015) Seasonal herding practices influence predation on domestic stock by African lions along a protected area boundary. Biol Conserv 191:546–554

Kulbhushansingh RS, Redpath SM, Bhatnagar YV et al (2017) Impact of wild prey availability on livestock predation by snow leopards. R Soc Open Sci 4(6):170026

Kumar P, Chandel S, Kumar V et al (2017) Leopard-human conflict led casualties and conservation awareness campaign in Shivalik Hills of Northern India. Proc Nat Acad Sci India Sect B Biol Sci 87:893–898

Kurtén B (1973) The age of mammals. Columbia University Press, New York

Kurtén B, Anderson E (1980) Pleistocene mammals of North America. Columbia University Press, New York

Kushnir H, Leitner H, Ikanda D et al (2010) Human and ecological risk factors for unprovoked lion attacks on humans in southeastern Tanzania. Hum Dimens Wildl 15:315–331

Kushnir H, Weisberg S, Olson E et al (2014) Using landscape characteristics to predict risk of lion attacks on humans in South-Eastern Tanzania. African J Ecol 52:524–532

Lagendijk DDG, Gusset M (2008) Human-carnivore coexistence on communal land bordering the greater Kruger area, South Africa. Environ Manag 42:971. https://doi.org/10.1007/s00267-008-9204-5

Landa AK, Gudvangen JE, Swenson JE et al (1999) Factors associated with wolverine *(Gulo gulo)* predation on domestic sheep. J Appl Ecol 36:963–973

Larson LR, Conway AL, Krafte KE et al (2016) Community-based conservation as a potential source of conflict around a protected area in Sierra Leone. Environ Conserv 43:242–252

Larue MA, Nielsen CK, Dowling M, Miller K et al (2012) Cougars are recolonizing the Midwest: analysis of cougar confirmations during 1990–2008. J Wildl Manag 76:1364–1369

Launay F, Scanlon JE (2018) We can stop the poaching of big cats. Here's how. https://www.nytimes.com/2018/03/12/opinion/conservation-poaching-tigers.html. Accessed 20 Apr 2018

Laundré JW (2010) Behavioral response races, predator-prey shell games, ecology of fear, and patch use of pumas and their ungulate prey. Ecology 91:2995–3007

Laurenson MK, Caro TM (1994) Monitoring the effects of non-trivial handing in free-living cheetahs. Anim Behav 47:547–557

Lee-Thorp J, Thackery F, Van der Merwe N (2000) The hunters and the hunted revisited. J Hum Evol 39:565–576

Lembo T, Hampson K, Haydon DT et al (2008) Exploring reservoir dynamics: a case study of rabies in the Serengeti ecosystem. J Appl Ecol 45:1246–1257

Lewis ME (1997) Carnivoran paleoguilds of Africa: implications for hominid food procurement strategies. J Hum Evol 32:257–288

Lewis ME (2017) Carnivore guilds and the impact of Hominin dispersals. In: Boivin N, Crassard R, Petraglia M (eds) Human dispersal and species movement: from prehistory to the present. Cambridge eBooks, UK, pp 29–61

Li J, Lu Z (2013) Human-snow leopard conflicts in the Sanjiangyuan region of the Tibetan plateau. Biol Conserv 166:118–123

Li J, Wang D, Yin H et al (2014) Role of Tibetan Buddhist monasteries in snow leopard conservation. Conserv Biol 28:87–94

Li G, Davis B, Eizirik E et al (2016) Phylogenetic evidence for ancient hybridization in the genomes of living cats (Felidae). Genome Res 26:1–11

Lichtenfeld L, Trout LC, Kisimir EL (2014) Evidence-based conservation: predator-proof bomas protect livestock and lions. Biodivers Conserv 24:483–491

Lindsey PA (2008) Trophy hunting in sub-Saharan Africa: Economic scale and conservation significance. In: RD Baldus, GR Damm, K-U Wollscheid (eds). Best practices in sustainable hunting- a guide to best practices around the world. CIC-International Council for Game and Wildlife Conservation, Technical Series Publication No 1, Budakeszi, pp 41–47

Lindsey PA, Alexander R, Frank LG et al (2006) Potential of trophy hunting to create incentives for wildlife conservation in Africa where alternative wildlife-based land uses may not be viable. Anim Conserv 9:283–291

Lindsey PA, Frank LG, Alexander R, Mathieson A, Romañach SS (2007) Trophy hunting and conservation in Africa: problems and one potential solution. Conserv Biol 21:880–883

Lindsey PA, Havemann CP, Lines R et al (2013a) Determinants of persistence and tolerance of carnivores on Namibian ranches: implications for conservation on southern African private lands. PLOS One 8:e52458

Lindsey PA, Balme GA, Funston P et al (2013b) The trophy hunting of African lions: scale, current management practices and factors undermining sustainability. PLoS One 8(9):9

Lindsey PA, Balme G, Becker M et al (2013c) The bushmeat trade in African savannas: impacts, drivers, and possible solutions. Biol Conserv 160:80–96. https://doi.org/10.1016/j.biocon.2012.12.020

Lindsey P, Balme G, Funston P et al (2015) Life after Cecil: channeling global outrage into funding for conservation in Africa. Conserv Lett 9(4):296–301. https://doi.org/10.1111/con.12224

Linnell JDC, Odden J, Smith ME, Aanes R, Swenson JE (1999) Large carnivores that kill livestock: do "problem individuals" really exist? Wildl Soc Bull 27:698–705

Löe J (2002) Large carnivore related deaths: a conservation issue. MSc Thesis, Norwegian University of Science and Technology, Trondheim, Norway

Löe J, Röskaft E (2004) Large carnivores and human safety: a review. Ambio 33:283–288

Loeffler IJ (1997) Letter to the editor. J Trauma 43:560

Lovari S, Minder I, Ferretti F et al (2013) Common and snow leopards share prey, but not habitats: competition avoidance by large predators? J Zool 291:127–135

Loveridge A, Searle A, Murindagomo F et al (2007) The impact of sport hunting on the population dynamics of an African lion population in a protected area. Biol Conserv 134:548–558

Loveridge AJ, Wang SW, Frank LG et al (2010) People and wild felids: conservation of cats and management of conflicts. In: MacDonald E, Loveridge A (eds) The biology and conservation of wild felids. Oxford Biology, Oxford, pp 161–195

Lynam A, Tun Khaing S, Maung Zaw K (2006) Developing a national Tiger action plan for the Union of Myanmar. Environ Mgmt 37:30–39

Macdonald C, Gallagher AJ, Barnett A, Brunnschweiler J, Shiffman DS et al (2017) Conservation potential of apex predation tourism. Biol Conserv 215:132–141

Macura B, Secco L, Pisani L (2016) All that glitters is not gold: the effect of top-down participation on conservation knowledge, attitudes and institutional trust in a central Indian tiger reserve. Regional Environ Change 16:125–140

Madden F (2004) Creating co-existence between humans and wildlife: global perspectives on local efforts to address human-wildlife conflict. Hum Dim Wildl 9:247–257

Madden F, McQuinn B (2014) Conservation's blind spot: the case for conflict transformation in wildlife conservation. Biol Conserv 178:97–106

Madhusudan MD (2003) Living amidst large wildlife: livestock and crop depredation by large mammals in the interior villages of Bhadra Tiger Reserve, South India. Environ Mgmt 31:466–475
Malkin CA (2003) Humans too tasty to resist. https://www.newscientist.com/article/mg17723821-700-humans-too-tasty-to-resist/
Mallick JK (2007) Ecology, status and aberrant behaviour of Bengal tigers in the Indian Sundarbans. In: Varma AK, Gupta VK (eds) Animal diversity, natural history and conservation, vol 2. Daya Publishing House, Delhi, pp 381–454
Marker L (2008) Cheetah conservation strategies in Namibia-A Model for the future. Afr Wildlife Conf Proc. 6–11. Zoo Dvur Kralove, Czech Republic
Marker LL, Boast LK (2015) Human-wildlife conflict 10 years later: lessons learned and their application to cheetah conservation. Hum Dimens Wildl 20:302–309
Marker LL, Dickman AJ (2005) Factors affecting leopard *(Panthera pardus)* spatial ecology, with particular reference to Namibian farmlands. So African J Wildl Res 35:105–115
Maskey TM, Bauer J, Cosgriff K (2001) Village children, leopards and conservation. Patterns of loss of human life through Leopards (Panthera pardus) In Nepal. Kathmandu, Nepal: Department of National Parks and Wildlife Conservation/Sustainable Tourism CRC
Massei G, Roger J, Quy A et al (2010) Can translocations be used to mitigate human-wildlife conflicts? Wildl Res 37:428–439
Mattern MY, McLennan DA (2000) Phylogeny and speciation of felids. Cladistics 16:232–253
Mattson D, Logan K, Sweanor L (2011) Factors governing risk of cougar attacks on humans. Hum-Wildl Interact 5:135–158
Mazak JH, Christiansen P, Kitchener AC (2011) Oldest known Pantherine skull and evolution of the Tiger. PLoS One 6(10):e25483. https://doi.org/10.1371/journal.pone.0025483
Mccain EB, Childs JL (2008) Evidence of resident jaguars *(Panthera onca)* in the Southwestern United States and the implications for conservation. J Mammal 89(1):1–10
Mccarthy T, Murray K, Sharma K et al (2010) Preliminary results of a long-term study of snow leopards in South Gobi, Mongolia. Cat News 53:15–19
McCombs ME, Shaw DL (1972) The agenda-setting function of mass media. Public Opin Quart 36:176–187
McDougal C (1987) The man-eating tiger in geographical and historical perspective. In: Tilson RL, Seal US (eds) Tigers of the world. Noyes Publications, Park Ridge
Mcgeachie J (1958) Isolation of Pasteurella septica from a lion bite wound and lion's mouth. J Pathol Bacteriol 75:467–470
McKee D (2003) Cougar attacks on humans: a case report. Wild Environ Med 14:169–173
McManus JS, Dickman AJ, Gaynor B et al (2014) Dead or alive? Comparing costs and benefits of lethal and non-lethal human-wildlife conflict mitigation on livestock farms. Oryx 49:687–695
Meachen-Samuels J, Van Valkenburgh B (2009) Craniodental indicators of prey size preference in the Felidae. Biol J Linnean Soc 96:784–799
Mech LD, Harper EK, Meier TJ et al (2000) Assessing factors that may predispose Minnesota farmers to wolf depredations on cattle. Wildl Soc Bull 28:623–629
Menotti-Raymond M, O'Brien SJ (1993) Dating the genetic bottleneck of the African cheetah. Proc Natl Acad Sci 90:3172–3176
Mercader J, Panger M, Boesch C (2002) Excavation of a chimpanzee stone tool site in the African rainforest. Science 296:1452–1455
Merola M (1994) A reassessment of homozygosity and the case for inbreeding depression in the Cheetah, *Acinonyx jubatus*: implications for conservation. Conserv Biol 8:961–971
Mettam RWM (1937) A short history of rinderpest with special reference to Africa. Uganda J 5:22–26
Michalski F, Boulhosa RL, Faria A et al (2006) Human-wildlife conflicts in a fragmented Amazonian forest landscape: determinants of large felid depredation on livestock. Anim Conserv 9:179–188

Miller JR (2015) Mapping attack hotspots to mitigate human-carnivore conflict: approaches and applications of spatial predation risk modeling. Biodivers Conserv 24:2887–2911

Miller J, Stoner KJ, Cejtin MR et al (2016a) Effectiveness of contemporary techniques for reducing livestock depredations by large carnivores. Wildl Soc Bull 40:806–815

Miller JRB, Jhala YV, Schmitz OJ (2016b) Human perceptions mirror realities of carnivore attack risk for livestock: implications for mitigating human-carnivore conflict. PLoS One 11(9):e0162685. https://doi.org/10.1371/journal.pone.0162685

Milner JM, Nilsen EB, Andreassen HP (2007) Demographic side effects of selective hunting in ungulates and carnivores. Conserv Biol 21:36–47

Minnie L, Boshoff AF, Kerley GIH (2015) Vegetation type influences livestock predation by leopards: implications for conservation in agro-ecosystems. Afr J Wildl Res 45:204–214

Miquelle DG, Smirnov EN, Quigley JB, Hornocker MG, Nikolayev IG et al (1996) Food habits of Amur tigers in the Sikhote-Alin Zapovednik and the Russian Far East, and implications for conservation. J Wildl Res 1:138–143

Mishra C, Madhusudan MD, Allen P et al (2003) The role of incentive programs in conserving the snow leopard. Conserv Biol 17:1512–1520

Mishra C, Redpath SR, Suryawanshi KR (2016) Livestock predation by snow leopards: conflicts and the search for solutions. In: McCarthy T, Mallon D, Nyhus PJ (eds) Snow leopards: biodiversity of the world: conservation from genes to landscapes. Academic Press, Cambridge, pp 59–67

Mloszewski MJ (1983) The behavior and ecology of the African buffalo. Cambridge University Press, Cambridge

Mol D, van Logchem W, de Vos J (2011) New record of the European jaguar, *Panthera onca gombaszoegensis* (Kretzoi, 1938), from the Plio-Pleistocene of Langenboom (The Netherlands). Cainozoic Res 8:35–40

Mondol S, Karanth KU, Ramakrishnan U (2009) Why the Indian Subcontinent holds the key to global tiger recovery. PLoS Genet 5(8):e1000585. https://doi.org/10.1371/journal.pgen.1000585

Montag J (2003) Compensation and predator conservation: limitations of compensation. Carnivore Damag Prev News 6:2–6

Montgomery S (2001) Spell of the tiger: the man-eaters of Sundarbans. Houghton Mifflin Harcourt, Boston

Morgan MS (1999) Tiger Bites. J Royal Soc Med 92:545

Mossaz A, Buckley RC, Castley JG (2015) Ecotourism contributions to conservation of African big cats. J Nat Conserv 28:112–118

Mosser A, Packer C (2009) Group territoriality and the benefits of sociality in the African lion, *Panthera leo*. Anim Behav 78:359–370

Moussaieff Masson J (2014) Beasts: what animals can teach us about the origins of good and evil. Bloomsbury USA, New York

Mukherjee S, Heithaus MR (2013) Dangerous prey and daring predators: a review. Biol Rev 88:550–563

Muriuki MW, Ipara H, Kiringe JW (2017) The cost of livestock lost to lions and other wildlife species in the Amboseli ecosystem, Kenya. Eur J Wildl Res 63:60

Nabi DG, Tak SR, Kangoo KA et al (2009) Injuries from leopard attacks in Kashmir. Injury 40:90–92

Naha D, Sathyakumar S, Rawat GS (2018) Understanding drivers of human-leopard conflicts in the Indian Himalayan region: Spatio-temporal patterns of conflicts and perception of local communities towards conserving large carnivores. PLoS One 13(10):e0204528. https://doi.org/10.1371/journal.pone.0204528

Naidoo R, Weaver LC, Diggle RW et al (2016) Complementary benefits of tourism and hunting to communal conservancies in Namibia. Conserv Biol 30:628–638

Naughton-Treves L, Grossberg LR, Treves A (2003) Paying for tolerance: rural citizen's attitudes toward wildlife depredation, wolf depredation and compensation. Conserv Biol 17:1500–1511

Neiburger EJ, Patterson BD (2000) The man-eaters with bad teeth. New York State Dental J 66(10):26–29. PMID 11199522
Nelson F, Lindsey PA, Balme G (2013) Hunting and lion conservation: a question of governance? Oryx 47:501–509
Neto MFC, Garrone Nato D, Haddad V (2011) Attack by jaguars *(Panthera onca)* on humans in Central Brazil: report of three cases, with observation of a death. Wild Environ Med 22(2):130–135
Newman P (2012) Tracking the Weretiger: supernatural man-eaters of India, China and Southeast Asia. McFarland & Company, Ind, London, pp 155–157
Newsome TM, Greenville AC, Cirovic D, Dickman CR, Johnson CN et al (2017) Top predators constrain mesopredator distributions. Nat Commun 8:15469. https://doi.org/10.1038/ncomms15469
Ng SJ, Dole JW, Sauvajot RM et al (2004) Use of highway undercrossings by wildlife in southern California. Biol Conserv 115:499–507
Nijman V, Shepherd CR (2015) Trade in tigers and other wild cats in Mong La and Tachilek, Myanmar-a tale of two border towns. Biol Conserv 182:1–7
Nowak RM, Paradiso JL (1983) Walker's mammals of the world, 4th edn. Johns Hopkins University Press, Baltimore
Nowell K, Jackson P (1996) Wild cats: status survey and conservation action plan. International Union for Conservation of Nature. Gland, Switzerland
Nyhus P (2016) Snow leopards. In: McCarthy T, Mallon D (eds) Biodiversity of the world: conservation from genes to landscapes. Academic Press, Elsevier, Amsterdam, pp 3–11
Nyhus P, Fischer H, Madden F et al (2006) Taking the bite out of wildlife damage: the challenges of wildlife compensation schemes. Conserv Pract 4(2). https://doi.org/10.1111/?j.1526-4629.2003.tb
Nyhus PJ, Dufraine CE, Ambrogi MC, Hart SE et al (2010) Human-tiger conflict over time. In: Tilson R, Nyhus PJ (eds) Tigers of the world: the science, politics and conservation of Panthera tigris, 2nd edn. Academic Press, Burlington, pp 132–135
O'Brien SJ, Johnson WE (2005) Big cat genomics. Ann Rev Genomics Hum Genet 6:407–429
O'Brien SJ, Johnson WE (2007) The evolution of cats. Sci Am July 207:68–75
O'Brien SJ, Wildt MBD (1986) The cheetah in genetic peril. Sci Am 254:68–76
Odden M, Wegge P (2005) Spacing and activity patterns of leopards Panthera pardus in the Royal Bardia National Park, Nepal. Wildl Biol 11:145–152
Odden J, Herfindal I, Linnell JDC, Andersen R (2008) Vulnerability of domestic sheep to lynx depredation in relation to roe deer density. J Wildl Mgmt 72:276–282
Odden M, Wegge P, Frdriksen T (2010) Do tigers displace leopards? If so, why? Ecol Res 25:875–881
Odden M, Arthreya V, Rattan S, Linnell DC (2014) Adaptable Neighbours: movement patterns of GPS-collared leopards in human dominated landscapes in India. PLoS One 9(11):e112044. https://doi.org/10.1371/journal.pone.0112044
Ogada MO, Woodroffe R, Oguge NO et al (2003) Limiting depredation by African carnivores: the role of livestock husbandry. Conserv Biol 17:1106–1117
Olsoy PJ, Zeller KA, Hicke JA et al (2016) Quantifying the effects of deforestation and fragmentation on a range-wide conservation plan for Jaguars. Biol Conserv 203:8–16
Ondaatje C (1992) The man-eater of Punanai. In: A journey of discovery to the jungles of old Ceylon. Harper Collins, Canada
Ordiz A, Bischof R, Swenson JE (2013) Saving large carnivores, but losing the apex predator? Biol Conserv 168:128–133
Ortolani A (1999) Spots, stripes, tail tips and dark eyes: predicting the function of carnivore colour patterns using the comparative method. Biol J Linnean Soc 67:433–476
Osmaston LS (1904) A man-eating panther. J Bombay Nat Hist Soc 15:135–138
Owen-Smith N, Mills MG (2008) Predator-prey size relationships in an African large mammal food web. J Animal Ecol 77:173–183

Packer C (2009) Rational fear. https://www.naturalhistorymag.com/0509/0509_feature.html. Accessed 22 Nov 2018

Packer C, Pusey AE (1997) Divided we fall: cooperation among lions. Sci Am 276:52–59

Packer C, Scheel D, Pusey AE (1990) Why lions form groups: food is not enough. Am Naturalist 136:1–19

Packer C, Ikanda D, Kissui B et al (2005) Conservation biology: lion attacks on humans in Tanzania. Nature 436:927–928

Packer C, Brink H, Kissui BM et al (2011a) Effects of trophy hunting on lion and leopard populations in Tanzania. Conserv Biol 25:143–153

Packer C, Swanson A, Ikanda D et al (2011b) Fear of darkness, the full moon and the nocturnal ecology of African Lions. PLoS One 6(7):e22285. https://doi.org/10.1371/journal.pone.0022285

Packer C, Loveridge A, Canney S et al (2013) Conserving large carnivores: dollars and fence. Ecol Lett 16:635–641

Padodara RJ, Jacob N (2014) Olfactory sense in different animals. Indian J Vet Sci 2:1–14

Pagnamenta R (2011) Leopard attacks. The Times, October 06, 2011

Palmeira FBL, Crawshaw PG, Haddad CM et al (2008) Cattle depredation by puma *(Puma concolor)* and jaguar *(Panthera onca)* in Central-Western Brazil. Biol Conserv 141:118–125

Pandey A. Had to shoot tigress Avni as she charged at us, says the shooter. https://www.indiatoday.in/india/story/had-to-shoot-the-tigress-as-she-charged-at-us-says-the-shooter-who-killed-tigress-avni-1382516-2018-11-05. Accessed 22 Nov 2018

Papadoupoulos M, Tubridy N (1999) Neurological symptoms 27 years after a tiger bite. J R Soc Med 92:303–304

Patel A, Boje E, Fisher C, Louis L, Lane E (2016) Quasi-steady state aerodynamics of the cheetah tail. Biol Open 5:1072–1076

Pathak H, Borkar J, Dixit P, Dhawane S (2013) Fatal tiger attack: a case report with emphasis on typical tiger injuries characterized by partially resembling stab-like wounds. Forensic Sci Intl 232:1–3

Patterson JH (1907) The man-eaters of Tsavo and other East African adventures. Macmillan Co, London

Patterson BD (2004) The lions of Tsavo: exploring the legacy of Africa's notorious man-eaters. McGraw Hill Professional Press, London, United Kingdom

Patterson B (2006) Ethics for wildlife conservation: overcoming the human-nature dualism. Bioscience 56:144–150

Patterson BD, Neiburger EJ, Kasiki SM (2003) Tooth breakage and dental disease as causes of carnivore-human conflicts. J Mammal 84:190–196

Patterson BD, Kasiki SM, Selempo E et al (2004) Livestock predation by lions *(Panthera leo)* and other carnivores on ranches neighboring Tsavo National Park, Kenya. Biol Conserv 119:507–516

Pawar S, Kshirsagar R, Raut P et al (2018) Maxillofacial injury from a leopard attack. Natl J Maxillofac Surg 9:96–98

Peebles KA, Wielgus RB, Maletzke BT et al (2013) Effects of remedial sport hunting on cougar complaints and livestock depredations. PLoS One 8(11):e79713. https://doi.org/10.1371/journal.pone.0079713

Penteriani V, Delgado MDM et al (2016) Human behavior can trigger large carnivore attacks in developed countries. Sci Rep 6. https://doi.org/10.1038/srep20552. Article number: 20552

Petter G (1973) Carnivores Pleistocènes du Ravin d'Olduvai (Tanzanie) In: SB Leakey, RLG Savage, SC Coryndon (eds) Fossil vertebrates of Africa 3. Academic Press, London, Pp. 43-100

Peterhans K, Julian C, Gnoske TP (2001) The science of 'man-eating' among lions *(Panthera leo)* with a reconstruction of the natural history of the "man-eaters of Tsavo". J East African Nat Hist 90(1):1–40. https://doi.org/10.2982/0012-8317

Pickering TR, Dominguez-Rodrigo M, Egeland CP (2004) Beyond leopards: tooth marks and the contribution of multiple carnivore taxa to the accumulation of the Swartkrans member 3 fossil assemblage. J Hum Evol 46:595–604

Pitman RT, Kilian PJ, Ramsay PM et al (2013) Foraging and habitat specialization by female leopards *Panthera pardus* in the Waterberg Mountains of South Africa. S Afr J Wild Res 43:167–176
Platt J (2010) Poaching, habitat loss taking toll on Bornean clouded leopard. Scientific American December 21
Polisar J (2000) Jaguars, pumas, their prey base, and cattle ranching: ecological perspectives of a management issue. Biol Conserv 209:297–310
Pooley S, Barua M, Beinart W et al (2016) An interdisciplinary review of current and future approaches to improving human-predator relations. Conserv Biol 31:513–523
Potgieter GC, Kerley GI, Marker LL (2016) More bark than bite? The role of livestock guarding dogs in predator control on Namibian farmlands. Oryx 50:514–522
Prasad A, Madan VS, Buxi TB (1991) Tiger assault: an unusual mode of head injury. Letter to the editor. Clin Neurol Neurosurg 93:171
Quamman D (2004) Monster of god: the man-eating predator in the jungles of history and the mind. WW Norton & Company, New York
Quigley H, Herraro S (2003) Characterization and prevention of attacks on humans. In: Woodroffe R, Thirgood S, Rabinowitz A (eds) People and wildlife: conflict or coexistence? Cambridge University Press, Cambridge, pp 27–48
Rabinowitz AR (1986) Jaguar predation on domestic livestock in Belize. Wildl Soc 14:170–174
Raheja NM (2015) In the search of man eating leopard of Uttrakhand. http://www.millenniumpost.in/in-the-search-of-man-eating-leopard-of-uttrakhand-110939. Accessed 15 Jan 2017
Raina P (2011) As humans draw nearer, leopards attack. New York Times. August 5, 2011
Ramesh T, Snehalatha V, Sankar K et al (2009) Food habits and prey selection of tiger and leopard in Mudumalai Tiger Reserve, Tamil Nadu, India. J Sci Trans Environ Technovation 2:170–181
Ramesh T, Kalle R, Sankar K et al (2019) To resettle or not? Socioeconomic characteristics, livelihoods, and perceptions toward resolving human-tiger conflict in the Nilgiri Biosphere Reserve, India. Land Use Pol 83:32–46
Rangarajan M (2001) India's wildlife history. Permanent Black, New Delhi
Rangarajan M (2013) Animals with rich histories: the case of the lions of Gir Forest, Gujarat, India. Hist Theory 52:109–127
Ravanelle J, Nyhus PJ (2017) Global patterns and trends in human-wildlife conflict compensation. Conserv Biol 31:1247–1256
Rawlinson G (1992) Herodotus. The Histories, The Everyman Library
Raza RH, Chauhan DS, Pasha MKS et al (2012) Illuminating the blind spot: a study on illegal trade in leopard parts in India (2001–2010). TRAFFIC India/WWF India, New Delhi
Redpath SM, Young J, Evely A et al (2013) Understanding and managing conservation conflicts. Trends Ecol Evol 28:100–109
Redpath SM, Bhatia S, Young J (2015) Tilting at wildlife: reconsidering human-wildlife conflict. Oryx 49:222–225
Reinhardt I, Rauer G, Kluth G et al (2012) Livestock protection methods applicable for Germany-a country newly recolonized by wolves. Hystrix 23:62–72
Riggio J, Jacobson A, Dollar L et al (2013) The size of savannah Africa: a lion's *(Panthera leo)* view. Biodivers Conserv 22:17–35
Ripple WJ, Estes JA, Beschta RL et al (2014) Status and ecological effects of the world's largest carnivores. Science 343(6167):1241484
Ripple WJ, Newsome TM, Kerley GIH (2016) Does trophy hunting support biodiversity? A response to Di Minin et al. Trends Ecol Evol 7:495–496
Rishi V. Man, mask and maneater. http://vinodrishi.blogspot.com/2011/05/man-mask-and-maneater.html. Accessed 18 Oct 2018
Roberge J, Angelstam P (2004) Usefulness of the umbrella species concept as a conservation tool. Conserv Biol 18:76–85
Roelke-Parker ME, Munson L, Packer C et al (1996) A canine distemper virus epidemic in Serengeti lions *(Panthera leo)*. Nature 379:441–445

Rollinger R (2012) Herodotus VII. Xerxes according to Herodotus. Encyclopaedia Iranica. Vol XII, Fasc.3, pp 270–276. Iranicaonline.org
Rollins C, Spencer D (1995) A fatality and the American mountain lion: bite mark analysis and profile of the offending lion. J Forensic Sci 40:486–489
Rosenberg DK, Noon BR, Meslow EC (1995) Towards a definition of wildlife corridor. Integrating People and Wildlife for a Sustainable Future. Proc First International Wildlife Congress, pp 436–439
Rosenblatt E, Becker MS, Creel S et al (2014) Detecting declines of apex carnivores and evaluating their causes: an example with Zimbabwean lions. Biol Conserv 180:176–186
Rostro-Garcia S, Kamler JF, Hunter LTB (2015) To kill, stay or flee: the effects of lions and landscape factors on habitat and kill site selection of cheetahs in South Africa. PLoS One. https://doi.org/10.1371/journal.pone.0117743
Rushby G (1965) No more the tusker. WH Allen, London
Saayman M, van der Merwe P, Saayman A (2018) The economic impact of trophy hunting in the South African wildlife industry. Global Ecol Conserv 16:e00510. https://doi.org/10.1016/j.gecco.2018.e00510
Saberwal VK, Gibbs JB, Chellam R et al (1994) Lion-human conflict in the Gir Forest, India. Conserv Biol 8:501–507
Saitoti TO (1980) Maasai. Abrams, New York
Sakai ST, Arsznov BM, Hristova AE et al (2016) Big cat coalitions: a comparative analysis of regional brain volumes in Felidae. Front Neuroanat 10:99. https://doi.org/10.3389/fnana.2016.00099
Sakamoto M, Ruta M (2012) Convergence and divergence in the evolution of cat skulls: temporal and spatial patterns of morphologic diversity. PLoS One 7(7):W39752. https://doi.org/10.1371/journal.pone.0039752
Saladié P, Rodríguez-Hidalgo A, Huguet R et al (2014) The role of carnivores and their relationship to hominin settlements in the TD6-2 level from Gran Dolina (Sierra de Atapuerca, Spain). Quat Sci Rev 93:47–66
Salles LO (1992) Felid phylogenetics: extant taxa and skull morphology (Felidae, Aeluroides). Am Mus Novitates 3047:67
Sanderson EW, Walston J (2019) Implications of the shared socioeconomic pathways for tiger *(Panthera tigris)* conservation. Biol Conserv 231:13–23
Sanderson E, Forrest F, Loucks C et al (2010) Setting priorities for the conservation and recovery of wild tigers: 2005–2015. In: Tilson R, Nyhus PJ (eds) Tigers of the world: the science, politics and conservation of Panthera tigris, 2nd edn. Academic Press, New York, pp 143–161
Scantlebury DM, Mills MG, Wilson RP et al (2014) Mammalian energetics. Flexible energetics of cheetah hunting strategies provide resistance against kleptoparasitism. Science 346(6205):79–81
Schaller GB (1967) The deer and the tiger: a study of wildlife in India. Chicago Press, Chicago
Schaller GB (1968) Hunting behaviour of the cheetah in the Serengeti National Park, Tanzania. African J Ecol 6:95–100
Schaller GB (1972) The Serengeti lion: a study of predator-prey relations. University of Chicago Press, Chicago
Schaller GB, Lowther GR (1969) The relevance of carnivore behavior to the study of early hominids. Southwestern J Anthropol 25:307–341
Schaller GB, Vaseconselos JMC (1978) Jaguar predation on capybara. Zeitschrift für Säugetierkunde 43:296–301
Scheel D (1993) Profitability, encounter rates, and prey choice of African lions. Behavior Ecol 4:90–97
Scheel D, Packer C (1991) Group hunting behaviour of lions: a search for cooperation. Anim Behav 4194:697–709
Schiess-Meier M, Ransauer S, Gabanpelo T et al (2007) Livestock predation-insights from problem animal control registers in Botswana. J Wildl Mgmt 71:1267–1274

Schmitz O (2017) Predator and prey functional traits: understanding the adaptive machinery driving predator-prey interactions. F1000 Res (6):1767. https://doi.org/10.12688/f1000research.11813.1

Schwartz M (2017) The lion and the cow: conservation, pastoralism, and conflict. http://voices.Nationalgeographic.com/2017/04/28/the-lion-and-the-cow-conservation-pastoralism-conflict

Seidensticker J (2010) Saving wild tigers: a case study in biodiversity loss and challenges to be met for recovery beyond. Integrat Zool 5:285–299

Seidensticker J, Lumpkin S (1990) Great cats. Weldon Owen Pty, Ltd., San Francisco

Shankar M (2008) Traditional Chinese medicine worries tiger, leopard activists. http://www.highbeam.com/doc/1P1-157120848.html. Accessed 20 Oct 2017

Shehzad W, McCarthy TM, Pompanon F et al (2012) Prey preference of snow leopard *(Panthera uncial)* in South Gobi, Mongolia. PLoS One 7(2):e32104. https://doi.org/10.1371/journal.pone.0032104

Shehzad W, Nawaz MA, Pomanon F et al (2015) Forest without prey: livestock sustain a leopard *Panthera pardus* population in Pakistan. Oryx 49:248–253

Shepherd SM, Mills A, Shoff WH (2014) Human attacks by large felid carnivores in captivity and in the wild. Wilderness Environ Med 25:220–230

Shivik JA, Treves A, Callahan P (2003) Nonlethal techniques for managing predation: primary and secondary repellents. Conserv Biol 17:1531–1537

Shrestha TK, Aryal A, Rai RK et al (2014) Balancing wildlife and human needs: the protected forest approach in Nepal. Nat Areas J 34:376–380

Shrestha B, Aihartza J, Kindlmann P (2018) Diet and prey selection by snow leopards in the Nepalese Himalayas. PLoS One 13(12):e0206310. https://doi.org/10.1371/journal.pone.0206310

Shrivastav AB, Singh KP (2017) Health management of big cats. In: Shrivastav AB, Singh KP (eds) Big Cats. IntechOpen, Rijeka. https://doi.org/10.5772/interchopen.72049

Sicuro FL (2011) Evolutionary trends on extant cat skull morphology (carnivore: Felidae): a three-dimensional geometrical approach. Biol J Linnean Soc 103:176–190

Simberloff D, Cox J (1987) Consequences and costs of conservation corridors. Conserv Biol 1:63–71

Singh HS (2005) Status of the leopard *(Panthera pardus fusca)* in India. CAT News 42:15–17

Singh AK (2019) Figuring out the enigma of leopard deaths in India: a complete analysis. Ranthambore National Park

Singh M, Raval PP, Dharaiya N et al (1999) Feeding niche of Asiatic lion *(Panthera leo persica)* and leopard *(Panthera pardus)* in the Gir protected area. Tigerpaper 26:12–15

Smilie S (2002) Killer cats hunted human ancestors. National Geographic News

Smith JA, Suraci JP, Clinchy M et al (2017) Fear of the human 'super predator' reduces feeding time in large carnivores. Proc Royal Soc B 284:20170433. https://doi.org/10.1098/rspb.2017.0433

Spassov N (2011) *Acinonyx pardinensis* remains from the middle Villafranchian locality of Varshets (Bulgaria) and the Plio-Pleistocene history of the cheetahs in Eurasia. Estud Geol 67:245–253

Srivastava KK, Bhardwaj AK, Abraham CJ et al (1996) Good habits of mammalian predators in Periyar Tiger Reserve, South India. Indian Forrester 122:877–883

Stahl P, Vandel JM (2001) Factors influencing lynx depredation on sheep in France: problem individuals and habitat. Carnivore Damag Prev News 4:6–8

Stander PE (1992) Cooperative hunting in lions: the role of the individual. Behavior Ecol Sociobiol 29:445–454

Stander PE, Llau K, Lui N et al (1997) Non-consumptive utilization of leopards: Community conservation and ecotourism in practice. In Proceedings of a symposium on lions and leopards as game ranch animals. Onderstepoort.

Stein AB, Athreya V, Gerngross P et al (2016) *Panthera pardus*. The IUCN Red List of Threatened Species. IUCN 15954/102421779. Doi:https://doi.org/10.2305/IUCN.UK.2016-1.RLTS.T15954A50659089.en

Steinberg MK (2016) Jaguar conservation in southern Belize: conflicts, perceptions and prospects among Mayan hunters. Conserv Soc 14:13–20
Sterndale RA (1877) Seonee or camp life on the Satpura Range: a tale of Indian adventure, 2nd edn. Sampson Low, Marston, Searle, & Rivington, London
Stiner MC (1990) The use of mortality patterns in archaeological studies of hominid predatory adaptations. J Anthropol Archaeol 9:305–351
Stiner MC (2012) Competition theory and the case for Pleistocene Hominin-carnivore co-evolution. J Taphonomy 2:129–145
Sullivan J (2018) Southeast Asian forest loss much greater than expected, with negative implications for climate. https://www.princeton.edu/news/2018/07/03/southeast-asian-forest-loss-much-greater-expected-negative-implications-climate. Accessed 13 Sep 2018
Sunde PK, Overskaug K, Kvam T (1998) Culling of lynxes (Lynx lynx) related to livestock predation in a heterogeneous landscape. Wildl Biol 4:169–175
Sunquist M (2010) What is a tiger? Ecology and behaviour. In: Tigers of the world: the science, politics and conservation of Panthera tigris, 2nd edn. Academic Press, London
Sunquist M, Sunquist F (2002) Snow leopard. In: Tilson R, Nyhus PJ (eds) Wild cats of the world. University of Chicago Press, Chicago, pp 377–394
Suryawanshi K, Sharma R, Timbadia R et al (2017) Reducing human snow leopard conflict in Upper Spiti Valley, India. Final Report: Conservation Leadership Report. Nature Conservation Project. Snow Leopard Trust. Project ID:2462
Swanepoel LH, Somers MJ, Dalerum F (2015) Functional responses of retaliatory killing versus recreational sport hunting of leopards in South Africa. PLoS One 10(4):e0125539. https://doi.org/10.1371/journal.pone.0125539
Swayne HCG (1899) The leopard in Somaliland. In: Anderson BH (ed) Great and small game of Africa. Rowland Ward, London, pp 575–579
Sweanor LL, Logan KA, Hornocker MG (2005) Puma responses to close approaches by researchers. Wildl Soc Bull 33:905–913
Sweeney SJ, Miller RS (2010) Free-ranging wildlife. In: Portacci K, Lombard J (eds) Assessment of pathways for the introduction and spread of Mycobacterium bovis in the United States, United States Department of Agriculture, Animal Plant Inspections Service, Fort Collins, pp 94–123
Taylor J (1959) Maneaters and marauders. Barnes and Co, New York
Taylor CR, Rowntree VJ (1973) Temperature regulation and heat balance in running cheetahs: a strategy for sprinters? Am J Phys 224:848–851
Teichman KJ, Cristescu B, Darimont CT (2016) Hunting as a management tool? Cougar-human conflict is positively related to trophy hunting. BMC Ecol 16:44
Terborgh J, Lopez P, Nunez M, Rao G et al (2002) Ecological meltdown in predator-free forest fragments. Science 294:1923
Terrell KA, Crosier AE, Wildt D et al (2015) Continued decline in genetic diversity among wild cheetahs *(Acinonyx jubatus)* without further loss of semen quality. Biol Conserv 200:192–199
Tessaro SV (1986) The existing and potential importance of brucellosis and tuberculosis in Canadian wildlife: a review. Can Vet J 27:119–132
Thaker M, Vanak AT, Owen CR et al (2011) Minimizing predation risk in a landscape of multiple predators: effects on the spatial distribution of African ungulates. Ecology 92:398–407
Thapa K, Manandhar S, Bista M et al (2018) Assessment of genetic diversity, population structure, and gene flow of tigers *(Panthera tigris tigris)* across Nepal's Terai Arc landscape. PLoS One 13(3):e0193495. https://doi.org/10.1371/journal.pone.0193495
Theile S (2003) Fading footprints: the killing and trade of snow leopards. Traffic Report. TRAFFIC International.
Therrien F (2005) Feeding behaviour and bite force of sabretoothed predators. Zool J Linnean Soc 145:393–426

Thinley P, Rajaratnam R, Lassoie JP et al (2018) The ecological benefit of tigers *(Panthera tigris)* to farmers in reducing crop and livestock losses in the eastern Himalayas: implications for conservation of large apex predators. Biol Conserv 219:119–125

Thorbahn PF (1979) Precolonial ivory trade of East Africa: Reconstruction of a human-elephant ecosystem. PhD Thesis. University of Massachusetts. Amherst

Tian Y, Wu J, Smith AT et al (2011) Population viability of the Siberian Tiger in a changing landscape: going, going and gone? Ecol Model 222:3166–3180

Titze IR, Fitch WT, Hunter EJ et al (2010) Vocal powers and pressure-flow relationships in excised tiger larynges. J Experim Biol 213:3866–3873

Torres SG, Mansfield TM, Foley JE et al (1996) Mountain lion and human activity in California: testing speculations. Wildl Soc Bull 24:457–460

Tortato FR, Layme VMG, Crawshaw PG Jr et al (2015) The impact of herd composition and foraging area on livestock predation by big cats in the Pantanal of Brazil. Anim Conserv 18:539–547

Treves A, Karanth U (2003) Human-carnivore conflict and perspectives on carnivore-management worldwide. Conserv Biol 17:1491–1499

Treves A, Naughton-Treves L (1999) Risk and opportunity for humans coexisting with large carnivores. J Hum Evol 36:275–282

Treves A, Jurewicz RR, Naughton-Treves L et al (2002) Wolf depredation on domestic animals: control and compensation in Wisconsin, 1976–2000. Wildl Soc Bull 30:231–241

Treves A, Wallace RB, Naughton-Treves L et al (2006) Co-managing human-wildlife conflicts: a review. Hum Dim Wildl 11:383–396

Treves A, Krofel M, McManus J (2016) Predator control should not be a shot in the dark. Frontiers Ecol Environ 14:380–388

Trivedi BP (2002) Female lions prefer dark-maned males. Behavior Ecol Sociobiol 52:303–307

Trophy Hunting by the Numbers: The United States Role in Global Trophy Hunting (2016) A report of the Humane Society of the United States & The Humane Society International.

Tseng ZJ, Wang X, Slater GJ et al (2013) Himalayan fossils of the oldest known pantherine establish ancient origins of big cats. Proc Biol Sci 281(1774):20132686. https://doi.org/10.1098/rspb.2013.2686

Tucker A (2009) The most ferocious man-eating lions. https://www.smithsonianmag.com/science-nature/the-most-ferocious-man-eating-lions-2577288/

Turnball-Kemp P (1967) The leopard. Howard Timmins, Cape Town

Turner A (1999) Evolution of the African Plio-Pleistocene mammalian fauna: correlation and causation. In: Bromage TG, Schrenk F (eds) African biogeography, climate change & human evolution. Oxford University Press, Oxford, pp 76–87

Turner A, Antón M (1997) The big cats and their fossil relatives. Columbia University Press, New York

Tyrrell P, Western D (2017) Seasonal movements of wildlife and livestock in a heterogeneous pastoral landscape: implications for coexistence and community based conservation. Global Ecol Conserv 12:59–72

Uphyrkina O, Johnson EW, Quigley H et al (2001) Phylogenetics, genome diversity and origin of modern leopard, *Panthera pardus*. Mol Ecol 10:2617–2833

Valeix M, Henson G, Loveridge AJ et al (2012) Behavioral adjustments of a large carnivore to access secondary prey in a human-dominated landscape. J Appl Ecol 49:73–81

Van Bommel L, Bij de Vaate MD, De Boer WF et al (2007) Factors affecting livestock predation by lions in Cameroon. African J Ecol 45:490–498

van der Meer F, Clancy E, Thomas A et al (2015) The Maasai need cows and the cows need Maasai, the use of a photovoice approach to assess animal health needs. Front Vet Sci 2:46–49

van der Merwe P, Saayman M, Rossouw R (2014) The economic impact of hunting: a regional approach. S Afr J Econ Manag Sci 17:379–379

van Eeden LM, Crowther MS, Dickman C, Macdonald DW, Ripple WJ et al (2017) Managing conflict between large carnivores and livestock. Conserv Biol 32:167–173 van Orsdol KG (1984) Foraging behaviour and hunting success of lions in Queen Elizabeth National Park, Uganda. Afr J Ecol 22:79–99

van Valkenburgh B (2001) The dog-eat-dog world of carnivores: a review of past and present carnivore community dynamics. In: Stanford C, Bunn H (eds) The early human diet: the role of meat. Oxford University Press, Oxford, pp 101–121

van Valkenburgh B, Ruff CB (1987) Canine tooth strength and killing behaviour in large carnivores. J Zool 212:379–397

van Valkenburgh B, Wayne RK (2010) Carnivores. Curr Biol 20:R915–R919

Viana M, Cleaveland S, Matthiopoulos J et al (2015) Dynamics of a morbillivirus at the domestic-wildlife interface: canine distemper virus in domestic dogs and lions. Proc Nat Acad Sci 112:1464–1469

Victor P (2017) Deforestation-a modern-day plague in Southeast Asia. https://theaseanpost.com/article/deforestation-modern-day-plague-southeast-asia. Accessed 23 Jan 2018

Viljoen IM, van Helden PD, Millar RP (2015) Mycobacterium bovis infection in the lion *(Panthera leo)*: current knowledge, conundrums and research challenges. Vet Microbiol 177:252–260

Volodina EV (2000) Vocal repertoire of the cheetah Acinonyx jubatus (Carnivora, Felidae) in captivity: sound structure and their potential for estimating the state of adult animals. Zoologischeskiī Zhurnal 79:833–843

Wagner KK, Schmidt RH, Conover MR (1997) Compensation programs for wildlife damage in North America. USDA National Wildlife Research Center. Staff Publications

Waldron A, Mooers AO, Miller DC et al (2013) Targeting global conservation funding to limit immediate biodiversity declines. Proc Natl Acad Sci U S A 110:144–148

Waldron A, Miller DC, Redding D, Mooers A, Kuhn TS et al (2017) Reductions in global biodiversity loss predicted from conservation spending. Nature 551:364–367

Walston J, Robinson JG, Bennett EL et al (2010) Bringing the tiger back from the brink-the six percent solution. PLoS Biol 8(9):e1000485. https://doi.org/10.1371/journal.pbio.1000485

Wassener B (2012) Leopard poaching on the rise, Group Warns. New York Times, September 28

Watson FG, Becker MS, Milanzi J et al (2015) Human encroachment into protected area networks in Zambia: implications for large carnivore conservation. Reg Environ Chang 15:415–429

Waylen KA, Fischer A, McGowan PJK et al (2010) Effect of local cultural context on the success of community-based conservation interventions. Conserv Biol 24:1119–1129

Weber DJ, Wolfson JS, Swartz MN et al (1984) Pasteurella multocida infections. Report of 34 cases and review of the literature. Medicine 63:135–154

Weilenmann M, Gusset M, Mills DR et al (2010) Is translocation of stock-raiding leopards into a protected area with resident co-species an effective management tool? Wildl Res 37:702–707

Weise FJ, Stratford KJ, van Vuuren RJ (2014) Financial costs of large carnivore translocations-accounting for conservation. PLoS One 9(8):e105042. https://doi.org/10.1371/journal.pone.0105042

Weise FJ, Lemeris J Jr, Stratford KJ et al (2015) A home away from home: insights from successful leopard *(Panthera pardus)* translocation. Biodivers Conserv 24:1755–1774

Weissengruber GE, Forstenpointner G, Peters G, Kübber-Heiss A, Fitch WT (2002) Hyoid apparatus and pharynx in the lion *(Panthera leo)*, jaguar *(Panthera onca)*, tiger *(Panthera tigris)*, cheetah *(Acinonyx jubatus)*, liger *(Panthera leo* x *Panthera tigris)*, Tigon *(Panthera tigris* x *Panthera leo)* and the domestic cat *(Felis silvestris, F. catus)*. J Anat 201:195–209

Werdelin L, Yamaguchi N, Johnson W, O'Brien SJ (2010) Phylogeny and evolution of cats (Felidae). In: MacDonald DW, Loveridge AJ (eds) Biology and conservation of wild felids. Oxford University Press, Oxford, pp 59–82

West PM, Packer C (2002) Sexual selection, temperature, and the Lion's mane. Science 298:1339–1343

West P, Igoe J, Brockington D (2006) Parks and peoples: the social impact of protected areas. Annu Rev Anthropol 35:251–277

Wiens MB, Harrison PB (1996) Big cat attack: a case study. J Trauma 40:829–831

Williams TM, Wolfe L, Davis T et al (2014) Instantaneous energetics of puma kills reveal advantage of felid sneak attacks. Science 346:81–85

Williams ST, Williams KS, Lewis BP et al (2017) Population dynamics and threats to an apex predator outside protected areas: implications for carnivore management. Royal Soc Open Sci 4:161090. https://doi.org/10.1098/rsos.161090. PMC 5414262

Wilson JW, Mills MGL, Wilson RP et al (2013a) Cheetahs *(Acinonyx jubatus)* balance turn capacity with pace when chasing prey. Biol Lett 9(50):20130620. https://doi.org/10.1098/rsbl.2013.0620. PMC 3971710

Wilson AM, Lowe JC, Roskilly K, Hudson PE, Golabek KA et al (2013b) Locomotion dynamics of hunting in wild cheetahs. Nature 498:185–189

Wilson AM, Hubel TY, Wilshin SD et al (2018) Biomechanics of predator-prey arms race in lion, zebra, cheetah and impala. Nature 554:183–188

Wilting A, Fischer F, Baker SA, Linsenmair KE (2006) Clouded leopards, the secretive top-carnivore of South East Asian rainforests: their distribution, status and conservation needs in Sabah, Malaysia. BMC Ecol 6:1–13

Wilting A, Christiansen P, Kitchener AC et al (2010) Geographical variation in and evolutionary history of the Sunda clouded leopard *(Neofelis diardi)* (Mammalia: Carnivora: Felidae) with the description of a new subspecies from Borneo. Mol Phylogenet Evol 58:317–328

Winterbach H, Winterbach C, Somers MJ, Hayward M (2013) Key factors and related principles in the conservation of large African carnivores. Mammal Rev 43:89–110

Wolf C, Ripple WJ (2016) Prey depletion as a threat to the world's large carnivores. R Soc Open Sci 3:160252. https://doi.org/10.1098/rsos.160252

Wolf C, Ripple WJ (2018) Rewilding the world's large carnivores. Royal Soc Open Sci 5(5):180514

Wolfe ML, Koons DN, Stoner DC, Terletzky P, Gese EM et al (2014) Is anthropogenic cougar mortality compensated by changes in natural mortality in Utah? Insight from long-term studies. Biol Conserv 182:187–196

Woodroffe R (2001) Strategies for carnivore conservation: lessons from contemporary extinctions. In: Gittelman JL, Funk SM, Macdonald DW, Wayne RK (eds) Carnivore conservation. Cambridge University Press, Cambridge, pp 61–92

Woodroffe R, Frank LG, Lindsey PA, Symon MK (2006) Livestock husbandry as a tool for carnivore conservation in African's community rangelands: a case-control study. Biodivers Conserv 16:1245–1260

Woolfrey BQ (1985) *Pasteurella multocida* in an infected tiger bite. Arch Pathol Lab Med 109:744–746

Yamaguchi N, Cooper A, Werdelin L, Macdonald DW (2006) Evolution of the mane and group-living in the lion *(Panthera leo)*: a review. J Zool Lond 263:329–342

Yeakel JD, Patterson BD, Fox-Dobbs K et al (2009) Cooperation and individuality among man-eating lions. PNAS 106:19040–19043

Yu L, Zhang YP (2005) Phylogenetic studies of pantherine cats (Felidae) based on multiple genes, with novel application of nuclear beta fibrinogen intron 7 to carnivores. Mol Phylogenet Evol 35:483–495

Zarco-Gonzalez MM, Monroy-Vilchis O, Alaniz J (2013) Spatial model of livestock predation by jaguar and puma in Mexico: conservation planning. Biol Conserv 159:80–87

Chapter 3
A Large Carnivore Among People and Livestock: The Common Leopard

Uzma Khan, Francesco Ferretti, Safdar Ali Shah, and Sandro Lovari

3.1 Introduction

Large carnivores are key components of ecosystems being involved in ecological processes, such as influencing prey abundance, distribution and/or behaviour and potentially other ecosystem components dependent on them, e.g. vegetation or smaller animals (e.g. Sinclair et al. 2003; Hebblewhite et al. 2005; Lovari et al. 2009; Ripple et al. 2014). In human-dominated areas, i.e. where wild, meso/large prey have disappeared or are scarce, carnivores also exploit livestock (e.g. Meriggi and Lovari 1996; Bagchi and Mishra 2006; Khorozyan et al. 2015). In turn, conflicts with humans are triggered, which often result in poaching/retaliatory killing of carnivores (e.g. Dinerstein et al. 2003; Treves and Karanth 2003; Packer et al. 2005; Khan et al. 2018).

Habitat destruction and/or fragmentation as well as poaching are expected to negatively affect prey abundance, resulting in a depletion of prey for large carnivores. Additionally, the absence of appropriate measures for livestock guarding, combined with the lack of adequate schemes that ensure a fast and sure financial

U. Khan
Research Unit of Behavioural Ecology, Ethology and Wildlife Management, Department of Life Sciences, University of Siena, Siena, Italy

WWF Pakistan, Lahore, Pakistan

F. Ferretti
Research Unit of Behavioural Ecology, Ethology and Wildlife Management, Department of Life Sciences, University of Siena, Siena, Italy

S. Ali Shah
Khyber Pakhtunkhwa Wildlife Department, Peshawar, Pakistan

S. Lovari (✉)
Research Unit of Behavioural Ecology, Ethology and Wildlife Management, Department of Life Sciences, University of Siena, Siena, Italy

Maremma Natural History Museum, Grosseto, Italy

F. M. Angelici, L. Rossi (eds.), *Problematic Wildlife II*,
https://doi.org/10.1007/978-3-030-42335-3_3

compensation for livestock losses, is expected to emphasise human-carnivore conflict. In turn, conflict can further endanger large carnivores because of retaliatory human responses.

In Pakistan, deforestation has been substantial over the last millennia, and currently, exploitation of forests by humans is ongoing, with a rate of destruction of *c.* 2%/year, i.e. 400 km^2/year, since 1990, i.e. the highest level of deforestation in Asia (FAO 2009). This deforestation level has been linked to a growing human population, which has shown a threefold increase in the last 50 years, i.e. *c.* 208 million in 2017, according to the last population census conducted by the Government of Pakistan (http://www.pbscensus.gov.pk). In turn, some iconic wildlife species have disappeared, e.g. the tiger *Panthera tigris*, and others have sharply declined, e.g. the leopard *Panthera pardus* (Roberts 1997). In Asia the historic range of the leopard has decreased by 80% (Jacobson et al. 2016). As for mountain-dwelling ungulates, the grey goral *Naemorhedus goral*, the musk deer *Moschus chrysogaster* and the barking deer *Muntiacus vaginalis* have declined or disappeared locally, which has narrowed the prey spectrum for large carnivores (Anwar et al. 2011; Shehzad et al. 2014).

The leopard is a large carnivore (body weight: female 28–60 kg, male 37–80 kg; Nowak 1999), and is classified as 'vulnerable' by the IUCN Red List of Threatened Species (Stein et al. 2016) but 'critically endangered' in Pakistan (Sheikh and Molur 2004). It is an ecologically flexible predator, which can be found in a wide range of habitats, from sub-desert to equatorial/boreal ones (Nowak 1999). This adaptable cat tends to use a wide spectrum of wild prey, mainly ungulates (up to >100 kg of weight, but especially prey belonging to a range of 2–40 kg; Lovari et al. 2013a, for Asia; Hayward et al. 2006, mainly for Africa). In the absence of wild ungulates, the leopard can prey on livestock or even domestic dogs/cats, as well as smaller wild prey (Shehzad et al. 2014; Athreya et al. 2016; Khan et al. 2018). In these cases, conflicts with human activities can escalate (Khan 2015; Khan et al. 2018) to dramatic consequences, i.e. lethal injuries to humans and human retaliatory killing of leopards (Treves and Naughton-Treves 1999; Athreya et al. 2010, 2013; Khan et al. 2018). In Pakistan, human-leopard conflicts are also emphasised by the absence of a policy framework that would aim to compensate livestock owners for their incurred economic losses due to depredation (Khan 2015). In turn, measures to mitigate conflict with humans should be identified through ad hoc studies that ideally carry out a multi-factorial approach.

Several aspects of leopard-human relationships were studied in three densely inhabited mountainous zones of northern Pakistan. Study areas differed in their relative extent of forest and management (i.e. the main refuge habitat for leopards, as well as suitable habitat for potential wild prey; see Study Area) but were similar in the fact that the leopard was the only large carnivore present (Khan 2015). Wild ungulates were virtually absent, and leopards survived out of domestic (goats, sheep, cattle and dog) and small wild prey (rhesus monkeys *Macaca mulatta*, small-to-meso carnivores, rodents and birds; Shehzad et al. 2014; Khan 2015; Khan et al. 2018). Across the areas, we compared (*i*) food habits of leopards, (*ii*) livelihood approaches of local communities and livestock depredation and (*iii*) forest loss in the last two decades, to ultimately propose effective mitigating solutions (cf. Dar et al. 2009; Kabir et al. 2014, for Pakistan).

3.2 Materials and Methods

3.2.1 *Study Areas*

Our study areas were located in a human-dominated portion of the Western Himalayan Ecoregion (Abbottabad District, Pakistan, *c.* 881,000 people over 1967 km^2, i.e. a density of *c.* 45,000 people/100 km^2) that showed varying levels of anthropogenic activities (in total, *c.* 32,800 ha, 1000–2800 m a.s.l.). The first area was the Ayubia National Park and its surroundings (ANP, in the Khyber Pakhtunkhwa Province; 10,800 ha). ANP is a protected area (*c.* 3300 ha) and is the largest forested sector within our chosen area of study (<10% of non-forested area, in 2011, Khan 2015). The other two areas of study were the Murree Forest (MF, in the Punjab Province, 12,200 ha; > 75% of non-forested area), and the Changla Gali-Koza Gali area, which is located in an intermediate position between Ayubia and Murree, hereafter called the *Transitional Area* (TA; 9800 ha; > 90% of non-forested area, Khan 2015). In comparison to ANP and surroundings, MF and TA showed heavier anthropogenic activities and smaller forest patches, interspersed with cultivated fields and human settlements (i.e. villages and hamlets) (Khan 2015). Overall, the forested areas included temperate, tropical and sub-tropical pine forests, with a few subalpine meadows. The main tree species were blue pine *Pinus wallichiana, chir* pine *Pinus roxburghii*, fir *Abies pindrow*, Himalayan yew *Taxus wallichiana*, deodar *Cedrus deodara*, morinda spruce *Picea smithiana*, horse chestnut *Aesculus indica*, oak *Quercus incana* and *Q. dilatata*, and *Rhododendron arboreum* (Khan 2015).

A growing human population is present in the Abbottabad District (ERRA 2007). Livestock, such as cattle, goats, sheep and equids, were abundant over the whole district (*c.* 269 livestock heads/km^2, AAVV 1998). Among wild ungulates, only the wild boar *Sus scrofa* and the grey goral were present during our study, but scantily (Khan 2015). The leopard was the only large predator, with the wolf *Canis lupus* being a rare visitor. Smaller carnivores were the leopard cat *Prionailurus bengalensis*, the golden jackal *Canis aureus*, the red fox *Vulpes vulpes*, the yellow-throated marten *Martes flavigula* and civets *Paguma larvata* and *Paradoxurus hermaphrodites*. Other potential prey of leopards were the rhesus monkey *Macaca mulatta* (min. 1–3 individuals/100 ha, Lovari S., unpublished data), the Indian crested porcupine *Hystrix indica*, the Kashmir flying squirrel *Eoglaucomys fimbriatus* and several species of gallinaceous birds. In this district, snowfall usually occurs from November to March and rainfall peaks during June–September and February–March with precipitations ranging between *c.* 1600 and 2600 mm/year (Khan 2015). December, January and February are the coldest months (minimum temperature down to −7.5 °C) and May, June and July are the warmest ones (maximum temperature up to 32 °C) (Khan 2015).

3.2.2 Food Habits

Diet composition of the leopard was assessed through analyses of food remains in scats, collected along trails (total length, ANP: 43.6 km; MF: 19.0 km; TA: 20.1 km), which were walked monthly (November 2010–October 2011) and bimonthly (November 2011–May 2013). Scats were thoroughly and conservatively selected on the basis of different features (e.g. smell, position, size, contents and presence of footprints in their vicinity: Lovari et al. 2009, 2015) to limit the risk of collecting scats of other carnivores. No feature alone is species specific, but a combination of them can be effective in identifying leopard scats (Lovari et al. 2009). Moreover, the absence of sympatric large predators reduced the probability of error in the identification of scats. Dubious cases were genotyped ($n = 130$); 82% of them were confirmed to come from leopards. Once collected, samples were put in paper bags, air-dried and stored for laboratory analyses. A sub-sample of each fresh scat was stored in silica, within a sterilised plastic bottle, for genetic tests. For laboratory analyses, once scats were air-dried, they were put at a temperature of 60 °C for at least 12 h in an oven. Later, they were soaked in water, separated using tweezers, washed and strained (through fine sieves with 1 and 3 mm meshes) to separate remains such as hair, bone fragments, teeth, nails/claws and scales/feathers. The remains were put in an oven on petri dishes, to dry out, for another 8–12 h.

Reference slides were made from hair/feather samples of potential prey species. For domestic prey species, samples were collected from the local communities and, for wild prey species, from the Donga Gali Information Centre, KP Wildlife Department and Pakistan Museum of Natural History collections. Slides were prepared according to Teerink (1991), Mukherjee et al. (1994) and De Marinis and Asprea (2006). Twenty individual hairs were collected from each scat and compared with reference slides and photographic keys to identify the prey by using clues such as appearance of hair, colour, cuticular and medullary patterns and relative length (Karanth and Sunquist 1995) through a 100–400 × microscope.

For each area and season (summer: April–September; winter: October–March), data were tabulated as absolute frequency of occurrence (AO, n. occurrences of each prey species/total n. scats × 100); relative occurrence (RO, n. occurrences of each prey species/total n. occurrences × 100); volume of each prey item, which was estimated by eye for each scat (estimated volume of each prey species/total estimated volume × 100, Kruuk 1989). Estimated volume, when present (%), was plotted versus the absolute frequency of occurrence (%) of the prey species to show the relative importance of prey items (e.g. Kruuk and Parish 1981; Kruuk 1989). Several uncertainties affect the methods used to calculate the biomass consumed, from scats (cf. Chakrabarti et al. 2016; Lumetsberger et al. 2017). Additionally, it is usually impossible to know (*i*) whether a young/subadult/male/female has been preyed upon (body mass is normally quite different in different age classes and/or sexes, especially of polygynous mammals), (*ii*) whether other carnivores participated in the usage of a carcass, or (*iii*) whether the carnivore fed alone on it or with conspecifics, e.g. a pair or a female with cubs. Thus, we limited our analyses to frequencies of occurrence and estimated volume (Kruuk and Parish 1981).

For each area and season, we determined the adequacy of sample size through the Brillouin diversity index (Hb, index range: 0–4.5; Brillouin 1956; cf. Glen and Dickman 2006; Hass 2009):

$$Hb = \left(\ln N! - \sum \ln n_i!\right) / N$$

where N = total n. individual prey taxa in all samples and n_i = *n*. individual prey taxa in the ith category. A diversity curve was calculated by sampling with replacement, in increments of two. Cumulative diversity (Hb) was plotted against *n*. scats and adequacy of sampling effort was determined by examining the asymptote reached in the diversity curves. We compared the diet of leopards across study areas through G-tests, both during the whole study period and seasonally. In particular, we compared – across areas – the total occurrences of food categories.

3.2.3 *Forest Cover Changes*

To evaluate recent changes in the suitable habitat for leopards, we assessed land cover in 1990 and 2011 (Khan 2015). We downloaded temporal satellite images (spatial resolution: 30 m, Landsat TM) from the United States Geological Survey (www.usgs.gov). Land cover classes were identified by applying Histogram Equalize and Standard Deviation Stretch. This enhanced the low contrast of satellite images and made them more interpretable for further processing. The features of the same colour tone were enhanced and differentiated through the brightness and contrast utilities, as well as different False Colour Composites (FCCs). Three land cover categories were defined: forest, human settlements/cultivations and other (i.e. bare rocks, open grassland and water channels). These classes were extracted by using Object-Based Image Analysis (OBIA) (Flack 1995). In this process, boundaries of the dominant objects were extracted at fine and coarse scales. For interpretation and processing, Digital Image Processing (DIP) software ERDAS Imagine 8.7® and Definiens Developer 7.0® were used. All the maps were developed in ArcGIS 10.2.2®.

3.2.4 *Socioeconomic Implications of Livestock Losses*

In May–July 2012, a household level survey was conducted in ANP and MF, which presumably were refuge areas to leopards during the day. A total of 1016 households were interviewed (ANP, *n* = 593; MF, *n* = 423). Data on the total number of

Table 3.1 Household-level survey profile and livestock ownership in ANP and Murree

Profile	ANP	Murree
No of union councils surveyed	7	6
Total villages	69	63
Number of households surveyed	593	423
Total respondents	593	423
Female	317	219
Male	276	204
Livestock owners	525	324
Total livestock owned by respondents	2875	2092
Reason for keeping livestock		
Domestic use	478	303
Sale of extra milk and occasional sale of an animal	47	17
Sale of livestock for livelihood/business		4

households present in each of the areas were obtained from the Union Councils of Abbottabad District. We selected 13 Union Councils (administrative units of a district), including 36,024 households (Table 3.1). We adopted a multistage sampling technique: a random sample of villages was selected from each union council, then, in each village a number of households proportional to the population were selected through systematic sampling, then, every fifth household was interviewed. Questions asked to the households included (*i*) type/number of livestock raised; (*ii*) livelihood approach; (*iii*) overall perception on the most frequent cause of livestock losses (leopard depredation, natural causes of mortality: disease and winter severity, other), i.e. to assess the general perception of factors of livestock losses; (*iv*) household livestock losses in the previous year; (*v*) type of household livestock lost; and (*vi*) cause of household livestock losses.

Livelihood approaches were analysed and the top three categories were used to calculate income brackets. The economic rates of various jobs were taken from the local markets and departments (see Khan 2015, for details); salary brackets were defined to calculate the economic losses and income of communities. For further details on interview surveys, see Khan et al. (2018).

3.3 Results

3.3.1 Food Habits

A total of 525 scats were analysed to assess food habits (ANP: $n = 300$; MF: $n = 84$; TA: $n = 141$). In each area/season, the sample size was greater than what is considered adequate to describe diet diversity in large cats ($n = 11$–44, Hass 2009; Lovari et al. 2013b, 2015; Lyngdoh et al. 2014; cf. below). Ten food categories were identi-

Table 3.2 Food habits of common leopards in the three study areas. In bold, broad food categories

PREY	ANP	TA	MF	Total occurrence	Absolute frequency of occurrence (%)	Relative frequency of occurrence (%)
Domestic goat (*Capra hircus*)	**184**	**85**	**49**	**318**	**60.6**	**54**
Canids						
Domestic dog (*Canis familiaris*)	28	20	14	62	11.8	10.5
Fox (*Vulpes vulpes*)	6	2	0	8	1.5	1.0
Jackal (*Canis aureus*)	3	2	2	7	1.3	1.4
***Bos* spp.**	**16**	**13**	**5**	**34**	**6.5**	**5.8**
Galliform (Phasianidae)	**6**	**2**	**3**	**11**	**2.1**	**1.9**
Small Carnivore						
Himalayan palm civet/ Masked civet (*Paguma larvata*) Common palm civet (*Paradoxurus hermaphrodites*)	4	8	1	13	2.5	2.2
Yellow-throated marten (*Martes flavigula*)	4	1	2	7	1.3	1.2
Leopard cat (*Prionailurus bengalensis*)	8	5	2	15	2.9	2.5
Rhesus (*Macaca mulatta*)	**14**	**10**	**3**	**27**	**5.0**	**4.6**
Small mammals	**12**	**1**	**0**	**13**	**2.5**	**2.2**
Plant matter	**28**	**4**	**7**	**39**	**7.4**	**6.6**
Others						
Wild boar (*Sus scrofa*)	3	1	1	5	1.0	0.8
Grey goral (*Naemorhedus goral*)	2	1	1	4	0.8	0.7
Flying squirrels	2	0	0	2	0.4	0.3
Indian crested porcupine (*Hystrix indica*)	3	0	0	3	0.6	0.5
Equid (donkey, mule, horse)	2	1	0	3	0.6	0.5
Sheep (*Ovis aries*)	1	1	0	2	0.4	0.3
Unidentified/degraded	**8**	**3**	**5**	**16**	**3.0**	**2.7**

fied, with one category being unidentified remains (Table 3.2). Major prey species were listed individually; minor prey items (i.e. occurring less than five times) were pooled together as 'others'. According to the Brillouin diversity index, our sample size was adequate to represent diet diversity for each season/area, because asymptotes were reached at *c.* 18–23 scats (Fig. 3.1).

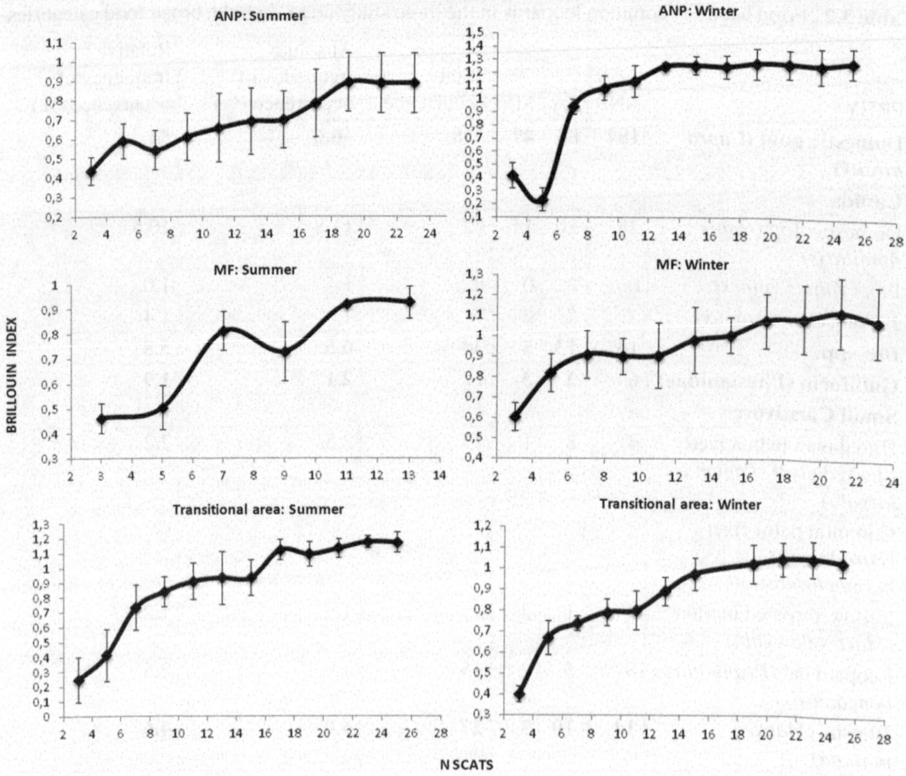

Fig. 3.1 Diversity curves (mean ± standard error) of food habits of the common leopard, estimated through the Brillouin index (Hb), in Ayubia National Park (ANP), Murree Forest (MF) and the Transitional Area, in summer and winter

In all of our areas of study, a single category (goats) dominated the relative volume and relative frequency of diet, building up more than 50% of the total volume and 58% of the frequency in diet (Figs. 3.2 and 3.3). In all the sites, livestock represented the staple of the leopard diet, with a frequency always greater than 80% (Table 3.2). In Fig. 3.3, we compared the absolute occurrence of different food categories with their relative volume in the total diet and with their relative volume when present. Again, the goat was not only the most frequently used food item, but it also constituted the large majority of the consumed volume (Fig. 3.3). Other meso-large prey (e.g. canids or *Bos* spp.) showed a remarkable volume when present, but constituted less than 15% of the overall diet (Fig. 3.3). Diet composition did not differ among areas of study (G-test: G = 38.298, df = 34, P = 0.281; Chi-square test: Chi-square: 38.627, df = 34, P = 0.269) or seasons (ANP: G-test:

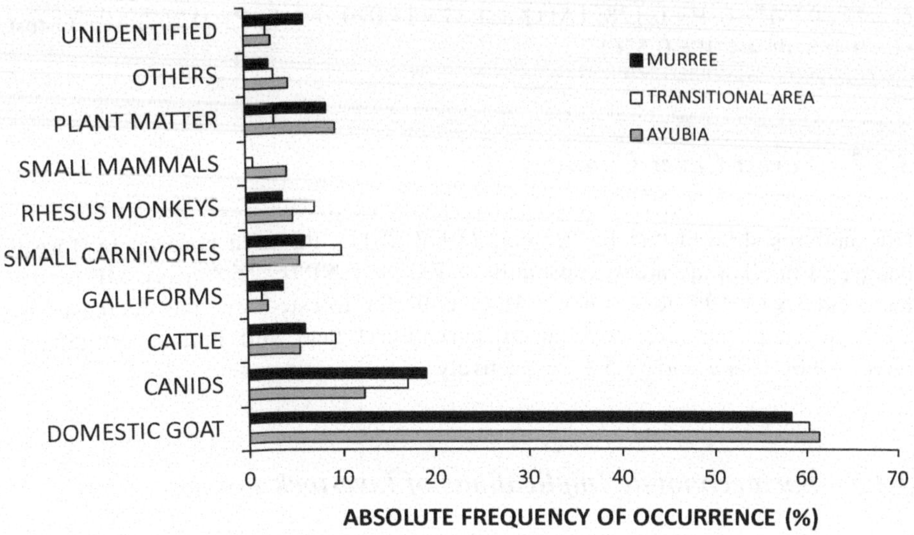

Fig. 3.2 Absolute frequency of occurrence of different prey in scats

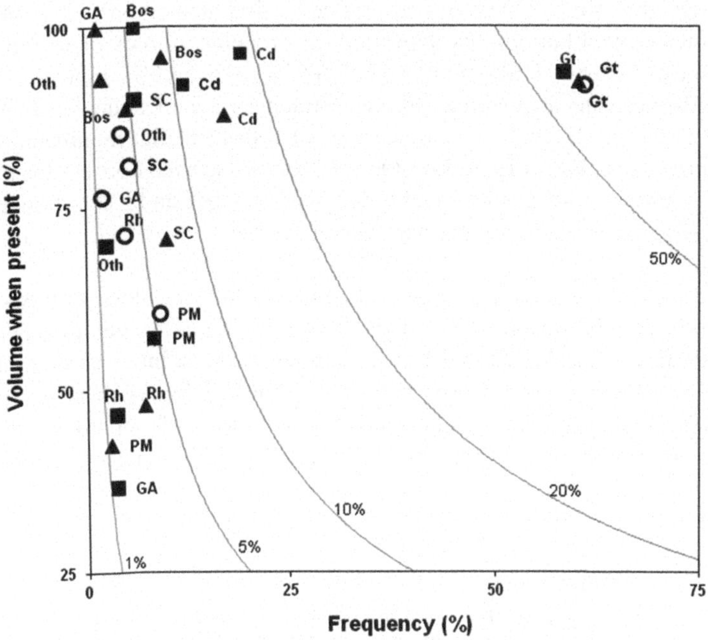

Fig. 3.3 Food habits of the common leopard, in terms of estimated volume when present (%) versus frequency of occurrence (%). Isopleths connect the equal relative volume (%). Goat (Gt), Canid (Cd), *Bos*, Galliform (GA), Rhesus monkey (Rh), Small carnivores (SC), Plant matter (PM) and Others (Oth). ○: ANP, ▲: Transitional Area, and ■: Murree

G = 12.592, df = 6, P = 0.128; TA: G-test: G = 11.070, df = 5, P = 0.267; MF: G-test: G = 9.488, df = 4, P = 0.556).

3.3.2 *Forest Cover Changes*

Our analyses showed that in 20 years (1992–2011), the area covered by forests decreased in all study areas, especially in TA, i.e. ANP (*c.* 15% loss), MF (*c.* 4% loss) and TA (*c.* 19% loss, Table 3.3). Overall, the forested area has decreased by 6.6%, whereas the area covered by agricultural land and human settlements increased by 15.4% and 81.5%, respectively.

3.3.3 *Socioeconomic Implications of Livestock Losses*

In both ANP and MF areas, goat owners mainly kept 2 (ANP: 32% out of 406 goat owners; MF: 33% out of 251 goat owners) or 1 goat (ANP: 26%; MF: 24%). Similarly, cattle owners mainly kept 1 (ANP: 51% out of 275 interviewed cattle owners; MF: 48% out of 275 cattle owners) or 2 individuals (ANP: 32%; MF: 37%).

Most interviewed households depended on irregular sources of income, with a salary range of 120–800 USD/month (ANP: 60.2%; MF: 53.6%; Table 3.4). Regular jobs, whether private or government, constituted 24.3% of sources of income in ANP and 24.6% in MF, with an income range of 100–150 USD/month and 150–200 USD/month, respectively (Table 3.4). Only 1% of interviewed households in Murree indicated livestock trade as their livelihood. However, 9% and 5% of households in ANP and MF, respectively, supplemented their income occasionally with livestock/milk trade.

According to the general perception of causes of livestock losses, in ANP, about 26% of households attributed livestock losses to leopards, but this percentage dropped to only 2% in MF (Table 3.5). In both areas, most interviewed households attributed livestock losses to disease (*c.* 55%; Table 3.5). As to actual livestock losses, 57% and 29% of livestock heads were lost because of leopard depredation in ANP and MF, respectively (Table 3.6). Additionally, 58% and 26% of interviewed

Table 3.3 Changes in forest cover in the three study areas between 1990 and 2011

Sub-area	Area size (km^2)	Forest cover in 1990 (km^2)	Forest cover in 2011 (km^2)	Forest loss in km^2	% Forest loss
ANP and surroundings	227.4	82.7	70.3	−12.4	15.0
Ta	148.3	14.8	11.9	−2.8	19.2
Mf	292.5	74.0	71.1	−2.9	3.9

Table 3.4 Livelihood approaches of the communities in the Ayubia National Park and surroundings, as well as in the Murree Forest area. NA, not available

#	Livelihood approaches	ANP		Murree		Income based on local market rates/month
		Frequency	%	Frequency	%	
1	Labour	169	28.5	122	29	80–120 USD
	Works in hotel	83	14	16	3.8	
	Driver	55	9.3	45	10.6	
	Works in factory	1	0.2	7	1.6	
	Security guards	8	1.3	11	2.6	
	Works in shop	40	6.7	17	4	
	Tailoring	1	0.2	8	2	
2	Other private jobs	55	9.3	32	7.6	100–150 USD
3	Government jobs	89	15	72	17	150–200 USD
4	Trading/business	22	3.7	34	8	–
5	Pension	17	2.9	15	3.5	–
6	Own property/agriculture produce	3	0.5	NA	NA	–
7	Resident abroad	11	2	NA	NA	–
8	None/doing nothing	39	6.5	38	9	–
9	Livestock dealing	NA	NA	4	0.9	–
10	Poultry farm	NA	NA	2	0.5	–
	Total	593	100	423	100.1	–

Table 3.5 General perceptions of reasons of livestock losses in the Murree Forest and in the Ayubia National Park and surroundings areas

Causes of domestic animal losses	Murree		Ayubia National Park	
	N Respondents	%	N Respondents	%
Disease	126	55.5	243	55.6
Winter severity	79	34.8	33	7.5
Falling from cliff	6	2.6	14	3.2
Other reasons	11	4.8	34	7.8
Leopard predation	**5**	**2.2**	**113**	**25.9**
Total	227	99.9	437	100

households who suffered livestock losses, in ANP and MF respectively, attributed them directly to leopards (Table 3.6).

Furthermore, the cost of domestic animals was assessed from the local market rates. The average value of livestock was 80 USD (goat), 450 USD (cow) and 600 USD (ox/bull[1]). Mean livestock losses were 2.3 (ANP) or 2.4 (MF) goats/year and 1.8 (ANP) or 1.0 (MF) cattle/year. A goat owner was likely to lose 18–29% (ANP) or 9–31% (MF) of their annual income, while a cattle owner lost 51–82% (ANP) or 46–51% (MF) of their annual income.

[1]. One United States Dollar (USD) = 100 Pakistani Rupees (Rs.) at the time of the study

Table 3.6 Losses of domestic animals with respect to the number of households (HH) in previous year in the Ayubia National Park and surroundings areas and in the Murree Forest

Domestic animal	Leopard depredation			Disease/winter severity/fall			Other reasons (stolen, starvation, snakebite)		
	Losses	% Loss	HH	Losses	% Loss	HH	Losses	% loss	HH
ANP									
Goat	283	60	125	151	32	57	39	8	15
Cattle	17	24.6	9	46	66.6	29	6	8.6	6
Dog	19	86.6	16	3	13.6	2	0	0	0
Equid	2	100	1	0	0	0	0	0	0
Total	321		151	200		88	45		21
Murree									
Goat	85	33	35	135	52.73	56	30	14	19
Cattle	1	2.5	1	28	71.8	24	10	25.6	4
Dog	1	100	1	0	0	0	0	0	0
Equid	0	0	0	0	0	0	0	0	
Total	87		37	163		80	46		23

3.4 Discussion

The leopard tends to concentrate its predation on meso- (i.e. 2–40 kg) and large mammals (Hayward et al. 2006, for Africa; Lovari et al. 2013a, for Asia). Livestock is rarely the staple prey of leopards (*c.* 5% out of 44 studies, Hayward et al. 2006; Lovari et al. 2013a; Athreya et al. 2016; Shehzad et al. 2014), but it may constitute its main prey when wild ones are not available (Shehzad et al. 2014; Athreya et al. 2016; Khan et al. 2018). In the Ayubia National Park, in the Murree Forest, and in the Transitional Area, larger wild preys are nearly absent. Instead, goats and dogs become the staple prey of leopards. Livestock tends to roam around freely, often unguarded; it is only corralled at night, but in poorly constructed sheds (Khan 2015). Not surprisingly, most predation events on goats were recorded to occur during the day; conversely, dogs were often preyed on during the night, from sheds or courtyards or even porches (Khan et al. 2018). In our study areas, no significant seasonal changes were reported in the diet of common leopards, which suggests a constant availability of livestock throughout the year (by contrast, Kabir et al. 2014 reported seasonal variation in the predation of livestock by leopards, for their study area). Although nomadic herders visit the area seasonally, with several hundred heads each (Khan 2015), most livestock owners are residents. Thus, livestock is largely available to leopards all year long. Moreover, livestock is confined to sheds only when/where heavy snowfall occurs, thus it usually roams freely (Khan 2015). In the Abbottabad District, livestock density is high (1998: 269 heads/km^2, AAVV 1998) and increasing (Khan 2015). In consequence, livestock depredation is expected to remain a viable solution for leopards.

Overall, nearly 20% of households lost livestock (13.4% of total heads) to leopards over a study period of 12 months. Across all of our areas of study, people tended to keep a small number of livestock heads. Although livestock owners did not depend on domestic animals for their livelihood, they did use them for domestic purposes, e.g. to get milk for the family. In turn, considering the economic value of livestock, the relevant economic loss was equivalent to 8–31% and 46–92% of annual income for goat or cattle owners, respectively, with negligible differences between areas (Khan et al. 2018). Not surprisingly, our interviews of the households indicated that the perception of human-leopard conflict was greater among respondents who did suffer losses from leopards than among people who did not suffer them (Khan et al. 2018). Moreover, the perception of human-leopard conflict was higher in ANP than in MF. The former is inclusive of a National Park, and the percentage of livestock losses caused by leopards was greater than in the latter. Khan et al. (2018) suggested that, across their sites of study, depredation level was intensive (i.e. over 13% of livestock losses) with respect to other areas holding large carnivores (cf. 2.3–4.5% of livestock losses: see Khan et al. 2018, for references). Despite the remarkable depredation intensity, financial compensations were very low to absent. Consequently, the absence of a sound and appropriate financial compensation scheme for livestock losses is likely to emphasise the conflict between people and leopards (Khan 2015). Other causes of mortality, such as disease, winter severity and other minor factors, represented a substantial proportion of losses (*c.* 40% in ANP and 70% in MF: Khan 2015). Improved management of livestock would reduce the frequency of alternative mortality factors. For example, suitable winter sheds are likely to limit mortality caused by winter rigours. Additionally, appropriate health management of livestock should reduce the incurrence of disease. We suggest that financial incentives from veterinary services or livestock insurance plans (Dar et al. 2009; Kabir et al. 2014; Mishra et al. 2016) need to be implemented as long-term solutions to limit losses. We suggest that a reduction in overall losses will, in addition, limit the negative perception of livestock owners to depredations by leopards.

The leopard is an elusive large predator, rarely seen in daylight. Telemetry data indicate that it can come very close to human settlements and remain undetected at very close distances from human habitations, i.e. within several tens of metres (Khan et al. 2018). However, predation on domestic animals and potential encounters with humans may ensue. Quantitative data on leopard attacks on humans are scanty. In a large area that encompasses our sites of study, 19 attacks on people by leopards were reported in 10 years (Khan et al. 2018). At least 17 attacks occurred in daylight and 6 within 100 m of houses (Khan et al. 2018). This frequency of attacks on people was relatively small (2 events/year), although the percentage of attacks leading to a lethal outcome for humans (*c.* 50%) was not negligible (Khan et al. 2018), and it was much greater than elsewhere (cf. 1.6–17 events/year; percentage of lethal attacks: 0–36%, Treves and Naughton-Treves 1999; Athreya et al. 2010, 2013). Immediate human response to these events usually includes retaliatory killing, which has caused the loss of around 2.5 leopards/year, in the Abbottabad

District alone, and at least 6 leopards/year at the national level (105 individuals from 1998 to 2015; see Khan et al. 2018). Attacks of large carnivores on livestock or people are often the consequence of ultimate determinants, which range from inadequate livestock protection measures, habitat manipulation by humans to, especially, depletion of wild prey communities (e.g. Meriggi and Lovari 1996; Bagchi and Mishra 2006; Athreya et al. 2013; Khorozyan et al. 2015). If the proximity of leopards and humans is determined by the absence of wild prey, loss of relevant suitable habitat and an increase in human settlements, the removal of individual leopards is not likely to solve – in the long term – issues related to livestock depredation or attacks on people. It may only postpone conflicts until a new leopard arrives in the area. Translocation of 'problematic' individuals to new areas should also be discouraged. This action may increase the risk of attacks on humans/livestock (Athreya et al. 2010), as well as lethal attacks by local territorial leopards on the translocated individuals. In fact, stress induced by the translocation process, post-release movements of the leopard through the unfamiliar, human-dominated landscapes, and loss of fear of humans during captivity have all been suggested as factors that increase the risk of human attacks by leopards when these are released in the new areas (Athreya et al. 2010). Financial compensation schemes for livestock losses may contribute to limit negative attitudes of people towards carnivores, but would not solve the ultimate determinants of livestock depredation, i.e. absence of wild, large prey and habitat loss.

The goral, the wild boar and the musk deer are all potential prey for leopards (Lovari et al. 2013a), but they are in very low density in our areas of study. In Pakistan, the goral is rated as 'vulnerable' (Sheikh and Molur 2004) and is threatened because of poaching, habitat loss, diseases and competition with livestock (Chaiyarat et al. 1999; Abbas et al. 2009; Perveen et al. 2013). The wild boar is very rare, and the musk deer has been extirpated from our study areas (Khan 2015). We suggest that reintroductions/re-stocking operations, in addition to legal protection, are urgently needed to restore a local guild of wild ungulates, but that these measures would only be effective if the size of the area under effective protection (e.g. from deforestation and/or poaching) is increased (Khan et al. 2018). Moreover, improved preventive measures should be implemented or the opposite effect may be achieved (Suryawanshi et al. 2017). Additionally, free livestock grazing should be restricted (Khan et al. 2018) to limit the potential for competition and transmission of diseases (e.g. Bagchi et al. 2004; Mishra et al. 2001; Ekernas et al. 2017).

In addition to prey depletion, habitat loss is a severe threat to leopards. A fast decline in forested area has been occurring in all three sites of study, especially in the Ayubia National Park and its surroundings and in the Transitional Area. The TA represents a potential ecological corridor between the two largest forest patches in the area. Forest loss has been a major consequence of human population growth (Khan 2015). In our study areas – leopard strongholds in Pakistan – human communities are highly dependent on natural resources, such as wood for fuel, fodder and pasture for livestock, medicinal plants, construction material, water and non-timber forest products. In mountainous areas of central Asia, the leopard is mainly linked to forested habitats (e.g. Karanth 2013; Lovari et al. 2013b). Khan et al. (2018) showed that, in their area of study, a tagged leopard selected forested sites over non-forested ones. The area used by this leopard was remarkably larger than what was reported in other study areas in Asia: (91.5 km^2, MCP95%; 72.6 Kernel 95%: Khan et al. 2018; vs. *c.* 17–50 km^2 for males, with different estimators/time scales, in Asia: Rabinowitz 1989; Grassman 1999; Karanth and Sunquist 2000; Odden and Wegge 2005; Simcharoen et al. 2008; Odden et al. 2014). This result may suggest that habitat manipulation – through forest loss – has led to a reduction in habitat quality for leopards, as home range size usually shows an inverse relationship with habitat quality (Harestad and Bunnell 1979). Khan et al. (2018) also showed that, in ANP and surroundings, depredation events occurred mainly within a 500 m distance from the border of the forest. One can predict that if forest loss is not stopped, the frequency of contacts between humans and leopards will increase, further emphasising the conflict. Nevertheless, food habits of leopards were consistent across our three study sites, which showed a different forest cover. In turn, prey depletion is probably a more important determinant of livestock depredation than forest exploitation. Additionally, in forested areas, livestock is regularly left free to roam, which, in turn, is expected to further increase the risk of depredation by leopards.

In summary, we propose that actions should be targeted to reduce both short-term and long-term determinants of human-leopard conflict (see also Khan et al. 2018). Actions to reduce short-term determinants should include (*i*) improvement of management practices for livestock (adequate protection, means to deter leopards and access to veterinary facilities) and (*ii*) livestock insurance and/or an adequate financial compensation scheme for losses. Actions to reduce long-term determinants include (*i*) habitat improvement (i.e. a better forest protection/restoration) and, in particular, (*ii*) reintroduction/re-stocking of wild prey. If so, the welfare of human communities could improve through (*i*) the reduced dependency of leopards on livestock for food, (*ii*) a decrease of human-leopard and leopard-livestock encounters and (*iii*) increased survival of livestock overall. Unless socioeconomic conditions of local communities are benefited – or are, at least, not damaged – by the presence of a predator, any attempt at their coexistence will fail.

Acknowledgements We are indebted to WWF Pakistan and Human Wildlife and Nature Conservation Society for financial support and to the Wildlife Departments in KP and Punjab for letting us conduct our work in Ayubia-Murree. Further financial support was given by the University of Siena, Department of Life Sciences, through a PhD study grant to Uzma Khan. M. Odden, Zia Ullah, Muhammad Asad and, in particular, Mohammad Waseem provided suggestions and help in trapping operations and/or at organisational levels. Muhammad Arif yielded precious advice on questionnaire sampling and surveys. Syed Ghulam Mohayuddin and Naeem Shahzad for GIS analysis and support. A large number of scats were collected by Mohammad Sajid, to whom we are grateful for his untiring assistance in the field. DNA analyses were conducted by J. Janecka. A. Ghoddousi and A. Kshettry kindly took the time to provide extensive suggestions to improve an earlier draft. M.C. Montanaro revised our English. This work would not have been conducted without the backing and encouragement of Ali Hassan Habib and Brig Mukhtar Ahmed.

References

AAVV (1998) District census report, 1998. Government of Pakistan

Abbas FI, Ruchstuhl KE, Main A, Akhtar T, Rooney TP (2009) Distribution, population size, and structure of Himalayan grey goral *Naemorhedus goral* bedfordi (Cetartiodactyla: Bovidae) in Pakistan. Mammalia 76:117–235

Anwar MB, Jackson R, Nadeem MS, Janecka JE, Hussain S, Beg MA, Muhammad G, Qayyum M (2011) Food habits of the snow leopard *Panthera uncia* (Schreber,1775) in Baltistan, Northern Pakistan. Eur J Wildl Res 57:1077–1083

Athreya V, Odden M, Linnell JDC, Karanth U (2010) Translocation as a tool for mitigating conflict with leopards in human-dominated landscapes of India. Conserv Biol 25:133–141

Athreya V, Odden M, Linnell JDC, Krishnaswamy J, Karanth U (2013) Big cats in our backyards: persistence of large carnivores in a human dominated landscape in India. PLoS One 8(3):e57872. https://doi.org/10.1371/journal.pone.0057872

Athreya V, Odden M, Linnell JDC, Krishnaswamy J, Karanth KU (2016) A cat among the dogs: leopard *Panthera pardus* diet in a human-dominated landscape in western Maharashtra, India. Oryx 50:156–162

Bagchi S, Mishra C (2006) Living with large carnivores: predation on livestock by the snow leopard (*Uncia uncia*). J Zool 268:217–224

Bagchi C, Mishra C, Bhatnagar YV (2004) Conflicts between traditional pastoralism and conservation of Himalayan ibex (*Capra sibirica*) in the Trans-Himalayan mountains. Anim Conserv 7:121–128

Brillouin L (1956) Science and information theory. Academic Press: New York

Chaiyarat R, Laohajinada W, Kutintara U, Nabhitabhata J (1999) Ecology of the goral in om koi wildlife sanctuary, Thailand. Nat Hist Bull Siam Soc 47:191–205

Chakrabarti S, Jhala YV, Dutta S, Qureshi Q, Kadivar RF, Rana VJ (2016) Adding constraints to predation through allometric relation of scats to consumption. J Anim Ecol 85:660–670

Dar NI, Minhas RA, Zaman Q, Linke M (2009) Predicting the patterns, perceptions and causes of human-carnivore conflict in and around Machiara National Park, Pakistan. Biol Conserv 142:2076–2082

De Marinis AM, Asprea A (2006) Hair identification key of wild and domestic ungulates from southern Europe. Wildl Biol 12:305–320

Dinerstein E, Loucks C, Wikramanayake E, Ginsberg J, Sanderson E, Seidensticker J, Forrest J, Bryja G, Heydlauff A, Klenzendorf S, Leimgruber P, Mills J, O'Brien TG, Shrestha M, Simons R, Songer M (2003) The fate of wild tigers. Bioscience 57:508–514

Ekernas LS, Sarmento WM, Davie HS, Reading RP, Murdoch J, Wingard GJ, Amgalanbaatar S, Berger J (2017) Desert pastoralists' negative and positive effects on rare wildlife in the Gobi. Conserv Biol 31:269–277

ERRA (2007) Earthquake reconstruction and rehabilitation authority, District Profile Abbottabad. www.erra.gov.pk

FAO (2009) The state of the world's forest. Food and Agriculture Organisation of the United Nations, Rome, p 111

Flack J (1995) Interpretation of remotely sensed data using guided techniques. PhD thesis, School of Computer Science, Curtin University of Technology, Western Australia

Glen AS, Dickman CR (2006) Diet of the spotted-tailed quoll (*Dasyurus maculatus*) in eastern Australia: effects of season, sex and size. J Zool 269: 241–248

Grassman LI Jr (1999) Ecology and behavior of the Indochinese leopard in Kaeng Krachan National Park, Thailand. Nat Hist Bull Siam Soc 47:77–93

Harestad AS, Bunnell FL (1979) Home range and body weight: a revaluation. Ecology 60:389–402

Hass CC (2009) Competition and coexistence in sympatric bobcats and pumas. J Zool 278:174–180

Hayward MW, Henschel P, Brien JO, Hofmeyr M, Balme G, Kerley GIH (2006) Prey preference of the leopard (*Panthera pardus*). J Zool 270:298–313

Hebblewhite M, White CA, Nietvelt CG, McKenzie JA, Hurd TE, Fryxell JM, Bayley SE, Paquet P (2005) Human activity mediates a trophic cascade caused by wolves. Ecology 86:2135–2144
Jacobson AP et al (2016) Leopard (*Panthera pardus*) status, distribution, and the research efforts across its range. PeerJ 4:e1974. https://doi.org/10.7717/peerj.1974
Kabir M, Goddousi A, Awan MS, Awan MN (2014) Assessment of human–leopard conflict in Machiara National Park, Azad Jammu and Kashmir, Pakistan. Eur J Wildl Res 60:291–296
Karanth KU (2013) Leopard *Panthera pardus*. In: Johnsingh AJT, Manjrekar N (eds) Mammals of South Asia, vol Ist. Universities Press, Hyderabad, pp 559–573
Karanth KU, Sunquist ME (1995) Prey selection by tiger, leopard and dhole in tropical forests. J Anim Ecol 64:439–450
Karanth KU, Sunquist ME (2000) Behavioural correlates of predation by tiger (*Panthera tigris*), leopard (*Panthera pardus*) and dhole (*Cuon alpinus*) in Nagarhole, India. J Zool 250:255–265
Khan U (2015) Ecology, socioeconomic implications and habitat change of the common leopard (*Panthera pardus*) in the Western Himalayan Ecoregion, Pakistan. PhD thesis, University of Siena, Italy
Khan U, Lovari S, Ali Shah S, Ferretti F (2018) Predator, prey and humans in a mountainous area: loss of biological diversity leads to trouble. Biodivers Conserv 27:2795–2813
Khorozyan I, Ghoddousi A, Soofi M, Waltert M (2015) Big cats kill more livestock when wild prey reaches a minimum threshold. Biol Conserv 192:268–275
Kruuk H (1989) The social badger. Ecology and behaviour of a group-living carnivore (*Meles meles*). Oxford University Press, Oxford/New York/Tokyo
Kruuk H, Parish R (1981) Feeding specialization of the European badger Meles meles in Scotland. J Anim Ecol 50:773–788
Lovari S, Boesi R, Minder I, Mucci N, Randi E, Dematteis A, Ale SB (2009) Restoring a keystone predator may endanger a prey species in a human-altered ecosystem: the return of the snow leopard to Sagarmatha National Park. Anim Conserv 12:559–570
Lovari S, Minder I, Ferretti F, Mucci N, Randi E, Pellizzi B (2013a) Common and snow leopards share prey, but not habitats: competition avoidance by large predators? J Zool 291:127–135
Lovari S, Ventimiglia M, Minder I (2013b) Food habits of two leopard species, competition, climate change and upper treeline: a way to the decrease of an endangered species? Ethol Ecol Evol 25:305–318
Lovari S, Pokheral CP, Jnawali SR, Fusani L, Ferretti F (2015) Coexistence of the tiger and the common leopard in a prey-rich area: the role of prey partitioning. J Zool 296:122–131
Lumetsberger T, Ghoddousi A, Appel A, Khorozyan I, Walter M, Kiffner C (2017) Re-evaluating models for estimating prey consumption by leopards. J Zool 203:201–210
Lyngdoh S, Shrotriya S, Goyal SP, Clements H, Hayward MW, Habib B (2014) Prey preferences of the snow leopard (*Panthera uncia*): regional diet specificity holds global significance for conservation. PLoS One 9:e88349. https://doi.org/10.1371/journal.pone.0088349
Meriggi A, Lovari S (1996) A review of wolf predation in southern Europe: does the wolf prefer wild prey to livestock? J Appl Ecol 33:1561–1571
Mishra C, Van Wieren SE, Ketner P, Heitkönig I, Prins HHT (2001) Competition between domestic livestock and wild bharal *Pseudois nayaur* in the Indian Trans-Himalaya. J Appl Ecol 41:344–354
Mishra C, Redpath SR, Suryawanshi KR (2016) Livestock predation by snow leopards: conflicts and the search for solutions. In: MCCarty T, Mallon D (eds) Snow leopards. Academic/Elsevier, London, pp 59–67
Mukherjee S, Goyal SP, Chellam R (1994) Standardisation of scat analysis techniques for leopards (*Panthera pardus*) in Gir National Park, Western India. Mammalia 58:139–143
Nowak RM (1999) Walker's mammals of the world, vol 1. The Johns Hopkins University Press, Baltimore
Odden M, Wegge P (2005) Spacing and activity patterns of leopards *Panthera pardus* in the Royal Bardia National Park, Nepal. Wildl Biol 11:145–151

Odden M, Atherya V, Rattan S, Linnell JDC (2014) Adaptable neighbours: movement pattern of GPS-collared leopards in human dominated landscape in India. PLoS One 9(11):e112044. https://doi.org/10.1371/journal.pone.0112044
Packer C, Ikanda D, Kissui B, Kushnir H (2005) Lion attacks on humans in Tanzania. Nature 436:927–928
Perveen F, Khan A, Shah AH (2013) Hunting and trapping pressure on the Himalayan Grey goral *Naemorhedus goral* (Hardwicke) (Artiodactyla Bovidae) in Kohistan, Pakistan. Am J Zool Res 1:5–11
Rabinowitz AR (1989) The density and behaviour of large cats in a dry tropical forest in Kha Khaeng Wildlife Sanctuary, Thailand. Nat Hist Bull Siam Soc 37:235–251
Ripple WJ, Estes JA, Beschta RL, Wilmers CC, Ritchie EG, Hebblewhite M, Berger J, Elmhagen B, Letnic M, Nelson MP, Schmitz OJ, Smith DW, Wallach AD, Wirsing AJ (2014) Status and ecological effects of the World's largest carnivores. Science 343(6167):1241484. https://doi.org/10.1126/science.1241484
Roberts TJ (1997) The mammals of Pakistan. Oxford University Press, Karachi, p 527
Shehzad W, Nawaz A, Pompanon F, Coissac E, Riaz T, Shah SA, Taberlet P (2014) Forest without prey: livestock sustain a leopard *Panthera pardus* population in Pakistan. Oryx 49:248–253
Sheikh KM, Molur S (2004) Status and Red list of Pakistan's Mammals, Based on Pakistan Conservation and Assessment Management Plan (CAMP), International Union for Conservation of Nature (IUCN), Pakistan
Simcharoen S, Barlow ACD, Simcharoen A, Smith JLD (2008) Home range size and daytime habitat selection of leopards in Huai Kha Khaeng Wildlife Sanctuary, Thailand. Biol Conserv 141:2242–2250
Sinclair ARE, Mduma S, Brashares JS (2003) Patterns of predation in diverse predator-prey system. Nature 425:288–290
Stein AB, Athreya V, Gerngross P, Balme G, Henschel P, Karanth U, Miquelle D, Rostro-Garcia S, Kamler JF, Laguardia A, Khorozyan I, Ghoddousi A 2016. *Panthera pardus*. The IUCN Red List of Threatened Species 2016: e.T15954A102421779. https://doi.org/10.2305/IUCN.UK.2016-1.RLTS.T15954A50659089.en
Suryawanshi KR, Redpath SM, Bhatnagar YV, Ramakrishnan U, Chaturvedi V, Smout SC, Mishra C (2017) Impact of wild prey availability on livestock predation by snow leopards. R Soc Open Sci 4(6):170026. https://doi.org/10.1098/rsos.170026. eCollection 2017 Jun
Teerink BJ (1991) Hair of west-European mammals. Cambridge University Press, Cambridge, UK
Treves A, Naughton-Treves L (1999) Risk and opportunity for humans coexisting with large carnivores. J Hum Evol 36:275–282

Chapter 4
Recent Changes in Wolf Habitat Occupancy and Feeding Habits in Italy: Implications for Conservation and Reducing Conflict with Humans

Alberto Meriggi, Elisa Torretta, and Olivia Dondina

4.1 Introduction

The wolf (*Canis lupus*), along with other species of large predators, is among the most controversial and challenging species to conserve in the human-dominated world (Mech and Boitani 2003). As a result of negative perception due to the species impact on some human activities (Treves and Karanth 2003), wolf populations historically have considerably been reduced all over Europe (Ripple et al. 2014a, b). At present, the European continent, taking into consideration all continental European countries, is succeeding in maintaining and to some extent restoring, viable wolf populations (Chapron et al. 2014). Wolves are currently, permanently, found in all countries, with the exception of Belgium and the Netherlands, where only occasional data has been recorded in recent years (Lelieveld et al. 2016). At present, the wolf is the most widespread large carnivore in Europe (Fig. 4.1), while it is the second most abundant species behind the brown bear (*Ursus arctos*), with an estimated total population of more than 12,000 individuals (Chapron et al. 2014). Most populations have increased or remained stable in recent years, although there are some exceptions, such as the Sierra Morena population (Spain), which is on the brink of extinction, with only one pack detected in 2010 (Kaczensky et al. 2013).

As the area of both occasional and stable wolf presence is increasing in Europe, a clear understanding of this expansion process on the continental level is needed to support actions aimed at population conservation (de Groot et al. 2016).

A. Meriggi (✉) · E. Torretta
Dipartimento di Scienze della Terra e dell'Ambiente, Università degli Studi di Pavia, Pavia, Italy
e-mail: alberto.meriggi@unipv.it

O. Dondina
Dipartimento di Scienze dell'Ambiente e della Terra, Università degli Studi di Milano-Bicocca, Milan, Italy
e-mail: olivia.dondina@unimib.it

F. M. Angelici, L. Rossi (eds.), *Problematic Wildlife II*,
https://doi.org/10.1007/978-3-030-42335-3_4

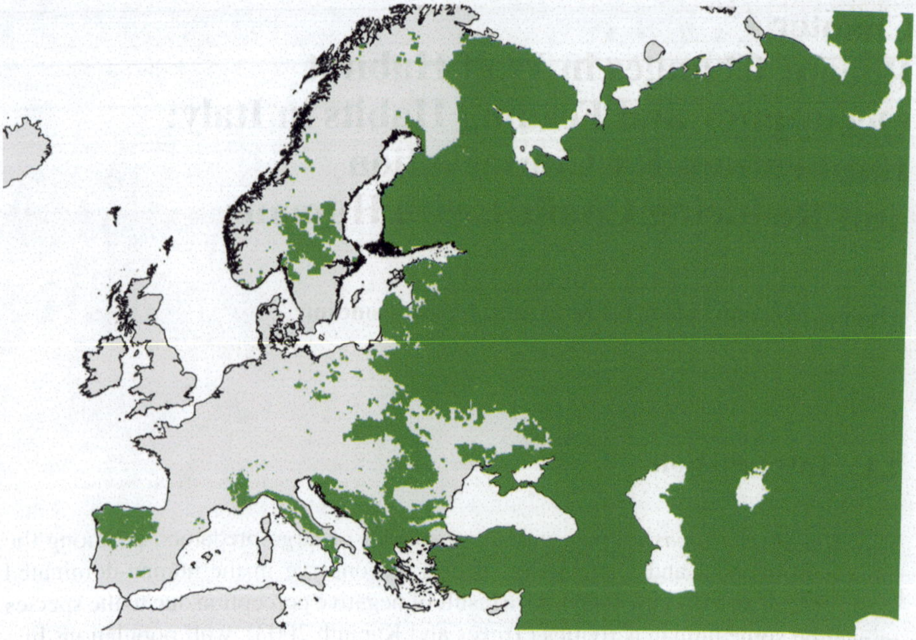

Fig. 4.1 Wolf distribution in Europe, obtained from the report on species of community interest compiled by ISPRA for the National Biodiversity Strategy referred to 2013 (Genovesi 2005) and the IUCN Red List of Threatened Species (Version 2017-2)

Unfortunately, even though the conservation of viable wolf populations needs to be planned and coordinated on a very large scale, which often spans over different intra- and international borders (Linnell and Boitani 2011), monitoring programmes usually cover relatively small areas. In fact, the legal responsibility to conserve wolves is handled independently in each country and, in the case of some federal countries, this responsibility has been devolved to an even lower level (regions, provinces or cantons; Linnell and Boitani 2011).

At an international level, the wolf is included in several conservation agreements. The most relevant European laws aimed at wolf conservation are the 1979 Bern Convention (Convention on the Conservation of European Wildlife and Natural Habitats), which includes the species in Appendix II (strictly protected species), and the 1992 Habitats Directive (Council Directive 92/43/EEC on the Conservation of Natural Habitats and of Wild Fauna and Flora), which includes the wolf in Appendix II (habitat conservation needs) and Appendix IV (fully protected). The overall goal is to maintain and to restore viable populations of wolves as an integral part of ecosystems in coexistence with people. Both the Bern Convention and the Habitats Directive have thus allowed some countries to make some exceptions or local modifications to the status of wolves, applying a number of measures, both preventive and reductive, including local removal of individuals (Boitani 2000).

In this study, we first summarized the dynamics of wolf distribution in Italy over the last 50 years. Subsequently, we traced changes that have occurred in Italy in the wolf diet, comparing it with what has been observed in other European countries, with the aim of understanding how it has affected the evolution of the human-predator conflict due to the impact on both livestock and wild ungulates. To provide specific insights about past and current distribution, feeding habits and conflict with humans characterizing the Italian wolf population, we present three case studies. All the studies were carried out in northern Italy, i.e. in the area where the dynamics of wolf packs could affect the future of the whole Italian wolf population.

4.2 Changes in Wolf Range in Italy Over the Last 50 Years

Historically, the wolf was widely distributed across Europe, but human persecution greatly reduced and fragmented its range. In Italy, the wolf was killed legally until it was totally eradicated from the Alps and almost totally from the Apennines (Breitenmoser 1998). At the beginning of the 70s, wolves in Italy were close to extinction, and here surviving at their, now historically, minimum size of less than 100 individuals in isolated areas in the central and southern Apennines (Zimen and Boitani 1975; Boitani 1992). Wolves dealt with the impoverishment of ecological conditions of the landscape, generally characterized by the destruction of natural habitats, and survived on various food sources of anthropogenic origin, such as livestock and garbage (Boitani 1982; Mattioli et al. 1995).

During the 1970s the first national laws aimed at species conservation were promulgated: in 1971 the wolf was excluded from the harmful species list and hunting activity was forbidden (Ministerial Decree 'Natali'), while in 1976 the wolf became a fully protected species (Ministerial Decree 'Marcora'). This large carnivore has also benefited from the socioeconomic changes that have taken place since the end of the Second World War, which have led to a general improvement in habitat quality. Indeed, human activities have decreased in many mountainous and hilly areas because of a widespread exodus from rural areas, and the associated abandonment of agricultural lands. This rural area abandonment has favoured the return of large mammals (Pereira et al. 2012), such as wild ungulates, which can sustain large carnivore populations. In this context, the surviving wolf populations quickly expanded into newly restored natural areas (Chapron et al. 2014). Such newly recolonized areas were characterized by a continuous and marked decrease in resident human population and a marked recovery of wild ungulates (Meriggi et al. 1991).

The species has re-colonized the northern Apennines (since the late 1980s) and reached the western Alps (from the early 1990s) through the ecological corridor of Liguria (a NW region of Italy) (Lucchini et al. 2002; Valière et al. 2003; Fabbri et al. 2007). A moderate bottleneck occurred during the re-colonization process, and gene flow between the Apennines and the Alps was moderated (1.25–2.50 wolves per generation; Fabbri et al. 2007). Bottleneck simulations showed that a total of 8–16 effective founders explained the genetic diversity observed in the Alps (Fabbri et al.

2007). Nowadays, the expansion of the wolf range is continuing towards the eastern Alps and Europe (Switzerland, Austria and France).

At present, the wolf is a protected species under National Law n. 157/1992 and the European legislation.

In Italy the wolf is one of most studied species, although unfortunately a comprehensive overview of the current population status at the national level is difficult to build, as many short-term and often small-scale monitoring projects have been conducted using heterogeneous methods (Galaverni et al. 2016). Most recent results have shown the presence of more than 300 packs corresponding to a minimum of ca. 1300 wolves in the period 2009–2013 in the Apennines (Galaverni et al. 2016) and ca. 190 wolves in the period 2015–2016 in the Alps (Marucco et al. 2017). Both in the Apennines and in the Alps, the population was double the size of previous estimates: in the Apennines the previous estimate was of about 600–800 wolves between 2006 and 2011 (Kaczensky et al. 2013), while in the Alps it was of 57–89 wolves in 2012 (Wolf Alpine Group 2014). Wolf range is continuous from the Aspromonte in the South (Calabria region) to the western Alps in the North (Piedmont and Aosta Valley regions), passing through the Liguria region. Moreover, there are scattered wolf packs in the central and eastern Alps: in 2015 a pack established itself in the Lombardy Prealps (across the province of Como and Switzerland) (Marucco et al. 2017), and since 2013 a pack representing the first case of natural mixing between the Italian and Balkan wolf populations occupies the Lessinia Regional Park (across the Veneto region and the province of Trento) (Ražen et al. 2016) (Fig. 4.2).

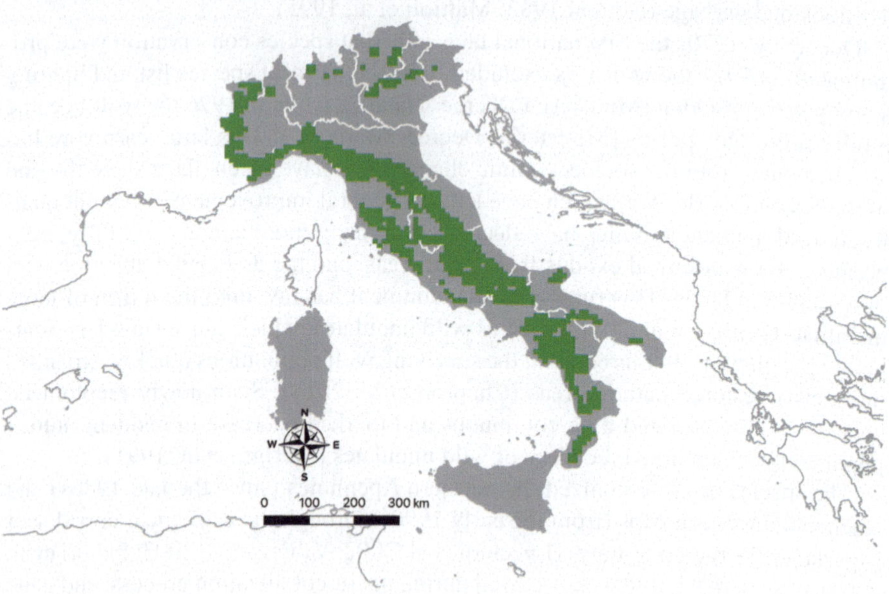

Fig. 4.2 Wolf distribution in Italy, obtained from the report on species of community interest compiled by ISPRA for the National Biodiversity Strategy referred to 2013 (Genovesi 2005) and updated with more recent results (Gaudiano et al. 2016; Marucco et al. 2018)

Overall, the current Italian wolf population size and trend seem favourable. Such a trend might be attributable to the colonization of suitable new areas, rather than to an increase in the number of individuals within already existing packs (Hayes and Harestad 2000; Galaverni et al. 2016). Interestingly, in many Apennine regions, wolves have expanded their range in both remote and nearly urbanized areas (Caniglia et al. 2012, 2014); furthermore, wolf observations are increasing in several parts of the Po plain were the presence of small packs is becoming stable (e.g. the provinces of Parma and Piacenza).

The recovery of the wolf population can positively influence ecosystem equilibria via the reconstruction of trophic cascades (Ripple et al. 2014a but see also Allen et al. 2017a, b). On the other hand, this wolf expansion is exacerbating conflicts with human activities, mainly livestock breeding and wild ungulate hunting, especially in areas where predators had been gone for a long time (Linnell and Boitani 2011).

4.3 Changes in Wolf Diet in Relation to the Increase in Wild Ungulate Species

The wolf is an opportunistic predator with high trophic adaptability and, consequently, a diet which changes according to the local abundance of and accessibility to food sources, even if at a local scale it is focused on few prey species (Ansorge et al. 2006; Gazzola et al. 2005; Barja 2009; Meriggi et al. 2011; Zlatanova et al. 2014; Newsome et al. 2016).

Similar to North America, in northern and eastern Europe the wolf mostly feeds on wild ungulates, generally, one or two main prey species characterize its diet (Peterson and Ciucci 2003). Conversely, during the re-colonization of southern Europe, the expanding wolf populations have locally adapted to feed on other, more available resources, such as livestock (Fico et al. 1993), small and medium-sized mammals (Castroviejo et al. 1975), fruits (Meriggi et al. 1991), and rubbish (Macdonald et al. 1980; Reig et al. 1985).

In the early 1970s, wild ungulates were rare in most of Italy, and the wolf diet showed high occurrences of rubbish and livestock (Boitani 1982; Meriggi et al. 2011). Subsequently, the diet of wolves evolved towards a higher consumption of large wild ungulates, becoming more and more similar to that of the wolves of North American and north-eastern European areas (Meriggi and Lovari 1996; Capitani et al. 2004; Meriggi et al. 2011, 2015a). This significant change in the wolf diet is certainly due to the increased availability of wild ungulates (Meriggi et al. 1991, 2011). Meriggi et al. (2015a) stated that, in Italy, the increase in wild ungulate abundance and distribution range was mainly due to the recovery of wild boar (*Sus scrofa*) (from 1900 individuals estimated in 1977 to 900,000 in 2010) and roe deer (*Capreolus capreolus*) (from 102,000 in 1977 to 457,794 in 2010) (Database Ungulates 2006–2010, ISPRA).

Currently, in Italy, wild ungulate communities are characterized by a high number of species with higher local population densities compared to northern Europe and North America; as a consequence, wolves have a wider range of prey choice to satisfy pack food requirements (Okarma 1995; Jędrzejewski et al. 2002; Peterson and Ciucci 2003; Melis et al. 2009). However, packs do not hunt according to prey abundance only, but there are a number of factors that can influence the selectivity of wolf predation on wild ungulates. The composition of the ungulate community, the degree of habitat overlap between predator and prey, accessibility, vulnerability and profitability of prey (encounter rate with prey and probability of a successful kill), previous hunting experience, and cultural transmission are the main factors affecting prey selection and, consequently, the predator diet (Endler 1991; Huggard 1993; Meriggi et al. 1996).

Overall, in Italy wild boar is the most important prey for wolves (Meriggi et al. 1996; Mori et al. 2017). This evidence agrees with the overall diet of wolves across Europe but it is in contrast with that of some European countries, where red deer (*Cervus elaphus*) is the preferred species of wolves, while wild boar represents the second species in terms of importance (Reig and Jędrzejewski 1988; Jędrzejewski et al. 1992; Smietana and Klimek 1993; Jędrzejewska et al. 1994; Okarma 1995; Meriggi and Lovari 1996; Gula 2004; Zlatanova et al. 2014; Newsome et al. 2016). The high occurrence of wild boar in the wolf diet in Italy probably depends on the wide distribution and remarkable densities of the species, as well as on its eco-ethological characteristics. Indeed, wild boar live in large groups, that are easily detectable by a predator (Meriggi et al. 2011). Moreover, sub-adults are forced to leave matriarchal groups when new births occur, thus becoming a highly vulnerable prey for wolves (Heck and Raschke 1980). The second species in terms of importance in the diet of the Italian wolf is the roe deer, which is widespread and characterized by high densities particularly in the northern Apennines and in the Alps (Meriggi et al. 2011). In general, the elusive behaviour of roe deer, together with their low aggregation rate, make this species a difficult prey for wolves. However, when present at high densities, roe deer can become a suitable prey for wolves mainly because of the high encounter rate and low handling time (Meriggi et al. 1996; Meriggi and Lovari 1996; Jędrzejewski et al. 2002). The red deer was the third most preyed species by the wolf in Italy, and the fourth after fallow deer (*Dama dama*) in the northern Apennines (Meriggi et al. 2011). This situation differs from what happens in other European countries, and it may be due to the clumped spatial distribution of red deer in Italy, which is the result of local reintroduction projects. Chamois (*Rupicapra rupicapra* and *Rupicapra pyrenaica*) and mouflons (*Ovis musimon*) generally represent accessory prey (Mori et al. 2017): chamois can climb quickly on steep cliffs, making them very difficult to catch for wolves (Poulle et al. 1997), while mouflons are rare and have a very scattered distribution. This is the general pattern of ungulate species consumption in Italy, but local exceptions exist. For example, Palmegiani et al. (2013) found that chamois was the most consumed species in the Gran Paradiso National Park (western Alps).

Meriggi et al. (2011) found that along with the increase of wild ungulates in the wolf diet, a significant reduction of livestock predation occurred. The authors concluded that when wolves have a choice between wild and domestic ungulate, they prefer the former. This result seems a constant of the wolf predatory behaviour in Europe, even if some local-scale cases depart from this general model (Okarma 1995; Cozza et al. 1996; Poulle et al. 1997; Zlatanova et al. 2014). Moreover, the authors have also highlighted that the breadth of the wolf diet, calculated in several studies, was negatively and significantly related to the occurrence of wild ungulates, suggesting that when wild ungulates are rare, wolves are forced to use alternative food sources, such as small mammals, lagomorphs, fruits and garbage, other than livestock.

This finding is of decisive importance from a conservation point of view. It suggests that to limit damage made by wolves on husbandry, which is the main cause of wolf persecution by humans, conservation measures should primarily restore and maintain a rich and abundant wild ungulate community. Indeed, when a well-structured ungulate community is present, if the abundance of one prey species drops, wolves can compensate with others to satisfy their food requirements, without focusing their predatory efforts on livestock (Meriggi and Lovari 1996).

4.4 Impacts on Husbandry and Wild Ungulate Populations

The impact of the wolf on livestock is different depending on the geographical context. Differences are mainly due to husbandry methods, environmental factors and wild ungulate community characteristics (Imbert et al. 2016; Pimenta et al. 2017). In areas characterized by a very low abundance of wild ungulates, such as in Portugal and Greece, wolf diet is mainly composed of livestock (Papageorgiou et al. 1994; Vos 2000; Migli et al. 2005; Torres et al. 2015). On the other hand, in areas where shepherds equip pastures with preventive methods and the availability of wild ungulate is high, such as in Germany, predation upon livestock is rare (Wagner et al. 2012). Conversely, in areas, such as France and northern Italy, were wolves are re-colonizing previously occupied areas, even if wild ungulates represent the main prey of wolves, the use of livestock is still noticeable (MEEDDAT–MAP 2008; Milanesi et al. 2012). For instance, a recent study on wolf trophic habits carried out in the northern Apennines (Meriggi et al. 2015a) showed that livestock still represents 20% of the predator's diet. This situation is probably due to the long absence of wolves from these areas, which has led to the loss of the cultural tradition of coexistence with the predator (Gazzola et al. 2005). Indeed, in areas of new re-colonization, many pastures lack any measures of attack prevention (e.g. wire or electric fences, sirens, flashing and strobe lights, guard dogs), making domestic ungulates the most vulnerable and profitable prey species in such areas. Another possible reason for which a non-negligible percentage of livestock occurs in the wolf diet in these areas is the high presence of dispersing lone wolves in areas of new re-colonization. Imbert et al. (2016) found that packs tended to consume more wild ungulates than dispersing wolves, which showed a greater use of livestock.

Even though the overall loss due to wolf attacks is around 20% at a regional level, several studies have demonstrated that predation events on livestock are generally concentrated on a few farms that consequently suffer remarkable damage (Fritts et al. 1992; Cozza et al. 1996; Ciucci and Boitani 1998; Gazzola et al. 2008; Dondina et al. 2014). This happens because different grazing areas are characterized by different degrees of predation risk depending, as mentioned above, on both management and environmental factors (Vos 2000; Espuno et al. 2004; Treves et al. 2004; Edge et al. 2011; Van Liere et al. 2013; Dondina et al. 2014; Pimenta et al. 2017). The type of breeding, the species being reared, and the age of grazing animals seem to be the two main factors affecting the frequency of wolf attacks (Meriggi et al. 2011). For instance, the frequencies of predation upon cattle depends on the availability of calves younger than 15 days old that are in grazing areas, because this is the age class most vulnerable to wolf attacks (Meriggi et al. 1991, 1996). Moreover, some management decisions, such as the lack of surveillance of free-grazing herds, the lack of preventive methods and the habit of allowing asynchronous births with a prolonged season directly in the grazing areas, definitely increases the risk of predation by wolves (Fritts et al. 1992; Paul and Gipson 1994; Espuno et al. 2004; Edge et al. 2011). In addition, some environmental characteristics, such as high percentage of forest and shrub cover within grazing areas, may increase pastures' vulnerability to attacks (Fritts et al. 1992; Mech et al. 2000; Kaartinen et al. 2009; Van Liere et al. 2013; Dondina et al. 2014).

Overall in Italy, sheep is the species most preyed upon, with a proportion ranging from 64% to 97%, followed by cattle and goats in similar proportions (Meriggi et al. 1996; Ciucci and Boitani 1998; Gazzola et al. 2008). However, there are differences among areas in the consumption of domestic ungulates, which are mainly due to the accessibility of grazing animals, in turn affected by the adoption of different husbandry methods (Meriggi et al. 1996).

Several authors have stressed that predation on livestock represents the main issue for the conservation of wolves in Italy as well as in Europe (Boitani 2000; Lovari et al. 2007; Imbert et al. 2016; Pimenta et al. 2017). Indeed, mortality due to illegal killing still represents one of the main risk factors for the species (Lovari et al. 2007). An effective management of the conflict between wolf and husbandry could thus be a crucial tool for a feasible long-term strategy for wolf conservation (Boitani 2000; Mech et al. 2000; Treves et al. 2004; Bradley and Pletscher 2005; Dondina et al. 2014).

The natural re-colonization of wolves has triggered serious conflicts with humans, indeed they consider large predators as competitors in hunting, because of predations on wild ungulates (Gazzola et al. 2007). Therefore, another crux regarding the management of the conflict between humans and the large carnivore lies in understanding to what extent predation by wolves is a regulatory or limiting factor acting on wild ungulates populations.

In a prey-predator system, regulation occurs when predation is density dependent, leading prey populations to a density equilibrium. Conversely, if predation is density independent, a limiting effect occurs, while if it is inversely density dependent, a depensatory effect occurs (Meriggi et al. 2011). In these last cases predation rate increases as prey density decreases. This situation mainly occurs when there are no refuges for prey, or when predators do not have an alternative prey source (Messier 1991; Marshal and Boutins 1999; Jędrzejewski et al. 2002; Wittmer et al. 2005; Sinclair et al. 2006). In the case of the wolf, several studies carried out in Europe showed that the predator typically preys on more abundant ungulate species that constantly increase despite the impact of wolf predation (Meriggi et al. 2015a). In Europe, wolves can have a limiting effect at a local scale only on red deer (Meriggi et al. 2011), for which predation represents 40% of total mortality (32% in the western Alps; Gazzola et al. 2007). Conversely, the mortality of wild boar and roe deer due to wolves accounts for a very small proportion of their annual mortality, and evidence of compensatory predation became obvious in central Europe (Milanesi et al. 2012). For these species, the main limiting factors were habitat quality, food resource availability, climatic conditions, hunting and traffic accidents (Okarma 1995; Jędrzejewski et al. 2002; Gazzola et al. 2007; Melis et al. 2009). The impact of wolf predation on the prey species community is likely to change with the composition of the multi-prey species community (Sand et al. 2016). Overall, in Italy, considering the high densities of wild ungulates and the high richness characterizing the ungulate community, it is unlikely that wolf predation can become a limiting effect. Indeed, the high productive habitats and the mild climate can reduce the negative impacts of predation, by enhancing the ability of ungulate species to compensate for predation losses by increasing its reproduction rate (Melis et al. 2009).

4.5 Management of Wolf-Human Conflicts

As stated above, one of the main issues concerning wolf conservation in Italy depends on wolf depredation on domestic ungulates, which leads to intense illegal killing in spite of the species' legal protection (Lovari et al. 2007). In addition, competition with hunters for game species can be considered a source of conflict.

Many wildlife management tools can be used to promote the coexistence between large carnivores and human activities; Carter and Linnell (2016) defined coexistence as a 'dynamic but sustainable state in which humans and large carnivores co-adapt to living in shared landscapes where human interactions with carnivores are governed by effective institutions that ensure long-term carnivore population persistence, social legitimacy, and tolerable levels of risk'. This concept suggests that managers should consider different measures to promote coexistence. On the one hand, there are many practical tools aimed at reducing the impact of wolves (and other large carnivores) on human activities, such as economic compensation and incentives, information campaigns, spatial zoning (e.g. habitat protection from human development), technical changes in livestock husbandry, restoration of wild prey populations and limited removal of individuals. On the other hand, there are efforts to engage diverse stakeholder groups, building trust and dialogue between groups with different viewpoints toward carnivores, and the adoption of novel decision-making structures that ensure participation and legitimacy (Carter and Linnell 2016).

Reducing the negative impact of wolf presence on human activities may involve both non-lethal and lethal measures; the former may sometimes be challenging to implement, but they are usually not very controversial per se. In contrast, the use of lethal measures, which involves killing wolves, can be highly controversial from ecological, social, and legal viewpoints (Linnell et al. 2017). Moreover, it seems that lethal control of wolves can be ineffective and that it can increase livestock depredation (Wielgus and Peebles 2014; Kompaniyets and Evans 2017).

As highlighted before, the European legislative agreements previously discussed have been instrumental in fostering the recovery of the wolf in Europe (Chapron et al. 2014; Fleurke and Trouwborst 2014). However, an emerging issue is to what extent international legislation provides constraints on the possibilities of single countries adopting lethal measures (Linnell et al. 2017). Linnell et al. (2017) analysed the arguments supporting wolf and other large carnivore hunting, which include reducing damage to livestock, reducing competition with hunters for game species, pleasing rural people who are able to directly influence their own interactions with wolves, adding a direct economic value to the presence of wolves (e.g. the sale of trophy hunting licenses) and reducing poaching. Nevertheless, considering the effects of wolf removal at a regional scale, a lot of research has pointed out that it did not decrease livestock depredations (Musiani et al. 2005; Harper et al. 2008; Wielgus and Peebles 2014). Specifically, Wielgus and Peebles (2014) found that the expected number of livestock killed increased by 4–6% each year despite the number of wolves culled the previous year, until wolf mortality exceeded 25% (data collected during the period 1987–2012 in Idaho, Montana and Wyoming). These findings were criticized by Poudyal et al. (2016), who, using the same dataset, reached opposite conclusions: they showed that more culling of wolves led to fewer killings of livestock in the following year than expected in the absence of culling, in particular considering the expected number of cattle and sheep killed (decreased by 1.9% and 3.4% respectively). Finally, Kompaniyets and Evans (2017) reanalysed the same dataset and confirmed the positive significant link between cattle depredation and the number of wolves killed found by Wielgus and Peebles (2014) but disagreed with their interpretation: these authors had claimed that removing wolves that depredate cattle would slow down the relative rate of cattle depredations; however, cattle depredations would increase until the wolf population reached stability. Thus, the effect of wolf removal on reducing cattle depredations only becomes evident when the wolf population growth closes in on its stability.

Considering the effects of wolf removal at a local scale, Bradley and Pletscher (2005) found that it reduced the recurrence of depredations, depending on the number of wolves remaining in the pack. In detail, partial pack removal was only slightly more effective in reducing depredation recurrence than no removal, only if it occurred within the first 7 days after the depredation event, while the removal of the entire pack decreased depredation recurrence the most (data collected during the period 1987–2008 in Idaho, Montana and Wyoming).

Managers making decisions in response to wolf depredation also need to consider biological issues. The removal of pack members can cause changes in the pack social structure and, sometimes, may lead to the breakup of the pack itself (Brainerd et al. 2008). Consequently, it can only lead to a temporary decrease in livestock attacks, because suitable empty areas are rapidly recolonized by dispersing individuals, which have a bigger consumption of livestock than packs (Imbert et al. 2016). Besides biological considerations, the costs related to wolf hunting might be considerable, and they could exceed the cost of the damage caused by wolf depredation. For example, Bradley and Pletscher (2005) argued that pack removal can be only effectively accomplished using radio-collaring and aerial gunning, two expensive management tools. Last but not least, it is important to underline that the Italian wolf population is considered 'vulnerable' on the national Red List of Threatened Species (Rondinini et al. 2013), as (i) it is still genetically isolated from other European populations (with the exception of the Lessinia pack), (ii) human-caused mortality remains high (about 20%; mainly poaching and accidents) and (iii) there is anthropogenic hybridization with dogs. This last issue, ignored for decades, deserves particular attention. Free-ranging or feral dogs (*C. l. familiaris*), which amount to an estimated value of 700,000 heads distributed especially in central and southern Italy, can successfully reproduce with wolves. A recent research showed that the first hybridization events might have occurred in the main population refugia in central and southern Italy (during and even before the 1970s), followed by more frequent events in the northern Apennines, and finally in human-dominated landscapes along the Tyrrhenian and Adriatic coasts, likely retracing the main population expansion wave around the 1990s (Galaverni et al. 2017). The current challenge is the collection of data on the behaviour and ecology of wolf-dog hybrids under free-ranging conditions, which are largely lacking at present.

Recently, another source of wolf-human conflict has emerged: since the perception of wolf presence seems to be increasing, both due to advances in technology (e.g. the non-professional use of camera traps), and because of the positive trend of wolf population, fear has become a frequent component of conflict. Finally, wolves have become symbols of social struggles, like the opposition between traditional rural and urban publics, or between local or regional, and national or international bodies (Linnell et al. 2017).

4.6 Case Studies in Northern Italy

To provide specific insights about the past and the current situation of the wolf in Italy, we present the results of different researches carried out in the northern part of the country, where the dynamics of wolf packs could affect the future of the entire Italian wolf population by regulating the dispersal flow from the Apennines to the Alps. In particular, we report studies carried out (i) in the provinces of Pavia and Piacenza, in the Lombardy and Emilia Romagna regions, respectively

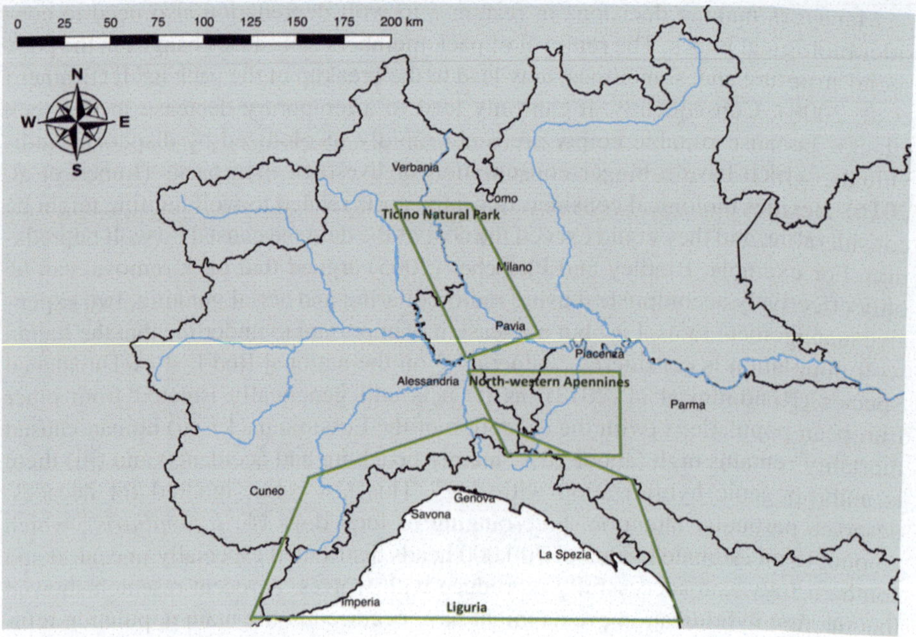

Fig. 4.3 The three study areas

(north-western Apennines); (ii) in the Liguria region; and (iii) in the Ticino Natural Park (lowland area of the Lombardy region) (Fig. 4.3).

4.6.1 North-Western Apennines

The north-western Apennines is a mountain area of about 1000 km^2. It can be divided into three altitudinal zones: (i) a low hill zone between 200 and 400 m. a. s. l. (urban areas 12.6%, arable lands 67.7%, woodlands 14.4%, pastures 2.8%); (ii) a medium hill zone between 400 and 700 m. a. s. l. (urban areas 4.3%, arable lands 39.7%, woodlands 47.3%, pastures 5.0%); and (iii) a mountain zone between 700 and 1700 m a. s. l. (urban areas 2.9%, arable lands 16.7%, woodlands 70.1%, pastures 7.5%). The community of wild ungulates inhabiting the study area includes, in order of abundance, wild boar, roe deer, fallow deer, and red deer, with the former two species widespread over the whole study area and the last two characterized by a clumped distribution (Meriggi et al. 2015a). Livestock (mainly cattle, but also goats, sheep and horses) graze freely from April to October on pastures located on the ridges of mountain chains.

The wolf population inhabiting the area has been monitored since 1987, 2 years after the first record of wolf reproduction. In 1988, two breeding pairs with pups were present in the part of the study area falling within the borders of Lombardy (about 300 km^2), plus an indeterminate number of non-breeding individuals (Meriggi et al. 1991). About 20 years later, between 2007 and 2008, four stable packs were present in the whole study area (Milanesi et al. 2012). Between 2007 and 2012 wolf range was limited to the medium hilly and mountainous zone of the study area covering about 480 km^2 (Lombardini 2009; Raviglione 2011; Perversi 2012), while in 2015 the total range covered about 800 km^2 (Fusari 2015). In 4 years, the total species range doubled, while the surface covered by core areas (200 km^2 in 2011 and 180 km^2 in 2015) remained almost the same, suggesting that the number of stable packs probably had not changed, while dispersing individuals and new pairs of wolves were colonizing new areas. The colonization of new hilly and lowland areas has been recently confirmed by several sightings occurring both in Lombardy (Fusari 2015) and Emilia Romagna (Wildlife Service of Emilia Romagna) regions.

4.6.1.1 Changes in Habitat Use

The first study performed on wolf habitat use in north-western Apennines (Meriggi et al. 1991) found that, overall, pastures and scrublands were the most used habitats all year round. The selection of these two habitat types probably depended on the feeding habits of wolves in the same period within the study area, mainly composed of livestock and *Rosaceae* fruits, particularly widespread in scrublands. A study carried out about 20 years later (Perversi 2012) in the same area highlighted that wolves tended to avoid elevations lower than 1000 m a. s. l., while they preferred areas above 1400 m a. s. l. This result agrees with several other studies carried out in Europe (Massolo and Meriggi 1998; Glenz et al. 2001; Ciucci et al. 2003; Jędrzejewski et al. 2005), and depends on the high forest cover characterizing such elevations. A high degree of forest cover was one of the habitat characteristics selected by wolves because it ensures the presence of several wild ungulate species and provides optimal areas for dens and rendezvous site locations. Other than woodlands, pastures were also selected by wolves, probably because of their importance in providing food resources during the grazing season. Finally, it turned out that wolves select uncultivated areas, which represent one of the preferred habitats of the roe deer. Interestingly, the changes in habitat use observed over 20 years in the study area follow the changes observed in the trophic habits of the predator in the same period. In fact, as described below, the diet of wolves shifted from a prevalent use of fruits and livestock to a prevalence of wild ungulates that typically inhabit woodlands and uncultivated areas. The most recent study carried out in the study area (Fusari 2015) confirmed that wolves select forest cover and an abundant and diversified community of wild ungulates, proving that this is one of the most important variables in determining the stable presence of the species (Massolo and Meriggi 1998; McLoughlin et al. 2004; Ronnenberg et al. 2017). Another interesting result from this last research was the avoidance of elevations lower than 200 m a. s. l. The avoidance of low elevation was also shown in the study carried out by Perversi (2012), however, in that study wolves avoided elevations lower than 1000 m a. s. l.

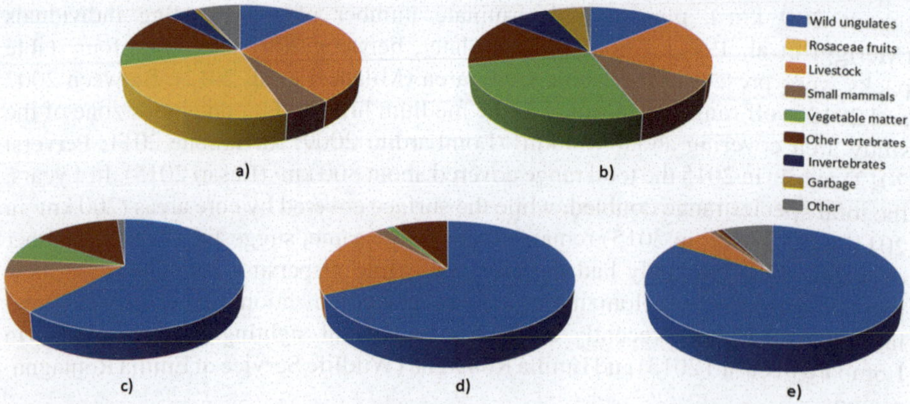

Fig. 4.4 Variation of mean percent volume of food categories in wolf diet in north-western Apennines (1987–2015). (**a**) Meriggi et al. (1991, 1987–1989). (**b**) Meriggi et al. (1996, 1987–1992). (**c**) Milanesi et al. (2012, 2007–2008). (**d**) Meriggi et al. (2015a, 2007–2012). (**e**) Procaccio (2015, 2014–2015)

This change was probably due to the saturation of all the optimum areas located at higher altitudes by the species and the consequent colonization of free areas located at lower altitudes.

4.6.1.2 Changes in Wolf Diet

During the late 1980s (1987–1989), the diet of wolves was composed of two main items: *Rosaceae* fruits and livestock, each accounting for about 25% of the total diet (Fig. 4.4a). Other important food items in terms of volume were wild ungulates, mainly wild boar (13.0%), which were the only species of ungulate available in the study area at the end of the 1980s; small rodents (5.4%); and other fruits and grasses (5.4%). Despite changes in diet composition during the year, only small variations were recorded in diet breadth calculated through the B-index (Feisinger et al. 1981) (winter: B = 0.53; spring: B = 0.61; summer: B = 0.51; autumn: B = 0.49).

A few years later (1987–1992), the wolf food habits were very similar to those described by Meriggi et al. (1991) (fruits 24.7%; livestock19.1%; wild ungulates 13.5%, with 11.9% represented by wild boar; other vertebrates 13.9%; small mammals 11.5%; invertebrates 7.6%; garbage 5.8% and other minor categories) (Fig. 4.4b), and the diet breadth was even higher (B = 0.79) (Meriggi et al. 1996). These results showed that 30 years ago wolves in the north-western Apennines were essentially opportunistic predators.

Conversely, the wolf diet analysed 20 years later (2007–2008) showed that the most used categories were wild ungulates (62.2%, with 47.2% represented by wild boar, 10.4% by roe deer, 3.8% by fallow deer and 1.0% by red deer), livestock (10.8%), vegetable matter (7.3%) and small mammals (4.4%) (Fig. 4.4c). The diet of wolves was thus mainly based on wild ungulates, causing a decrease in the diet breadth value (B = 0.47) (Milanesi et al. 2012). The marked increase of the occurrence of wild boar and roe deer in the wolf diet can be explained by the dramatic increase in the abundance of both species over the same period, as demonstrated by the number of hunted wild boars and by the results from roe deer monitoring studies (Milanesi et al. 2012).

Enlarging the study period to 2007–2012, some differences emerged. The category mainly used were wild ungulates (68.0%, composed of wild boar 59.9%, roe deer 23.1%, fallow deer 3.6% and red deer 13.4%) followed by livestock (20.2%), small mammals (2.2%) and vegetable matter (0.4%) (Meriggi et al. 2015a) (Fig. 4.4d). This research showed, even more markedly, that the wolf diet in the north-western Apennines is typical for a predator of large herbivores. Indeed, wild ungulates cover almost the whole range of trophic categories, while the others have dropped to negligible levels, as highlighted by the lower values observed for diet breadth (B = 0.33). Even if the frequency of the occurrence of wild ungulates was similar to that reported for the period 2007–2008, there are some differences regarding the species consumed; the use of wild boar slightly decreased, while the use of roe deer and red deer increased, probably due to the recent increase in the abundance of these two species in the study area (Milanesi et al. 2012; Meriggi et al. 2015a).

The most recent research (Procaccio 2015) confirmed the high selectivity which characterizes the wolf diet, with an almost exclusive consumption of wild ungulates (83.8%, represented by wild boar 25.2%, roe deer 65.9% and fallow deer 3.4%) (Fig. 4.4e) and a very low value of diet breadth (B = 0.17). The most important result of this last study was the significant increase in the use the roe deer, which probably depends on the recent expansion of wolves towards lower elevation hilly zones. The study showed that the roe deer was used according to its availability in mountainous areas (Manly Index ũFC; = 0.30, CI 95% = 0.10–0.85; Manly 2006), while it was selected in the hilly part of the study area (Manly Index ũFC; = 0.83, CI 95% = 0.68–0.98), where wolves occur as dispersal individuals, breeding pairs or small packs, and predation upon roe deer can satisfy their food requirements, also allowing complete consumption of the prey in a short time (Thurber and Peterson 1993; Schmidt and Mech 1997; Hayes and Harestad 2000; Jędrzejewski et al. 2002; Sand et al. 2016).

The consumption of livestock by wolves in the north-western Apennines did not change significantly between the periods 1987–1989 and 2007–2012, passing from a percentage of consumption of 25% to 20%. Conversely, the most recent research carried out in the study area (Procaccio 2015) showed a significant decrease (6%) in livestock consumption by wolves. Moreover, some changes in species selection occurred over the last 30 years. In 1987–1989 and 1987–1992 the species most preyed upon was sheep (55.6% and 51.3%, respectively) (Meriggi et al. 1991; Meriggi et al. 1996), in 2007–2008 goats (62.0%) (Milanesi et al. 2012) and in 2007–2012 calves (43.0%) (Meriggi et al. 2015a), while in 2014–2015 the wolf returned to using mostly sheep (34.4%). These differences probably do not depend on variations in the number of grazing animals of different species (Milanesi et al. 2012), but on changes in their availability due to management decisions. In general, the not negligible use of domestic prey by wolfs in the study area for a long period, probably depends on the high accessibility of livestock due to ineffective preventive measures against wolf attacks. Indeed, Dondina et al. (2014) analysed data on official predation events occurring during the period 2005–2012 in the study area, and showed that 56.6% of the pastures of the whole area are potentially exposed to wolf predation risk. In particular, the authors suggested that the pastures that suffered predation during the study period were those in which cattle births occur directly on pasture areas, those that had at least one period of free grazing during the year and cattle farms that were lacking any preventive methods. Moreover, the risk of wolf attacks seems to increase with the increasing complexity of pasture shape and the decrease in the percentage of coniferous forest.

4.6.2 *The Liguria Region*

Liguria is a 5343 km^2 region located in north-western Italy (Fig. 4.3); it is characterized by a broad altitude range, from 0 to 2200 m a. s. l. Forests cover 63.7% of the whole area. Pastures and scrublands cover 5.3% and 10.3%, respectively. Cultivated lands (12%) are localized along main valleys, and permanent crops are dominated by olive trees and vineyards. Urban areas (6.3%) are concentrated near the coasts and along flat and wide valleys.

The wild ungulate community includes wild boar and roe deer, widely distributed with high densities, and fallow deer and chamois, with scattered distribution (Imbert et al. 2016; Torretta et al. 2017). Hunters harvest these ungulate species annually. Moreover, the red deer has a sporadic presence along the boundaries with Piedmont and Emilia Romagna regions.

Sheep and goat farms were more than double the cattle farms. Seventy-five percent of livestock farms practiced a free grazing period from April to October on mountain pastures or in areas surrounding the folds.

A wolf monitoring project was carried out from 2007 to 2014 along the whole region, using non-invasive survey methods. Non-invasive genetic analysis estimated a minimum of 5 wolf packs, with an average pack size of 4.2 ± 0.8 individuals (mean ± SD). Twenty-two dispersing wolf resamples showed an average distance between subsequent sampling of 25.6 km (SD = 38.2) with a maximum distance of 138.1 km.

4.6.2.1 Wolf Diet: Differences Between Packs and Dispersing Individuals

Wild ungulates were the main food category consumed by wolves (64%), followed by domestic ungulates (27%). Other food categories (small mammals, medium-sized mammals, grasses, invertebrates, fruits and garbage) were consumed occasionally (Fig. 4.5a). Among wild ungulates, the most consumed were wild boar (33.5%) and roe deer (22.6%); the other species were consumed less. Among

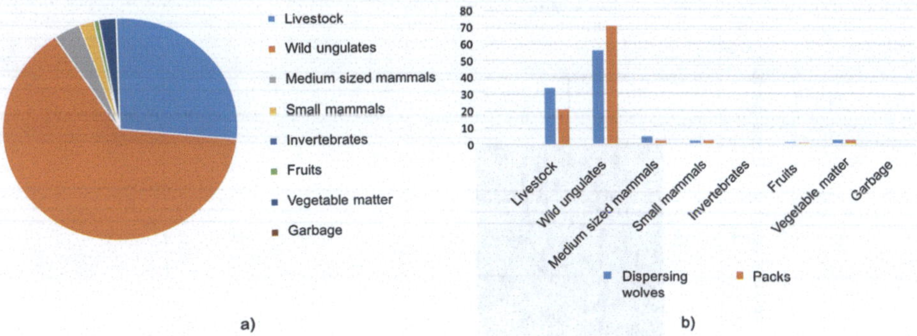

Fig. 4.5 Mean percent volume of food categories in wolf diet in Liguria (2007–2014) (**a**) and differences in mean percent volume of food categories between dispersing wolves and packs (**b**)

livestock species, wolves chiefly consumed goats (63.8%), followed by sheep (26.9%) and cattle (3.7%).

Comparing the diets of packs and dispersing wolves, a higher consumption of livestock and medium-sized mammals emerged for dispersing wolves (34% and 4.4%, respectively, vs. 21.2% and 1.8%, respectively, for packs), whereas the consumption of wild ungulates was higher for individuals belonging to packs (70.5% vs. 55.9% for dispersing wolves; Fig. 4.5b). The consumption of livestock was therefore negatively related to the presence of packs. Indeed, structured packs hunt on their territory and know where to find wild prey, whereas dispersing individuals, new to the area, do not know the distribution of wild prey and prey on livestock, which is easier to detect. Moreover, the consumption of livestock was negatively related with the adoption of prevention methods (nocturnal shelter, presence of shepherds and dogs, electric fences) and with roe deer abundance, the second most consumed potential prey species for wolves. Finally, the extent of deciduous woods decreased livestock consumption, probably in relation to the great density of wild boar and deer that can be found in this habitat type (Imbert et al. 2016).

4.6.2.2 Livestock Depredation and Prevention Measures

From 2002 to 2014, 349 attacks occurred on 82 pastures (out of 303) and wolves killed 808 heads. The attack frequency had a seasonal trend, with cattle mainly depredated in June and sheep and goats in September.

A predictive model of predation risk, performed on pastures with predation vs. pastures without predation, which considered as predictive variables both habitat and pasture characteristics, was used to classify each pasture. The best models obtained classified 40.3% of pastures (n = 122) as being at high predation risk (Meriggi et al. 2013, 2015b).

This analysis was performed each year on the basis of the attacks recorded the previous year and the results were used to identify pastures at high depredation risk.

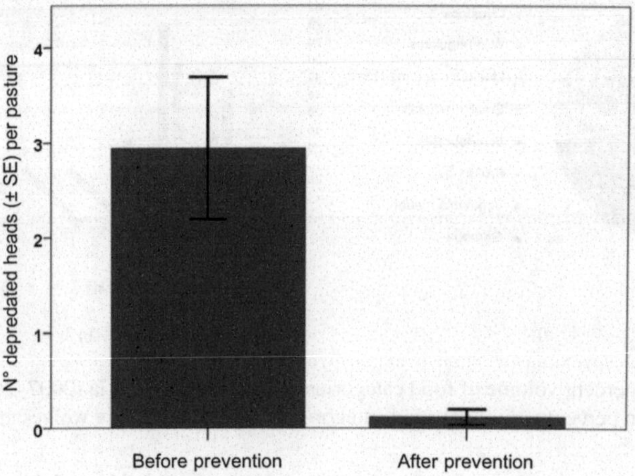

Fig. 4.6 Depredated heads per pasture before and after the adoption of preventive measures

These pastures were supported with prevention measures, such as fencing, shelters for births and night-time, and acoustic alarm devices. From 2008 to 2014, 25 pastures were supported with one or two preventive measures. By analysing the number of depredated heads, a significant decrease was observed between pre- and post-preventive measures adoption (Fig. 4.6).

4.6.2.3 Wolves and Wild Ungulates Interactions

In the Ligurian Alps the most consumed species among wild ungulates were the wild boar (47.3%) and, secondly, the roe deer (42.5%). The other species were less consumed (fallow deer: 5.9%; chamois: 1.2%). Considering the most consumed species, it emerged that the wild boar was selected only during the non-denning season (from September to March; Manly Index ũFC; = 0.41, CI 95% = 0.30–0.55), while it was consumed in proportion to its abundance during the denning season (from April to August; Manly Index ũFC; = 0.20, CI 95% = 0.13–0.27), whereas the roe deer was always selected (non-denning season: Manly Index ũFC; = 0.42, CI 95% = 0.29–0.56; denning season: Manly Index ũFC; = 0.52, CI 95% = 0.39–0.65).

In this area, wolf range encompasses all wild ungulate ranges, but it tended to most closely match the roe deer and wild boar ranges, selecting and avoiding similar habitats, with few differences between seasons.

Considering the species' use of the diel cycle, a key factor influencing the encounter rate between wolves and prey, a low temporal overlap between the wolf and the roe deer patterns during the denning season was observed. This might be evidence of the anti-predator strategy of the roe deer trying to avoid the predator during the reproductive period. Regarding the non-denning season, the temporal overlap between the wolf and the roe deer patterns was moderate. The wolf and the wild boar revealed very similar nocturnal activity patterns, with coinciding peaks of activities, especially during the non-denning season, when the two species had a high temporal overlap.

Spatial and temporal results suggested that the roe deer and the wild boar represent the most accessible prey species for wolves in the Ligurian Alps, because predator habitat use and activity rhythms overlap with those of these ungulate species.

To successfully hunt prey species, wolves need to share their range, searching for them in the most suitable habitat types and in the periods of the diel cycle during which they are mainly active. Fallow deer and chamois consumption was low in the Ligurian Alps, and wolves showed also relatively low overlap with these species. These results suggested that wolves might be primarily specialized in wild boar predation, as they showed substantial spatial and temporal overlap with this species, and secondly on roe deer predation, especially during the denning season when they probably take advantage of the presence of fawns (Torretta et al. 2016).

During 2007–2016, 23 roe deer (37.1%), 18 fallow deer (29%), 16 wild boars (25.8%) and 5 chamois (8.1%) were localized on 62 kill sites. The main environmental and human-related factors influencing the distribution of wild ungulate kill sites were analysed, and the results showed a negative effect of mixed forests, urban areas and road density, the latter without statistical significance, and a positive effect of steep slopes (> 60°) and open areas. Prey are more vulnerable to predators under certain conditions and predators are able to select for these conditions. Wolves select certain habitats in which to kill their prey: they preferred steep, open habitats far from human presence, where wild ungulates are more easily detected and chased (Torretta et al. 2017).

4.6.3 *Ticino Natural Park*

The Ticino Natural Park is a 220 km^2 wide protected area, which comprises the residual continuous forests of the lowland area of north-western Italy (Lombardy region).

In May 2017 a female wolf was camera-trapped in the central part of the Park. From morphological characteristics, it is an individual that can be ascribed to the Italian subspecies (*Canis lupus italicus* Altobello 1921). This is the first finding of a living wolf inhabiting the Ticino Valley after 150 years. The presence of wolves in the Ticino Natural Park was confirmed by a second observation of a young male obtained by camera-trapping. Based on landscape characteristics and on the distribution of wolf presence signs collected in the Park in summer 2017, wolves probably reached the woodland areas of the Ticino Valley starting from the hilly areas of north-western Apennines, which are about 10 km^2 from the Park boundaries. It is a finding of exceptional importance, especially if it is considered in the wider context of the wolf distribution dynamic in Italy. While the occupation of the Apennines and western Alps seems to have been completed and stabilized, the wolf currently finds it difficult to expand in the eastern Alps and to therefore join the Dinaric-Balkan population. The re-colonization of the eastern Alps would thus be a fundamental step for wolf conservation in Italy as well as in Central Europe (Genovesi 2002), since it would increase the chances of originating mixed packs and of increasing the local genetic diversity (Randi 2011; Fabbri et al. 2014). In this context, the importance of wolf presence within the Ticino Natural Park is evident, as it suggests that the habitats of the Park are characterized by a level of quality that guarantees the effectiveness of the Ticino Natural Park in playing the role of an ecological corridor between the Apennines and the Alps for this large predator. In particular, given its strategic position, the Park could play a key role within the wolf distribution dynamic in Italy, creating a direct shortcut that could facilitate the movement of individuals from the Apennines towards the North, directly in correspondence with the colonization front in the Alps.

4.7 Conclusions

4.7.1 How to Mitigate Conflicts with Husbandry and Hunting?

The increase in the wolf population in Europe and the enlargement of its range in highly populated areas have enhanced the negative perception of the predator presence, and increased conflict with husbandry and hunting. Several researches carried out in Europe have shown that the impact on livestock farms mainly depends on the rearing methods and pasture location and characteristics. In particular, it is known that: (i) among livestock species, sheep and goats are selected; (ii) predation on cattle occurs mainly upon calves younger than 15 days; (iii) free-ranging livestock farms are more exposed to predation risk; (iv) the presence of an abundant, rich and diversified community of wild ungulates can reduce the attacks on livestock; and (v) isolation of pasture areas, their size and the presence of more livestock species increases predation risk. In the light of these findings it seems that the damage to husbandry can be reduced by the adoption of different prevention methods, in particular: (i) livestock sheltering at night, (ii) surveillance of livestock during grazing, (iii) using livestock guard dogs, (iv) electric fences and (v) visual and acoustic deterrents. These measures can be effective but, in some circumstances, they fail because of the wolf habituation. In France, for example, despite the wide and increasing use of prevention methods, the number of depredated livestock heads increased exponentially from 1992 to 2016, reaching 10,000 (Meuret et al. 2017). Moreover, the adoption of prevention measures is very expensive, for example in France, in 2016, € 27.7 million was spent on prevention and € 3.2 million for the reimbursement of losses (Meuret et al. 2017). For these reasons it is necessary that prevention measures are targeted at farms with a high predation risk, which can be easily identified by predictive models of risk probability. In this way, prevention can be adopted before the occurrence of the attack, effectively lowering the number of depredated heads.

As far as the perceived impact of wolves, on wild ungulate populations, is concerned, and the consequent conflict with hunting, the problem can be solved only by detailed and widespread information about the true effect of wolf predation on the population dynamics of wild herbivores, while at the same time promoting the establishment of rich and diversified communities of wild ungulates.

4.7.2 *Perspectives on the Management of Wolf Populations*

Making a budget of costs and benefits of the presence of the wolf is extremely difficult. On the one hand, the presence of the wolf in a geographic area has economic costs that in some cases can be very high. They can be summarized as follows: (i) expenditure for verification surveys, including the costs staff of public administrations, veterinary services and experts who have to ascertain and quantify the damage; (ii) reimbursement for depredated heads, which should include both the immediate damage and the reduction of the farm yield because of fertility lowering, abortion increase and decrease in milk production (Meuret et al. 2017), and that should also consider that the economic value of a depredated animal could be much lower than that of an animal at the end of the breeding cycle (e.g. lambs and calves); (iii) charges for the disposal of carcasses of killed animals; and (iv) prevention costs, including changes in breeding methods.

On the other hand, the presence of wolves has several benefits that are much more difficult to quantify than costs. They include: (i) the natural regulation of populations of wild ungulates that, if uncontrolled, can cause damage to agriculture, livestock farming by competition and forestry (reduction of renewal, reduction of the growth of forestry plants, etc.); (ii) the health prophylaxis, consuming carcasses of wild and domestic animals that have died from natural causes and preying on individuals debilitated by diseases; (iii) maintaining and/or restoring the equilibrium in prey communities (density-dependent predation guarantees the presence of all prey species of a community and reduces the risk of extinction by preventing some species from taking advantage of others by competition); (iv) natural selection of wild ungulate species (the coexistence between predators and prey lead to the fixing of eco-ethological characteristics of prey species, which is lost if the selective pressure exerted by wolves is lacking); (v) increased local biodiversity and system stability; and (vi) maintaining the human cultural aspects linked to the presence of predators.

A definitive resolution to the conflict between humans and wolf is unrealistic, mainly because it seems difficult to completely eliminate damage to husbandry. Nevertheless, it should be considered that livestock farming in mountain areas is rarely a self-sustaining activity. In fact, almost all farms benefit from economic aid from the countries and the European Union. Why not think that such aid might be a compensation for the wolf predation? If this could be accepted, important resources could be economized. It does not mean, however, that all possible actions for the reduction of damage should not be implemented, in particular the indirect methods for the restoration of the wild ungulate communities, the increase in the density of prey populations and the maintenance of stable packs in suitable areas. In this framework, the numerical control of the wolf cannot be considered an effective method for solving depredation of livestock. Considering the economic issue, the most effective management tools are too expensive to be employed (Bradley and Pletscher 2005), especially for many European countries. Considering the biological issue, wolf removal would act primarily on packs, which are more easily contactable, possibly fragmenting their members (Brainerd et al. 2008), and thus increasing the number of dispersing wolves, with consequent increase in livestock consumption (Imbert et al. 2016).

Acknowledgements The collection and the analysis of data within the Liguria region was supported by the project 'Il Lupo in Liguria' (2012–2014), funded by the Regional Administration of Liguria (ROP/ERDF funds), coordinated by the Antola Regional Park and supervised by the University of Pavia.

Wolf monitoring in the Ticino Natural Park involves two regional parks (Parco Regionale Lombardo della Valle del Ticino and Ente di Gestione delle Aree Protette del Ticino e del Lago Maggiore) and two universities (University of Milano-Bicocca and University of Pavia).

We thank A. Brangi, N. Brielli, L. Caviglia, F.F. Codina, G. D'Alessandro, V. Dagradi, C. Delfoco, S.M. Fusari, F. Gatti, C. Imbert, M. Lombardini, M. Magliano, P. Milanesi, A. Pastorino, M. Perversi, E.L. Procaccio, F. Puopolo, S. Raviglione, A. Repossi, L. Schenone, M. Serafini, D. Signorelli and M. Zanzottera for their field work collaboration.

References

Allen BL, Allen LR, Andrén H, Ballard G, Boitani L, Engeman RM, Fleming PJS, Ford AT, Haswell PM, Kowalczyk R, Linnell JDC, Mech LD, Parker DM (2017a) Can we save large carnivores without losing large carnivore science? Food Webs 12:64–75

Allen BL, Allen LR, Andrén H, Ballard G, Boitani L, Engeman RM, Fleming PJS, Ford AT, Haswell PM, Kowalczyk R, Linnell JDC, Mech LD, Parker DM (2017b) Large carnivore science: non-experimental studies are useful, but experiments are better. Food Webs 13:49–50

Ansorge H, Kluth G, Hahne S (2006) Feeding ecology of wolves *Canis lupus* returning to Germany. Acta Theriol 51:99–106

Barja I (2009) Prey and prey-age preference by the Iberian Wolf (*Canis lupus signatus*) in a multiple-prey ecosystem. Wildl Biol 15:147–154

Boitani L (1982) Wolf management in intensively used areas of Italy. In: Harrington FH, Paquet PC (eds) Wolves of the world. Noyes Publications, Park Ridge, pp 158–172

Boitani L (1992) Wolf research and conservation in Italy. Biol Conserv 61:125–132

Boitani L (2000) Action plan for the conservation of wolves in Europe (*Canis lupus*), No. 18–113. Council of Europe, Strasbourg

Bradley EH, Pletscher DH (2005) Assessing factors related to wolf depredation of cattle in fenced pastures in Montana and Idaho. Wildl Soc Bull 33:1256–1265

Brainerd SM, Andrén H, Bangs EE, Bradley EH, Fontaine JA, Hall W, … Mack CM (2008). The effects of breeder loss on wolves. J Wildl Manag 72:89–98

Breitenmoser U (1998) Large predators in the Alps: the fall and rise of man's competitors. Biol Conserv 83:279–289

Caniglia R, Fabbri E, Cubaynes S, Gimenez O, Lebreton JD, Randi E (2012) An improved procedure to estimate wolf abundance using non-invasive genetic sampling and capture-recapture mixture models. Conserv Genet 13:53–64

Caniglia R, Fabbri E, Galaverni M, Milanesi P, Randi E (2014) Noninvasive sampling and genetic variability, pack structure, and dynamics in an expanding wolf population. J Mammal 95:41–59

Capitani C, Bertelli I, Varuzza P, Scandura M, Apollonio M (2004) A comparative analysis of wolf (*Canis lupus*) diet in three different Italian ecosystems. Mamm Biol 69:1–10

Carter NH, Linnell JDC (2016) Co-adaptation is key to coexisting with large carnivores. Trends Ecol Evol 31:575–578

Castroviejo J, Palacios F, Garzon J, Cuesta L (1975) Sobre la alimentacion de los canides ibericos. Proceeding XII Congress International Union Game Biologists, Lisboa, pp 39–46

Chapron G, Kaczensky P, Linnell JD, Von Arx M, Huber D, Andrén H, …, Balčiauskas L (2014) Recovery of large carnivores in Europe's modern human-dominated landscapes. Science 346:1517–1519

Ciucci P, Boitani L (1998) Wolf and dog depredation on livestock in Central Italy. Wildl Soc Bull 26:504–514

Ciucci P, Lucchini V, Boitani L, Randi E (2003) Dewclaws in wolves as evidence of admixed ancestry with dogs. Can J Zool 81(12):2077–2081

Cozza K, Fico R, Battistini ML, Rogers E (1996) The damage-conservation interface illustrated by predation on domestic livestock in central Italy. Biol Conserv 78:329–336

de Groot GA, Nowak C, Skrbinšek T, Andersen LW, Aspi J, Fumagalli L, …, Marucco F (2016) Decades of population genetic research reveal the need for harmonization of molecular markers: the grey wolf *Canis lupus* as a case study. Mammal Rev 46:44–59

Dondina O, Meriggi A, Dagradi V, Perversi M, Milanesi P (2014) Wolf predation on livestock in an area of northern Italy and prediction of damage risk. Ethol Ecol Evol 27:200–219

Edge JL, Beyer DE, Belant JL, Jordan MJ, Roell BJ (2011) Adapting a predictive spatial model for wolf Canis spp. Predation on livestock in the Upper Peninsula, Michigan, USA. Wildl Biol 17:1–10

Endler JA (1991) Interactions between predators and prey. In: Davies NB, Krebs JR (eds) Behavioural ecology. An evolutionary approach. Blackwell Scientific Publications, Oxford, pp 169–196

Espuno N, Lequette B, Poulle ML, Migot P, Lebreton JD (2004) Heterogeneous response to preventive sheep husbandry during wolf recolonization of the French Alps. Wildl Soc Bull 32:1195–1208

Fabbri E, Miquel C, Lucchini V, Santini A, Caniglia R, Duchamp C, Weber JM, Lequette B, Marucco F, Boitani L, Fumagalli L, Taberlet P, Randi E (2007) From the Apennines to the Alps: colonization genetics of the naturally expanding Italian wolf (*Canis lupus*) population. Mol Ecol 16:1661–1671

Fabbri E, Caniglia R, Kusak J, Galov A, Gomercic T, Arbanasic H, Huber D, Randi E (2014) Genetic structure of expanding wolf (*Canis lupus*) populations in Italy and Croatia, and the early steps of the recolonization of the Eastern Alps. Mamm Biol 79:138–148

Feisinger P, Spers EE, Poole RW (1981) A simple measure of niche breadth. Ecology 62:27–32

Fico R, Morosetti G, Giovannini A (1993) The impact of predators on livestock in the Abruzzo region of Italy. Rev Sci Tech OIE 12:39–50

Fleurke FM, Trouwborst A (2014) European regional approaches to the transboundary conservation of biodiversity: the Bern Convention and the EU Birds and Habitats Directives. In: Kotze L, Marauhn T (eds) Transboundary governance of biodiversity. Martinus Nijho Publishers, Leiden/Boston, pp 128–162

Fritts SH, Paul WJ, Mech LD, Scott DP (1992) Trends and management of wolf-livestock conflicts in Minnesota, Resource publication 181. U.S. Department of the Interior Fish and Wildlife Service, Washington, DC

Fusari S (2015) Fattori influenzanti la distribuzione e la frequentazione del lupo (*Canis lupus*) in Oltrepò Pavese. Master thesis, University of Pavia, Italy

Galaverni M, Caniglia R, Fabbri E, Milanesi P, Randi E (2016) One, no one, or one hundred thousand: how many wolves are there currently in Italy? Mamm Res 61:13–24
Galaverni M, Caniglia R, Pagani L, Fabbri E, Boattini A, Randi E (2017) Disentangling timing of admixture, patterns of introgression, and phenotypic indicators in a hybridizing wolf population. Mol Biol Evol 34:2324–2339
Gaudiano L, Sorino R, Corriero G, Frassanito AG, Strizzi C, Notarnicola G (2016) Stato delle conoscenze del lupo *Canis lupus* in Puglia. X Congresso Italiano di Teriologia (Acquapendente, VT), 20–23 Aprile 2016
Gazzola A, Bertelli I, Avanzinelli E, Tolosano A, Bertotto P, Apollonio M (2005) Predation by wolves (*Canis lupus*) on wild and domestic ungulates of the western Alps, Italy. J Zool 266:205–213
Gazzola A, Avanzinelli E, Bertelli I, Tolosano A, Bertotto P, Musso R, Apollonio M (2007) The role of the wolf in shaping a multi-species ungulate community in the Italian western Alps. Ital J Zool 74:297–307
Gazzola A, Capitani C, Mattioli L, Apollonio M (2008) Livestock damage and wolf presence. J Zool 274:261–269
Genovesi P (2002) Piano d'azione nazionale per la conservazione del lupo (*Canis lupus*). Quad Cons Nat 13:1–94
Genovesi P (2005) Mitigation of conflicts and Italian policies for large carnivore conservation. In: Ciucci P, et al (eds) Grandi Carnivori e Zootecnia tra conflitto e coesistenza. Ozzano dell'Emilia: Istituto Nazionale della Fauna Selvatica 'Alessandro Ghigi'. Biologia e Conservazione della Fauna 115. pp 21–25
Glenz C, Massolo A, Kuonen D, Schlaepfer R (2001) A wolf habitat suitability prediction study in Valais (Switzerland). Landsc Urban Plan 55(1):55–65
Gula R (2004) Influence of snow cover on wolf *Canis lupus* predation patterns in Bieszczady Mountains, Poland. Wildl Biol 10:17–23
Harper EK, Paul WJ, Mech LD, Weisberg S (2008) Effectiveness of lethal, directed wolf depredation control in Minnesota. J Wildl Manag 72:778–784
Hayes RA, Harestad AS (2000) Demography of a recovering wolf population in the Yukon. Can J Zool 78(1):36–48
Heck L, Raschke G (1980) Die Wildsauen. Parey, Hamburg
Huggard DJ (1993) Prey selectivity of wolves in Banff National Park. I. Prey species. Can J Zool 71:130–139
Imbert C, Caniglia R, Fabbri E, Milanesi P, Randi E, Serafini M, Torretta E, Meriggi A (2016) Why do wolves eat livestock?: factors influencing wolf diet in northern Italy. Biol Conserv 195:156–168
Jędrzejewski W, Jędrzejewska B, Okarma H, Ruprecht A (1992) Wolf predation and snow cover as mortality factors in the ungulate community of the Bialowieza National Park, Poland. Oecologia 90:27–36
Jędrzejewska B, Okarma H, Jędrzejewski W, Milkowski L (1994) Effects of exploitation and protection on forest structure, ungulate density and wolf predation in Bialowieza Primeval Forest, Poland. J Appl Ecol 31:664–676
Jędrzejewski W, Niedzialkowska M, Mysłajek RW, Nowak S, Jędrzejewska B (2005) Habitat selection by wolves Canis lupus in the uplands and mountains of southern Poland. Acta Theriol 50(3):417–428
Jędrzejewski W, Schmidt K, Theuerkauf J, Jędrzejewska B, Selva N, Zub K, Szymura L (2002) Kill rates and predation by wolves on ungulate populations in Białowieza Primeval Forest (Poland). Ecology 83:1341–1356
Kaartinen S, Luoto M, Kojola I (2009) Carnivore-livestock conflicts: determinants of wolf (*Canis lupus*) depredation on sheep farms in Finland. Biodivers Conserv 18:3503–3517
Kaczensky P, Chapron G, von Arx M, Huber D, Andrén H, Linnell J (2013) Status, management and distribution of large carnivores – bear, lynx, wolf and wolverine – in Europe (Report to the EU Commission, Part 1 and Part 2)

Kompaniyets L, Evans MA (2017) Modeling the relationship between wolf control and cattle depredation. PLoS One 12(10):e0187264
Lelieveld G, Beekers B, Kamp J, Klees D, Linnartz L, van Norren E, Polman E, Vermeulen R (2016) The first proof of the recent presence of wolves in the Netherlands. Lutra 59:23–31
Linnell JD, Boitani L (2011) Building biological realism into wolf management policy: the development of the population approach in Europe. Hystrix It J Mamm 23:80–91
Linnell JDC, Trouwborst A, Fleurke FM (2017) When is it acceptable to kill a strictly protected carnivore? Exploring the legal constraints on wildlife management within Europe's Bern Convention. Nat Conserv 21:129–157
Lombardini M (2009) Ecologia del lupo (*Canis lupus*) in un'area dell'Appennino settentrionale. Master thesis, University of Pavia, Italy
Lovari S, Sforzi A, Scala C, Fico R (2007) Mortality parameters of the wolf in Italy: does the wolf keep himself from the door? J Zool 272:117–124
Lucchini V, Fabbri E, Marucco F, Ricci S, Boitani L, Randi E (2002) Noninvasive molecular tracking of colonizing wolf (*Canis lupus*) packs in the western Italian Alps. Mol Ecol 11:857–868
Macdonald DW, Boitani L, Barrasso P (1980) Foxes, wolves and conservation in the Abruzzo mountains. Biogeographica 18:223–235
Manly BFJ (2006) Randomization, bootstrap and Monte Carlo methods in biology. Manly editorial text in statistical science, 3rd edn. Chapman and Hall/CRC, Boca Raton, 480 p
Marshal JP, Boutins S (1999) Power analysis of wolf-moose functional responses. J Wildl Manag 63:396–402
Marucco F, Avanzinelli E, Bassano B, Bionda R, Bisi F, Calderola S, Chioso C, Fattori U, Pedrotti L, Righetti D, Rossi E, Tironi E, Truc F, Pilgrim K, Engkjer C, Schwartz M (2018) La popolazione di lupo sulle Alpi Italiane 2014–2018. Relazione tecnica, Progetto LIFE 12 NAT/IT/00080 WOLFALPS – Azione A4 e D1
Massolo A, Meriggi A (1998) Factors affecting habitat occupancy by wolves in northern Apennines (northern Italy): a model of habitat suitability. Ecography 21(2):97–107
Mattioli L, Apollonio M, Mazzarone V, Centofanti E (1995) Wolf food habits and wild ungulate availability in the Foreste Casentinesi National Park, Italy. Acta Theriol 40:387–402
McLoughlin PD, Walton LR, Cluff HD, Paquet PC, Ramsay MA (2004) Hierarchical habitat selection by tundra wolves. J Mammal 85(3):576–580
Mech LD, Boitani L (2003) Wolf social ecology. In: Mech LD, Boitani L (eds) Wolves: behavior, ecology, and conservation. University of Chicago Press, Chicago, pp 1–34
Mech LD, Harper EK, Meiert J, Paul WJ (2000) Assessing factors that may predispose Minnesota farms to wolf depredation on cattle. Wildl Soc Bull 28:623–628
MEEDDAT–MAP (2008) Plan d'action national sur le loup 2008–2012, dans le contexte français d'une activité importante et traditionnelle d'élevage. Ministère de l'Ecologie, de l'Energie, du Développement Durable et de l'Aménagement du Territoire et Ministère de l'Agriculture et de la Pêche, Paris, France
Melis C, Jędrzejewska B, Apollonio M, Barton KA, Jędrzejewski W, Linnell JDC, Kojola I, Kusak J, Adamic M, Ciuti S, Delehan I, Dykyy I, Krapinec K, Mattioli L, Sagaydak A, Samchuk N, Schmidt K, Shkvyrya M, Sidorovich VE, Zawadzka B, Zhyla S (2009) Predation has a greater impact in less productive environments: variation in roe deer, *Capreolus capreolus*, population density across Europe. Glob Ecol Biogeogr 18:724–734
Meriggi A, Lovari S (1996) A review of wolf predation in southern Europe: does the wolf prefer wild prey to livestock? J Appl Ecol 33:1561–1571
Meriggi A, Rosa P, Brangi A, Matteucci C (1991) Habitat use and diet of the wolf in Northern Italy. Acta Theriol 36:141–151
Meriggi A, Brangi A, Matteucci C, Sacchi O (1996) The feeding habits of wolves in relation to large prey availability in northern Italy. Ecography 19:287–295
Meriggi A, Brangi A, Schenone L, Signorelli D, Milanesi P (2011) Changes of wolf (*Canis lupus*) diet in Italy in relation to the increase of wild ungulate abundance. Ethol Ecol Evol 23:195–210

Meriggi A, Dagradi V, Dondina O, Perversi M, Milanesi P, Lombardini M, Raviglione S, Repossi A (2015a) Short-term responses of wolf feeding habits to changes of wild and domestic ungulate abundance in Northern Italy. Ethol Ecol Evol 27:389–411

Meriggi A, Milanesi P, Scehnone L, Signorelli D, Serafini M, Torretta E, Puopolo F, Caniglia R, Fabbri E, Randi E (2015b) La presenza del lupo in Liguria: approccio integrato per la gestione dei conflitti. Relazione tecnica, Progetto "Il lupo in Liguria"

Meriggi A, Milanesi P, Serafini M, Torretta E, Imbert C, Caniglia R, Fabbri E, Randi E, Scehnone L, Signorelli D, Puopolo F, Zanzottera M, Magliano M (2013) Status ed ecologia del lupo in Liguria Dal monitoraggio alla gestione dei conflitti. Relazione tecnica, Progetto "Il lupo in Liguria"

Messier F (1991) The significance of limiting and regulating factors on the demography of moose and white-tailed deer. J Anim Ecol 60:377–393

Meuret M, Garde L, Moulin CH, Nozières-Petit MO, Vincent M (2017) Élevage et loups en France: historique, bilan et pistes de solution. INRA Prod Anim 30:1–23

Migli D, Youlatos D, Iliopoulos Y (2005) Winter food habits of wolves in Central Greece. J Biol Res (Thessaloniki) 4:217–220

Milanesi P, Meriggi A, Merli E (2012) Selection of wild ungulates by wolves *Canis lupus* (L. 1758) in an area of the Northern Apennines (North Italy). Ethol Ecol Evol 24:81–96

Mori E, Benatti L, Lovari S, Ferretti F (2017) What does the wild boar mean to the wolf? Eur J Wildl Res 63:9

Musiani M, Muhly T, Gates C, Callaghan C, Smith ME, Thsoni E (2005) Seasonality and reoccurrence of depredation and wolf control in western North America. Wildl Soc Bull 33:876–887

Newsome TM, Boitani L, Chapron G, Ciucci P, Dickman CR, Dellinger JA, López-Bao JV, Peterson RO, Shores CR, Wirsing AJ, Ripple WJ (2016) Food habits of the world's grey wolves. Mammal Rev 46:255–269

Okarma H (1995) The trophic ecology of wolves and their predator role in ungulate communities of forest ecosystems in Europe. Acta Theriol 40:335–386

Palmegiani I, Gazzola A, Apollonio M (2013) Wolf diet and its impact on the ungulates community in a new recolonized area of Western Alps: gran Paradiso National Park. Folia Zool 62:59–66

Papageorgiou N, Vlachos C, Sfougaris A, Tsachalidis E (1994) Status and diet of wolves in Greece. Acta Theriol 39:411–416

Paul WJ, Gipson PS (1994) Wolves. In: Hyngstrom SE et al (eds) Prevention and control of wildlife damage. University of Nebraska Cooperative Extension, Institute of Agriculture and Natural Resource, Lincoln, p C–129

Pereira HM, Navarro LM, Martins IS (2012) Global biodiversity change: the bad, the good, and the unknown. Annu Rev Environ Resour 37:25–50

Perversi M (2012) Selezione dell'habitat da parte del Lupo (*Canis lupus* L. 1758) nell'Appennino pavese-piacentino e previsione dell'espansione della popolazione. Master thesis, University of Pavia, Italy

Peterson RO, Ciucci P (2003) The wolf as a carnivore. In: Mech LD, Boitani L (eds) Wolves: behavior, ecology and conservation. University of Chicago Press, Chicago/London, pp 104–130

Pimenta V, Barroso I, Boitani L, Beja P (2017) Wolf predation on cattle in Portugal: assessing the effects of husbandry systems. Biol Conserv 207:17–26

Poudyal N, Baral N, Asah ST (2016) Wolf lethal control and livestock depredations: counter-evidence from respecified models. PLoS One 11:e0148743

Poulle ML, Carles L, Lequette B (1997) Significance of ungulates in the diet of recently settled-wolves in the Mercantour mountains (southeastern France). Rev Ecol (Terre Vie) 52:357–368

Procaccio EL (2015) La dieta del lupo (*Canis lupus*) in Oltrepò Pavese in relazione all'abbondanza di grandi erbivori domestici e selvatici. Master thesis, University of Pavia, Italy

Randi E (2011) Genetics and conservation of wolves *Canis lupus* in Europe. Mammal Rev 41:99–111

Raviglione S (2011) Distribuzione ed ecologia alimentare del Lupo (*Canis lupus* L, 1758) nell'Appennino pavese-piacentino. Master thesis, University of Pavia, Italy

Ražen N, Brugnoli A, Castagna C, Groff C, Kaczensky P, Kljuin F, Knauer F, Kos I, Krofel M, Luštrik R, Majić A, Rauer G, Righetti D, Potočnik H (2016) Long-distance dispersal connects Dinaric-Balkan and Alpine grey wolf (*Canis lupus*) populations. Eur J Wildl Res 62:137–142
Reig S, Jędrzejewski W (1988) Winter and early spring food of some carnivores in Bialowieza National Park, eastern Poland. Acta Theriol 33:57–65
Reig S, Cuesta L, Palacios F (1985) The impact of human activities on the food habits of red fox and wolf in Old Castille, Spain. Rev Ecol (Terre Vie) 40:151–157
Ripple WJ, Beschta RL, Fortin JK, Robbins CT (2014a) Trophic cascades from wolves to grizzly bears in Yellowstone. J Anim Ecol 83:223–233
Ripple WJ, Estes JA, Beschta RL, Wilmers CC, Ritchie EG, Hebblewhite M, Berger J, Elmhagen B, Letnic M, Nelson MP, Schmitz OJ, Smith DW, Wallach AD, Schmitz OJ (2014b) Status and ecological effects of the world's largest carnivores. Science 343:1241484
Rondinini C, Battistoni A, Peronace V, Teofili C (2013) Lista rossa IUCN dei vertebrati italiani. Comitato Italiano IUCN e Ministero dell'Ambiente e della Tutela del Territorio e del Mare, Roma, 56
Ronnenberg K, Habbe B, Gräber R, Strauß E, Siebert U (2017) Coexistence of wolves and humans in a densely populated region (Lower Saxony, Germany). Basic Appl Ecol 25:1–14
Sand H, Eklund A, Zimmermann B, Wikenros C, Wabakken P (2016) Prey selection of Scandinavian wolves: single large or several small? PLoS One 11(12):e0168062
Schmidt PA, Mech LD (1997) Wolf pack size and food acquisition. Am Nat 150:513–517
Sinclair ARE, Fryxell JM, Caughley G (2006) Wildlife ecology conservation and management, 2nd edn. Blackwell Science, Oxford. xii + 469 p
Smietana W, Klimek A (1993) Diet of wolves in the Bieszczady Mountains, Poland. Acta Theriol 42:241–252
Thurber JM, Peterson RO (1993) Effects of population density and pack size on the foraging ecology of gray wolves. J Mammal 74:879–889
Torres RT, Silva N, Brotas G, Fonseca C (2015) To eat or not to eat? The diet of the endangered Iberian wolf (*Canis lupus signatus*) in a human-dominated landscape in Central Portugal. PLoS One 10(6):e0129379
Torretta E, Serafini M, Milanesi P, Imbert C, Meriggi A (2016) Wolf and wild ungulates in the Ligurian Alps (Western Italy): prey selection and spatial-temporal interactions. Mammalia 81:537–551
Torretta E, Caviglia L, Serafini M, Meriggi A (2017) Wolf predation on wild ungulates: how slope and habitat cover influence the localization of kill sites. Curr Zool 64(3):271–275. https://doi.org/10.1093/cz/zox031
Treves A, Karanth KU (2003) Human-carnivore conflict and perspectives on carnivore management worldwide. Conserv Biol 17:1491–1499
Treves A, Naughton-Treves L, Harper K, Mladenoff DJ, Rose RA, Sickley TA, Wydeven P (2004) Predicting human-carnivore conflict: a spatial model derived from 25 years of data on wolf predation on livestock. Conserv Biol 18:114–125
Valière N, Fumagalli L, Gielly L, Miquel C, Lequette B, Poulle ML, Weber JM, Arlettaz R, Taberlet P (2003) Long-distance wolf recolonization of France and Switzerland inferred from non-invasive genetic sampling over a period of 10 years. Anim Conserv 6:83–92
Van Liere D, Dwyer C, Jordan D, Premik-Banič A, Valenčič A, Kompan D, Siard N (2013) Farm characteristics in Slovene wolf habitat related to attacks on sheep. Appl Anim Behav Sci 144:46–56
Vos J (2000) Food habits and livestock depredation of two Iberian wolf packs (*Canis lupus signatus*) in the north of Portugal. J Zool 251:457–462
Wagner C, Holzapfel M, Kluth G, Reinhardt I, Ansorge H (2012) Wolf (*Canis lupus*) feeding habits during the first eight years of its occurrence in Germany. Mamm Biol 77:196–203
Wielgus RB, Peebles KA (2014) Effects of wolf mortality on livestock depredations. PLoS One 9:1–16

Wittmer HU, Sinclair ARE, Mclellan BN (2005) The role of predation in the decline and extirpation of woodland caribou. Oecologia 144:257–267

Wolf Alpine Group (2014) Wolf population status in the Alps: pack distribution and trend up to 2012. 6 p. Available at http://www.lcie.org. 2014 June

Zimen E, Boitani L (1975) Number and distribution of wolves in Italy. Z Säugetierkd 40:102–111

Zlatanova D, Ahmed A, Valasseva A, Genov P (2014) Adaptive diet strategy of the wolf (*Canis lupus* L.) in Europe: a review. Acta Zool Bulg 66:439–452

Part III
Urban Wildlife Conflicts Are an Emerging Problem

The interactions between wildlife and urban environment are an extremely current and constantly evolving topic (Luniak 2004; Adams 2016). In fact, the urban and peri-urban environment is constantly expanding with the increase in human population and anthropization, and this implies the need to study the phenomenon in depth (Soulsbury and White 2015).

This part begins with a review of some cases of wildlife urbanization illustrating 'positive' and 'negative' impacts (Perry et al. 2020).

The urban environment offers many possibilities for wildlife because it is a quite stable habitat with few natural predators (Adams 2016). However, there are also negative scenarios for human safety, the possible introduction of zoonoses, or the danger that certain species represent to humans and their activities (e.g. Mackenstedt et al. 2015; Landy et al. 2018). On the other hand, even the presence of dangerous species may have positive implications, such as in the case of the urban leopard *Panthera pardus* in India, which preying on stray dogs reduces the risk of rabies transmission and could save up to 90 human lives per year in highly urbanized area (Braczkowski et al. 2018).

Following is the chapter by Seiler and Bardwaj (2020): a review about the impact of road traffic on wildlife which also deals with the analysis of data obtained from studies conducted by the authors in Sweden. The problem is global and, with the incessant construction of new roads and highways, even in areas of the world particularly important for wildlife and biodiversity, represents a continuous challenge to be faced (Glista et al. 2009; Bennett 2017).

The last chapter (Benussi and Fraissinet 2020) concerns a species of sea bird now invasive in many cities: the Western yellow-legged Gull (*Larus michahellis*). It offers an introductory review of the phenomenon of urban seagulls worldwide (e.g. Belant 1997) and then analyses in particular the case of a coastal city in Northeast Italy and the solutions adopted to try to solve, or at least mitigate, the problem. The ever-increasing presence of breeding pairs of this species creates many problems, especially on balconies and roofs, also because of its territorial behaviour and aggressiveness even towards humans.

References

Adams CE (2016) Urban wildlife management, 3rd edn. CRC Press, Boca Raton

Belant JL (1997) Gulls in urban environments: landscape-level management to reduce conflict. Landsc Urban Plan 38:245–258

Bennett VJ (2017) Effects of road density and pattern on the conservation of species and biodiversity. Curr Landsc Ecol Rep 2:1–11

Benussi E, Fraissinet M (2020) The colonization of the Western Yellow-legged Gull (*Larus michahellis*) in an Italian city: evolution and management of the phenomenon. In: Angelici FM, Rossi L (eds) Problematic wildlife II. New conservation and management challenges in the human-wildlife interactions. Springer, Cham, pp 191–212

Braczkowski AR, O'Bryan CJ, Stringer MJ et al (2018) Leopards provide public health benefits in Mumbai, India. Front Ecol Environ 16:176–182

Glista DJ, DeVault TL, DeWoody A (2009) A review of mitigation measures for reducing wildlife mortality on roadways. Landsc Urban Plan 91:1–7

Landy F, Rodary E, Calas B (2018) Why did leopards kill humans in Mumbai but not in Nairobi? Wildlife management in and around urban national parks. In: Landy F (ed) From urban national parks to natured cities in the global south. Springer Nature, Cham, pp 157–179

Luniak M (2004) Synurbization – adaptation of animal wildlife to urban development. In: Shaw WW, Harris LK, VanDruff L (eds) Proceedings of the 4th international symposium on urban wildlife conservation (May 1–5, 1999), Tucson, pp 50–55

Mackenstedt U, Jenkins D, Romig T (2015) The role of wildlife in the transmission of parasitic zoonoses in peri-urban and urban areas. Int J Parasitol Parasites Wildl 4:71–79

Perry G, Boal C, Verble R et al (2020) "Good" and "bad" urban wildlife. In: Angelici FM, Rossi L (eds) Problematic wildlife II. New conservation and management challenges in the human-wildlife interactions. Springer, Cham, pp 141–170

Seiler A, Bardwaj M (2020) Wildlife and traffic – an inevitable but not unsolvable problem? In: Angelici FM, Rossi L (eds) Problematic wildlife II. New conservation and management challenges in the human-wildlife interactions. Springer, Cham, pp 191–212

Soulsbury CD, White PCL (2015) Human–wildlife interactions in urban areas: a review of conflicts, benefits and opportunities. Wildl Res 42: 541–553

Chapter 5
"Good" and "Bad" Urban Wildlife

Gad Perry, Clint Boal, Robin Verble, and Mark Wallace

5.1 Urban Environments and Wildlife

In an age when most natural habitats are in decline, urbanization stands out as an increasing, if novel, environment, though exactly what constitutes "urban" can be hard to define (e.g., McIntyre et al. 2008). Although greater attention is paid to multimillion resident metropolises such as Calcutta or São Paulo, most urban centers are much smaller and the vast majority of urbanites live in smaller cities (Montgomery 2008). According to United Nations (2015a) estimates, by 2014, over half of the world's population lived in urban areas, compared to about 30% in 1950 (World Bank data depicted in Fig. 5.1 are only slightly different). By 2050, some two thirds of the world's population will live in urban areas (United Nations 2015a). This trend is especially strong in Europe, North America, and Latin America, where over 70% of the population is already urban, but more than half the residents of Africa and Asia are also projected to live in urban areas by 2050 (United Nations 2015a).

Urbanization is a leading cause of habitat degradation and extinctions (reviewed in McKinney 2002), leading to the urban setting being termed a "landscape of fear" (Stillfried et al. 2017). However, studies of urban ecology have been uncommon until relatively recently (Fig. 5.2A). Although urban areas show lowered biodiversity (McKinney 2008), particularly of highly specialized species (Soulsbury and White 2015), the common view of cities as "biological deserts" not worth studying (e.g., Sukopp 1998, p. 3; Ditchkoff et al. 2006) is overly simplistic. Urban areas provide a home for a range of species that are, for various reasons, tolerant of human

G. Perry (✉) · C. Boal · M. Wallace
Department of Natural Resource Management, Texas Tech University, Lubbock, TX, USA
e-mail: Gad.Perry@ttu.edu

R. Verble
Department of Natural Resource Management, Texas Tech University, Lubbock, TX, USA

Missouri S&T Ozark Research Field Station, Lubbock, TX, USA

F. M. Angelici, L. Rossi (eds.), *Problematic Wildlife II*,
https://doi.org/10.1007/978-3-030-42335-3_5

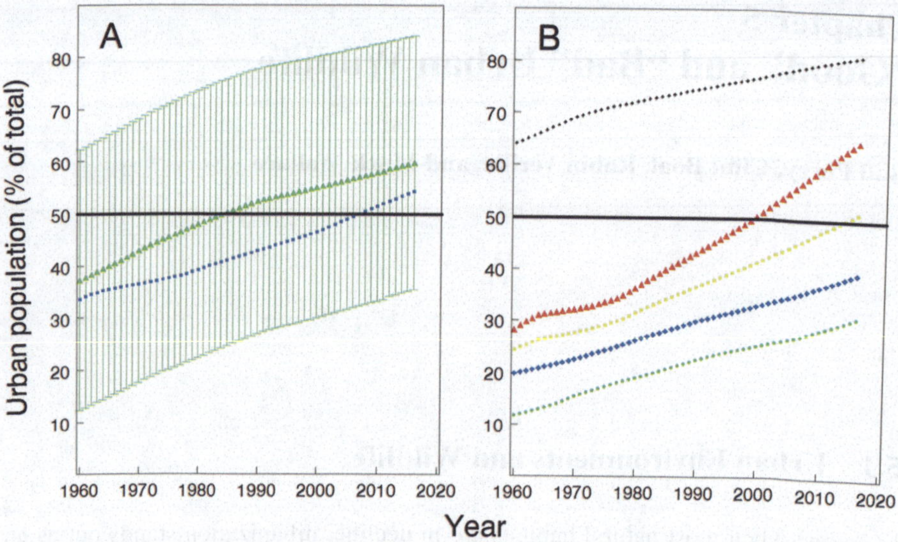

Fig. 5.1 Global urban population patterns, 1980–2016. (**A**) Worldwide trends. Blue squares: total world population; Green triangles: Average of all nations represented in the World Development Indicators database (https://data.worldbank.org/indicator/SP.POP.TOTL) from which all data are taken. Note the consistently broad standard deviation error bars, indicating ongoing differences between nations. Also note that both lines have already crossed the 50% horizontal line (black), indicating that a majority of the world's population is now urban. (**B**) Patterns according to national economic position. Green circles: low income nations; Blue diamonds: lower middle income; Yellow squares: middle income; Red triangles: Upper middle income; Black plusses: High income nations. Although all lines show a steady increase in urbanization, populations in the two lowest-income categories remain mostly rural

presence or activity (McKinney 2008). Many of them are nonnative and globally widespread. In fact, some species, including ones that are of conservation concern in other settings, can reach particularly high densities in urban areas (Fuller et al. 2009). The discipline of urban ecology focuses on organisms located in and around cities, their distributions, abundances, and interactions with the urban biotic (human and otherwise) and abiotic environment (Pickett et al. 2001). Yet it was not until the 1990s that the term "urban wildlife" emerged as a description of species found in proximity to humans (Adams 2016), and studies of urban wildlife still remain underrepresented in the scientific literature (Magle et al. 2012). Based on topics proposed by members of the British Ecological Society, Sutherland et al. (2013) identified "100 fundamental ecological questions." Questions 73 through 89 were contained in the "human impacts and global change" section, which "recognized that current ecological dynamics and ecosystem function occurs within the context of a human-dominated planet." However, although "[h]uman impacts on ecosystems include direct impacts on habitats such as land conversion …," the effects of urbanization were not explicitly included even once by Sutherland et al. (2013).

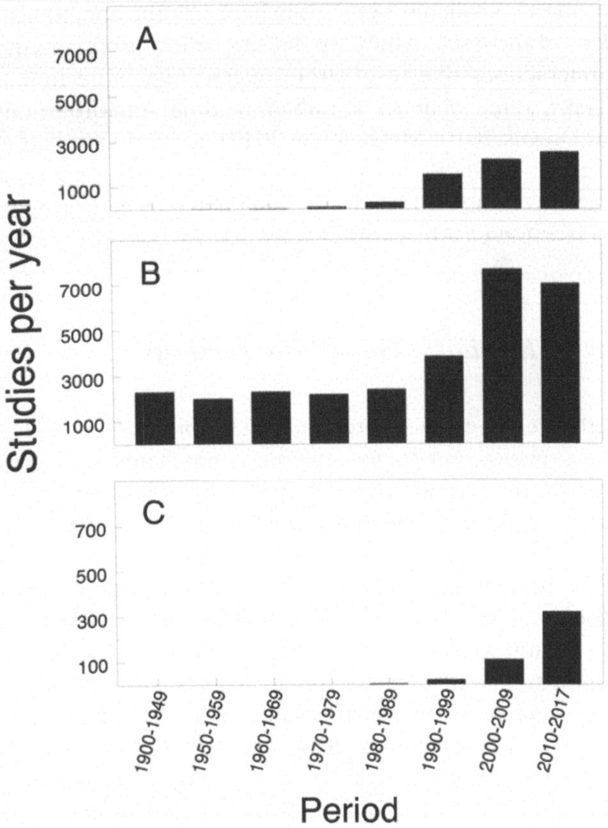

Fig. 5.2 Annualized number of hits returned by Google Scholar for the terms "urban ecology" (**A**), "human wildlife conflict" (**B**), and "human dimensions of wildlife – conflict" (**C**). Note that the Y axis of C is an order of magnitude smaller than those of A and B

Human urban residents interact with nonhuman ones on a regular basis. Some of these are usually categorized as "bad" interactions. For example, few people like rats (*Rattus* sp.) or cockroaches (e.g., *Periplaneta americana*). This is perhaps not surprising, given that some rat and cockroach species are near-global in their distribution and also associated with disease and refuse. Studies of such conflicts have been common for quite some time (Fig. 5.2B). A relatively new insight (Fig. 5.2C) is that other interactions are "good," in the sense that people seek them out, or at least enjoy them. For example, many people insist on keeping pets in highly built environments and consider amenities such as green-belts an improvement to their quality of life (e.g., Briffett et al. 2004). In some cases, the perception of the interaction varies. As other chapters point out, urban cats (*Felis silvestris catus* or *Felis catus*) are highly cherished as pets, but they are also commonly viewed as pests when feral. Urban pigeons (*Columba livia*) are a treasured symbol of highly touristy locations such as Trafalgar Square in London (UK) and Piazza San Marco, Venice

(Italy), but are often considered pests elsewhere (Jerolmack 2008). In addition to these two "sides" of the issue, which we discuss below, urban environments have a set of unique interactions that result from intersecting large human and wildlife populations. These often channel to urban wildlife rehabilitation centers (e.g., Wimberger and Downs 2010; McGaughey 2012), a poorly studied topic which we also address below. In addition, we separately provide taxonomically oriented reviews of issues involving reptiles, birds, and mammals and provide a section that looks at the intersection of urban wildlife and disease issues.

5.1.1 Cities as Habitat: "Good" Interactions

Much of the urban ecology literature emphasizes biogeochemical cycles and the biology of urban species, but socioeconomic components are increasingly being incorporated into the discussion (McIntyre et al. 2008; Pickett et al. 2011). (An even larger literature deals with populations considered to be pests and is reviewed below.) With humans increasingly incorporated into the analyses, a distinct subsection examines the benefits that natural environments provide to urban populations (Hines 2017), although much of that discussion focuses on vegetation rather than wildlife. In this section we review the literature that addresses the positive interactions between humans and nonhuman animals found in urban settings.

A number of recent studies explicitly evaluated the benefits of wildlife presence on the feelings of well-being among urban dwellers. For example, Mumaw et al. (2017) reported "wellbeing benefits including strengthened connections with nature, place and community." Shanahan et al. (2015) reported similar findings. Perhaps because of this sense of enhanced welfare, or perhaps because features leading to homeowner happiness also benefit wildlife, Farmer et al. (2011) found that urban dwellers were willing to pay higher prices for houses in neighborhoods characterized by relatively high indices of bird diversity. More recently, Clucas et al. (2015) estimated the economic value to residents of enjoying native urban songbirds at approximately 120 million USD/year in Seattle, USA, and 70 million USD/year in Berlin, Germany. These and other studies are consistent with the idea of Kellert and Wilson (1995) that humans exhibit "biophilia," an innate need, or at least desire, for being surrounded with biological nature. Additional human benefits can accrue from unintended consequences, such as reduced vehicular collisions resulting from predator control of urban herbivores (Gilbert et al. 2017). Creating wildlife-friendly environments such as green roofs provides habitat for at least some wildlife species (Williams et al. 2014), but different development strategies benefit different types of urban wildlife (Soga et al. 2014a, b).

The presence of wildlife in the urban setting also benefits conservation of natural habitats and the species dwelling in them. Most urbanites encounter only urban species, which therefore come to represent all wildlife (Lunney and Burgin 2004) and affect perceptions of wildlife in general. For example, Hosaka et al. (2017a, b) found that childhood exposure to wildlife makes urbanites somewhat more tolerant of nuisance wildlife encountered later in life. Conversely, the so-called extinction of experience that comes with lack of exposure to wildlife can lead to declining support for conservation (Miller 2005; Soga and Gaston 2016). Thus, ensuring exposure of urban dwellers to even a limited biodiversity palette is likely to improve the chance that nonurban species and populations would survive.

The final area we explore here is also the best studied, and receives additional taxonomically organized attention below (Sect. 5.2): urban environments as habitats for wildlife. In some ways, cities and towns offer foreign, often hostile environments. Urban settings have different thermal profiles, strong nightlights, persistent noise, and other states of natural conditions that are highly divergent or greatly exaggerated (Longcore and Rich 2004; Ditchkoff et al. 2006; Francis and Barber 2013). Nonetheless, they "might buffer some species against wholesale regional population depletion," given that exurban environments are themselves often highly impacted by other human activities, such as agriculture (Fuller et al. 2009).

Animals can respond to human encroachment in three main ways. Some, often termed "avoiders," stay away from areas where conditions are too different from what they normally experience in nature (McKinney 2002; Fischer et al. 2015). "Exploiter" species, many of them ubiquitous and widely distributed, are found at the other extreme of the spectrum and may be encountered in almost any human-dominated environment (Blair 1996; Lowry et al. 2013). Anthropophilic species, such as those that readily adapt to the nightlight niche, are particularly suited to invading new urban environments after being transported around the globe, often in association with economic activity or the international pet trade (Perry et al. 2008; Powell et al. 2011). That cities share many common features even when they are far apart and in drastically different climactic regions (Soulsbury and White 2015) also helps. In between these extremes, "adapter" species can be found in urban environments and may even do well, but are rarely found in the densest areas of cities (McKinney 2002; Fischer et al. 2015).

Whether native or introduced, anthropophilic species are the ones most likely to become problematic for humans (see Sect. 5.1.2, below). In between are species that show some ability to become habituated to human presence. Sometimes known as "adjusters," such species may increase their presence over time, either because particular individuals are immediately better suited to urban settings or because evolution allows tolerance to increase in magnitude and frequency (Whittaker and Knight 1998).

5.1.2 Wildlife as Urban Problems: "Bad" Interactions

The story of the Pied Piper of Hamlin emerged in Europe in the thirteenth century and recounts the story of a rat catcher during a plague, whose nature is still being debated (Dirckx 1980). Thus, the interaction of humans with urban wildlife, and particularly the black rat and the Norwegian rat, is far from new (Soulsbury and White 2015). In fact, some records exist going back to ancient Egypt (Dixon 1989).

Despite this, there is still new research to show, for example, that black rats benefit from the addition of refuge spaces (Price and Banks 2017). An extensive literature on human-wildlife conflict in urban environments and on the management strategies that have evolved as a result already exists (Adams 2016). We do not attempt to replicate it all here, but rather to provide a brief summary and an update.

As with other habitats (Morrison et al. 2006), management of urban wildlife populations is often closely related to management of wildlife habitat. To an even greater degree than in other habitats (Riley et al. 2002), urban wildlife management is deeply interlaced with addressing human wants, needs, and values. In other words, what makes a species problematic is rarely intrinsic. Rather, it is that its presence is in some way inconvenient to humans, at least in the densities it displays, and that its removal or reduction is considered acceptable and preferable to continued friction.

Common reasons for considering urban wildlife problematic include damage to food, as with rats, or other property, as with animal-automobile accidents in North America (Lee and Miller 2003); creating noise or smell nuisance, as with gulls (*Larus* spp.) in northern latitudes (Belant 1997); threatening humans, pets, or ornamental vegetation, as by carnivores (Bateman and Fleming 2012) or beavers (Loven 1985); being disease vectors, as with rats and mosquitoes; and other economic damage, as with brown treesnakes (*Boiga irregularis*) causing power outages on Guam (Fritts 2002). Common responses include lethal control, relocation, and habitat modification to discourage continued presence. In some cases, all three may be used with a single species, as with urban deer (*Odocoileus* spp.; Messmer et al. 1997). In other cases, as with urban cats, some strategies may be much more acceptable to the public than others (e.g., Loyd and Miller 2010).

5.1.3 *Urban Wildlife Rehabilitation Centers*

The combination of well-intentioned "rescue" of wildlife, actions taken to remedy perceived nuisance situations, and accidental encounters often results in unwanted urban wildlife needing a home. Few dedicated options were available until relatively recently (Fitzwater 1988), when urban wildlife rehabilitation centers began stepping in to address the growing need. In the USA, some 500–1000 certified wildlife rehabilitators are thought to operate (Southeastern Outdoors 2010; Wildlife Rehabilitation Information Directory 2016). Thousands are similarly involved in Australia (Tribe and Brown 2000). Nor is this a uniquely developed-nation phenomenon. For example, Vyas (2013) describes practices at several urban wildlife rescue centers in the state of Gujarat, India. As Karesh (1995) points out, however, rehabilitation centers in developing countries face some unique challenges.

Of the growing literature on wildlife rehabilitation centers, a large preponderance focuses on nonurban centers dealing with charismatic vertebrates, most often birds or mammals. A relatively small number deal with urban rehabilitation centers, the species admitted into them, the disposition of those animals, and the makeup and motivation of their staffs. In an early study from North America, Beringer et al. (2004) reported that most rehabilitated white-tailed deer (*Odocoileus virginianus*) died following release. Furthermore, many of those that survived remained near human dwellings, where they had the potential to become a nuisance. In Durban, South Africa, Wimberger and Downs (2010) showed that annual rehabilitation center intakes included well over 2500 animals each year, some 90% of them birds. In Belo Horizonte, Minas Gerais, Brazil, hundreds of calls were made to the environmental police to remove unwanted white-eared opossums (*Didelphis albiventris*) between 2002 and 2007 (Souza et al. 2012). Many of those were later released into city forest fragments or sent to veterinary clinics or government facilities. Unfortunately, recent budget cuts have resulted in many of those centers closing.

One of the better datasets comes from a series of studies conducted at the South Plains Wildlife Rehabilitation Center (SPWRC) in Lubbock, Texas, USA. McGaughey et al. (2011) showed that ornate box turtles (*Terrapene ornata*) are the most common species of reptile brought in to the SPWRC. Turtles were also the most commonly seen group at a similar facility in Illinois (Rivas et al. 2014). Sosa and Perry (2015) examined the outcomes of releases of those turtles received by the SPWRC once they are let go, following examination and/or treatment in an urban center. They found that few of those settle and successfully establish a new home range (Sosa and Perry 2015).

Over two decades, SPWRC admitted over 12,000 birds representing over 200 volunteer-identified species from 43 families, many of them only represented by one or two individuals (M. Jones, K. McGaughey, and G. Perry, unpublished data). The family represented by the most individuals was Columbidae, with the largest number of species in the order Passeriformes. The most common reported reason for admission was that individuals were sick or wounded (Fig. 5.3), with juveniles brought in during spring being the most common age group and season for admission. The most common outcome was death from natural causes or by euthanasia

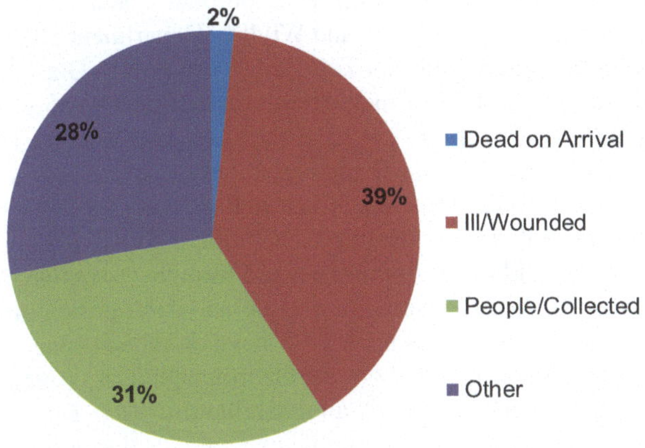

Fig. 5.3 Reason for admission of birds to the South Plains Wildlife Rehabilitation Center, 1991–2010

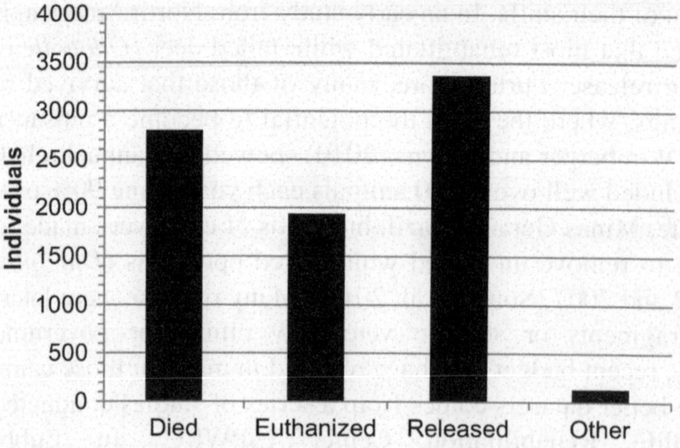

Fig. 5.4 Disposition of birds admitted to the South Plains Wildlife Rehabilitation Center, 1991–2010

(Fig. 5.4), though post-release follow-up was nearly nonexistent and it is quite possible that many of the individuals let go following rehabilitation also died (M. Jones, K. McGaughey, and G. Perry, unpublished data; Boal 2008).

Over the same period (1991–2010), SPWRC accepted nearly 6600 individual mammals belonging to 57 (volunteer-identified) species, the vast majority of them natives (McGaughey 2012). About 80% of these belonged to three common species: eastern cottontails (*Sylvilagus floridanus*), eastern gray squirrels (*Sciurus carolinensis*), and Virginia opossums (*Didelphis virginiana*). Most admissions were young individuals found on the ground by well-meaning residents, and the largest segment, about 40%, were eventually released, though information on post-release survival is virtually nonexistent (McGaughey 2012).

McGaughey (2012) also attempted to survey all wildlife rehabilitators in the state of Texas – over 200 permitted wildlife rehabilitators and an unknown number of unpermitted assistants (Texas Parks and Wildlife Department 2017) – in order to characterize them and gauge relevant knowledge and beliefs. Based on 222 completed surveys returned (not all of them from urban centers), almost 90% of rehabilitators are female and many do not have full time employment. They primarily rehabilitate for a mixture of personal and emotional reasons, about a third of them spending at least 20 h/week doing so. About half of them have 5 years or less of experience, but some 80% of the permitted rehabilitators running the centers require volunteers to undergo additional training beyond formal classes they may have formerly taken. Most respondents show moralistic and ecologistic attitudes (Kellert 1979), characterized by responses such as "I have an ethical concern for animals and nature" and "I am concerned about the environment as a system, and the wild species within natural habitats." Nearly all rehabilitators (97%) believe that wildlife rehabilitation has a positive impact on individual animals that are released. A large majority (84%) believe their actions benefit populations of species in the wild, and

(75%) help ensure the continued persistence of species. A significant difference was found between unpermitted volunteers and permitted leaders, however, with the latter more likely to accept that rehabilitated wildlife might not survive being released. Unfortunately, the majority of rehabilitators did not consider "the education of the public" as a top reason for their activities (McGaughey 2012). These results are overall similar to those few reported previously, which also demonstrate that professional biologists often view rehabilitators as overly optimistic and even naïve (Siemer and Brown 1993; Add et al. 1996; Tribe and Brown 2000). However, Siemer et al. (1991) found much greater interest in public education among New York rehabilitators than McGaughey (2012) reported from Texas.

5.2 Taxonomic Focus

Although urban environments tend to be ecologically impoverished, most taxonomic groups have at least some metropolitan representation. Here we focus on the three groups that are generally represented in this book: birds, mammals, and reptiles. We also discuss zoonotic diseases, those that can be transferred from animals to humans or between urban wildlife species. As these are often transmitted by invertebrate vectors such as insects, this very large taxonomic group also receives some representation in this section.

5.2.1 Reptiles

Studies of urban reptiles are relatively new. A Google Scholar search for the terms "urban ecology + reptile," "urban ecology + lizard," "urban ecology + snake," and "urban ecology + turtle" returned no relevant papers published between 1900 and 1970. Two decades ago, Rodda et al. (1999) summarized what remains perhaps the most expansive case of human-reptile conflict, that of the brown treesnake on Guam (and, to a lesser degree, elsewhere). A decade later, Mitchell et al. (2008) captured the state of the knowledge of urban reptiles at the time. Here we briefly review urban reptile conflicts and focus on recent studies not already presented in those two volumes.

5.2.1.1 Turtles

Innocuous, relatively slow moving, and often small, turtles are not generally hated or feared in the way that many other reptiles are. In many cases, they are actually liked (e.g., Sosa and Perry 2015) and can be prominent fixtures at urban wildlife rehabilitation centers (McGaughey et al. 2011). There are more studies on the negative impacts of urban pollution and water body management (e.g., Vyas 2015) or addressing the problems that light pollution causes sea turtles (e.g., Perry et al. 2008) than ones showing impacts by turtles. Nonetheless, turtles can infrequently be considered a nuisance. For example, Teixeira et al. (2016a) report multiple turtle-caused calls to local police in Belo Horizonte, Brazil. Ironically, these were mostly caused by freshwater species (*Trachemys* spp.), many of them presumably released pets displaced by rainfall events and wandering into yards were they were not wanted (Teixeira et al. 2016). In addition, urban turtles can serve as reservoirs for disease-causing organisms that might affect people (Dezzutto et al. 2017).

5.2.1.2 Crocodilians

Large and carnivorous, several species of crocodilians are capable of killing and consuming humans, and others are persecuted because of the perception of danger. Most crocodilian populations are found well outside of urban centers, but a few do occur near human settlements. Vyas (2010) documented cases of attack on humans by muggers (*Crocodylus palustris*) and the management actions taken in response to their presence in Gujarat, India. Saltwater crocodiles (*C. porosus*) are known to encroach on urban areas in Sri Lanka (Amarasinghe et al. 2015) and Australia (Fukuda et al. 2014). Although the Nile crocodile (*C. niloticus*) is a well-documented cause of human and cattle mortality in Africa, most conflict there appears to occur in rural settings. Unexpectedly, of the small number of terrestrial reptiles reported to collide with aircraft on Brazilian runways, crocodilians had the largest proportion (Novaes et al. 2016).

5.2.1.3 Lizards

Although common in urban settings, lizards are rarely a cause of conflict. For example, calls to the Environmental Police of Belo Horizonte, Brazil, between 2002 and 2008 rarely involved the large populations of wild lizards found in the area (Teixeira et al. 2016). In Puerto Rico, invasive green iguanas (*Iguana iguana*) found on the runways increasingly pose a risk for incoming aircraft (Engeman et al. 2005). Some lizard species, and particularly nonnative ones, find urban environments to be especially productive (e.g., Hall and Warner 2017), whereas others show little effect (Becker and Buchholz 2016) or reduced fitness (e.g., Lazić et al. 2017). A few studies show that urban lizards can have beneficial effects by reducing the populations of invertebrate pests (Canyon and Hii 1997; Ramires and Fraguas 2004; Kwenti 2017). However, other species can become problematic by feeding on ornamental vegetation and damaging infrastructure (Engeman and Avery 2016; Sementelli et al. 2008).

5.2.1.4 Snakes

Of the 469 calls to the Environmental Police of Belo Horizonte, Brazil, reviewed by Teixeira et al. (2016), 60% involved snakes. About half of the calls involved venomous snakes, most of them sighted in green belts and urban parks rather than in densely urbanized cores that form most of the study area (Teixeira et al. 2016). The preponderance of snake calls is a reflection of the facts that some are venomous, most are suspect as such, and few are appreciated in modern cultures (see Perry et al. chapter 18 in this volume). Venomous snakes are particularly likely to be killed or, at best, removed (Pitts et al. 2017), usually to locations where their chances of survival are low (Devan-Song et al. 2016). In addition, urban environments often present inhospitable environments to those snakes that do remain (e.g., Wolfe et al. 2017). Finally, snakes originating in the pet trade and released or escaped into the wild are some of the most damaging reptiles reported (e.g., Dove et al. 2011).

5.2.2 Birds

Birds are popular among many urban residents who intentionally attract them to backyards for viewing enjoyment (Horn and Johansen 2013). For example, an estimated $ US 5 billion was spent in 2011 for bird seed, bird houses, and other associated items in the USA (U.S. Fish and Wildlife Service 2012). Expenditures for similar items in the UK were estimated as high as £290 million ($ US 387 million; Jones and Reynolds 2008). Indeed, the enjoyment of viewing birds translates to an economic value of millions of dollars to local municipalities (Clucas et al. 2015). Thus, when considering birds in urban settings, conflict is likely not the first thing that comes to mind. Conflict between humans and birds in urban settings, however, is quite common, and at times the two issues may be conflated. For example, the earliest study of urban ecology we have located, Matheson (1944), discusses predation of urban birds by roaming cats associated with humans. Below we review the more common situations under which conflicts may occur between avifauna and humans in urban settings.

5.2.2.1 Nocturnal Roosts

Perhaps one of the greatest sources of conflict between urban birds and human city dwellers are the large nocturnal roosts that some species will form. Examples include rainbow lorikeets (*Trichoglossus haematodus*) creating nocturnal roosts exceeding one thousand individuals in cities across Australia (Chapman 2005); roosts of house crows (*Corvus splendens*) in Singapore (Peh and Sodhi 2002); communal urban roosts numbering in the tens of thousands of individuals of great-tailed grackle (*Quiscalus mexicanus*), brown-headed cowbirds (*Molothrus ater*), and introduced European starling (*Sturnus vulgaris*) in cities across North America

(Coon and Arnold 1977; Caccamise 1990); roosts of over one hundred black vultures (*Coragyps atratus*) in South American cities (Novaes and Cintra 2013); and more. Primary complaints are the volume of noise such congregations make, especially at dusk and dawn; the mess and health concerns associated with the excrement that can accumulate under large roosts; and the occasional presence of dead or dying birds following storms or exceptionally cold winter nights (e.g., Hall and Harvey 2007). An additional concern is damage to fruit tree or crops (Chapman 2005), but this primarily occurs at foraging sites away from roost locations.

5.2.2.2 Heronries and Other Rookeries

Herons form multispecies roosting aggregations and will also nest semi- to fully colonially. For example, in the arboretum at the University of California, Davis, USA, Hattori (2009) found 866 active nests of four heron and egret species. Such aggregations can cause substantial conflict, primarily due to accumulated feces and human health concerns. Compounding these concerns, these heronries can be located in places of extremely high urbanization (Telfair et al. 2000; Hattori 2009). For example, the Riverwalk in San Antonio, Texas, USA, is a popular tourist destination visited by millions of guests annually. What most visitors do not realize when seated for dinner along the River Walk is that the umbrellas over their tables are more to protect them from the excrement of yellow-crowned night herons (*Nyctanassa violacea*) nesting and roosting overhead than to provide shade (Boal 2011). In Israel, Mendelssohn and Yom-Tov (1999) reported cattle egret (*Bubulcus ibis*) concentrations so dense that the trees they roost on "become painted white from the birds' urine and later die from that." Moreover, "the noise and discarded food," which can include regurgitated used condoms collected on foraging trips, "disgust the people of the area," who kept petitioning authorities for removal of the nuisance. In the Indian state of Kerala, nearly half of the heronries are located in urban settings (Subramanya 2005). Members of such rookeries are less likely to flush when humans approach, benefit from the nearness of anthropogenic food sources such as fish markets, and appear to enjoy reduced numbers of predators (Roshnath and Sinu 2017a, b).

5.2.2.3 Geese and Greenspace

An increasingly common wildlife management issue in North America is the burgeoning numbers of Canada geese (*Branta canadensis*) occupying urban lakes and adjacent greenspaces. Large aggregations of sometimes-aggressive geese damage lawns by grazing, cause an accumulation of feces on areas used for family relaxation, create noise, and degrade water quality (Forbes 1993; Smith et al. 1999). The conflict has been prevalent enough to lead to the development of numerous strategies to attempt to alleviate the problem (Smith et al. 1999).

5.2.2.4 Pigeons

Perhaps no bird is more closely associated with humans and urbanization, more typifying of the "exploiter" species, than rock doves (*Columba livia*), more commonly known as pigeons. Urban pigeons, the feral descendants of a species domesticated over 5000 years ago (Johnston and Janiga 1995), are found in cities around the globe (del Hoyo et al. 2005). Quite frequently, they are considered a pest species because they nest on horizontal surfaces of buildings and their excrement accumulates on ledges and the ground below, presenting a human health risk. Pigeons are known to carry several diseases that are transmittable to humans (Giunchi et al. 2012). Additionally, damage to buildings has been attributed to salts, pH level of solutions when water interacts with pigeon excrement, and fungal growth facilitated by excrement (Giunchi et al. 2012).

5.2.2.5 Predatory Birds

Hawks and owls, as predators, can elicit strong emotional responses from people, both of excitement and of fear and concern. Urban landscapes can provide abundant trees for nests, and buildings offer cliff-like habitats that replace native sites used before dense human habitation (e.g., Boal and Mannan 1999; Mendelssohn and Yom-Tov 1999). The primary conflict of urban residents with predatory birds is a fear of threat to their pets. Other concerns are when predatory birds are unwelcome guests hunting songbirds at backyard feeders. Some raptors are successful "adapters," but will nonetheless become protective of their nests and young. This can lead to diving swoops at people and pets that venture close to nest trees (Morrison et al. 2006). Although these can be frightening, physical contact is rarely made.

5.2.2.6 Airports

Undoubtedly, worldwide, the most serious conflict between birds and humans in urban areas is the threat posed to aircraft. Nearby garbage dumps have attracted large numbers of birds, most often gulls, to areas near busy airports (Davidson et al. 1971; Mendelssohn and Yom-Tov 1999). Even the strike of small birds can lead to substantial damage to aircraft and risk to passengers. Larger birds such as geese and hawks can lead to loss of human life and aircraft (Washburn et al. 2015; Dolbeer et al. 2016), as was dramatically demonstrated in the crash of US Airways Flight 1549 into the Hudson River in January 2009. Considerable management efforts have been made toward reducing such conflict, including recent attempts to use drones made to look and fly like predatory birds to scare birds away from airports (Rosenberg 2017).

5.2.3 Mammals

A Google Scholar search of the terms "urban ecology + mammal" prior to 1950 returns a single relevant article, Matheson (1944), dealing with urban cats (the topic is covered in some detail in Chap. 13). Conducted in the UK, the study used surveys "to obtain an approximate ratio of domestic cats to people in big cities" – about 13%, most of them "house-kept" (Matheson 1944). We now know these numbers were gross underestimates (Marra and Santella 2016). For the two decades that follow, Google Scholar also returns a single relevant paper, a study of introduced gray squirrels (*Sciurus carolinensis*) in a British Columbia, Canada, urban park (Robinson and Cowan 1954). Publications picked up in the 1970s, and especially the later years of the decade, with papers focusing on both ecology (e.g., Harris 1977) and interactions between humans and urban greenspaces (e.g., Laurie 1979). Work further accelerated thereafter, with a series of studies originating in England, with Dickman (1987; Dickman and Doncaster 1987, 1989) and Harris (1981; Harris and Smith 1987; Trewhella and Harris 1988) being notable tone-setters.

5.2.3.1 Mammals Declining near Humans

"Avoider" mammal species which are most sensitive to urbanization include large animals with correspondingly large movement patterns. Species such as bison (*Bison bison*) and elk (*Cervus canadensis*) in North America or red deer (*Cervus elaphus*) in Europe and across Asia, and elephants (*Loxodonta* spp.) in Africa frequently come in conflict with humans as they attempt to use human crops or simply move through human settlements (McCleery 2010). Predators, particularly species like gray wolf (*Canis lupus*), mountain lion (*Felis concolor*), or lion (*Panthera leo*), that came in conflict with rural pastoralist societies were persecuted. Although individual humans may be injured or killed in these confrontations, invariably, management solutions lead to eradication of offending individual animals or whole populations. Local eradication can be followed by creation of barriers which result in relocation of seasonal ranges, migratory routes, or populations of wildlife. However, many species of mammals are associated with human habitations, with species diversity dropping off as urban intensity increases (Ordeñana et al. 2010).

5.2.3.2 Co-Adaptation of Humans and Nonhuman Mammals

Although the majority of mammal species cannot survive in urban habitats, a surprising subset of mammals have habituated to living with and around humans. Several species have adapted, through evolution or by human-driven artificial selection with varying degrees of success, to live with humans. Perhaps first of those was the dog (*Canis familiaris*) followed by the house cat – the epitome of "exploiter" species. In developed countries, humans value them highly and their numbers can be almost as high as those of people: roughly 70 million dogs and 74 million cats in the USA in 2012 (American Veterinary Medical Association 2012), about 10.5 million of each in England in 2007 (Murray et al. 2010), and approximately 2.7 million cats and 4.8 million dogs in Australia in 2016 (Animal Medicines Australia 2016). While both species are now considered domesticated, both show a propensity to successfully revert to a nearly wild or "feral" state and can live well and with high reproductive rates on human refuse or in its absence. Abandonment of pet cats and dogs has resulted in "feral" populations requiring control in areas, including urban and peri-urban areas (e.g., Home et al. 2017). Feral cats are thought to be primarily responsible for the extinction of 33 species of birds (Crooks and Soule 1999), and the presence of feral and free-ranging dogs or cats makes some otherwise suitable locations for reintroduction of other wild species unsuitable. In many parts of the world dogs are a ubiquitous problem driven by the perceptions of people. Perhaps where these populations are best controlled is where dogs are still a regular part of the human diet (Ekanem et al. 2013).

5.2.3.3 "Adapter" Mammals

Some mammals, such as the black tufted marmoset (*Callithrix penicillata*) in Brazil, mostly persist by avoiding dense human concentrations in favor of larger urban parks (Teixeira et al. 2016b). Urban adapters adjust their risk perceptions in ways that allow them to tolerate the "landscape of fear" posed by the novel urban environment (Stillfried et al. 2017). Adapters use food resources available around human developments, including ornamental plants, gardens, garbage, and supplemental food. Common urban adapted mammals include some burrowing species – moles (family Talpidae), groundhogs (*Marmota monax*), armadillos (*Dasypus novemcinctus*) – which find refuge from humans in their burrows, often under porches and houses. Medium-size omnivores and carnivores, such as opossums (*Didelphis virginiana*), raccoons (*Procyon lotor*), foxes (*Vulpes* spp.), and coyotes (*Canis latrans*), have also been successful urban adapters, in part because of the elimination of large predators (Crooks and Soulé 1999). In fact, urban Virginia opossums are roughly one third heavier than those living in rural environments (Wright et al. 2012). Beaver (*Castor canadensis*) do well enough in some urban settings to become pests and merit control efforts because of their impact on ornamental vegetation (Loven 1985). Even larger species such as wild boar (*Sus scrofa*) have learned to exploit major urban centers such as Berlin, Germany (Stillfried et al. 2017), and overabundant urban white-tailed deer (*Odocoileus viginianus*) populations have become a major management issue in cities in North America (Doerr et al. 2001). However, most large mammals, particularly predators such as mountain lions (*Puma concolor*), which decline the closer one gets to urban centers in California, USA (Ordeñana et al. 2010), become uncommon or disappear in proximity to dense human populations. Urban exploiters are usually omnivorous and may become dependent on human resources, with little reliance on local vegetative communities (Nilon and Vandruff 1987; McKinney 2002). They can also become highly adept at using human food resources, such as trash, gardens, and bird seed (McKinney 2002; Yirga et al. 2016). Urban rodents commonly found in cities around the world include house mice (*Mus musculus*), black rats, and Norway rats.

As modern societies abandon rural life in favor of cities, previously persecuted predators such as brown bears (*Ursus arctos*), gray wolves, lynx (*Lynx lynx*), and wolverine (*Gulo gulo*) are making a comeback in Europe (Chapron et al. 2014). Urban black bears (*Ursus americanus*) feed on high-calorie garbage and have larger litter sizes than rural populations, giving birth to up to three times more cubs than exurban bears (Beckmann and Berger 2003; Prange et al. 2003).

Several canid species are also thriving in human-dominated landscapes. Red foxes (*Vulpes vulpes*) are notoriously successful in urban settings (Harris 1977, 1981), and coyotes (*Canis latrans*) and golden jackals (*Canis aureus*) do well around human settlements in North America, Asia, and Africa (Sillero-Zubiri et al. 2004). Spotted hyena (*Crocuta crocuta*) in Ethiopia also utilize trash located on the outskirts of towns (Yirga et al. 2016). Commensal rodents likewise enjoy reproductive advantages. Urban fox squirrel (*Sciurus niger*) populations have litter sizes comparable to rural populations, but unlike rural individuals, urban females have more than one litter annually (McCleery 2009). Similarly, urban striped field mice (*Apodemus agrarius*) extend their breeding season later into autumn in the urban environment and may reach sexual maturity earlier in urban areas (Andrzejewski et al. 1978; Gliwicz et al. 1994). Garbage dumps can provide additional food and shelter for indigenous chipmunks (*Tamias striatus*) and mice (*Peromyscus* spp.; Courtney and Fenton 1976), among others, though those species are found throughout urban areas within their range. Their presence near humans allows various types of conflicts to develop, ranging from mild nuisance through damage to structures to risk to human life.

5.2.3.4 Nocturnal Roosts

Just as urban bird roosts can become noise and smell nuisances (see Sects. 5.2.2.1 and 5.2.2.2), so can urban bat roosts. To date, 24 of the 45 bat species known from the USA have been recorded roosting under bridges or in culverts (Keeley and Tuttle 1999). In Texas, large colonies of Mexican free-tailed bats (*Tadarida brasiliensis*) can be found roosting in city centers on buildings, bridges, and stadiums (Scales and Wilkins 2007). Although such insectivorous bats can often be found in urban settings (e.g., Threlfall et al. 2011), it is the larger frugivorous "flying foxes" which are most often problematic. The best documented cases of this kind come from Australia. As Tidemann and Vardon (1997) pointed out, members of the genus *Pteropus* play multiple roles in the lives of humans in Australia: they are traditionally eaten by Aborigines, kept as pets in some urban centers, destroyed as pests in orchards and cities, especially after they were shown to transmit Hendra (equine morbillivirus), which is transmissible and sometimes lethal to humans. The complexities involved in their management in urban settings, and the human conflicts revolving around it, were reviewed by Perry and Perry (2008). Despite the species declining to the point of being listed as vulnerable (Commonwealth of Australia

2017), such conflict is ongoing, with local governments calling on the Australian Federal government to better manage the species (Fernbach and Gordon 2017). This kind of problem is not unique to Australia, however. In Malaysia, surveys showed that people hold negative attitudes towards the local species (*P. hypomelanus*) and do not recognize its ecological importance in pollination and seed dispersal (Aziz et al. 2017). Similarly, the Indian flying fox (*P. giganteus*) roosts in urban environments in Pakistan (Javid et al. 2017).

5.2.3.5 Urban Structures as Dens

Small- and medium-sized mammals, such as squirrels, brushtail possums (*Trichosurus vulpecula*), stone martins (*Martes fonica*), and raccoons, commonly use attics and other unused spaces in buildings for denning (Adams 2016; Hill et al. 2007). Some species, like urban red foxes, skunks (Mephitidae), Virginia opossums, Eurasian badgers (*Meles meles*), woodchucks (*Marmota momax*), and armadillos (Dasypodidae) regularly den under houses (Adams 2016; Davison et al. 2008; Harris 1981; Lariviere et al. 1999). Noise and smell nuisance, as well as concerns about disease transmission and attacks, are common reasons for removal of such individuals, when they are detected.

In some cases, urban-adapted wildlife almost exclusively use anthropogenic structures for denning. For example, Herr et al. (2009) found that 97% of urban stone martin dens were found in buildings, and inhabited buildings were selected for during winter months, presumably for warmth. Culverts and garbage are common features used by urban mammals for shelter. Coyotes, kit foxes (*Vulpes velox*), raccoons, skunks, black bears, and even spotted hyenas (*Crocuta crocuta*) are known to use culverts for dens in both urban and exurban locations (Barnes and Bray 1966; Reese et al. 1992; Adams 1994; Pokines and Kerbis Peterhans 2007; Grubbs and Krausman 2009). In England, native small mammals used building refuse and garbage for nesting (Dickman 1987), and North American woodrats (*Neotoma* spp.) are commonly found nesting in trash and building materials (McCleery et al. 2006). Noise, smell, and disease concerns are common where such dens occur near humans.

5.2.3.6 Urban Mammals and Roads

Roads are an important urban feature. They have been shown to restrict the movements of small mammals, such as hedgehogs (Erinaceinae; Rondinini and Doncaster 2002), woodrats (*Neotoma* spp.; McCleery et al. 2006), and white-footed mice (*Peromyscus leucopus*; Merriam et al. 1989) in urban and exurban settings. Roads present less of a barrier to medium-sized mammals, but their movements may still be restricted by larger highways (Prange et al. 2003; Gehrt 2005; Riley 2006). Even large mammals such as white-tailed deer appear to avoid highways with more than eight lanes, and dense development (Etter et al. 2002).

Road kill is commonly the greatest cause of mortality for many urban mammals. Collision with vehicles is a major cause of mortality for urban white-tailed deer (Etter et al. 2002; Lopez et al. 2003), raccoon (Prange et al. 2003), fox squirrel (McCleery et al. 2008), coyote (Gehrt 2007), and fox populations (Gosselink et al. 2007). Many of the mammals brought to urban wildlife rehabilitation centers (Sect. 5.1.3) were injured on urban roads.

5.3 Urban Wildlife and Disease

"Most of us did not embark on a career in wildlife ecology, management, or disease to study in the urban environment," states Gibbs (2007) in her review of the original edition of Adams (2016). "It is an unfortunate reality, however," she continued, "that we must understand the issues surrounding urban wildlife populations in order to make appropriate decisions about wildlife diseases." Although we disagree with the use of the word "unfortunately" by Gibbs (2007), we agree that understanding urban wildlife is an essential component of management of wildlife disease.

Urban arthropod ecology traditionally focuses on three topics: pests of human homes and gardens (e.g., spiders, cockroaches, stink bugs, thrips), maintenance or creation of habitats for pollinators and beneficial insects (bees, butterflies), and vectors of human illnesses (e.g., fleas, ticks, mosquitoes). Indeed, over three quarters of human diseases are zoonotic and are linked to wildlife and domestic animals (Taylor et al. 2001), and this is one of two aspects we address in this chapter.

5.3.1 Urban Wildlife as Vectors of Human Disease

Changes in human land use patterns can alter the distribution and abundance of wildlife-borne infectious diseases (Patz et al. 2004), and wildlife-pathogen dynamics shift in response to urban land use (Bradley and Altizer 2007). Patz et al. (2004) compiled a list of vector-borne and zoonotic diseases with the strongest links to urbanization and land use changes. Globally, these diseases include malaria, dengue, Lyme disease, yellow fever, Rift Valley fever, Japanese encephalitis, onchocerciasis, trypanosomiasis, plague, filariasis, meningitis, rabies, leishmaniasis, Kyasanur Forest fever, hantavirus, and Nipah virus (Patz et al. 2004).

One of the more well-studied arthropod-vectored diseases, with respect to urbanization, is Lyme disease (*Borrelia burgdorferi*). In the northeastern USA, prevalence of Lyme disease is tied to decreases in biodiversity, fragmentation of forests, and urbanization (Schmidt and Ostfeld 2001). White-footed mice exhibited greater densities of ticks infected with Lyme disease in forest fragments associated with suburban development (LoGiudice et al. 2003). While these links are direct, the ecology of these systems is often considerably more complex. For example, in a European study, urban Norway rats with Lyme disease were more infectious than urban mice (*Apodemus flavicollis*) with the disease (Matuschka et al. 1996). Urban development provides increased food sources and shelters for wild rodents, which also increases human plague (*Yersenia pestis*) risk (Mann et al. 1979). Urban plague generally results from wildlife host entries into human habitats, rather than from human entry into wild environments (Gibbons and Humphreys 1941; Mann et al. 1979). Finally, although rarely considered or encountered in developed countries, the trade in dog meat for human consumption (Ekanem et al. 2013) and especially bites by feral dogs (e.g., Sudarshan et al. 2006) can be important sources of rabies infections in humans.

5.3.2 *Vectors of Urban Wildlife Disease*

Disease outbreaks are a common cause of wildlife mortality in urban areas, likely due to the high densities of many urban populations of mammals (Gosselink et al. 2007; Prange et al. 2003) and other taxa. Agents of urban wildlife disease are often less thoroughly examined. Here, we review trends in urban wildlife disease ecology, with special emphasis on arthropod-vectored diseases. Direct linkages between urban land use changes and wildlife disease have been established in avian taxa. For example, Bradley and Altizer (2007) observed that pathogenic nematodes only influenced wading birds in coastal Florida, where stream engineering and nutrient fluxes had occurred. In addition, passerine birds had higher prevalence of potentially lethal West Nile virus (*Flavivirus* sp.) in densely populated areas (Ezenwa et al. 2006). The importance of biodiversity is apparent, for example, in studies showing that passerine bird diversity is associated with lower West Nile virus infection rates in both mosquitoes and humans (Ezenwa et al. 2006).

These trends are driven by habitat changes resulting from urbanization. However, they are likely exacerbated by insecticide use and changes in public health policy, resulting in new urban disease threats for wildlife (Gubler 1998) – and perhaps also in more novel zoonoses that urban wildlife could carry. An increased awareness and understanding of the prevalence and impacts of urban wildlife disease will "improve our understanding of the ecological drivers behind spatial variation in pathogen occurrence" (Bradley and Altizer 2007) and can improve our ability to combat and treat their vectors.

Finally, we need to consider the relationship between human population trends and globalization of trade and the growth in prevalence of emerging zoonoses (e.g., Brown 2004). Many diseases use wildlife as a reservoir, primary, secondary, or temporary host. Certainly, factors such as consumption of "bushmeat" contribute greatly to human exposure to such diseases, but many of them fall outside the scope of this chapter. In the context of urbanization, we are particularly concerned with the growth of the pet trade in both volume and diversity of available species, many of them wild-caught, and, to a lesser extent, with access to petting zoos and similar access at events such as county fairs (Chomel et al. 2007; Smith et al. 2017). Thus, the potential for growth in new wildlife and human diseases arriving via the pet and exotic food trade is substantial and appears not to have received sufficient policy consideration.

5.4 Urban Wildlife, Global Change, and the Future

As reviewed above, the unique microclimates provided by urban settings help buffer the wildlife dwelling in cities against extreme cold and drought, allowing species to expand their ranges, either by themselves (e.g., beavers now expanding into the Arctic tundra, both benefitting from and affecting changes in local water cycles (Rozell 2017) or with human-aided transport (i.e., invasive species). On the other hand, urban heat islands, enhanced illumination, and other changes keep other species away or negatively affect them. In some cases, as with the reported subsidence of some major urban centers as a result of over-drawing of underlying aquifers (e.g., Ruggerio 2017), the impacts on wildlife are not clear-cut. What does the future hold, then, under realistic scenarios?

A variety of predictions of human population have been published by organizations such as the United Nations over the years. Many of these have predicted a stabilization in the human population during the current century. This, it turns out, was probably optimistic (Gerland et al. 2014). Human populations are still growing, and will continue to do so. Human needs are growing apace (Crist et al. 2017), as are the land use consequences of land degradation and fragmentation (Minelli et al. 2017). At the same time that human numbers, life expectancy, and needs are increasing, human standard of living is also improving, with global poverty and hunger declining drastically since 1990 (United Nations 2015b; but see Hickel 2016; Pingali 2016). In the words of Tilman et al. (2017), "[f]uture population growth and economic development are forecasted to impose unprecedented levels of extinction risk on many more species worldwide, especially the large mammals of tropical Africa, Asia and South America."

As already mentioned, however, urban environments are continually expanding (Fig. 5.1) – perhaps the only type of habitat to do so. We therefore predict that species that do well in or near urbanization will fare proportionally much better in the decades ahead. What will it take to do so? Clearly, species that are tolerant of human presence, noise, nightlight, and other current activities will be winners. Urban environments already offer suitable habitat for species endangered elsewhere, such as amphibians (Chovanec 1994), peregrine falcons (*Falco peregrinus*; Cade and Bird 1990), and other taxa (Alvey 2006). However, we predict rapid evolution of traits that allow additional species, both common and not, to inhabit urban settings as a result of the strong artificial selection involved, and there is evidence that this is already occurring (Alberti et al. 2017; Johnson and Munshi-South 2017). In addition, we expect that other changes already being noticed to increasingly create un welcome wildlife conflict.

Finally, future cities will differ from current ones. One change that is emerging is the development of self-driving vehicles. Will those be programmed to avoid wildlife as they do humans? If so, will they respond differently to a large bear than to a small turtle attempting to cross the road? Another trend that is already apparent is the increase in the number of drones of various sizes filling the skies (Murray and Chu 2015). What will urban birds experience when delivery of parcels is largely airborne? In the absence of solid data, we can imagine a range of plausible scenarios. Lastly, there is overwhelming agreement among scientists that climate is changing. Although the exact scenarios are still being debated, and will greatly depend on the policy options adopted, planners are already busy devising ways of adapting cities to likely scenarios (e.g., Bulkeley and Betsill 2005). Given the increasing importance of urban environments for the survival of future wildlife, will planners take the welfare of nonhuman species into account as they design or modify future cities? Conservation biologists, who have generally treated urban environments with disdain, might want to consider these questions in order to contribute their insights to future trends.

5.5 Summary

Despite being long neglect, studies of urban environments and their ecological value are becoming more common (Fig. 5.2), perhaps because urban ecosystems are becoming increasingly dominant (Fig. 5.1). Human dimensions of wildlife are also receiving more attention, and not only in the area of human-wildlife conflict that has long been the dominant subdiscipline. For example, the substantial value of wildlife in the tourism industry – itself a major source of global economic value (Scott and Gössling 2015) – is increasingly explicitly recognized (e.g., Mustika et al. 2013). Our review shows that a focus on conflict remains common in an urban context, with reptiles, birds, and mammals all responsible for a wide range of impacts. However, all three groups are also increasingly valued as urban co-inhabitants, with numbers on the costs people willingly incur to feed birds in their yards being both well documented and in the tens of millions of US dollars. Another reflection of the eagerness for wild animal companionship is the expansion of the international pet trade, which has grown just as international trade in general has increased. The increasing movement of humans and nonhuman animals across the globe has, in turn, increased the transport (potential and realized) of diseases to novel locations. Many of those disease, sometimes unique to humans or other animals and sometimes shared, are carried or transmitted by invertebrates, a group barely touched upon in this chapter. Finally, this chapter briefly points out that urban environments and their human inhabitants provide an increasingly important habitat for some animals that are at growing risk of decline – sometimes already formally considered threatened – elsewhere. Exposure of urbanites to animals, especially at a young age, may provide an essential encouragement for conservation of wildlife elsewhere. Thus, the interactions between humans and nonhumans in urban environments are multifaceted and in some cases still developing. Perhaps not surprisingly, policy and management responses to this fluid situation are lagging behind the complex reality and would benefit from careful reexamination.

References

Adams LW (1994) Urban wildlife habitats: a landscape perspective. University of Minnesota Press, Minneapolis

Adams CE (2016) Urban wildlife management, 3rd edn. CRC press, Boca Raton

Add AH, Kidd RM, Zasloff RL (1996) Characteristics and motives of volunteers in wildlife rehabilitation. Psychol Rep 79:227–234

Alberti M, Marzluff J, Hunt VM (2017) Urban driven phenotypic changes: empirical observations and theoretical implications for eco-evolutionary feedback. Philos Trans R Soc B 372:20160029

Alvey AA (2006) Promoting and preserving biodiversity in the urban forest. Urban For Urban Green 5:195–201

Amarasinghe AT, Madawala MB, Karunarathna DS et al (2015) Human-crocodile conflict and conservation implications of Saltwater Crocodiles *Crocodylus porosus* (Reptilia: Crocodylia: Crocodylidae) in Sri Lanka. J Threat Taxa 7:7111–7130

American Veterinary Medical Association (2012) US pet ownership and demographics sourcebook. American Veterinary Medical Association, Schaumburg

Andrzejewski R, Babinska-Werka J, Gliwicz J, Goszczynski J (1978) Synurbanization process in a population of *Apodemus agrarius*. Acta Theriol (Warsz) 23:341–358

Animal Medicines Australia (2016) Pet ownership in Australia, 2016, Animal Medicines Australia. Australian Capital Territory, Australia, Barton

Aziz SA, Clements GR, Giam X et al (2017) Coexistence and conflict between the island flying fox (*Pteropus hypomelanus*) and humans on Tioman Island, peninsular Malaysia. Hum Ecol 45:377–389

Barnes VG Jr, Bray OE (1966) Black bears use drainage culverts for winter dens. J Mammal 47:712–713

Bateman PW, Fleming PA (2012) Big city life: carnivores in urban environments. J Zool 287:1–23

Becker M, Buchholz S (2016) The sand lizard moves downtown–habitat analogues for an endangered species in a metropolitan area. Urban Ecosyst 19:361–372

Beckmann JP, Berger J (2003) Using black bears to test ideal-free distribution models experimentally. J Mammal 84:594–606

Belant JL (1997) Gulls in urban environments: landscape-level management to reduce conflict. Landsc Urban Plan 38:245–258

Beringer J, Mabry P, Meyer T et al (2004) Post-release survival of rehabilitated white-tailed deer fawns in Missouri. Wildl Soc Bull 32:732–738

Blair RB (1996) Land use and avian species diversity along an urban gradient. Ecol Appl 6:506–519

Boal CW (2008) Hacked nestling Mississippi kite survives six years and demonstrated philopatry. Bull Tex Ornithol Soc 41:66–68

Boal CW (2011) Colonial nesting of Yellow-crowned Night Herons on the San Antonio River Walk. Bull Tex Ornithol Soc 43:45–48

Boal CW, Mannan RW (1999) Comparative breeding ecology of Cooper's hawks in urban and exurban areas of southeastern Arizona. J Wildl Manag 63:77–84

Bradley CA, Altizer S (2007) Urbanization and the ecology of wildlife disease. Trends Ecol Evol 22:95–102

Briffett C, Sodhi N, Yuen B, Kong L (2004) Green corridors and the quality of urban life in Singapore. In: Proceedings of the 4th international urban wildlife symposium, pp 56–63

Brown C (2004) Emerging zoonoses and pathogens of public health significance – an overview. Rev Sci Tech Off Int Epiz 23:435–442

Bulkeley H, Betsill MM (2005) Cities and climate change: urban sustainability and global environmental governance, vol 4. Routledge, New York

Caccamise DF (1990) Communal starling roosts: implications for control. In: Davis LR, Marsh RE (eds) Proceedings of the 14th vertebrate pest conference. University of California, Davis

Cade TJ, Bird DM (1990) Peregrine falcons, *Falco peregrinus*, nesting in an urban environment: a review. Can Field Nat 104:209–218

Canyon DV, Hii JLK (1997) The gecko: an environmentally friendly biological agent for mosquito control. Med Vet Entomol 11:319–323

Chapman T (2005) The status and impact of the rainbow lorikeet (*Trichoglossus haematodus moluccanus*) in Southwest Western Australia, Report 04/2005. Department of Agriculture and Food, Western Australia, Perth

Chapron G, Kaczensky P, Linnell JDC et al (2014) Recovery of large carnivores in Europe's modern human-dominated landscapes. Science 346:1517–1519

Chomel BB, Belotto A, Meslin FX (2007) Wildlife, exotic pets, and emerging zoonoses. Emerg Infect Dis 13:6–11

Chovanec A (1994) Man-made wetlands in urban recreational areas—a habitat for endangered species? Landsc Urban Plan 29:43–54

Clucas B, Rabotyagov S, Marzluff JM (2015) How much is that birdie in my backyard? A cross-continental economic valuation of native urban songbirds. Urban Ecosyst 18:251–266

Commonwealth of Australia (2017) Living with fruit bats: inquiry into flying-fox management in the eastern states. House of Representatives Standing Committee on the Environment and Energy. Available from http://parlinfo.aph.gov.au/parlInfo/download/committees/reportrep/024037/toc_pdf/Livingwithfruitbats.pdf;fileType=application%2Fpdf. Accessed 12 Dec 2017

Coon DW, Arnold KA (1977) Origins of Brown-headed Cowbird populations wintering in central Texas. N Am Bird Bander 2:7–11

Courtney PA, Fenton MB (1976) The effects of a small rural garbage dump on populations of *Peromyscus leucopus* rafinesque and other small mammals. J Appl Ecol 13:413–422

Crist E, Mora C, Engelman R (2017) The interaction of human population, food production, and biodiversity protection. Science 356:260–264

Crooks KR, Soule ME (1999) Mesopredator release and avifaunal extinctions in a fragmented system. Nature 400:563

Davidson GR Jr, Degeare GV Jr, Sorg TJ, Clark RM (1971) Land disposal sites near airports reporting bird/aircraft hazards: a survey for the inter-agency bird hazard committee, US Environmental Protection Agency Division of Technical Operations/Open-File Report (TSR 1.6.004/0). EPA, Washington

Davison J, Huck M, Delahay RJ, Roper TJ (2008) Urban badger setts: characteristics, patterns of use and management implications. J Zool (Lond) 275:190–200

del Hoyo J, Elliot A, Christie D (2005) Handbook of the birds of the world, vol 10. Lynx Ediciones, Barcelona

Devan-Song A, Martelli P, Dudgeon D et al (2016) Is long-distance translocation an effective mitigation tool for white-lipped pit vipers (*Trimeresurus albolabris*) in South China? Biol Conserv 204:212–220

Dezzutto D, Barbero R, Canale G et al (2017) Detection of *Leptospira* spp. in water turtle (*Trachemys scripta*) living in ponds of urban parks. Vet Sci 4(51). https://doi.org/10.3390/vetsci4040051

Dickman CR (1987) Habitat fragmentation and vertebrate species richness in an urban environment. J Appl Ecol 24:337–351

Dickman CR, Doncaster CP (1987) The ecology of small mammals in urban habitats. I. Populations in a patchy environment. J Anim Ecol 56:629–640

Dickman CR, Doncaster CP (1989) The ecology of small mammals in urban habitats. II. Demography and dispersal. J Anim Ecol 58:119–127

Dirckx JH (1980) The pied Piper of Hamelin: a medical-historical interpretation. Am J Dermatopathol 2:39–46

Ditchkoff SS, Saalfeld ST, Gibson CJ (2006) Animal behavior in urban ecosystems: modifications due to human-induced stress. Urban Ecosyst 9:5–12

Dixon DM (1989) A note on some scavengers of ancient Egypt. World Archaeol 21:193–197

Doerr ML, McAninch JB, Wiggers EP (2001) Comparison of 4 methods to reduce white-tailed deer abundance in an urban community. Wildl Soc Bull 2001:1105–1113

Dolbeer RA, Weller JR, Anderson AL, Begier MJ (2016) Wildlife strikes to civil aircraft in the United States 1990–2015, Serial Report Number 22. Federal Aviation Administration, National Wildlife Strike Database, Washington, DC

Dove CJ, Snow RW, Rochford MR, Mazzotti FJ (2011) Birds consumed by the invasive Burmese python (*Python molurus bivittatus*) in Everglades National Park, Florida, USA. Wilson J Ornithol 123:126–131

Ekanem EE, Eyong KI, Philip-Ephraim EE et al (2013) Stray dog trade fuelled by dog meat consumption as a risk factor for rabies infection in Calabar, southern Nigeria. Afr Health Sci 13:1170–1173

Engeman R, Avery L (2016) Prioritizing management and research actions against invasive reptiles in Florida: a collaboration by an expert panel. USDA National Wildlife Research Center – Staff Publications 1862

Engeman RM, Smith HT, Constantin B (2005) Invasive iguanas as airstrike hazards at Luis Munoz Marin International Airport, San Juan, Puerto Rico. J Aviat Aerosp Edu Res 14(3):Art. 5

Etter DR, Hollis KM, Deelen TRV et al (2002) Survival and movements of white-tailed deer in suburban Chicago, Illinois. J Wildl Manag 66:500–510

Ezenwa VO, Godsey MS, King RJ, Guptill SC (2006) Avian diversity and West Nile virus: testing associations between biodiversity and infectious disease risk. Proc R Soc Lond 273:109–117

Farmer MC, Wallace MC, Shiroya M (2011) Bird diversity indicates ecological value in urban home prices. Urban Ecosyst 16:131–144. https://doi.org/10.1007/s11252-011-0209-0

Fernbach N, Gordon K (2017) LGAQ calls on governments to take control of flying fox issue. ABC News, 28 February 2017. Available online at http://www.abc.net.au/news/2017-02-28/flying-fox-inquiry-local-government-association-responds/8310516. Accessed 12 Dec 2017

Fischer JD, Schneider SC, Ahlers AA, Miller JR (2015) Categorizing wildlife responses to urbanization and conservation implications of terminology. Conserv Biol 29:1246–1248

Fitzwater WD (1988) Solutions to urban bird problems. Proc Vertebr Pest Control Conf 13:254–259

Forbes JE (1993) Survey of nuisance urban geese in the United States. In: Proceedings of the 11th great plains wildlife damage control workshop, Kansas City, Missouri, USA, pp 92–101

Francis CD, Barber JR (2013) A framework for understanding noise impacts on wildlife: an urgent conservation priority. Front Ecol Environ 11:305–313

Fritts TH (2002) Economic costs of electrical system instability and power outages caused by snakes on the island of Guam. Int Biodeterior Biodegradation 49:93–100

Fukuda Y, Manolis C, Appel K (2014) Management of human-crocodile conflict in the Northern Territory, Australia: review of crocodile attacks and removal of problem crocodiles. J Wildl Manag 78:1239–1249

Fuller RA, Tratalos J, Gaston KJ (2009) How many birds are there in a city of half a million people? Divers Distrib 15:328–337

Gehrt SD (2005) Seasonal survival and cause-specific mortality of urban and rural striped skunks in the absence of rabies. J Mammal 86(6):1164–1170

Gehrt SD (2007) Ecology of coyotes in urban landscapes. In: Proceedings of the 12th wildlife damage conference. University of Nebraska, Lincoln, pp 303–311

Gerland P, Raftery AE, Ševčíková H et al (2014) World population stabilization unlikely this century. Science 346:234–237

Gibbons RJ, Humphreys FA (1941) Plague surveys in western Canada. Can Public Health J 32:24–28

Gibbs SE (2007) Urban Wildlife Management. J Wildl Dis 43:321–321

Gilbert SL, Sivy KJ, Pozzanghera CB et al (2017) Socioeconomic benefits of large carnivore recolonization through reduced wildlife-vehicle collisions. Conserv Lett 10:431–439

Giunchi D, Albores-Barajas YV, Baldaccini NE et al (2012) Feral pigeons: problems, dynamics and control methods. In: Soloneski S (ed) Integreated pest management and pest control – current and future tactics. InTech, Rijeka, pp 215–240

Gliwicz J, Goszczynski J, Luniak M (1994) Characteristic features of animal populations under synurbanization-the case of the blackbird and of the stripped field mouse. Mem Zool 49:237–244

Gosselink TE, Van Deelen TR, Warner RE, Mankin PC (2007) Survival and cause specific mortality of red foxes in agricultural and urban areas of Illinois. J Wildl Manag 71:1862–1873
Grubbs SE, Krausman PR (2009) Use of urban landscape by coyotes. Southwest Nat 54:1–12
Gubler DJ (1998) Resurgent vector-borne diseases as a global health problem. Emerg Infect Dis 4:442–450
Hall DW, Harvey TM (2007) Mortality at a night roost of great-tailed grackles and European starlings during a spring hail storm. Wilson J Ornithol 119:309–312
Hall JM, Warner DA (2017) Body size and reproduction of a non-native lizard are enhanced in an urban environment. Biol J Linn Soc 122(4):860–871. https://doi.org/10.1093/biolinnean/blx109
Harris S (1977) Distribution, habitat utilization and age structure of a suburban fox (*Vulpes vulpes*) population. Mammal Rev 7:25–38
Harris S (1981) An estimation of the number of foxes (Vulpes vulpes) in the city of Bristol, and some possible factors affecting their distribution. J Appl Ecol 18:455–465
Harris S, Smith GC (1987) Demography of two urban fox (*Vulpes vulpes*) populations. J Appl Ecol 24:75–86
Hattori M (2009) Adaptive management of urban wildlife: resolving human-heron conflicts. MS thesis, University of California, Davis, California, USA
Herr J, Schley L, Engel E, Roper TJ (2009) Den preferences and denning behaviour in urban stone martens (*Martes foina*). Mamm Biol 75:138–145
Hickel J (2016) The true extent of global poverty and hunger: questioning the good news narrative of the Millennium Development Goals. Third World Q 37:749–767
Hill NJ, Carbery KA, Deane EM (2007) Human–possum conflict in urban Sydney, Australia: public perceptions and implications for species management. In: Human dimensions of wildlife. Taylor and Francis, Oxford, pp 101–113
Hines RK (2017) Review of literature: potential benefits of urban nature exposure and stream corridor recreation. Illuminare A Stud J Recreation Parks Tour Stud 15:33–46
Home C, Bhatnagar YV, Vanak AT (2017) Canine conundrum: domestic dogs as an invasive species and their impacts on wildlife in India. Anim Conserv 21:275. https://doi.org/10.1111/acv.12389
Horn DJ, Johansen SM (2013) A comparison of bird-feeding practices in the United States and Canada. Wildl Soc Bull 37:293–300
Hosaka T, Sugimoto K, Numata S (2017a) Effects of childhood experience with nature on tolerance of urban residents toward hornets and wild boars in Japan. PLoS One 12(4):e0175243
Hosaka T, Sugimoto K, Numata S (2017b) Childhood experience of nature influences the willingness to coexist with biodiversity in cities. Palgrave Commun 3:17071. https://doi.org/10.1057/palcomms.2017.71
Javid A, Gulariz TL, Ashraf M et al (2017) Proximate analysis of Indian flying fox's (*Pteropus giganteus*) natural food, with a note on its roost variations in urban areas of Lahore, Pakistan. Turk J Zool 41:714–721
Jerolmack C (2008) How pigeons became rats: the cultural-spatial logic of problem animals. Soc Probl 55:72–94
Johnson MT, Munshi-South J (2017) Evolution of life in urban environments. Science 358:eaam8327
Johnston RF, Janiga M (1995) Feral pigeons. Oxford University Press, Oxford
Jones DN, Reynolds SJ (2008) Feeding birds in our towns and cities: a global research opportunity. J Avian Biol 39:265–271
Karesh WB (1995) Wildlife rehabilitation: additional considerations for developing countries. J Zoo Wildl Med 26:2–9
Keeley BW, Tuttle MD (1999) Bats in American bridges, Resource Publication 4. Bat Conservation Int, Austin
Kellert SR (1979) Public attitudes toward critical wildlife issues. U.S. Government Printing Office, Washington, DC
Kellert SR, Wilson EO (eds) (1995) The biophilia hypothesis. Island Press, Washington, DC

Kwenti TE (2017) Biological control of parasites. In: Khater H, Govindarajan M, Benelli G (eds) Natural remedies in the fight against parasites. InTech, Rijeka. https://doi.org/10.5772/68012

Lariviere S, Walton LR, Messier FO (1999) Selection by striped skunks (Mephitis mephitis) of farmsteads and buildings as denning sites. Am Midl Nat 142:96–101

Laurie I (1979) Nature in cities: the natural environment in the design and development of urban open space. Wiley, New York

Lazić MM, Carretero MA, Živković U, Crnobrnja-Isailović J (2017) City life has fitness costs: reduced body condition and increased parasite load in urban common wall lizards, *Podarcis muralis*. Salamandra 53:10–17

Lee ME, Miller R (2003) Managing elk in the wildland-urban interface: attitudes of Flagstaff, Arizona residents. Wildl Soc Bull 2003:185–191

LoGiudice K, Ostfeld RS, Schmidt KA, Keesing F (2003) The ecology of infectious disease: effects of host diversity and community composition on Lyme disease risk. Proc Natl Acad Sci 100:567–571

Longcore T, Rich C (2004) Ecological light pollution. Front Ecol Environ 2:191–198

Lopez RR, Vieira MEP, Silvy NJ et al (2003) Survival, mortality, and life expectancy of Florida key deer. J Wildl Manag 67:34

Loven JE (1985) Reported beaver damage and control methods used in Texas. Great plains wildlife damage control workshop proceedings, vol. 170. Available online from http://digitalcommons.unl.edu/gpwdcwp/170. Accessed 20 Dec 2017

Lowry H, Lill A, Wong B (2013) Behavioural responses of wildlife to urban environments. Biol Rev 88:537–549

Loyd KA, Miller CA (2010) Factors related to preferences for Trap–Neuter–Release management of feral cats among Illinois homeowners. J Wildl Manag 74:160–165

Lunney D, Burgin S (2004) Urban wildlife management: an emerging discipline. In: Proceedings of a forum held by the Royal Zoological Society of New South Wales at Taronga Zoo on 20 October. 2001. Available online at http://researchdirect.westernsydney.edu.au/islandora/object/uws%3A7286/. Accessed 11 Nov 2017

Magle SB, Hunt VM, Vernon M, Crooks KR (2012) Urban wildlife research: past, present, and future. Biol Conserv 155:23–32

Mann JM, Martone WJ, Boyce JM et al (1979) Endemic human plague in New Mexico: risk factors associated with infection. J Infect Dis 140:397–401

Marra PP, Santella C (2016) Cat wars: the devastating consequences of a cuddly killer. Princeton University Press, Princeton

Matheson C (1944) The domestic cat as a factor in urban ecology. J Anim Ecol 13:130–133

Matuschka F-R, Endepols S, Richter D et al (1996) Risk of urban Lyme disease enhanced by the presence of rats. J Infect Dis 174:1108–1111

McCleery RA (2009) Reproduction, juvenile survival and retention in an urban fox squirrel population. Urban Ecosyst 12:177–184

McCleery R (2010) Urban mammals. In: Urban ecosystem ecology, Agronomy Monograph 55. ASA, CSSA, SSSA, Madison, pp 87–102

McCleery RA, Lopez RR, Silvy NJ et al (2006) Population status and habitat selection of the endangered Key Largo woodrat. Am Midl Nat 155:197–209

McCleery RA, Lopez RR, Silvy NJ, Gallant DL (2008) Fox squirrel survival in urban and rural environments. J Wildl Manag 72:133–137

McGaughey EK (2012) Wildlife rehabilitation centers: survey of rehabilitators' attitudes, motivations, and knowledge & study of animal admittance to the south plains wildlife rehabilitation center. Unpublished MS thesis., Texas Tech University

McGaughey EK, Wallace M, Perry G (2011) Herpetofauna admitted to the South Plains wildlife rehabilitation center (Lubbock, Texas): a two-decade perspective. IRFC Reptiles Amphib 18:18–23

McIntyre NE, Knowles-Yánez K, Hope D (2008) Urban ecology as an interdisciplinary field: differences in the use of "urban" between the social and natural sciences. Urban Ecol 2008:49–65

McKinney ML (2002) Urbanization, biodiversity, and conservation. Bioscience 52:883–890
McKinney ML (2008) Effects of urbanization on species richness: a review of plants and animals. Urban Ecosyst 11:161–176
Mendelssohn H, Yom-Tov Y (1999) A report of birds and mammals which have increased their distribution and abundance in Israel due to human activity. Isr J Zool 45:35–47
Merriam G, Kozakiewicz M, Tsuchiya E, Hawley K (1989) Barriers as boundaries for metapopulations and demes of *Peromyscus leucopus* in farm landscapes. Landsc Ecol 2:227–235
Messmer TA, George SM, Cornicelli L (1997) Legal considerations regarding lethal and nonlethal approaches to managing urban deer. Wildl Soc Bull 25:424–429
Miller JR (2005) Biodiversity conservation and the extinction of experience. Trends Ecol Evol 20:430–434
Minelli S, Erlewein A, Castillo V (2017) Land degradation neutrality and the UNCCD: from political vision to measurable targets. In: International yearbook of soil law and policy. Springer, Cham, pp 85–104
Mitchell JC, Jung Brown RE, Bartholomew B (2008) Urban herpetology. Herpetological conservation, vol 3. Society for the Study of Amphibians and Reptiles, Salt Lake City
Montgomery MR (2008) The urban transformation of the developing world. Science 319:761–764
Morrison JL, Terry M, Kennedy PL (2006) Potential factors influencing nest defense in diurnal North American raptors. J Raptor Res 40:98–110
Mumaw LM, Maller C, Bekessy S (2017) Strengthening wellbeing in urban communities through wildlife gardening. Cities Environ (CATE) 10(1):Article 6. Available at: http://digitalcommons.lmu.edu/cate/vol10/iss1/6
Murray CC, Chu AG (2015) The flying sidekick traveling salesman problem: optimization of drone-assisted parcel delivery. Transp Res C Emerg Technol 54:86–109
Murray JK, Browne WJ, Roberts MA et al (2010) Number and ownership profiles of cats and dogs in the UK. Vet Rec 166:163–168
Mustika PLK, Birtles A, Everingham Y, Marsh H (2013) The human dimensions of wildlife tourism in a developing country: watching spinner dolphins at Lovina, Bali, Indonesia. J Sustain Tour 21:229–251
Nilon CH, Vandruff LW (1987) Analysis of small mammal community data and applications to management of urban greenspaces. Proc Natl Symp Urban Wildl 2:53–59
Novaes WG, Cintra R (2013) Factors influencing the selection of communal roost sites by the black vulture *Coragyps atratus* (Aves: Cathartidae) in an urban area in Central Amazon. Zoologia (Curitiba) 30:607–614
Novaes WG, Grossmann NV, Pimentel DS, Prada M (2016) Terrestrial mammal and reptile hazards in an airport in the Brazilian Amazon. Hum Wildl Interact 10:122–127
Ordeñana MA, Crooks KR, Boydston EE et al (2010) Effects of urbanization on carnivore species distribution and richness. J Mammal 91:1322–1331
Patz JA, Daszak P, Tabor GM et al (2004) Unhealthy landscapes: policy recommendations on land use change and infectious disease emergence. Environ Health Perspect 112:1092–1098
Peh KSH, Sodhi NS (2002) Characteristics of nocturnal roosts of house crows in Singapore. J Wildl Manag 66:1128–1133
Perry D, Perry G (2008) Improving interactions between animal rights groups and conservation biologists. Conserv Biol 22:27–35
Perry G, Buchanan BW, Fisher RN et al (2008) Effects of artificial night lighting on reptiles and amphibians in urban environments. In: Jung RE, Mitchell JC (eds) Urban herpetology. Herpetological conservation, vol 3. Society for the Study of Amphibians and Reptiles, Salt Lake, pp 239–256
Pickett STA, Cadenasso ML, Grove JM et al (2001) Urban ecological systems: linking terrestrial ecological, physical, and socioeconomic components of metropolitan areas. Annu Rev Ecol Syst 32:127–157
Pickett ST, Cadenasso ML, Grove JM et al (2011) Urban ecological systems: scientific foundations and a decade of progress. J Environ Manag 92:331–362

Pingali PL (2016) Opinion: the hunger metrics mirage: There's been less progress on hunger reduction than it appears. Proc Natl Acad Sci 113:4880–4883

Pitts SL, Hughes BD, Mali I (2017) Rattlesnake nuisance removals and urban expansion in Phoenix, Arizona. West N Am Nat 77:309–316

Pokines JT, Kerbis Peterhans JC (2007) Spotted hyena (*Crocuta crocuta*) den use and taphonomy in the Masai Mara National Reserve, Kenya. J Archaeol Sci 34:1914–1931

Powell R, Henderson RW, Farmer MC et al (2011) Introduced amphibians and reptiles in the greater Caribbean: patterns and conservation implications. In: Hailey A, Wilson BS, Horrocks JA (eds) Conservation of Caribbean Island herpetofaunas, vol 1, pp 63–143

Prange S, Gehrt SD, Wiggers EP (2003) Demographic factors contributing to high raccoon densities in urban landscapes. J Wildl Manag 67:324–333

Price CJ, Banks PB (2017) Habitat augmentation for introduced urban wildlife: the use of piles of railway sleepers as refuge for introduced black rats *Rattus rattus*. Aust Zool 39:513. https://doi.org/10.7882/AZ.2017.029

Ramires EN, Fraguas GM (2004) Tropical house gecko (*Hemidactylus mabouia*) predation on brown spiders (*Loxosceles intermedia*). J Venomous Anim Toxins Incl Trop Dis 10:185–190

Reese EA, Standley WG, Berry WH (1992) Habitat, soils, and den use of San Joaquin kit fox (*Vulpes velox macrotis*) at Camp Roberts Army National Guard Training Site, California (No. EGG-10617-2156). EG and G Energy Measurements, Inc., Goleta, CA (United States). Santa Barbara Operations

Riley SPD (2006) Spatial ecology of bobcats and gray foxes in urban and rural zones of a national park. J Wildl Manag 70:1425–1435

Riley SJ, Decker DJ, Carpenter LH et al (2002) The essence of wildlife management. Wildl Soc Bull 2002:585–593

Rivas AE, Allender MC, Mitchell M, Whittington JK (2014) Morbidity and mortality in reptiles presented to a wildlife care facility in Central Illinois. Hum Wildl Interact 8:78–87

Robinson DJ, Cowan IM (1954) An introduced population of the gray squirrel (*Sciurus carolinensis* Gmelin) in British Columbia. Can J Zool 32:261–282

Rodda GH, Sawai Y, Chiszar D, Tanaka H (eds) (1999) Problem snake management: the Habu and the Brown Treesnake. Cornell University Press, Ithaca/New York

Rondinini C, Doncaster CP (2002) Roads as barriers to movement for hedgehogs. Funct Ecol 16:504–509

Rosenberg T (2017) Where birds and planes collide, a winged robot may help. New York Times, 28 November 2017. Available from https://www.nytimes.com/2017/11/28/opinion/birds-planes-robird.html?action=click&pgtype=Homepage&clickSource=story-heading&module=opinion-c-col-right-region&region=opinion-c-col-right-region&WT.nav=opinion-c-col-right-region&_r=0. Accessed 28 Nov 2017

Roshnath R, Sinu PA (2017a) Nesting tree characteristics of heronry birds of urban ecosystems in peninsular India: implications for habitat management. Curr Zool 63:599–605

Roshnath R, Sinu PA (2017b) Are the heronry birds adapting to urbanization? Zoo's Print 32:27–33

Rozell N (2017) 'Tundra be dammed': Beavers head north, leaving their mark on the Arctic. Anchorage Daily News, 18 November 2017. Available online at https://www.adn.com/alaska-news/wildlife/2017/11/18/tundra-be-dammed-beavers-head-north-leaving-their-mark-on-the-arctic/. Accessed 20 Dec 2017

Ruggerio A (2017) The ambitious plan to stop the ground from sinking. BBC, 1 December 2017. Available online at http://www.bbc.com/future/story/20171130-the-ambitious-plan-to-stop-the-ground-from-sinking. Accesses 21 Dec 2017

Scales JA, Wilkins KT (2007) Seasonality and fidelity in roost use of the Mexican free-tailed bat, *Tadarida brasiliensis*, in an urban setting. West N Am Nat 67:402–408

Schmidt KA, Ostfeld RS (2001) Biodiversity and the dilution effect in disease ecology. Ecology 82:609–619

Scott D, Gössling S (2015) What could the next 40 years hold for global tourism? Tour Recreat Res 40:269–285

Sementelli A, Smith HT, Meshaka WE Jr, Engeman RM (2008) Just green iguanas? The associated costs and policy implications of exotic invasive wildlife in South Florida. Public Works Manag Policy 12:599–606
Shanahan DF, Fuller RA, Bush R, Lin BB, Gaston KJ (2015) The health benefits of urban nature: how much do we need? Bioscience 65:476–485
Siemer WF, Brown TL (1993) Key perceptions held by wildlife managers and wildlife rehabilitators in New York: an assessment of communication coorientation, an executive summary. Human Dimensions Research Unit, Department of Natural resources, Cornell University, Ithaca/New York
Siemer WF, Brown TL, Martin PP, Stumvoll RD (1991) Tapping the potential of the wildlife rehabilitation community for public education about wildlife damage management. Proc East Wildl Damage Control Conf 5:143–147
Sillero-Zubiri C, Hoffmann M, Macdonald DW (2004) Canids: foxes, wolves, jackals and dogs. status survey and conservation action plan. IUCN/SSC Canid Specialist Group, Gland/Cambridge. x + 430 pp
Smith AE, Craven SR, Curtis PD (1999) Managing Canada geese in urban environments, Jack Berryman Institute Publication 16. Cornell Cooperative Extension, Ithaca/New York
Smith KM, Zambrana-Torrelio C, White A et al (2017) Summarizing US wildlife trade with an eye toward assessing the risk of infectious disease introduction. EcoHealth 14:29–39
Soga M, Gaston KJ (2016) Extinction of experience: the loss of human-nature interactions. Front Ecol Environ 14:94–101
Soga M, Yamaura Y, Koike S, Gaston KJ (2014a) Land sharing vs. land sparing: does the compact city reconcile urban development and biodiversity conservation? J Appl Ecol 51:1378–1386
Soga M, Yamaura Y, Koike S, Gaston KJ (2014b) Woodland remnants as an urban wildlife refuge: a cross-taxonomic assessment. Biodivers Conserv 23:649–659
Sosa JA, Perry G (2015) Site fidelity, movement, and visibility following translocation of ornate box turtles (*Terrapene ornate ornata*) from a wildlife rehabilitation center in the high plains of Texas. Herpetol Conserv Biol 10:255–262
Soulsbury CD, White PC (2015) Human–wildlife interactions in urban areas: a review of conflicts, benefits and opportunities. Wildl Res 42:541–553
Southeastern Outdoors (2010) Wildlife rehabilitators directory – USA. Available online at http://www.southeasternoutdoors.com/wildlife/rehabilitators/directory-us.html. Accessed 12 Nov 2017
Souza CSA, Teixeira CP, Young RJ (2012) The welfare of an unwanted guest in an urban environment: the case of the white-eared opossum (*Didelphis albiventris*). Anim Welf UFAW J 21:177–183
Stillfried M, Gras P, Börner K et al (2017) Secrets of success in a landscape of fear: urban wild boar adjust risk perception and tolerate disturbance. Front Ecol Evol 5:157. https://doi.org/10.3389/fevo.2017.00157
Subramanya S (2005) Heronries of Kerala. Malabar Trogon 3:1–2
Sudarshan MK, Mahendra BJ, Madhusudana SN et al (2006) An epidemiological study of animal bites in India: results of a WHO sponsored national multi-centric rabies survey. J Commun Dis 38:32–39
Sukopp H (1998) Urban ecology—scientific and practical aspects. In: Breuste J, Feldman H, Uhlmann O (eds) Urban ecology. Springer, Berlin, pp 3–16
Sutherland WJ, Freckleton RP, Godfray HCJ et al (2013) Identification of 100 fundamental ecological questions. J Ecol 101:58–67
Taylor LH, Latham SM, Woolhouse ME (2001) Risk factors for human disease emergence. Philos Trans R Soc Lond 356:983–989
Teixeira CP, Passos L, Goulart VD, Hirsch A, Rodrigues M, Young RJ (2016) Evaluating patterns of human–reptile conflicts in an urban environment. Wildl Res 42(7):570–578
Teixeira CP, Passos L, Goulart VD et al (2016a) Evaluating patterns of human–reptile conflicts in an urban environment. Wildl Res 42:570–578

Teixeira B, Hirsch A, Goulart VDLR et al (2016b) Good neighbours: distribution of black-tufted marmoset (*Callithrix penicillata*) in an urban environment. Wildl Res 42:579–589

Telfair RC II, Thompson BC, Tschirhart L (2000) Nuisance heronries in Texas: characteristics and management, PWD BK W7000-134 (1/00), 2nd edn. Wildlife Department's Wildlife Diversity Program., Texas Parks and Wildlife Department, Austin

Texas Parks and Wildlife Department (2017) How to contact wildlife rehabilitators in Texas. Available online at https://tpwd.texas.gov/huntwild/wild/rehab/. Accessed 12 Nov 2017

Threlfall C, Law B, Penman T, Banks PB (2011) Ecological processes in urban landscapes: mechanisms influencing the distribution and activity of insectivorous bats. Ecography 34:814–826

Tidemann CR, Vardon MJ (1997) Pests, pestilence, pollen and pot roasts: the need for community based management of flying foxes in Australia. Aust Biol 10:77–83

Tilman D, Clark M, Williams DR et al (2017) Future threats to biodiversity and pathways to their prevention. Nature 546:73–81

Trewhella WJ, Harris S (1988) A simulation model of the pattern of dispersal in urban fox (*Vulpes vulpes*) populations and its application for rabies control. J Appl Ecol 25:435–450

Tribe A, Brown PR (2000) The role of wildlife rescue groups in the care and rehabilitation of Australian fauna. Hum Dimens Wildl 5:69–85

U.S. fish and Wildlife Service (2012) 2011 national survey of fishing, hunting, and wildlife-associated recreation. National overview. U.S. Fish and Wildlife Service, Washington, DC

United Nations (2015a) World urbanization prospects: the 2014 revision, ST/ESA/SER.A/366. Department of Economic and Social Affairs, Population Division. Available online at https://esa.un.org/unpd/wup/Publications/Files/WUP2014-Report.pdf

United Nations (2015b) The millennium development goals report 2015. United Nations Development Program, New York

Vyas R (2010) Mugger (*Crocodylus palustris*) population in and around Vadodara city, Gujarat state, India. Russ J Herpetol 17:43–50

Vyas R (2013) Snake diversity and voluntary rescue practice in the cities of Gujarat State, India: an evaluation. Reptile Rap 15:27–39

Vyas R (2015) Status of Ganges soft-shell turtle *Nilssonia gangetica* amidst deplorable scenarios in urban Wetlands of Central Gujarat State, India. Newsl South Asian Reptile Netw 17:3–12

Washburn BE, Begier MJ, Wright SE (2015) Collisions between eagles and aircraft: an increasing problem in the airport environment. J Raptor Res 19:192–200

Whittaker D, Knight RL (1998) Understanding wildlife responses to humans. Wildl Soc Bull 26:312–317

Wildlife Rehabilitation Information Directory (2016) Available online at http://wildliferehabinfo.org/. Accessed 12 Nov 2017

Williams NS, Lundholm J, Scott MacIvor J (2014) Do green roofs help urban biodiversity conservation? J Appl Ecol 51:1643–1649

Wimberger K, Downs CT (2010) Annual intake trends of a large urban animal rehabilitation centre in South Africa: a case study. Anim Welf 19:501–513

Wolfe AK, Bateman PW, Fleming PA (2017) Does urbanization influence the diet of a large snake? Curr Zool 64(3):311–318. https://doi.org/10.1093/cz/zox039

Wright JD, Burt MS, Jackson VL (2012) Influences of an urban environment on home range and body mass of Virginia opossums (*Didelphis virginiana*). Northeast Nat 19:77–86

Yirga G, Leirs H, De Iongh DD et al (2016) Spotted hyena (*Crocuta crocuta*) concentrate around urban waste dumps across Tigray, northern Ethiopia. Wildl Res 42:563–569

Chapter 6
Wildlife and Traffic: An Inevitable but Not Unsolvable Problem?

Andreas Seiler and Manisha Bhardwaj

6.1 Introduction

Problems between wildlife and traffic are well acknowledged and on the rise in many countries. Not only is there ample evidence for the negative effect of traffic and infrastructure on wildlife, but the presence and movement of animals impose a growing traffic safety issue. These conflicts appear inevitable and will require more effective mitigation, given the anticipated expansion of transportation infrastructure. However, can these conflicts be resolved?

Countries, societies, and economies depend heavily on a safe, efficient, and well-connected transport system, and this demand is constantly growing. Every day, more than 2 billion vehicles are estimated to travel on the 64 million kilometers of road that currently transect our globe (Sperling and Gordon 2009; CIA 2013). European transport policies (e.g., The EU White Book on Transport, European Commision 2011) as well as international transportation strategies (e.g., OECD/ITF 2017 or the Belt and Road Initiative, Cai 2017; The Economist 2017) aim at unifying transportation corridors across countries and even continents. These designated corridors aim to ensure efficient mobility with fast intermodal transport and demounted legal, technical, and physical barriers. Although these strategies mainly focus on transport along roads, the use and demand of rail transportation is growing as well (Stewart et al. 2014; Upton 2016). According to the International Transport Forum, passenger mobility on railroads may increase by a staggering 200–300% by 2050, and freight activity may increase by as much as 150–250% (OECD/ITF 2017; UN 2016), which will require 335 thousand kilometers of railway (Dulac 2013). In order to support this anticipated increase in traffic, the world's road network should

A. Seiler (✉) · M. Bhardwaj
Department of Ecology, Grimsö Wildlife Research Station, Swedish University of Agricultural Sciences, Uppsala, Sweden
e-mail: andreas.seiler@slu.se; manisha.bhardwaj@live.ca

F. M. Angelici, L. Rossi (eds.), *Problematic Wildlife II*,
https://doi.org/10.1007/978-3-030-42335-3_6

increase by 60% by 2050 (from 2010), requiring an additional 25 million kilometers of road (Dulac 2013).

Transportation networks threaten the goals of international policies on biodiversity conservation. European legal frameworks and policies such as the *Habitat Directive* (Directive 92/43/EEC) and the *Bern Convention* (Convention on the Conservation of European Wildlife and Natural Habitats, ETS no. 104) state that species of community interest shall be kept at a favorable conservation status and that biodiversity values and ecosystem services shall not be diminished (European Parliament 2012). Thus, we are supposed to maintain viable and thriving wildlife populations that facilitate healthy ecosystem services. This requires that populations are of a sufficient size and have sufficiently connected habitats so that stochastic variations in local fecundity and survival can be compensated for by immigration from neighboring populations (Hodgson et al. 2009; Nicholsson and Ovaskainen 2009). However, roads and railways can threaten this connectivity and movement.

Connected habitats are necessary for biodiversity preservation; however, few landscape around the globe remain unfragmented and "roadless" (Ibisch et al. 2016). More than 97% of the terrestrial surface of the United States is within 5 km of a road (Riitters and Wickham 2003). In Sweden, one of the least fragmented countries in Europe, more than 70% of the managed forests are within 500 m to the nearest road access (Seiler and Folkeson 2006). In southern parts of the country, the average mesh size (i.e., the size of the areas not subdivided by roads) within the network of public roads is less than 4 km^2 (Seiler et al. 2016), while it drops down to 1 km^2 in the Netherlands (van Langevelde et al. 2009) and 0.25 km^2 in Belgium (Trocmé et al. 2003). This leaves little or no space for most animals to establish home ranges that are not dissected by roads. Species with large home ranges, such as carnivores (e.g., wolf, *Canis lupus* (Mattisson et al. 2013); red fox, *Vulpes vulpes* (Walton et al. 2017)) and large ungulates (red deer, *Cervus elaphus* (Jerina 2012); moose, *Alces alces* (Olsson et al. 2010)), will have to cross roads regularly during their daily activities and face a significant risk of collision.

Road mortality is one of the main negative impacts of roads on wildlife (Forman et al. 2003). In some species, road mortality exacerbates other factors that are already causing a decline in population size. For example, road mortality accounts for 17% of the annual mortality of the endangered Iberian lynx (*Lynx pardinus*) in Spain (Ferreras et al. 1992), 35% of the annual mortality of the endangered Florida panther (*Puma concolor coryi*) in the USA (Taylor et al. 2002), and for approximately 50% of the known mortality in badgers (*Meles meles*) in Sweden (Seiler et al. 2003). Traffic is speculated to overtake hunting as the leading cause of vertebrate mortality on land by people (Forman and Alexander 1998). Expanding transportation networks will further fragment landscapes, reduce the size of roadless habitats, increase mortality, isolate habitats, and disturb the areas adjacent to the road, ultimately threatening the persistence of wildlife species in the landscape (Ibisch et al. 2016).

Not all impacts of roads are detrimental to wildlife and not all species are equally sensitive (Rytwinski and Fahrig 2015). Some species may even benefit from new habitats or food provided by road verges. If managed appropriately, the vegetated verges along roads and railways can sustain a variety of flora and small fauna despite being polluted and disturbed (Bellamy et al. 2000; Milton et al. 2015; Rosell Pagès et al. 2016; Villemey et al. 2018). Species that can benefit from these verge habitats include not only opportunistic species and pioneers but also some endangered species that may otherwise no longer be able to survive in the surrounding anthropogenic landscape (Havlin 1987; Helldin et al. 2015; Spooner 2015; Bernes et al. 2016). Infrastructure verges can also perform many other functions, such as producing seeds for adjacent landscapes; filtering traffic noise, light, and runoff water into the surroundings; providing a safety buffer for traffic; or being of aesthetic value for road users (Milton et al. 2015). However, such verges need an optimal design and regular maintenance. If not well maintained, verge habitats may constitute ecological traps that drain wildlife populations from the surrounding landscape rather than supporting them (Clevenot et al. 2018), may aid the spread of weeds and invasive species (Kalwij et al. 2008), or attract larger wildlife with palatable forage or road salt, obscure their detectability, and thus increase collision risks (Planillo et al. 2018). The potentials and the limitations of road and rail corridors to contribute to the ecological, "green infrastructure" of the landscape is now well recognized (Benedict and McMahon 2006; European Commission 2013). The open challenge to road administrations, as proclaimed by the international expert network of road ecologists, is to find a balance between the positive and the hazardous effects of road verges (IENE declaration 2016).

We can resolve the growing conflict between developing efficient transportation systems and achieving biodiversity conservation goals by addressing these conflicts head-on (Alamgir et al. 2017; Laurance et al. 2014). Current international policies in the field of nature conservation, such as the *Aichi Biodiversity Targets* under the UN Convention on Biological Diversity (CBD 2014) and the EU-wide *Strategy on Green Infrastructure* (European Commission 2013), recognize the transport sector and transportation facilities as important players in the endeavor toward a greener and sustainable future. Yet implementation of these policies appears to be lagging behind. Failing to incorporate nature conservation into transportation design does not appear to be due to a lack of technical or biological knowhow, but rather due to political challenges. People are the main end-users given priority in transportation projects. All other "factors" are secondary and are often not given any priority at all. We urge road agencies to look at the issue from a different perspective. Instead of maintaining that the problem is that "the deer crosses the road," we must remember that it is in fact "the road that crosses the forest." Moving forward, designs for sustainable transportation infrastructure should not only meet the needs of the immediate users and customers but also be integrated into the process of enhancing the coexistence between humans and wildlife. The incorporation of wildlife provisions into infrastructure planning from the beginning can help prevent irreversible damage, enhance potentially positive effects, and save money for expensive repairs in the end.

In this chapter, we focus on two significant aspects in the conflict between wildlife and traffic: barrier effects and traffic mortality. The conflict between wildlife and traffic requires special and growing attention, but can be mitigated by an updated, direct, approach.

6.2 Identifying the Conflict: Disruptions to Movement

The intersection of transportation infrastructure and wildlife habitat is often detrimental for wildlife and humans alike. On one hand, roads and railways disrupt natural flows and processes (such as wildlife movements across the landscape), kill animals as they attempt to cross, and ultimately threaten the persistence of wildlife populations (Rytwinski and Fahrig 2015; Torres et al. 2016). On the other hand, collisions between vehicles and large animals can be fatal for drivers and passengers or at least produce costly material damage and long delays in traffic. In both cases, the movement of both subjects, humans and wildlife, is affected, and individuals may face mortality. Identifying these disruptions to movement and mitigating them early in the planning process is essential if we intend to provide a safe and efficient transport network for all.

6.2.1 Disrupting the Movement of Wildlife

The barrier effect of transportation infrastructure to wildlife results from a combination of physical hindrances, behavioral avoidance effects, and traffic mortality (Fig. 6.1, see review in Seiler 2003; Rytwinski and Fahrig 2015). Not only large mammals but also many other wildlife species including arthropods, small mammals, and reptiles experience physical difficulties when attempting to cross fences, gullies, road embankments and the road surface itself (Mader and Pauritsch 1981; Swihart and Slade 1984; Mader et al. 1990; Richardson et al. 1997; Anderson et al. 2002; Andrews and Gibbons 2005; Kornilev et al. 2006). Many individuals also refrain from entering the open road corridor (Rost and Bailey 1979; van der Zande et al. 1980; Abbott et al. 2015). Those species that still dare to move onto the road face the risk of collision with vehicles. Therefore, only a small fraction of the animals attempting to cross an infrastructure corridor will eventually succeed. As a rule of thumb, roads with more than 10,000 vehicles per day should be considered absolute barriers to most wildlife, while roads with intermediate traffic volumes may be considered as a significant source of mortality (Müller and Berthoud 1995; Iuell et al. 2003; Jacobson et al. 2016).

Barriers and mortality affect two key factors of wildlife population viability: dispersal and survival (Fahrig 2003; Jaeger et al. 2005). Barrier effects impede mobility, reduce the access to necessary foraging and breeding habitat, and limit the exchange of individuals and genes between populations. Depending on the species, the strength of the barrier, and the size of the isolated habitat/population, these impacts can occur in a single breeding season or take generations to manifest (Rytwinski and Fahrig 2012). Road mortality reduces population survival rates and can drain local populations (Rhodes et al. 2014; Torres et al. 2016). Species with slow reproduction rates, sparse populations, high mobility, and extensive spatial requirements are particularly exposed and sensitive to mortality on transportation

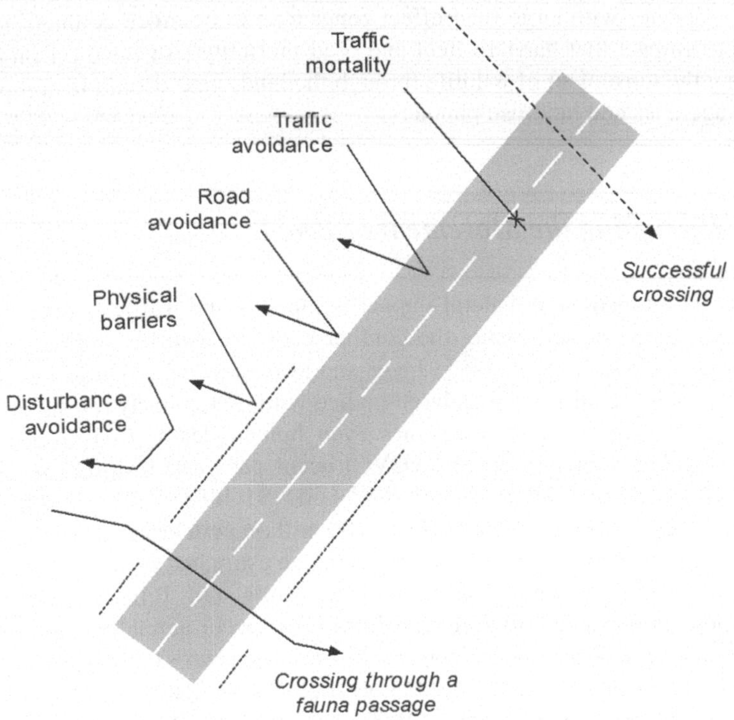

Fig. 6.1 Components of the barrier effect of infrastructure on wildlife. Through the interplay of behavioral and physical barriers with the risk of death in traffic, only a fraction of those individuals that expect to move across the barrier will succeed

networks (Coffin 2007; Rytwinski and Fahrig 2015; Seiler et al. 2016). By addressing and mitigating the barrier effect and road mortality impacts of roads on animals, we can improve dispersal and the survival of wildlife populations in our landscapes.

Disturbances from transportation corridors and traffic can have far-reaching consequences on the environment (Mader et al. 1990; Fahrig et al. 1995; Coffin 2007; Shepard et al. 2008). Light, noise, and chemicals can spread from the road or railway into the surrounding landscape, degrading the ecological and biological capacity of the adjacent habitats, and hindering animals from approaching or crossing the transportation corridor (Forman et al. 2003). Chemical pollutants, like nitrogen or road salt, usually affect flora and fauna in close proximity to infrastructure, while noise from vehicles and groundwater effects may spread much further (Reck and Kaule 1993; Forman and Alexander 1998). The extent of the spread of the impacts of roads is known as a "road-effect zone" (Forman and Alexander 1998) and, depending on the mobility and spatial requirements of the species, it can extend up to 1 km for some species of birds and to 5 km for some species of mammals (see review in: Benitez-Lopez et al. 2010). The road-effect zone can exceed the physical imprint of roads and railways more than tenfold (Forman and Alexander 1998; Forman and Deblinger 2000) and can amplify the barrier effect of roads on wildlife – species with large road-effect zones tend to be more isolated. Given the deleterious impact that barrier effect and road mortality can have on biodiversity, improving the movement of wildlife through the landscape should be a top concern for road agencies and land-use planners.

6.2.2 Disrupting the Movement of People

Reducing the immediate collateral impact to humans and traffic is one of the primary motivations for addressing the conflict between wildlife and transportation infrastructure (Conover et al. 1995; Joyce and Mahoney 2001; Huijser et al. 2009; Seiler et al. 2016). Collisions with large-bodied wildlife regularly results in material damage, human injury, and sometimes even human death. In Europe, at least 500,000 ungulate-vehicle collisions (UVC) occur per year, resulting in over 300 human deaths, 30,000 injuries, and over USD $1 billion in material damage (Bruinderink and Hazebroek 1996). Over 100 million vertebrate road kills per year occur in Spain (Caletrio et al. 1996), producing an estimated annual economic cost of about 105 million euros (Sáenz-de-Santa-María and Tellería 2015). In the Netherlands, an estimated minimum of 5500 UVC occur annually, resulting in 80 human injuries and an overall societal cost of 17 million euros (Ooms 2010). Finally, in Britain, estimates point up to 74,000 UVC per year with an average cost of about 25,000 euros per incident (Langbein 2007; Langbein et al. 2011). Although already quite high, all of these UVC estimates are conservative, as many collisions go unreported (Hesse and Roy 2016); thus it is important to regard these figures as *minimum estimates* and not absolute numbers (Seiler and Jägerbrand 2016). Although discussed less commonly, the impact of wildlife on railway traffic is increasingly recognized. Train collisions with large-bodied animals can cause expensive repairs and extensive delays, especially on high-speed rail systems (Fig. 6.2, Seiler et al. 2014; Borda-de-Água et al. 2017). In Sweden, approximately 2700 UVC with trains are registered annually, producing a societal bill estimated to 150 million euros per year (Seiler et al. 2014; Seiler and Olsson 2017).

Swedish car drivers are legally obliged to report any accident with ungulates and large carnivores to the police, allowing for a comprehensive and reliable data record. During the late 1960s, around 1800 deer-vehicle collisions were recorded in Sweden, evoking national concern and initiating a comprehensive research and prevention program (Almkvist et al. 1980). This led to the fencing of over 7000 km of public roads (Swedish Transport Administration in 2016). Despite these efforts, the number of UVC quickly increased as ungulate populations grew, reaching 27,000 UVC in 1999 (Seiler 2004). By 2016, there were more than 60,000 accidents (NVR 2018), which carried an estimated societal cost of over 250 million euros (Seiler et al. 2016). The greatest concern was accidents with moose; due to the large body mass and long legs of the animal, collisions inflict serious neck and head trauma, if not fatal injuries to the driver and passengers of the car (Jägerbrand et al. 2018). Luckily, thanks to improved vehicle technologies and a strong traffic safety policy

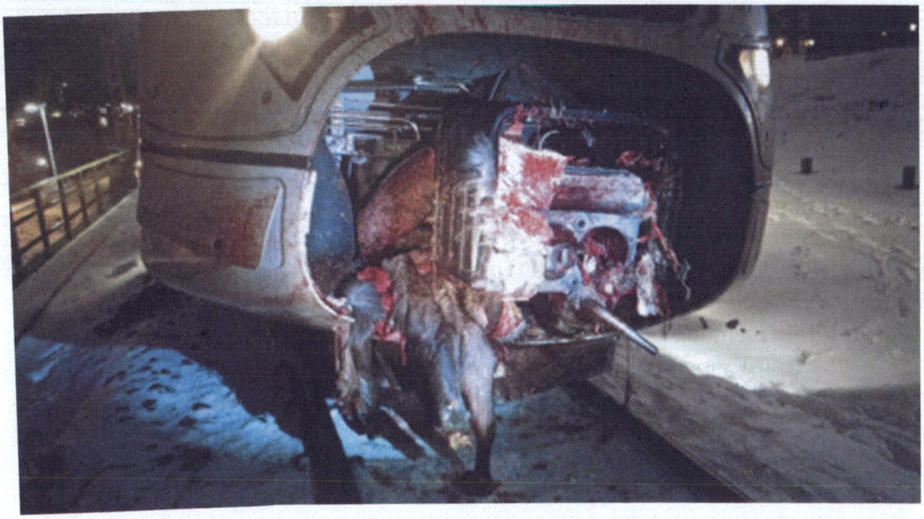

Fig. 6.2 Result of a collision between a moose and a train travelling 200 km/h in Sweden. The train's front cover was completely destroyed and there was extensive damage to sensitive electronics. The passengers had to be transferred to a bus and the train had to be pulled into station for repair. (Photo: Jimmy Nilsson, Swedish Railways AB)

on preventing human fatalities (Trafikverket 2012), the trend in severe human injuries and fatalities is decreasing despite an increasing number of incidents overall (Seiler, unpublished data; Strandroth 2015). If traffic volume and ungulate populations continue to rise, and their conflicts are not properly addressed and mitigated now, it is easy to see that UVCs may grow to an unmanageable extent, to the detriment of both humans and wildlife.

6.3 Addressing and Mitigating the Conflict

Successful mitigation of wildlife-traffic conflicts depends on a working cooperation between transportation agencies, ecologists, economists, sociologists, engineers, landowners, and the general transport customer, i.e., drivers. Local policies and strategies must reflect international and national concern, and local actions must be based on quantifiable goals and measureable targets (Seiler and Sjölund 2005; van der Ree et al. 2015b). This often means aiming higher than the minimum requirement for species conservation or habitat protection described in laws and standards. Although it may seem like a daunting, insurmountable challenge, only through purposeful planning and detail-oriented execution of mitigation strategies can we alleviate and minimize the conflict between wildlife, people, and transportation.

To minimize and avoid the conflict between wildlife and linear infrastructure we must change the way we approach the issue. Too often, we consider human mobility and the mobility in wildlife as separate issues, despite the necessity of both, humans and wildlife, to travel through the landscape. While humans create a web of physical "gray" infrastructures, wildlife use a more obscure and diffuse network of "green infrastructures". Where gray and green infrastructures intersect, we must develop ways to separate the movements of humans and wildlife while maintaining connectivity in both (IENE 2012). Successful examples of such endeavors for existing transportation networks can be found in the *Dutch defragmentation plan* (Rijkswaterstaat 2004), the *Swiss defragmentation plan* (Trocmé 2005), and the *German plan for wildlife and roads* (Hänel and Reck 2011). In Sweden, a special *Guideline for Landscapes* has been published, setting standards for mitigation for wildlife while planning new infrastructure projects (Trafikverket 2015).

The three basic approaches in addressing the impact of transportation infrastructure and traffic on wildlife are avoidance, mitigation, and compensation (Iuell et al. 2003). Avoiding damage is better than repairing it, but where an impact cannot be prevented, mitigation measures must be installed to limit the negative effects. When mitigation is insufficient or impossible, implementation of compensation measures is necessary, but only as a last resort. Identifying which approach is the correct one to take can be challenging since humans and wildlife utilize the landscape on different scales. For example, we may avoid the destruction of a breeding pond for amphibians by rerouting a new road outside the critical habitat, but still cause high mortality in animals that move to and from the breeding habitat or isolate the local population due to unresolved barrier effects caused by the road. Destroyed or disturbed local breeding habitats may be compensated by creating new habitats elsewhere, but as long as the road is present in the landscape, the fragmentation effect resides. For large species, barrier and mortality effects are inevitable and cannot entirely be avoided nor compensated for. Instead, they require direct mitigation on site. Thus, understanding the scales at which impacts are detrimental for wildlife is essential to reduce these impacts, and the identification of appropriate methods, to address these problems.

6.3.1 Reducing the Impact Begins with Planning

Planning of transport infrastructure at the landscape scale is essential and requires specific attention, especially when hitherto unfragmented and unexploited areas are in question (Laurance et al. 2006; Laurance and Balmford 2013; Laurance 2015). Where it can threaten the survival of a population, construction of transport infrastructure should be avoided entirely and these areas should remain unfragmented (DeVelice and Martin 2001; Selva et al. 2015). This is particularly true for landscapes that support endemic or endangered populations, landscapes that are part of an important migration corridor, or landscapes with important or sensitive habitat that cannot be replaced or replicated (Selva et al. 2011). It is important to critically think of the project and determine if new "gray" infrastructure is indeed the only option to meet the growing transportation needs (Fig. 6.3). The only true way to

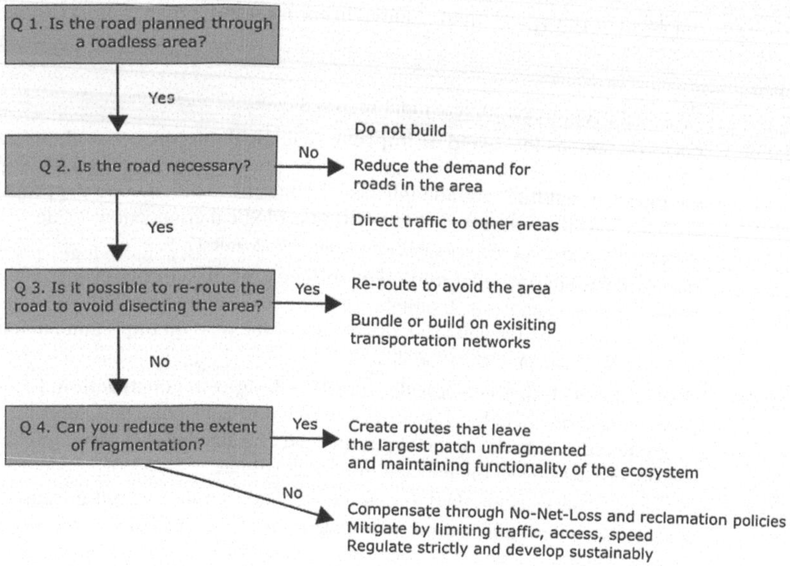

Fig. 6.3 Four main questions to ask when planning a road project in roadless or low-traffic areas. (Adapted from Selva et al. 2015)

ensure no net impact on wildlife is to avoid construction of new roads and railways where possible.

Where new transportation infrastructure is necessary, every stage of the project, from strategic planning to operation, must incorporate mitigation strategies for wildlife (Table 6.1, Seiler and Eriksson 1997; Roberts and Sjölund 2015). The planning stage must include clearly defining tangible goals for wildlife management, such as percentage of intended road mortality reduction or connectivity increase, with realistic targets to work towards (van Der Grift et al. 2016). This will help develop rules, standards, and effective solutions at both strategic and practical levels. Recent handbooks on infrastructure ecology (e.g., Iuell et al. 2003; van der Ree et al. 2015b; Borda-de-Água et al. 2017; van der Grift et al. 2017) provide general support and advice in this process. These guidelines should be adapted to fit the needs of national transport authorities and to be able to be incorporated within their infrastructure development "toolkit."

6.3.2 Mitigating Road Mortality and Barrier Effects

To date, the most effective, albeit expensive, solution to mitigate wildlife and traffic conflicts is the combination of exclusion fences and crossing structures (Iuell et al. 2003; Putman et al. 2004; Huijser et al. 2008; Bissonette and Rosa 2012; van der Ree et al. 2015a). Exclusion fencing will keep wildlife from entering the road and

Table 6.1 Type and detail of ecological input required in each stage of a road project (Roberts and Sjölund 2015)

Stages in a road project	Type and detail of ecological input required
Strategic planning	Focus on options that **avoid or improve ecological outcomes** based on strategic environmental assessment. Examples of key questions include: Can the impact on important wildlife migration/dispersal routes be avoided? Can the project enhance wildlife connectivity by restoring ecological connections? Can developing a project in areas without roads be avoided?
Physical planning	Focus on road designs that **minimize, mitigate, or offset impacts** based on detail ecological analysis. Examples of key questions include: Where should fauna crossings be located? Can the impact of the road on important habitat be reduced by modifying the road design?
Construction	Focus on **translating ecologically sensitive designs to construction**. Examples of key questions include: Has the design of wildlife crossing structures met the key required standards for the target species? Has the detailed drainage design considered the impact on adjacent important habitat?
Operation	Focus on ongoing **ecological management, maintenance of mitigation measures, review, and adaptive management**. Examples of key questions include: Is there a plan for post-construction monitoring and maintenance in place to ensure that crossing structures remain effective over time? Does a management plan for the areas with important habitat adjacent to the road project exist to ensure that the indirect impacts of the operation of the road itself do not degrade these habitats?

lead them toward safe crossing facilities, provided the fences are well-designed and maintained. Fences need to be sufficiently tall (the species of interest should not be able to climb over), robust, and continuous, leaving no holes or gaps where animals could sneak through, or crawl under. Crossing structures can provide effective connection through the landscape for wildlife, as long as they appear safe to use by wildlife and are well-positioned in landscape (e.g., following known migration routes or connecting two patches of necessary habitat; Beckmann et al. 2010; van der Ree et al. 2015a; Bhardwaj et al. 2017). Although commonly implemented, fencing and crossing structures need to be adapted to the target species (e.g., Clevenger and Waltho 2005; Ascensão and Mira 2006; Seiler and Olsson 2009; Smith et al. 2015; Bhardwaj et al. 2017). Solutions for amphibians will necessarily be different from solutions for ungulates, but both can potentially be combined in a multiple-species approach. Standards and guidelines for the design of fences and passages are proposed and published in handbooks such as the *Handbook of Road Ecology* (van der Ree et al. 2015b) and the *European Handbook on Wildlife and Traffic* (Iuell et al. 2003, and its updated online version: https://handbookwildlifetraffic.info).

A variety of solutions are available, but which design may be most appropriate or how many crossing structures are needed will differ from case to case and depend on the overall mitigation objectives. As proposed in the Austrian prioritization approach (Woess et al. 2002), expensive investments in a single large wildlife crossing structure (e.g., overpass or ecoduct) may be necessary and justified where infrastructure crosses important interregional wildlife corridors. In cases of more local significance, mitigation objectives may be met by constructing a series of smaller, species-specific solutions. Understanding the landscape and the needs of the target species is essential to mitigate the intersection between gray and green infrastructure and ensure continued wildlife movement and traffic safety (IENE 2012).

6.3.3 Identifying the Problem Helps Identifying the Solution

Mitigating the impacts of roads on wildlife is not always straightforward. The discussion often involves infrastructure planners choosing between installing fencing that keep animals off the roadway, investing in crossing structures to allow for continued movement, or simply accepting a certain level of isolation and mortality. Economic costs and benefits of these options can be compared (Huijser et al. 2008, 2009; Seiler et al. 2016), but it may be unclear how much mitigation is actually needed to comply with species conservation needs, legal requirements, and policy objectives or where to employ which type of solution. This is often because there are many case-specific factors that influence these decisions. The type of transportation corridor and its features (e.g., size, traffic volume etc.), the landscape through which it is constructed and the response of the animals needs to guide decision-making.

Barrier effects and road mortality are often present together; however, their relative importance differs between species. While some animals (such as large carnivores or ungulates) may be able to recognize vehicles as a potential threat and refrain from crossing busy roads or railways, others (such as amphibians) may be entirely ignorant of the danger (Jacobson et al. 2016; Seiler et al. 2016). As traffic increases, the first group of species will experience elevated barrier effects, while the latter will pay an increasing death toll. Therefore, it is important to understand the ecosystem in which the road is constructed, to be able to plan mitigation that appropriately targets the species of concern.

Geospatial data on wildlife mortality or wildlife-vehicle collisions can help to locate hotspots where mitigation may be needed (Bíl et al. 2013; Rea et al. 2014; Gunson and Teixeira 2015). This data may be difficult to come by, but recent citizen-science projects such as the Austrian project *Roadkill* (www.roadkill.at, Heigl et al. 2016) or the *California Roadkill Observation System* (http://www.wildlifecrossing.net, Shilling and Waetjen 2015) suggest a promising approach that not only provides data but also involves and educates the public. Other valuable sources, though available for fewer species, may be police records (e.g., Seiler 2004; Hothorn et al. 2012), hunter reports (e.g., Heigl et al. 2016), train driver reports (e.g., Seiler and Olsson 2017), or records made by road maintenance services (e.g., Shilling and Waetjen 2015). Sufficient data is the foundation for implementing effective mitigation strategies in appropriate places in the landscape.

Identifying barrier effects can be more challenging than identifying mortality hotspots, as there are fewer "easy-to-measure" variables for barrier effects than there are for mortality (Zimmermann Teixeira et al. 2017). High accident frequencies may indicate high barrier effects if most animals approaching to cross the road are killed. However, it may also indicate that animals are frequently moving across this road but some get hit. Low mortality rate, on the other hand, may suggest that animals are capable of avoiding collisions when crossing the road, or it may indicate a strong barrier effect that animals are deterred from approaching the road. For some species, radiotelemetry data can be used to map their movements and predict where crossings are likely to occur (Kämmerle et al. 2017; Neumann et al. 2012). In other species, it may be possible to obtain their probable movement path from expert knowledge on habitat preferences and general movement characteristics (Zeller et al. 2015). Even though it is difficult, it is essential to identify barrier effects of transportation on wildlife, to best plan installation of appropriate mitigation strategies.

Different processes may cause barrier or mortality impacts and understanding the mechanisms of these impacts in different species is important when choosing an appropriate mitigation strategy (Jacobson et al. 2016). If the animals are mostly threatened or deterred by traffic, crossing structures in combination with leading fences, light screens, or noise protection walls may prove effective (Beckmann et al. 2010; van der Ree et al. 2015a). Traffic rerouting or traffic calming may offer other powerful alternatives (Jaarsma and Willems 2002; van Langevelde and Jaarsma 2009). For animals, such as deer, that respond to vehicles but are less afraid of traffic itself, level crossings with driver warning systems may provide a cost-effective mitigation – albeit only on minor to intermediate roads (Huijser et al. 2015). In situations where the avoidance of street lighting is the cause of the barrier, mitigation strategies may include the removal of the light, modifying light fixture location, emission spectrum and intensity, or light shielding techniques to reduce the spill of light into the surrounding habitat (Blackwell et al. 2015). Roads may be built of a substrate that is not easily traversed by some species such as mice (Brehme et al. 2013), and turtles and amphibians may be incapable of climbing over rails causing them to be trapped on railways (Kornilev et al. 2006). In both cases, minor adjustments to the barrier surface may provide effective means for mitigation. Some arboreal species, such as squirrel gliders (*Petaurus norfolcensis*), may be incapable of crossing a road or railway if the gap in the canopy is too large and there are not enough trees to enable arboreal movement (van der Ree et al. 2010). Here, crossing structures like rope bridges or glider poles that reconnect forest canopies may overcome barrier problems for these species (van der Ree 2006; Soanes et al. 2013). Finally, on railways, wildlife warning devices that manage to scare animals from the train corridor shortly before a train approaches while allowing free passage during train-free intervals may be effective (Babińska-Werka et al. 2015; Seiler and Olsson 2017). Each of these strategies deliver a positive outcome; however how effective they are depends on how well they target the true mechanism(s) of the barrier effect and mortality for the focal species.

Successful mitigation of roads and railways on wildlife is possible, but it requires early planning and a thorough understanding of the environment through which the transportation corridor will transect. This includes identifying important landscape features, the true cause(s) of road impacts, the needs of the focal species, and the needs of the people using the roads. Transportation system that promotes the connectivity of, both, humans and wildlife can only be created through open collaboration among all stakeholders and through careful consideration of wildlife management throughout transportation infrastructure planning. Although conflicts between wildlife and traffic seem inevitable, they do not need to be unresolvable.

6.4 Conclusion: Conflict Resolution Begins by Changing the Way We View the Problem

The conflicts between wildlife and transportation infrastructure are not a mystery – where roads and railways exist, wildlife is impacted. Although the conflict is well known, it still does not receive adequate attention in infrastructure design. This needs to change. Although many (inter)national jurisdictions and policies require the protection of biodiversity, including species and habitats, the current legal framework is not sufficient, as it often merely marks the ultimate bottom line to prevent total extinction of a species. There is great potential to achieve much better outcomes for both humans and wildlife alike. Current knowledge and present technology can turn the conflict with wildlife into an opportunity to create safe and sustainable infrastructure. This ambitious target can be achieved if we: face the numerous challenges through interdisciplinary cooperation, develop common standards and rules, and integrate mitigation strategies throughout every stage of roads and railways design, including early stages. Reducing the conflict between wildlife and transportation infrastructure is possible, but it requires a change in perspective: "it is not the deer that crosses the road but the road that crosses the forest."

Acknowledgments We are most grateful to Francesco Maria Angelici and Lorenzo Rossi for giving us the opportunity to publish this work in their book on problematic wildlife. This chapter has been financially supported by the Swedish Transport Administration and the Marie Claire Cronstedts foundation. Thanks to Filippo Favilli and Anke Benten providing comments to an early draft of this manuscript.

References

Abbott IM, Berthinussen A, Stone E, Boonman M, Melber M, Altringham J (2015) Bats and roads. In: van der Ree R, Smith DJ, Grilo C (eds) Handbook of road ecology. Wiley, pp 290–299. https://doi.org/10.1002/9781118568170.ch34

Alamgir M, Campbell MJ, Sloan S, Goosem M, Clements GR, Mahmoud MI, Laurance WF (2017) Economic, socio-political and environmental risks of road development in the tropics. Curr Biol 27:R1130–R1140. https://doi.org/10.1016/j.cub.2017.08.067

Almkvist B, André T, Ekblom S, Rempler SA (1980) Slutrapport Viltolycksprojekt (VIOL), vol TU146:1980-05. Vägverket Borlänge, Sweden

Anderson P, Sherwood B, Cutler D, Burton JA (2002) Roads as barriers. In: Wildlife and roads. Imperial College Press, London, pp 169–184

Andrews KM, Gibbons JW (2005) How do highways influence snake movement? Behavioral responses to roads and vehicles. Copeia 2005:772–782

Ascensão F, Mira A (2006) Factors affecting culvert use by vertebrates along two stretches of road in southern Portugal. Ecol Res 22:57–66. https://doi.org/10.1007/s11284-006-0004-1

Babińska-Werka J, Krauze-Gryz D, Wasilewski M, Jasińska K (2015) Effectiveness of an acoustic wildlife warning device using natural calls to reduce the risk of train collisions with animals. Transp Res Part D: Transp Environ 38:6–14. https://doi.org/10.1016/j.trd.2015.04.021

Beckmann JP, Clevenger AP, Huijser MP, Hilty JA (2010) Safe passages: highways, wildlife, and habitat connectivity. Island Press, Washington, DC

Bellamy PE, Shore RF, Ardeshir D, Treweek JR, Sparks TH (2000) Road verges as habitat for small mammals in Britain. Mammal Rev 30:131–139

Benedict MA, McMahon ET (2006) Green infrastructure: smart conservation for the 21st century. https://trid.trb.org/view.aspx?id=643710

Benitez-Lopez A, Alkemade R, Verweij PA (2010) The impacts of roads and other infrastructure on mammal and bird populations: a meta-analysis. Biol Conserv 143:1307–1316. https://doi.org/10.1016/j.biocon.2010.02.009

Bernes C, Bullock JM, Jakobsson S, Rundlof M, Verheyen K, Lindborg R (2016) How are biodiversity and dispersal of species affected by the management of roadsides? A systematic map protocol. Environ Evid 5:4. https://doi.org/10.1007/s10980-015-0329-7

Bhardwaj M, Soanes K, Straka TM, Lahoz-Monfort JJ, Lumsden LF, van der Ree R (2017) Differential use of highway underpasses by bats. Biol Conserv 212:22–28. https://doi.org/10.1016/j.biocon.2017.05.022

Bíl M, Andrášik R, Janoska Z (2013) Identification of hazardous road locations of traffic accidents by means of kernel density estimation and cluster significance evaluation. Accid Anal Prev 55:265–273. https://doi.org/10.1016/J.Aap.2013.03.003

Bissonette JA, Rosa S (2012) An evaluation of a mitigation strategy for deer-vehicle collisions. Wildl Biol 18:414–423. https://doi.org/10.2981/11-122

Blackwell BF, DeVault TL, Seamans TW (2015) Understanding and mitigating the negative effect of road lighting on ecosystems. In: van der Ree R, Smith DJ, Grilo C (eds) Handbook of road ecology. Wiley, Chichester, pp 143–150

Borda-de-Água L, Barrientos R, Beja P, Pereira HM (2017) Railway ecology. Springer International Publishing. https://doi.org/10.1007/978-3-319-57496-7_1

Brehme CS, Tracey JA, McClenaghan LR, Fisher RN (2013) Permeability of roads to movement of scrubland lizards and small mammals. Conserv Biol 27:710–720. https://doi.org/10.1111/cobi.12081

Bruinderink GWTAG, Hazebroek E (1996) Ungulate traffic collisions in Europe. Conserv Biol 10:1059–1067. https://doi.org/10.1046/j.1523-1739.1996.10041059.x

Cai P (2017) Understanding China's belt and road initiative, vol 2018. vol 2018-06-01. Lowy Institute for International Policy, Sydney. https://www.lowyinstitute.org/publications/understanding-belt-and-road-initiative

Caletrio J, Fernandez JM, Lopez J, Roviralta F (1996) Spanish national inventory on road mortality of vertebrates. Global Biodivers 5:15–18

CBD (2014) Global biodiversity outlook 4. Secretariat of the Convention on Biological Diversity, Montréal. www.cbd.int/GBO4

CIA (2013) The world fact book 2013–14. Central Intelligence Agency, Washington, DC

Clevenger AP, Waltho N (2005) Performance indices to identify attributes of highway crossing structures facilitating movement of large mammals. Biol Conserv 121:453–464. https://doi.org/10.1016/j.biocon.2004.04.025

Clevenot L, Carré C, Pech P (2018) A review of the factors that determine whether stormwater ponds are ecological traps and/or high-quality breeding sites for amphibians. Front Ecol Evol 6:40. https://doi.org/10.3389/fevo.2018.00040
Coffin AW (2007) From roadkill to road ecology: a review of the ecological effects of roads. J Transp Geogr 15:396–406. https://doi.org/10.1016/j.jtrangeo.2006.11.006
Conover MR, Pitt WC, Kessler KK, Dubow TJ, Sanborn WA (1995) Review of human injuries, illnesses, and economic losses caused by wildlife in the United States. Wildl Soc Bull 23:407–414
DeVelice RL, Martin JR (2001) Assessing the extent to which roadless areas complement the conservation of biological diversity. Ecol Appl 11:1008–1018. https://doi.org/10.1890/1051-0761(2001)011[1008:ATETWR]2.0.CO,2
Dulac J (2013) Global land transport infrastructure requirements. International Energy Agency, Paris, 20, 2014
European Commision (2011) Accompanying paper to the white paper on transport. Commission staff working document. In: White paper on transport – roadmap to a single European transport area: towards a competitive and resource efficient transport system, vol SEC(2011) 391 final. European Commision, Brussles
European Commission (2013) Green infrastructure (GI) – enhancing Europe's natural capital, vol COM (2013) 249 final. European Commission, Brussels. http://eur-lex.europa.eu/resource.html?uri=cellar:d41348f2-01d5-4abe-b817-4c73e6f1b2df.0014.03/DOC_1&format=PDF
European Parliament (2012) Our life insurance, our natural capital: an EU biodiversity strategy to 2020. European Parliament resolution of 20 April 2012 (2011/2307(INI)). http://ec.europa.eu/environment/nature/biodiversity/comm2006/pdf/EP_resolution_april2012.pdf
Fahrig L (2003) Effects of habitat fragmentation on biodiversity. Annu Rev Ecol Evol Syst 34:487–515. ISI: 000220102000018
Fahrig L, Pedlar JH, Pope SE, Taylor PD, Wegner JF (1995) Effect of road traffic on amphibian density. Biol Conserv 73:177–182. https://doi.org/10.1016/0006-3207(94)00102-V
Ferreras P, Aldama JJ, Beltran JF, Delibes M (1992) Rates and causes of mortality in a fragmented population of Iberian Lynx Felis-Pardina Temminck, 1824. Biol Conserv 61:197–202. https://doi.org/10.1016/0006-3207(92)91116-A
Forman RT, Alexander LE (1998) Roads and their major ecological effects. Annu Rev Ecol Syst 29:207–231
Forman RT, Deblinger RD (2000) The ecological road-effect zone of a Massachusetts (USA) suburban highway. Conserv Biol 14:36–46. https://doi.org/10.1046/j.1523-1739.2000.99088.x
Forman RT et al (2003) Road ecology – science and solutions. Island Press, Washington, DC
Gunson K, Teixeira FZ (2015) Road–wildlife mitigation planning can be improved by identifying the patterns and processes associated with wildlife-vehicle collisions. In: van der Ree R, Smith DJ, Grilo C (eds) Handbook of road ecology. Wiley, pp 101–109. https://doi.org/10.1002/9781118568170.ch13
Hänel K, Reck H (2011) Bundesweite Prioriteten zur Wiedervernetzung von Ökosystemen: Die Überwindung strassenbedingter Barrieren vol Naturschutz und Biologische Vielfalt Heft 108. Bundesamt für Naturschutz, BfN, Bonn, Bad Godesberg. http://www.bfn.de/0306_zerschneidung.html
Havlin J (1987) On the importance of railway lines for the avifauna in agrocoenoses. Folia Zool 36:345–358
Heigl F, Stretz C, Steiner W, Suppan F, Bauer T, Laaha G, Zaller J (2016) Comparing road-kill datasets from hunters and citizen scientists in a landscape context. Remote Sens 8:832. https://doi.org/10.3390/rs8100832
Helldin J-O, Wissman J, Lennartsson T (2015) Abundance of red-listed species in infrastructure habitats – "responsibility species" as a priority-setting tool for transportation agencies' conservation action. Nat Conserv 11:143–158. https://doi.org/10.3897/natureconservation.11.4433
Hesse G, Roy RV (2016) Quantifying wildlife vehicle collision underreporting on northern British Columbia highways (2004–2013). Unpublished report prepared for the BC Ministry

of Transportation and Infrastructure, Canada. http://www.wildlifecollisions.ca/docs/quantifying_wvc_underreporting_april6-2016finalpdf:53

Hodgson JA, Moilanen A, Thomas CD (2009) Metapopulation responses to patch connectivity and quality are masked by successional habitat dynamics. Ecology 90:1608–1619. https://doi.org/10.1890/08-1227.1

Hothorn T, Brandl R, Müller J (2012) Large-scale model-based assessment of deer-vehicle collision risk. PLoS One 7:e29510. https://doi.org/10.1371/journalpone.0029510

Huijser MP, McGowen P, Clevenger AP, Ament R (2008) Wildlife-vehicle collision reduction study: best practices manual, vol DTFH61-05-D-00018. U.S. Department of Transportation, Federal Highway Agency. http://ebookbrowse.com/huijser-et-al-2008-fhwa-manual-binder11-1-pdf-d29701169

Huijser MP, Duffield JW, Clevenger AP, Ament RJ, McGowen PT (2009) Cost-benefit analyses of mitigation measures aimed at reducing collisions with large ungulates in the United States and Canada: a decision support tool. Ecol Soc 14:15

Huijser MP, Mosler-Berger C, Olsson M, Strein M (2015) Wildlife warning signs and animal detection systems aimed at reducing wildlife-vehicle collisions. In: van der Ree R, Smith DJ, Grilo C (eds) Handbook of road ecology. Wiley, pp 198–212. https://doi.org/10.1002/9781118568170.ch24

Ibisch PL et al (2016) A global map of roadless areas and their conservation status. Science 354:1423–1427. https://doi.org/10.1126/science.aaf7166

IENE (2012) IENE 2012 declaration: overcome barriers – Europe-wide and now. IENE. http://www.iene.info/declarations/overcome-barriers-europe-wide-now/. Accessed 2018-07-10

IENE (2016) IENE 2016 declaration: ecological design and maintenance for habitats related to transport infrastructures. IENE. http://www.iene.info/declarations/ecological-design-maintenance-habitats-related-transport-infrastructures/. Accessed 2018-07-10

Iuell B et al (2003) Wildlife and traffic – a European handbook for identifying conflicts and designing solutions. Prepared by COST 341 – habitat fragmentation due to transportation infrastructure. Ministry of Transport, Public Works and Water Management, Road and Hydraulic Engineering division, Delft. http://www.iene.info/wp-content/uploads/COST341_Handbook.pdf

Jaarsma C, Willems GPA (2002) Reducing habitat fragmentation by minor rural roads through traffic calming. Landsc Urban Plan 58:125–135. https://doi.org/10.1016/S0169-2046(01)00215-8

Jacobson S, Bliss-Ketchum L, De Rivera C, Smith WP (2016) A behavior based framework for assessing barrier effects to wildlife from vehicle traffic volume. Ecosphere 7:e01345. https://doi.org/10.1002/ecs2.1345

Jaeger JA, Fahrig L, Haber W (2005). Reducing habitat fragmentation by roads: a comparison of measures and scales. UC Davis: Road Ecology Center. Retrieved from https://escholarship.org/uc/item/3dd9z5r1

Jägerbrand AK, Gren I-M, Seiler A, Johansson Ö (2018) Uppdatering och nya effektsamband i effektmodellen för viltolyckor. Calluna AB, Linköping. https://doi.org/10.13140/RG.2.2.15211.23849

Jerina K (2012) Roads and supplemental feeding affect home-range size of Slovenian red deer more than natural factors. J Mammal 93:1139–1148. https://doi.org/10.1644/11-MAMM-A-136.1

Joyce TL, Mahoney SP (2001) Spatial and temporal distributions of moose-vehicle collisions in Newfoundland. Wildl Soc Bull 29:281–291. ISI: 000168179900038

Kalwij JM, Milton SJ, McGeoch MA (2008) Road verges as invasion corridors? A spatial hierarchical test in an arid ecosystem. Landsc Ecol 23:439–451. https://doi.org/10.1007/s10980-008-9201-3

Kämmerle JL, Brieger F, Kroschel M, Hagen R, Storch I, Suchant R (2017) Temporal patterns in road crossing behaviour in roe deer (*Capreolus capreolus*) at sites with wildlife warning reflectors. PLoS One 12:e0184761. https://doi.org/10.1371/journalpone.0184761

Kornilev YV, Price SJ, Dorcas ME (2006) Between a rock and hard place: responses of Eastern Box Turtles (*Terrapene carolina*) when trapped between railroad tracks. Herpetol Rev 37:145–148

Langbein J (2007) National deer-vehicle collisions project. In: Deer initiative research report 07/1. The Deer Initiative, Wrexham, pp 215–259

Langbein J, Putman R, Pokorny B (2011) Traffic collisions involving deer and other ungulates in Europe and available measures for mitigation. In: Putman R, Apollonio M, Andersen R (eds) Ungulate management in Europe: problems and practices. Cambridge University Press, Cambridge, pp 215–259

Laurance WF (2015) Bad roads, good roads. In: van der Ree R, Smith DJ, Grilo C (eds) Handbook of road ecology. Wiley, pp 10–15. https://doi.org/10.1002/9781118568170.ch2

Laurance WF, Balmford A (2013) Land use: a global map for road building. Nature 495:308–309. https://doi.org/10.1038/495308a

Laurance WF et al (2006) Impacts of roads and hunting on central African rainforest mammals. Conserv Biol 20:1251–1261. https://doi.org/10.1111/j.1523-1739.2006.00420.x

Laurance WF et al (2014) A global strategy for road building. Nature. Advance Online Publication. https://doi.org/10.1038/nature13717

Mader HJ, Pauritsch G (1981) Nachweis des Barriere-Effektes von verkehrsarmen Straßen und Feldwegen auf Kleinsäuger der Waldbiozönose durch Markierungs- und Umsetzversuche. Natur und Landschaft 56:451–454

Mattisson J, Sand H, Wabakken P, Gervasi V, Liberg O, Linnell JD, Rauset GR, Pedersen HC (2013) Home range size variation in a recovering wolf population: evaluating the effect of environmental, demographic, and social factors Oecologia 173:813–825. https://doi.org/10.1007/s00442-013-2668-x

Mader HJ, Schell C, Kornacker P (1990) Linear barriers to arthropod movements in the landscape. Biol Conserv 54:209–222

Milton SJ, Dean WRJ, Sielecki LE, van der Ree R (2015) The function and management of roadside vegetation. In: van der Ree R, Smith DJ, Grilo C (eds) Handbook of road ecology. Wiley, pp 373–381. https://doi.org/10.1002/9781118568170.ch46

Müller S, Berthoud G (1995) Fauna/Traffic safety: Manual for engineers. Ecole Polytechnique Fédérale de Lausanne, Département de génie civil. Laboratoire des voies de circulation (LAVOC).

Neumann W, Ericsson G, Dettki H, Bunnefeld N, Keuler NS, Helmers DP, Radeloff VC (2012) Difference in spatiotemporal patterns of wildlife road-crossings and wildlife-vehicle collisions. Biol Conserv 145:70–78. https://doi.org/10.1016/j.biocon.2011.10.011

Nicholsson E, Ovaskainen H (2009) Conservation prioritization using metapopulation models. In: Moilanen A, Wilson KA, Possingham H (eds) Spatial conservation prioritization: quantitative methods and computational tools. Oxford University Press, Oxford

NVR (2018) Viltolycksstatistik (in Swedish). Nationella Viltolycksrådet – Swedish National Council for Wildlife Vehicle Accidents. www.viltolycka.se. Accessed 2018-07-10

OECD/ITF (2017) ITF transport outlook 2017. OECD Publishing, Paris. https://doi.org/10.1787/9789282108000-en

Olsson M, Cox JJ, Larkin JL, Widén P, Olovsson A (2010) Space and habitat use of moose in southwestern Sweden. Eur J Wildl Res 57:241–249. https://doi.org/10.1007/s10344-010-0418-5

Ooms JW (2010) Wildlife accidents – preventive measures and their application (In Dutch: Wildongevallen – Preventieve maatregelen en hun toepassingsgebied). NOVI Verkeersacademie, Utrecht. http://www.verkeersacademie.com

Planillo A, Mata C, Manica A, Malo JE (2018) Carnivore abundance near motorways related to prey and roadkills. J Wildl Manag 82:319–327. https://doi.org/10.1002/jwmg.21384

Putman RJ, Langbein J, Staines BW (2004) Deer and road traffic accidents: a review of mitigation measures: costs and cost-effectiveness. Report for the Deer Commission for Scotland, Contract RP23A. http://www.snh.org.uk/pdfs/publications/dcs/mitigation%20measures.pdf

Rea RV, Johnson CJ, Emmons S (2014) Characterizing moose–vehicle collision hotspots in northern British Columbia. J Fish Wildl Manag 5:46–58. https://doi.org/10.3996/062013-jfwm-042

Reck H, Kaule G (1993) Strassen und Lebensräume: Ermittlung und Beurteilung strassenbedingter Auswirkungen auf Pflanzen, Tiere und ihre Lebensräume. Forschung, Straßenbau und

Straßenverkehrstechnik. Bundesministerium für Verkehr, Abteilung Strassenbau, Bonn-Bad Godesberg, Germany
Rhodes JR, Lunney D, Callaghan J, McAlpine CA (2014) A few large roads or many small ones? How to accommodate growth in vehicle numbers to minimise impacts on wildlife. PLoS One 9:e91093. https://doi.org/10.1371/journalpone.0091093
Richardson JH, Shore RF, Treweek JR, Larkin SC (1997) Are major roads a barrier to small mammals? J Zool 243:840–846
Riitters KH, Wickham JD (2003) How far to the nearst road? Front Ecol Environ 1:125–129. https://doi.org/10.1890/1540-9295(2003)001[0125:HFTTNR]2.0.CO;2
Rijkswaterstaat (2004) MJPO – Meerjarenprogramma Ontsnippering. Ministerie van Verkeer en Waterstaat. Ministerie van Landbouw, Natuur en Voedselkwaliteit, Ministerie van Volkshuisvesting, Ruimtelijke Ordening en Milieubeheer, DenHaag. www.mjpo.nl
Roberts K, Sjölund A (2015) Incorporating biodiversity issues into road design: the road agency perspective. In: van der Ree R, Smith DJ, Grilo C (eds) Handbook of road ecology. Wiley, Chichester, pp 27–31
Rosell Pagès C, Helldin JO, Reck H, Navas F, Cama A, O'Brien E (2016) Road maintenance guidelines to improve wildlife conservation and traffic safety. SAFEROAD technical report 5. Conference of European Directors of Roads (CEDR), Brussels. https://www.saferoad-cedr.org/upload_mm/9/6/4/80f84c02-1d65-44c0-bed9-60c67e090a7e_CEDR%20Call%202013_SAFEROAD_Technical%20Report%205.pdf
Rost GR, Bailey JA (1979) Distribution of mule deer and elk in relation to roads. J Wildl Manag 43:634–641
Rytwinski T, Fahrig L (2012) Do species life history traits explain population responses to roads? A meta-analysis. Biol Conserv 147:87–98. https://doi.org/10.1016/j.biocon.2011.11.023
Rytwinski T, Fahrig L (2015) The impacts of roads and traffic on terrestrial animal populations. In: van der Ree R, Smith DJ, Grilo C (eds) Handbook of road ecology. Wiley, pp 237–246. https://doi.org/10.1002/9781118568170.ch28
Sáenz-de-Santa-María A, Tellería JL (2015) Wildlife-vehicle collisions in Spain. Eur J Wildl Res 61:399–406. https://doi.org/10.1007/s10344-015-0907-7
Seiler A (2003) Effects of infrastructure on nature. In: Trocmé M et al (eds) COST 341 – habitat fragmentation due to transportation infrastructure: the European review. Office for Official Publications of the European Communities, Luxembourg, pp 31–50
Seiler A (2004) Trends and spatial patterns in ungulate-vehicle collisions in Sweden. Wildl Biol 10:301–313. https://doi.org/10.2981/wlb.2004.036
Seiler A, Eriksson IM (1997) New approaches for ecological consideration in Swedish road planning. In: Canters K, Piepers A, Hendriks-Heersma A (eds) Proceedings of the international conference on "Habitat fragmentation, infrastructure and the role of ecological engineering" Maastricht & DenHague 1995. Ministry of Transport, Public Works and Water Management, Road and Hydraulic Engineering division, Delft, pp 253–264
Seiler A, Folkeson L (2006) Habitat fragmentation due to transportation infrastructure: COST 341 national state-of-the-art report Sweden. VTI report 530 A, Linköping
Seiler A, Jägerbrand AK (2016) Mörkertal i viltolycksstatistik – tolkning av viltolycksenkät och databasanalyser (Hidden statistics in wildlife-vehicle accident statistics – interpretation of drivers surveys and database comparisons). Triekol. Trafikverket Publikation 2016:055, Borlänge
Seiler A, Olsson M (2009) Are non-wildlife passages effective passages for wildlife? Paper presented at the ICOET 2009, international conference on ecology & transportation, Duluth, Minnesota, Sept 13–17, 2009. Available at: http://www.icoet.net/ICOET_2009/downloads/proceedings/ICOET2009-Proceedings-Complete.pdf
Seiler A, Olsson M (2017) Wildlife deterrent methods for railways – an experimental study. In: Borda-de-Água L, Barrientos R, Beja P, Pereira HM (eds) Railway ecology. Springer International Publishing, Cham, pp 277–291. https://doi.org/10.1007/978-3-319-57496-7_17
Seiler A, Sjölund A (2005) Target orientation for ecologically sound road management in Sweden. Gaia 14:178–181. https://doi.org/10.14512/gaia.14.2.24

Seiler A, Helldin JO, Eckersten T (2003) Road mortality in Swedish Badgers (*Meles meles*). Effect on population. The toll of the automobile: Wildlife and roads in Sweden. Ph.D. thesis at the Department for Conservation Biology, SLU, Uppsala
Seiler A, Söderström P, Olsson M, Sjölund A (2014) Costs and effects of deer-train collisions in Sweden. Paper presented at the IENE 2014 international conference on ecology and transportation, Malmö. Available at: http://iene2014.iene.info
Seiler A, Olsson M, Rosell C, Van Der Grift EA (2016) Cost-benefit analyses for wildlife and traffic safety. SAFEROAD technical report 4. Conference of European Directors of Roads (CEDR), Brussels. https://www.saferoad-cedr.org/upload_mm/4/d/1/2346434d-206d-463e-a2ea-3136722a2e1f_CEDR%20Call%202013_SAFEROAD_Technical%20Report%204.pdf
Selva N et al (2011) Roadless and low-traffic areas as conservation targets in Europe. J Environ Manag 48:865–877. https://doi.org/10.1007/s00267-011-9751-z
Selva N, Switaliski A, Kreft S, Ibisch P (2015) Why keep areas road-free? The importance of roadless areas. In: van der Ree R, Smith DJ, Grilo C (eds) Handbook of road ecology. Wiley, Chichester, pp 16–26
Shepard DB, Kuhns AR, Dreslik MJ, Phillips CA (2008) Roads as barriers to animal movement in fragmented landscapes. Anim Conserv 11:288–296. https://doi.org/10.1111/j.1469-1795.2008.00183.x
Shilling FM, Waetjen DP (2015) Wildlife-vehicle collision hotspots at US highway extents: scale and data source effects. Nat Conserv 11:41–60. https://doi.org/10.3897/natureconservation.11.4438
Smith DJ, van der Ree R, Rosell C (2015) Wildlife crossing structures: an effective strategy to restore or maintain connectivity across roads. In: van der Ree R, Smith DJ, Grilo C (eds) Handbook of road ecology. Wiley, Chichester, pp 172–183
Soanes K, Lobo MC, Vesk PA, McCarthy MA, Moore JL, van der Ree R (2013) Movement re-established but not restored: inferring the effectiveness of road-crossing mitigation for a gliding mammal by monitoring use. Biol Conserv 159:434–441. https://doi.org/10.1016/j.biocon.2012.10.016
Sperling D, Gordon D (2009) Two billion cars: driving toward sustainability. Oxford University Press, New York
Spooner PG (2015) Minor rural road networks: values, challenges, and opportunities for biodiversity conservation. Nat Conserv 11:129–142. https://doi.org/10.3897/natureconservation.11.4434
Stewart C, Luebkeman C, Morell M, Goulding L (2014) Future of rail 2050. ARUP.com, London. https://www.arup.com/publications/research/section/future-of-rail-2050
Strandroth J (2015) Analysis of road safety trends 2014. Management by objectives for road safety work towards the 2020 interim targets, vol 2015:103. Swedish Transport Administration 2015:103, Borlänge
Swihart RK, Slade NA (1984) Road crossing in Sigmodon hispidus and Microtus ochrogaster. J Mammal 65:357–360. https://doi.org/10.2307/1381184
Taylor SK, Buergelt CD, Roelke-Parker ME, Homer BL, Rotstein DS (2002) Causes of mortality of free-ranging Florida panthers. J Wildl Dis 38:107–114. https://doi.org/10.7589/0090-3558-38.1.107
The Economist (2017) What is China's belt and road initiative? The Economist: 2–3. https://doi.org/10.1080/0163660X.2017.1302743
Torres A, Jaeger JA, Alonso JC (2016) Assessing large-scale wildlife responses to human infrastructure development. Proc Natl Acad Sci U S A 113:8472–8477. https://doi.org/10.1073/pnas.1522488113
Trafikverket (2012) Säker trafik – Nollvisionen på väg. Swedish Transport Administration (Trafikverket), Borlänge
Trafikverket (2015) Guideline landscape (In Swedish: Riktlinje landskap) Swedish Transport Administration (Trafikverket) TDOK 2015-0323, Borlänge
Trocmé M (2005) The Swiss defragmentation program–reconnecting wildlife corridors between the Alps and Jura: an overview. UC Davis: Road Ecology Center. Retrieved from https://escholarship.org/uc/item/8w66w88s

Trocmé M et al (2003) COST 341 – habitat fragmentation due to transportation infrastructure: the European review. Office for Official Publications of the European Communities, Luxembourg
UN (2016) Mobilizing sustainable transport – analysis and policy recommendations from the United Nations Secretary-General's High-Level Advisory Group on Sustainable Transport, United Nations. https://sustainabledevelopment.un.org/index.php?page=view&type=400&nr=2375&menu=35
Upton L (2016) 9 major factors shaping the future of the rail industry. https://www.smartrailworld.com/9-major-factors-shaping-the-future-of-the-rail-industry
van der Grift EA, Seiler A, Helldin JO (2016) Guidelines for outcome-based specifications in road mitigation. SAFEROAD technical report 2. Conference of European Directors of Roads (CEDR), Brussels. https://www.saferoad-cedr.org/upload_mm/c/0/4/3e3d1f2a-b53c-40bf-9872-f68b6473f72e_SAFEROAD_Technical%20Report%202.pdf
van der Grift EA, Seiler A, Rosell C, Simeonova V (2017) Safe roads for wildlife and people. SAFEROAD final report. CEDR, Brussels. https://www.saferoad-cedr.org/upload_mm/4/6/b/80eb4659-f6e6-4327-8358-f692b98f3419_CEDR_finalreport.pdf
van der Ree R (2006) Road upgrade in Victoria a filter to the movement of the endangered Squirrel glider (Petaurus norfolcensis): results of a pilot study. Ecol Manage Restor 7:226–228
van der Ree R, Cesarini S, Sunnucks P, Moorre JL, Taylor A (2010) Large gaps in canopy reduce road crossing by a gliding mammal. Ecol Soc 15:35
van der Ree R, Gagnon JW, Smith D (2015a) Fencing: a valuable tool for reducing wildlife-vehicle collisions and funnelling fauna to crossing structures. In: van der Ree R, Smith D, Grilo C (eds) Handbook of road ecology. Wiley, pp 159–171. https://doi.org/10.1002/9781118568170.ch20
van der Ree R, Smith DJ, Grilo C (2015b) Handbook of road ecology. Wiley, Chichester. https://doi.org/10.1002/9781118568170
van der Zande AN, Ter Keurs WJ, van der Weijden WJ (1980) The impact of roads on the densities of four bird species in an open field habitat – evidence of a long-distance effect. Biol Conserv 18:299–321
van Langevelde F, Jaarsma CF (2009) Modeling the effect of traffic calming on local animal population persistence. Ecol Soc 14(2):39. http://www.ecologyandsociety.org/vol14/iss2/art39/
van Langevelde F, van Dooremalen C, Jaarsma CF (2009) Traffic mortality and the role of minor roads. J Environ Manage 90:660–667. https://doi.org/10.1016/j.jenvman.2007.09.003
Villemey A et al (2018) Can linear transportation infrastructure verges constitute a habitat and/or a corridor for insects in temperate landscapes? A systematic review. Environ Evid 7. https://doi.org/10.1186/s13750-018-0117-3
Walton Z, Samelius G, Odden M, Willebrand T (2017) Variation in home range size of red foxes Vulpes vulpes along a gradient of productivity and human landscape alteration. PLoS One 12:e0175291. https://doi.org/10.1371/journalpone.0175291
Woess M, Grillmayer R, Voelk FH (2002) Green bridges and wildlife corridors in Austria. Zeitschrift für Jagdwissenschaft 48(Suppl 1):25. https://doi.org/10.1007/BF02192389
Zeller KA, McGarigal K, Cushman SA, Beier P, Vickers TW, Boyce WM (2015) Using step and path selection functions for estimating resistance to movement: pumas as a case study. Landsc Ecol 31:1319–1335. https://doi.org/10.1007/s10980-015-0301-6
Zimmermann Teixeira F, Kindel A, Hartz SM, Mitchell S, Fahrig L, Cadotte M (2017) When road-kill hotspots do not indicate the best sites for road-kill mitigation. J Appl Ecol 54:1544–1551. https://doi.org/10.1111/1365-2664.12870

Chapter 7
The Colonization of the Western Yellow-Legged Gull (*Larus michahellis*) in an Italian City: Evolution and Management of the Phenomenon

Enrico Benussi and Maurizio Fraissinet

7.1 Introduction

Following the strong numerical increase of the yellow-legged gull *Larus michahellis* and herring gulls *L. argentatus* in Europe, starting in the early twentieth century and mainly due to an increase in food sources of anthropic origin, there has been a progressive saturation of usual breeding sites (Burger and Lesser 1980; Burger and Gochfeld 1983; Vincent 1985). As a result, there has been an expansion of the reproductive area and the colonization of new environments, including urban ones (Cadiou 1997). In this environment the gulls have found very favourable breeding conditions. In fact, man-made constructions offer a wealth of suitable sites, which greatly reduce intraspecific predation and eliminate the problem of terrestrial predators. Moreover, urban waste is a source of additional food, which, in contrast to natural food sources, is abundant and constant. In many cities, populations of yellow-legged gull have now reached such dimensions as to cause considerable disturbances to the public, mainly through noise, dirt and aggressiveness of territorial pairs towards people wishing to use terraces and balconies, which the gulls have chosen as a reproductive site. This has prompted various local administrations to study attempts to reduce, or at least contain, gull populations. These interventions are often not completely successful, due in part to the limited involvement of ornithologists. In Italy, where the phenomenon has been monitored for

E. Benussi (✉)
Studio Immagine Natura, Trieste, Italy

M. Fraissinet
Wildlife Research Unit, Department of Agriculture, Federico II University of Napoli, Portici (Napoli), Italy

Association of Southern Italian Ornithological Studies – ASOIM, San Giorgio a Cremano (Napoli), Italy
e-mail: mfraissinet@tiscali.it

F. M. Angelici, L. Rossi (eds.), *Problematic Wildlife II*,
https://doi.org/10.1007/978-3-030-42335-3_7

years, there has been a close cooperation between the National Association of Italian Municipalities (ANCI) and ornithological experts that study the phenomena (Fraissinet 2015) to provide information to local administrators and the public.

7.2 The Colonization of Cities in Europe

The phenomenon of gulls colonizing urban areas is worldwide. In US cities, the glaucous-winged gull *Larus glaucescens* and the ring-billed gull *L. delawarensis* are commonly found, while in Argentinian and New Zealand cities, it is the kelp gull *L. dominicanus*. In European cities, the yellow-legged gull *L. michahellis*, the herring gull *L. argentatus*, the Caspian gull *L. cachinnans* and lesser black-backed gull *L. fuscus* predominate, and, to a lesser extent, the great black-backed gull *L. marinus*, common gull *L. canus* and black-headed gull *Chroicocephalus ridibundus* do. Lastly, in Barcelona city's port, in the years 2013, 2016 and 2017, a colony of Audouin's gull *Ichthyaetus audouinii* nested. This particular event presumably followed the decline in nest numbers that occurred in a large nesting colony found in the Ebro Delta (Anton et al. 2017).

In Europe, the colonization of urban centres began in the twentieth century with its first instances reported in the United Kingdom as far back as the 1940s. The phenomenon manifested itself in other European countries, in a clear and widespread manner, in the 1970s. In France, urban breeding first began in 1970 and in Spain in 1975 at the Barcelona Zoo. In Italy, the first case of urban nesting dates back to 1971, with a pair that successfully breeded on an artificial rock in the Zoo of Roma. It is interesting to note that both first urban nesting in London, which dates back to 1966, and in Barcelona that dates back to 1975 took place in similar settings (Zoological Gardens) (Garcia Petit et al. 1986; Oliver 1997).

France, due to its geographical location, hosts four similar size breeding species of gulls: the herring, yellow-legged, lesser black-backed and the great black-backed gulls. It would be interesting to examine whether interspecific differences have been observed in the colonization of urban environments by gulls. Cadiou (1997) reports data for the 1980; the first half of the 1990s links the increase of French urban breeding with the increase in nesting pairs along the coasts after the beginning of the twentieth century. At the end of the 1980s, in the French cities, 1800 pairs of herring gull were breeding in 25 colonies and by 1996 the number of urban pairs increased to 7200–7500 (perhaps 8000–10,000) in about forty colonies, of which seven numbered more than 500 pairs. The average annual growth rate of urban populations was 20%. For a good number of cities, the first pairs nested on the roofs of fish markets. The urban population of the lesser black-backed gull in France in the late 1980s was about 80 pairs distributed throughout six colonies. This represented 0.3% of the French breeding population. In the second half of the 1990s, the population increased to 700–800 pairs spread across twenty colonies. This represented about 3% of French nesting pairs, with the species establishing themselves in cities that had pre-existing herring gull colonies with spontaneous colonization by the species being unknown. The first hybrid lesser black-backed gull x herring gull pair was recorded in Rennes. In France, as in Great Britain, there are no urban colonies where the lesser black-backed gull predominates. At most colonies of the lesser black-backed gull represent microcolonies within larger colonies of gulls. At the end of the 1980s, there were dozens of urban great black-backed gull breeding pairs in France, this is equivalent to 0.4% of the French population and by the mid-1990s, 26–29 pairs were nesting in urban areas at a dozen sites, i.e. about 1% of the French breeding population for that species. The great black-backed gulls settle on large roofs already colonized by other gulls, placing their nests in a central position in the colony, as normally happens in non-urban sites. The status of the yellow-legged gull in France in the 1990s is interesting. Urban nesting on the Atlantic coast was rare, but it was much more widespread on the Mediterranean coast where even though that by the second half of the 1990s urban pairs numbered only 70–100, but this figure was probably underestimated although urban breeding was an exceptional event in Corsica. In addition, it was the only species to nest in the French mainland cities of Paris, Toulouse and Saint-Girons, in the 1990s (Cadiou 1997). In 2010, Paris had three species of nesting gulls: yellow-legged with three pairs and the lesser black-backed with three pairs, both located within herring gull colonies totalling 50 pairs. In 2006 a mixed pair of yellow-legged gull × herring gull was also observed (Malher 2010).

The herring gull, in particular, increased sharply in numbers in the United Kingdom and Ireland from the late 1930s onwards and this led to the need for them to search for new nesting sites which were found on the roofs of abandoned buildings. From the 1970s, the growth of urban populations was quite rapid, passing from 1250 pairs in 1970 to 3000 in 1976 and on to 16,900 pairs in 1994. By the end of the twentieth century, 8% of herring gulls and 4% of lesser black-backed gulls in Great Britain and Ireland were urban breeders. In South Shields, in north-eastern England, the number of urban pairs of herring gull increased from 5 in 1963 to 209 in 1976, with an annual increase of more than 20%. Compared to natural breeding populations on the British and Irish coasts, it has been observed that urban herring gulls suffer less juvenile mortality. The reason for this is attributed to lower rates of cannibalism. It has also been ascertained that urban gulls originate from natural colonies, where the young, once mature and unable to find a suitable site for reproduction, moves to urban centres (Monaghan 1979).

In Bulgaria, the colonization of urban centres on the Black Sea coast is even older, dating back to the late nineteenth century. By 1992, two thirds of the population of yellow-legged gull was urban with 95.5% of the breeding pairs found in the coastal towns of the Black Sea and 4.5% nesting in the inland cities (Nankinov 1992). At the time of Nankinov's article, however, there was no separation between the two species – *Larus michahellis* and *L. cachinnans* – so it is not known which species started the colonization process. At the moment both species are present in coastal urban areas of the Black Sea. In the Romanian city of Constance there is a clear prevalence of *L. michahellis.*

In the Spanish coastal cities, the presence of yellow-legged gull is particularly frequent and in the city of Barcelona, in 2017, 500 nesting pairs were estimated,

150–200 of which were on roofs and terraces and 15–30 at the Zoo (Anton et al. 2017).

7.2.1 *The Colonization of Italian Cities*

In Italy, as already mentioned, the first urban breeding dates back to 1971 when a pair nested on an artificial rock at the Zoo of Roma (now called 'Bioparco'). The pair nested for several years, but they remained an isolated case for a long time, because it was only from the 1980s onwards that there was an increase in the number of urban nests of yellow-legged gulls. The first cities to be affected, besides Roma (which by 1984 had four pairs), were Sanremo (north-west) with a pair in 1982, Livorno (central) in 1984 and Trieste (north-east) in 1987. In some cities, the population growth, as observed in other European urban centres, was very fast, with an exponential trend. In Trieste, for example, in the period between 1988 and 2000 there was an annual increase equal to 28.9% with an average annual growth of 4.3% between 2001 and 2018, while in Napoli (south) a figure of 22% per year was measured between 1990 and 2014 (Table 7.1). At present, nesting is known to occur in about fifty urban centres with more than 10,000 inhabitants (Fig. 7.1). Figure 7.2 shows the progress of the colonization process. There is a sharp growth after 2000 and a further increase after 2010. It should be noted that the colonization of the Adriatic cities along the Italian coast with the exception of Venezia and Trieste has only recently started, in most cases from 2015 onwards. There have been no cases

Table 7.1 The evolution of the number of breeding pairs in certain Italian cities

City	No. pairs and period	Source
Torino	1 pair in 2007 – 12/15 in 2011	Di Rienzio A. EBN Italia
Genova	1 pair in 1986 – 78 in 2014	Milia L., unpublished data
Cremona	10/15 pairs in 1987 – 30/40 in 1998	Allegri M. (1999)
Venezia	22 pairs in 2003 – 200/250 in 2018	Sartori A. estimate
Trieste	1 pair in 1987 – 571 in 2018	Benussi E., unpublished data
Cesenatico	160 pairs in 2004 – 400 in 2014	Brina S., unpublished data
Sesto Fiorentino	45 pairs in 2013 – 72 in 2014	Del Sere M., Malfatti L., Puglisi L., unpublished data
Livorno	16 pairs in 1999 – 240 in 2013	Franceschi A., unpublished data
Piombino	2 pairs in 1994 – 40 in 2007	Franceschi A., unpublished data
Roma	1 pair in 1971 – 1000/1500 in 2014	Fraticelli F., unpublished data
Napoli	14 pairs in 1990 – 300/350 in 2018	Fraissinet M., unpublished data
Portici	2 pairs in 1999 – 25/30 in 2014	Fraissinet M., unpublished data

Fig. 7.1 Italian cities with a population exceeding 10,000 inhabitants with nesting yellow-legged gull *Larus michahellis*. Map updated to 2016

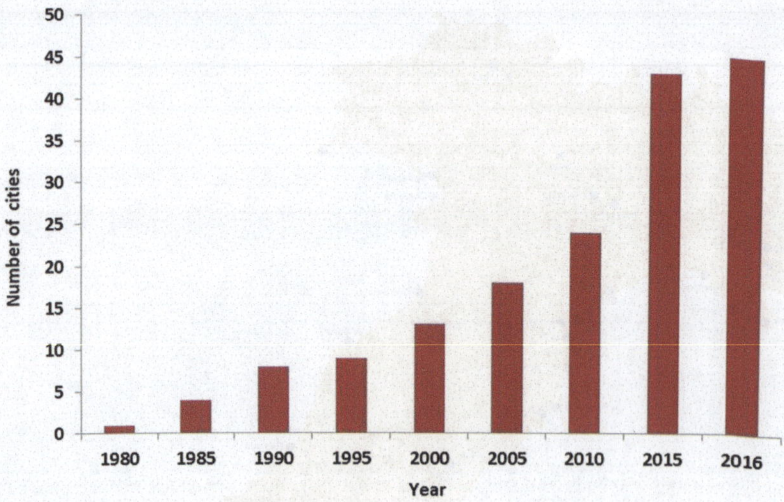

Fig. 7.2 Growth in the number of Italian cities with a population of more than 10,000 inhabitants with nesting yellow-legged gull *Larus michahellis* updated to 2016

of colonies disappearing so far. An estimate of the breeding population in urban centres with more than 10,000 inhabitants leads to a figure of at least 4000 pairs, equal to about 8% of the Italian breeding population. Regarding the choice of nest sites in cities, the species mainly uses man-made structures, nesting on terraces, tiled roofs, bell towers and on eaves. Ground nesting has also been observed recently in Roma and Trieste.

7.3 The Process of Colonization and the Ecology of the Species in Urban Areas

There may be many causes that lead to the process of colonization of urban centres by yellow-legged gulls, a species originally widespread in coastal environments. First of all, the strong numerical recovery of European populations since the mid-twentieth century should be taken into consideration, with the need, therefore, over time, to find new breeding sites. Added to this are more specific reasons: in urban centres, food is abundant and easily available, allowing even the first-year birds, still inexperienced, to feed easily. There are also some favourable ecological conditions, such as a higher average temperature than that in areas out of town, and lower rates of predation, although this positive aspect has been decreasing over the years as a result of the entry into the city of species which predate eggs and nestlings such as the hooded crow *Corvus cornix*, carrion crow *C. corone*, jackdaw *C. monedula*, magpie *Pica pica* and even raven *C. corax*. Many studies highlight the distance between the nests in the city being greater than that of the natural colonies and this might lead to the supposition that urban pairs are components of a single colony widely distributed across an area. There are, however, exceptions and some colonies established on roofs of industrial warehouses or near the coast have distances between nests similar to natural ones (Benussi and Bembich 1998; Fraissinet and De Rosa 2012).

The key to the success of the yellow-legged gull is also to be found in its high ecological plasticity that allows it, in particular, to modify its diet, adapting it to the most abundant and easily available resources present in urban centres or in the vicinity. This includes eating food waste of anthropic origin and carrying out an active predatory behaviour on both insects and small mammals but also birds, including especially the feral pigeon (*Columba livia* var. *domestica*), abundant in cities (Fig. 7.3). The process of colonization in the reproductive period has also been facilitated by the fact that, in the wild, the yellow-legged gull is a species that nests in rocky environments and has therefore found a surrogate of those environments in cities and building a nest on a cliff or on the roof of a building (Fig. 7.4) does not make much difference, especially if the roof is rarely accessed and also hosts spontaneous vegetation.

In the city of Napoli, an urban population and a natural population coexist within a short distance of each other, the former breeding on buildings of the city, the latter on the tuff cliffs along the Posillipo and Nisida coast. It was therefore possible to carry out some interesting comparisons between the two populations, to study their ecology and reproductive biology and verify possible differences between them.

In the 3 years between 2005 and 2007, the two populations were monitored. In this case as well, as mentioned earlier, we noticed a greater distance between the nests in urban pairs than natural ones, with the sole exception of the nesting colony on the roof of the Royal Palace, a historic building located a short distance from the

Fig. 7.3 The predation of feral pigeon is a rather common fact and is carried out by certain 'specialist' individuals (M. Fraissinet)

Fig. 7.4 Nesting in the urban centre of Trieste (E. Benussi)

sea. As far as their reproductive biology was concerned, there were no differences in the average number of eggs nor the average number of fledged chicks. Differences were found however in the fledging dates, where the nesting pairs in the natural context exhibited a certain heterogeneity in fledging dates, with respect to those nesting on man-made structures and in an urban context. This phenomenon has been observed in other Italian urban settings (C. Soldatini, personal communication). Further differences are found in diet composition, with the nesting pairs on the Posillipo and Nisida coasts having a more fish-based diet compared to those in urban environments that eat more birds, as shown in Fig. 7.5 (Fraissinet and De Rosa 2012).

Further monitoring of the breeding pairs carried out in Napoli brought another interesting piece of data on the growth of the urban breeding population compared to that in natural settings. Since 2006, the latter has shown no growth (Fig. 7.6) (Fraissinet 2016) and it can be noted that the rapid urban population growth coincides with stasis in the natural one which has evidently reached a saturation point in the available sites, in sharp contrast to the urban population which is still a long way from arriving at this point.

The yellow-legged gull should also be seen as an element of the urban ecosystem, in which it plays the role of secondary and tertiary consumer, depending on the prey, but also that of necrophage, having been observed to feed on carrion of other animals (even rats), as well as human food waste. At the same time, through its broods, it offers trophic resources to other predators such as corvids.

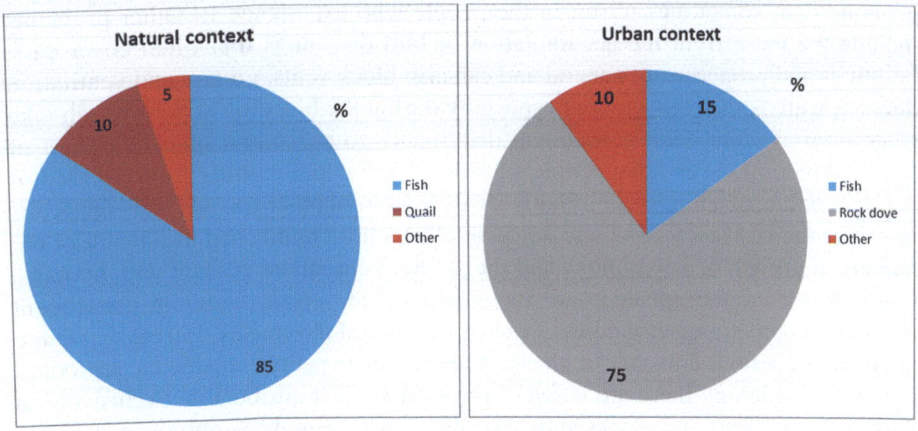

Fig. 7.5 Differences in diet composition observed in Napoli's two populations, found in a natural context and in an urban context

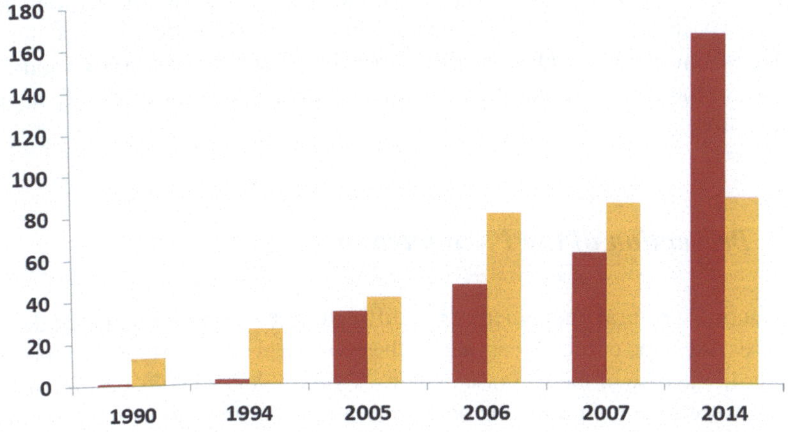

Fig. 7.6 Trends in the number of breeding pairs in urban and in natural contexts in Napoli. Red: pairs breeding in urban context. Yellow: pairs breeding in natural context

7.4 Problems Created by Urban Gulls to Humans

The presence of large numbers of nesting yellow-legged gull pairs in a city can cause problems of coexistence with the inhabitants. The difficulties, especially felt during the breeding season and in all European cities where the gulls nest, are manifold. Perhaps the largest impact is the aggressive behaviour of adults towards people – intruders from the gulls' perspective! – who, inadvertently, approach a nest with chicks; such behaviour known as 'dive bombing' is the highest level of aggression that the species can manifest towards humans and although it is 'no more' than flying just over the head of the victim, without any physical contact, it is enough to generate fear, sometimes terror, in the people who experience it. Other problems include the mess from the accumulation of bird droppings that often cover cars, buildings and properties in general and can also block vents, gutters and ventilation ducts as well as the calls emitted especially during the breeding season which take place even at night, with consequent disturbance to peace and quiet, the theft of food – sometimes even from people's hands – a behaviour coming from the klepto-parasitic habits of a bird that normally steals food from other species and conspecifics. Over the last few years these negative effects have multiplied as a result of the numerical growth of populations and the greater consequent contact with humans. Often, however, the episodes are magnified by the press, with the consequent increase in sometimes groundless or even unjustified psychoses. In reality, we are not aware of serious episodes of physical aggression to people outside the reproductive period although there are cases of physical contact attributable exclusively to the defence of nests and chicks at the breeding sites. Even the health risks, deriving from the fact that gulls can be carriers of some pathogenic bacteria to humans, must be traced to particular contexts and circumstances and to date, within the limits of Italian urban contexts, these have never been verified.

It should be remembered that the problems are partly offset by the ecological role that gulls play in the urban ecosystem as predators, as well as the pleasure that may be derived from the observation of their 'free flight' and the 'natural vivacity' that they bring to the cities. Recently, in some cities, people have increasingly begun to feed the gulls.

7.4.1 Prevention of the Phenomenon

Counteracting a natural phenomenon is a difficult, often impossible, undertaking. In some cases, it can be curbed by acting on the ecological causes triggering and maintaining it. In the case of the increase in pairs of yellow-legged gulls nesting in a city, it is necessary to act, first of all, on prevention. This prevention must be implemented both by tackling the ecological causes and behavioural aspects of the species. Prevention, moreover, must take place both, on a large scale, and be organized and financed by the municipal administrations. At the level of individual homes, they could be helped and coordinated by the municipal administration but finance it themselves.

In this context, it is worth mentioning the initiative of the National Association of Italian Municipalities – ANCI – and the Municipality of Naples, which, with the involvement of some ornithologists, have produced a brochure that contains information for local authorities of Italian cities affected by the presence of yellow-legged gulls on how to prevent and limit the phenomenon, in particular during the breeding season, as well as providing advice to citizens on how to behave in case of forced coexistence with nesting pairs (Fraissinet 2015).

The local authorities are called on to carry out two types of preventive action: the reduction and removal of constant food sources, correct information to the public (civic education) repeated over time. Although the species is very mobile and is able to obtain food from sometimes dozens if not hundreds of kilometres away from a city, it is also appropriate to reduce the opportunities for an urban supply. Therefore, commercial businesses, fishmongers, fruit and vegetable shops, butchers and restaurants of various kinds should be encouraged to avoid disposal of their waste in the area. Many yellow-legged gulls are known to have memorized the closing hours of fishmongers or local markets and are ready and waiting to pick the 'leftovers' up from the ground. In addition to the commitment from the individual shopkeepers, there must also be that from local authorities whom should be prompted to clean the areas. Another important food source present in the city are rubbish bins for urban waste, although in many cities, these are progressively disappearing to make way for new and more modern disposal techniques based on separated rubbish collection, with organic waste being left out only on certain days, in closed containers, and removed quickly. This is certainly a valid method for reducing food sources available to the gulls. Controlling spontaneous feeding by some people is more difficult to implement and enforcing prohibitions of this activity have regularly failed in the European cities in which it has been attempted, due to the difficulty of its application and the social implications for sometimes problematic human subjects. Another area in which administrations need to take action is the maintenance of roofs and terraces. A roof or a terrace that is not used and is in a state of neglect creates the ideal conditions to entice a pair to nest. It is therefore of great importance that, in the month of February, municipal authorities issue a reminder which states that from the following month (March) yellow-legged gulls will resume breeding activity and will also begin the search for suitable nesting sites. This reminder should suggest that from the period between February and March it would be appropriate for homeowners to access roofs and terraces frequently in order to show the gulls that, due to continuous human presence, the site is unsafe. The reminder, moreover, may also provide information regarding the different ways that individuals may adopt to prevent the phenomenon, remembering in any case that this is a species protected by various laws and therefore no cruel or violent action of any kind may be carried out towards these animals. These interventions, on the other hand, might have no effect considering the noteworthy ability of the species to respond to adversity.

For individuals who do not appreciate the presence of breeding pairs in or near their home, as well as the managers of public areas, industrial warehouses, shopping centres or anywhere else that may be negatively affected by the presence of the species during their reproductive period, the information booklet, issued to the town mayors, suggests various types of interventions and reiterates the need to access, if possible, the areas where the gulls have already nested in the past or that could attract them in the future such as roofs, terraces, balconies, chimneys and gutters. In case there are difficulties in carrying this out frequently, it would seem that only two techniques have given positive feedback. The first – an alternative to the 'anti-intrusion net', which is expensive, ugly and not always effective – consists in spreading over the terrace, above head height, a simpler net formed of taut, parallel wires with a mesh of between 50 cm and 3 m, although a distance between 1.50 and 2 m is usually also sufficient. The wires must be strong (with a diameter of at least

3 mm), of a material which is resistant to sunlight and corrosion, such as stainless steel or, even better, fishing line. This method has proven itself effective for flat surfaces, even large ones (such as car parks, squares, etc.), as it acts on the mechanical impossibility of animals to get their wings through the net and the considerable insecurity that it generates in the breeding pair for the raising of chicks. Furthermore, it is not cruel and, above all, does not generate 'familiarity' over time; i.e. the birds do not become accustomed to its presence and therefore do not lose their sense of insecurity (Blokpoel and Tessler 1984). The second technique, which is also a mechanical obstruction, can be used to prevent nesting on sloping, tiled roofs or near chimneys and other roof projections on which the gulls nest. It consists in the use of 'support spikes' similar to the ones used for pigeons, but of a different shape and orientation; i.e. they must be inclined wedges and rather close together, with a height that is not less than 15 cm, but it should be borne in mind that in some cases, longer lengths may have to be used, sometimes as much as 30 cm. Furthermore, it should be underlined that this method requires precise installation to prevent damage to other wild birds that might land on the roof and continuous maintenance over the years. There are different types of such dissuaders and therefore the identification of those that are best suited to the type and need of each roof is necessary. A wrong choice (elements not curved but straight, not very tall, etc.) may even promote nesting in certain circumstances!

Ultimately, the promotion of scientific research is of fundamental importance. Multi-year projects (at least 5 years) involving the marking of birds with legible rings, for example, are able to provide indispensable information on population dynamics.

7.4.2 Strategies to Be Adopted in Case of Nesting

The presence of important urban populations of yellow-legged gulls can often lead to real or presumed conflictual situations with people. The most frequently made complaints can be summarized as follows:

- Assaults
- Predation on pets
- Public disturbance
- Dirt and droppings on roofs, monuments and statues
- Health risks
- Damage to structures

Each of these situations presents various levels of criticality, shown below, which must be carefully evaluated if specific action is to be contemplated.

- *Assaults*. Even if it is true that, while rearing its chicks, the yellow-legged gull applies a behavioural model of defence to its nest that is particularly aggressive towards those, including humans, getting too close, we must consider that, with the exception of isolated incidents, it is mostly just a display of aggression. However, there are more and more frequent cases in which particularly aggressive individuals even engage in physical contact, attacking in flight and striking the heads of intruders with their feet and beak. Regardless of the level of real risk, this behaviour puts a question mark on the everyday use of spaces (balconies or terraces) where a nest is. It can also represent a real element of risk for those who, in the course of their work, must visit the tops of buildings with situations where balance is vital (construction workers, installers of aerials, chimney sweeps). The period of greatest aggressiveness is limited to some extent to incubation but particularly emerges in the weeks after the hatching of the eggs until the young birds fledge. Considering that nesting can last several months in the spring and summer, this type of disturbance is most keenly felt by people.

- *Predation*. In most cases, the attack or active defence (mobbing) behaviour connected with the defence of the nest, or the attempt to steal food put down for dogs and cats, sometimes aggressively, are erroneously interpreted as predation attempts on pets. However, real attempts at predation cannot be ruled out, especially with regard to small animals left free to wander on terraces or in courtyards, such as dwarf rabbits or guinea pigs. Cases of predation have already been extensively documented for newborn kittens.
- *Disturbance.* Yellow-legged gulls are particularly vocal and annoying at night when they engage in collective flights in large numbers. The meaning of this behaviour is still unclear and probably has a social function. The undoubted disturbance to peace and quiet, especially on the upper floors of buildings near nests, is not however limited to these flights, but also to the constant loud calling of the pairs and their chicks, especially during the weaning period when the latter begin to fly.
- *Dirt and droppings.* Gulls use many types of materials in the construction of the nest, even if in limited quantities (Figs. 7.7 and 7.8). They also carry a range of objects to the tops of buildings, perhaps to play with. These materials, often helped by the slope of the roofs, when driven by rain, tend to accumulate at the mouth of the drainpipes and guttering, causing blockages and leaks (Figs. 7.9 and 7.10). In addition, there are numerous cases of monuments and bronze statues being used as perches that are soiled by the bird droppings which causes corrosion, potentially damaging the structures seriously and even irreversibly. In addition, the remains of predated pigeons and feathers may accumulate on roofs.
- *Health risks.* Gulls' excrement is particularly liquid and therefore does not tend, even if deposited by substantial numbers of individuals, to form layers of guano as happens in the case of urban pigeons. Given that the yellow-legged gull is also a predator of other vertebrates, remains of fish, birds or other food waste in a state of putrefaction near the nest may be found. To date, there are no reports of infection directly transmissible from gulls to humans, but in these situations it is advisable to maintain the correct precautions and strive to ensure suitable hygiene conditions in areas frequented by people, as the international scientific community, to date, has no doubts about the possible zoonotic risk linked to the presence of gulls in urbanized areas.

Figs. 7.7 and 7.8 The supply of miscellaneous material for the construction of the nest and an unusual location between the plants of a window box on a balcony (E. Benussi)

Figs. 7.9 and 7.10 Nests built in gutters can obstruct the flow of rainwater causing overflows and considerable consequent damage (E. Benussi)

- *Damage*. The yellow-legged gull chicks, before starting to fly and moving away from the nesting site, spend several days walking about on the infrastructure near the nest. During these excursions, apparently demonstrating playful behaviour – but probably practising using their beak for their future independence – they tend to fiddle with various objects such as window fittings and dormers, electric or television cables, roof covering materials and so forth. This behaviour can cause serious damage.
- Multiple stratagems have been proposed with a view to stopping, or at least containing, the damage and disturbance caused by this species but none has given satisfactory results or is applicable in all situations. The following list refers exclusively to methodologies that have, if nothing else, a rational approach.
- *Distress call*. The alarm or distress call, recorded and amplified, can function as a means of removal for some species. In the management of urban populations of starling *Sturnus vulgaris*, this has given positive results allowing the removal, or at least, the breaking up of flocks that concentrate in roosts. Using this approach with the yellow-legged gull, however, has not produced the same results but has rather generated aggregative responses in a sort of mutualistic behaviour that brings together the individuals present in the vicinity of the site.

- *Ultrasound*. Although there are many ultrasound systems available, which are sold to deter annoying birds, their effectiveness is nil because birds do not have the ability to hear sounds above 20 kHz, in other words ultrasound.
- *Olfactory repellents*. Despite the sense of smell having recently been reconsidered in many species of birds, in the yellow-legged gull, the use of products with strong repellent odours did not result in their relocation.
- *Laser lights*. Even this system, which consists in flashing laser light beams at the gulls as they perch, has not provided any result. It is also necessary to consider the real risks to public health that this technique can cause.
- *Protection nets*. Covering the roofs of buildings or terraces with nets can provide positive results especially on small surface areas. Above larger spaces, these structures are not always able to fulfil their purpose. The potential advantages should be weighed against the installation costs and the limitations posed by making the surfaces affected inaccessible to humans. This does not, of course, include the suspended wires mentioned earlier as they can also be placed at heights that allow human access.
- *Spikes*. These consist of plastic or stainless steel points that are designed to prevent gulls from landing or perching on surfaces or building nests. This system, used in many urban centres to limit the presence of pigeons along the edges of buildings and monuments, does not give appreciable results for gulls, unless the forms and types described above are adopted (Figs. 7.11 and 7.12). Moreover, using the spikes intended for pigeons – besides the obvious difficulty of intervening on all surfaces with often extremely irregular shaped material – may even prove to be counterproductive as the spikes could constitute a support for nest material especially on sloping surfaces where the nest would otherwise slip in the absence of such intervention.

Regarding the following techniques, it must be borne in mind that the yellow-legged gull, in many countries, is a species protected by law and, consequently, any action involving it or its nests must have a prior permission from the authorities. The

Figs. 7.11 and 7.12 Spikes and metal nets sometimes serve to better anchor the nest and might not prevent nesting (E. Benussi)

authorization process must be undertaken directly by the individual's public administration, while any planned actions must be carried out by personnel with clear and certified ornithological skills. From a biological point of view, these interventions can be meaningful only in the presence of critical situations related to single individual birds. However, they are totally ineffective as forms of management of the species as a whole, because given that there are so many gulls, local numerical reduction work would be immediately thwarted by the arrival of new birds from neighbouring areas.

- *Culls.*Apart from ethical reservations, the effect of euthanizing one of the gull of a breeding pair is nullified in a short time because of the large number of individuals that have failed to pair or which have not found a site suitable for reproduction and thus a new partner is found immediately. Furthermore, the elimination of large numbers of individuals in urban environments would involve techniques that could represent a real risk to the public.
- *Egg removal and nest destruction.* This does not produce results unless it is carried out on a large scale and repeated several times a year on the replacement clutches. This technique can mainly solve the problem on single buildings used as homes, preventing reproduction and therefore removing the source of irritation and the cause of damage to the structures that the presence of breeding pairs may bring. In many cases, the destruction of the nest must be continued for several years because gulls tend to return to the same sites year after year and even when the gulls leave, the problem merely moves elsewhere. In this case laws that protect the species must as well be taken into consideration.
- *Sterilization of eggs.* The technique consists in the drilling of the eggs or in their waterproofing with paraffin, both aimed at interrupting embryonic development. These interventions have proved to be inadequate because within a short time the breeding pair perceives the absence of an embryo in development and lays a replacement clutch. This forces the repetition of the operation several times because there may be multiple replacement clutches. Furthermore, the operational difficulties that this technique entails, the considerable commitment of people involved and the high costs must also be considered. For example, in central Roma alone, fewer than 5% of the nests can be reached, in an area of maximum reproductive density for yellow-legged gull, without the use of rock-climbing techniques (F. Fraticelli, personal communication).
- *Relocation of nests.* This technique involves physically moving the nest from high impact areas to areas of less impact. Beyond the obvious operational limits similar to the previous intervention, this technique requires that alternative sites for relocating the nest exist. It is therefore understandable that at most this can be applied in a limited number of cases.
- *Administration of baits treated with contraceptive chemicals* This experimentation has thrown up difficulties in the preparation and dosage of baits, such as difficulties in their systematic distribution and controls on consistent and balanced intake as well as high costs which thus makes it inapplicable.

- *Sterilization of adults*. This is not an effective method of control, particularly if performed on sexually immature subjects, as it cannot reduce the size of the population, unless 100% of the territorial males are castrated (or 100% of the females are sterilized), the decrease in the population arriving through the natural mortality of adults. The interventions on juveniles are useless considering the high mortality they undergo in the first 2 years of life (over 50%) and the marked erraticism that distinguishes all subjects until sexual maturity in their third year, which allows for only a very low percentage, and predominantly males, to return to reproduce at the natal nest site. Castrating/unnecessarily sterilizing a low percentage of sexually mature subjects without actually altering population dynamics represents an error in planning, but above all is a pointless episode of animal cruelty.

In conclusion, considering the ineffectiveness of the methods described above, the only really effective action is the prevention of the phenomenon. Another fundamental step is the communication to the public of the simple rules of coexistence that may be summarized as follows:

- *Do not feed the gulls*. When checking many situations of presumed risk, it became evident that the same people who had thought themselves threatened had previously offered food to the gulls (Fig. 7.13), thus conditioning their behaviour (Figs. 7.14 and 7.15).
- *Do not make pet food available*. Avoid feeding dogs and cats on terraces and balconies and if necessary remove leftover food.
- *Reduce alternative trophic sources*. In urban centres, avoid leaving rubbish outside bins and limit open landfills outside towns.
- *Protect small pets*. Don't leave them outdoors.

Finally, it should be added that in some cities (e.g. in Italy, Trieste, Cesenatico and Napoli) groups of experts have come forward both from the scientific and the conservation world who, when called on by people, intervene to find solutions that – while guaranteeing the safety of the birds – are also able to meet the expectations of people.

Acknowledgements The following collaborated by providing data and information: Enrico Borgo, Sandro Brina, Marcello Bruschini, Barbara Cursano, Mauro Del Sere, Luciano Del Gaudio, Flavio Ferlini, Alessandro Fioretti, Alessio Franceschi, Fulvio Fraticelli, Loris Galli, Angelo Genovese, Piero Giovacchini, Salvo Grenci, Davide Grimaldi, Maria Grotta, Giovanni Leonardi, Raffaele Luise, Luigi Malfatti, Bruno Massa, Angelo Meschini, Laura Milia, Salvatore Mirra, Giampaolo Necco, Giuseppe Nuovo, Emily Prall, Fabio Procaccini, Giacomo Radi, Alessandro Sartori, Alessio Usai, and Nereo Verginella.

Thanks to Danilo Russo for the critical reading of the text and to Emily Prall and Paul Tout for correcting the English texts.

Fig. 7.13 'Domesticating' gulls by providing them systematically with food causes the species to become more dangerous towards humans, especially during the breeding season when the most aggressive individuals can physically attack people in defence of the offspring (E. Benussi)

Figs. 7.14 and 7.15 In some cases, particularly tame gulls nest in conditions that allow for them to be approached without problem but greater aggressiveness can occur however, following hatching (E. Benussi)

Appendix: Experiments for the Management of the Species in the City of Trieste

The species in the urban area of Trieste has been breeding since 1987 and is annually monitored using standardized census methods.

The number of pairs has progressively increased and in 2018 the breeding pairs counted totalled 571 (estimated 600–620), with an average annual increase of 14.6%

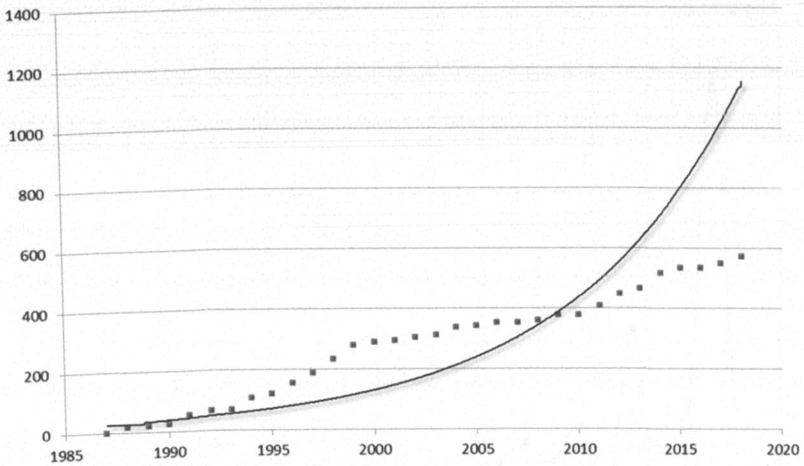

Fig. 7.16 Trend of the breeding population in Trieste

recorded in the period 1988–2018 (Fig. 7.16), levelling off at 4.3% in the period 2000–2018 (Fig. 7.17).

Since the end of the 1990s, the municipal administration has undertaken an experimental campaign aimed at limiting the breeding population, carried out on sample pairs, through birth control (1999–2004) (Benussi 2005) and removal from reproductive sites (2014–2018).

Recorded below is the information obtained in a total of 11 years of experimentation using different types of intervention:

(a) *Drilling of eggs* (1999, 2000, 2001, 2004) (Fig. 7.18). A single intervention carried out on 60 sample pairs: 29 (48.3%) continued incubation for a greater number of days and then definitively abandoned the site and moved elsewhere; 26 (43.3%) laid again on a new nest built in the immediate vicinity, successfully hatching their eggs; and 5 (8.3%) immediately abandoned the clutch and moved elsewhere from the site; the data obtained shows that, often but not always, pairs leave the nest to build another in a new site, often nearby, and thus breed successfully. The experimentation was considered ineffective for the purposes intended.

(b) *Removal of eggs and destruction of the nest* (2000, 2001, 2004, 2014–2018). Out of 210 sample pairs 132 (62.8%) laid again within 15 days at the same site in a new nest but generally fewer eggs (63% one egg, 31% two eggs, 6% three eggs), demonstrating, however, a high percentage of clutch replacement (up to three times) and strong attachment to the site; 78 pairs (37.1%) permanently abandoned the site, probably moving elsewhere. This experimentation was to be considered effective if carried out with interventions which have to be repeated periodically until the pair leaves the roof. The problem at the site in question is resolved for one season (aggressiveness, disturbance, damage) but the gulls presumably move elsewhere and it is quite possible that they will reoccupy the site the following season. Such interventions do not favour a decrease and a numerical control of the species.

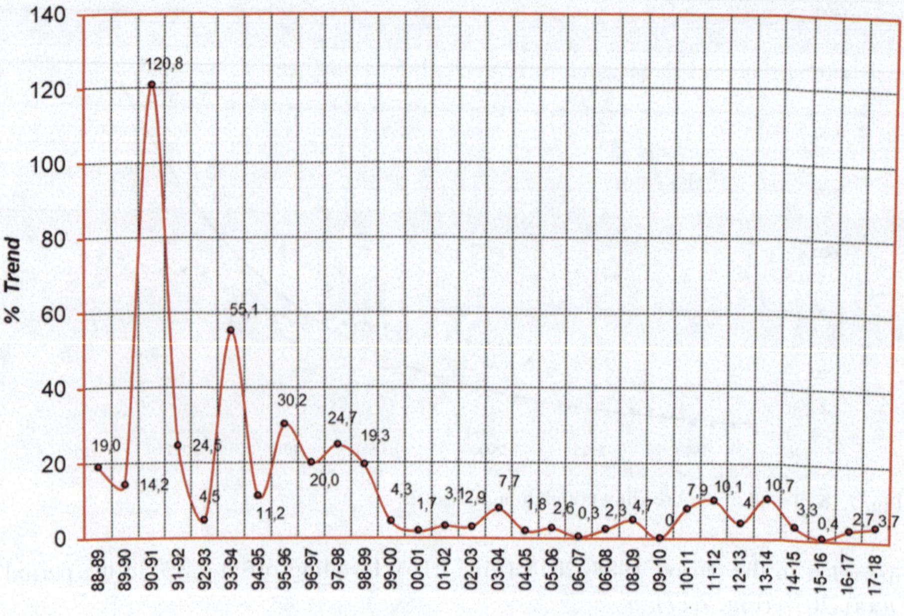

Fig. 7.17 Percentage trend of the growth of the breeding population in Trieste

Fig. 7.18 Drilling eggs did not provide the expected results in limiting nesting owing to the difficulties in reaching all urban nests (E. Benussi)

(c) *Administration of bait treated with the contraceptive substance* 'Glisol-T' (1999). This partial experimentation revealed difficulties in the preparation and dosage of bait, difficulties in the distribution and control of its balanced consumption and high costs. Out of six ringed pairs, four (66.6%) did not breed successfully (infertile eggs), two (33.4%) reproduced normally but four out of six eggs in the nest were infertile (probably as a result of insufficient intake of the drug). This experimentation is not to be considered widely applicable on a large scale.

From the results obtained, it can be stated that the interventions implemented with the aim of containing the urban population have proven to be ineffective. It is therefore objectively impossible to prevent urbanization and nesting by gulls using cruelty-free methods.

Any containment through the reduction of hatching, if carried out on a substantial part of the urban population, may represent a way to slow down (but not stop) the process of population growth. This avoids drastic high impact measures such as euthanizing juveniles and adults, which targeted management would recommend.

The management of a population characterized by a marked dynamism such as that of the yellow-legged gull must plan for continuous action over a period of time (5–8 years). Occasional interventions are ineffective and nullify the results obtained and constitute a waste of public resources.

Combined actions are essential for the drastic reduction of available trophic sources from anthropic sources such as landfills, rubbish bins and food provided directly by humans ('adopted' gulls). Therefore, civic education projects are as important as public awareness campaigns.

It is also difficult to predict the evolution of the city colonies due to the difficulty of quantifying the potential capacity of the urban ecosystem. A further limitation is caused by the impossibility of predicting the immigration of other gulls from surrounding areas.

References

Allegri M (1999) Nesting of Gull *Larus cachinnans* and lesser spotted woodpecker *Picoides minor* in Cremonese. Avocetta 23:142

Anton M, Herrando S, Garcia D, Ferrer X, Cebrian R (2017) Atles des ocells nidifcants de Barcelona. Ajuntament de Barcelona, ICO, UB, Barcelona Zoo

Benussi E (2005) The management of *Larus cachinnans* in the urban area of Trieste. First results on methods of census and containment of nesting. Acts XIII Conv Ital Ornithol Avocetta 29:135

Benussi E, Bembich L (1998) Characteristics, status and evolution of the urban colony of *Larus cachinnans michaellis* in the city of Trieste. Annales 13:67–74

Blokpoel H, Tessler GD (1984) Overhead wires and monofilament lines exclude Ring-billed Gulls from public place. Wildl Soc Bull 12:55–58

Burger J, Gochfeld M (1983) Behavioral responses to human intruders of Herring Gulls (*Larus argentatus*) and Great black-backed Gulls (*L. marinus*) with varying exposure to human disturbance. Behav Process 8:327–344

Burger J, Lesser F (1980) Nest side selection in an expanding population of Herring Gulls. J Field Ornithol 51(3):270–280

Cadiou B (1997) La reproduction des Goélands en milieu urban: historique et situation actuelle en France. Alauda 65(3):209–227

Fraissinet M (ed) (2015) The colonization of Italian urban centers by the Gull (*Larus michahellis*). To know the phenomenon, to prevent it, to manage it. ANCI and Councilor for the Environment of the Comuna di Napoli eds, Naples, p 32
Fraissinet M (2016) Monitoring of the population of Yellow-legged Gull *Larus michahellis* nesting in the city of Naples. Picus 42(81):35–39
Fraissinet M, De Rosa D (2012) Distribution and ecology of the Gull *Larus michahellis* nesting in Naples in the three-year period 2005–2007. Picus 38:5–12
Garcia Petit J, Marti Gabernet M, Javier F, Gimeno T, Carrera Galissà E, (1986) Nesting of Gull *Larus cachinnans* in the city of Barcelona (Spain). In: Proceedings of the first symposium on Mediterranean seabirds. Editions of the Sun, Alghero, pp 483–485
Malher F (2010) Oiseaux nicheurs de Paris. Corif. Délachaux et Niestlé ed, Paris
Monaghan P (1979) Aspects of the breeding biology of Herring Gulls *Larus argentatus* in urban colonies. Ibis 121:475–481
Nankinov DN (1992) The nesting by the Herring Gull (*Larus argentatus*) in the Towns and Villages of Bulgaria. Avocetta 16:93–94
Oliver PJ (1997) The breeding birds of Inner London, 1966–1994. Br Birds 90:211–225
Vincent T (1985) Les Goelands argentes (*Larus argentatus*), leur expansion en center ville and les moyens de lutte permant de s'y opposer. Bull Trim Soc Géol. Normandie et Amis Muséum du Havre 72(1–2):79–89

Part IV
Hunting and Ecotourism: Possible Mechanisms for Conservation and Coexistence?

This part of the book deals with a rather controversial topic: the possibility of coexistence between hunting and conservation. In addition, the theme of ecotourism as a modern resource that can improve conservation of animal species and habitats (see Krüger 2005) is also discussed.

The first chapter (Perco 2020), starting from worldwide case studies, continues by discussing – both philosophically and in practice – the possibility of coexistence between hunting activities and species conservation, with benefits for the hunted species but also for their preys and/or predators. The second chapter (Giménez-Anaya et al. 2020) is a very thorough review of the status of the wild boar in the Iberian Peninsula. It deals with the topics of hunting and its compatibility with the conservation of the species as a resource, human-boar interactions (at various levels), acceptance from population, economic implications, etc. (see Giacomelli et al. 2018).

Finally, the chapter about ecotourism (Esposito et al. 2020) addresses the topic from a global perspective and shows how this activity represents a significant economic source that must be used for the conservation of the interested areas generating jobs for the local population. The important point here is that there should not be overexploitation by eco-visitors, in order to avoid negative effects of disturbance and deterioration of habitats (Libosada Jr 2009).

References

Esposito EM, Palumbo D, Lucidi P (2020) Travelling in a fragile world: the value of ecotourism. In: Angelici FM, Rossi L (eds) Problematic wildlife II. New conservation and management challenges in the human-wildlife interactions. Springer, Cham, pp 273–355

Fischer A, Sandström C, Delibes-Mateos M et al (2013) On the multifunctionality of hunting– an institutional analysis of eight cases from Europe and Africa. J. Environ. Plan. Manag 56:531–552

Giacomelli S, Gibbert M, Viganò R (2018) Community empowerment for managing wild boar: a longitudinal case study of northern Italy 2001–2018. Ecol Soc. https://doi.org/10.5751/ES-10353-230412

Giménez-Anaya A, Bueno CG, Fernández-Llario P et al (2020) What do we know about wild boar in Iberia? In: Angelici FM, Rossi L (eds) Problematic wildlife II. New conservation and management challenges in the human-wildlife interactions. Springer, Cham, pp 251–271

Knezevic I (2009) Hunting and environmentalism: conflict or misperceptions. Hum Dimen Wildl 14:12–20

Krüger O (2005) The role of ecotourism in conservation: panacea or Pandora's box? Biodivers Conserv 14:579–600

Libosada Jr CM (2009) Business or leisure? Economic development and resource protection-concepts and practices in sustainable ecotourism. Ocean Coast Manage 52:390–394

Lindsey P, Alexander R, Balme G et al (2012) Possible relationships between the South African captive-bred lion hunting industry and the hunting and conservation of lions elsewhere in Africa. S Afr J Wildl Res 42:11–22

Loveridge AJ, Reynolds JC, Milner-Gulland EJ (2007) Does sport hunting benefit conservation? In: Macdonald DW, Service K (eds) Key topics in conservation biology. Blackwell, Oxford, pp 224–240

Perco F (2020) How hunting and wildlife conservation can coexist. Review and cases study. In: Angelici FM, Rossi L (eds) Problematic wildlife II. New conservation and management challenges in the human-wildlife interactions. Springer, Cham, pp 215–250

Chapter 8
How Hunting and Wildlife Conservation Can Coexist: Review and Case Studies

Franco Perco

8.1 Introduction

Hunting is the activity of killing or trapping wildlife or feral animals or tracking them with the intent of doing so.

This is the current definition but a more precise one can be "the set of activities that aim to take possession of a wild or feral animal, in order to have it permanently but for various purpose (food, economic, recreational, amateur, scientific etc.)."

Stressing that this represents the exploitation use of a renewable resource, hunting and related activities are an art of the ecosystem services (ES), i.e., in a certain way, the services that nature is able to provide to society. These services are implicit (i.e., they exist in themselves, such as atmospheric CO_2 regulation by vegetation) or explicit (i.e., they depend on human activities of various kinds, such as hunting) (Buckley and Mossaz 2015; Di Minin et al. 2013b; Crosmary et al. 2015a).

The ownership of wildlife is normally attributable to the state or to the owner of the land where the animal usually lives (or finds itself in that particular moment). In some cases, wild animals can be considered *res nullius*, that is, not belonging to anyone until their capture or killing (this applies to all traditional or subsistence survival types of hunting). Game may also take the form of collective, feudal, or religious property or belong to public bodies other than the state or even to private bodies other than the state (as in the case of concessions issued by the state to private collective organizations involved in hunting management).

The problem of property arises, however, with regard to species of a certain importance, such that they constitute a genuine resource for which rules are necessary to regulate their appropriation.

In conclusion, hunting always requires the deprivation of liberty or the death for the hunted subject as well as a change of ownership or its reinforcement (if the

F. Perco (✉)
Montelupone (MC), Italy

F. M. Angelici, L. Rossi (eds.), *Problematic Wildlife II*,
https://doi.org/10.1007/978-3-030-42335-3_8

hunter is also the owner of the fauna), even if such cases are largely conditioned by various regulations.

Poaching is illegal or non-permitted hunting.

Hunting is licit only if codified by written rules. The same hunting may be licit, if it is allowed under traditional and cultural behaviors. This is the case for many traditional hunting methods, as well as for those few populations that do not have legislation in this regard.

The species that are hunted are referred to as "game" and are usually mammals or birds.

However, we talk about hunting even in the case of specific reptiles (in particular the crocodile in Africa and Australia; see Webb et al. 2004). Furthermore, the sport hunting of snakes is not unknown (the Python Challenge in the Everglades, promoted in Florida by the Fish and Wildlife Conservation Commission, a challenge with the aim of capturing as many pythons as possible). Other reptiles and amphibians are normally hunted by all native populations.

In the case of fish and other organisms living in an aquatic environment (with the exception of marine mammals) such as mollusks, crustaceans, etc., we talk about "fishing" (Giles 1978), even if basically it is the same activity.

8.2 Conservation of Fauna (Wildlife *Stricto Sensu* and Its Ecosystem Services)

Wildlife (WL) is a renewable resource. In this sense, its sustainable exploitation, that is to say for as long as possible, is a guarantee of its preservation to the extent that the benefits brought about by the exploitation itself involve a large number of subjects. These benefits may include provisioning services (indicators: meat from hunting, value of game, etc.), regulatory services (indicators: pest control, number of IAS, invasive alien species, etc.), and cultural services (indicators: number of hunters, ecotourism operators, etc.) (Andersson et al. 2007; Bateman et al. 2013; Bauer et al. 2009; Freeman 2003; Jeffers et al. 2015; Mace et al. 2012; Tallis et al. 2008; Western 1984).

The evaluation of the WL, in several aspects (not just hunting), allows an appreciation of the natural capital, and this should be considered by the different states in their decision-making processes. The Second Session of the IUCN World Conservation Congress (Wiseman and Hopkins 2001) again promoted the sustainable use of WL as a way of protecting biodiversity and supporting the development of rural communities (Jeffers et al. 2015). Generally, it is the maintenance of high levels of biodiversity that guarantee the conservation of natural faunal balances, which entails the conservation or the reconstitution of adequate faunistic biodiversity values at subspecific levels as well (Hawksworth and Bull 2007; Zunino and Zullini 2004).

If the case is restricted to a group of species or just to a species, it is very different in the case of the large carnivores, or, e.g., the Siberian tiger (*Panthera tigris altaica*), because their protection implies the preservation of an entire ecosystem (Quammen 2003). However, in the case of the threatened subspecies of the red deer (*Cervus elaphus*), it does not mean that the natural ecosystem is preserved, but only the "ecological-managerial" aspects that are favorable for their continued survival.

Conserving species at the top of the food chain (key species) or focal species in a broader sense is the norm for the conservation of an ecosystem, but it must also be said that there are hybrid situations such as that of the case of the brown bear (*Ursus arctos*) in Romania (Quammen 2003; Stringham 1989), which is widely "artificially fed" and yet "preserved," even at the expense of naturalness. In a certain sense, even the provision of carrion dumps for scavengers (e.g., griffon vulture, *Gyps fulvus*) falls into this category. This certainly applies to all endangered species that, in any case, take advantage of environmental transformations triggered by human activity.

I would therefore propose to distinguish three forms of WLC:

1. Type A conservation of all the "natural balances," implying a great limitation of human activities.
2. Type B conservation versus a partial naturalness and enabled by substantial human-engineered assistance: e.g., Ursus arctos, Gyps fulvus, etc.
3. Type C includes species at the base of the food chain including *Cervus* sp. and mountain sheep *Ovis* spp., namely, species which may be a priority – although this is not always the case – and involves only certain interventions such as their reintroduction or the imposing of limitations on their harvesting, etc.

A paradox is that the conservation of type C may require the control of key species in terms of conservation of the type A. A case of this kind took place, for example, in the National Park of Abruzzo, Lazio, and Molise (Italy) with the hypothesis of reintroducing the lynx (*Lynx lynx*) (whose autochthony was in any case questionable) where, however, the Apennine chamois (*Rupicapra pyrenaica ornata*) might be threatened, when at that time there were less than 500 individuals of this endemic threatened subspecies.

Hunting can make a contribution to reintroduction of the huntable species (Perco 1997) as well as to the control of problematic species such as wild boar (Meriggi et al. 2016; Perco 2017), red deer (Ferretti et al. 2015; Perco et al. 2001), elephant (Le Bel et al. 2016; Schloles and Smart 2008; Slotow et al. 2008), and IAS (Dahl et al. 2000).

However, WL can be managed from a range of perspectives and with different purposes (Di Minin et al. 2013b), dependent on the context and the institution charged with their management.

The management of the WL in a protected area is different from that to be conducted in an urban setting or on a hunting reserve. Not only according to the institutional context but also to the stakeholders (Varner 1998), the geographical area, the dominant culture, and the historical period (Buijs et al. 2009; Giles 1978; Gossow 1983; Leopold 1949; ISPRA 2011; Kiss 1990; Mace et al. 2012; Saunders 2013). In short, the management of WL can respond to the following priority purposes: protection, hunting, economy, research, leisure, observation, sensitivity training, or ideal or ethical values (Bookbinder et al. 1998; Perco 1997; Reimoser 2018b; Ripple 2016; Ulrich 1983; Williams et al. 2000). Each of these purposes has the obligation to preserve and guarantee the satisfaction of the ES.

8.3 Coexistence

As previously mentioned, all management aims should allow, or rather consolidate, the conservation of the species managed while taking into account the priorities.

In a hunting area, the priority is the game bag, but without threatening the protected species. In some speciale cases, such as the conservation of the gray wolf outside a protected area, we cannot ignore the interests of farmers (economic option, Mech 2017). Naturally all this requires careful balancing and a positive relationship with stakeholders (Giles 1978; Kiss 1990; Leader-Williams and Hutton 2005).

Coexisting means "existing together," at the same time, or in the same place without regard to the differences. Therefore, coexisting does not imply active conservation.

The crux of the matter of compatibility (coexistence) between hunting and conservation must be tackled in two respects: when hunting is inherently incompatible with conservation or if its compatibility depends on how it is carried out.

It is compatible if it is carried out "correctly"; it is not if it is carried out in an inappropriate fashion. The problem must be faced in an objective manner, regardless of ideologies. If there are cases in which hunting preserves biodiversity, irrespective of the number of cases and the difficulties that exist, this means that it is compatible with conservation.

8.4 Types of Coexistence

We can distinguish four types of coexistence between hunting and WLC:

1. Non-impactful Hunting (NIH). These are hunts that are not "impactful"; that is, the harvesting and related activities (disturbance) are minimal and without significant or permanent consequences for the resource.
2. Impactful and Eliminatory Hunting (IEH). This hunting is both impactful and "eliminatory." An example of this kind includes the megafauna of the Pleistocene (outside Africa) which was destroyed by Homo sapiens. This type of hunting exists also today, e.g., Steller's sea cow Hydrodamalis gigas (1768+), the passenger pigeon Ectopistes migratorius (1914+) and the Tasmanian tiger Thylacinus cynocephalus (1936+), etc.

3. Impactful but Resilient Hunting (IRH). This hunting is impactful but WL exhibits "resilience" (the capacity of the resource to recover from exploitation), restoring its losses (quantity, structure, other characteristics) over variable timescales.
4. Impactful but Contributory Hunting (ICH). This hunting, although impactful, brings some advantages to the WLC in question.

8.5 Non-impactful Hunting (NIH) and Its Effects on Conservation

Neglecting a logical but only theoretical discourse with regard to the number of hunters (ten hunters in the USA would be non-impactful) (but see also Martin 1984: 367, in Martin and Klein 1984) involves forms of hunting that could be termed "primitive" (i.e., "natural" or "native") and therefore very different from modern ones, with firearms.

These "primitive" hunts, for survival or subsistence, can be distinguished by their great variety of forms.

They may be practiced without tools (e.g., collecting eggs, capturing nestlings or juvenile and injured individuals) or with tools of various types including poisons, intoxicants or adhesive substances (birdlime), snares, nets, traps, leghold traps, and up to bladed weapons, bows and arrows, or spears.

Today this type of hunting is part of indigenous peoples' traditional livelihood, not yet under the influence by so-called "civilization," and could be considered marginal as they are still used only by the few surviving tribes and groups of natives (in the Amazon Forest and the Fayu in New Guinea), Pygmies, Khoe-San, Inuit, indigenous Australians, etc. Despite this, they are not at all unknown in developed countries, as they are part of the illegal methods of hunting (poaching).

Traditional hunts are involuntarily conservative depending on the abundance of prey, the ability to obtain them, the presence of alternative prey, and the lack of technology and hunting techniques.

Therefore, the level of awareness is generally absent, as it is linked to necessity and the fauna is seen as an endless resource (Corlett 2007; Duda 2017; Festa-Bianchet et al. 2011; Gunn 1994; Harrison et al. 2016).

This does not mean that there is or there has been a certain awareness that the resource should not be exhausted, in order to be able to use it again, but this only occurs when it is not subordinate to immediate requirements. In regards to the rules governing primitive hunting, this is not a question about true norms but involves shared rituals. These are not infrequently more compulsory than written rules. This said, regulated hunting activities are, *par excellence*, the only ones able to achieve a certain level of compatibility, not least because they are subject to modification as a result of an improvement in knowledge, awareness, and sensitivity.

A distinction must be made between survival hunting and subsistence hunting. The latter is necessary to live (Cahill 1998), while the former is an exclusive activity in which Homo sapiens has no resource other than hunting and/or harvesting.

Survival hunting exists beside other productive activities, such as agriculture and livestock rearing. Subsistence hunting implies a minimum level of resources which is constant albeit extremely poor, while in the survival hunting, the hunted resource is the only resource available for survival (Donaldson 1988; Harrington 1981; Kenny et al. 2018; Kobayashi Issenman 1997). Subsistence hunting may in fact represent a prelude to a hunting economy (Peres et al. 2006; Vizina and Kobei 2017), passing through a phase of sustainable subsistence hunting, that is to say hunting which slowly ceases to be a forced requirement and then able to plan the exploitation of animal resources.

This also occurs because agriculture and livestock rearing bring about a change in thinking; in fact both are based on a certain level of planning and forethought, a sense of saving (of seeds, food, and breeding stock) and, therefore, of a "forward-looking" use of the resource. A change with regard to the awareness of traditional hunting is made difficult, by the weight of lore and the related rituals which conflict with the need to change, even in the case of excessive withdrawals and environmental and climate change. Culture and tradition are a powerful curb on change and not only in this field.

Changing a traditional hunt with deep cultural roots related to the economy is a very difficult job (e.g., the poaching of European honey buzzard, *Pernis apivorus*, in Calabria (Italy) (Agostini et al. 1999). It is simpler to prohibit it, at least apparently, but there is a risk of a profound resistance including revenge poaching. When hunting takes advantage of modern technologies (e.g., seal hunting with carbines by the Inuit), violations are not only the norm but take on an even greater danger for the resource in question. Some ancient hunts are strongly ritualized and inserted into a popular or noble tradition and have been progressively emptied of their cruelest aspects (e.g., foxhunting on horseback or the Gran Venerié with hounds used for hunting red deer or the falconry, etc.; Poplin 1987, Dickson et al. 2009).

To a greater or lesser extent all these forms of hunting possess an entertainment element, cultural ritualization, a sense of belonging, and physical exercise but their contribution to conservation, however, is lacking.

In conclusion, if they do, survival and subsistence hunting preserve only involuntarily, regardless of whether they are ritualized, cultural, or not.

8.6 Hunting That Is Both Impactful and Eliminatory: Impactful and Eliminatory Hunting (IEH)

All "primitive" populations (Martin 1984; Diamond 1999, 2006) have destroyed many species of mammals and birds who were not used to the presence of the humankind. The only populations of megafaunal communities to have escaped this extinction are those that have coevolved with early humans (in Africa and perhaps part of Asia, according to Martin 1984 and Diamond 2006).

But even in very recent times many other species have been brought to extinction by hunting. It is also true that in these cases other causes have often contributed and that in some cases hunting has proved to be less important than in others. Environmental destruction, the domestication of the species (the aurochs, *Bos primigenius*, and the tarpan, *Equus ferus ferus*; Hainard 1949) with economic or social and cultural needs, are the main causes of disappearance, but it must be said that hunting has often dealt the coup de grace. Today, however, it seems that hunting is almost irrelevant while other causes appear much more important, except in cases where hunting is closely connected with economic needs such as the Asiatic lion (*Panthera leo persica*) (Quammen 2003) or cultural one such as a status symbol (horns of black rhinoceros).

8.7 Impactful but Resilient Hunting (IRH)

Hunts of this kind allow a certain increase or a natural change in the ecosystem but it is necessary to distinguish two separate scenarios: (1) unaware (unaware conservation, UC) due to a lack of technology or the extent of pressure and (2) aware (aware conservation, AC) for different reasons which will be examined. UC is typically involuntary but may have lasting effects which can also be positive even from a psychological perspective, in the sense that it may end up motivating individuals or human groups who are interested in continuing to hunt noninvasively (though their reasons may vary) and they therefore undertake a range of actions, the simplest of these being to exclude other subjects from harvesting the resource and to therefore operate a sort of surveillance that may be informal or even professional.

In addition to these activities, which are the simplest and the most common, there may be others concerning the environment, its general management, and that of other species that ranges from monitoring to control. In short, and not without reserve and caution, this represents a passage from an unconscious "conservative" hunting to a conscious and programmed one, but with the warning that in many cases the objective is partial and specific (i.e., red deer, *Cervus elaphus*) and which therefore entails the elimination of species perceived as detrimental (e.g., gray wolf, *Canis lupus*).

However, the conservation of one species to the detriment of another is not real conservation.

It is worth pointing out that this culling is not hunting but represents a set of actions aimed at eliminating the (from a human perspective) "harmful" effects of an animal population (Scholes and Smart 2008). Professionals or even volunteers may carry out the control, but the merely recreational part of the hunting is absent or reduced, even if it may also be present in some forms, especially when someone is volunteering. The control itself aims at efficiency and effectiveness while the hunting is aimed at personal satisfaction. When hunting is mainly for direct economic purposes (the production of consumable meat) it becomes close to control. A very clear difference, however, does not seem to exist; in fact there are several intermediate forms, especially because of the operator's subjectivity, and a good definition should be based on the extremes, in this case entertainment versus the elimination of damage.

Regulated hunts have at least one common basic idea: the protection of some species or their non-extinction, which does not mean they represent the same thing. Professionals (e.g., wildlife managers, WM) and hunters themselves therefore distinguish between hunts that are "well carried out" from those that are not (Fukuda et al. 2011).

In ancient times, with the beginning of agriculture and animal husbandry, the first Homo sapiens felt the need to furnish themselves with rules either because hunting was reserved for powerful people (and they reserved it for themselves through severe punishments) or because, in hunting, those rules of prescient use were applied, having been borrowed from their own agro-zootechnical practices and obviously dependent on the various cultures and requirements. Hunting was a male activity representing virility training and might have been preparatory for war as well as representing a sphere of freedom and therefore one with as few rules as possible. According to its virile and warlike rules, hunting was more important the more a trophy was coveted and dangerous. Hence the interest of the ruling classes (rulers) as a symbol of their powerful status, but even in a such context hunting, although reserved for a select few, did not coexist with conservation. This said, sometimes in Europe single species were preserved from massive harvesting and humankind witnessed the enactment of edicts, true and proper regulations, aimed at ensuring that hunting could still continue.

Since hunts using poor technology coexist quite well with conservation, it is worth remembering a tendency, particularly prevalent in the United States, to voluntarily renounce certain advantages that the hunter has over the prey which includes hunting with muzzle-loading rifles and bowhunting, equipment that, in fact, requires much more skill on the part of the hunter and in some ways gives more chance for the prey to escape (see Case Study 3 - Oregon).

8.8 Special Situations Involving IRH

In Europe, particularly in countries that make reference to a Central European tradition, hunting has a managerial character that is explained by the phrase "Jagd ist Hege" (Hunting is management) (Silva-Tarouca 1927, in Raesfeld 1958). The meaning is that hunting is an "entertainment" economy and part of the overall management of natural resources, in which the forest (and forestry) plays the major role. In this sense, the hunter does not simply buy a permit (as in the USA) and neither is he interested above all in the result (as an expression of great skill), but believes he has obtained the prey because, in a sense, he has taken care of it, "loving it" (Hegen in German, Reimoser 2001 and 2018a).

This approach directly involves the hunter – often only the owner of the hunting rights – in the entire set of management activities including censuses, environmental interventions, etc., and is particularly the case in German-speaking countries and surrounding areas (Linder 1978). It has profoundly influenced an attitude toward all the activities connected with hunting, starting with cynophilia.

An approach of this kind is very important in cultural terms and represents a subjective attitude toward nature (with some deviations, which we will address later). We can distinguish three types of approach to hunting. These are the hunter (*venator*) *dominus*, the *venator socius*, and the *venator emptor*.

The *venator dominus* is the one that directly manages and organizes hunting because the law allows him to exclude other subjects. Generally, he is the landowner or the owner of the hunting rights and runs and/or rules them alone or at most with a few other colleagues, friends, or relatives, never more than 3–5 people with a range of responsibilities. The list of hunts of this type covers the royal and noble hunts (Howe 1981; Ortega y Gasset 1972) to those of landowners or even owners in *consortium*, a formal association of contiguous landowners operating as a single unit (such as the private reserves, *cotos* in Spain, etc.).

The *venator socius* is associated, i.e., registered, to a specific district and hunts along with others in an ongoing enterprise (with social hunting objectives) and pays a periodic fee, usually annual. Depending on personal inclination, time constraints, and individual resources, he (for he is usually but not exclusively male) is involved to a greater or lesser extent in the management, but the situation is not always obvious. The contexts greatly vary at the global level and it ranges from associates who know each other (with a maximum of about 40 individuals involved), this is because the objective is restricted, and therefore a certain degree of self-regulation is facilitated, to other cases in which the members barely know each other at all, because they hunt on areas covering tens of thousands of hectares.

A smaller number of members still allows for some sort of self-management. Naturally this is a matter that applies to a group of hunters who are known to each other, often living close to one another and who share language, family ties, and social relationships and a context in which social control, very strict and personal, is particularly effective (Dunbar 2004, 2010).

The *venator emptor* often hunts abroad or in any case far from his place of residence, buying hunting rights from time to time, thus investing large amounts of money. This category fits very well with the category of trophy hunters (Gunn 2001).

There are also mixed approaches in which *venatores domini* are often also *emptores*, mainly because they can afford to, while ***socii*** are more rarely emptores or in some cases – and not in Europe – are associated de facto; in other words they are forced to hunt in a given territory for a range of circumstances, generally due to financial constraints or other types of necessities.

NIH-type hunts can evolve positively toward impactful contributory ones (ICH), even though the opposite is also often the case, with the degradation of an ICH toward a "flat" coexistence with conservation, of a passive type. For example, sometimes the Central European hunters tend to see gray wolves and/or lynx as an obstacle to good management results (Treves 2009).

A particular case is that of trophy hunting (TH). TH implies the hunt of animals with specific characteristics which can be stored as a memento of the adventure including antlers, horns, tusks, skulls, skins, or even the entire stuffed specimen. The characteristics of these trophies have been codified internationally, initially with the "Records of Big Game" (1895), then with the "Records of North American Big Game" (1932), and, finally, with the rules of the Conseil International de la Chasse, CIC (1934), currently still in force and which unify the Anglo-Saxon, North American, and European systems.

This kind of revolution (definable, because TH establishes the quality of trophies with objective criteria, which allow for a comparison between them) was made possible by the progress of war and hunting weapons in the late nineteenth century (smokeless gunpowders, progressive and special optics), which allowed the adoption of sniper rifles.

Unlike the Anglo-Saxon/North American school, the Mitteleuropean school has made its "livestock" its flag (Raesfeld 1957; Ückermann 1952) being based, at least at the beginning, on concepts derived directly from animal husbandry and more precisely on the idea that it is possible to improve the quality of a species (from a trophy perspective) by eliminating subjects with undesirable characteristics. Basically, a sort of special selection, so much so that some authors had adopted the slogan of "breeding with the rifle" (Rehwildhege mit der Büchse – Roe Deer Management with the rifle – Wagenknecht 1976). This criterion was called into question after 1960 (Bubenik 1970; Elsmann 1971; Hennig 1961; Jelinek 1989; Kurt 1991; Stubbe and Passarge 1979) but especially post-2000 (Apollonio et al. 2010) and at present is considered non realiable.

Nowadays, with the new opportunities for mobility, TH has assumed global dimensions. With reference to the following chapter (impactful contributory hunts, ICH) and above all to an extra-European examine, it can be argued that this form of hunting can be considered compatible on condition that some essential biological rules are respected:

- The harvesting of endowed subjects takes place in very low percentages, thus avoiding affecting the average quality of the population (Perco 2014).
- The avoidance of harvesting during the mating season and before.
- Harvesting even during the mating season but only a modest fraction of the qualitatively important subjects.
- Harvesting only very old subjects at the end of their reproductive careers. This is a case of sound (biological) effectiveness for Bovidae but not practicable for Cervidae.

It should however be remembered that rare species (and subspecies) should be managed with particular caution. The trophies of rare species or subspecies have a greater value than the most common ones and this could encourage excessive harvesting (Palazy et al. 2012).

8.9 Impactful but Contributory Hunting (ICH)

We now address the positive contribution of ICH to conservation and for which it is possible to make a distinction depending on the fields of intervention and the modalities for which there is conservation or not. WL is preserved, more precisely:

By interventions on wildlife:

1. Reintroductions
2. Control
3. Eradication

By environmental improvements:

1. Habitat restoration or habitat maintenance
2. Supplementary feeding

By interventions on (local) communities:

1. Through the activation of anti-poaching surveillance
2. Through monitoring
3. Through the improvement of the social conditions of local residents
4. Through an improvement in the attitude toward wildlife (nonresidents)

8.10 Interventions on Wildlife

1. *Reintroductions*

   In this case, we mean the reintroduction of species that will be huntable in the future and for which the intervention is explicitly or at least indirectly done for this reason. Coexistence and awareness are high and so, obviously, these correspond to an effectiveness if the planning and operations have been carefully carried out. One risk is to make the community in question, of hunters, but not exclusively suspect that a reintroduction may serve as a pretext or opportunity to create a protected area with obvious consequences (including hunting prohibitions and limitations of various types). Therefore good communication is necessary.
2. *Control*

   By this term we mean the control of antagonistic species, excluding domestic ones (and not part of wildlife) in a numerical or structural sense. This may prove a very contradictory measure, as it is aimed at strengthening species that can be hunted to the detriment of other species and is generally quite effective, but the level of coexistence/awareness is very low, as it tends to label species as useful or harmful and is thus recommended only for "opportunist" and/or those of least concern (LC) including magpie (*Pica pica*), crows (*Corvus* spp.), and wild boar (*Sus scrofa*) or European rabbit (*Oryctolagus cuniculus*) in certain contexts. However, the numerical reduction of some highly competitive species (*Cervus elaphus*, *Loxodonta Africana*, etc.) may also prove necessary or indispensable for environmental reasons (Perco 2001; Reimoser 2001, 2018b; Scholes and Smart 2008; Slotow et al. 2008). In reality there are quite a few species that, in certain contexts (environmental, social, etc.), can be considered "problematic" and that require a sort of control carried out, sometimes using lethal means (Angelici 2016).

3. *Eradication*

   The complete elimination of a species in a given area is necessary for IAS (Invasive alien species), such as coypu, *Myocastor coypu*, and gray squirrel, *Sciurus carolinensis*, in Europe. In very particular circumstances, eradication is also recommended for some native species (gray wolf, according to Mech 2017), when it comes to highly problematic species in very delicate socioeconomic or cultural contexts. In some cases, the collaboration of hunting organizations was essential for the success of project management (raccoon *Procyon lotor*, Dahl et al. 2000; muskrat *Ondatra zibethicus*, Gosling and Baker 1989).

8.11 Environmental Improvements

1. *Habitat Restoration or Habitat Maintenance*

   For example:

   Advanced forestry management (high forest introduction, naturalistic cutting of coppice, extirpation of allochthonous species, thinning, etc.)

   Restoration and care of natural meadows and pastures

   Controlled fires

   Care of wetlands (flooding, reed cutting, grazing, etc.)

   Intentional and programmed abandonment of open areas

   The level of coexistence/awareness is high but is directed toward certain (huntable) species even if it can indirectly favor many other species. The effects are positive.

2. *Supplementary Feeding*

Wildlife crops. These are specific cultivations aimed at strengthening certain (huntable) species or reducing damage by the species in the nearby areas. The level of coexistence/awareness is good but oriented (see above) and the effectiveness is comparable with environmental improvements. A contraindication is that these interventions can sometimes increase certain species in an abnormal fashion.

Direct provision of food. This includes several food attractants in addition to water and salt or fodder treated with fragrant essences, etc. The resultant increase in huntable species (e.g., ungulates) has a clear purpose, with greater harvests in loco (Raesfeld 1958; Reimoser 2001). Initially, this was carried out for ungulates (in Europe), but now it involves many other species including ducks and geese (France, Italy, Argentina, etc.) and brown bear (Romania). In the case of ducks, it allows excessive and never-before-seen bags and leads to the culling of protected species (Boyd 1990; Perco Fa and Perco Fr 1992). Although it has clear management approaches, the level of coexistence/awareness is ambiguous or modest and the overall effects may be negative for conservation. Open-pit carrion dumps, fixed or movable, to attract carnivores or scavenging bird species, are useful when they are not associated with hunting, even if they have been subjected to severe criticisms in the case of certain mammals such as the grizzly subspecies of brown bear (Stringham 1989; Craighead 1998).

8.12 Interventions on the Local Community

1. *Activation of anti-poaching surveillance.* This is mainly aimed at protecting huntable species but its effectiveness is nevertheless high because it concerns all species (Roe et al. 2017). The level of coexistence/awareness is very good with only one criticality to be suspected, or worse accused, of acting against the interests and/or traditions of local communities (Corry 2011, 2015). The effectiveness remains high.
2. *Performing monitoring.* Unlike the previous intervention, this is effective only for the species monitored, therefore those being hunted, even if for these the level of coexistence/awareness is high. Indirectly it may also require monitoring other species but this is very unusual. One critical aspect is that if it is not carried out with the collaboration of local stakeholders, then their interest (see also the following points) in the conservation of the resource will decline. The case described applies to venator emptor and not for members who put themselves forward to carry out the monitoring, which is not always in a reliable fashion if they are not supported and/or organized by professional technicians. The case of venator domini is different because it can be convenient to involve local actors, on large areas; but this situation is very rare in Europe and elsewhere; it only applies if the dominus is represented by large landowners (Africa) or institutions (governmental and not) (M. Fabris pers. comm. 2018). Anyway, a minimum and essential level of monitoring is generally a duty for all states (Selier and Di Minin 2015).
3. *Improvement of Residents' Social Conditions*

   Economic improvements. These can be divided into three categories: compensation or reimbursement for damage, income integration, and targeted improvements in traditional productive activities (agriculture and livestock rearing). Furthermore, to the extent that the natural capital has been retained, payments for ecosystem services ensure benefits and well-being.

   (http://ec.europa.eu/environment/integration/research/newsalert/pdf/30si_en.pdf)

   Compensation or reimbursement. Compensation presupposes responsibility while reimbursement represents only a pro bono pacis contribution, but not always, according to rates and/or appraisals. Without going into the context of the various very different disputes which depend on the juridical regime of the matter and the "ownership" of wildlife, it should be noted that the level of coexistence/awareness is modest or absent and the efficacy is rarely appreciable as the person who has suffered damage feels underestimated and generally cheated with regard to the size, promptness, and the method of compensation. These are procedures to be used only when there is a lack of better initiatives, as they do not create "alliances." In addition, wildlife is always perceived as damage and "it would be better if it wasn't there."

Income integration. This is such only if it takes place before any incidents or requests and could almost be considered a type of rent paid in advance to local communities. The level of coexistence/awareness, always meaning from those paying it (the venatores), is average or even good but everything depends upon the way in which it is paid. An appreciable personal relationship between the parties can at least be established, but often it is only provided for the purpose of "keeping them quiet" so that they do not demand subsequent compensation. In theory it is an effective method but it depends very much on the venator's ability to establish a relationship with the potentially damaged parties (Nguinguiri et al. 2017).

Targeted improvements in traditional productive activities (agriculture and livestock rearing). These are by far the best in creating coexistence/awareness on both sides and are also highly effective in establishing good relationships, the key being to make people understand that wildlife is a resource that helps the local community if certain rules are followed. A problem, and not a small one, is balancing agriculture and/or animal rearing with wilderness or naturalness, because if we support additional activities (see over) they can, merely by their spread, jeopardize the very conservation they sought to obtain, and this process can therefore be a double-edged sword.

4. *Cultural Improvements*

These include the improvement of education and culture and therefore also knowledge about wildlife as a resource. This is an excellent way to obtain good results in the future and the degree of coexistence/awareness is definitely high. In one way it accomplishes the improvement in social conditions and the life of local communities and represents a particularly effective and long-term solution if individuals deal with problems, related to education, integration, gender equality, and demography, that is to say with respect for culture and local sensibilities. One risk, which is not unimportant, may prove to be a possible (probable) increase in the local population with obvious consequences on conservation. It is true that a greater number of residents aware of the economic and cultural value of wildlife is much better than the opposite, but nevertheless there are limits to man-made density beyond necessity, which may mean the disappearance and/or the reduction in biodiversity that we were seeking to protect. It is therefore similar to the previous case (that of economic improvements) of a poorly managed instrument that may become a double-edged sword. At least theoretically, cultural improvements are more effective even over the long term than economic ones, which may lead, in not a few marginal situations, to phenomena of hoarding and exploitation, where reliable public authorities and/or effective social controls are lacking (Bocci 2011; Di Minin et al. 2015; Leader-Williams and Hutton 2005).

8.13 Creating an Improvement in the Approach to Fauna (Nonresidents)

Tourism, or, even better, aware tourism, can reduce the use of wildlife through hunting when it is carried out in the same areas. Nevertheless, mixed uses of wildlife should not be ruled out (tourism, especially but not only photographic tourism and hunting, etc.) for which there may be a temporal or (partial) geographical displacement of the two activities. The first consists of a type of tourism outside the hunting season, while the second one includes protected areas on the edge of areas used for hunting. These are possible and manageable hypotheses even for hunting managers and for this reason we mention that for iconic species (especially in Africa) tourism requires, however, "[...] political stability, proximity to good transport links, minimal disease risks, high density wildlife populations to guarantee viewing, scenic landscapes, high-capital investment, infrastructure (hotels, food and water supply, waste management), and local skills and capacity" (IUCN 2016: 8), and this creates many difficulties, even if they may be considered highly complementary activities. Wetlands and areas suitable for ungulates may play an effective role by allowing the use in part of the structures planned for hunting as observation points, during moratoria periods. The difficulties involved in these compromises should not be underestimated as the tourists' approach is generally contrary to hunting and to the hunters themselves although their organizations (tourism, bird-watching, etc.) prefer not to come to terms with the hunting associations. However, these are possibilities that should be explored with the aim of increasing mutual awareness and tolerance, combined with sound knowledge; moreover, there are some educational projects to be borne in mind to at least explore.

8.14 The Problem of Trophy Hunting Especially for the *Venator Emptor*

Trophy hunting is a recent development around the middle of the twentieth century, so well after the invention of the "rifle." Hemingway (1935) in North America and Raesfeld (1958) in Europe mention it in their writings.

Trophy hunting is based on choice and is therefore typically selective, using the excellent optical instruments available. Certainly its subjective motivations are the

Table 8.1 Hunters in the world (estimate)

Continent	Population (2016)	Area	Density	Hunters ([a])	% hunters/ population	Hunter density
Europe	743,100,000	993,525,900	75	7,950,000	1.1%	0.80
Asia	4,393,296,000	4,461,522,000	98	17,500,000	0.4%	0.39
Africa	1,216,000,000	3,022,153,200	40	3000,000	0.2%	0.10
North America	579,000,000	2,470,900,000	23	23,500,000	4.1%	0.95
South America	422,500,000	1,884,100,000	22	4,250,000	1.0%	0.23
Oceania	36,100,000	852,598,900	4	550,000	1.5%	0.06
Total	**7,389,996,000**	**13,684,800,000**	**54**	**56,750,000**	**0.77%**	**0.41**

[a]Source: FACE, WIKI, expert estimation and personal estimate

proof of great skill or luck, a symbol of status which is nice to share with a few friends and some enthusiasts, albeit in a very personal way.

The search for the most "beautiful" or greatest value trophy, following a score validated by an international commission, the CIC, is constant. Local and world exhibitions of trophies are important events for the experts in the subject and may represent real competitions with a slightly sporty flavor, even if hunting is not a sport in the proper sense.

Only a fraction, 10–20%, according to T. Terzi (2018) (pers. comm.), of world hunters are interested in trophy hunting. Therefore, they vary between 5.7 and 11.3 million. It is believed that the percentage of hunters in the EU is 20–25% higher than this total, namely, around 1.3 million (a personal estimate). A significant part of these (more than 60%) mainly hunt in their own country (Table 8.1).

TH has often been criticized, especially recently, when a beloved lion named Cecil was killed by an American dentist in Zimbabwe (2015), sparking fierce criticism from animal rights groups and fueling a backlash that set social media ablaze with condemnation and death threats. Criticisms are of different kinds: technical, economic, social, and ethical (Aryal et al. 2015; Buckley et al. 2015; Cohen 2014; Coltman et al. 2003; Crosmary et al. 2015a, b; Dahles 1993; Di Minin et al. 2016; Festa-Bianchet et al. 2014; Kaltenborn et al. 2013; Kurt 1991; Leader-Williams et al. 2005; Lindberg et al. 2003; Lindsey et al. 2005, 2006, 2007a, b, 2012; Mysterud 2011; Nelson et al. 2013; Norton 1984; Palazy et al. 2012; Ripple et al. 2016; Treves 2009).

In particular, they are critical of the following elements:

- *From a technical point of view,* the excessive numbers culled, the culling of subjects at the peak of their reproductive powers, the destructuring and disturbance when widespread (e.g., in Europe), the vulnerability of certain species, the absence of specific studies, and the fact that there are no forecasts looking at consequences on the ecosystem.

- *From an economic point of view,* the distribution of profits, the absence, or the low sums destined for reinvestment in conservation projects, while it is suggested that aware ecotourism may be an alternative option.
- *From a social point of view,* corruption and/or political instability, the possibility of circumventing the rules, the scarcity of controls, and the charismatic value of some species. It is also assumed that TH is a boost to poaching.
- *From an ethical point of view,* the existence of animal rights, hunting versus canned animals, the fact that TH is elitist and/or confirming a certain white supremacy, and an indicator of colonialism. Corry (2015) supports the notion that even the parks in the USA are a colonial construct, a judgment that is certainly exaggerated.

The criticism of TH from this perspective is often a condemnation of hunting in general, sometimes from very particular points of view (ecofeminism: Kheel 1996) with nods toward anthropology and traditions (Howe 1981), to the relationship with the landscape and prey (Reis 2009) and to respect for the prey (Taylor Ang 1996): "There are circumstances in which the killing of a human being may be justified, but to mount this person's head on a wall is not usually [!] considered acceptable."

Remaining with the point of view of ethics, Gunn states (2001): "I assume animals have interests, and that we have an obligation to take some account of those interests: roughly, that we are entitled to kill animals only in order to promote or protect some nontrivial human interest and where no reasonable alternative strategy is available."

Others however consider the hunt from the moral point of view (Causey 1989; Loftin 1984, 1988; Leopold 1949; Ortega y Gasset 1972; Shepherd 1973; Taylor Ant 2004, 2009; Vitali 1990) with greater or lesser intensity.

Cohen (2014), again quoting Gunn (2001) argues that: "None of the Gesinnungsethik (Weber 1919) approaches, is completely consequential in its prioritization of animal life; all concede that hunting animals might become ethically permissible under certain circumstances…The Verantworungtsethik (again Weber 1919) representatives [on the other hand] of an environmental ethics in fact sought to define these circumstances."

Apart from the problem of the morality of hunting, it should be noted that the definition of TH (according to Gunn 2001, Cohen 2014, and others already quoted) is rather reductive and incomplete. In them, they tend to believe that the motivations of the trophy hunters (THs) are all the same or very similar and therefore formulate a condemnation or especially ethical perplexity, based on the concept that people conducting TH are very rich people who pay just to kill for fun or who enjoy killing defenseless/innocent animals. THs would mostly be shooters, within game parks or farms, but if we analyze the phenomenon it is not so because THs include a lot of domestic hunters (both in the USA, in Canada, and in Europe). But the relative appreciation of the aforementioned authors for the sport's hunter that implements a kind (kind of) "fair chase" with the prey is also very debatable and even negative, at least for the hunter-manager of the Mitteleuropean school (Bubenik 1984; Elsmann

1971), that is to say the hunters who are directly involved in the management of their own hunting (see over, Raesfeld 1958).

Above all for the ethical reasons mentioned above, the proposals include its replacement with ecotourism, strong restrictions or a total ban, and a ban on the import of trophies (a request of some members of the European Parliament to the EU, 2016).

Other authors are much more drastic (Harrison et al. 2007). In addition Corlett (2007) says: "Over the last 50 yr, the importance of hunting for subsistence has been increasingly outweighed by hunting for the market. The hunted biomass is dominated by the same species as before, sold mostly for local consumption, but numerous additional species are targeted for the colossal regional trade in wild animals and their parts for food, medicines, raw materials, and pets. …. Most of this hunting is now illegal, but the law enforcement is generally weak. However, examples of successful enforcement show that hunting impacts can be greatly reduced where there is sufficient political will. Ending the trade in wild animals and their parts should have the highest regional conservation priority." But their findings concern Asia and suggest rather a hunting moratoria and/or systems which are more aware.

Technical, economic, and, in part, social criticisms are generally accepted, but the overwhelming majority of authors above quoted, and technicians claimed, that these faults do not cancel out the usefulness of TH.

In particular, in 2016 the International Union for Conservation of Nature issued a document entitled "Informing decisions on Trophy Hunting. A Briefing Paper for European Union Decision-makers regarding potential plans for the restriction of imports of hunting trophies," which highlights the substantial opportunity of TH.

- Not only the IUCN (2016) but other authors as well (Cooney et al. 2017; Di Minin et al. 2013a, 2016; Festa-Bianchet 2017; Leader-Williams and Hutton 2005; Leader-Williams et al. 2005; Lindsey et al. 2006, 2012; Roe et al. 2017; Webb et al. 2004) argue that, even if there are unpleasant situations, there are also several reasons to support a coexistence with, and indeed the usefulness of, TH, including that:
- A total ban would see important financial resources lost (at all).
- These resources are indispensable for conservation and for the fight against poaching (Roe et al. 2017).
- The resources, taken out the operators' remuneration, generally go 50% to the local communities and for the rest to the government agencies (in Namibia 90% goes to the local communities, IUCN 2016).
- The resources are fundamental in order to help local communities understand that it is worth to preserve wildlife through reintroduction projects (Makombe 1993; Webb et al. 2004).
- The resources are also in the form of "consumable meat" (bushmeat) and as such they are "essential" for populations living in total destitution (Kupiers et al. 2016; M. Pani pers. comm. 2018).
- Ecotourism is not practicable everywhere, particularly for security reasons (Kiss 2004; IUCN 2016).

- Ecotourism requires many investments (IUCN 2016) and "customers" would also be easy to please (Di Minin et al. 2013a; Goodwin and Leader-Williams 2000; Krüger 2005; Novelli et al. 2006).
- Many hunters prefer poorly accessible areas and to enjoy the wilderness of these places (Lindsey et al. 2006).
- TH (in Africa) has a lower ecological footprint than tourism (Di Minin et al. 2013a, 2016).
- There are several cases in which TH has had positive effects on conservation (Kock 1996). The IUCN document (2016) cites ten case studies of TH having positive conservation and livelihood benefits: rhinos in South Africa and Namibia, argali in Mongolia, bighorn sheep *Ovis canadensis* in North America, private wildlife lands in Zimbabwe, communal conservancies in Namibia, markhor *Capra falconeri* and urial *Ovis orientalis vignei* in Pakistan, and markhor in Tajikistan. Moreover it has benefits to nontarget threatened species and revenues for government wildlife agencies, including for anti-poaching and polar bears in Canada (IUCN 2016).
- Where there has been a total ban on hunting, the wildlife situation not only has not improved but has even worsened, even if the reasons for this vary (Western et al. 2009).
- Interventions of similar importance are not feasible, on an ongoing and non-episodic basis (IUCN 2016).
- There is no need to renounce these activities for reasons, above all the ethical ones, in exchange for proposals that are not feasible everywhere: it is not good to throw out the baby with the bathwater (Cooney et al. 2017).

8.15 Case Studies of Hunting Related to Positive or Critical Consequences to Conservation

8.15.1 Case Study 1 Italy

Italy covers an area of 302,073 km^2, of which 23.2% is mountain, 41.6% is hill, 35.2% is intensely cultivated plains, and about 8.000 km^2 of which is wetlands (ISPRA 2011). There are roughly 6,5 million inhabitants with a density of 200/km^2, among the highest in the EU. Italy has a very high wildlife biodiversity. The area covered by hills and mountains is suitable for the ungulates with at least 20,000 roe deer present in lowland areas in various populations.

Ungulates have increased about 200-fold since 1945, for environmental reasons (abandonment of the uplands) and currently (2017) there are estimated to be almost 2 million animals (mostly roe deer and wild boar; see Perco 2014). There have been several reintroductions of roe deer, red deer, and chamois both in hunting areas and in protected areas (60%). The gray wolf and the brown bear (the latter in two sub-species) are estimated at about 2000 and 110 individuals, respectively. The hunters' contribution to the management of ungulates is positive where the hunting is selective (about 75% of the territory occupied by roe deer) and negative or neutral elsewhere. Wild boar is hunted mainly through drives with several hounds. 85% of the territory is suitable for the species (Carnevali et al. 2009; Pedrotti et al. 2001; Perco 2014).

10.5% of the Italian territory is within protected areas (MATTM 2010), 7,2% in urban areas (ISPRA 2015), and 6,9% in other non-huntable areas (ISTAT 2007), while the huntable area covers about 222,800 sq. km (about 75% of the national surface area). Hunting is regulated at the national level by a framework law with regional application laws except in the regions and provinces with special status (5/20). Eighteen of the 20 Italian regions are divided into territorial hunting areas (THA), of a social nature and on a provincial basis (one or more per province), each of at least 1000 square kilometers, with an unlimited number of hunters. Trentino-Alto Adige (and the autonomous provinces of Trento and Bolzano: Flaim and Brugnoli pers. comm. 2018) and Friuli Venezia Giulia (100% of the territory: Colombi pers. comm. 2018) as well as other provinces in the Eastern Alpine area (four, part of their territory) instead have social hunting reserves on a municipal basis. Private hunting reserves occupy around 5% of the area in question.

In Italy, fauna is an "unavailable state resource" and the landowner cannot prevent hunting on the land he/she owns, an almost unique situation in Europe. There are 48 huntable species, 13 mammal species, and 35 bird species.

The 580,000 Italian hunters spend an average of USD 500 (€425)/year for membership fees, regional taxes, and national licenses. The figure, which is significant (USD 290 million/€255 million), goes into the general financial management of the regions and/or the state and it does not concern conservation.

In order to become a hunter, a regional examination is required, with a differentiated difficulty depending on the region in question. A special course and exam are compulsory for selective hunting. Most hunters form part of the group of "migratory hunters" who do not contribute directly to the maintenance of habitats, with some exceptions (but taking large hunting bags) in the private hunting reserves of the Venetian lagoon. Poaching on all species is very high (Merli et al. 2017; Perco and Forconi 2016) especially in central southern Italy, in Sicily and Sardinia, but also in other regions (birds). Annual estimates of hunted species as well as bags are not available. The largest hunting association (Italian Hunting Federation, FIDC, covering about 50% of the Italian hunters) is carrying out some wildlife research.

The current national hunting law does not oblige the carrying out of censuses or even annual management reports, a situation that is judged as outdated together with the absence of the owner's consent for the use of their own land for hunting purposes.

In addition, due to the shortcomings of the law, we would therefore like to suggest that only in the province of Trento and in that of Bolzano does hunting coexist with conservation, owing to the specific managerial methods provided under their autonomous system. This is also the case but with some exceptions for several Alpine municipal hunting reserves. In just a few districts of Emilia Romagna and, very locally, in Tuscany, hunting does not seem to openly be in conflict with conservation itself. As a trend, hunting management would be deemed, at least, passable in the region of Friuli Venezia Giulia, the only region in which the closed number of hunters is established by law (7923 on 6724 sq. km, divided into 247 municipal hunting reserves).

8.15.2 Case Study 2 Hunting, Wetland Restoration, Game Conservancy

In North America (United States and Canada) "Ducks Unlimited" is very active (Boyd 1990). It collects and invests huge funds for the re-naturalization and restoration of wetlands. Ducks Unlimited is an association that has operated since 1937, mainly in North America (Canada, United States, and Mexico, but also in Australia, New Zealand, etc.) particularly in the field of wetland restoration and reactivation.

It collects funds not only from its members (over 1000,000) but also through donations and research activities of various kinds. Since its foundation it has contributed to the re-naturalization of about 14 million acres (more than 5.6 million ha.) for waterfowl in general. An important component of DU is represented by hunters, as well as breeders of Anatidae, birdwatchers, etc. Again in the USA important re-naturalization work has been carried out by the army, some quite close to New York and elsewhere and covering thousands of hectares.

The Wildfowl and Wetlands Trust is a similar organization, founded in 1946 by Peter Scott under the name of the Severn Wildfowl Trust, which currently runs several wetland centers in the UK and promotes the field of protection and restoration of wetlands and its associated fauna worldwide. The strategy of conservation followed in this case is based on hunting areas or full protection, extended as a fragmented patchwork which represents the only truly effective model for the management of migratory or wintering waterbirds that are also game species. In this case hunters have also been, and still are, an active component of the organization since its inception (Lampio 1974, 1982).

Wetlands International (WI, formerly the International Waterfowl Research Bureau) is the most internationally active European organization in the field of research with projects aimed at conserving and increasing biodiversity, organizing technical meetings and conferences, etc. WI provides consultancy to wetland managers, including state offices as well as within the framework of the International Convention of Ramsar and the European Natura 2000 regulations (which do not exclude hunting).

Furthermore, other organizations of this kind exist at the international level. In Europe, but not only, hunting is often carried out on properties managed by private individuals. Therefore, many wildlife management organizations provide advice, even if they use state funds to some extent. For example "ICONA" (Instituto Nacional por la Conservation de la Naturaleza) in Spain is an important example where individuals deal with all kinds of wildlife. Similar institutions and organizations, with similar aims and to a large extent financed by private individuals, exist in other European countries, for example, in France, Germany, Austria, and Italy.

Particular attention should be paid to the large areas falling within the wetland areas of the Upper Adriatic and similar "private game reserves" elsewhere in Italy, where the owners of fishing valleys have largely moved from fish production to hiring out annual hunting pegs as a more profitable enterprise (USD 62,000, €50,000, or even more for the rent of a barrel per year). These initiatives are based on massive feeding programs and allow game bags sometimes numbering more than 200 ducks per day (Perco Fa pers. comm. 2018).

Since there is no real control, even on the killing of protected species or on the use of nontoxic ammunition (not containing lead), these activities are largely incompatible with conservation (Perco Fa and Perco Fr 1992).

In the case of huge game bags, in fact, the degree of involvement of the hunting world seems ambiguous, because although it is true that in the past there were significant efforts to preserve wetlands in the most suitable environmental conditions for hosting the typical fauna, in the rented or privately owned areas, today there is a growing tendency to favor intensive winter feeding of game species (Anatidae) to facilitate huge individual daily game bags. This tendency increases the habit of considering migratory or wintering wildlife essentially as moving targets for entertainment shooting (Bell and Owen 1990; Bregnballe and Madsen 2004; Fox and Madsen 1997). A remedial or mitigation action must be found in line with international experiences, with the establishment or maintenance of large areas of tranquility, reducing intensive winter feeding and increasing controls. The recent use of unmanned aerial vehicles (UAVs) to police these areas in the Po Delta is an interesting development in this respect.

Again based in Great Britain, but with important initiatives at the international level, we should finally cite and appropriately highlight the activity of the "Game & Wildlife Conservation Trust" (formerly the "Game Conservancy"), an authoritative organization that operates in various sectors especially (though not exclusively) involving small game of interest for hunting by tradition and with economic implications, especially through interventions in agriculture, both in relation to wetlands and in cultivated or uncultivated areas (Keddy 2010; Tamisier 1985).

8.15.3 Case Study 3 Oregon (USA)

Oregon is one of the 50 states (in addition to the federal district of Washington) that make up the United States. It is on the west coast and it covers 255,000 square kilometers with about 4 million inhabitants (15.5/q km).

It is one of the states with the greatest environmental diversity, that is to say volcanoes, abundant water bodies, dense evergreen and mixed forests, as well as high deserts and semiarid shrub lands.

The "big game" hunting fauna consists of wapiti *Cervus canadensis* (two subspecies), three species of deer (*Odocoileus* spp.), black bear *Ursus americanus*, pronghorn *Antilocapra americana*, mountain goat *Oreamnos americanus*, bighorn sheep (two subspecies), and puma *Puma concolor*. Game bird hunting includes a variety of Anseriformes (ducks and geese) and upland birds such as wild turkey *Meleagris gallopavo* (two subspecies), grouse (six species), quail (two species), pheasant *Phasianus colchicus*, gray partridge *Perdix perdix*, and chukar *Alectoris chukar*, these last three being introduced.

Hunting is managed by a state agency, the Department of Fish and Wildlife (ODFW), which oversees all aspects of wildlife hunting and research. The decision-making and executive power is a matter for a commission, the Fish and Wildlife Commission, composed of seven members appointed by the governor of the state which determines hunting, regulations, fees, and license costs, the latter diversified by age (veteran, senior, junior, etc.). The enforcement of rules and regulations is ensured by a special section of the Oregon State Police – assisted locally, when necessary, by other agencies.

The US federal agencies have jurisdiction only on the lands for which they are responsible. In the case of many Western states (Oregon, Idaho, Nevada, Washington, etc.) almost half or more of their territory is owned by the National Forest Service or the Bureau of Land Management. Some other federal laws regulate possession and use of weapons, the use of lead in hunting, movement of motor vehicles, access and use, etc.

It is worth pointing out that in the USA each state has its own particular structure with substantial differences:

- In some states landowners are involved in hunting programs.
- In others, the owner can "manage in his own way" the game that lives on his land or comanage it with the authorities.
- In other cases (gated property), they have an almost exclusive powers of control.

The licenses for the culling of some wild animals (over-the-counter tags) can be purchased directly (together with the annual hunting license) at the point of sale or online. For other wild animals, the hunter purchases a "lottery" ticket and the culling licenses are assigned by extraction (at random). Some permits are linked to particular territories (hunting units); other permits depend on the type of game, meaning a hunter can hunt in larger areas or even the entire state. A "preference point" system gives a percentage advantage to those hunters who have not won in previous years: 14–16 points for the pronghorn and 2–3 for the mule deer (*Odocoileus hemionus*).

The cost of the license ranges from USD 160 (nonresidents), USD 32 (residents) to USD 6 (pioneers, i.e., 50 years of residence in Oregon) and even free for disabled war veterans. Young people pay USD 10 and the minimum age to get the license is 9–12 years old but they must be accompanied by an adult aged 21 or over (unless they hunt on lands owned by a parent or a legal guardian).

Oregon has about 200,000 hunters (5% of total population). It is worth noting that the percentage of hunters is lower in urban areas, while it is much higher in rural areas.

A course is required, for individuals that are 18 years of age and younger, to purchase the yearly hunting license. This hunter education course has an average duration of 14–16 hours and requires both classroom time and a "live fire" test in the field. In recent years, several school districts have allowed hunting education courses to be offered, as after-school programs. These courses are taught in school by volunteers certified as hunter education instructors who are often school teachers.

The hunting seasons vary depending on the species and the type of weapon (bow, pistol, muzzle-loading rifle, shotgun, or carbine) which is determined based on the lethality of the mean: more lethal, e.g., carbine = less time. The maximum period concerns migratory game as well as ducks and geese (about 90–120 days), black bear (3 months), and the puma (11 months).

In Oregon, hunting activities and wildlife conservation practices coexist in a delicate context of increased urbanization and a slow decrease of participation in hunting activities. However, despite these general challenges, hunters are still strongly committed to earning their hunting privileges by paying out of their own pockets (by purchasing hunting licenses, tags, P&R Act, and other fees) for education, habitat restoration, law enforcement, wildlife research, and management programs. This model (the North American Wildlife Management and Conservation System) of direct involvement and support has made good contributions to the preservation and protection of the wildlife patrimony and of hunting heritage (Mahoney 1995).

8.15.4 Case Study 4 Safari Club International

Safari Club International (SCI) is an international organization of hunters that was founded in 1972 in the USA.

The aim of SCI is to protect the freedom to hunt and therefore to promote wildlife conservation. Today SCI has about 50,000 members (84% USA, 6% Canada, 5% Europe, 4% other countries) and 200 local branches.

The Safari Club International Foundation, a branch of SCI, promotes international projects to restore wildlife conservation and to fight against poaching and believes that wildlife conservation would not be possible without hunters (Safari Club 2017).

Between 2000 and 2016, SCI Foundation invested USD 60 million in conservation, education, and humanitarian services and funded over 80 wildlife conservation projects in more than 27 countries. (Safari Club International Foundation, pers. comm).

For example, between 2008 and 2016:

- North America. The investment was more than USD 3 million including 21 species in 16 US states, four Canadian provinces, and Mexico. Featured projects: Michigan predator-prey, Alaska wood bison reintroduction, Missouri black bear, Newfoundland woodland caribou, and Arizona desert bighorn sheep reintroduction.
- Africa. The investment is more than USD 3.5 million including 12 species in 14 countries. Featured projects: African Wildlife Consultative Forum (AWCF), Tanzania lion, Zambia lion, Namibia leopard, and Tanzania Ruaha buffalo.
- Asia. The investment is more than USD 1 million including ten species in five countries. Featured projects: Tajikistan argali sheep and snow leopard projects in Mongolia, Pakistan, and Russia.

SCI published the *Safari Club International Record Book*, which is the largest record keeping system of this kind, in the world. Trophies are measured and listed according to the size of horns and antlers, to where and how they are taken, and to their specific quality and form.

Most SCI members are interested in big game animals but they are not the majority of trophy hunters. A rough estimate of this category is about 10–20% (8.5 million) of hunters around the world (T. Terzi pers. comm. 2018).

SCI has been often criticized for supporting the hunting in high fenced game ranches where, for example, less valuable species are replaced by more valuable species and predators are therefore persecuted. In addition, fencing on game ranches fragments wildlife populations, leading to the disruption of dispersal and migratory movements (Ripple et al. 2016).

Other criticalities are the giving of awards for hunting leopards, elephants, lions, rhinos, and buffalos in Africa. SCI responds to this second problem by saying hunting provides needed funds for habitat preservation and enhancement and that SCI has enhanced the propagation or survival of several species by rescuing these species from near extinctions and providing the founder the stock necessary for reintroduction.

The first problem is actually not a simple question because SCI members are debating "if and how" to evaluate the trophies from "canned wildlife" (animals in enclosure). Therefore, recently SCI has been opposing the hunting of African lions bred in captivity.

SCI supports the thesis of "use it or lose it" talking about wildlife conservation, i.e., hunting and opposition to poaching. "If hunting tourism is suspended, instead of having legal hunting, there will be illegal hunting," says Dr. Adelhelm Meru, Tanzania's Permanent Secretary for the Ministry of Natural Resources and Tourism.

8.15.5 *Case Study 5 Trophy Hunting in Sub-Saharan Africa*

This activity covers about seven countries that extend over about 5.5 million km^2 (18% of Africa) with about 180 million inhabitants (15% of Africa's current population) and a density of 30 inhabitants per km^2, less than that of Africa as a whole continent (around 40/km^2) (See Table 8.2).

Table 8.2 Human population estimate in some African countries

Countries	Area (km²)	Total population (million)	Density (people per km²)
Botswana	581,726	2.25	3.87
Mozambique	801,590	28.83	35.96
Namibia	825,418	2.03	2.46
South Africa	1,221,037	57.16	46.82
Tanzania	947,303	55.60	58.69
Zambia	752,618	16.60	22.06
Zimbabwe	390,757	16.15	41.33
Total	**5,520,449**	**178.62**	**30.17**

The protected areas cover about 29% (ranging from 6% in South Africa to 43% in Namibia) while game ranches cover 17.5% (ranging from 10.5% in Mozambique to 26% in Tanzania). In this part of Africa (about a fifth of the total area) the annual income deriving from trophy hunting for the trophy itself, generally the big four (elephant, buffalo, lion, and leopard) and other species as impala (*Aepyceros melampus*), warthog (*Phacochoerus africanus*), greater kudu (*Tragelaphus strepsiceros*), hippo (*Hippopotamus amphibius*), zebra (*Equus quagga*), chacma baboon (*Papio ursinus*), Nile crocodile (*Crocodylus niloticus*), and lechwe (*Kobus leche*), from 2009 to 2013 amounted to USD 217.2 million, with USD 40 per km2 (ranging from USD 5 to 60, Zambia and Tanzania) (Di Minin et al. 2013a; Di Minin 2016).

Considering that the minimum wage in this part of Africa is about USD 3000 per year, the figures would seem quite important, especially in Namibia and Botswana, precisely to support the local economy, even because it is believed that in general within the Campfire projects about 75% of the income (of the hunting for trophies) goes to local populations, according to credible local sources. Botswana, however, banned hunting in 2014, with negative results for conservation due to the increase in poaching and culling through PAC [problem animal control] (on lions attacking livestock and elephants for damage to agriculture or for breaking fences in order to enter farms).

In general, the illegal market of parts of the body of endangered and charismatic species has assumed incredible dimensions with rhinoceros (*Diceros bicornis* and *Ceratotherium simum*) horn dust estimated on the black market to about USD 60–100,000 per kg (Goga and Salcedo Albarán 2017). Worked and inlaid ivory on the Asian market costs up to USD 10,000 per kg according to Gao and Clark (2014).

The problem of the fight against poaching (Roe et al. 2017) as a control of species, which are harmful to agriculture and other communities' traditional activities, is at a turning point, in the sense that they are part of the human-wildlife conflict (HWC), which is not a recent concern in Africa. Resources are necessary for the repression of poaching, but if local populations believe that wildlife is a source of damage (and often it is so, from their point of view), then every battle for conservation is lost (Di Minin et al. 2015).

There are a significant number of trophy hunting programs where indigenous or local communities have freely chosen to use trophy hunting as a tool to provide the incentives and revenue to help them conserve and manage their wildlife and/or improve their livelihoods (IUCN 2016).

Furthermore, the communities benefit directly from TH through the concessions or other investments deriving from it, which can be invested in schools or clinics, as well as in employment directly dependent on the organization of TH including guides, game guards, wildlife managers, and other hunting-related employment. In addition, the hunting product is also a consumable resource (meat) and in many communities there are no alternatives (Lindsey et al. 2006).

For these above reasons, all the coexistence with the conservation of TH is essential and given some critical issues, Di Minin et al. (2016) suggest a set of proposals:

1. Mandatory levies should be imposed on safari operators by governments so that they can be invested directly into trust funds for conservation and management.
2. Eco-labeling certification schemes could be adopted for trophies coming from areas that contribute to broader biodiversity conservation and respect animal's welfare concerns.
3. Mandatory population viability analyses should be done to ensure that harvests cause no net population declines.
4. Post-hunt sales of any part of the animals should be banned to avoid illegal wildlife trade.
5. Priority should be given to fund trophy hunting enterprises run (or leased) by local communities.
6. Trust funds to facilitate equitable benefit sharing within local communities and promote long-term economic sustainability should be created.
7. Mandatory scientific sampling of hunted animals, including tissue for genetic analyses and teeth for age analysis, should be enforced.
8. Mandatory 5-year (or more frequent) reviews of all individuals hunted and detailed population management plans should be submitted to government legislators to extend permits.
9. There should be full disclosure to public of all data collected (including the sums levied).
10. Independent government observers should be assigned randomly and without forewarning on safari hunts whenever they take place.
11. Trophies must be confiscated and permits must be revoked when illegal practices are disclosed.
12. Backup professional shooters and trackers should be present for all hunts to minimize welfare concerns.

We might agree with these opinions but they should be completed with a strategy that could provide cultural, social, and economic improvement to the existing communities where it would take place.

8.16 Conclusion

Coexistence/compatibility between hunting and conservation may depend upon the type of management. If the harvest rates are conservative, scrupulously monitored by wildlife authorities and abided by the hunting operators or by the managers themselves (venatores socii), most of the problems described can be avoided.

However, this requires good organizational skills, political stability, and constant and accurate research (long-term monitoring systems that integrate the social and financial benefits of trophy hunting for local communities and consider the costs and benefits of different conservation alternatives) (Crosmary et al. 2015b), but these circumstances are not guaranteed everywhere in the world.

It should also be remembered that hunters can use their experience and ability to solve problems such as the management and control of problematic species and of invasive alien species. In a world more and more modified by humankind, this social role played by hunters may essentially be for the conservation of nature, of biodiversity, of ecosystem services, and of the future itself of hunting.

The awareness that WL is an important resource must be the main objective of management (hunting), fighting poaching and illegal trade of body parts of rare and endangered species, and know the objectives of criminal markets, this especially being the case in Africa and Asia.

Criticisms of an ethical nature directed at TH, but also to hunting in general, should be listened to very carefully but also placed in the right context. One may wonder, for example, if it is better supporting ethics which, despite various compromises, manage to achieve important conservation goals, rather than ethics that place their own values first and simply say "Oh well" to reduced conservation, as a result of the achievement of the moral high ground.

Personally, I am for practical ethics, of the first type.

Conservation is not just a technical and ethical problem. Its implications are economic, social, and cultural. Interventions that are not effective from an economic point of view have profound negative consequences regardless of their technical correctness, leaving aside ethical considerations.

When wildlife management contrasts with the culture, habits, and expectations of local populations, it is impossible not to reflect on the social problems which must be resolved, at the same time or even a priori. This requires a great deal of attention by considering wildlife management as a discipline that cannot do without social studies and communication.

Inevitably, there will be compromises, with the aim of an "ethically" winning result: that is, however, conservation, and that's it.

At this point, collaboration between a range of groups with different cultures, habits, and values is necessary and the world of research and conservation cannot refuse to work with them, even if their ethics do not always correspond to their own.

On the other side, another similar cultural problem is conceiving sustainable hunting as a way of giving more value to biodiversity, wildlife, and habitats – i.e., the nature capital itself. Conceiving hunting as part of the ecosystem services potentially allows to increase, and in any case to enhance, some natural resources. Otherwise, in our modern world, these resources may risk to be trivialized, exactly because of the lack of an adequate and widespread nature conservation culture.

It goes without saying that the slogan dear to Webb et al., (2004), "use or lose it," can be applied to every situation, although it is necessary, however, to accept that conservation problems cannot be solved with moral positions for or against hunting. The same is also stated by Makombe (1993) "Use or non-use is not the issue; sustainable use is."

The increase in human population in Africa (and not only) will seal the fate of many species. In 2050 Africa is expected to reach 2.5 billion inhabitants (update to 2018 is 1.29 billion) with an 11-fold increase compared to 1950. For a long time the natural environment has been in a state of suffering due to environmental changes (Blackburn et al. 2016; Niamir-Fuller et al. 2012; Said et al. 2006; Wronski et al. 2015; Yurco 2017) caused above all by pastoralism, agriculture, deforestation, the increase in traffic and urban centers, etc.

Gunn (2001) is also of the same opinion which also has no sympathy for TH: "As Africa's population continues to grow, and habitat shrinks, pressure on wildlife will increase. Africans, like Western environmentalists, are entitled to a materially adequate standard of life. They cannot and should not be expected to protect wildlife if it is against their interests to do so. The only feasible strategy to protect the interests of both wildlife and people is one that integrates conservation and development, as in Zimbabwe. Whatever we may think of trophy hunting, and I share the distaste of serious sports hunters for it, at present it is a necessary part of wildlife conservation in Southern Africa."

This does not have to mean "letting down your guard" on improper and harmful hunting methods. The problems must also be tackled to the extent that they can be solved and the extension of good hunting practices is always opportune, or rather a duty.

However, I believe that we would be making a big mistake if the abovementioned best practices did not also deal with supporting education, equal rights, and an improvement of the cultural conditions of the communities in question (Bocci 2011). This is not a new brand of Western colonialism, but above all a counseling action, by taking into account existing local traditions.

Governments, bodies, and hunting organizations have the responsibility to include it in a process of social maturity, which is not only economic but also cultural and educational, for a new environmental awareness. And if, as I believe, the greatest danger is the population increase on the planet, it is highly likely that improvements in culture and education can at least slow this growth.

However, it is true that, from an individual perspective, it is easier and more reassuring to hunt in uncontaminated places, after paying a sum, and then having the trophy (venator emptor) delivered to the house, uninterested in anything that is not a trophy, completely ignoring, with utter superficiality, the social and environmental conditions of the area where it was hunted with such great effort. But an uninterested hunter such as this is an enemy of conservation.

The alternative is scraping the bottom of the barrel, looking for more and more striking hunting results (venator dominus and venatores socii).

In any case, the aware hunters' burden must be something else. He must contribute to conservation preserving the ecosystem services which are essential for him, and he must also worry about the economic, social, and cultural difficulties of those who live in the areas in which he hunts.

Not just as a lasting testament but also because this is the right way to go about it.

Acknowledgments I am particularly grateful to Fabio Perco (Co.Na., Trieste, Italy), Paul Tout (Italy), Maurizio Valerio (Oregon, US), Tiziano Terzi and Elena Fileppo (SCI Italian Chapter), Marco Festa-Bianchet (Sherbrooke University, Canada), Mauro Fabris (Safaris & Co Pty Ltd), Marco Pani (IUCN CEESO/SSC), and Giacomo Nicolucci (EPS, Italy). I would also like to thank Piero Genovesi (ISPRA, Italy), Sandro Lovari (University of Siena, Italy), Silvano Toso (formerly ISPRA, Italy), Sandro Flaim (UNCZA, Italy), Alessandro Brugnoli (ACT, Trento, Italy), Dario Colombi (Regional Authority Friuli Venezia Giulia Italy), Michele Sorrenti, and Marco Ramanzini (FIdC, Italy), Federico Liboi Bentley, Guenther Mittenzwei, Vittorio Taveggia (Italy), and especially Eleonora Stortoni (Italy).

References

Agostini N, Logozzo D, Colero C (1999) The orientation/navigation hypothesis: an indirect evidence in migrating honey buzzards, *Pernis apivorus*. Riv ital Orn, Milano 69:153–159

Andersson E, Barthel S, Ahrne K (2007) Measuring social-ecological dynamics behind the generation of ecosystem services. Ecol Appl 17:1267–1278

Angelici FM (ed) (2016) Problematic wildlife. A cross-disciplinary approach. Springer, Cham

Apollonio M, Andersen R, Putman R (eds) (2010) European ungulates and their management in the 21st century. Cambridge University Press

Aryal A, Dhakal M, Panthi S et al (2015) Is trophy hunting of bharal (blue Sheep) and Himalayan tahr contributing to their conservation in Nepal? Hystrix Ital J Mamm 26:85–88. https://doi.org/10.4404/hystrix-26.2-11210

Bateman IJ, Harwood AR, Mace GM et al (2013) Bringing ecosystem services into economic decision-making: land use in the United Kingdom. Science 341:45–50

Bauer N, Wallner A, Hunziker M (2009) The change of European landscapes: human-nature relationships, public attitudes towards rewilding, and the implications for landscape management in Switzerland. J Environ Manag 90:2910–2920

Bell DV, Owen M (1990) Shooting disturbance - a review. In: Matthews GVT (ed) Managing waterfowl populations. Proc. IWRB Symp. Astrakhan 1989. IWRB Spec. Publ. 12, Slimbridge, pp 159–171

Blackburn S, Hopcraft JGC, Ogutu JO et al (2016) Human–wildlife conflict, benefit sharing and the survival of lions in pastoralist community-based conservancies. J Appl Ecol 53:1195–1205

Bocci A (2011) La caccia al trofeo come strumento di gestione e conservazione. Cacciare a palla (VIII) 1:12–19

Bookbinder MP, Dinerstein E, Arun R et al (1998) Ecotourism's support of biodiversity conservation. Conserv Biol 12:1399–1404

Boyd H (1990) Hunting and the reported kill of duck and geese in the USA and Canada. In: Matthews GVT (ed) Managing waterfowl populations. Proc. IWRB Symp. Astrakhan 1989. IWRB Spec. Publ. 12, Slimbridge

Bregnballe T, Madsen J (2004) Tools in waterfowl reserve management: effects of intermittent hunting adjacent to a shooting-free core area. Wildl Biol 10:261–268. https://doi.org/10.2981/wlb.2004.031

Bubenik AB (1970) Rehwildhege und Rehwildbiologie. Der Deutsche Jäger 5–8: 109–112, 137–141, 168–172, 193–196, 389–394

Bubenik AB (1984) Ernährung, Verhalten und Umwelt des Schalenwildes. BLV, München Wien Zuerich

Buckley R, Mossaz A (2015) Hunting tourism and animal conservation. Anim Conserv 18:133–135

Buijs AE, Elands BHM, Langers F (2009) No wilderness for immigrants: cultural differences in images of nature and landscape preferences. Landsc Urban Plan 91:113–123

Cahill GF Jr (1998) Survival in starvation. Am J Clin Nutr 68:1–2

Carnevali L, Pedrotti L, Riga F, Toso S (2009) Banca Dati Ungulati. Status, distribuzione, consistenza, gestione e prelievo venatorio delle popolazioni di Ungulati in Italia. Rapporto 2001–2005. Biol. Cons. Fauna, 117:168

Causey AS (1989) On the morality of hunting. Environ Ethics 11:327–343

Cohen E (2014) Recreational hunting: ethics, experiences and commoditization. Tour Recreat Res 39:3–17

Cooney R, Freese C, Dublin H et al (2017) The baby and the bathwater: trophy hunting, conservation and rural livelihoods. Unasylva 249:3–15

Corlett RT (2007) The impact of hunting on the mammalian Fauna of tropical Asian forests. Biotropica 39:292–303

Corry S (2011) Tribal peoples for tomorrow's world. Freeman Press, Alcester

Corry S (2015) The colonial origins of conservation. http://www.truth-out.org/opinion/item/32487-the-colonial-origins-of-conservation-the-disturbing-history-behind-us-national-parks. Accessed 8 May 2019

Craighead JJ (1998) Status of the Yellowstone Grizzly Bear population: has it recovered, Should it be delisted? Ursus 10:597–602

Crosmary WG, Côté SD, Fritz H (2015a) The assessment of the role of trophy hunting in wildlife conservation. Anim Conserv 18:36–137

Crosmary WG, Côté SD, Fritz H (2015b) Does trophy hunting matter to long-term population trends in African herbivores of different dietary guilds? Anim Conserv 18:117–130

Dahl F, Åhlén PA, Swartström J, et al (2000) Management of the invasive Raccoon Dog *(Nyctereutes procyonoides)* in the north-European countries LIFE09 NAT/SE/000344. Swedish Association for Hunting and Wildlife Management, Öster-Malma, 611 91 Nyköping

Dahles H (1993) Game killing and killing games: an anthropologist looking at hunting in a modern society. Soc Anim 1:169–184

Di Minin E, Fraser I, Slotow R, MacMillan DC (2013a) Understanding heterogeneous preference of tourists for big game species: implications for conservation and management. Anim Conserv 16:249–258

Di Minin E, Macmillan DC, Goodman PS et al (2013b) Conservation businesses and conservation planning in a biological diversity hotspot. Conserv Biol 27:808–820

Di Minin E, Laitila J, Montesino Pouzols F et al (2015) Identification of policies for a sustainable legal trade in rhinoceros horn based on population projection and socioeconomic models. Conserv Biol 29:545–555

Di Minin E, Leader-Williams N, Bradshaw CJA (2016) Banning trophy hunting will exacerbate biodiversity loss. Trees 31:99–102

Diamond J (1999) Guns, germs, and steel: the fates of human societies, 1st edn. W. W. Norton & Company, New York City

Diamond J (2006) Collapse. How societies choose to fail or succeed. Penguin, New York

Dickson B, Hutton J, Adams WA (eds) (2009) Recreational hunting, conservation and rural livelihoods: science and practice. Blackwell Publishing Ltd, Hoboken, New Jersey

Donaldson JL (1988) The economic ecology of hunting: A case study of the Canadian Inuit. Dissertation, Harvard University

Duda R (2017) Ethnoecology of hunting in an empty forest practices, local perceptions and social change among the Baka (Cameroon). Dissertation, Universitat Autònoma de Barcelona
Dunbar RIM (2004) Gossip in Evolutionary Perspective. Rev Gen Psychol 8:100–110
Dunbar RIM (2010) How many friends does one person need? Dunbar's number and other evolutionary quirks. Faber and Faber, London
Elsmann H (1971) Rehwildhege und das Knopfbockproblem. F.C. Mayer, München
Ferretti F, Corazza M, Campana I et al (2015) Competition between wild herbivores: reintroduced red deer and Apennine chamois. Behav Ecol 26:550–559. https://doi.org/10.1093/beheco/aru226
Festa-Bianchet M (2017) When does selective hunting select, how can we tell, and what should we do about it? Mamm Rev 47:76–81
Festa-Bianchet M, Ray JC, Boutin S, Côté SD et al (2011) Conservation of caribou *(Rangifer tarandus)* in Canada: an uncertain future. Can J Zool 89:419–434. https://doi.org/10.1139/z11-025
Festa-Bianchet M, Pelletier F, Jorgenson JT et al (2014) Decrease in Horn size and increase in age of Trophy Sheep in Alberta Over 37 Year. J Wildl Manag 78:133–141
Fox AD, Madsen J (1997) Behavioural and distributional effects of hunting disturbance on waterbirds in Europe: implications for refuge design. J Appl Ecol 34:1–13. https://doi.org/10.2307/2404842
Freeman AM (2003) The measurement of environmental and resource values. Theory and methods. RFF press, Washington, DC
Fukuda Y, Webb G, Manolis C et al (2011) Recovery of saltwater crocodiles following unregulated hunting in tidal rivers of the Northern Territory, Australia. J Wildl Manag 75:1253–1266
Gao Y, Clark SG (2014) Elephant ivory trade in China: Trends and drivers. Biol Conserv 180:23–30
Giles RH Jr (1978) Wildlife management. Freeman & Co, S. Francisco
Goga K, Salcedo Albarán E (2017) Background on South Africa, rhino poaching and rhino horn trafficking. The global observatory of transnational criminal networks - research Paper No. 8. Vortex Working Papers No. 22. Vortex Foundation, Bogotà, 18 p
Goodwin H, Leader-Williams N (2000) Tourism and protected areas – distorting conservation towards charismatic megafauna? In: Entwhistle A, Dunstone N (eds) Priorities for the conservation of mammalian diversity: has the panda had its day? Cambridge University Press, Cambridge, pp 257–275
Gosling LM, Baker SJ (1989) The eradication of muskrats and coypus from Britain. Biol J Linn Soc. https://doi.org/10.1111/j.1095-8312.1989.tb01561.x
Gossow H (1983) Zur geschichtlichen Entwicklung der Beziehungen zwischen Jagd und Waldwirtschaft. Centralblatt für das gesamte Forstwesen 100:191–207
Gunn AS (1994) Environmental ethics and tropical rainforests: should greens have standing? Environ Ethics 16:21–40
Gunn AS (2001) Environmental ethics and trophy hunting. In: Stephens G (ed) Ethics and the environment, vol 6, Number 1. Indiana University Press, Bloomington, Indiana, pp 68–95
Hainard R (1949) Les Mammifères sauvages d'Europe, II. Neuchatel, Paris, p 274
Harrington R (1981) The Inuit – Life as it was. Hurtig, Edmonton. ISBN 0–88830–205-3
Harrison RD, Sreekar R, Brodie JF, Brook S, Luskin M, O'Kelly H, Rao M, Scheffers B, Velho N (2016) Impacts of hunting on tropical forests in Southeast Asia. Conserv Biol 30:972–981
Hawksworth DL, Bull AT (2007) Vertebrate Conservation and Biodiversity. Springer International, Cham
Hennig R (1961) Das Rehwildproblem in neuer Sicht. P. Haupt, Bern, p 64
Howe J (1981) Fox hunting as ritual. Am Ethnol 8(2):278–300
ISPRA (2011) Contributi per la tutela della biodiversità delle zone umide. Rapporto 153, p 461
ISPRA (2015) Il consumo del suolo in Italia. Rapporto 259, p 290
ISTAT Istituto Nazionale di Statistica (2007) Agricoltura e Zootecnia. Caccia. Tavv K03 and K04

IUCN (2016) Informing decisions on trophy hunting. A briefing paper for European Union decision-makers regarding potential plans for restriction of imports of hunting trophies. IUCN Report: 1–19

Jeffers ES, Nogué S, Willis KJ (2015) The role of palaeoecological records in assessing ecosystem services. Quat Sci Rev 112:17–32

Jelinek R (1989) Hege und Bewirtschaftung des Rehwildes in der Kulturlandschaft. Leopold Stocker Verlag, p 145

Kaltenborn BP, Andersen O, Linnell JDC (2013) Is hunting large carnivores different from hunting ungulates? Some judgments made by Norwegian hunters. J Nat Conserv 21(5):326–333. https://doi.org/10.1016/j.jnc.2013.05.004

Keddy PA (2010) Wetland ecology: principles and conservation. Cambridge University Press, Cambridge

Kenny TA, Fillion M, Simpkin S, Wesche SD, Chan HM (2018) Caribou (Rangifer tarandus) and inuit nutrition security in Canada. EcoHealth 15(3):590–607. https://doi.org/10.1007/s10393-018-1348-z

Kheel M (1996) The killing game: an ecofeminist critique of hunting. J Philos Sport 23:30–44

Kiss A (1990) Living with wildlife: wildlife resource management with local participation in Africa. Washington, DC: World Bank Technical Paper No. 130, Africa Technical Department Series

Kiss A (2004) Is community-based ecotourism a good use of biodiversity conservation funds? Trends Ecol Evol 19:232–237

Kobayashi Issenman B (1997) Sinews of survival – the living legacy of inuit clothing. UCB Press, Vancouver

Kock MD (1996) Zimbabwe: a model for the sustainable use of wildlife and the development of innovative wildlife management practices. In: Taylor VJ, Dunstone N (eds) The exploitation of mammal populations. Springer, Dordrecht. https://doi.org/10.1007/978-94-009-1525-1_1

Krüger O (2005) The role of ecotourism in conservation: panacea or Pandora's box? Biodivers Conserv 14:579–600

Kurpiers LA, Schulte-Herbrüggen B, Ejotre I, Reeder DM (2016) Bushmeat and emerging infectious diseases: lessons from Africa. In: Angelici FM (ed) Problematic wildlife. Springer, Cham, pp 507–551. https://doi.org/10.1007/978-3-319-22246-2_24

Kurt F (1991) Das Reh in der Kulturlandschaft: Sozialverhalten und Ökologie eines Anpassers. Parey, Hamburg und Berlin, p 284

Lampio T (1974) Hunting rationalisation studies. Riistatieteellisia Julkaisuja, Finnish Game Research Helsinki, 34–60

Lampio T (1982) Improvement of methods and practices in waterfowl hunting. Wetland Management, I.W.R.B.:302–307

Le Bel S, La Grange M, Czudek R (2016) Managing human–elephant conflict in Zimbabwe: a boundary perspective rather than a problematic species issue. In: Angelici (ed) Problematic wildlife. a cross-disciplinary approach. Springer, Cham, pp 123–142. https://doi.org/10.1007/978-3-319-22246-2_7

Leader-Williams N, Hutton JM (2005) Does extractive use provide opportunities to reduce conflicts between people and wildlife? In: Woodroffe R, Thirgood SJ, Rabinowitz A (eds) People and WILDLIFE: CONflICT OR COEXISTENCE. Cambridge University Press, Cambridge, pp 140–161

Leader-Williams N, Milledge S, Adcock K, Brooks M, Conway A, Knight M, Mainka S, Martin EB, Teferi T (2005) Trophy Hunting of Black Rhino: Proposals to Ensure Its Future Sustainability. J Int Wildl Law Policy 8(1):1–11. https://doi.org/10.1080/13880290590913705

Leopold A (1949) Game Management. 1996 reprint by Natraj Publisher, p 481

Lindberg K, James B, Goodman P (2003) Tourism's contribution to conservation in Zululand: an ecological survey of private reserves and public protected areas. In: Aylward B, Lutz E (eds) Nature tourism, conservation, and development in KwaZulu-Natal, South Africa. The World Bank, Washington, DC, pp 203–240

Lindner K (1978) Geschichte des deutschen Weidwerks (2013) De Gruyter. Reprint first ed. (1937) p 441
Lindsey PA, Alexander RR, du Toit JT, Mills MGL (2005) The potential contribution of ecotourism to African wild dog Lycaon pictus conservation in South Africa. Biol Conserv 123:339–348
Lindsey PA, Alexander R, Frank LG, Mathieson A, Romañach SS (2006) Potential of trophy hunting to create incentives for wildlife conservation in Africa where alternative wildlife-based land uses may not be viable. Anim Conserv 9:283–291
Lindsey PA, Frank LG, Alexander R, Mathieson A, Romañach SS (2007a) Trophy hunting and conservation in Africa: problems and one potential solution. Conserv Biol 21(3):880–883
Lindsey PA, Roulet PA, Romañach SS (2007b) Economic and conservation significance of the trophy hunting industry in sub-Saharan Africa. Biol Conserv 134:455–469
Lindsey PA, Balme GA, Booth VR, Midlane N (2012) The significance of African lions for the financial viability of trophy hunting and the maintenance of wild land. PLoS One 7:e29332
Loftin RF (1984) Morality of Hunting. Environ Ethics 6:241–250
Loftin RF (1988) Plastic Hunting and Real Hunting. Behav Polit Anim Stud 1:317–323
Mace GM, Norris K, Fitter AH (2012) Biodiversity and ecosystem services: a multilayered relationship. Trends Ecol Evol 27:19–26
Mahoney SP (1995) The role of hunters in the conservation of wildlife. Paper presented at the fourth annual governor's symposium on North America's hunting heritage, August, 1995, Green Bay, WI, USA. 73–80
Makombe K (ed) (1993) Sharing the land: wildlife, people and development in Africa. IUCN/ROSA Environmental Series No. 1. Harare, IUCN/ROSA, and Washington DC, IUCN/SUWP, p 36
Martin PS, Klein RG (eds) (1984) Quaternary extinctions: a prehistoric revolution. University of Arizona Press, Tucson, p 892
MATTM (2010) Censimento delle aree protette in Italia. Ministero dell'Ambiente e della Tutela del Territorio e del Mare
Mech LD (2017) Where can wolves live and how can we live with them? Biol Conserv 210:310–317
Meriggi A, Lombardini M, Milanesi P, Brangi A, Lamberti P, Giannini F (2016) Management of Wild Boar in protected areas: the case of Elba Island. In: Angelici FM (ed) Problematic Wildlife. Springer, Cham, pp 229–251. https://doi.org/10.1007/978-3-319-22246-2_11
Mysterud A (2011) Selective harvesting of large mammals: how often does it result in directional selection? J Appl Ecol 48:827–834
Nelson F, Lindsey PA, Balme GA (2013) Trophy hunting and lion conservation: A question of governance? Oryx 47(4):501–509. https://doi.org/10.1017/S003060531200035X
Nguinguiri JC, Czudek R, Julve Larrubia C, Ilama L, Le Bel S, Angoran EJ, Trebuchon JF, Cornelis D (2017) Managing human–wildlife conflicts in central and southern Africa. Unasylva 249 668:39–44
Niamir-Fuller M, Kerven C, Reid R, Milner-Gulland E (2012) Co-existence of wildlife and pastoralism on extensive rangelands: competition or compatibility? In: Pastoralism: research, policy and practice. Springer, pp 1–14. https://doi.org/10.1186/2041-7136-2-8
Norton BG (1984) Environmental ethics and weak anthropocentrism. Environ Ethics 6:131–148
Novelli M, Barnes JI, Humavind M (2006) The other side of the ecotourism coin: consumptive tourism in Southern Africa. J Ecotour 5(1&2):62–79
Ortega Y, Gasset J (1972) Meditations on Hunting. Charles Scribner's and Sons, New York
Palazy L, Bonenfant C, Gaillard JM, Courchamp F (2012) Rarity, trophy hunting and ungulates. Anim Conserv 15:4–11
Pedrotti L, Duprè E, Preatoni D, Toso S (2001) Banca Dati Ungulati: status, distribuzione, consistenza, gestione, prelievo venatorio e potenzialità delle popolazioni di Ungulati presenti in Italia. Biol Cons Fauna 109(p):132
Perco Fa, Perco Fr (1992) Waterfowl hunting pressure and regulations in Italy with special reference to the lagoons of the Upper Adriatic. In: Managing Mediterranean wetlands and their birds. IWRB Special Publication 20:23–27

Perco Fr (1997) Problemi gestionali delle popolazioni immesse. In: Spagnesi M, Toso S, Genovesi P (eds), Atti del III Convegno Nazionale dei Biologi della Selvaggina. Suppl. Ric. Biol. Selvaggina XXVII: 123–53

Perco Fr (2014) Il Capriolo. Vol I e vol II. CTO, Vicenza, p 210 and 115

Perco Fr (2017) Il controllo faunistico con particolare riferimento al Cinghiale nelle Aree Protette. In: Achille G (ed) Scritti in onore di Franco Pedrotti. Collana Natura e aree protette Temi, Trento 32:136–171

Perco Fr, Forconi P (2016) Seasonal pattern of the wolf (Canis lupus) population in Italy. III National Congress Problematic Wildlife, Cesena 24–26 ott 2016

Perco Fr, Carmignola G, Pasolli C, Pedrotti L (2001) Progetto Cervo. Programma triennale di indagine e sperimentazione per una gestione del Cervo nel Parco Nazionale dello Stelvio - rapporto conclusivo. Consorzio del Parco Nazionale dello Stelvio, p 150

Peres CA, Hilton S, Nascimento HS (2006) Impact of game hunting by the Kayapo´ of south-eastern Amazonia: implications for wildlife conservation in tropical forest indigenous reserves. Biodivers Conserv 15:2627–2653. https://doi.org/10.1007/s10531-005-5406-9

Poplin F (1987) La découpe et le partage du cerf en Vénerie françois. Anthropozoologica, 1987, Premier Numéro Spécial, pp 19–20

Quammen D (2003) Monster of God: the man-eating predator in the jungles of history and the mind. W. W. Norton. Pimlico, New York, p 528

Raesfeld F (1957) Das Rotwild. Parey, Hamburg und Berlin, p 385

Raesfeld F (neubearbeitete Auflag von Frevert W) (1958) Das deutsche Waidwerk: Lehr- und Handbuch der Jagd. Parey, Hamburg und Berlin, p 497

Reimoser F (2001) Waldbau, Wildökologie und Jagdstrategien. In: Deutscher Forstverein e.V. (ed) Ein Wald für alle Fälle. Kongressbericht, Dresden, pp 424–430

Reimoser F (2018a) Die Jagd als wirtschaftlicher Faktor. In: Dietlein J, Froese J (eds), Jagdliches Eigentum, Band 17 der „Bibliothek des Eigentums', Springer-Verlag

Reimoser F (2018b) Wildschadensproblem und Forst-Jagd-Konflikt im Alpenraum – Hintergründe, Entwicklungen, Perspektiven. Game-damage problem and forest-hunting conflict in the Alpine region – background, developments, perspectives. In: Verein zum Schutz der Bergwelt e.V. (ed) Jahrbuch Verein zum Schutz der Bergwelt, München, Germany, 83, 2, pp 61–116

Reis AC (2009) More than the kill: hunters' relationships with landscape and prey. Curr Issue Tour 12, 5/6:573–587. https://doi.org/10.1080/13683500903042881

Ripple WJ, Newsome TM, Kerley G (2016) Does Trophy Hunting Support Biodiversity? A response to De Minin et al. Trees 2093:1–2

Roe D, Cooney R, Dublin H, Challender D, Biggs D, Skinner D, Abensperg-Traun M, Ahlers N, Melisch R, Murphree M (2017) First line of defence: engaging communities in tackling wildlife crime. Unasylva 249 668:33–38

Safari Club International Foundation (2017) Conservation Highlights. Department of Conservation Fiscal Year 2015–2016, p 20

Said MY, Ogutu JO, Kifugo SC, Makui O, Reid RS, de Leeuw J (2006) Effect of extreme land fragmentation on wildlife and livestock population abundance and distribution. J Nat Conserv 34:151–164

Saunders FP (2013) Seeing and doing conservation differently: a discussion of landscape aesthetics, wilderness, and biodiversity conservation. J Environ Dev 22:3–24

Scholes R, Smart K (eds) (2008) Elephant management: a scientific assessment for South Africa. Wits University Press, Johannesburg

Selier SAJ, Di Minin E (2015) Monitoring required for effective sustainable use of wildlife. Anim Conserv 18:131–132

Shepherd P (1973) The tender carnivore and the sacred game. Charles Scribner's Sons, New York

Slotow R, Whyte I, Hofmeyr M, Kerley GHI, Conway T, Scholes RJ (2008) Lethal management of elephants. In: Scholes R, Smart K (eds) Elephant management: a scientific assessment of South Africa. Wits University Press, pp 370–405

Stringham SF (1989) Demographic consequences of bears eating garbage at dumps: an overview. In: Bromley M (ed) Bear-People conflicts: proceedings of a symposium on management strategies. Northwest Territories Dep. Renewable Resour, Yellowknife, pp 35–42
Stubbe C, Passarge H (1979) Rehwild. J. Neumann-Neudamm Melsungen – Berlin – Basel – Wien, p 432
Tallis H, Kareiva P, Marvier M, Chang A (2008) An ecosystem services framework to support both practical conservation and economic development. Proc Natl Acad Sci USA 105:9457–9464
Tamisier A (1985) Hunting as a key environmental parameter for the Western Palearctic duck populations. Wildfowl 36:95–103
Taylor Ang (1996) Animal rights and human needs. Environ Ethics 18:249–260
Taylor Ant (2004) Pig-sticking princes: royal hunting, moral outrage, and the republican opposition to animal abuse in nineteenth- and early-twentieth-century Britain. History 89(293):30–48
Taylor Ant (2009) Animals and ethics - third edition. Broadview Press, Calgary, p 240
Treves A (2009) Hunting for large carnivore conservation. J Appl Ecol 46:1350–1356. https://doi.org/10.1111/j.1365-2664.2009.01729.x
Ückermann E (1952) Rehwild und Standort. Der Anblick, Graz, 7 (5) p 31
Ulrich RS (1983) Aesthetic and affective response to natural environment. In: Altman I, Wohlwill JF (eds) Human behavior and environment: advances in theory and research, New York, pp 85–125
Varner GE (1998) In Nature's Interests? Interests, Animal Rights, and Environmental Ethics. Oxford University Press, New York
Vitali T (1990) Sport Hunting: Moral or Immoral? Environ Ethics 12:69–82
Vizina Y, Kobei D (2017) Indigenous peoples and sustainable wildlife management in the global era. Unasylva 249, vol 668: 27–32
Wagenknecht E (1976) Rehwildhege mit der Büchse. Neumann, Lipsia-Radenbeul. p 207
Webb G, Brook B, Whitehead P, Manolis C (2004) Wildlife management principles and practices in crocodilian conservation and sustainable use. In: Crocodiles. Proceedings of the 17th working meeting of the Crocodile Specialist Group, IUCN - The World Conservation Union, Gland/Cambridge UK, pp 84–91
Weber M (1919) Politik als Beruf. München (Vortragsmitschrift mit Nachwort von Dahrendorf R. Reclam, Ditzingen DBR, reprint 1992)
Western D (1984) Amboseli National Park: human values and the conservation of a savanna ecosystem. In: Jeffrey A, McNeely, Miller KR (eds) National Parks, conservation, and development: the role of protected areas in sustaining society: proceedings of the World Congress on National Parks, Bali, Indonesia, 11–22 October 1982. Smithsonian Institution Press, Washington, D.C, pp 93–99
Western D, Russell S, Cuthill I (2009) The status of wildlife in protected areas compared to non-protected areas of Kenya. PLoS One 4:e6140
Williams PH, Burgess ND, Rahbek C (2000) Flagship species, ecological complementarity and conserving the diversity of mammals and birds in sub-Saharan Africa. Anim Conserv 3:249–260
Wiseman R, Hopkins L (eds) (2001) Sowing the seeds for sustainability: agriculture, biodiversity, economics and society. Proceedings of the eighth interactive session held at the second IUCN
Wronski T, Bariyanga JD, Apio A, Plath M (2015) Interactions between wildlife, humans and cattle: activity patterns of a remnant population of impala on the degraded Mutara Rangelands, Rwanda. Rangel J 37:357–365
Yurco K (2017) Herders and herdsmen. The remaking of pastoral livelihoods in Laikipia, Kenya. Pastoralism: Research, Policy and Practice. In: Pastoralism: research, policy and practice pastoralism. Springer, pp 1–17
Zunino M, Zullini A (2004) Biogeografia. La dimensione spaziale dell'evoluzione. CEA, Milan, p 374

Chapter 9
What Do We Know About Wild Boar in Iberia?

Alberto Giménez-Anaya, C. Guillermo Bueno, Pedro Fernández-Llario, Carlos Fonseca, Ricardo García-González, Juan Herrero, Carlos Nores, and Carme Rosell

9.1 Introduction

Wild boar *Sus scrofa* L. has one of the largest distributions among terrestrial mammals (Oliver and Leus 2008), which include the entire Iberian Peninsula (Bosch et al. 2012) where it has had a significant impact on ecosystems (Barrios-Garcia and Ballari 2012). Both in its native range and in areas where it has been introduced the wild boar can act as an ecosystem engineer (Barrios-Garcia and Ballari 2012;

A. Giménez-Anaya (✉) · J. Herrero
Area of Ecology, Department of Agrarian and Environmental Sciences, Technical School, University of Zaragoza, Huesca, Spain
e-mail: herreroj@unizar.es

C. G. Bueno
Institute of Ecology and Earth Sciences, Department of Botany, University of Tartu, Tartu, Estonia

P. Fernández-Llario
Innovación en Gestión de Ungulados SL, Cáceres, Spain

C. Fonseca
CESAM and Departamento de Biologia, Universidade de Aveiro, Aveiro, Portugal
e-mail: cfonseca@bio.ua.pt

R. García-González
Pyrenean Institute of Ecology (CSIC), Jaca, Spain
e-mail: rgarciag@ipe.csic.es

C. Nores
INDUROT, University of Oviedo, Oviedo, Spain
e-mail: cnores@uniovi.es

C. Rosell
Animal Biology Department, University of Barcelona, Barcelona, Spain

Minuartia, Barcelona, Spain
e-mail: crosell@minuartia.com

F. M. Angelici, L. Rossi (eds.), *Problematic Wildlife II*,
https://doi.org/10.1007/978-3-030-42335-3_9

Ballari and Barrios-García 2014), modifying habitat conditions for other organisms (Puigdefábregas 1980; Rosell et al. 2004; Sforzi and Tonini 2004; Engeman et al. 2007; Herrero et al. 2008; Bueno 2011) and, in some cases, affecting the viability and conservation of sensitive species (Canut 2006; Bertolero 2007; Lecomte 2007; Giménez-Anaya et al. 2008; Santaeufèmia 2008; Navàs et al. 2010). For instance, wild boar can modify the productivity and human use of natural, semi-natural, and human-made ecosystems (Markina 1999; Mayer et al. 2000; Cahill et al. 2003; Cahill and Llimona 2004; Malo et al. 2004; Fernández-Bou et al. 2006; Gortázar et al. 2006; Gómez and Hódar 2008; Fillat et al. 2008; Rosell et al. 2008; Santos et al. 2009; Bueno et al. 2010; Langbein et al. 2011; Cahill et al. 2012; Colino et al. 2012; Duarte et al. 2012; Lagos et al. 2012; Rodríguez-Morales et al. 2012; García-Jiménez et al. 2013; Rosell et al. 2013; Burrascano et al. 2014; Burrascano et al. 2015; Torrellas 2015; García-Jiménez et al. 2016; Torrellas et al. 2016). Despite the importance of wild boar, the knowledge about the species is limited, which is the current case in Mediterranean areas. In recent decades, wild boar has gone from a huntable species that lives in populations that range widely in density to the wild ungulate that has the largest distribution and density throughout the Iberian Peninsula (Bosch et al. 2012). This represents an important conflict with human activities (Markina 1999; Mayer et al. 2000; Cahill et al. 2003; Malo et al. 2004; Gortázar et al. 2006; Gómez and Hódar 2008; Rosell et al. 2008; Santos et al. 2009; Bueno et al. 2010; Colino et al. 2012; Lagos et al. 2012; Matilde 2012; Rodríguez-Morales et al. 2012; García-Jiménez et al. 2013; Torrellas 2015; Giménez-Anaya et al. 2016; Torrellas et al. 2016) and biodiversity conservation (Rosell et al. 2004; Sforzi and Tonini 2004; Engeman et al. 2007; Muñoz and Bonal 2007; Fernández-Bou et al. 2006; Herrero et al. 2008; Bueno et al. 2010; Rosell et al. 2016). Those changes have made applied research a priority, which should focus on minimizing the conflicts that the species has already posed.

This review focused on gathering the information available on wild boar in the Iberian Peninsula (Portugal and Spain), including historical aspects of the research, ecology (behaviour, distribution, habitat use, expansion, reproduction, demography, and diet), environmental impact, and management. To be able to evaluate the effects of wild boar population fluctuation, the species' impact on ecosystems and human activities related to global trends in climate, and to followed changes in biodiversity, we identified priority for future fields of research.

9.2 Study Area

Continental Portugal and Spain cover 582,459 km^2. Andorra was not included in the review because no research was reported on wild boar in this country. Portugal comprises 18 districts, and Spain has 15 regions (Fig. 9.1). The average elevation of the Iberian Peninsula is 600 m (maximum = 3482 m). The Pyrenees act as a natural barrier on the north of the peninsula and with its geographic position, its

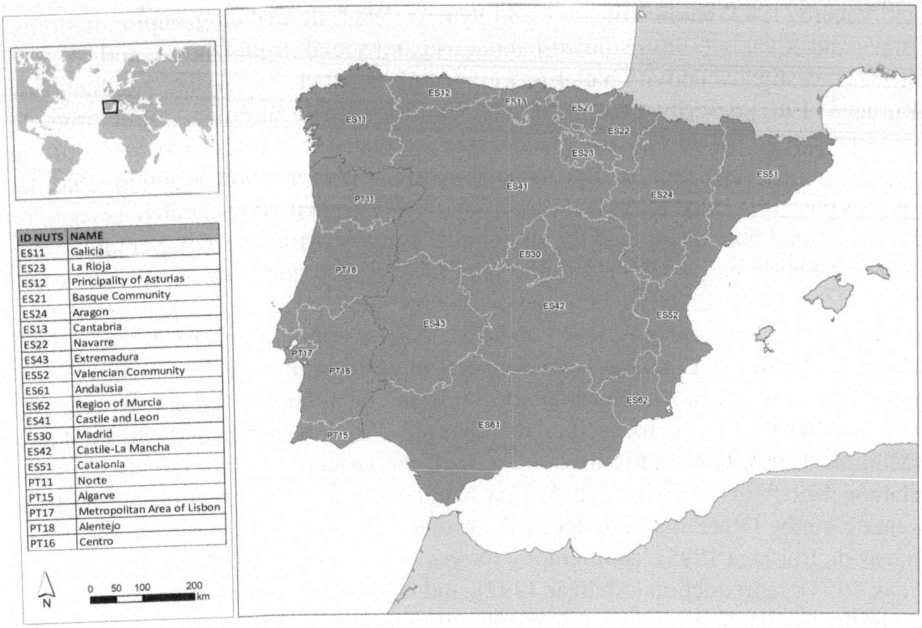

Fig. 9.1 Regions (Spain) and departments (Portugal)

mountainous topography, and environmental characteristics, it distinguishes itself from elsewhere in Europe.

The human population density was about 93 inhabitants per km^2, and most of it was concentrated in large cities, while large areas had densities of <5 km^2 because of human abandonment. This, in turn, has promoted the recovery of wild ungulate, particularly wild boar, populations.

9.3 History of Research

The first scientific research on wild boar in the peninsula, a description of the Iberian subspecies, was published in the early twentieth century (Cabrera 1914). Scientists did not return to it until the late 1970s and early 1980s, which included studies on rooting (Puigdefábregas 1980), reproduction (Vericad 1971, 1983), diet (Rodríguez Berrocal et al. 1982), estimation of abundance and habitat use (Rogers and Myers 1980), social behaviour in captivity (Martínez-Rica 1980), and natural history (Morais 1979; Vericad 1971).

In the 1980s, Serôdio (1985) and Sáez-Royuela (1989) did their PhD on natural history of wild boar (Saez–Royuela and Telleria 1986, 1987, 1988; Tellería and Sáez-Royuela 1984, 1985, 1986). Magalhães (1983) and Bugalho et al. (1984) described the historical populations of wild boars in Portugal. Garzón et al. (1983)

and Venero (1983) studied the diet, and Venero (1984) described grouping patterns. Braza and Álvarez (1989) studied habitat use and social organization, and Cuartas (1987) investigated activity patterns. Leránoz (1983) focused on diet, and Sáenz de Buruaga (1987) described the historical evolution of wild ungulate populations. In the 1990s, regional administrations became interested in wild boar because of the population increase and the new management and conservation problems that the species posed. This led to several reports and a more practical research approach. In Portugal, wild boar has spread throughout the country (Morais 1979; Serôdio 1985; Santos 1994; Fonseca 1999), and the hunt has become generalized. Research has focused on the interaction between wild boar and red partridge*Alectoris rufa*, demography, diet, and physical condition (Santos 1994; Fonseca 1996, 1999). Herrero's (2003) PhD research investigated the effects of wild boar battues on a relict population of brown bears *Ursus arctos* and on agricultural damage and culling. Another PhD thesis focused on demography (Markina 1998) and car accidents (Markina 1999). Other PhD thesis described the species ecology (Abaigar 1990; Garzón 1991; Rosell 1998) and Arroyo Nombela et al. (1990) published the first genetic study. Other research focused on diet (Abáigar 1993; Valet et al. 1994; Sáenz de Buruaga 1995), demography (Nores et al. 1995), caused damages (Nores et al. 1994), reproduction (Abáigar 1992), and habitat uses (Abáigar et al. 1994).

In the twenty-first century, research focused on habitat use (Santos et al. 2004; Fernández-Llario 2005; Rosell et al. 2006, 2010, 2012b; Bueno 2011), reproduction and demography (Fernández-Llario and Mateos-Quesada 2003; Uzal and Nores 2004; Herrero et al. 2008; Fonseca et al. 2004b, 2011), population management (Fonseca 2006), genetics (Ferreira et al. 2009; Pérez-González et al. 2014), diet (Giménez-Anaya et al. 2008), range expansion (Santos et al. 2004; González et al. 2013), and long-term population monitoring in protected areas (PA) and game reserves (Rosell et al. 2007; Herrero et al. 2008, Pita 2012). Moreover, the first analysis on the evolution of hunted wild boars and hunters in Europe, including Portugal and Spain, was done (Massei et al. 2015). The last decade has been marked by the conflicts caused by wild boars, such as car accidents, urban wild boars, culling in PA, and caused damage to agriculture and pastures (Putman et al. 2014; Rosell et al. 2004, 2008; Herrero et al. 2006; Bueno et al. 2011a; Cahill et al. 2012; Colino et al. 2012; Duarte et al. 2012; Lagos et al. 2012; Rodríguez-Morales et al. 2012; Sáenz de Santamaria and Tellería 2015). Further, an effort was made to standardize methods used to estimate population size (Acevedo et al. 2007; Nores et al. 2010; Gonçalves et al. 2013; Nores 2013).

Among the infectious diseases that can affect wild boar, research on tuberculosis *Mycobacterium tuberculosis*(TB) has been particularly of interest with an emphasis on the role of the species as a reservoir and transmitter to livestock, especially cattle (García-Jiménez et al. 2013; García-Jiménez et al. 2016). Focusing on livestock management but also evaluating other risk factors, such as co-infection with other pathogens or parasites, TB in wild boar was studied from multiple perspectives (Vieira-Pinto et al. 2011; Risco et al. 2013; Risco et al. 2014). In addition to TB other diseases were described in wild boar which traditionally were rather associated with domestic pigs, like swine erysipelas (Risco et al. 2011), salmonellosis

Salmonella sp. (Navarro-González et al. 2012), Glasser disease *Haemophilus parasuis* (Cuesta-Gerveno et al. 2013), virosis (Vicente et al. 2002; Ruiz-Fons et al. 2006), and parasitosis (García-Sánchez et al. 2009; García-González et al. 2013; Navarro-González et al. 2013). Common diseases that affect intensive breeding of domestic pigs also affect wild boar populations irrespective of population density (Risco 2011).

9.4 Population Development

The wild boar was considered a relatively scarce species during the nineteenth century and early twentieth century (Magalhães 1983; Bugalho et al. 1984; Gortázar et al. 2000; Carranza 2010). In recent decades, however, it is the native wild ungulate whose population has expanded the most in Iberia (Saez–Royuela and Telleria 1986; Gortázar et al. 2000; Fonseca et al. 2008). Its range currently covers the whole Iberian Peninsula (Rosell and Herrero 2007) except for some coastal, semi-desert, and high-elevation areas. In summer, wild boars were even found to occur at elevations of >2000 m (Rosell et al. 2016). Further, they even expanded into large towns (Fonseca and Correia 2008; Carranza 2010; Cahill et al. 2012) which was a recent phenomenon also elsewhere in Europe (Ferreira et al. 2009; Bosch et al. 2012; Stillfried et al. 2016).

In Iberia, wild boar is the main big-game species as reflected by the number of animals harvested each year (Lopes and Borges 2004). The estimated annual hunting bag was about 240,000 animals in 2010 (Carranza 2010; Vingada et al. 2010; Massei et al. 2015), although the actual number probably is much higher. Several interrelated factors have contributed to the increase in abundance including, among others: rural abandonment, increase in shrub and forest cover (Fonseca 2004); ageing of rural human populations (Nores et al. 1995); climate change, expressed through milder winters (Carranza 2010); captive breeding (Fernández-Llario et al. 1996); adaptability to habitats and foods (Rosell et al. 2001); and an exceptionally high reproductive potential (Table 9.1), which allows them to reproduce even in their first year of life (Herrero et al. 2008; Rosell et al. 2012). Easy year-round access to food through feeding and intensified agriculture realized this high growth potential, and the control of population abundance and expansion became a major challenge (Giménez-Anaya et al. 2016). In addition, polygyny, female parental care, and parturition in matriarchal groups increase piglet survival (Fernández-Llario and Mateos-Quesada 1998), which was reported to contribute to the problem.

In northern Iberia, wild boar densities typically ranged between 1 and 10 animals per km^2 and are influenced by climate, orography, forest cover, and land use (Nores 2010). In recent years, however, in some areas, density increased to >15 individuals per km^2 (García-Jiménez et al. 2013; Risco et al. 2013). In southern Iberia, management practices, sometimes including supplementary feeding within hunting grounds, and highly favourable habitats, have even led to densities of 40 individuals per km^2 (Table 9.2).

Table 9.1 Wild boar litter size in Iberia

Locality	Average	References
Empordà, Catalonia	5,0	Rosell et al. (2012)
Valle del Ebro, Aragon	4,5	Herrero et al. (2008)
Burgos, Castile and Leon	4,3	Saez–Royuela and Telleria (1987)
Western Pyrenees, Aragon	3,3	Vericad (1983)
Western Pyrenees, Aragon	4,2	Herrero et al. (2008)
Portugal	4,2	Fonseca et al. (2004a, b)
Portugal, Central Region	4,2	Fonseca et al. (2011)
Andalusia	4,1	Abáigar (1992)
Castile-La Mancha	3,9	Ruiz Fons et al. (2006)
Extremadura	3,9	Garzón (1991)
Montseny, Catalonia	3,8	Rosell et al. (1998)
Extremadura	3,7	Fernández-Llario and Mateos-Quesada (2005)
Andalusia	3,0	Fernández-Llario y Carranza (2000)

Table 9.2 Iberian wild boar densities

Locality	Density km^{-2}	References
Western Pyrenees, Aragon	3,3	Herrero (2003)
Navarre	2,6–3,0	Leránoz and Castién (1996)
Burgos, Castile and Leon	1,9–4,2	Tellería and Sáez-Royuela (1986)
León, Castile and Leon	1,7–11,4	Purroy et al. (1988)
Extremadura	3	Garzón (1991)
Álava, Basque Country	0,4–6,1	Markina (1998)
Asturias	3,8–10	Nores, own data
Pre-Pyrenees, Aragon	7,0	Marco et al. (2011)
Garrotxa, Catalonia	3,6–8,5	Rosell et al. (2001)
Alt Empordà, Catalonia	7–12,5	Rosell et al. (2001)
Catalonia	2–17	Departament Agricultura, Ramaderia i Pesca (2016)
Extremadura	10	Fernández-Llario et al. (1996)
Extremadura	6,5–32[a]	Risco et al. (2013)
Extremadura	>40	Gonçalves (in prep)

[a]Fenced hunting grounds with supplementary feeding

9.5 Diet and Environmental Impacts

Diet has been studied through analyses of stomach contents, typically collected during the hunting season, which roughly goes from September to February (Herrero et al. 2006). However, in population control programs, samples could be collected year-round (Herrero et al. 2005; Giménez-Anaya et al. 2008). Results indicate a generalist, opportunistic, and mainly phytophagous diet (Giménez-Anaya et al. 2008). Wild boars consume a wide variety of trophic resources, which was influenced by food abundance and availability (Herrero et al. 2005; Giménez-Anaya et al. 2008). As an omnivore, it also consumes animal matter, but the basis of its diet was hard mast of Fagaceae acorns *Quercus* sp., beechnuts *Fagus sylvatica*, and chestnuts *Castanea sativa*, as well as agricultural products (Irizar et al. 2004; Herrero et al. 2006, 2008). In some areas and periods cultivated plants were in fact the main food (Herrero et al. 2006; Giménez-Anaya et al. 2008), which also represents the main basis for the conflict with human interests.

The underground parts of plants (roots, bulbs, rhizomes) and, to lesser extent, insect larvae, small vertebrates, and worms can be an important part of the diet as well (Herrero et al. 2006, Giménez-Anaya et al. 2008). If hard mast was scarce, the proportion of the diet that is aboveground plant matter increased and included agricultural (Leránoz 1983) and wild plants (Herrero et al. 2005).

The impact of wild boar on biodiversity was based on the species capacity to act as an ecosystem engineer by transforming habitats, mostly through rooting (Barrios-Garcia and Ballari 2012; Ballari and Barrios-García 2014). Those effects cannot always be classified as damages, because, in some cases, they only make up a small amount of the many factors that cause changes in the composition and structure of living communities (Muñoz and Bonal 2007; Fernández Bou et al. 2008; Bueno et al. 2010; Rosell et al. 2016).

Rooting can be an important perturbation that acts on multiple scales (Bueno 2011). For instance, on alpine and subalpine pastures, wild boar looked for areas that had dense vegetation, deep soils, and specific pastoral use (Bueno et al. 2009). At that scale, rooting modified the structure of vegetal communities (Burrascano et al. 2014, 2015). It reduced diversity and heterogeneity between communities, but increased it within them (Bueno 2011; Fillat et al. 2008; Fernández-Bou et al. 2006). Furthermore, rooting can modify the physical and chemical structure of soils (Bueno et al. 2013) and their biotic components, like the seed bank (Bueno et al. 2011a), the bulb community (Palacio et al. 2013), and earthworms (Bueno and Jiménez 2014). In addition, rooting can alter the relationships between species, like mycorrhizae (Puigdefábregas 1980). This can affect important ecosystem interactions, alter the regeneration capacity of the affected communities, and lead to a recovery of vegetal diversity within 1 year following rooting (Bueno 2011). In subalpine environments, recovery after rooting can take as long as 3 years (García-González, unpublished data). Generally rooting can be an annual recurrence, especially in sensitive habitats that are not adapted to soil perturbations and where natural revegetation is ineffective (Bueno et al. 2011a). The intensity of rooting fluctuated and appeared to be influenced by factors associated with wild boar abundance, aerial food availability, the kind of forest, and its proximity (García-González et al. 2003; Bueno 2011). In addition, the pastoral value of subalpine meadows that have a high cattle density, and therefore tend to be dominated by matgrass *Nardus stricta* that is avoided by cattle, can be improved through rooting, as it creates new opportunities for other plant species to colonize these patches (Fernández-Bou et al. 2006).

In other cases, wild boar was a threat to the conservation of some endangered species, especially in wetland areas (Herrero et al. 2006). For instance, ground breeding bird species like the purple swamphen *Porphyrio porphyrio* or the Eurasian bittern *Botaurus stellaris* were especially sensitive to the impact of wild boar (Giménez-Anaya et al. 2008). Another example were endangered orchids, from which wild boar like to consume the bulbs and for which there are already recovery plans in action (Lecomte 2007; Santaeufèmia 2008; Navàs et al. 2010). Other research found negative effects of wild boar, on eggs and juveniles of Mediterranean tortoise *Testudo hermanni* (Bertolero 2007), ptarmigan *Lagopus muta* (Canut 2006), and brown bear (Nores et al. in prep). More abundant species that might be affected include rabbit *Oryctolagus cuniculus* (Carpio et al. 2014b), red partridge (Carpio et al. 2014c), and invertebrates (Carpio et al. 2014a). Thus, wild boar was perceived as a threat to biodiversity in some protected areas and subsequently monitored and culled intensively, especially in some wetlands (Rosell et al. 2004; Sforzi and Tonini 2004; Engeman et al. 2007; Herrero et al. 2008; Giménez-Anaya et al. 2016).

9.6 Behaviour and Habitat Use

In part because it is a nocturnal species, only few studies have investigated the behaviour of the wild boar in Iberia. New techniques such as camera trapping and drone-based thermal cameras helped to increase the knowledge about certain unknown aspects of the wild boar's biology (Sarmento et al. 2010; Casas-Díaz et al. 2011; Cahill et al. 2012).

Social organization was studied based on diurnal direct sighting and indicated a matriarchal system (Braza and Álvarez (1989; Fernández-Llario et al. 1996). Changes in group composition occurred during rut, as during this time adult and expelled young males joined the sounders (Fernández-Llario et al. 1996). After the rut, males abandoned the matriarchal groups again (Fernández-Llario et al. 1996).

Night cameras provided a means of quantifying activity rhythms of individual groups and their behaviour (Gonçalves, pers. comm., Rosell et al. in prep.). Males change the timing of their activity to adapt to the activity of a matriarchal group that has 5–6-month-old piglets; i.e. they are less active in the night hours and become more active at dawn and dusk (Gonçalves, pers. comm.). Another study has shown that hunting causes wild boars to concentrate in areas where there is no hunting during the hunting period (Rosell et al. in prep.). Molecular genetic studies showed that more than 20% of the litters had more than one father (Delgado et al. 2008).

Activity rhythms were also reflected in habitat use. In the south of the peninsula, wild boar movements were based on food availability; specifically, animals focus mainly on areas that have acorns and avoid or select sunny or shaded areas, depending on the season. In summer and early autumn, shaded areas that had water available were preferred as resting and feeding areas. In winter, the animals moved to sunny, high-slope areas, which allowed them to avoid soils where water had accumulated (Fernández-Llario 2004). In the north, in addition to the factors described above, hunting activity influenced the activity rhythms of the animals (Nores 2010; Rodrigues et al. 2016).

9.7 Conflicts with Human Activities

At moderate densities, and if limited to forest areas, the social and economic conflict potential was low. Conflicts mainly occurred, if they exceeded social carrying capacity, in agricultural lands, in urbanized areas, or in sensitive habitats such as high pastures or wetlands (Giménez-Anaya et al. 2016).

Among the social and economic conflicts, agricultural damage to field crops and pastures caused the most complaints and damage claims. Rooting in pastures can affect extensive grazing by livestock (Bueno et al. 2010) and its environmental and productive value (Bueno et al. 2011a). Among agricultural products, corn *Zea mays*, fruit trees, and vegetables are the most frequently damaged. No global estimates are available; however, in Asturias, in 2008, the regional government paid €1,215,573 in compensation, which is equivalent to € 115 per km^2. In addition, wild boars caused damage to forest regeneration and, particularly, to forest restoration, based on seed and seedling consumption, which obliged a new plantation and the use of proper protection (Mayer et al. 2000; Gómez and Hódar 2008).

In the case of livestock breeding, high density of wild boar can influence disease transmission (Gortázar et al. 2006; Santos et al. 2009; García-Jiménez et al. 2013). In the south, some infectious diseases, particularly bovine tuberculosis (TB), have become one of the main threats to the sustainability of extensive livestock farms because the wild boar is the main reservoir of the disease (García-Jiménez et al. 2016). Some populations showed this TB prevalence at >50%, and the species role in the transmission of bacillus to other species has been demonstrated (Hermoso de Mendoza et al. 2006; Santos et al. 2009; Rodríguez Campos 2013). Interbreeding with domestic pigs can occur, but it did not have a significant impact on wild boars. This however might occur with the interbreeding with Vietnamese pigs *Sus scrofa domestica*, which have produced crossed populations (Delibes–Mateos and Delibes 2013).

The presence of wild boars on the outskirts of and within large cities is another important source of conflicts (Cahill et al. 2003). These conflicts were difficult to control if there was no rapid action to eliminate habituated wild boars. These boars were no longer 'suited' to living freely (Cahill and Llimona 2004; Cahill et al. 2012; Duarte et al. 2012). Given that a city's outskirts are not huntable areas, other methods to avoid their presence have to be used, e.g. non-lethal battues (Tolon 2010), trapping, and anaesthetic captures, together with proper protocols. These methods have to be sustainable, and they require neighbourhood training, regarding the behaviour of these animals, whom can also be seen during the day, searching for food in dustbins and gardens, or receiving food directly from humans, which causes potentially dangerous situations.

Car accidents caused by wild boars have increased in recent years. Although they are a small proportion of the damages caused by ungulates in Europe (Langbein et al. 2011), in Spain, the amount of damage increased by 30% between 2007 and 2011 (Matilde 2012). In Catalonia, > 85% of all accidents that involved wildlife (about 1000 accidents, annually) involved wild boars (Departament de Territori i

Sostenibilitat 2015). That report estimated an average damage of €3787 per car although others have reported per capita amounts of €2700 (Colino et al. 2012) and €9119 (Sáenz de Santamaria and Tellería 2015). The complexity of the management of the problem and the complications derived from the damage claims that do not include insurance companies often blame the presence of hunting grounds for the accident (Markina 1999; Malo et al. 2004; Rosell et al. 2008; Colino et al. 2012; Lagos et al. 2012; Rodríguez-Morales et al. 2012; Torrellas 2015; Torrellas et al. 2016).

The frequency of car-boar accidents is highest between September and January and between dusk and twilight. The measures taken to prevent car crashes alter the behaviour of wild boar, and certain aspects must be considered, such as perimetral closing, which needs fauna passages (Rosell et al. 2008). In Catalonia, the map with the locations of car crashes involving ungulates and the concentration of hotspots has recently been updated, and a plan to reduce these accidents has been implemented. This plan includes fences and fauna passes on highways and other measures used on conventional roads with accord with the temporality shown by these events (Rosell et al. 2013).

9.8 Hunting Management

In northern Iberia, wild boar have been the cornerstone of big-game hunting, which brings together the hunting crew. In the north, it has been the most hunted and the most expensive species to manage. In Asturias, for example, the damages caused by huntable and protected species were paid by the regional government, and wild boars are the one that caused the most damage, more than wolf *Canis lupus*, red deer *Cervus elaphus*, or brown bear *Ursus arctos*. In 2008, the cost of damage was over €1.5 million, which is equivalent to €152/km^2. The cost of damage per wild boar within the hunting grounds was €25.5 per animal, and the cost per hunted wild boar was €37–187.

In the north, where the species has existed historically (Leránoz and Castién 1996; Markina 1998; Herrero 2003, Nores unpublished data), most hunts have occurred within small 10–600 ha battues (Table 9.3), covering an average size of 131–150 ha, with 11–16 hunters, 2–7 beaters, and 5–13 dogs. Hunting was in open grounds, which guaranteed gene flow and natural movements of predators. In Asturias, each year 20–25% of the total population was harvested (C. Nores unpublished data). There was no management beyond hunting and protection of field crops, and population density was more strongly influenced by environmental and intrinsic factors, rather than by human intervention (Uzal and Nores 2004; Nores et al. 2010).

In the south, the most popular hunting method were and are large battues, in which the hunter waits for the wild boars to appear, after they have been forced to move by packs of at least 20 dogs who are guided by one beater. Legally, the area must be more than 500 ha and, except in special circumstances, can be hunted only

Table 9.3 Characteristics of hunting battues with dogs in Iberia. Average values

Locality	Surface (ha)	Hunters	Beaters	Dogs	Efficiency (%)	References
Western Pyrenees, Aragon	168	9,5	2,5	5,4	25	Herrero (2004)
Navarre	151	15,4		8,1	31	Leránoz and Castién (1996)
Burgos, Castile and Leon	75	19,7	16	15,4	15,4–18	Saez–Royuela and Telleria (1988)
Leon, Castile and Leon	122,7		9,8			Sáenz de Buruaga et al. (1987)
Asturias	131	12,6	6,4	7,3	21	Nores et al. (2010)
Alava, Basque Country	142	12	6	10		Markina (1998)
Middle Ebro Valley, Aragon		6	1	10	39	Giménez-Anaya et al. (2016)
Central Region, Portugal	163	64	7.6	21		Fonseca (unpublished data)

once during the hunting season. About 30 hunters stay in fire lines around the beaten area. Effectiveness has been about 30%, which is similar to the effectiveness of battues in the north (Table 9.3). In the south, to achieve these results, management developed all year long. Control by rangers involved artificial feeding (Fernández-Llario and Mateos-Quesada 1998) and increasing the availability of water points (Fernández-Llario 2004), which promoted an increase of wild boar density.

Canned hunt (fenced grounds) of private areas is widespread in the South of Spain and is associated with watering, artificial feeding, high densities of wild boar, and high prevalence of several diseases (Fernández-Llario and Mateos-Quesada 1998; Fernández-Llario 2004; Risco et al. 2011).

In Portugal, battues were the most common traditional and widespread hunting method. Dog packs averaged 25 animals were driven by more than one beater. The area covered could be of <500 ha, and the number of hunters was very variable (i.e. dozens to hundreds). Night waiting during full moon was another traditional method, which could be used year-round, particularly in the period when agricultural damage was worst.

9.9 Survey Methods

The population dynamics of wild boar is complex because of its demographic adaptability, which makes it difficult to estimate population size (Barrett 1982), the factors that influence populations (Uzal and Nores 2004), and to obtain information from population change models, the data series must be several decades long (Turchin 2003).

The main methods used to monitor wild boar, in regard to its demography on the long term, are battues (Rosell et al. 1998; Giménez-Anaya et al. 2016), direct sightings (Nores et al. 2000), and hunting bags (Bosch et al. 2012). For sanitary purposes necropsies and sera analysis are performed (Risco et al. 2014). For short periods of time, radiotelemetry (Santos et al. 2004) and faecal counts (Rodrigues et al. 2016) are used. Most populations are not monitored and hunting bags are not a reliable method as they are based on the voluntary declaration by the owners of their hunting rights.

9.10 The Future

Wild boar has been living in the Iberian Peninsula, with more or less abundant populations, since the Middle Pleistocene (Groves 1981). It has, however, never been an important economic resource, possibly because of the limited resources that have been dedicated to the research and management of the species. That is in contrast to the important and diverse problems posed by the species (social, economic, and for biodiversity conservation) in many areas of the peninsula, because of the increase in the extent and size of wild boar populations. For that reason more interest and demand for a better understanding of the species and more research on management and control are needed.

A species with such a reproductive and ecosystem transformation potential should receive particular attention and monitoring within the context of natural area management. The species' extraordinary capacity to adapt to a wide variety of natural and cultural environments has allowed it to become established in urban areas, as well as to colonize agricultural lands which have higher productivity compared to forest environments (Herrero 2003). The basis for an agile and effective adaptive management must be created. Delayed response and low efficacy has caused high socioeconomic costs. There has been an increase in the frequency of interactions between humans and wild boars. Problems that have arisen will probably worsen in the near future (Cahill et al. 2012) and should be addressed by actions taken on wild boar, but also through the education of the public, to avoid risky situations. The response to the increase in wild boar in urban areas and the risk of car accidents should include environmental education, which has not been needed, until now.

There are several aspects of the species ecology that are still unknown and many questions that will have to be addressed in the near future. In Iberia, there has been long-term population research, which has shed light on population knowledge. Some important aspects are trends (Leránoz and Castién 1996), environmental impacts, foods particularly in summer (outside the hunting season) (Herrero et al. 2006; Giménez-Anaya et al. 2008), rooting recurrence and its effect on the capacity of ecosystems to recover (Bueno 2011), trend in management and hunt (hunting efficiency) to control populations and reduce damages (Giménez-Anaya et al. 2016), as well as cross-breeding with domestic breeds (Delibes–Mateos and Delibes 2013). We emphasize that, in the context of change in the composition and distribution of biodiversity caused by climate change, research on ecosystem engineer species should be a priority for evaluating the species' future impact on ecosystems and on human-affected productive activities. Accordingly, a clear commitment for research on wild boar ecology is essential and urgent for the development of sound proactive management measures that should identify possible irreversible perturbations on ecosystems.

Acknowledgments This chapter is dedicated to Carlos Sáez Royuela, '*Cartucho*' (Shotgun Shell), whose PhD thesis, the first on wild boar in Iberia, was a remarkable example and a source of inspiration to all of us. Our thanks go to the countless collaborators who helped us with our field and lab work, who were motivated by their interest in learning more about this fascinating species. Regional administrations and several scientific institutions contributed financing to local studies, exhibited interest towards the issues that we have studied, and helped to increase the understanding of the role of wild boar in Iberian ecosystems. Bruce MacWhirter reviewed the English.

References

Abaigar T (1990) Características biológicas y ecológicas de una población de jabalíes (Sus scrofa L. 1758) en el SE Ibérico. PhD thesis, Universidad de Navarra, Spain

Abáigar T (1992) Paramètres de la reproduction chez le sanglier (*Sus scrofa*) dans le sud-est de la Péninsule Iberique. Mammalia 56:245–250

Abáigar T (1993) Régimen alimentario del jabalí (*Sus scrofa*, L. 1758) en el sureste ibérico. Doñana Acta Vertebrata 20:35–48

Abáigar T, Del Barrio G, Vericad JR (1994) Habitat preference of wild boar (*Sus scrofa* L., 1758) in a Mediterranean environment. Indirect evaluation by signs. Mammalia 58:201–210

Acevedo P, Vicente J, Höfle U, Cassinello J, Ruiz-Fons F, Gortázar C (2007) Estimation of European wild boar relative abundance and aggregation: a novel method in epidemiological risk assessment. Epidemiol Infect 135:519–527

Arroyo Nombela JJ, Rodríguez Murcia C, Abáigar T, Vericad JR (1990) Cytogenetic analysis (GTG, CBG and NOR bands) of a wild boar population (*Sus scrofa scrofa*) with chromosomal polymorphism in the South-East of Spain. Genet Sel Evol 22:1–9

Ballari SA, Barrios-García MN (2014) A review of wild boar *Sus scrofa* diet and factors affecting food selection in native and introduced ranges. Mammal Rev 44:124–134

Barrett RH (1982) Habitat preferences of feral hogs, deer, and cattle on a Sierra foothill range. J Range Manag 35:342–346

Barrios-Garcia MN, Ballari SA (2012) Impact of wild boar (Sus scrofa) in its introduced and native range: a review. Biol Invasions 14(11):2283–2300

Bertolero A (2007) Cens i distribució de la tortuga mediterrània a la serra de l'Albera. Departament de Medi Ambient i Habitatge, Barcelona. Unpublished report

Bosch J, Peris S, Fonseca C, Martínez M, De la Torre A, Iglesias I, Muñoz MJ (2012) Distribution, abundance and density of the wild boar on the Iberian Peninsula, based on the CORINE program and hunting statistics. Folia Zool 61:138–151

Braza F, Álvarez F (1989) Utilisation de l'habitat et organization sociale du sanglier (*Sus scrofa* L.) Doñana (Sud-Ouest de l'Espagne). Can J Zool 67:2047–2051

Bueno CG (2011) Las perturbaciones del jabalí en los pastos alpinos del Pirineo Central: una aproximación multiescalar. PhD thesis, Universidad de Zaragoza, p 191. Zaragoza, Spain

Bueno CG, Jiménez JJ (2014) Livestock grazing activities and wild boar rooting affect alpine earthworm communities in the Central Pyrenees (Spain). Appl Soil Ecol 83:71–78

Bueno CG, Alados CL, Gómez-García D, Barrio IC, García-González R (2009) Understanding the main factors in the extent and distribution of wild boar rooting on Alpine grasslands. J Zool 279:195–202

Bueno CG, Barrio IC, García-González R, Alados CL, Gómez-García D (2010) Does wild boar rooting affect livestock grazing areas in alpine grasslands? Eur J Wildl Res 56:765–770

Bueno CG, Barrio IC, García-González R, Alados CL, Gómez-García D (2011a) Assessment of wild boar rooting on ecological and pastoral values of alpine Pyrenean grasslands. Pirineos 166:51–67

Bueno CG, Azorín J, Gómez-García D, Alados CL, Badía D (2013) Occurrence and intensity of wild boar disturbances, effects on the physical and chemical soil properties of alpine grasslands. Plant Soil 373(1–2):243–256

Bugalho JF, Carvalho JC, Borges JF (1984) Situation du sanglier au Portugal Symp. Sur le sanglier du Cons. Intern. de la Chasse, 5–11 July, 16 pp

Burrascano S, Giarrizzo E, Bonacquisti S, Copiz R, Del Vico E, Fagiani S, Mortelliti A, Blasi C (2014) Quantifying *Sus scrofa* rooting effects on the understorey of the deciduous broadleaf forests in Castelporziano Estate (Italy). Rendiconti Lincei 26:317–324

Burrascano S, Copiz R, Del Vico E, Fagiani S, Giarrizzo E, Mei M, Mortelliti A, Sabatini FM, Blasi C (2015) Wild boar rooting intensity determines shifts in understorey composition and functional traits. Community Ecol 16:244–253

Cabrera A (1914) Fauna ibérica. Mamíferos. Museo Nacional de Ciencias Naturales, Madrid, España

Cahill S, Llimona F (2004) Demographics of a wild boar Sus scrofa Linnaeus, 1758 population in a metropolitan park in Barcelona. Galemys 16:37–52

Cahill S, Llimona F, Gracia J (2003) Spacing and nocturnal activity of wild boar *Sus scrofa* in a Mediterranean metropolitan park. Wildl Biol 9:3–13

Cahill S, Llimona F, Cabañeros L, Calomardo F (2012) The increasing dilemma of wild boar (*Sus scrofa*) habituation to urban areas: traits from Collserola Park (Barcelona) and comparison with this problem in other cities. Anim Biodivers Conserv 35(2):221–233

Canut J (2006) Depredación. In: Afonso I, Aniz M, Arribas O, Canut J, Marty E, Mossoll M, Piqué J, Racionero C (eds) El lagópodo alpino (*Lagopus muta pyrenaica*) en el Parque Nacional de Aigüestortes i Estany de Sant Maurici. OAPN. Ministerio de Medio Ambiente, Madrid, pp 77–84

Carpio AJ, Castro–López J, Guerrero–Casado J, Ruiz–Aizpurua L, Vicente J, Tortosa FS (2014a) Effect of wild ungulate density on invertebrates in a Mediterranean ecosystem. Anim Biodivers Conserv 37(2):115–125

Carpio AJ, Guerrero-Casado J, Ruiz-Aizpurua L, Vicente J, Tortosa FS (2014b) The high abundance of wild ungulates in a Mediterranean region: is this compatible with the European rabbit? Wildl Biol 20(3):161–166

Carpio AJ, Guerrero-Casado J, Tortosa FS, Vicente J (2014c) Predation of simulated red-legged partridge nests in big game estates from South Central Spain. Eur J Wildl Res 60:391–394

Carranza J (2010) Ungulates and their management in Spain. In: Apollonio M, Andersen R, Putman R (eds) Ungulate management in Europe in the XXI century. Cambridge University Press, Cambridge, pp 419–440

Casas-Díaz E, Peris A, Serrano E, Closa-Sebastià F, Torrentó J, Miño A, Casanovas R, Marco I, Lavín S (2011) Estima de la densidad de una población de jabalí (Sus scrofa) mediante trampeo fotográfico: estudio piloto en Cataluña. [Wild boar (Sus scrofa) density estimation by mark resighting: a pilot study in Catalonia]. Galemys Span J Mamm 23:99–104

Colino V, Bosch J, Reoyo MJ, Peris S (2012) Influence of new irrigation croplands on wild boar (*Sus scrofa*) roadkills in NW Spain. Anim Biodivers Conserv 35(2):247–252

Cuartas P (1987) Estructura social y reparto temporal de actividades del jabalí (*Sus scrofa*) en Doñ*ana*. BSc thesis, University of Oviedo, Spain

Cuesta-Gerveno JM, Risco D, Gonçalves P, García-Jiménez W, Gil M, Fernandez-Llario P, Hermoso de Mendoza J, Gómez LJ (2013) Fatal infection due to *Haemophilus parasuis* in a young wild boar (*Sus scrofa*). J Vet Diagn Investig 25:297–300

Delgado R, Fernández-Llario P, Azevedo M, Beja-Pereira A, Santos P (2008) Paternity assessment in free-ranging wild boar (*Sus scrofa*) – Are littermates full-sibs? Mamm Biol 73(3):169–176

Delibes–Mateos M, Delibes A (2013) Pets becoming established in the wild: free–living Vietnamese potbel-lied pigs in Spain. Anim Biodivers Conserv 36(2):209–215

Departament d'Agricultura, Ramaderia, Pesca i Alimentació (2016) Programa de seguiment de les poblacions de senglar (Sus scrofa) a Catalunya. Informe inédito elaborado por Minuartia, 71 pp

Departament de Territori i Sostenibilitat (2015) Estudi de l'accidentalitat provocada per animals en llibertat a la xarxa de vies de carreteres de la Generalitat de Catalunya. Asistencia técnica Minuartia. Informe inédito, 69pp + Annexes and maps

Duarte J, Farfán MA, Vargas MJ (2012) Ungulados en las nuevas zonas urbanas de la Costa del Sol (Málaga). III Reunión sobre Ungulados Silvestres Ibéricos 'Ungulados, biodiversidad y actividades humanas: gestión de conflictos' 4-5 Octubre de 2012. Castelló d'Empúries (Girona)

Engeman RM, Constantin BU, Schwiff SA, Smith HT, Woolard J, Allen J, Dunlap J (2007) Adaptive and economic management methods for feral hog control in Florida. Human - Wildlife Conflicts 1:178–185

Fernández-Bou M, Rosell C, Giménez A, Guardiola M, Navàs F, Espelta JM (2006) Selecció d'hàbitats i efectes de l'activitat del senglar en prats del Parc Nacional d'Aigüestortes i Estany de Sant Maurici. En: La investigació al Parc Nacional d'Aigüestortes i Estany de Sant Maurici. Setenes Jornades sobre Recerca al Parc Nacional d'Aigüestortes i Estany de Sant Maurici. Parc Nacional d'Aigüestortes i Estany de Sant Maurici. pp. 285–300. Departament de Medi Ambient i Habitatge, Lleida.

Fernández-Bou M, Rosell C, Giménez-Anaya A, Guardiola M, Navàs F, Espelta JM (2006) La població de senglar (*Sus scrofa*) al Parc Nacional d'Aigüestortes i Estany de Sant Maurici. 2006. VII Jornades sobre Recerca al Parc Nacional d'Aigüestortes i Estany de Sant Maurici. Barruera, 25–27 de October 2006

Fernández-Llario P (2004) Enviromental correlates of nest site selection by wild boar Sus scrofa. Acta Theriol 49:383–392

Fernández-Llario P (2005) The sexual function of wallowing in male wild boar (*Sus scrofa*). J Ethol 23:9–14

Fernández-Llario P, Mateos-Quesada P (1998) Body size and reproductive parameters in the wild boar Sus scrofa. Acta Theriol 43:439–444

Fernández-Llario P, Mateos-Quesada P (2003) Population structure of the wild boar (*Sus scrofa*) in two Mediterranean habitats in the western Iberian Peninsula. Folia Zool 52:143–148

Fernández-Llario P, Carranza J, Hidalgo De Trucios SJ (1996) Social organization of the wild boar (*Sus scrofa*) in Doñana National Park. Miscel·lània Zoològica 19:9–18

Fernández-Llario P, Carranza J (2000) Reproductive performance of the wild boar in a Mediterranean ecosystem under drought conditions. Ethol. Ecol. Evol 12:335–343

Fernández-Llario P, Mateos-Quesada P, Silvério A, Santos P (2003) Habitat effects and shooting techniques on two wild boar (*Sus scrofa*) populations in Spain and Portugal. Z Jagdwiss 49:120–129

Fernández-Llario P, Mateos-Quesada P (2005) Influence of rainfall on the breeding biology of wild boar (Sus scrofa) in a Mediterranean ecosystem. Folia Zool 54:240–248

Ferreira E, Souto L, Soares AM, Fonseca C (2009) Genetic structure of the wild boar population in Portugal: evidence of a recent bottleneck. Mamm Biol – Z Säugetierkd 74:274–285

Fillat F, García-González R, Gómez D, Reiné R (2008) Pastos del Pirineo. Departamento de Publicaciones del CSIC, Madrid, Spain

Fonseca C (1996) Aspectos da Biologia de Javali (*Sus scrofa* L.). Relatório final de estágio científico. Departamento de Zoologia da Faculdade de Ciências e Tecnologia da Universidade de Coimbra. Policopiado, 30 pp

Fonseca C (1999) Ecologia do javali (*Sus scrofa* L, 1758) no Centro de Portugal. MSc thesis, University of Coimbra, Coimbra, Portugal

Fonseca C (2004) Population dynamics and management of wild boar (Sus scrofa L.) in Central Portugal and South- Eastern Poland. PhD thesis, Universidade de Aveiro, Aveiro, Portugal

Fonseca C (2006) Gestão das populações de javali (*Sus scrofa*) e dos seus habitats. Gestão das populações cinegéticas e dos seus Habitats. Federação de caça e Pesca da Beira Litoral, Coimbra, Portugal, pp 109–126

Fonseca C, Correia F (2008). O Javali. Colecção Património Natural Transmontano. João Azevedo Editor (1.ª Edição). Mirandela, 168 pp

Fonseca C, Herrero J, Luís A, Soares AMVM (2004a) Wild boar research 2002. A selection and edited papers from the 4th International wild boar symposium. Galemys 16 Special Issue, Málaga, Spain

Fonseca C, Santos P, Monzón A, Bento P, Alves da Silva A, Alves J (2004b) Reproduction in the wild boar (*Sus scrofa*, Linnaeus, 1758) populations of Portugal. Galemys 16:53–65

Fonseca C, Cancela JH, Silva A, Vingada J, Soares AMVM (2008) Libro Perspectivas em Gestão de Fauna Selvagem Páginas 45–58 Editor Waves Portugal

Fonseca C, Alves da Silva A, Alves J, Vingada J, Soares AMVM (2011) Reproductive performance of wild boar females in Portugal. Eur J Wildl Res 57:363–371

García-González R, Marinas A, Sanz G, Gómez D, Herrero J, Valadon A, Arthur CP (2003) Valoración del impacto de las hozaduras de jabalí en los pastos supraforestales del Parc National des Pyrénées. Tercer informe parcial Informe inédito. Parc National des Pyrénées (Tarbes, Francia) and Instituto Pirenaico de Ecología, (CSIC), Jaca, Spain

García-González AM, Pérez-Martín JE, Gamito-Santos JA, Calero-Bernal R, Alonso MA, Carrión EMF (2013) Epidemiologic study of lung parasites (*Metastrongylus* spp.) in wild boar (*Sus scrofa*) in southwestern Spain. J Wildl Dis 49:157–162

García-Jiménez WL, Fernández-Llario P, Benítez-Medina JM, Cerrato R, Cuesta J, García-Sánchez A, Gonçalves P, Martínez R, Risco D, Salguero FJ, Serrano E, Gómez L, Hermoso-de-Mendoza M (2013) Reducing Eurasian wild boar (*Sus scrofa*) population density as a measure for bovine tuberculosis control: Effects in wild boar and a sympatric fallow deer (*Dama dama*) population in Central Spain. Prev Vet Med 110:435–446

García-Jiménez WL, Cortés M, Benitez-Medina JM, Hurtado I, Martínez R, García-Sanchez A, Risco D, Cerrato R, Sanz C, Hermoso-de-Mendoza M (2016) Spoligotype diversity and 5-year trends of bovine tuberculosis in Extremadura, southern Spain. Trop Anim Health Prod:48(8):1533–1540

García-Sánchez RN, Nogal-Ruiz JJ, Manzano-Lorenzo R, Díaz JMA, López GP, Ruano FJS, Casas AR, Bascn CC, Bols-Fernández F, Martínez-Fernández AR (2009) Trichinellosis survey in the wild boar from the Toledo mountains in south-western Spain (2007-2008): Molecular characterization of *Trichinella* isolates by ISSR-PCR. J Helminthol 83:117–120

Garzón P (1991) Biología y ecología del jabalí (*Sus scrofa* L, 1758) en el Parque Natural de Monfragüe. PhD thesis, Universidad Autónoma, Madrid, Spain

Garzón P, Palacios F, Ibáñez C (1983) Primeros datos sobre la alimentación del jabalí (*Sus scrofa* baeticus Thomas, 1912) en el Parque Nacional de Doñana. II Reunión Iberoamer Cons Zool Vert, Cáceres, Spain, pp 466–475

Giménez-Anaya A, Herrero J, Rosell C, Couto S, García-Serrano A (2008) Food habits of wild boars (*Sus scrofa*) in a Mediterranean coastal wetland. Wetlands 28:197–203

Giménez-Anaya A, Herrero J, García-Serrano A, García-González R, Prada C (2016) Wild boar battues reduce crop damages in a protected area. Folia Zool 65(3):214–220

Gómez JM, Hódar JA (2008) Wild boars (*Sus scrofa*) affect the recruitment rate and spatial distribution of holm oak (*Quercus ilex*). For Ecol Manag 256:1384–1389

Gonçalves P (in prep.) Biological indicators of wild boar management in Mediterranean ecosystems. PhD Thesis

Gonçalves P, Fernández-Llario P, Risco D, Gil M, Varela E, García-Jiménez WL, Cerrato R, Martínez R, Hermoso de Mendoza J, Mateos C (2013) Novel method to estimate wild boar

populations based on food intake. VIII International Symposium on Wild Fauna. Book of abstracts. León, Spain, 2013
González J, Herrero J, Prada C, Marco J (2013) Changes in wild ungulate populations in Aragon, Spain between 2001 and 2010. Galemys 25:51–57
Gortázar C, Herrero J, Villafuerte R, Marco J (2000) Historical examination of the status of large mammals in Aragon, España. Mammalia 64:411–422
Gortázar C, Acevedo P, Ruiz-Fons F, Vicente J (2006) Disease risks and overabundance of game species. Eur J Wildl Res 52:81–87
Groves CP (1981) Ancestors for the pigs: taxonomy and phylogeny of the genus Sus. Aust Natl Univ Dep Prehist Tech Bull 3:196
Hermoso de Mendoza J, Parra A, Tato A, Alonso RJM, Peña J, García-Sánchez A, Larrasa J, Teixidó J, Manzano G, Cerrato R, Pereira G, Fernández-Llario P, Hermoso de Mendoza J (2006) Bovine tuberculosis in wild boar (*Sus scrofa*), red deer (*Cervus elaphus*) and cattle (*Bos taurus*) in a Mediterranean ecosystem (1992–2004). Prev Vet Med 74:239–247
Herrero J (2003) Adaptación funcional del jabalí *Sus scrofa* L. a un ecosistema forestal y a un sistema agrario intensivo en Aragón. Consejo de Protección de la Naturaleza de Aragón, Zaragoza, España
Herrero J, Irizar I, Laskurain A, García-Serrano A, García-González R (2005) Fruits and roots: wild boar foods during the cold season in the southwestern Pyrenees. Ital J Zool 72:49–52
Herrero J, García-Serrano A, Couto S, Ortuño VM, García-González R (2006) Diet of wild boar *Sus scrofa* L. and crop damage in an intensive agroecosystem. Eur J Wildl Res 52:245–250
Herrero J, García-Serrano A, García-González R (2008) Differences in reproductive and demographic parametres in two Iberian wild boar *Sus scrofa* L. populations. Acta Theriol 53(4):355–364
Irizar I, Laskurain NA, Herrero J (2004) Wild boar Frugivory in the Atlantic Basque Country. Galemys 16(N.S.):125–134
Lagos L, Picos J, Valero E (2012) Temporal pattern of wild ungulate-related traffic accidents in northwest Spain. Eur J Wildl Res 58(4):661–668
Langbein J, Putman R, Pokorny B (2011) Traffic collisions involving deer and other ungulates in Europe and available measures for mitigation. In: Putman R, Apollonio M, Andersen R (eds) Ungulate management in Europe. Cambridge University Press, Cambridge/Reino Unido, pp 215–259
Lecomte T (2007) Causes, impact et gestion d'une surpopulation de sanglier á la réserve naturrelle des Mannevilles (Marais Vernier – EURE). Colloque Modalités de Gestion du Sanglier, Marzo de 2007
Lеránoz I (1983) Sobre la relación del jabalí (*Sus scrofa* L.) con la agricultura, en Navarra septentrional XV Congreso Internacional de Fauna Cinegética y Silvestre, Trujillo, Spain, pp 639–645
Leránoz I, Castién E (1996) Evolución de la población del jabalí (*Sus scrofa* L., 1758) en Navarra (N península ibérica). Miscel·lània Zoològica 19:133–139
Lopes F, Borges J (2004) Wild boar in Portugal. Galemys 16(special issue):243–251
Malgalhaes C (1983) Javali (Notas de Biologia, idade e troféu). Direcção Geral das Florestas – Direcção de Serviços de Caça. Internal publication, Portugal, 25 pp
Malo J, Suárez F, Díez F (2004) Can we mitigate animal-vehicle accidents using predictive models? J Appl Ecol 41:701–710
Marco J, Herrero J, Escudero MA, Fernández-Arberas O, Ferreres J, García-Serrano A, Giménez-Anaya A, Labarta JL, Monrabal L, Prada C (2011) Veinte años de seguimiento poblacional de unguladosen Aragón. Pirineos 166:135–153
Markina F (1998) Estudio de las poblaciones de corzo (*Capreolus capreolus*) y jabalí (*Sus scrofa*) y análisis de su explotación cinegética en el Territorio Histórico de Álava. PhD thesis, University of León, León, Spain
Markina F (1999) Accidentes de carretera con ungulados cinegéticos en el territorio histórico de Álava. In: Asociación Técnica de Carreteras (ed) Fauna y Carreteras. Asociación Técnica de Carreteras, Madrid, pp 129–138

Martínez-Rica JP (1980) Nuevos datos sobre conducta social del jabalí, en condiciones de semilibertad. Publicaciones del Centro Pirenaico de Biología Experimental 12:19–54

Massei G, Kindberg J, Licoppe A, Gačić D, Šprem N, Kamler J, Baubet E, Hohmann U, Monaco A, Ozoliņš J, Cellina S, Podgórski T, Fonseca C, Markov N, Pokorny B, Rosell C, Náhlik A (2015) Wild boar populations up, numbers of hunters down? A review of trends and implications for Europe. Pest Manag Sci 71(4):492–500

Matilde E (2012) Contributions to the prediction of road-collision accidents with wild boar (*Sus scrofa*). MSc thesis, Universidade de Évora, Évora, Portugal

Mayer JJ, Nelson EA, Wike LD (2000) Selective depredation of planted hardwood seedlings by wild pigs in a wetland restoration area. Ecol Eng 15:79–85

Morais JLP (1979) Introduction to the wild boar *(Sus scrofa L., 1758)* biology study in Portugal graduation report. Faculdade de Ciências de Lisboa, Lisboa, Portugal

Muñoz A, Bonal R (2007) Rodents change acorn dispersal behaviour in response to ungulate presence. Oikos 116:1631–1638

Navarro-González N, Mentaberre G, Porrero CM, Serrano E, Mateos A, López-Martín JM, Lavín S, Domínguez L (2012) Effect of cattle on salmonella carriage, diversity and antimicrobial resistance in free-ranging wild boar (*Sus scrofa*) in Northeastern Spain. PLoS One 7:e51614

Navarro-González N, Fernández-Llario P, Pérez-Martín JE, Mentaberre G, López-Martín J, Lavín S, Serrano E (2013) Supplemental feeding drives endoparasite infection in wild boar in Western Spain. Vet Parasitol 196:114–123

Navàs F, Rosell C, Romero de Tejada S, Santaeufèmia X, de Dalmases I (2010) Does the presence of wild board in protected wetlands affect sensitive species?. In: 8th international symposium on wild boar and Other Suids. York, 1–4 de Septiembre de 2010

Nores C (2010) Wid boar abundance and altitude. In: 8th international symposium on wild boar and other suids. Book of Abstracts. York (UK) 1–4 September 2010: 46

Nores C (2013) Manual para la gestión de poblaciones de mamíferos. INDUROT-Universidad de Oviedo, Mieres, 88 pp

Nores C, Fernández A, García E, Rodriquez R (1994) Informe sobre los daños del jabalí en el Principado de Asturias en 1993. Análisis de la variabilidad espacial del jabalí y su incidencia en las haciendas campesinas. Consejería de Medio Ambiente y Urbanismo, Principado de Asturias and University of Oviedo, Oviedo. Unpublisehd report. Oviedo, Spain

Nores C, González F, García P (1995) Wild boar distribution trends in the last two centuries: an example in Northern Spain. Ibex J Mt Ecol 3:137–140

Nores C, Fernández A, Corral N (2000) Estimación de la población de jabalí (*Sus scrofa*) por recuento de grupos familiares. Naturalia Cantabricae 1:53–59

Nores C, Rodríguez O, Segura A (2010) Wild boar counting on drives: testing a method. In: 8th international symposium on wild boar and other suids. Book of Abstracts. York (UK) 1–4 September 2010: 41

Nores C, Martín A, Ballesteros F, Blanco JC, García-Serrano A, Herrero J, Palomero G (in prep.). Wild boar increase limits brown bear reproduction in the Cantabrian Mountains

Oliver W, Leus K (2008) *Sus scrofa*. The IUCN red list of threatened species 2008: e.T41775A10559847

Palacio S, Bueno CG, Azorin J, Maestro M, Gomez-Garcia D (2013) Wild-boar disturbance increases nutrient and C stores of geophytes in subalpine grasslands. Am J Bot 100:1790–1799

Pérez-González J, Costa V, Santos P, Slate J, Carranza J, Fernández-Llario P, Zsolnai A, Monteiro NM, Anton IN, Buzgó J (2014) Males and females contribute unequally to offspring genetic diversity in the polygynandrous mating system of wild boar. PLoS One 9:e115394

Pita M, Casas S, Herrero J, Prada C, García-Post R (2012) Game reserves in Spain: the public management of hunt. Forest Systems, 21(3):398–404

Puigdefábregas J (1980) Observaciones sobre la hozadura del jabalí en ambiente forestal. Publicaciones del Centro Pirenaico de Biología Experimental 12:7–16

Purroy FJ, Clevenger AP, Costa L, Saenz de Buruaga M (1988) Demografía de los grandes mamíferos (jabalí, corzo, ciervo, lobo y oso) de la Reserva Nacional de Caza de Riaño: análisis de la prelación e incidencia en la ganadería. Biología Ambiental, 1 pp. 375–387

Putman R, Langbein J, Watson P, Gree P, Cahill S (2014) The management of urban populations of ungulates. In: Putman R, Apollonio M (eds) Behaviour and management of European ungulates. Whittles Publishing, Dunbeath, pp 148–177

Risco D, Llario P, Velarde R, García W, Benítez J, García A, Bermejo F, Cortés M, Rey J, de Mendoza J (2011) Outbreak of swine erysipelas in a semi intensive wild boar farm in Spain. Transbound Emerg Dis 58(5):445–450

Risco D, Fernández-Llario P, García-Jiménez WL, Gonçalves P, Cuesta JM et al (2013) Influence of porcine circovirus type 2 infections on bovine tuberculosis in wild boar populations. Transbound Emerg Dis 60:121

Risco D, Serrano E, Fernández-Llario P, Cuesta JM, Gonçalves P, García-Jiménez WL, Martínez R, Cerrato R, Velarde R, Gómez L (2014) Severity of bovine tuberculosis is associated with co-infection with common pathogens in wild boar. PLoS One 9:e110123

Rodrigues P, Herrero J, García-Serrano A, Prada C, Giménez-Anaya A, Ayala R, Fernández-Arberas O, Fonseca C (2016) Habitat use by wild boar *Sus scrofa* in Moncayo Nature Park, Spain. Pirineos 171:1–7

Rodríguez Berrocal J, Zamora Lozano M, Aaedina Blanco M (1982) Hábitos alimenticios del jabalí: un condicionante para su explotación cinegética racional Arch. Zootaxa 31:149–154

Rodríguez Campos S (2013) Epidemiología molecular de mycobacterium bovis y mycobacterium caprae en España. PhD thesis, Universidad Complutense de Madrid

Rodríguez-Morales B, Marey MF, Díaz-Varela ER (2012) Caracterización temporal de los accidentes de tráfico con corzo y jabalí en la provincia de Lugo. III Reunión sobre Ungulados Silvestres Ibéricos 'Ungulados, biodiversidad y actividades humanas: gestión de conflictos' 4-5 Octubre. Castelló d'Empúries (Girona)

Rogers PM, Myers K (1980) Animal distribution, landscape classification and wildlife management, Coto Doñana, Spain. J Appl Ecol 17:545–565

Rosell C (1998) Biologia i ecologia del senglar *(Sus scrofa L, 1758)* a dues poblacions del nord-est ibèric. Aplicació a la gestió. PhD thesis, Departament de Biologia Animal Universitat de Barcelona, España

Rosell C, Herrero J (2007) *Sus scrofa* Linnaeus, 1758. In: Palomo LJ, Gisbert J, Blanco JC (eds) Atlas y libro rojo de los mamíferos de España. Dirección General para la Biodiversidad-SECEM -SECEMU, Madrid, pp 348–351

Rosell C, Carretero MA, Bassols E (1998) Seguimiento de la evolución demográfica del jabalí (*Sus scrofa*) y efectos del incremento de presión cinegética en el Parque Natural de la Zona Volcánica de la Garrotxa. Demografía del jabalí y presión de caza. Galemys 10:59–73

Rosell C, Fernández-Llario P, Herrero J (2001) El jabalí (*Sus scrofa* Linnaeus, 1758). Galemys 13:1–26

Rosell C, Navàs F, Romero S, De Dalmases I (2004) Activity patterns and social organization of wild boar (*sus scrofa*, l.) in a wetland environment. preliminary data on the effects of shooting individuals. Galemys 16:153–162

Rosell C, Fernández Bou M, Giménez-Anaya A, Guardiola M, Navàs F, Espelta JM (2006) Seasonal variation in habitat use of wild boar in Pyrenean National Park. In: 6th international symposium on wild boar (*Sus scrofa*) and on sub-order suiformes. Monastery of Kykkos, Chipre, 25–28 October, 2006

Rosell C, Navàs F, Casanovas R (2007) Wild boar monitoring program in Catalonia (NE Spain). A basis for adaptive management. International Union of Game Biologists. XVIII Congreso. Uppsala, Suecia, 13–18 August 2007

Rosell C, Navàs F, Fernández-Pou M, Carol Q (2008) Wild boar-vehicle collisions. Spatial and temporal patterns and measures for the mitigation of the conflict. In: 7th international symposium on wild boar (*Sus scrofa*) and on Sub-order Suiformes. Sopron, Hungary 28–30 August de 2008

Rosell C, Navàs F, Carol J, Fernández M, Romero de Tejada S (2010) Use of fauna passages by wild boar: some change is observed. In: 8th international symposium on wild boar and other Suids. York, 1–4 September 2010

Rosell C, Navàs F, Romero S (2012) Reproduction of wild boar in a cropland and coastal wetland area: implications for management. Anim Biodivers Conserv 35:209–217

Rosell C, Fernández-Bou M, Camps F, Boronat C, Navàs F, Martínez M, Sorolla A (2013) Animal-vehicle collisions: a new cooperative strategy is needed to reduce the conflict. In: Proceedings ICOET 2013 international conference on ecology and transportation. Scottsdale, Arizona, pp 23–27

Rosell C, Muñoz A, Navàs F, Fernández-Bou M, Sunyer P, Bonal R, Espelta JM (2016) Principals resultats del projecte "Impactboar" de seguiment de l'activitat del senglar al Parc Nacional d'Aigüestortes i Estany de Sant Maurici. In: X Jornades sobre Recerca al Parc Nacional d'Aigüestortes i Estany de Sant Mauric. Departament de Territori i Sostenibilitat – Parc Nacional d'Aigüestortes i Estany de Sant Maurici, pp 209–218

Ruiz-Fons F, Vicente J, Vidal D, Höfle U, Villanúa D, Gauss C, Segalés J, Almería S, Montoro V, Gortázar C (2006) Seroprevalence of six reproductive pathogens in European wild boar (*Sus scrofa*) from Spain: the effect on wild boar female reproductive performance. Theriogenology 65:731–743

Sáenz de Buruaga M (1987) Evolución de las áreas de distribución de corzo, ciervo y jabalí en la Montaña Leonesa. Actas de las Jornadas de Estudios sobre la Montaña, Urz, León, pp 315–322

Sáenz De Buruaga M (1995) Alimentación del jabalí (*Sus scrofa*) en el norte de España. Ecología 9:367–386

Sáenz de Buruaga M, Costa L, Purroy F (1987) Distribution and abundance of three wild ungulates in the Cantabrian Mountains. 18 Congres of the IUGB, Cracovia, Poland, pp 627–630

Sáenz de Santamaria A, Tellería JL (2015) Wildlife-vehicle collisions in Spain. Eur J Wildl Res 61:399–406

Sáez-Royuela C (1989) Biología y Ecología del jabalí (*Sus scrofa*). PhD thesis collection, INIA 78. Ministerio de Agricultura Pesca y Alimentación. Instituto Nacional de Investigaciones Agrarias, Madrid, Spain

Saez–Royuela C, Telleria JL (1986) The increased population of the Wild Boar (*Sus scrofa* L.) in Europe. Mammal Rev 16:97–101

Saez–Royuela C, Telleria JL (1987) Reproductive trends of the wild boar (*Sus scrofa*) in Spain. Folia Zool 36:21–25

Saez–Royuela C, Telleria JL (1988) Las batidas como método de censo en especies de caza mayor: aplicación al caso del jabalí (*Sus scrofa* L.) en la provincia de Burgos (Norte de España). Doñana. Acta Vertebrata 15:215–223

Santaeufèmia FX (2008) Estat actual de les orquídies del Remolar-Filipines. Consorci per a la Protecció i Gestió dels Espais Protegits del delta del Llobregat. Unpublished report

Santos P (1994) Interacção entre o javali, *Sus scrofa L. 1758,* e a perdiz-vermelha, *Alectoris rufa (L.)* 1758: dinâmica populacional e ordenamento cinegético. University of Évora, Évora, Portugal

Santos P, Mexía-de-Almeida L, Petrucci-Fonseca F (2004) Habitat selection by wild boar *Sus scrofa* L. In Alentejo, Portugal. Galemys 16(Special Issue):167–184

Santos N, Correia-Neves M, Ghebremichael S, Källenius G, Svenson SB, Almeida V (2009) Epidemiology of Mycobacterium bovis infection in wild boar (*Sus scrofa*) from Portugal. J Wildl Dis 45:1048–1061

Sarmento P, Cruz J, Fonseca C, Eira C (2010) Occupancy of wild boars in a non-hunting area. A camera trapping approach. Booklet of the 8th international symposium on wild boar and other Suids, York, UK

Serôdio HM (1985) Alguns aspectos da Biologia do Javali (Sus scrofa L., 1758) em Portugal. Relatório de estágio. Faculdade de Ciências de Lisboa. 257 pp

Sforzi A, Tonini L (2004) Le catture di cinghiali nel Parco Regionale della Maremma: risultati, riflessioni critiche e spunti gestionali. Comunicación presentada al "Seminario tecnico sulle catture di cinghiale nelle aree protette". Badia Prataglia, Italy, 18–19 Novembre 2004

Stillfried M, Fickel J, Börner K, Wittstatt U, Heddergott M, Ortmann S, Kramer-Schadt S, Frantz AC (2016) Do cities represent sources, sinks or isolated islands for urban wild boar population structure? J Appl Ecol 54(1):272–281

Tellería JL, Sáez-Royuela C (1984) The large mammals of central Spain. Mammal Rev 14:51–56

Tellería JL, Sáez-Royuela C (1985) L'évolution démographique du sanglier (*Sus scrofa*) en Espagne. Mammalia 49:195–202

Tellería JL, Sáez-Royuela C (1986) El uso de la frecuencia en el estudio de la abundanciade grandes mamíferos. Acta Oecologica Oecologia Applicata 7:69–75

Tolon V (2010) Impacts of refuges on wild boar – hunters spatial interactions and wild boar mortality. In: Oral communication. 8th international symposium on wild boar (Sus scrofa) and on Sub-order Suiformes. York, England

Torrellas M (2015) Ungulate-vehicle collisions in Catalonia: Identifying the main landscape and road-related variables describing the most hazardous road locations. Màster d'Ecologia Terrestre i Gestió de la Biodiversitat. Especialitat en Ecologia Terrestre. Universitat Autònoma de Barcelona, Barcelona, Spain

Torrellas M, Rosell C, Fernández-Bou C, Andrášik R, Sedoník J, Bíl M, Camps F, Seiler A, Rodà F (2016) Main landscape and road-related variables describing ungulate-vehicle collisions hazardous locations in Catalonia. IENE International Conference 2016 (accepted comunication)

Turchin P (2003) Complex population dynamics: a theoretical/empirical synthesis. Princeton University Press, Princeton. 456 pp

Uzal A, Nores C (2004) Endogenous and exogenous constraints in the population changes of wild board (*Sus scrofa* Linnaeus, 1758). Galemys 16(Special issue):83–98

Valet G, Rosell C, Chayron L, Fosty P, Spitz F (1994) Regime alimentaire automnal du sanglier (*Sus scrofa*) en Ariege, France, et en Catalogne, Espagne. Gibier Faune Sauvage 11:313–326

Venero JL (1984) Dieta de los grandes fitófagos silvestres del Parque Nacional de Doñana. Doñana. Acta Vertebrata 2:19–130

Venero JL (1983) Gregarismo del jabalí (Sus scrofa) en el Parque Nacional de Doñana, España. XV° Cong Int Fau Cineg Silv 1:331–339 Trujillo (Cáceres).

Vericad JR (1971) Estudio faunístico y biológico de los mamíferos del Pirineo. Publicaciones del Centro Pirenaico de Biología Experimental

Vericad JR (1983) Estimación de la edad fetal y períodos de concepción y parto del jabali (*Sus scrofa* L.) en los Pirineos occidental Proceedings of the IUGB XV congress, Trujillo, España

Vicente J, León-Vizcaíno L, Gortázar C, Cubero MJ, González M, Martín-Atance P (2002) Antibodies to selected viral and bacterial pathogens in European wild boars from southcentral Spain. J Wildl Dis 38:649–652

Vieira-Pinto M, Alberto J, Aranha J, Serejo J, Canto A, Cunha MV, Botelho A (2011) Combined evaluation of bovine tuberculosis in wild boar (*Sus scrofa*) and red deer (*Cervus elaphus*) from Central-East Portugal. Eur J Wildl Res 57:1189–1201

Vingada J, Fonseca C, Cancela J, Ferreira J, Eira C (2010) Ungulates and their management in Portugal. In: Apollonio M, Andersen R, Putman RJ (eds) European ungulates and their management in the 21st century. Cambridge University Press, Cambridge

Chapter 10
Traveling in a Fragile World: The Value of Ecotourism

Ernesta Martina Esposito, Davide Palumbo, and Pia Lucidi

10.1 Introduction

What is ecotourism and why is it a current topic more than ever? Tourism represents a significant economic sector and is further forecasted to grow at a global level—exceeding in the first few months of 2018 the most optimistic growth expectations according to WTTC (2018a).

According to the International Ecotourism Society, ecotourism can be defined as a form of responsible tourism that prefers natural areas and that focuses its attention and commitment to conserve the environment and sustain the well-being of local people through interpretation (of heritage, traditions) and education (habitats, animals, cultures).

Ecotourism has the potential to contribute, directly or indirectly, to all the objectives set by the 2030 Agenda for Sustainable Development, which establishes ambitious global targets for people, the planet, prosperity, and peace through partnerships (UNWTO 2019; WCED 1987). Moreover, it provides the opportunity to preserve natural areas, through natural resource management and increasing environmental awareness and eco-friendly practices; provide sustainable economic growth of local communities in countries like Nepal, Costa Rica, or Ecuador; preserve indigenous

E. M. Esposito (✉)
Universita' degli studi di Teramo, facoltà di Medicina Veterinaria, Piano d'Accio,
Tassarnas Djurklinik, Göteborg, Sweden
e-mail: ernestamarti.esposito@studenti.unite.it

D. Palumbo
Museo dell'Ecologia di Cesena, Cesena, Italy

P. Lucidi
Facoltà di Bioscienze e Tecnologie Agro-Alimentari e Ambientali, Università degli Studi di Teramo, Polo Silvio Spaventa, Teramo, Italy
e-mail: plucidi@unite.it

F. M. Angelici, L. Rossi (eds.), *Problematic Wildlife II*,
https://doi.org/10.1007/978-3-030-42335-3_10

culture and tradition through educational programs; and reinvest money for conservation efforts like the protection of species or reforestation.

However, critics to ecotourism emphasize the negative impacts that this industry has on local people and environments as a consequence of long distance travel (like the pollution generated by planes) and the negative impacts deriving from the presence of tourists in delicate environments (and the related increase of waste or pollution).

This paper aims to provide a general examination of the available data about ecotourism activities on a global level, presenting examples from representative countries worldwide.

We consider both the qualitative and quantitative aspects of this industry, trying to focus on what it represents in terms of its impacts and benefits for the country's natural resources, communities, and economy.

In the first part of the manuscript, we compare ecotourism to other forms of natural resource uses such as trophy hunting and mass tourism, trying to evaluate whether these represent a preferable alternative or not in terms of sustainability and economic benefits.

The next section is divided by regions: Africa, the Americas, Europe, and Australia.

For the study, we used statistics made available mainly by the World Travel and Tourism Council (WTTC) and the UN Environment World Conservation Monitoring Centre (UNEP-WCMC), together with information from specialized international literature.

10.2 Ecotourism: Definition, Principles, and Practices

Though the term "ecotourism" does not have a univocal definition, the common denominator of all the definitions is the fact that it is mainly linked to nature with minimal environmental impact and therefore on a small scale (Edwards et al. 2003). All the definitions also start from some basic assumptions concerning the form of "responsibility" linked to the principles of social and economic respect and respect for environments and cultures. The concept of *education* is one of the fundamental keys of ecotourism. Although virtually all tourism that takes place in natural environments always involves a certain degree of learning, ecotourism produces a different experience. It lies precisely in the *education* and above all in the *interpretation* of the natural environment and of every cultural-associated manifestation. It involves a conscious process based on defined learning objectives and the use of specific learning procedures, with the aim of revealing meanings and correlations (Blamey 2001).

Whatever the definition, ecotourism should produce at least three objectively verifiable outcomes: (1) be a positive force for conservation, emphasizing the protection and perpetuation of naturalistic scenarios and attractive local peculiarities; (2) bring economic benefit to the host communities and ensure that the people who must endure the social and environmental impacts of tourism development also share in rewards; and (3) promote environmental awareness both among tourists and local communities (Edwards et al. 2003). This requires the planning, selection, implementation, and evaluation of tourism sustainability indicators, and "in this respect, ecotourism is the segment with the highest growth in the world and with the greatest potential to integrate the dimensions of sustainability. Ecotourism is an activity that, in the best of its meanings, produces a minimal impact on the environment, and relates aspects of learning to conservation, understanding and appreciation for the environment and for the cultures visited, it is usually established in virgin or well-conserved areas or in landscape where the presence of human beings is minimal The tourist is motivated to be educated and is sensitized from the environmental and cultural point of view through the experience with the nature" (Camacho-Ruiz et al. 2016).

10.3 Ecotourism and the Value of Natural Capital

Because ecotourism is often grouped together with other forms of naturalistic tourism, it is difficult to grasp its real entity and the rate of its growth. In the absence of some traditional measurement tools used to calculate the expansion of general tourism, other indicators should be used to measure its growth. For example, growth in ecotourism education, international recognition and regional support, international funding opportunities, and growth in tourism eco-certification and eco-label programs, each of these indicators provides a look beyond traditional statistical analysis, both on the current position of ecotourism and expected future growth (Hawkins and Lamoureaux 2001).

Nature-based tourism thrives in many countries, but the target of this type of tourism, namely, wildlife and habitats, is increasingly threatened by human demographic expansion, economic activity, illegal poaching, and lack of funding (Twining-Ward et al. 2018).

In August 2018, the total number of protected areas registered in the World Database on PAs (WDPA) amounted to 235,536. Worldwide, protected areas cover almost 21% of inland waters, 20% of natural forests, 19% of mountainous surfaces, 17% of islands, and 13% of arid areas on the planet (UNEP-WCMC and IUCN 2016).

To evaluate the potential benefits of tourism to the economy of protected areas is not an easy task. There is an emerging evidence base that PAs do work, especially when they are well managed, but understanding what constitutes good management is an ongoing challenge (Coad et al. 2015).

WTTC, recognizing that the measure of the direct contribution of T&T to GDP is not enough to assess the real revenue due to the sector, aims to capture its indirect and induced impacts through continuous research.

Successful nature-based tourism experiences are emerging across the world, especially in Southern and Eastern Africa, Southeast Asia, Latin America, and the Caribbean. Well-planned, sustainably run tourism operations also enhance the perceived value of live animals, reduce poaching, and increase investments in protected areas and reserves and can also provide opportunities for rural communities to improve their livelihoods through tourism-related jobs, revenue-sharing arrangements, and co-management of natural resources (Twining-Ward et al. 2018).

Wildlife tourism is often focused on one or more charismatic animal species, such as big carnivores (tiger, jaguar, leopard, lion, panther), elephants, giraffes, apes (gorilla, orangutan), cetaceans (whale, dolphins), coral reef ecosystems, etc. Many travelers, indeed, choose their destination based on the presence of their beloved animal in a country or in an area. In that case, the tourists' expenditure can be obtained from the park's/reserve's entry fees.

A few examples are given in Table 10.1 (data from Twining-Ward et al. 2018).

Sometimes ecotourism can become such a tool for financial development and economical security to drive people to abandon other works and become agro-eco farmers for a sustainable use and protection of local resources, which in turn attract ecotourism. This is the case of Cuba, where many areas, before becoming agro-ecotourism enterprises, were vacant and overrun with weeds and trash, and now doctors, engineers, and former government ministers have left their previous job to work on these farms for better salaries and, as a result of more revenue, the farms can sustain new positions (Duffy et al. 2016).

The economy of protected areas becomes essential for sustainable development to occur in poor regions and is a key tool to achieve many sustainable development goals (SDGs). For example, the Protected Place 2016 Report (UNEP-WCMC and IUCN 2016) described a few benefits whose value is higher than money revenue: More than one billion people depend on protected areas for a significant percentage of their livelihood, and physical activity within Victoria Parks in Australia has resulted in health cost savings of about 200 million Australian dollars. Between 2000 and 2005, unprotected humid tropical forests lost about twice as much carbon for deforestation compared to the equivalent protected forest areas. Protected areas provide a significant proportion of the drinking water for 1/3 of the world's 105 largest cities (Stolton et al. 2010).

Protected areas also provide services, such as pollination, that are vital to produce foodstuff. In Costa Rica, the economic value of the pollination services provided by feral bees to coffee planters has been assessed by measuring the variation in yield in areas adjacent to, proximate to, and distant from patches of forest that

Table 10.1 Some of the key species that attract tourists as highlighted in the report of Twining-Ward et al. 2018

Species	Countries	Visitors n.	Expenditure (USD)	Time
Tigers	India	324,000	300,000	2013–2014
Tigers	India	78,235	430,383	2016–2017
Jaguar	Brazil	nd	6.8 million/year	nd
Orangutan	Borneo	nd	23 million	2011
Mountain gorilla	Rwanda	20,000	15 million	2014
Elephant	Africa	nd	1.6 million/elephant	2013

provided habitat to pollinating bees. The increased yield value of the price received by planters provided an estimate of the added producers' surplus generated by the free pollination service of the bees (Freeman III et al. 2014). The annual pollination value was estimated to be between USD 120 billion and USD 200 billion globally; some estimates suggest that over 75% of crops rely on pollination by animal vectors (Stolton et al. 2010).

Marine protected areas play an important role in conservation and fishing as they can help fish stocks to replenish and restock waters beyond the boundaries of the area; they also help conserve important habitats for aquatic life, promote the development of natural biological communities, and are generally more resilient to threats such as climate change (Stolton et al. 2010). It has been estimated that the conservation of 20–30% of global oceans in marine protected areas could create one million jobs, support a fishery worth 70–80 billion USD per year, and provide ecosystem services with a gross value of roughly 4.5–6.7 trillion USD/year (UNEP-WCMC and IUCN 2016).

In general, the average global value of wetlands for flood control and storm mitigation has been estimated (in 2000 figures) at USD 464 per ha per year (Stolton et al. 2010).

The importance of forests to the planet has been highlighted in the 2030 Agenda, which places the forest management and sustainability into the international development framework and emphasizes the importance of these objectives in both developing and developed country. In fact, the need to combat climate changes has resulted in increased attention to the role that forests play: Forests indeed capture and store carbon and dwindle the large quantity of CO_2 derived from human activities (i.e., up to 22–26% since the turn of the century) (Morrison-Métois and Lundgren 2016). According to the Intergovernmental Panel on Climate Change (IPCC, 2013), deforestation and forest degradation are the second leading human cause of CO_2 emissions contributing to global warming. The creation of forests PAs has been used by different governments as a method to safeguard the forests from deforestation and degradation. However, the creation of a PA itself can be poorly effective in reducing deforestation, while participation of local people and forest dwellers in the forest management can make PAs more successful and avoid conflict (Assunção et al. 2012; Porter-Bolland et al. 2012; Nolte et al. 2013; Nepstad et al. 2014; Morrison-Métois and Lundgren 2016). But, it has been demonstrated that PAs management, account too: Areas that allow sustainable use are more effective than strictly protected areas, and indigenous areas are the most effective of all (Morrison-Métois and Lundgren 2016). Among locals, the elderly, together with people with high education level, seem to demonstrate a stronger conservation behavior, the former because they often grew up in close contact with the forest—depending upon it for food, health, and livelihood (Stem et al. 2003). The participation of local communities thus seems to be the real breakthrough for a successful conservation. Ecotourism, in this regard, by ensuring a regular income to forests dwellers and an accurate and updated educational support, can help local communities or private owners to not shift to other commercial activities which can be in competition with the protection of forests. Some ecologists, in fact, argue that protected areas which are surrounded by inimical land uses cannot avoid long-term extinction of the species they purport to protect (Weller 2017).

Across the world, areas with high or important biodiversity are often located within indigenous peoples' and local communities' conserved territories and areas (ICCAs), and traditional and contemporary systems of stewardship embedded within cultural practices enable the conservation, restoration, and connectivity of ecosystems, habitats, and specific species in accordance with indigenous and local world views (Jonas et al. 2012). Indigenous peoples have always protected their lands and the rich resources they hold, managing their resources through customary laws and traditional practices. Some of the best protected biodiversity-rich areas are those in the lands and territories of indigenous peoples (Walker Painemilla et al. 2010).

The literature suggests that the costs for the effective management of protected areas (in areas with similar socioeconomic, institutional, and management contexts) of tropical developing countries can be on average between 185 and 644 USD/km^2/year (Emerton et al. 2015). It has been estimated that protected areas globally receive eight billion visits per year (sustainable wildlife tourism), generating as much as USD 600 billion of tourism expenditure annually, in contrast to the cost to protect these areas, which accounts for less than USD 10 billion per year (Twining-Ward et al. 2018).

The financial resources needed to globally implement the Aichi Biodiversity Targets have been estimated at 150–440 billion USD/year (Convention on Biological Diversity, the 11th Conference of the Parties meeting (COP 11), held in India in 2012). According to Balmford et al. (2015), altogether, the terrestrial PA (excluding only the smallest ones) roughly receives eight billion visits/y—which generates approximately USD 600 billion/year in direct in-country expenditure and US $250 billion/year in consumer surplus (Marshall 1920, quoted by Freeman III et al. 2014). These figures dwarf current, typically inadequate, spendings on PAs conservation.

If one considers that over one billion people, almost a sixth of the world's population, depend on protected areas for a significant percentage of their livelihoods, whether it be food, fuel, or support of economic activity, one can see how the efficient functioning of protected areas can offer significant returns (TEEB 2009). In fact, we do not lose biodiversity and ecosystems primarily for lack of conservation funding but also due to poor governance, wrong policies, perverse incentives, and other factors (Berghöfer et al. 2017). Globally, it is estimated that ecosystems within protected areas can deliver a value of USD 100 for every dollar invested in management for the maintenance and improvement of their ecosystem services (Stolton et al. 2010).

Since conservation spending has proved to be able to support economic activities such as forestry, fishing, wildlife, and tourism, governments have to ask if short-term solutions, such as budgetary cuts to environmental programs in favor of other defined social priority, could be a useful strategy to overcome poverty, unemployment, and other social challenge in the long run, because these choices will put a heavy weight on the future generations (Adelaja et al. 2008).

10.4 Trophy Hunting

The hunting of wildanimals as trophies—which consists of tourists paying to kill and retain the body of the animal they shot or part of it—due to the ethical implications that it entails has attracted a wide attention in recent years. The targets of this kind of hunting (rhinos, elephants, hippo, giraffes, buffalo, kudu, bush pig, blue sheep, lions, leopard, cheetahs, caracal, genet, bears, wolf, etc.) are relegated—after being commoditized, killed, and dismembered—to the sphere of mere things and turned into souvenirs, oddities, and collectibles, a practice which is morally indefensible (Batavia et al. 2018).

There have been significant trophy hunting industries in South Africa, Mozambique, Namibia, Tanzania, Zambia, and Zimbabwe and smaller industries in Ethiopia and various West and Central African countries, and Botswana has banned all lion trophy hunting since 2008 and all trophy hunting in public areas since 2014 (Lindsey et al. 2015; MacDonald 2016; Price 2017).

The International Fund for Animal Welfare (IFAW) estimates that as many as 1.7 million hunting trophies, of which at least 200,000 trophies of threatened taxa (an average of 20,000 trophies/year), could have been traded between nations in just 10 years (2004–2014).

Although a decreasing number of trophies should be expected to occur along the gradient of extinction risks, the attribution of a IUCN category seems to have no significant effect on the number of trophies recorded in the Safari Club International (SCI); on the contrary, hunting pressure increases on the most threatened species, with the occurrence of an anthropogenic Allee effect, i.e., the disproportionate valorization and exploitation of rare species (Palazy et al. 2012). In particular, a study has demonstrated that trophy hunting can constitute an underestimated threat to fragile felids species, since the value of rarity makes them disproportionately desirable to hunters (Palazy et al. 2011).

According to IFAW, SCI boasts approximately 50,000 members and 150 chapters. In 2015, it collected approximately \$3.6 million in membership service fees, product sales, dues, and subscriptions, and approximately \$14.4 million more is raised from its annual hunting convention. SCI members only have killed more than 2000 lions, 1800 leopards, almost 800 elephants, and 93 black rhinos over the past 60 years (IFAW 2016).

IUCN stresses the exercise of wildlife hunting requires urgent action and reform because, apart from the more publicized cases, there are many examples of weak governance, corruption, lack of transparency, excessive quotas, illegal hunting, inadequate monitoring, and other problems threatening wildlife worldwide (IUCN 2016). For example, during an investigation of the League Against Cruel Sports, the director of the E J Churchill Sporting Agency "admitted that '90% of the trophy fee goes straight into some Nigerian's pocket or African politician or whatever it is" (Rodriguez 2004). In fact, because of the large amount of money involved, criminal activity often accompanies trophy hunting (Milner-Gulland and Leader-Williams 1992, quoted by Naevdal et al. 2012). However, in the name of conservation, the

IUCN supports trophy hunting with a series of arguments; this consideration is validated by the fact that IUCN 2016 report explicitly sends the reader back to the *Guiding Principles for Trophy Hunting as a Tool for Conservation Incentives*. In 2016 again, IUCN claims that trophy hunting has been unfairly charged to lead to the decline of iconic species, which is not, and that trophy hunting is not a significant threat to any species. Moreover, it considers that photographic tourism cannot alternatively replace it because this tourism is only viable over a limited percentage of the areas currently used for hunting. Thus, what people know about trophy hunting are—for IUCN—just misinformation. Instead, IUCN considers trophy hunting a practice able to provide a substantial revenue flow from developed to developing countries, with local community benefit as high as 100%, reduce human-wildlife conflicts, and reduce illegal killing. However, at least the IUCN recognizes that because trophy hunting takes place in a great variety of countries, with different administrations, management, and ecological contexts, its impact on conservation varies enormously, from negative to neutral to positive (IUCN 2016).

Justifying a "good" trophy hunting opposed to a bad or neutral hunting is a limited version of utilitarianism, which judges the merits of an action by its positive and negative outcomes (MacDonald 2016). Yet conservation policies also involve judgments reflecting values, which, being very different to price, cannot be appraised by a mere economic point of view: Values can be sacrosanct (Can and MacDonald 2018).

In the absence of human effect, the mortality rate of adult lions is low, and trophy hunting is often the most common cause of death (Creel et al. 2016). Moreover, trophy hunting has serious genetic implications too, since the hunters' targets are primarily males with the largest manes or biggest horns, i.e., the stronger and healthier animals; thus, anthropogenic selection pushes traits away from their naturally selected optima (Coltman et al. 2003; Rodriguez 2004; Wilfred 2012; Coulson et al. 2017; Hariohay et al. 2018). Heavily hunted populations generate more conflict, because their age structure shifts toward younger inexperienced predators, who may turn to predictable but risky foods like livestock (Treves and Naughton-Treves 2005; Teichman et al. 2016). An uncommon effect of hunting was recorded in South Africa in the 1990s, where more than 40 white rhinos were killed by young orphaned African elephants experiencing premature musth (Slotow et al. 2000 quoted by Rodriguez 2004).

10.4.1 Trophy Hunting Incomes

According to Lindsey et al. (2007) and MacDonald et al. (2016), trophy hunting accounts for more than 200 million USD/year in gross revenue across sub-Saharan Africa, in other words a very modest contribution to the GDP (IUCN/PACO 2009) considering that this figure has been criticized for the weak and unverifiable sources, at least for the Lindsey paper (Price 2017).

Among big-game hunting, lion hunts attract the highest mean prices than any other trophy (USD 24,000–71,000), and it generates 5–17% of the national hunting income at the national level, especially in Mozambique, Tanzania, and Zambia (Lindsey et al. 2012). However, Zambia's revenue from trophy hunting is very little (MacDonald 2016), and the country's government shifted their policy on trophy hunting according to its lack of revenue, after banning the hunting on big cat in 2013, and later Zambia's national authorities reversed their decision because of concerns over a lack of income (MacDonald 2016). In some case, trophy hunting can only survive through subsidies, in absence of which it is not a financially viable business venture, as in the case of CAMPFIRE program in Zimbabwe. The program, promoted as a model for employing local communities in trophy hunting, only survived because it was given massive subsidies by the US government through USAID (USD eight million from 1989 to 1996, USD 20.5 million in 1997–2000). With only USD 2.5 million a year in program revenue, the trophy hunting project was making a massive loss (Rodriguez 2004).

Yet the contribution of trophy hunting to national economies is insignificant, the average contribution to the GDP being 0.59‰ (IUCN/PACO 2009), and only Namibia and Botswana have better results (4.52 and 1.85‰) on a per hectare basis. However, Botswana decided that more value can be earned from promoting safari tourism than hunting and closed the Okavango Delta area to hunting in 2009 (IUCN/PACO 2009).

The modest income generated by trophy hunting becomes clearer if we compare hunting with photographic tourism that accounts for 80% (and growing) of trips sold to go to Africa as wildlife watching tourism (WTO 2014).

Just to give an example, the total inbound tourism expenditure for wildlife watching at the Serengeti-Ngorongoro southern circuit, in Tanzania, was estimated in 2009 at USD 500 million/year (WTO 2014), of which USD 100 million/year considered pro-poor. It is also important to note that the majority of the inbound tour operators and tourism providers are owned by Tanzanians (WTO 2014). In Namibia, it has been estimated that the total expenditure in 2010 made by "photographic" tourists in protected areas was of the order of USD 154 million, against the 6.4 million (2018 exchange rates) spent by tourists attracted by the hunting concessions (Berghöfer et al. 2017). From 1990 to 2016, Namibia's successful model of nature-based tourism generated approximately 488 million USD (2018 exchange rates) to the net national income and created 5147 jobs, promoting community conservation, according to a World Bank report (Twining-Ward et al. 2018). In Zimbabwe, tourism is 6.4% of its GDP, as compared to 0.2% from trophy hunting (Markarian 2015). In South Africa, a large game ranch conducting ecotourism can generate nearly 4–4.5 times more revenue per year than the same area devoted to hunting: R 9–12 million vs. 2–3.5 million per annum (Taylor et al. 2016). In Zambia, the 176,000 visitors in 2005 realized an export value of USD 194 million, generating fiscal revenues of USD 5–8 million, which by far exceeded the USD one million in fund allocated to Zambia Wildlife Authority in the same year (WTO 2014).

Another study assessed that ecotourism on private game reserves generated "more than 15 times the income of livestock or game rearing or overseas hunting. Eco-tourism lodges in Eastern Cape Province produce almost 2000 rand (£180) per hectare. Researchers also noted that more jobs were created, and staff received extensive skills training" (Rodriguez 2004).

Finally, for the 145 tour operators (international and Africa-based) of the 34 countries surveyed by WTO, annual revenues of wildlife watching tours represented 80% of the total annual revenues of trips to Africa (2014).

Photographic ecotourism undoubtedly generates greater gross revenue than trophy hunting in Africa, with the number of tourists being large and generating more employment opportunities for local people compared to hunting (Lindsey et al. 2007). "For instance, an adult male lion in Amboseli National Park, Kenya, will draw $515,000 in foreign exchange revenue for wildlife watching, compared with $8,500 for sport hunting or $1,324 for a commercial skin (Thresher 1981). An elephant herd for viewing in Amboseli is worth $610,000 per year, hunting would result in less than 10% of this value" (Western and Henry 1979, quoted by Hoyt and Hvenegaard 2002).

Also in the 2016 IUCN report, it is said that photographic tourism, in comparison to hunting, can be a very valuable option in many places and has generated huge benefits for conservation, although it is only viable in a very limited proportion of currently managed wildlife territories for trophy hunt [it requires political stability, good transport links, minimal risks of disease, high density of wild animals to ensure observations, scenic landscapes, high capital investments, infrastructure, local skills and competences]. Yet some author supports the concept that—opposed to ecotourism—trophy hunting has a smaller footprint in terms of carbon emission, infrastructure development, and personnel and can generate more revenue from a lower volume of tourists, who are even interested in maintaining good-quality habitat in order to harvest high-quality individuals (Di Minin et al. 2016).

10.5 Whale Watching

There are only two areas in the oceans where whales are safe, i.e., the whale sanctuaries designated by the International Whaling Commission (IWC): One, established in 1979, is located in the Indian Ocean south to 55°S, and the second is in the Southern Ocean around Antarctica (south of 40°S), defined since 1994. The third sanctuary in the South Atlantic Ocean has never seen light: Although its realization has been repeatedly advocated by the Australian and New Zealand governments, it has been repeatedly rejected, as its application did not reach the needed votes, and this decision was reiterated at the 67th IWC meeting on the 11th of September 2018 at Florianopolis, Brazil (International Found for Animal Welfare 2018). Japan voted against, followed by other IWC new-entry countries, leading the International Fund for Animal Welfare (IFAW) to claim that there is a strong correlation between the compositions and voting patterns of countries that support pro-Japan membership in the IWC and the disbursement of aid from Tokyo under the Grant Aid for Fisheries program of overseas development aid (Mulvaney and Taylor 2013).

Moreover, a Pelagos Sanctuary for Mediterranean Marine Mammals has been established in the Ligurian basin of Mediterranean Sea in 1991 by the Italian Environment Ministry and became an international PA in 2002 when Italy, France, and Monaco princedom agreed to its recognition (Notarbartolo di Sciara et al. 2008).

Whaling is permitted in all other international sea waters, where three types of catching are allowed:

- Aboriginal subsistence (AS), a multispecies whaling which supports indigenous communities (e.g., Inuit).
- Commercial whaling, which has been subjected to a moratorium in 1986, and it is overtly conducted by Iceland and Norway today and can only be carried out by those countries under objection or reservation to the current moratorium (O'Connor et al. 2009).
- Scientific (special permit) whaling, whose license is issued by individual countries according to the international law on whaling, with IWC having an advisory role only, was previously carried out by different countries (Korea, Norway, Iceland, and Japan), but from 2008 onward, only Japan practices these catches (according to IWC), throughout Antarctic, NW Pacific, Japan, and Iceland sea.

Whaling killed almost three million whales in the twentieth century alone, with some populations estimated to have been reduced by 99% of their pristine abundance, both from the failure of regulatory efforts and for the large-scale illegal whaling by the former Soviet Union and Japan (Ivanshenko et al. 2013; Ivashchenko and Clapham 2015; Rocha et al. 2015; Clapham 2016). The credibility of IWC indeed has been seriously undermined by this unbelievable lack of control, leading to the assertion that "the future of the International Whaling Commission is tenuous and might be rescued by a 'whale conservation market" (Costello et al. 2012). Overall, whaling has more than doubled in the past 20 years (Gerber et al. 2014).

In addition to hunting, whales have to face other threats and conditions, mostly human-related, such as opportunistic killing or nontarget catch, by-catch (the accidental capture in fishing gear), collision with boats, noise (from ships, seismic surveys, sonars, military operations), pollution (PBC, oil, biotoxins, plastic), climate change, ocean acidification, overfishing, genetic risks associated with small populations size, and infectious disease (Perez 2003; Reeves et al. 2003; Read 2008; Knowles and Campbell 2011; Leaper and Miller 2011; Knowlton et al. 2015; Clapham 2016; Thomas et al. 2016; Porter and Lai 2017; Desforges et al. 2018; Ford et al. 2018). Given all the threats whales are exposed to and given their value in the ecosystem (discussed later), whales should deserve a supernational commission devoted to their protection rather than their hunting.

In fact, whales, as other threatened, endangered, and rare species, also have an intrinsic economic value, i.e., a nonconsumptive value, which consists of nonconsumptive use values such as viewing (as opposed to consumptive use values, such as harvesting) and nonuse values apart from on-site active use, which are usually attributed to bequest and existence values (Lew 2015).

According to a so-called stakeholder theory, ecotourism's promotion, by conferring to natural resources a sufficient economic value, could help to incentivize locals to preserve those resources rather than extract them, thus helping conservation strategies in situ (Fletcher and Neves 2012).

Whale watching is a globally recognized use of cetacean resources, and since 1955, it has been an activity of growing economic importance, leading the IWC to consider this issue since 1975 (IWC 2011). The International Fund for Animal Welfare (IFAW) has undertaken, since 1994, many efforts for the promotion of responsible whale watching worldwide as an alternative to commercial whaling, alongside with the attempt to make whale watching a source of scientific information for IWC as an alternative to "scientific whaling" carried out by Japan (O'Connor et al. 2009).

In contrast to whale watching, a generous estimate of the total annual profit from all global commercial whaling activity is suggested to be at around 31 million USD (Costello et al. 2012). However, whaling relies heavily upon government subsidies, at least in Japan (Cunningham et al. 2011), where these average around USD 9.78 million/year (Mulvaney and Taylor 2013). "The good people of Japan are paying billions to support a dying industry [...] Whaling is an economic loser in the 21st century" (Patrick R. Ramage, IFAW Whale Program Director, quoted by Mulvaney and Taylor 2013).

Whale watching is no more a young gamble; it has today become a strong and mature sector in T&T industry. Just 10 years ago, in 2008, it accounted for annual revenue of USD 2.1 billion and supported more than 13,000 employees in 119 countries, with approximately 13 million people participating in whale observation around the world (O'Connor et al. 2009). Most of the revenue came from indirect tourist expenditure (USD 1.2 of 2.1 billion, half of which originated from US whale watchers) that was likely to undervalue the total contribution of whale watching to the local economies (O'Connor et al. 2009).

Whale watching is a story of strong and effective conservation policy delivering economic and development opportunities around all corners of the globe, with the population of wild cetaceans recovering (O'Connor et al. 2009). Whale-watch tourists tend on average to be "higher-end" tourists with higher levels of expenditure than average inbound tourists (O'Connor et al. 2009).

10.5.1 Whale Watching-Related Issues

Whale watchingrecognizes some problems that raise concerns about the welfare of the target species. In fact, despite the practice going hand in hand with a population growth of whales worldwide, for example, in the case of the gray whale (*Eschrichtius robustus*), which had been very close to extinction, and the humpback whale (*Megaptera novaeangliae*) (Ivashchenko and Clapham 2015; Miller et al. 2015; Amerson and Parsons 2018; Pallin et al. 2018), the disturbance caused to the whales (not always confirmed; see Di Clemente et al. 2018), the failure to respect regulations, countries' different guidelines, and the ineffectiveness of some regulations can reverse the benefits or even be unsustainable from the point of view of whale protection (Parson 2012; Amerson and Parson 2018; Di Clemente et al. 2018). Cetaceans exhibit changes in behavior to whale-watching boat traffic such as change in surfac ing, diving, rail slapping and beaching, acoustic, group size and cohesion, and swimming speed and direction and altered feeding and resting (Parsons 2012). The most prevalent disruption to Eastern North Pacific gray whale behavior as a result of whale-watching activity is that they relocate and swim further offshore from their known preferred costal habitat (Parsons and Brown 2017; Amerson and Parson 2018). The review by Amerson and Parson (2018) reported that in gray whale nursing lagoons (Mexico), whales are known to approach small vessels with their calves and tourists are allowed to touch them, whereas along migratory routes and in feeding grounds, gray whales are not known to approach vessels. This difference in behavior, in Mexican waters in contrast to migratory routes, has led to the development of different international strictures. The problem is that sometimes there is a high level of noncompliance with the guideline strictures for sustainable whale watching, and in other cases, the guidelines are too complicated or not effective at all.

There is sufficient evidence that humpbacks have also changed their behavior: They are shifting migration times and changing resting and delivery sites (Ramp et al. 2015, cited by Meynecke et al. 2017). Being meteorological models become less and less predictable and being the whale-watching industry very vulnerable to climate change, especially when the observations are related to migratory species, the lack of sighting by too many tourists or the continuous movement from one place to another can in fact compromise the success of the activity.

The recent shift toward mass wildlife tourism, particularly in the whale and dolphin tourism sector, can make running tours very difficult in terms of tourist satisfaction, as these tourists can have a very narrow fixation on specific charismatic species and high expectations which are impossible to guarantee (Curtin, 2013). Strangely, these customers can be more difficult to satisfy than the serious wildlife watching tourists, which tend to have a broader interest and understand the chance factors of wildlife watching (Curtin 2013).

10.5.2 *Total Value of Whales*

1. Specifically, baleen whales can provide clues as to the nature, direction, and mechanisms of ecosystem shifts, such as where, when, and how "new" NPP is cycled (Moore 2016). Whales likely play a significant role in carbon sequestration, and restoring their populations to pre-whaling levels would potentially help mitigate global warming (Clapham 2016).
2. Baleen whales also act as ecosystem engineers, and their recovering numbers may actually buffer the marine ecosystem from destabilizing stresses associated with rapid change: With more baleen whales recycling nutrients vertically and

horizontally and with increasing numbers of bowhead and gray whales resuspending sediments during epibenthic and benthic foraging, the Pacific Arctic marine ecosystem will probably continue to change in ways now difficult to predict (Moore 2016).

3. Education: Changing human behavior is quite difficult, but experience of sighting a whale or a whale family has proven to support, in whale watchers, more significant comprehension of cetacean conservation and marine environment themes, through the evoking of emotions and feelings of responsibility rather than only providing information about whale biology and ecology (Garcia-Cegarra and Pacheco 2017).
4. Whale-watch operators have contributed millions of dollars toward whale research in terms of raw data, substantial free boat time for researchers, as well as financial contributions; the best operators have created "floating classrooms" from which to learn about the sea (Hoyt and Iñígues 2008). Whale and Dolphin Tracker has proven to be a reliable method for citizen science data collection, contributing to the protection and enhanced understanding of the species they are watching (Currie et al. 2018).
5. "Yet, an even greater indication of true value, perhaps, comes from the degree to which many of us depend on whales for our sense of wonder — our hope about the future of the sea itself. We want a world alive with possibility, a world in which whales swim free in the sea. Even for those of us who may never see whales, we want to reserve the possibility that we could see them one day — something that economists seek to measure as the elusive but important so-called 'existence value' and 'option value" (Afterword of Erich Hoyt in O'Connor et al. 2009).

10.5.3 How Much Does Whaling Cost? How Much Money Do NGOs Spend to Save Whales?

Whalers spend millions of dollars to harvest whales, many of which are then sold on global markets, with a per-whale profit in the ballpark of 13,000–85,000 depending on the species, respectively, minke whale and fin whale (Costello et al. 2012). A conservative estimate of the amount spent annually by nonprofit organizations on anti-whaling (Greenpeace USA, Greenpeace International, Sea Shepherd Conservation Society, WWF International, and WWF UK) is instead equal to USD 25 million. Knowing that, Costello et al. (2012) suggested that rather than supporting anti-whaling protests and movements (and their accompanying carbon footprint), the money spent by NGOs could be used to purchase whales, arguably with the same or better effect.

Similarly, Gerber et al. (2014) studied a mixed biological and economical model relying on (i) whale population dynamics, (ii) demand for whales (both by conservationists and whalers), and (iii) rule allocation for quota shares. From their calculation and "for all case studies, conservationists are made better off by the increased whale population and whalers are made better off by more efficiently allocating the quota between selling to conservationists and harvesting the whales to sell in the market." The estimated cost to save whales is roughly USD 114 million for 20 years, to save a total of 8424 whales (Gerber et al. 2014), and by multiplying the 25 million dollars spent in 1 year of anti-whaling campaign by 20 years, the final figure would reach USD 500 million alone that would be able to stop whaling. Similar conclusion can be found in the paper of Huang et al. (2017), whose calculation demonstrated that the conservationist could employ 14–17% of their current annual expenditure to potentially drive to zero the whaling market.

10.6 African Countries

10.6.1 Botswana

Number of protected areas: 22
Protected land areas coverage: 29.14%
Protected marine area coverage: 0.0%

2017 contribution of T&T to GDPa	**% GDP**	**USD (billion)**	**2018 forecast**
Direct	3.8	0.7	↑
Total	11.5	2.1	↑
2017 contribution of T&T to employment[a]	**% jobs**	**'000 jobs**	
Direct	2.6	25.9	↑
Total	7.6	75.8	↑
	% of total export	**USD (billion)**	
2017 visitor exports[a]	7.4	0.7	↑
	% total investment	**USD (billion)**	
2017 investment	8.3	0.4	↑
GDP international country 2017[b]**(current USD)**		17.4 bn	

[a]World Travel and Tourism Council: Economic Impact 2018, March 2018. All rights reserved. Licensed under the Attribution, Non-Commercial 4.0 International Creative Commons Licence
[b]World Development Indicators (https://data.worldbank.org/products/wdi)

In terms of the country's economy, tourism ranks second in Botswana after the diamond industry. Tourism and travel contributed to 11.5% of the country's GDP in 2017, with an amount of USD 2072.9 mn (WTTC 2018). In particular, Botswana's spectacular abundance of wildlife attracts millions of tourists every year, making it among one of the most popular countries for safaris.

Ecotourism in Botswana is said to contribute 4–5% toward the country's GDP (Yasukawa and Pisa 2016).

Ecotourism projects found in northern Botswana, where there is an abundance of wildlife resources, generate more income than those found in the western, central, and eastern parts of the country where there are less wildlife resources (Mbaiwa 2008). The reinvestment of funds from ecotourism into other economic activities happening in rural villages is an important aspect of community development; in this sense, ecotourism can be described as one of the tools that promotes economic development in rural areas of Botswana (Mbaiwa 2008).

According to the World Database of Protected Areas, some 169,370 km^2 or 29.14% of Botswana's surface area has been allocated to protected areas, which are distinguished as national parks, game reserves, forest reserves, bird sanctuaries, and game sanctuary. Botswana National Parks and Game Reserves are run by the government through the Department of Wildlife and National Parks (DWNP) which employs more than 1200 people. National Conservation Policies were first formulated in the 1980s under the principles of "low-volume, high-value" wildlife tourism, with high entry fees for protected areas and a limitation on the number of visitors for any lodge in a national park or game reserve. This aimed to avoid mass tourism and to maintain an exclusive, quality product. "Raise the price, reduce the numbers" was the thinking, and this philosophy still holds. Revenues all accrue to the central government, which funds an operating budget of over US$1 million. In addition, donors support some research and development initiatives. Revenues have been increasing during the years, and in 2002, they were estimated at US$2 million a year, with around 170,000 annual visitors (Main and Warburton-Lee 2002).

Even though tourism in Botswana has long been based on trophy hunting, the government banned it in 2014, allowing for the development of ecotourism activities like photographic tourism (Siyabona Africa 2017).

Other ecotourism activities in Botswana involve game viewing, bushwalks, safari hunting, camping, lodging, boat driving, mekoro (dug-in canoe) safaris, storytelling, dancing, and many others (Mbaiwa 2008).

Botswana charges the same entrance fee for all protected areas and has rates for citizens of Botswana, residents of Botswana, and nonresidents (Spenceley et al. 2017). User fees charged by the Department of Wildlife and Natural Park in Botswana range from 1.41 to 2.81 and 5.62–11.24 USD (respectively, for resident/nonresident and children/adult) per day plus vehicle fees/day (4.68–140.54) and camping fees/night (1.41–2.81) (Maun Self Drive 4x4 Hire For Camping, Game Viewing and Off road Travel 2015).

Climate change is emerging as one of the biggest threats to the Okavango Delta as it disrupts the seasonal flooding of the plains (the International Ecotourism Society). Declines in the number of wildlife impact the ability of Wildlife Management Areas to generate income from community-based tourism initiatives, with the possibility for income from hunting licenses being removed entirely by the national hunting ban that came into effect in early 2014. The government supports measures that continue to incentivize the livestock sector in the southern Kalahari leading to land degradation, raising concerns of bush encroachment and dune activation, as well as wildlife declines (Dougill et al. 2016).

10.6.1.1 The Case of Kgalagadi Transfrontier Park

The Kgalagadi Transfrontier Park is one of South Africa's unique attraction for national and international tourists and one of the largest wilderness area in the world with a surface of more than 3.6 million hectares. It is an amalgamation of South Africa's former Kalahari Gemsbok National Park and Botswana's Gemsbok National Park. With the official opening of the park in 2000, it represents the first formally declared transfrontier park in Africa (SANParks 2012) that aims at returning ecosystems into their natural state by overcoming political borders (Hanks 2003). Besides the location and the magnitude of the park, its flora and fauna augment the uniqueness and make the area of special value to conservation (SANParks 2012).

10.6.2 Kenya

Number of protected areas: 411
Protected land areas coverage: 12.36%
Protected marine area coverage: 0.8%

2017 contribution of T&T to GDP[a]	**% GDP**	**USD (billion)**	**2018 forecast**
Direct	3.7	2.8	↑
Total	9.7	7.4	↑
2017 contribution of T&T to employment[a]	**% jobs**	**'000 jobs**	
Direct	3.4	429	↑
Total	9	1137	↑
	% of total export	**USD (billion)**	
2017 visitor exports[a]	18.1	1.9	↑
	% of total investment	**USD (billion)**	
2017 investment[a]	5.7	0.8	↑
GDP international country 2017[b](current USD)		74.9 bn	

[a]World Travel and Tourism Council: Economic Impact 2018 – March 2018. All rights reserved. Licensed under the Attribution, Non-Commercial 4.0 International Creative Commons Licence
[b]World Development Indicators (https://data.worldbank.org/products/wdi)

According to Tourism and Wildlife Cabinet Secretary Najib Balala, Kenya's tourism sector is leaving an important improvement. In fact, it registered earnings of Sh120 billion in 2017, a 20.3% growth compared with the Sh99.69 billion earned in 2016 (Kimanthi 2018).

The 2018 Economic Survey by Kenya National Bureau of Statistics shows that international arrivals increased by 8.1% reaching 1.4 million in 2017 from 1.3 million of the previous year (Njugunah 2018). Moreover, tourism contribution to the country's GDP was USD 7432.9 mn (9.7% of GDP) in 2017, and it is forecasted to rise in the next years (WTTC 2018e).

In Kenya, wildlife conservation is greatly funded by the revenue generated from this industry (Okello et al. 2008a).

According to the World Database on Protected Areas, Kenya's territory includes a total of 411 protected areas, of which 39 have management effectiveness evaluations. These areas cover about 12.36% of the total land area and about 0.8% of the marine areas and are organized under different national designations such as national parks, community nature reserves, national reserves, etc. As noted above, the protected areas in Kenya are categorized as either parks or reserves. The difference between these two categories is that while parks are characterized by complete protection of natural resources and only tourism or research-related activities are allowed, in reserves, human activities are allowed, under specific conditions, such as for firewood collection or fishing. Moreover, a lot of Kenya's wildlife lives outside protected areas, due to the fact that protected areas are not fully fenced, and animals are free to move in and out of these areas in search of pasture and water. This is causing human-wildlife conflicts that require strategic collaboration between the local communities living next to these areas and the Kenya Wildlife Service (KWS), which is responsible for the protection and conservation of these territories (Kenya Wildlife Service 2019). According to a study conducted on 50 protected areas (parks and reserves) around the country in 2007, the KWS and its preceding institutions responsible for the protection of the country's biodiversity have not taken significant initiatives in order to implement conservation strategies of the country. This lack of government regulations and of a strategic plan for wildlife conservation in Kenya's protected areas leads to an increase of pressure and threats to the country's wildlife and biodiversity (Mwale 2000, cited after Kiringe and Okello 2007).

However, a remarkable initiative is the Maasai Olderkesi community, who is given approximately USD 10,000 (KES one million) per month and additional rewards, in exchange of information that lead to the capture of poachers, guns, and ivory stocks (Cottar 2015).

Among the main threats affecting Kenya's protected areas, the study identifies those activities that are directly related to actual killing of wildlife (bushmeat, poaching for international commercial purposes, and human-wildlife conflicts). Also, tourism is identified as a threat factor: Its negative impacts on protected areas are mainly due to the unequal distribution of the tourists flux, which is heavily concentrated in certain areas, and a lack of appropriate management of tourist behavior and tourist facilities (Kiringe and Okello 2007).

Kenya's protected area system needs to be implemented, and the diversification of tourism attraction together with an increase of the entrance fees in order to limit tourist traffic could help reduce the pressure on the country's protected areas (Kiringe and Okello 2007).

10.6.2.1 The Masai Mara Nature Reserve

The Masai Mara Nature Reserve hosts the Kichwa Tembo Masai Mara Tented Camp. This camp attracts many tourists every year because of its year-round concentration of wildlife due to the camp's location on the route of the Great Migration. As it is a private concession land, bushwalks and night drives are allowed. The tented camp can be considered as an example of ecotourism since with its revenue, it supports local schools, reforestation, environmental education, health, and anti-AIDS programs.

The camp employs 200 employees, 70% of whom are locals from the Masai Mara region, and can host a maximum number of 80 guests. About 60% of farm products, consumed by tourists, are obtained from local suppliers. The annual revenue generated by the camp is estimated at US$ 8–10 million, of which US$ 1.5 million is paid directly to local communities for the lease fee, salaries, and purchases of local products (Lengefeld 2013).

10.6.3 Namibia

Number of protected areas: 148
Protected land areas coverage: 37.08%
Protected marine area coverage: 1.71%

2017 contribution of T&T to GDP[a]	**% GDP**	**USD (billion)**	**2018 forecast**
Direct	2.9	0.4	↑
Total	13.8	1.8	↑
2017 contribution of T&T to employment[a]	**% jobs**	**'000 jobs**	
Direct	3.2	22.8	↑
Total	14.0	98.2	↑

	% of total export	USD (billion)	
2017 visitor exports[a]	6.2	0.3	↑
	% total investment	**USD (billion)**	
2017 investment[a]	12.0	0.3	↑
GDP international country 2017[b]**(current USD)**		13.2 bn	

[a]World Travel and Tourism Council: Economic Impact 2018 – March 2018. All rights reserved. Licensed under the Attribution, Non-Commercial 4.0 International Creative Commons Licence
[b]World Development Indicators (https://data.worldbank.org/products/wdi)

Namibia's tourism industry has increasingly become an important contributor to the gross domestic product, making it the third largest economic sector after mining and fisheries (Kimaro et al. 2015) and one of the most competitive tourist destinations in Africa by 2017 as measured by the World Economic Forum Travel and Tourism Competitiveness Index (World Economic Forum 2017).

According to the statistics, tourism and travel contributed for a total of 13.8% of the country's GDP in 2017, with an amount of USD 1778.2 mn (WTTC 2018d).

Ecotourism activities in Namibia include adventure tours like mountain biking, horseback or walking safaris, self-drive safaris, research expeditions, and cultural workshops (Prokosch 2015b)

Namibia is one of the only countries in the world that addresses conservation and environmental protection directly in its constitution (Prokosch 2015b).

According to the World Database of Protected Areas, in Namibia, there are 148 protected areas which include national parks, communal conservancies, and community forests, as well as private reserves and tourism concessions. Protected areas have a significant value to the national economy; however, they continue to experience substantial underfunding. Three main sources of funding for protected areas in Namibia are government, donor, and park revenues channeled via Game Product Trust Fund (GPTF).

Revenues generated by the parks directly go to the central government with only a portion of these revenues being reinvested into the management of national parks. Funding for the management of protected areas is currently in the order of N$215 million (TEEB 2017). Moreover, the estimated annual recurrent expenditure for park management is N$275 million (TEEB 2017).

Namibia's national park entry fees have remained unchanged since 2005 and are among the lowest in Africa, corresponding to half of what is charged in Botswana, a third of Zimbabwe, South Africa, and Zambia. The total amount generated by park entry fees in the country during 2014/2015 was N$56.4 mn.

Among protected areas, communal conservancies, recognized by the Ministry of Environment and Tourism (MET), generated N$111,232,053 for local communities and created about 5116 jobs in 2015. In order to encourage increased conservation efforts by conservancy members, stronger incentives are recommended (TEEB 2017).

Among the main threats to maintaining biodiversity and ecosystems values in Namibia, there are poaching, overstocking (artificial water holes), excessive disturbance and off-road driving by tourists, mining, and climate change as well as lack of financial resources (as indicated above), lack of capacity, and persistent poverty outside PAs (TEEB 2017).

10.6.3.1 The Case of Etosha National Park (ENP)

Etosha National Park is the second largest national park after Namib-Naukluft Park (Berry 1997) and one of the oldest in the country. The size of the park has been reduced considerably since it was first proclaimed in 1907, but it still remains larger than several European countries.

Etosha is the most visited national park, with approximately 220,000 visitors per year, and the third most visited place in the country (MET 2013) after Windhoek and Swakopmund.

10.6.4 Rwanda

Number of protected areas: 10
Protected land areas coverage: 9.11%
Protected marine area coverage: 0.0%

2017 contribution of T&T to GDP[a]	% GDP	USD (billion)	2018 forecast
Direct	5.2	0.5	↑
Total	12.7	1.1	↑
2017 contribution of T&T to employment[a]	**% jobs**	**'000 jobs**	
Direct	4.4	132.2	↑
Total	11.1	333.6	↑
	% of total export	**USD (billion)**	
2017 visitor exports[a]	30.5	0.5	↑
	% total investment	**USD (billion)**	
2017 investment[a]	8.6	0.2	↑
GDP international country 2017[b](current USD)		9.1 bn	

[a]World Travel and Tourism Council: Economic Impact 2018 – March 2018. All rights reserved. Licensed under the Attribution, Non-Commercial 4.0 International Creative Commons Licence
[b]World Development Indicators (https://data.worldbank.org/products/wdi)

Despite its difficult postconflict situation, the country has succeeded in establishing the right strategies and instruments to maintain conservation as one of its priorities. In addition, tourism has been a significant contributor to the improved image of the country, and it has also been seen as a tool to reduce poverty and involve the communities (Nielsen and Spenceley 2010).

Today, tourism is one of the fastest-growing sectors, with a total contribution to GDP of USD 1129.9 mn in 2017 (WTTC 2018f). According to the African Wildlife Foundation, one of the main reasons for its success is that Rwanda is fortunate to be one of only three countries where tourists can visit the endangered mountain gorillas, successfully promoted worldwide.

Rwanda's ecotourism includes a large variety of activities such as volcano climbing, golden monkey trekking, bird watching, and cultural visits among others. Gorilla tours remains one of the most popular ecotourist activities which attracts more and more visitors each year. According to the Rwanda Development Board (2018) (Tashobya 2018), 10% of the income derived from gorilla, safari, and other tourist permits, as well as park fees, is spent in partnership with local communities to change lives for the better.

Rwanda has a total of 10 protected areas (UNEP-WCMC 2018o) which include national parks (Akagera, Nyungwe, and Volcanoes National Park) and forest reserves (Gishwati, Iwawa Island, and Mukura forest reserves). Besides those forests with a legal status of protected areas, there are other forests of cultural importance (Busaga Forest in Muhanga District) and other remnant natural forests which are more or less protected by law (REMA (Rwanda Environment Management Authority) 2009).

Among the environmental issues facing Rwanda, there's the fact that it is one of the most densely populated countries in all of Africa with more than 11 million, most of which live in poverty. This generates pressure to take away protected areas and convert it into farmland. It also causes deforestation, as people exploit forest products like bamboo or firewood for fuel (African Wildlife Foundation 2019a).

10.6.4.1 The Case of Volcanoes National Park and the Mountain Gorilla

The Volcanoes National Park was established in 1925. It is Africa's oldest national park and lies along the Virunga Mountains and borders with the Virunga National Park in Congo and the Mgahinga Gorilla National Park in Uganda. It spans 160 km^2 covered in rain forest and bamboo (Volcanoes National Park Rwanda 2019).

The park is best known for the mountain gorilla (*Gorilla beringei beringei*), but it is also home to other mammals such as the golden monkey, the black-fronted duiker, buffalo, spotted hyena, and bushbuck. The park also harbors 178 bird species including at least 29 endemics to Rwenzori Mountains and the Virungas.

Gorilla tourism is the key pillar of Rwanda's tourism and conservation industry, bringing in over 50% of the country's GDP.

In the last 9 years, gorilla tourism has generated over 107 million UD dollars to Rwanda. Records show that over 298,000 tourists have visited the VNP between 2006 and 2017.

Recently, the Rwanda Development Board (RDB) received land from the African Wildlife Foundation (AWF) in order to expand the VNP. According to RDB, the 27.8 hectares donated by AWF will be added to the existing 16,000 hectares that is the current size of the park (Tashobya 2018).

The major aim of the yet to start project, which is estimated to cost over $200 million, is to promote wildlife conservation providing more space for the increasing population of mountain gorillas as well as to promote gorilla tourism and improve the well-being of local people. According to the plans, local people living adjacent to the park will not be relocated outside of Musanze, the area where they have been living for over 100 years, but will remain to coexist with wildlife in the park.

RDB also allocates 10% of the total revenue collected from gorilla tourism in VNP to communities around the park and assured for new infrastructures (like roads, schools, and hospitals) and employments opportunities (Volcanoes National Park Rwanda).

10.6.5 South Africa

Number of protected areas: 1544
Protected land areas coverage: 8.0%
Protected marine area coverage: 12.06%

2017 contribution of T&T to GDP[a]	**% GDP**	**USD (billion)**	**2018 forecast**
Direct	2.9	10.2	↑
Total	8.9	31.0	↑
2017 contribution of T&T to employment[a]	**% jobs**	**'000 jobs**	
Direct	4.5	726.6	↑
Total	9.5	1530.3	↑
	% of total export	**USD (billion)**	
2017 visitor exports[a]	9.2	9.5	↑
	% total investment	**USD (billion)**	
2017 investment[a]	8.2	5.3	↑
GDP international country 2017[b] **(current USD)**		349 bn	

[a]World Travel and Tourism Council: Economic Impact 2018 – March 2018.
[b]World Development Indicators (https://data.worldbank.org/products/wdi)

Tourism is one of the country's greatest sources of revenue, contributing to 2.9% of the total GDP in 2017 (World Travel and Tourism Council).

South Africa's natural resources form the basis of the tourism industry, attracting millions of local and international ecotourists every year. Ecotourism in South Africa is based on adventure, culture, history, and, of course, wildlife.

Moreover, because of its rich bird population, South Africa is a popular destination for bird-watching activities. A quantitative study on avitourism conducted in 1997 in South Africa conservatively estimated that the country received between 11,400 and 21,200 bird watchers per year which contributed US$ 12–26 million to the South African economy (Turpie and Ryan 1998 cited by Biggs et al. 2011). Since 1997, there has been a significant increase in bird-watching tourism in South Africa, which has led to an increase in the number of tour operators specializing in this type of ecotourism. Currently, there are more opportunities for small business development along birding routes, which contributes to the creation of jobs for local communities (e.g., local birding guides) and supports conservation (Biggs et al. 2011).

In South Africa, the revenues generated by ecotourism have been studied on privately owned properties: According to Taylor et al. (2016), although in the Eastern Cape region there is a relatively small number of ecotourism-based game farms, the earning power of these private properties had a significant impact on the economy of Eastern Cape, with the amount of a gross income of approximately ZAR 2000/ha/year in 2004 (meaning a contribution equal to USD 5.8 million/year), which in 2010 was achieved with the contribution of nine reserves only. In these reserves, 87% of revenues from the different uses of wildlife land were ascribable to wildlife watching and accommodation (Taylor et al. 2016). Small ranches too, although lacking money for gross investments, had better returns (sometimes four times higher) from ecotourism than from other land use activities such as hunting, game sales, agriculture, or commercial livestock farming (Taylor et al. 2016). The only Nambiti private game reserve (< 10,000 ha) reports an income turnover of approximately ZAR 5.5 million/year, excluding the income generated by the ten-individual lodge, but only considering the reserve's operation (Taylor et al. 2016).

South Africa's conservation network is made of 22 protected areas accounting for 37,000 km^2 or 4% of the nation's total land area (an additional 4.9% is formally protected as provincial parks and public and private game reserves). It has an annual budget of close to R1 billion ($65 million), and 80% of this is self-generated by thriving tourism activities (Prokosch 2016).

Moreover, this revenue allows SANParks, the leading conservation organization in South Africa and also the largest provider of ecotourism experiences in the country (de Witt et al. 2011), to expand the conservation estate year after year, with 700,000 ha added since the advent of democracy in 1994 (Prokosch 2016).

10.6.5.1 The Case of Kruger National Park

The Kruger National Park (KNP), in Afrikaans *Nasionale Krugerwildtuin,* is the largest natural reserve of South Africa with a surface of 7580 miles^2 (19,633 km^2), bordering Mozambique in the east and Zimbabwe in the north (South African National Parks 2019).

KNP was formally declared a national park on December 10, 1926, although portions had already enjoyed conservation status for considerably longer. It is South Africa's number one wildlife reserve and encompasses 14 different ecosystems. The extreme north of KNP is unique due to its diverse assemblage of rock formations. Giant baobab, fever, and marula trees tower above the savanna, thornveld, and woodland landscape of the park.

The Big Five, buffalo, elephants, leopards, lions, and rhinos, all reside here, as well as Nile crocodiles, hippos, and rare birds like southern ground hornbills and lappet-faced vultures.

What's more, Kruger's Marula and Nxanatseni regions house the Albasini and Masorini ruins, where Portuguese colonists and members of the indigenous Ba-Phalaborwa ethnic group once traded metal products, beads, clothes, and more (US News and World Report).

KNP's patterns of geology, soil, fire, and rainfall are local factors which are vital attributes that make the park unique (South African National Parks Annual Report 2016/17).

Kruger National Park is ranked as:

- #5 in Best Places to Visit in Africa and the Middle East
- #9 in Best National Parks in the World
- #16 in Best Summer Vacations

10.6.6 Tanzania

Number of protected areas: 839
Protected land areas coverage: 38.15%
Protected marine area coverage: 3.02%

2017 contribution of T&T to GDP[a]	% GDP	USD (billion)	2018 forecast
Direct	3.8	2	↑
Total	9	4,7	↑
2017 contribution of T&T to employment[a]	**% jobs**	**'000 jobs**	
Direct	3,3	446	↑
Total	8,2	1092	↑

	% of total export	USD (billion)	
2017 visitor exports[a]	26	2,2	↑
	% of total investment	**USD (billion)**	
2017 investment[a]	8.7	0,3	↑
GDP international country 2017[b]**(current USD)**		52 bn	

[a]World Travel and Tourism Council: Economic Impact 2018 – March 2018.
[b]World Development Indicators (https://data.worldbank.org/products/wdi)

Tourism is now Tanzania's leading economic sector, earning US$1 billion a year and thus overtaking agriculture and growing at a steady rate for the past 7 years (Thome 2015). The number of international arrivals in Tanzania has been increasing by 12.9%, from 1,284,279 in 2016 to 1137,182 tourists in 2015. Moreover, revenues from tourism have been increasing year after year with USD 2 billion in 2016 against USD 1.9 billion in 2015 (Tanzaniainvest 2017).

Ecotourism becomes an important sector in Tanzania as it contributes much to the national economy (currently about 17.2% of GDP), and it has directly and indirectly supported about 400,000 jobs, nearly one job for every additional tourist (Carlson 2009 cited by Pasape and Mujwiga 2017). Most of Tanzania's visitors come as ecotourists, as it is estimated that at least 90% of tourists follow nature-based tourism (Anderson 2010 cited after Pasape and Mujwiga 2017).

The country provides wide range of opportunities for recreational activities such as hunting, hiking, canoeing, wildlife viewing, bird watching, mountain hiking, camping, and walking safaris associated with photographic tourism.

Tanzania has allocated 38.15% of its terrestrial areas and 3.02% of its marine areas to protected areas. The present network of wildlife protected areas is comprised of national parks, game reserves, the Ngorongoro Conservation Area, game-controlled areas, partial game reserves, and forest reserves (UNEP-WCMC 2018a, b, c, d, e, f, g, h, i, j, k, l, m, n, o, p, q, r, s).

Some studies have considered reviewing entrance prices and activity fees for Tanzania's protected areas in order to be more competitive with other PAs and to raise revenues to pay for management costs. For example, it was predicted that increasing at USD60 the Serengeti conservation fee over several years could raise an additional USD14.8 million by 2020 (equivalent to increasing the park's revenue by 57%). However, the study also established that this increase was park specific and not recommended for all the parks (like Kilimanjaro National Park where fees were already considered to be high) (Bruner et al. 2015, cited after Spenceley et al. 2017). However, the current entrance fees for noncitizen adults in protected areas in Tanzania correspond to USD30/USD50 in game reserves and USD10 in Wildlife Management Areas (excluding conservation fees) (Spenceley et al. 2017).

According to the International Ecotourism Society, challenges to conservation include poverty, education, population growth, governance issues, development pressures, and lack of financial resources.

In particular, key financial challenges are inadequate capital for the marketing of tour activities which affects the success and sustainability of tour operations, business, and entrance fec to the attraction sites and inadequate support from the existing support from local financial institution (Pasape and Mujwiga 2017). These financial challenges are being faced by the Tanzanian Association of Tour Operators network.

10.6.6.1 The Case of the Serengeti-Ngorongoro Circuit

According to a study conducted in 2009, the southern circuit at Serengeti-Ngorongoro receives 300,000 tourists per annum. The total inbound tourism expenditure generated here is US$ 500 million per year, which is more than half of Tanzania's foreign exchange earnings from tourism. The price of a typical wildlife watching package is US$ 1600 for 6 days/5 nights (US$ 320 per day). Additionally, tourists spend an average US$ 226 out of pocket (US$ 37/day). Along the safari circuit, there are about 3500 crafts and souvenir stalls that employ 7000 sellers and 21,000 crafters. About US$100 million per year (19% of the earnings) reach local people via wages and tips when they are employed by tourism providers. Furthermore, local small producers provide about half of the food consumed in the circuit. The local population obtains indirect benefits from tourism, through allocated funds by the protected area management to the communities (Steck/ODI (2009), cited in UNWTO 2014a, b).

10.6.7 Uganda

Number of protected areas: 712
Protected land areas coverage: 16.06%
Protected marine area coverage: 0.0%

2017 contribution of T&T to GDP[a]	**% GDP**	**USD (billion)**	**2018 forecast**
Direct	3.0	0.5	↑
Total	7.1	1.2	↑
2017 contribution of T&T to employment[a]	**% jobs**	**'000 jobs**	
Direct	1.7	27.4	↑
Total	4.4	69.2	↑
	% of total export	**USD (billion)**	

2017 visitor exports[a]	4.7	0.2	↑
	% total investment	**USD (billion)**	
2017 investment[a]	4.7	0.1	↑
GDP international country 2017[b]**(current USD)**		17.8 bn	

[a]World Travel and Tourism Council: Economic Impact 2018 – March 2018.
[b]World Development Indicators (https://data.worldbank.org/products/wdi)

Uganda's economy today relies primarily on small-scale agriculture. However, tourist arrivals in Uganda increased to 1449 thousand in 2017 from 1323 thousand in 2016. It averaged 648.79 thousand from 1990 until 2017, reaching an all-time high of 1449 thousand in 2017 and a record low of 69 thousand in 1990 (Trading Economics 2019a).

According to the World Bank Group, $1 of expenditure by a foreign tourist generates, on average, $2.5 of GDP, and the total impact includes the indirect value added along the supply chain plus the induced effects of households spending the wages generated (The World Bank Group 2013).

In particular, the number of tourists visiting the country's national park has grown from 246 thousand in 2016 to 286 thousand in 2017. This is a significant growth, compared with the number of tourists visiting the country's natural areas, 110 thousand in 2006 (Uganda Bureau of Statistics 2018).

The ecotourism destinations in Uganda include all the ten national parks, wildlife and game reserves, forest reserves, events/cultural centers, community wetlands, theme parks, resorts, and important bird areas. With the development of a wide range of eco-friendly activities, Uganda has taken significant steps in shifting from traditional tourism to responsible travel to ecologically sensitive areas (Uganda Eco Tours 2011). Indeed, wildlife safari (39%), gorilla viewing (26%), and adventure tourism (25%) are the most popular trip activities (The World Bank Group 2013).

Managed by Uganda Wildlife Authority (UWA), the ten national parks present in the country offer savanna safaris, boat tours, forest hikes, mountain climbing, and wildlife research activities (UWA (Uganda wildlife Authority) 2019).

However, Uganda's ecotourism development has to deal with several problems such as frequent human-wildlife conflicts which represent some of Uganda's biggest conservation challenges. Poachers illegally hunting wildlife for bushmeat, people clear-cutting forests to make charcoal, forest clearance, and disease spread by humans are serious issues (African Wildlife Foundation 2019b).

Referring to the tourism sector, local transport in Uganda, insufficient visitor information, and the quality of customer services are the most frequently cited sources of dissatisfaction and suggested areas for improvement (The World Bank Group 2013).

Going forward, the World Bank Group highlights the need for more policy and government investment essential to attract more tourists to visit Uganda. In particular:

- Marketing initiatives: use of new media, strengthening links with travel agencies in source markets, and attracting high-profile foreign operators.
- Increasing supply: private sector investments, new policies and regulations specific to the tourism sector, and reforms of the concession policy for tourism operators.
- Investing in natural assets: investments in park infrastructure, machinery, and equipment and investments for the protection and management of wildlife and for the improvement of staff skills.

10.6.7.1 The Case of Bwindi Forest National Park

The Bwindi Forest National Park is located in southwestern Uganda on the edge of the Rift Valley. It was declared as a national park in 1991 and a UNESCO Natural World Heritage Site in 1994.

The national park is home to one of Uganda's oldest and most biologically diverse rain forests, which contains almost 400 species of plants. There are 120 mammals, including several primate species such as baboons and chimpanzees, as well as elephants and antelopes, and around 350 species of birds, including 23 Albertine Rift endemics. Moreover, the Bwindi National Park also protects an estimated 400 mountain gorillas which represents roughly half of the world's population (Uganda Wildlife Authority 2018). Mountain gorilla families that are accustomed to humans can be visited by small tourist groups for one hour with a special guide. The permit to visit a gorilla family costs between US$ 500 and 700 per person. The visits to a single gorilla family that attracts an average of ten tourists a day generate between US$ 5000 and 7500 per day. Over a year's time, visits to this same family can generate up to about US$ 500,000 per year (visits are not made every day). The total income of gorilla visits in the Bwindi Forest National Park is about US$ 15 million per year. Additionally, a similar amount is spent by the tourists on accommodation, transport, and other services (Lengefeld 2013).

10.6.8 Zimbabwe

Number of protected areas: 232
Protected land areas coverage: 27.21%
Protected marine area coverage: 0.0%

2017 contribution of T&T to GDP[a]	% GDP	USD (billion)	2018 forecast
Direct	3.0	0.5	↑
Total	7.1	1.2	↑
2017 contribution of T&T to employment[a]	**% jobs**	**'000 jobs**	

Direct	1.7	27.4	↑
Total	4.4	69.2	↑
	% of total export	**USD (billion)**	
2017 visitor exports[a]	4.7	0.2	↑
	% total investment	**USD (billion)**	
2017 investment[a]	4.7	0.1	↑
GDP international country 2017[b]**(current USD)**		17.8 bn	

[a]World Travel and Tourism Council: Economic Impact 2018 – March 2018.
[b]World Development Indicators (https://data.worldbank.org/products/wdi)

The country's political instability combined with the lack of infrastructures has affected the tourism sector in Zimbabwe over the years. However, its natural beauty remained a focus of touristic attraction even during the difficult times. The Victoria Falls and the Zambezi River remain the most visited places by tourists.

In recent years, the number of international tourists visiting Zimbabwe steadily grew: In 2017, Zimbabwe received a total of 2,422,930 tourist arrivals and an increment of 12% compared with 2016 (Trading economics 2018). The tourism industry represented 7.1% of the country's GDP in 2017, generating USD 1199.8 mn (WTTC 2018i).

Zimbabwe has recently made some progress in the tourism infrastructure area: In December 2015, a new terminal at the Victoria Falls International Airport was opened to passengers. This is expected to attract 1.5 million passengers per year, a huge jump from the previous 176,000.

In 2015, the killing of a famous lion named Cecil, by an American tourist, exposed the country to international attention. The Zimbabwean authorities responded to the case showing a growing concern toward increasing measures that would avoid the repeat of such a tragic event, gaining positive international recognition. This paves the road to more sustainable and environmentally friendly ecotourism (Veras 2017).

The protected areas (PAs) network of Zimbabwe covers 28% of the land. The ZPWMA, established by an Act of Parliament in 2001, is mandated responsible for the entire wildlife population in Zimbabwe.

It is estimated that protected areas management budget is US$40 million annually. According to the statistics, in 2012, about US$31 million was directly invested in protected area management by the protected area agencies, local authorities and communities (through CAMPFIRE), and donors, and the estimated direct revenue from protected areas was US$382.5 million.

N.B.: Criticisms to CAMPFIRE in Trophy Hunting." See section "Trophy Hunting" for conflicting opinion".

Protected areas in Zimbabwe include national parks (constituting 13%), gazetted forests (3%), conservancies and private game parks (1.9%), and the Communal

Areas Management Program for Indigenous Resources (CAMPFIRE) covering 11.9% (Ministry of Environment, Water and Climate 2014).

The CAMPFIRE program consists in involving communities in the co-management of wildlife in communal areas. According to studies, this system has supported various ecotourism projects and benefitted several communities throughout Zimbabwe (Gandiwa et al. 2012). Moreover, it benefits more than million households and generated US$2.5 million in 2012 (Madzara 2013 cited after Ministry of Environment, Water and Climate 2014).

The economic benefits generated by the CAMPFIRE depend mostly on safari hunting and elephants, which are the major species producing the revenue stream that creates these benefits. As a consequence, the local communities are encouraged to value these animals instead of viewing them as a dangerous nuisance. The US importation ban in 2014 has not only negatively affected investment for the protection of wildlife but has also removed direct incentives, at the community level, to protect elephants. Disgruntled CAMPFIRE communities will turn to pastoralism and agriculture, thereby reducing wildlife habitat (Jonga and Pangeti 2015).

Also, gazetted forests form important habitats for commercially valuable wildlife (through sport hunting) and generate about $1,112,400 annually (Madzara 2013 cited after Ministry of Environment, Water and Climate 2015).

In Zimbabwe, trophy hunting is believed to work as a conservation tool, offering incentives for habitat protection and rural development. However, more effort needs to be put toward law enforcement as illegal hunting activities have occurred in some parts of Zimbabwe.

A total of US$39 million was generated by CAMPFIRE between 1994 and 2012, of which US$22 million was allocated to communities and used for resource management (22%), household benefits (26%), and community projects (52%) (Gandiwa et al. 2014, cited after V.K. Muposhi et al. 2016).

According to the Zimbabwe's Fifth National Report in 2015 to the Convention on Biological Diversity (Ministry of Environment, Water and Climate Republic of Zimbabwe 2015), successful nature conservation efforts have given rise to growth in the ecotourism industry. According to the available data, the revenue generated from the use of PAs was $ 6,507,297 during 2012. Twenty-seven percent of the 493,327 national tourists' arrivals visited PAs and generated a total of $ 214.2 million from PAs-related activities.

The major threats to the integrity of the protected areas network are major drought, poverty, a growing population, and a lack of fuel which have all led to massive deforestation (African Wildlife Foundation 2019a, b, c).c

There are costs associated with these threats: An estimated US$ 1.1 million is lost to poaching of bushmeat, and the government spends US$ 50 million annually to fight invasive alien plant species and livestock diseases emanating from wildlife areas. Poaching of major species in parks estates between 2009 and 2012 led to an estimated cumulative loss of US$ 47.5 million (ZPWMA).

The main reasons for poaching are poverty, food insecurity, dissatisfaction over the distribution of revenue and poor staffing by the ZPWMA, and the demand of illegal markets for ivory and rhino horns (Ministry of Environment, Water and Climate Republic of Zimbabwe 2015).

10.6.8.1 The Case of Gonarezhou National Park (GNP)

GNP is located in southeast Zimbabwe and covers an area of 5053 km^2. It was established in the early 1930s as a game reserve and transformed into a national park under the Parks and Wildlife Act of 1975 (Gandiwa, 2011). GNP has been part of the Great Limpopo Transfrontier Conservation Area since 2002 together with Limpopo National Park in Mozambique and Kruger National Park in South Africa (Mutanga et al. 2017).

It is the second largest national park in Zimbabwe, after Hwange National Park, and is widely known for the wilderness experience and its exceptional landscapes which include Chilojo Cliffs and Red Hills (Mutanga et al. 2017).

GNP is home to a wide variety of large carnivores, such as lion and spotted hyena, and herbivore species such as African buffalo, giraffe, waterbuck, roan antelope, sable, Burchell's zebra, blue wildebeest, African elephant, and hippopotamus (Zisadza et al. 2010).

There have been some variations in the temporal visit of tourist in the northern part of GNP. The increase in the number of visitors between 2009 and 2014 in northern GNP could be attributed to a number of factors which include the wildlife-based land reform which started in 2004 resulting in a peaceful environment between protected area staff and local communities and an improved economic and political environment in Zimbabwe (Mutanga et al. 2017) and the establishment of the conservation partnership arrangement which led to improved infrastructure which facilitated enhanced access into the park and increased accommodation facilities, i.e., tented camps. Moreover, and finally the improvement of law enforcement and a boundary or veterinary control fence established in the northern part of the park (Gandiwa et al. 2012; Mutanga et al. 2016). This teaches us that successful conservation alone is not enough to attract tourists, especially after natural and social disasters, e.g., political instability and economic crises (Mutanga et al. 2017).

10.7 Asian Countries

10.7.1 India

Number of protected areas: 231
Protected land areas coverage: 5.97%
Protected marine area coverage: 0.17%

2017 contribution of T&T to GDP[a]	% GDP	USD (billion)	2018 forecast
Direct	3.97	91.3	↑
Total	9.4	234.0	↑

2017 contribution of T&T to employment[a]	% jobs	'000 jobs	
Direct	5.0	26148.1	↑
Total	8.0	41622.5	↑
	% of total export	USD (billion)	
2017 visitor exports[a]	5.8	27.3	↑
	% total investment	USD (billion)	
2017 investment[a]	6.3	41.6	↑
GDP international country 2017[b](current USD)		2597 bn	

[a]World Travel and Tourism Council: Economic Impact 2018 – March 2018. All rights reserved. Licensed under the Attribution, Non-Commercial 4.0 International Creative Commons Licence
[b]World Development Indicators (https://data.worldbank.org/products/wdi)

Tourism in India has a strong relevance to economic development, cultural growth, and national integration. According to the 2018 economic impact report by the World Travel and Tourism Council (WTTC 2018a). India is expected to establish itself as the third largest travel and tourism economy by 2028 in terms of direct and total GDP. The WTTC report also said India will add nearly 10 million jobs in the tourism sector by 2028 and that the total number of jobs dependent directly or indirectly on the travel and tourism industry will increase from 42.9 million in 2018 to 52.3 million in 2028 (Mathur 2018). Moreover, India has moved up 13 positions, from 65th to 52nd, in tourism and travel competitive index.

In India, wildlife tourism has grown by 15% annually (Bindra and Karanth 2013 cited after Tortato et al. 2017), and the main reasons for visiting protected areas are opportunities to see nature in general, tigers (*Panthera tigris*), and scenic beauty (Karanth and DeFries 2011 and Karanth et al. 2012 cited after Tortato et al. 2017). Ecotourism needs to be promoted so that India's natural and cultural environments will be preserved and sustained. Among the positive consequences of this tourism growth indeed, there is the fact that the industry helped preserve several places around the country by declaring them as heritage sites for their historical importance (some examples are the famous Taj Mahal, entitled one of Seven Wonders of the World, and the Qutub Minar, the tallest minaret in the world made up of bricks), which would have been decayed and destroyed had it not been for the efforts taken by the Tourism Department to preserve them. Moreover, the tourism sector development also helps in conserving the natural habitats and with it the life of many endangered species (Dayananda 2016).

On the other hand, among the adverse effects of tourism on the environment, there is the increase in transport and construction activity which led to large-scale deforestation and destabilization of natural landforms and the increase in solid waste dumping as well as water depletion and fuel resources caused by the flow of tourists. Moreover, noise pollution, water pollution, vehicular emissions, untreated sewage, etc. also have direct effects on the natural environment (Dayananda 2016).

The government efforts to protect the country's biodiversity are shown in the establishment of 231 protected areas including 39 national parks, 190 wildlife sanctuaries, 1 game reserve, and 1 national marine park (UNEP-WCMC 2018h). All these protected areas are notified by the State Governments and protected by the Forest Departments. A national park is defined as a protected area which is reserved for the conservation of only animals, where no human interference in any form is allowed. Wildlife sanctuary differs for the fact that human activities like harvesting of timber, collecting minor forest products, and private ownership rights are allowed as long as they do not interfere with the well-being of animals (Edake 2016).

Certain national parks and wildlife sanctuaries which support a good tiger population have been redesignated as tiger reserves and enjoy a special status with the highest level of protection (Edake 2016). These kinds of reserves are very important to the country, not only because they support more than half of the global tiger population and are cornerstones of biodiversity conservation but also because they provide a wide range of economic, social, and cultural benefits in the form of ecosystem services (Verma et al. 2016). Moreover, many wildlife conservationists and respected ecotourism operators advocate that the presence of tourists in these tiger reserves can help save this iconic predator from activities like poaching, revenge killings, pelt sales, and deforestation which have been considered for years some of the main causes of tiger's population loss.

Toward this, the last census of 2014 had estimated India tiger's population at 2226, up from 1706 in 2010 (Jhala et al. 2015). Most experts expect the growth trend to continue. The results are likely to be announced early next year (Bhattacharya 2018).

"That is why we want to create more tiger reserves," said the environment minister Prakash Javadekar, highlighting how this strategy is leading India to the right direction in terms of biodiversity conservation (Agence France-Presse 2015).

According to the scientists, conserving the tigers currently present in India is equivalent to keeping a secure capital of USD 230 billion. This means that each tiger has a flow benefit of about USD 2.19 million per year which can be viewed as the interest one earns annually by conserving each tiger. The total annual expenditure to maintain these six tiger reserves was just about Rs. 23 crore. The return on investment for saving each tiger is of the order 356 times the investment (Bagla 2017). Saving tigers makes good economic sense!

However, there is an urgent need to establish and enforce regulations to manage ecotourists, resource use, and land use change around Indian protected areas (Karanth and DeFries 2011 cited after Tortato et al. 2017). In 2012, lack of control in tiger tourism resulted in India's Supreme Court temporarily banning tourism access to core areas of tiger reserves. This measure increased the risk of tigers being killed by poachers and reduced tourism revenue for park management (Buckley and Pabla 2012 cited after Tortato et al. 2017).

10.7.1.1 The Corbett National Park

Corbett National Park is India's first national park, established in 1936. Famous for its wealthy biodiversity which includes more than 25 species of reptiles, 585 kinds of birds, and over 50 varieties of mammal species, the main attraction of tourists to this park is the presence of the rare species of tigers that are facing extinction (Uttarakhand Tourism Development Board (2019).

10.7.2 Malaysia

Number of protected areas: 717
Protected land areas coverage: 19.12%
Protected marine area coverage: 1.54%

2017 contribution of T&T to GDP[a]	**% GDP**	**USD (billion)**	**2018 forecast**
Direct	4.8	15.2	↑
Total	13.4	41.9	↑
2017 contribution of T&T to employment[a]	**% jobs**	**'000 jobs**	
Direct	4.6	669.8	↑
Total	11.8	1704.5	↑
	% of total export	**USD (billion)**	
2017 visitor exports[a]	8.3	18.5	↑
	% total investment	**USD (billion)**	
2017 investment[a]	6.7	5.3	↑
GDP international country 2017[b]**(current USD)**		314.5 bn	

[a]World Travel and Tourism Council: Economic Impact 2018 – March 2018.
[b]World Development Indicators (https://data.worldbank.org/products/wdi)

Tourism is an important source of income in Malaysia. In fact, the tourism sector ranks second after oil and gas in Malaysia (Kaur 2007). According to the Malaysia Tourism Promotion Board (MTPB), in 2017, Malaysia received a total of 25,948,459 international tourists and recorded a 0.1% growth in tourist receipts, thus contributing RM82.2 billion to the country's revenue (approximately 19.86 US dollar). Meanwhile, the average length of stay during the same year for foreign tourists decreased to 5.7 nights from 5.9 nights in the previous year.

Although tourist arrivals dropped by 3%, in terms of numbers, Malaysia was the second most-visited South East Asian country after Thailand, which had 35.3 million tourists in 2017.

The Tourism Malaysia Integrated Promotional Plan 2018–2020 has also been formulated and implemented to tackle existing challenges and improve Malaysia's tourism performance. Meanwhile, the kick-off of Visit Malaysia 2020 campaign in various markets this year, targeting 36 million tourists and RM168 billion in tourist receipts by 2020, is also expected to revive Malaysia's position as a holiday destination choice (MTPB 2018).

Plenty of those tourists coming to Malaysia were drawn, at least in part, by the natural wonders of this country. And so ecotourism plays an important part in the country's economy, especially in biodiverse states like Sabah and Sarawak. In 2016, Sabah had a record of 3.43 million visitors who brought in RM7.25 billion in tourism revenues (CleanMalaysia.com 2017). According to Global Data consumer survey, in fact, Malaysia is among the countries that present the biggest interest in ecotourism with a percentage of 76%, followed by China (67%) and Turkey (65%).

Despite its great ecotourism potential, the country's ecotourism hot spots need to be better protected from the depredations of mass tourism. Sadly, often their very popularity is a threat to some of Malaysia's unique areas and landscapes (CleanMalaysia.com 2017). Many efforts to better understand ecotourism as well as to improve its planning, management, and marketing techniques have been carried out, but efforts still seem to be inadequate (Kaur 2007).

In order to safeguard its biodiversity, several networks of protected areas (PAs) have been established around the country. At present, according to the World Database on Protected Areas, Malaysia consists of 717 protected areas covering 19.12% of the country's land and 1.54% of the water area. These are organized by a system of 25 different national designations that include 29 national parks, 33 wildlife reserves, 42 protection forest reserves, etc.

But PAs under different networks are governed by different laws with varying degrees of protection status and different procedures. In Peninsular Malaysia alone, for example, there are at least four PA networks covering a total area of 2.98 million ha, managed by different agencies including the Federal Department of Wildlife and National Parks, Johor National Parks Corporation, Perak State Parks Corporation, and the respective state forestry departments.

The fact that Malaysia does not yet have a uniform system of national PAs under a common umbrella leads to suboptimally managed and severely underfinanced PAs and makes it impossible to achieve biodiversity conservation goals. The insufficient understanding of the economic value of the PAs and for the key role they play to national development explains the lack of investments by state government and the insufficient capacity within PA system management (PA Financing Project).

The federal government says it wants to change that by creating a total of 110 eco-parks as major new tourist attractions around Malaysia. Many of these new ecotourism hot spots include recreational forests near urban areas, and several of them are state forest parks managed by the Peninsular Malaysia Forestry Department. They boast scenic natural landscapes and a diversity of flora and fauna (CleanMalaysia.com 2017). Moreover, in August 2018, Malaysia's Terengganu state government announced that it has designated 25,664 acres of land formerly slated for logging as a new protected area for wildlife, home to 18 highly threatened mammal species, including the Asian elephant *Elephas maximus*, Sunda pangolin *Manis javanica*, Malayan tapir *Tapirus indicus*, dhole *Cuon alpinus*, white-handed gibbon *Hylobates lar*, and the critically endangered Malayan tiger *Panthera tigris jacksoni*. "This new protected area not only brings more key wildlife habitat under protection, but also protects vital forested watersheds that provide important ecosystem services to the people of Terengganu," said Dr. Sheema Abdul Aziz, President of Rimba (Wiltse-Ahmad 2018).

One of the main environmental issues that Malaysia is facing is deforestation. From 2001 to 2017, Malaysia lost 7.29 Mha of tree cover, equivalent to a decrease of 25% since 2000 and 784 Mt. of CO_2 emissions (Global Forest Watch 2019). The biggest drivers of deforestation are conversion of land for oil palm production and illegal logging for timber (Boucher et al. 2011). According to some studies, oil palm is responsible for carbon dioxide emissions 10 times the magnitude of fossil fuels about 10% of greenhouse gas emissions in the country (Wetlands International 2016). Malaysia is listed as one of the top 6 countries for wood product exports (Kaplinsky et al. 2003), and timber exports make up to 2% of total exports (Boucher et al. 2011). Furthermore, it is estimated that 25% of the country's overall log production is illegal (Lawson 2010). Most of Malaysia's remaining primary forest is found on the island of Borneo.

10.7.2.1 The Case of Redang Island Marine Park

Redang Island Marine Park (RIMP) is situated in the South China Sea. It comprises the main island of Redang Island, which is associated with eight islets. The park is maintained by the Malaysian government and also works as a research facility. It is one of the most attractive places to visit around the Redang Island as it also provides snorkeling and diving activities (Redang Island Malaysia Official 2019). RIMP is becoming an increasingly important ecotourism destination in Malaysia. For example, in 1990, RIMP was visited by just a few hundred people, and this has increased on a yearly basis, and in 2005, it received more than 120,000 visitors. According to a study conducted for the *International Journal of Economics and Management*, the total revenue obtained by the ecotourism operators in RIMP for 2003 was RM32.33 million. Furthermore, the development of ecotourism in RIMP created employment opportunities to local people, both through the involvement with ecotourism operators and the involvement in other sectors which are related to ecotourism industries (Shuib et al. 1994). The finding of the study was that the total employment created in RIMP from ecotourism was 937 employees, with 765 direct employments with ecotourism operators, 12 indirect employments in related sectors, and 160 induced employments (Mohd Rusli et al. 2008).

10.7.3 Nepal

Number of protected areas: 14
Protected land areas coverage: 23.63%
Protected marine area coverage: 0.0%

2017 contribution of T&T to GDP[a]	% GDP	USD (billion)	2018 forecast
Direct	4.0	1.0	↑
Total	7.8	1.9	↑
2017 contribution of T&T to employment[a]	**% jobs**	**'000 jobs**	
Direct	3.2	497.7	↑
Total	6.6	1027.1	↑
	% of total export	**USD (billion)**	
2017 visitor exports[a]	28.0	0.7	↑
	% total investment	**USD (billion)**	
2017 investment[a]	2.3	0.2	↑
GDP international country 2017[b](current USD)		24.4 bn	

*World Travel and Tourism Council: Economic Impact 2018 – March 2018.
[a] World Travel and Tourism Council: Economic Impact 2018 – March 2018.
[b]World Development Indicators (https://data.worldbank.org/products/wdi)

As one of the world's poorest countries, Nepal's economy relies heavily on aid and tourism (BBC News 2018).

The number of tourists has been increasing over the years, with a decreasing period registered in 2000, because of the internal conflicts in Nepal, and in 2015, when a 7.8 Richter scale earthquake took place in the country (Chapagain 2017).

According to government statistics, the total number of tourists visiting Nepal in 2017 was 940,218, from 753,002 in 2016, which represents an increase of 25%. Moreover, out of the total number of tourists that arrive in Nepal, tourists from five countries occupy more than 50%. The proportion of tourists from these countries are India, 17.1%; China, 11.1%; United States, 8.4%; United Kingdom, 5.4%; and Sri Lanka, 4.8% (Ministry of Culture, Tourism and Civil Aviation 2018). These visitors spent ±US $1.9 billion in 2017, about 7.8% of Nepal's national GDP, and directly and indirectly employ more than one million Nepalese (WTTC 2018k).

As Nepal is rich in ecological, social, cultural, and ethnic diversity, there is a great possibility for developing ecotourism. In particular, adventure tourism and wild tourism are important attractions for visitors.

Lonely Planet's Annual "Best in Travel List" says Nepal is the Best Value Destination to travel in the year 2017. "It remains a fabulous choice for budget-conscious travelers, who can access the best of its world-famous trekking routes and underrated wildlife for well south of US$ 50 a day," Lonely Planet writes (Lonely Planet 2017). Also, ShermansTravel has listed Nepal on its list of "Top 10 Ecotourism Destination" in 2013 (Shermans Travel 2013).

According to the Ministry of Foreign Affairs, Nepal's major tourist activities include climbing, trekking, hiking, bird watching, mountain flights, jungle safaris, and rafting/kayaking/canyoning.

The natural resources of this country are the major attractions for the foundation and acceleration of tourism industry in Nepal, which helps in environmental conservation, social enhancement, and economic development of a particular area where ecotourism is being promoted (Anup 2017).

Poverty alleviation, rural development, agricultural transformation, and community enrichment are promoted by ecotourism in Nepal (Anup 2017).

Community-based tourism in protected areas and outside protected areas in different regions of Nepal had supported livelihood of local communities (Nepal 1997; Acharya and Halpenny 2013).

As a result, the government of Nepal has developed protected areas and cultural heritage sites for conserving wildlife, preserving culture, and enhancing ecotourism (Baral et al. 2012).

Many efforts have been applied for the conservation of biodiversity. The various efforts include protected areas, zoos, different types of law, conventions, nongovernmental organizations (NGOs), local and national authorities, national and international organizations, etc. Among them, the protected areas are the main ones (Thapa 2010). Nowadays, the protected areas cover 34,898 km^2, 23.63% of the total area of Nepal. They include 1 hunting reserve, 11 conservation areas, 10 national parks, 3 wildlife reserves, 9 national park-buffer zones, and 3 wildlife reserve-buffer zones (UNEP-WCMC 2018l).

The Department of National Parks and Wildlife Conservation (DNPWC), established in 1980 after the National Parks and Wildlife Conservation Act (1973), is responsible for conserving and managing ecological systems, wildlife, and its habitat and for promoting ecotourism without any negative consequences (Acharya 2014). Although Nepal remains one of the poorest countries in the world (Malik 2013), it is now a model for successful biodiversity conservation (Heinen and Kattel 1992; Heinen and Shrestha 2006; Heinen and Yonzon 1994). In fact, Nepal's protected areas are significant for many reasons.

First, they developed collaboration between different figures (such as the government, nongovernmental organizations, local people, and the conservation authorities) to support conservation. Second, they support community development activities, including education, roads, drinking water, health and sanitation, income generation and capacity building, and wildlife damage relief at the local level. Third, tourism became a major contributor to protected area incomes. Fourth and most importantly, programs created by protected areas developed a network of conservation-related community-based organizations (such as buffer zone management committees, conservation area management council, functional groups, etc.) that have achieved global recognition by demonstrating practical and effective ways to achieve positive outcomes for conservation and local communities (Thakali et al. 2018).

Nepal's ecotourism is a consequence of the country's conservation success which leads to important economic benefits. The government records show that 70.36% and 51.49% of total international visitors to Nepal spent time in a protected area or other nature-based tourism activities in 2015 and 2016. Over the period from 2006 to 2015, the number of international visitors to protected areas increased on average by 10% annually and revenue increased by 13% (Samarth-NMDP 2016 cited after Thakali 2018).

Sustainable uses of forest products make important contributions to Nepal's economy. Some 40% of all Nepali families are dependent upon forest products. One study estimates that forestry directly contributes 9.45% to Nepal's GDP and provides full-time equivalent jobs for 9.23% of the economically active population (Pant 2016).

10.7.3.1 The Chitwan National Park and the Baghmara Buffer Zone Community Forest

In 1973, the National Park and Wildlife Conservation Act was promulgated, and that same year, the Chitwan National Park (CNP) was officially established as the first national park of the country.

The Royal Chitwan National Park is famous for its unique diversity of flora and fauna. UNESCO declared the park a World Heritage Site in 1983. While the number of animals in the park increased as a result of an effective conservation effort, their habitat decreased due to succession, erosion, and encroachment.

In 1989, the Nepal Conservation Research and Training Centre (NCRTC) initiated a buffer zone plantation program with the aim to increase habitat for the endangered wildlife while providing fodder, fuel wood, and timber for the local people. The project was launched in the Baghmara Forest located on the northeast boundary of the park.

As a result, community-managed ecotourism in the Baghmara has been able to generate local guardianship in the conservation of biodiversity of the area, the animal habitats have increased, and the economic benefits created from tourism helped to decrease pressure on the park because when people are economically well-off, they will be able to afford alternatives.

For these reasons, the Baghmara project has become known as a model of sustainable community forest conservation (CSD-7: Sustainable Development Success Stories).

Moreover, in 2018, a total of 133,171 tourists visited CNP and buffer zone areas in the first 10 months of the year. During the period, Rs. 215 million in revenue from tourism was collected against Rs. 225 million in total revenue (Devkota 2018).

10.8 American Countries

10.8.1 Canada

Number of protected areas: 7642
Protected land areas coverage: 9.69%
Protected marine area coverage: 0.87%

2017 contribution of T&T to GDP[a]	% GDP	USD (billion)	2018 forecast
Direct	2.0	32.0	↑
Total	6.5	106.5	↑
2017 contribution of T&T to employment[a]	**% jobs**	**'000 jobs**	
Direct	4.0	739.3	↑
Total	8.6	1588.4	↑
	% of total export	**USD (billion)**	
2017 visitor exports[a]	3.3	17.2	↑
	% total investment	**USD (billion)**	
2017 investment[a]	3.6	13.4	↑
GDP international country 2017[b] (current USD)		1653 bn	

[a]World Travel and Tourism Council: Economic Impact 2018 – March 2018. All rights reserved. Licensed under the Attribution, Non-Commercial 4.0 International Creative Commons Licence
[b]World Development Indicators (https://data.worldbank.org/products/wdi)

In 2017, a record of 6.7 million travelers from overseas (countries other than the United States) arrived in Canada, up by 6.6% from 2016. This marked the eighth consecutive year of increased travel to Canada by overseas residents, with volumes increasing by close to 60% since 2009 (The Daily 2017). Tourism in Canada is the largest service and represents more than 2% of the country's GDP. Moreover, one in 11 Canadian jobs (more than 1.7 million) depends on the tourist economy, and tourism is the number one employer of youth and is an important provider of employment for new Canadians (Government of Canada 2017a).

The country's protected areas represent a significant contribution to the protection of global biodiversity. Although parks are the most common type, protected areas in Canada are divided among more than 100 different types, ranging from ecological reserves and wilderness areas to community parks and conservation zones. Environment Canada's network of protected areas totals over 12.4 million hectares, with more than 85% of the network classified as wilderness areas under the IUCN protected areas management system. The Environment Canada's network of protected areas consists of 54 national wildlife areas and 92 migratory birds areas (Government of Canada 2017b). Parks Canada is responsible for both protecting the ecosystems of these magnificent natural areas and managing them for visitors to understand, appreciate, and enjoy in a way that doesn't compromise their integrity.

Recently, the government of Canada has made a significant investment in terms of nature conservation: The 2018 federal budget includes a $1.3 billion allocation to meet Canada's international commitment to protect 17% of its lands and 10% of waters by 2020 and points to the importance of indigenous partnerships in achieving its conservation goals (Indigenous leadership Initiatives 2018).

10.8.1.1 The Case of the Spirit Bear Lodge in the Great Bear Rainforest

The Spirit Bear Lodge (SBL) is a community-based ecotourism venture owned and operated by the Kitasoo/Xaixais First Nation. It is located in the Great Bear Rainforest (GBR) of British Columbia, Canada, and it is the largest intact temperate rain forest in the world. Since 2002, guests from around the world have traveled to the remote coastal village of Klemtu, in the heart of the rain forest, to experience a unique wildlife tour with local Kitasoo guides. With an abundance of wildlife such as spirit bears, grizzly and black bears, wolves, whales, and dolphins, SBL is an example of a successful, profitable cultural ecotourism business which aims to provide economic benefits and sustainable local employment to residents of Klemtu. The business is now an integral part of the conservation economy in the GBR, now recognized as a globally significant conservation model, and, according to National Geographic, "the wildest place in North America." SBL is a showcase community tourism business in the GBR and a recognized best practice model for indigenous community-based tourism in Canada (Prokosch 2018). SBL has become a successful model for conservation-based ecotourism. The lodge has helped strengthen economic, conservation, and cultural well-being in the community of Klemtu (Coast Funds 2018). In 2014, the Center for Responsible Travel (CREST) conducted the first study aiming to compare the economic value of bear viewing and trophy hunting for both grizzly and black bears in the Great Bear Rainforest (GBF). The study concluded that bear viewing in the GBF generates far more value to the economy, both in terms of total visitor expenditures and GDP, and provides greater employment opportunities and returns to government than does bear hunting. In 2012, bear-viewing companies in the GBF generated more than 12 times in visitor spending than bear hunting: Viewing expenditures were $15.1 million, while guided nonresident and resident hunters combined generated $1.2 million. The study also finds that organized bear-viewing activities are generating over 11 times more in direct revenue for the BC government than bear hunting carried out by guide outfitters: GDP is $7.3 million for bear viewing and $660,500 for nonresident and resident hunting combined. Nonresident grizzly hunting has a higher economic contribution rate than does resident grizzly hunting ($244,600 in nonresident grizzly GDP for four kills or $61,000 per kill compared to $60,000 for resident grizzly GDP for six kills or $10,000 per kill). Further, bear-viewing companies are estimated to directly employ 510 persons per year, while guide outfitters generate only 11 jobs per year in the GBF. In addition, bear viewing is attracting many more visitors to the GBF than is bear hunting (CREST 2014).

10.8.2 *United States of America (USA)*

Number of protected areas: 34,075
Protected land areas coverage: 12.99%
Protected marine area coverage: 41.06%

2017 contribution of T&T to GDP[a]	**% GDP**	**USD (billion)**	**2018 forecast**
Direct	2.6	509.4	↑
Total	7.7	1501.9	↑
2017 contribution of T&T to[a]	**% jobs**	**'000 jobs**	
Direct	3.4	5285.7	↑
Total	8.9	13668.0	↑
	% of total export	**USD (billion)**	
2017 visitor exports[a]	8.6	200.7	↑
	% total investment	**USD (billion)**	
2017 investment[a]	4.6	176.3	↑
GDP international country 2017[b]**(current USD)**		19,390 bn	

[a]World Travel and Tourism Council: Economic Impact 2018 – March 2018.
[b]World Development Indicators (https://data.worldbank.org/products/wdi)

Global tourism keeps growing, a 7% increase was registered in 2017, which is the strongest result in 7 years, and the United States is on pace for its sharpest drop in foreign travelers since the wake of the recession (Josephs 2018).

In the first 7 months of 2017 indeed, the United States took in 41 million international visitors, a 4 percent decline from the previous year, according to the Commerce Department. That follows a 2 percent drop from a year earlier. The US Travel said this equals to the loss of $4.6 billion and 40,000 jobs (Josephs 2018).

While data is showing a decline in tourist numbers, US President Donald Trump's policy raises concerns over America's image as a welcoming country. The travel ban blocking visitors from nations including Iran, Libya, and Syria is an example. "Politics is not helping us," said Tilo Krause-Duenow, owner of German tour operator CANUSA, which specializes in trips to North America (Sheahan 2018).

President Trump's new policy is raising concerns also in terms of wildlife protection and conservation: In June 2018 indeed, he signed a new executive order detailing a revised US oceans policy which revokes the 2010 oceans policy issued by then-President Barack Obama and replaces it with a markedly different template for what the government should focus on in managing the nation's oceans, coastal waters, and Great Lakes (Malakoff 2018).

Basically, the US policy toward the ocean is moving from conservation to moneymaking (Housman 2018).

Furthermore, according to the National Wildlife Federation (The National Wildlife Federation 2019), the country is living a wildlife crisis, and there is an urgent need for a change in the way the country funds conservation:

State fish and wildlife agencies have identified roughly 12,000 species in need of proactive conservation efforts in the United States, and the number of species petitioned for listing under the Endangered Species Act has increased by 1000 percent in less than a decade.

Habitat loss, invasive species, and severe weather are the main reasons why all types of wildlife are declining across the country.

The dramatic decline of so many species of wildlife and the habitats they depend on threatens Americans' quality of life, as well as their outdoor economy. Today, the outdoor recreation industry annually contributes $887 billion to its national economy, creates 7.6 million direct jobs, and generates $124.5 billion in federal, state, and local tax revenue, according to the Outdoor Industry Association (The National Wildlife Federation 2019).

According to the federation, the current levels of funding are less than 5% of what is necessary, and the Recovering America's Wildlife Act is the solution as it is a twenty-first-century model of conservation funding that will proactively, and cost-effectively, address these widespread drops in our wildlife populations (The Recovering America's Wildlife Act Factsheet).

A remarkable step in terms of international conservation efforts involving the United States is the US-China Nature Conservation Protocol: established in 1986. With this protocol, the two countries agreed to establish and manage protected natural areas for conservation of wildlife and habitats, protect migratory birds, regulate trade of endangered species, and participate in cooperative research and management projects (U.S. Fish and Wildlife Service 2019).

The most recent US-China work plan, known as Annex 12, negotiates the activities to be carried out during 2014–2016 in terms of nature reserve management, wildlife conservation, habitat protection, and other topics (Annex 12 2014).

As a large continental country with vast and diverse natural resources, the United States has a long tradition of ecotourism on public and private lands and waters from coast to coast.

While there are no consolidated data on all ecotourism activities taking place in the United States, it is estimated that Americans spend billions of dollars annually visiting national forests, parks, protected areas, and wildlife refuges and reserves. National parks attract millions of visitors each year, owing to their vast capacity for outdoor recreation—an estimated 331 million people visited national parks in the United States in 2017. National park tourism benefits both the national and local economies. In 2017, national park visitor's spending reached over 17 billion US dollars, and around 30% of this was spent on accommodation and 20% in restaurants and bars (Statista.com 2018a).

As of 2018, the United States has a total number of 34,075 protected areas, covering almost 13% of the land area and an additional 41.04% of the total marine area of the country (UNEP-WCMC 2019c).

These protected areas are managed by an array of different federal-, state-, and local-level authorities and receive widely varying levels of protection. The highest levels of protection, as described by the International Union for Conservation of Nature, are Level I (strict nature reserves and wilderness areas) and Level II (national parks).

But according to two new studies, few national parks are large enough to contain ecosystems: America's national parks and other protected public lands are too small and fragmented to sufficiently preserve the nation's biodiversity. Often missing are conservation corridors linking islanded protected areas (Sainato 2015).

Ecotourism activities in the United States may be subject to regulation at various levels of government depending on their location and status. However, the US government (USG) has several major land and water management agencies that support and promote ecotourism, including the National Park Service (NPS), National Forest Service, US Fish and Wildlife Service (FWS), Bureau of Land Management (BLM), and National Marine Fisheries Service (NMFS) (US Department of State Archive 2003).

10.8.2.1 The Case of Yellowstone National Park in Wyoming

The Yellowstone National Park is America's first national park, located in the State of Wyoming where travel and tourism make up a significant portion of the economy (National Parks Conservation Association 2019).

More than 8.7 million overnight visitors traveled to Wyoming in 2017, spending approximately $3.5 billion. This is equivalent to approximately $9.5 million dollars per day, an 8.9% increase over the previous year (Wyoming Office of Tourism 2018).

The Yellowstone National Park was established in 1872 and named after the river that runs through it. Within the park's massive boundaries, visitors can find mountains, rivers, lakes, waterfalls, and some of the most concentrated geothermal activity in the world. The park has 60% of the world's geysers, as well as hot springs and mud pots (National Parks Conservation Association 2019). Yellowstone National Park is widely considered to be the finest megafauna wildlife habitat in the United States. There are 67 species of mammals in the park, including the gray wolf, the threatened lynx, and the grizzly bear. Visitors in Yellowstone engage in a wide variety of activities throughout the year. With over 85% of visitors in all seasons, wildlife watching is one of the top reasons people choose to visit Wyoming. In particular, wolves watching represents a significant activity for the tourists coming to the park. When asked, 44% of the visitors said that watching wolves was their main reason for visiting the park, and 325,000 park visitors saw wolves in 2005. In total, it is estimated that the presence of wolves in Yellowstone National Park in 2005 led to an increase of visitors that resulted in an additional net economic value of between $18.3 and $30.6 million. Approximately 45–50% of visitors year-round participated in wildlife photography while in the park, with the highest percentage of participation in the spring (Duffield et al. 2006).

In particular, the gray wolves of Yellowstone have an interesting history:

- Eradication of wolves in 1872–1926. By the end of the 1920s, the gray wolf had been hunted to eradication, predominantly by ranchers protecting their livestock, becoming among the first species to be listed as endangered.
- Reintroduction of wolves in 1995. Seventy years later, the park began the reintroduction of these animals with 31 wolves trucked into the park from Canada. This decision was made not only in order to protect the gray wolf population but also because of the consequences that their eradication was having on the other animals in the park. In particular, the elk population exploded, and they grazed their way across the landscape killing young brush and trees.
- The effects of the reintroduction of wolves. Since the reintroduction, the wolf population has continued to grow and stabilize (Robbins 2017). But the reintroduction of wolves in the park had big consequences on the big-game numbers, in particular on the elk population. There are two generally opposing views. The first view is compensatory: That is, wolves have only had a little impact on northern Yellowstone elk population since they primarily take elk that would normally succumb to winter kill, disease, or old age (Vucetich et al. 2005, cited after Duffield 2016). The second view is that wolf predation of elk is largely additive and that it substantially increased the rate of recent declines in elk population (White and Garrott 2005, cited after Duffield 2006). A third view is that elk populations have decreased due to wolf predation, but not fully to the extent that would be predicted from the number of elk killed by wolves (Varley and Boyce 2006, cited after Duffield 2006).

According to the 2017 Montana Gray Wolf Program Annual Report, population estimates suggest there are approximately 900 wolves in Montana today, and approximately $380,000 was generated for wolf conservation and management by wolf license sales.

In 2017, livestock depredation by wolves was approximately 25% of what it was in 2009, when it was at a peak. The US Department of Agriculture's Wildlife Services confirmed 80 livestock losses to wolves in 2017. During the same year, the Montana Livestock Loss Board paid $64,133 for livestock which Wildlife Services confirmed as probable or certain wolf kills (Montana.gov 2018).

Based on the amount of money spent in the entire area around Yellowstone National Park, visitors who specifically wanted to see or hear wolves generated approximately $35.5 million in 2005, compared to the $27.74 million in 1991 (Duffield 2006).

10.8.3 Brazil

Number of protected areas: 2299
Protected land areas coverage: 29.42%
Protected marine area coverage: 26.62%

2017 contribution of T&T to GDP[a]	% GDP	USD (billion)	2018 forecast
Direct	2.9	59.6	↑
Total	7.9	163.0	↑
2017 contribution of T&T to employment[a]	**% jobs**	**'000 jobs**	
Direct	2.6	2337.0	↑
Total	7.3	6591.3	↑
	% of total export	**USD (billion)**	
2017 visitor exports[a]	2.3	19.7	↑
	% total investment	**USD (billion)**	
2017 investment[a]	6.1	6.0	↑
GDP international country 2017[b](current USD)		19,390 bn	

[a]World Travel and Tourism Council: Economic Impact 2018 – March 2018. All rights reserved. Licensed under the Attribution, Non-Commercial 4.0 International Creative Commons Licence
[b]World Development Indicators (https://data.worldbank.org/products/wdi)

Brazil's Ministry of Tourism showed that in 2017 Brazil received the largest number of international tourists in the country's history: In total, 6,588,770 international tourists landed in Brazil last year, beating previous years' records (Belen 2018).

With its reputation as having one of the most impressive ecosystems in the world, Brazil is a top destination for adventurous people who are interested in practicing ecotourism. Recently, many studies have drawn *attention to the forest-based ecotourism products, such as iconic wild animals* (pink river dolphins *Inia geoffrensis*, brown-throated sloth *Bradypus variegatus*, caimans *Caiman* spp., *Melanosuchus niger*, anacondas *Eunectes* spp., squirrel monkeys *Saimiri* spp., etc.), natural landscapes, indigenous culture, and heritage (D'Cruze et al. 2017). However, the country's enormous potential to be an ecotourism destination is not translated into big numbers for this market, as it accounted for only USD 515 million in revenue in 2012 while attracting 3.7 million tourists. In fact, its potential is even less inspiring when taking into account that ecotourism had global revenue of USD 260 billion in 2012, with Brazil being responsible for only 0.027% of this market (Bruha 2015).

The country hosts 76 national parks, 88 natural biological reserves, and 94 areas that have environmental protection managed by the National System of Protected Areas (SNUC in Portuguese), which was established in 2000 in order to ensure the creation, management, and consolidation of protected areas in Brazil (WWF 2014). However, SNUC does not have a long-term strategy toward its consolidation and financial sustainability, and even with its large number of protected areas of different sorts, some important areas of the country are still not protected enough (WWF 2014).

Recently, the Brazilian government announced the decision to promote ecotourism in Brazilian protected areas: The Ministries of the Environment and Tourism, ICMBio (Chico Mendes Institute for the Conservation of Biodiversity), and Embratur (the Brazilian Tourism Board) signed cooperation agreement which aims not only to develop joint actions in federal protected areas but also to increase the numbers of ecotourists visiting the country's protected areas and to promote sustainable development through ecotourism (BrazilGovNews 2017).

Moreover, according to Thomson Reuters Foundation, in March 2018, the Brazilian president declared that Brazil will protect two vast marine areas, meaning a quarter of the country's oceans, in a bid to help preserve its biodiversity (Hares 2018).

Despite these and other initiatives undertaken by the Brazilian government, Brazil is renowned for the destruction of its natural environment. Amazonia, which holds most of the country's biological diversity, has long been in a state of environmental crisis. Current measurements indicate that legal deforestation has been steadily reducing and is now significantly lower than the 2004 peak (27,772 km^2), although the number, however, indicates an increase in 2012 (Brazil Ministry of the Environment 2015). WWF estimated instead that forest losses in the Amazon averaged 1.4 million hectares/year between 2001 and 2012, with a total loss of 17.7 million hectares (177,000 Km^2), mostly in Brazil, Peru, and Bolivia. WWF also foresees that, at the actual trend, more than a quarter of the Amazon biome will be without trees by 2030 (WWF 2017).

10.8.3.1 The Brazilian Pantanal

Part national park, part UNESCO World Heritage Site, the Pantanal is the world's largest tropical wetland, covering over 70,000 km^2 in the smack-dab center of South America. Wildlife is highly visible and abundant in the Brazilian Pantanal, which quietly boasts the highest concentration of wildlife on the continent. This includes jaguars, giant anteaters, piranha, howler and capuchin monkeys, and green anacondas—the world's largest snakes (Stonich 2016). For this reason, the Pantanal is widely considered to be the most important wildlife tourism cradle in Latin America (Chardonnet et al. 2002 cited after Tortato et al. 2017).

In particular, Jaguar ecotourism represents a significant source of income: According to the data, the total annual revenue that jaguars represented for seven local lodges across the Brazilian Pantanal in 2015 was USD 6,827,625, and values per lodge ranged from USD 81,000 to USD 3,105,000 (Tortato et al. 2017).

Illegal hunting represents the main threat to the population of jaguars in the Pantanal and is supposedly mainly driven by economic losses induced by jaguars on cattle herds in private ranches (Zimmerman et al. 2005, and Boulhosa and Azevedo 2014, cited after Tortato et al. 2017). These losses range widely from USD 350 to USD 22,400 per year within a single ranch in the Pantanal (Tortato et al. 2015 cited after Tortato et al. 2017). Considering the aggregate costs of jaguar depredation on livestock within the same area, the authors estimated that the resident jaguar population would induce a hypothetical damage of only USD 121,500 per year in bovine cattle losses. In terms of observed costs and benefits of providing safe habitat to a jaguar population, the benefits accrued from ecotourism far outweighed the costs of cattle losses in private ranches. Moreover, the cattle losses induced by jaguars could be compensated by a system of voluntary donations from tourists, ensuring that both traditional livestock husbandry and ecotourism can coexist promoting landscape-scale jaguar conservation. In fact, 80% of all tourists visiting the study area suggested that they would be willing to donate on average of USD 84 over a 3-day stay, demonstrating the financial feasibility of a relatively doable compensation scheme through wildlife tourism, which would create a "win-win" scenario for both of these main economic activities (Tortato et al. 2017).

10.8.4 Costa Rica

Number of protected areas: 187
Protected land areas coverage: 27.6%
Protected marine area coverage: 0.83%

2017 contribution of T&T to GDP[a]	**% GDP**	**USD (billion)**	**2018 forecast**
Direct	5.0	2.9	↑
Total	12.9	7.5	↑
2017 contribution of T&T to employment[a]	**% jobs**	**'000 jobs**	
Direct	5.1	104.3	↑
Total	12.5	254.3	↑
	% of total export	**USD (billion)**	
2017 visitor exports[a]	20.3	3.9	↑
	% total investment	**USD (billion)**	
2017 investment[a]	4.0	0.4	↑
GDP international country 2017[b]**(current USD)**		57 bn	

[a]World Travel and Tourism Council: Economic Impact 2018 – March 2018.
[b]World Development Indicators (https://data.worldbank.org/products/wdi)

Once famous as a producer of agricultural goods such as coffee and bananas, today Costa Rica is a famous tourist destination.

With more than two million visits per year, tourism is today Costa Rica's leading sector and one of the most important sources of money. According to the Costa Rican Tourism Board (ICT 2017), the secret for the development of Costa Rica as a tourism destination is in its particular tourism supply which focuses on a wide range of services and products, including ecotourism, sun and beach, adventure, rural tourism, meetings, and more.

In the last 10 years, tourism has generated approximately USD 28 billion, with the exception of 2009 and the post-stock market crash global crisis, and that amounts to roughly 10% of the nation's GDP per year. Moreover, 23.9 million visitors from around the world arrived in Costa Rica in the last decade, spending an average of USD 1200 per person (Luty 2018).

Ecotourism in Costa Rica is a robust industry because the country was one of the earliest adopters of connecting nature/wildlife conservation with responsible travel. The country is considered an ideal introduction to the rain forests for its biodiversity, its excellent and accessible natural park system, and its relative safety for tourists. In some areas, tourism has proved a little too much for the environment, and some parks now have restrictions on the number of visitors allowed at any given time.

Moreover, Costa Rica has been ranked among the world's best ecotourism destinations on more than one occasion and became the first country in the Americas to ban hunting in 2012 Cota 2012 (Embajada de Costa Rica 2008). Approximately half of Costa Rica's international tourists visit a protected area (Kogan 2010, cited after Ferraro and Hanauer 2014).

Costa Rica is recognized worldwide as one of the countries at the forefront of conservation and particularly by its system of protected areas. Costa Rica boasts 28 national parks, 73 national wildlife refuges, 12 wetland areas, 9 forest reserves, and 8 biological reserves among others (UNEP-WCMC 2018f).

Since the 1990s, the Costa Rican government has developed a mixed economic-regulatory policy to protect the forests and ecosystems they host. They support a Payment for Environmental Services (PES), which is a program that rewards forest owners for the ecosystem services their forests provide: conserving wild species, protecting water sources, capturing and storing atmospheric carbon, and landscape beauty (Stolton et al. 2010; Porras et al. 2013). By 2005, 451,420 ha had been protected as part of the scheme, and the program had made payments to more than 4400 farmers and forest owners; it has been estimated that each hectare of forest is worth between 40 and 100 dollars for the services given to protect the river basins (Stolton et al. 2010). In 2012, the annual payment ranged between 41 and 294 USD/ha and 0.43 and 0.65 per tree. Despite the voluntary and short-term nature of the contracts (which clash with the long-term ideal of nature conservation), the program has helped to conserve nearly one million hectares of forest by payments for protection (90%), reforestation (6%), sustainable management (3%), and more recently regeneration of abandoned pastures (1%) (Porras et al. 2013). The PES program benefits people directly, through subsidies and possibly new jobs, and indirectly, i.e., by promoting a healthier ecosystem, which in turn is essential for the attraction of ecotourism operators. Here again, the level of education was significant for the adhesion to the program (Porras et al. 2013). At the time of Porras' and colleagues' report, the total forests' area amounted to 52% of the country vs. 21% registered in the 1980s.

Moreover, a study published in the journal of the National Academy of Sciences reveals that ecotourism has a positive impact on poverty reduction and has improved the quality of life of Costa Ricans living in areas close to parks and protectedareas by 16% (Ferraro and Hanauer 2014).

10.8.4.1 The Case of Osa Peninsula and the Corcovado National Park

The promotional materials of the Costa Rican Institute for Tourism (ICT 2018) prominently promote the Osa Peninsula as a place where "ecotourism features as the main product."

The Osa Peninsula, in fact, is considered not only the country's last wilderness frontier but also one of the most biodiverse places left on earth: It includes 13 ecosystems, such as forests, lakes, mangroves and freshwater wetlands, beaches, and coral reefs. It is home to more than 375 species of birds, 124 species of mammals, 40 species of freshwater fish, about 8000 species of insects, and 117 species of reptiles and amphibians (Driscoll et al. 2011).

Much of the Peninsula is protected by Corcovado National Park, covering over 42,570 hectares in its terrestrial area and 5375 hectares of marine protected area (Global Conservation 2018). In 2017, the Costa Rican government inaugurated new infrastructure at Corcovado National Park, investing USD2.4 million dollars of the Sustainable Tourism Program. The purpose of the investment is not only to raise the number of tourists but also to improve the conditions of the local communities and to improve the working condition of the park rangers (Alvarado 2017). Threats to the park include illegal mining, illegal logging, and commercial poaching of wildlife, which had decimated the jaguar population (now down to less than 30 individuals in Corcovado National Park) and its prey, the paca, a popular game meat for restaurants and families in Costa Rica (Global Conservation 2018).

10.8.5 Ecuador

Number of protected areas: 83
Protected land areas coverage: 21.69%
Protected marine area coverage: 13.55%

2017 contribution of T&T to GDP[a]	**% GDP**	**USD (billion)**	**2018 forecast**
Direct	2.2	2.2	↑
Total	5.4	5.4	↑
2017 contribution of T&T to employment[a]	**% jobs**	**'000 jobs**	

Direct	2.2	156.2	↑
Total	5.1	363.1	↑
	% of total export	**USD (billion)**	
2017 visitor exports[a]	10.6	1.2	↑
	% total investment	**USD (billion)**	
2017 investment[a]	4.9	2.1	↑
GDP international country 2017[b]**(current USD)**		103 bn	

[a]World Travel and Tourism Council: Economic Impact 2018 – March 2018. All rights reserved. Licensed under the Attribution, Non-Commercial 4.0 International Creative Commons Licence
[b]World Development Indicators (https://data.worldbank.org/products/wdi)

Despite this slowdown in general tourism trends (UN-ECLAC 2018), Ecuador has the fortunate distinction of being the easiest gateway to the Galapagos Islands, which is well appreciated by ecotourists for its distinct wildlife: Along with a booming community-based tourism trade, there is a highly regarded ecotourism certification program which focuses on environmental education and tourist waste management though there are some issues with deforestation (Lane 2018).

The country hosts 83 protected areas scattered throughout the whole national territory, among them 12 national parks, 5 biological reserves, 9 ecological reserves, 1 geobotanic reserve, 11 wildlife refuges, 5 marine reserves, and 6 national recreational areas, which make Ecuador the perfect place for ecotourism activities (UNEP-WCMC 2018g). The Protected Areas National System of Ecuador (Sistema Nacional de Áreas Protegidas, SNAP) is responsible for all of the PA and manages an area that covers more than 20% of the country's surface, with the aim to ensure the coverage and connectivity of important ecosystems at the terrestrial, marine, and coastal marine levels, of their cultural resources, and of the main water sources (MAE 2006, cited by Ministerio del Ambiente 2015).

With all of its national parks, protected areas, and heritage sites, Ecuador has become a leader in the ecotourism movement. By focusing on education as a fundamental element in the future of conservation, preservation, and sustainable tourism management, Ecuador has created a truly sustainable path for ecotourism (Lindsontheroad 2011). Furthermore, the government is working to support local communities to develop responsible and sustainable tourism industries that stimulate rural economies and protect the environment and local cultures. Ecuador is the first country in South America to become part of the Global Council of Sustainable Tourism, and the ministries are working with local communities to help them meet quality standards (Scherffius 2015). Despite their enormous value, the biodiversity and ecosystems in Ecuador have been threatened over the years by several environmental factors: Deforestation, water pollution, and soil contamination are the top three issues that negatively affect the environment in Ecuador. According to a report from the Ministerio del Ambiente (2014), between 2000 and 2008, the Ecuadorian

Amazon lost 109,000 hectares of natural forest annually. In September 2014, the New York Declaration on Forests (NYDF) outlined ten goals with the aim of achieving net zero deforestation by 2030. But according to a report from 2017, more finance is required in order to achieve the goals. For example, the USD 20 billion invested since 2010 for "providing support for the development and implementation of strategies to reduce forest emissions" (Goal 8) and "reward countries and jurisdictions that, by taking action, reduce forest emissions–particularly through public policies to scale-up payments for verified emission reductions and private-sector sourcing of commodities" (Goal 9) is considered insufficient and does not reflect the importance of forests as part of the climate solution (Climate Focus 2017).

In order to prevent deforestation and drilling for oil in one of the most biodiverse areas in Ecuador—the Yasuni Biosphere Reserve—in 2007, Ecuadorian President Rafael Correa launched the so-called Yasuni-ITT project in the Ishpingo-Tambococha-Tiputini oil field (Rickerby et al. 2016). This area is indeed estimated to contain 20–30% of the country's entire oil reserve, with an estimated amount of 846 million to 1.5 billion barrels. The Yasuni-ITT proposed that the global community contribute roughly USD350 million annually for 10 years to fund renewable energy while respecting biodiversity and social equality. In return, Ecuador would provide "Yasuni Guarantee Certificates" corresponding to the avoided CO_2 emissions (Bucaram et al. 2017).

The Yasuni-ITT initiative failed by 2013 due to economic and political reasons, as it was only able to collect USD13 million, i.e., 0.36% of the required total.

In an attempt to alleviate some of the inevitable damage (e.g., oil spills, disruption of tribes, etc.), the government of Ecuador has taken the responsibility to invest in the latest (and thus safest) extraction technology. This, however, is unreliable and not enough to minimize damage within Yasuni National Park. Evidence for this can be seen in the recent oil spill disaster caused by the American company Texaco, which resulted in catastrophic damage to both the environment and the health of many indigenous peoples (Bucaram et al. 2017).

Moreover, in 2018, Ecuador's state oil company has begun a second phase of drilling inside a new field of the Yasuni national park, the Tambococha-2 well. This decision has triggered fierce criticism from conservationists who say it is likely to accelerate deforestation, hunting, and conflicts with the local tribes (Rickerby et al. 2016; Vidal 2016; Watts 2018).

10.8.5.1 The Case of Galapagos National Park

The Galapagos National Park is one of the greatest natural treasures in the world. It consists of 13 large islands, 17 small islands, and 40 rocks, and it covers some 90% of the entire land surface area in this ocean region. Despite this, there are 54 different sites that are open to tourists (Ecuador.com 2019).

The Galapagos National Park is at the forefront of developing and promoting ecotourism.

Indeed, the two organizations which are responsible to protect the park and its many delicate resources (the Galapagos National Park Service and the Charles

Darwin Research Station) introduced a number of changes to the way the islands manage tourism:

- Setting carrying capacity limits for different sites and restricting the number of visitors allowed in some areas.
- Planning boat routes so that areas are not overwhelmed by visitors at any one time.
- Introducing entrance fees for visitors to the national park.
- Creating a ruling that no tourist is allowed to explore the islands without being accompanied by a guide who will not only educate the various visitors but help to enforce laws and protect the park environment.
- Developing educational opportunities for visitors and local people.
- Requiring boat licenses. These have now come into place for both tour operators and fishing boats and help the marine reserve police the waters far more carefully.
- Establishing urban development zones (Galapagos Conservation Trust 2019).

10.9 European Countries

10.9.1 Italy

Number of protected areas: 3899
Protected land areas coverage: 21.54%
Protected marine area coverage: 8.79%

2017 contribution of T&T to GDP[a]	**% GDP**	**USD (billion)**	**2018 forecast**
Direct	5.5	106.8	↑
Total	13.0	253.5	↑
2017 contribution of T&T to employment[a]	**% jobs**	**'000 jobs**	
Direct	6.5	1490.5	↑
Total	14.7	3394.7	↑
	% of total export	**USD (billion)**	
2017 visitor exports[a]	7.4	44.9	↑
	% of total investment	**USD (billion)**	
2017 investment[a]	3.4	11.6	↑
GDP international country 2017[a]**(current USD)**		1934 bn	

[a]World Travel and Tourism Council: Economic Impact 2018 – March 2018. All rights reserved. Licensed under the Attribution, Non-Commercial 4.0 International Creative Commons Licence
[b]World Development Indicators (https://data.worldbank.org/products/wdi)

Tourism represents an extraordinary engine of economic development for Italy. In fact, Italy has an abundance of high-quality natural and cultural heritage, which constitutes unique resources for tourism development.

The industry contributed a total of USD 253.5 bn, 13.0% of GDP in 2017 (WTTC 2018b).

Together with general tourism, also nature-related tourism is growing around the country.

According to the Annual Report on Nature Tourism from 2016, tourism in protected areas during 2015 registered a total of 105 million presences, corresponding to a 2.3% growth compared to the previous year, which generated 13,942 milliards USD (Gazzetta del Turismo 2016).

According to the World Database of Protected Areas (UNEP-WCMC 2019b), the country hosts 3899 protected areas of which 27 with management effectiveness evaluation. These include among others 365 regional/provincial nature reserves, 27 natural marine reserves and natural protected marine areas, 24 national parks, and 134 regional/provincial nature parks.

According to the type of area, the management body for a protected area may be an independent public organization, at a national or regional level, a consortium, a municipal administration, or an association. These areas are funded by public sources managed by national and regional administrations.

The Italian Federation of Parks and Nature Reserves aims to promote and implement projects and plans for the conservation and the modern management of protected natural areas supporting measures to make tourism in protected natural areas more sustainable.

Protected areas system in Italy faces several issues due to lack of legislative clarity, lack of financial resources, territorial institutional role not fully recognized, and lack of national coordination and support programs (Parks.it2018).

With a strategic plan for tourism (MiBACT), regarding a time period of 6 years (2017–2022), the Italian government is reshaping its plan for the tourism economy, implementing a lasting and sustainable approach to the environmental and cultural heritage of the country. The plan aims to broaden Italy's tourism supply and make it more sustainable and competitive also through a responsible use of natural settings such as nature parks and marine parks, mountains, and rural areas (MiBACT).

10.9.1.1 The Case of the Abruzzo Region's Parks

Abruzzo is considered the Europe's greenest region, with a third of its territory set aside in protected areas: 3 national parks, 1 regional park, and more than 30 nature reserves. This is the reason why this region is considered a leader in ecotourism or also "green tourism." The region is of national and international importance as 75% of all Europe's flora and fauna species are represented there. The region's integrated network of parks and nature reserves has allowed the conservation and protection of a huge number of resident and migrant species. These include a few unique species, such as the Abruzzo's chamois, the Apennine wolf, and the Marsican brown bear, whose survival has depended totally on the local mountains (Martella et al. 2003; Ciucci and Boitani 2008; Ciucci et al. 2017; Masseti and Salari 2017; Mattioli et al. 2018).

10.9.2 Norway

Number of protected areas: 2998
Protected land areas coverage: 17.11%
Protected marine area coverage: 0.83%

2017 contribution of T&T to GDP[a]	**% GDP**	**USD (billion)**	**2018 forecast**
Direct	3.7	14.9	↑
Total	9.0	33.8	↑
2017 contribution of T&T to employment[a]	**% jobs**	**'000 jobs**	
Direct	6.5	176.1	↑
Total	12.7	335.6	↑
	% of total export	**USD (billion)**	
2017 visitor exports[a]	4.6	6.8	↓
	% total investment	**USD (billion)**	
2017 investment[a]	6.4	3.5	↑
GDP international country 2017[b]**(current USD)**		398.8 bn	

[a]World Travel and Tourism Council: Economic Impact 2018 – March 2018. All rights reserved. Licensed under the Attribution, Non-Commercial 4.0 International Creative Commons Licence
[b]World Development Indicators (https://data.worldbank.org/products/wdi)

There were approximately 1.3 billion arrivals across national borders in 2017. International arrivals have been steadily increasing since 2009 (Statista.com 2018b). The rise in the number of trips across national borders demonstrates the robustness of the industry. Figures from the World Travel and Tourism Council show that the tourism industry was an important contributor to the country's economy, with a total contribution of travel and tourism to the country's GDP of USD 36.3 bn (9.0%) in 2017 (WTTC 2018c).

Due to its many natural wonders, Norway is the best place for nature-related activities, and ecotourism is very common. According to a report from 2016, nature outdoor experiences ranked highly among travelers that showed interest in participating in activities such as experiencing the magnificent scenery and natural phenomena such as the midnight sun or northern lights, hiking, cycling, skiing, snowboarding, and similar (Innovation Norway 2016).

Norway has a long conservation history establishing its first nature conservation area in 1884. Today, about 17% of the country is protected under the Nature Diversity Act, which aims to safeguard a representative selection of Norwegian habitats and landscapes for future generations and to protect areas of special value for plants and animals. According to the World Database on Protected Areas, currently there are 2998 protected areas around the country, organized in 2189 natural reserves, 175 wildlife conservation areas, 39 national parks, 194 protected landscapes, etc. (UNEP-WCMC 2018m).

Moreover, according to the statistics, a total of 9.7% of mainland Norway is national parks (Statistics Norway 2018).

Protected areas play an important role in maintaining viable populations of plants and animals, but protection alone is not enough. In fact, according to the Norwegian Environment Agency (NEA 2016), even when all the current conservation plans have been implemented, some habitat types will not be adequately represented, and most of the country will not be protected under the terms of the Nature Diversity. Moreover, Fredman and Haukeland (2016) identify five main challenges for the future of ecotourism development in Scandinavia: urbanization, increased mobility, changing demography, new lifestyles, and climate changes.

The Norwegian Red List for Ecosystems and Habitat Types assesses the status of 80 habitat types in the country. There is no documentation that any habitat types have been lost completely, but 40 habitat types are listed as threatened: 2 as critically endangered, 15 as endangered, and 23 as vulnerable (CBD 2014b).

10.9.2.1 Protected Areas in Svalbard

Located in the Arctic Ocean, the archipelago of Svalbard is part of the Kingdom of Norway.

Spitsbergen is the largest island of this archipelago and also the place where most of the human activities take place.

The total land area is 61,022 km^2, which correspond approximately to 16% of Norway's total land area.

Svalbard is one of the most sparsely populated regions in the world, with 0.04 inhabitants per km^2 (2650 people in total).

Once considered a terra nullius, a land over which no single state held sovereignty, after the Svalbard Treaty of 1920 (The Svalbard Treaty 1920), this region becomes part of the Kingdom of Norway. Since then, Norway is responsible for managing the archipelago and to safeguard its species and habitats.

Approximately 60% of the archipelago's land area is covered by glaciers. Altogether, 98% of Svalbard's land area is natural wilderness.

Today, Svalbard region is considered as one of the best-protected wilderness areas, and in comparison with the Norwegian mainland, much more of Svalbard's area is protected, including large marine areas: There are 7 national parks and 21 nature reserves, all protected under the 2002 Svalbard Environmental Protection Act

(The Governor of Svalbard 2019). They cover 65% of the area of the islands and about 87% of the territorial waters out to the 12-nautical-mile territorial limit.

Svalbard's biodiversity includes about 30 species of birds (all of them nest annually in Svalbard, but only a few remain here in winter) and few species of mammals of which just 3 live on land: the Svalbard reindeer, the Arctic fox, and the sibling vole. The polar bear is regarded as a marine mammal, along with 15/20 species of seals and whales.

Since 1607, when the British explorer Henry Hudson discovered the potential of Svalbard's natural resources, this has been the land for exploration and researches, but also for hunting and mining. Over the years, the animal populations have been severely reduced while the mining industry reached a great economic value.

In recent years, the turnover in the growing industries of tourism and culture overtook the turnover in the mining and quarrying industry, which investments have declined drastically, from USD404 million in 2008 to only USD 6.9 million in 2015.

Today, tourism is the leading industrial sector in Svalbard. According to the statistics, 450 out of the 1650 full-time equivalent jobs that are carried out by Svalbard's inhabitants are employed in tourism and culture (Statistics Norway 2016, Norwegian Directorate for Nature Management, Bergquist 2016).

The increasing number of tourists coming to Svalbard by plane or ship (estimated at 60,000) is required to pay a fee which is placed into the conservation fund used every year to promote education, information, nature conservation, and research projects for the management of tourism and protected areas in the region. Moreover, some tour operators are doing a great job educating tourists about the values and importance of Svalbard's nature and its protection. These initiatives, together with the joined political action of tour operators and conservation NGOs (which resulted in the creation of new national parks), are the main reasons why Svalbard is considered a good example of linking tourism and conservation (Prokosch 2015a).

10.9.3 Poland

Number of protected areas: 3067
Protected land areas coverage: 39.65%
Protected marine area coverage: 22.57%

2017 contribution of T&T to GDP[a]	**% GDP**	**USD (billion)**	**2018 forecast**
Direct	1.9	10.2	↑
Total	4.5	23.9	↑
2017 contribution of T&T to employment[a]	**% jobs**	**'000 jobs**	
Direct	2.0	332.0	↑
Total	4.5	738.2	↑
	% of total export	**USD (billion)**	
2017 visitor exports[a]	4.6	13.0	↑

	% total investment	USD (billion)	
2017 investment[a]	3.0	2.8	↑
GDP international country 2017[b]**(current USD)**		524.5 bn	

[a]World Travel and Tourism Council: Economic Impact 2018 – March 2018.
[b]World Development Indicators (https://data.worldbank.org/products/wdi)

Poland ranks among the world's ten most visited countries, with the bulk of visitors coming from the neighboring nations. According to the statistics, in 2017, there were 83.8 million foreigners, including 18.3 million tourists and 65.5 million same-day visitors, who came to Poland (Statistics Poland 2018). Moreover, the expenditures of foreigners visiting Poland (made before the trip in their home country and in Poland) amounted to ca. 56.7 billion PLN (approximately 15.41 billion USD), which was 3.6% more than in 2016 (Statistics Poland 2018).

Poland has undoubtedly plenty of resources to develop ecotourism or nature tourism's products. However, ecotourism is still a new trend in Poland. The ecotourism precursor in the country is the European Centre for Ecological Agriculture and Agro-Touristic-Poland (ECEAT), a nongovernmental organization working since 1993 to protect the environment and preserve the traditions and culture of the Polish countryside. On the Polish market, several travel agencies are also specialized in ecotourism, but according to a recent study, the development of ecotourism in the country still lacks an overall, coherent strategy both at a state and regional level (Oniszczuk-Jastrząbek et al. 2017). Some of the suggested activities include training of employees who understand the importance of local natural and cultural values and officials and decision-makers who can make the right environmental decisions, promoting marketing materials and sustainable/ecological profile of the region, and a wide range of activities organized for visitors which promote and respect nature and culture (Oniszczuk-Jastrząbek et al. 2017).

In recent years, Poland has made an attempt at a valuation of economic benefits from the use of forests which corresponds to a total value of USD 1.15 billion (CBD 2014a, b).

According to data from 2013, the number of visitors to national parks is estimated at about 11 million people per year, which accounts for nearly 30% of the Polish population (CBD 2014a, b).

A study conducted in the stork village of Zywkowo in the Masurian Lake estimates that the total annual benefit from storks *Ciconia ciconia* as a tourist attraction corresponds to 0.18 million USD per year.

This proves the importance of tourist and cultural ecosystem services (Czajkowski et al. 2014).

For many years, Poland has been engaged in nature protection efforts, including wildlife and biodiversity conservation measures. As a result of the low industrialization, many areas, ecosystems, and species have been preserved in good condition. The establishment of a system of protected areas with the aim of conserving the biological diversity is one of the priority directions in Poland's nature conservation. The national system of protected areas in Poland is based on the Nature Conservation Act (FAO 2019) and is organized by four different national designations: 23 national parks, 1488 nature reserves, 122 landscape parks, and 401 protected landscape areas (UNEP-WCMC 2018a, b, c, d, e, f, g, h, i, j, k, l, m, n, o, p, q, r, s).

Moreover, natural habitats and rare or endangered species of animals and plants are protected also by Natura 2000 (European Commission 2015), a network of nature protection areas of the European Union. In Poland this network overlaps with national forms of protection, including some of the reserves, all the national parks, and about half of the landscape parks. It covers 81 natural habitats, 40 plant species, 90 animal species, 74 bird species nesting in Poland, and 83 species of migratory birds. Also, 17 marine Natura 2000 sites have been established: 8 bird areas, 8 habitat areas, and 1 site—Lawica Slupska—which is both a bird and a habitat area (CBD 2014a).

Data from the Convention of Biological Diversity (CBD 2014a, b) a suggest that Poland's protection and conservation efforts have many achievements but still face many challenges. The biggest challenge is the insufficient linking of biodiversity with the economic growth of the country.

Agricultural changes (intensification or setting aside of land) and some aspects of forest management have a negative impact on 70% of habitats and 61% of animal species. Also fishery and hunting management have an impact on 31% of species. Other threats to biodiversity include pollution of the country's water through dumping of chemicals and seeping in of pesticides, habitat loss due to intense development pressure such as the construction of new road networks and a not-fully efficient land use planning system, environmental pollution, and spreading of alien species. The threats are mainly caused by a low efficiency of the environmental procedures and an insufficient use of legislation (CBD 2014a, b; Sawe 2017).

10.9.3.1 The Case of the Bialowieza Forest

The Bialowieza Forest is the largest and best preserved area of primeval forest in Europe. It is characterized by very high biological diversity which includes about 12,000 species of animals, more than 1000 species of plants, and 20 forest complexes. It is also home to the world's largest population of free European bison *Bison bonasus*. The forest is listed as a UNESCO World Heritage site. Currently, about 15% of the forest's area has the status of a national park, despite more than 20 years of efforts made by environmental organizations to extend the park to cover the entire forest. The issue of expanding the national park raises controversy due to a divergence of interests between the timber industry, the local population, the environmental objectives to increase biodiversity, and the recreational use of the forest (CBD 2014a).

How much is the Bialowieza Forest worth?

A debate on the most effective way to use the forest has raged for many years. Therefore, various attempts at assessing its value have been made:

The annual revenue from 110 to 150 thousands m^3 of wood from logging in the Bialowieza Forest reaches USD 1.1–1.6 million. The value of the forest as a recreation area has been estimated at USD 3.53 million a year with 110,000 people visiting the forest annually (CBD 2014a).

Finally, economists considered also the nonproductive values of the forest, which include the value that consumers can use or draw satisfaction from, even if they don't use it in a direct way. For example, people can be satisfied and thus benefit from the fact that the forest is protected, even if they never plan to visit it. In practice, this means that they may be willing to pay for the protection of strict nature reserves they will never be allowed to enter.

To estimate both the productive and nonproductive values of the protection of the Bialowieza Forest, a study was carried out in 2009 with the use of methods based on declared preferences (Giergiczny 2009; Czajkowski et al. 2009). Based on the obtained results, it was possible to estimate how much money a more extensive protection of the forest is worth. The sum stood at USD 22.5 a year per household. For about 12 million households in Poland, this means a total of USD 270 million per year. The obtained sum far exceeds the current income from logging (USD 1.1–1.6 million).

Such studies are an important contribution to the debate on changing the current use of the Bialowieza Forest from a combination of protection and economic use to an increased level of protection and preservation of it, in order to restrain further degradation (Giergiczny 2009; Kalinka 2003; Czajkowski et al. 2009).

10.10 Oceanian Countries

10.10.1 Australia

Number of protected areas: 11,045
Protected land areas coverage: 19.27
Protected marine area coverage: 40.56%

2017 contribution of T&T to GDP[a]	**% GDP**	**USD (billion)**	**2018 forecast**
Direct	3.0	41.7	↑
Total	11.0	151.4	↑
2017 contribution of T&T to employment[a]	**% jobs**	**'000 jobs**	
Direct	4.3	531.7	↑
Total	12.2	1501.6	↑
	% of total export	**USD (billion)**	
2017 visitor exports[a]	7.8	23.4	↑

	% of total investment	USD (billion)	
2017 investment[a]	5.4	18.0	↑
GDP international country 2017[b] **(current USD)**		1323 bn	

[a]World Travel and Tourism Council: Economic Impact 2018 – March 2018. All rights reserved. Licensed under the Attribution, Non-Commercial 4.0 International Creative Commons Licence
[b]World Development Indicators (https://data.worldbank.org/products/wdi)

Tourism is Australia's largest service export industry. Australia's proximity to Asia, its natural assets, and its high standard of living make it a desirable holiday location for those traveling overseas.

But Australia's isolation from the other continents makes it a difficult destination for travelers living in the northern hemisphere. The long distances and cost of flights can act as a deterrent, and in fact, Australia is ranked 40th in the world for number of incoming international visitors (Budgetdirect.com 2019).

According to the statistics, in 2017, Australia registered a number of 8.8 million arrivals, with a total of USD 41.3 billion spent by total international visitors. Tourism Australia estimates that in 2026–2027, there will be a 50% increase in USD spent in Australia on tourism which is roughly USD 151.4 billion, 15 million international visitors, and 14.8 million outbound trips from Australian residents. By 2020, Tourism Australia expects to see over USD 115 billion in overnight spent. It is anticipated that Chinese tourism will grow to 25.7% share of the market, while tourism from New Zealand, the United States, the United Kingdom, and Singapore will all experience below-average visitor growth (Budgetdirect.com 2018).

Australia has been described as a nature wonderland, thereby a great destination for ecotourism.

Even though entry to most national parks is free or at a modest charge, there is considerable economic activity due to spending of ecotourists in the region of the national parks. In fact, nearly one-quarter of all tourism expenditure in Australia in 2007 was made by tourists who visited national (or state) parks as part of their trip, corresponding to USD 15.4 billion, which was 21% of USD 73.6 billion that tourism spent in total in 2007 (Tourism Research Australia 2008a, b).

Today, the nature-based tourism industry is experiencing positive and sustainable growth in some regions, increasing 4% per annum since 2010. Year 2015 saw significant growth in the number of international visitors to state and national parks, increasing 13% since 2014. Moreover, in the year ending June 2016, 68% (or 5.0 million) of international visitors engaged in some form of nature-based activity.

This said the potential of ecotourism growth is yet to be fully realized: Only selected regions are receiving the benefits from this sector, and there still is a lack of a collaborative and nationwide approach.

Ecotourism is significant and represents a great opportunity to assist regions suffering from declining resources sector jobs and investment. In fact, ecotourists generate higher yield on average, spending more and staying longer compared to those who did not undertake nature-based activities: International nature-based visitors spend USD 5548 per trip compared to USD 3621 per trip by other international visitors (Ecotourism Australia 2016).

The country's unique biodiversity is legally protected through the Declaration of Protected Areas at the national, state, territory, and local government levels. Australia's protected areas system covers both terrestrial and marine areas. The first protected area was the Royal National Park in New South Wales, established in 1879 and commonly referred to as the second declared national park in the world, after Yellowstone National Park in the United States, established in 1872 (Boer and Gruper 2010).

According to the World Database on Protected Areas (UNEP-WCMC 2019a) Australia includes 11,045 protected areas of which 1502 with management effectiveness evaluation. PAs are organized with 68 different national designations including among others 20 regional parks, 57 commonwealth marine reserves, 49 marine parks, 94 forest reserves, 259 nature conservation reserves, and more.

In the State of Queensland, the current state government capital budget for "green infrastructure," the strategic growth of national parks, is USD 1.4 million. By contrast, the departmental capital budget is USD 49.5 million, while the state budget for built infrastructure is over USD 10 billion (7000 times greater), and yet visitors to national parks in Queensland spend about USD 5.6 billion a year, at least USD 952 million of which can be attributed to the existence of national parks (WWF Australia 2018).

According to the Australian Government, over the last 200 years, Australia has suffered the largest documented decline in biodiversity of any other continent. Despite the efforts to manage threats and pressures to biodiversity in Australia, biodiversity is still in decline due to the following factors:

- Loss, fragmentation, and degradation of habitat.
- Spread of invasive species.
- Unsustainable use of natural resources.
- Climate change.
- Inappropriate fire regimes.
- Changes to the aquatic environment and water flow.

All levels of Australian Government have enacted legislation to protect biodiversity, and Australia has made good progress in increasing the extent of the National Reserve System since 2011. Conservation efforts within Australia have increased, and the Australian Government monitors progress in achieving biodiversity outcomes over time through regular reporting by delivery agents (CBD, Australian Government 2019a).

10.10.1.1 The Great Barrier Reef Marine Park

The Great Barrier Reef Marine Park was created in 1975 through the Great Barrier Reef Marine Park Act (Australian Government 2019b). The Park itself extends south from the northern tip of Queensland, in northeastern Australia, to just north of

Bundaberg. It ranges between 60 and 250 km in width and has an average depth of 35 m in its inshore waters. On the outer reefs, continental slopes extend to depths of more than 2000 m.

The Great Barrier Reef Marine Park Act created guidelines through which visitors can interact with the reef, as well as guidelines regarding which areas can be visited and at what times of the year. This heavy level of regulation is necessary to give the reef adequate protection.

It is known worldwide as one of the most biologically diverse areas on earth, as well as one of the most ecologically sensitive ones. It is the world's largest World Heritage area, consisting of reefs, mangroves, islands, and ocean waters. It is composed of 3000 individual reef systems, 600 tropical islands, and about 300 coral cays (WWF 2018).

But all of this is at risk. The reef is indeed under pressure from climate change, poor water quality from land-based runoff, impacts from coastal development, and illegal fishing (Great Barrier Reef Marine Park Authority).

10.11 Conclusions

According to the WTTC (2018a), the global economic impact of travel and tourism (T&T) in 2017 represented 10.4% of global GDP (gross domestic product) and 9.9% of total employment (one in every ten jobs on the planet). Nowadays, the activity created by tourism at a global level is very solid and advanced so much that the WTTC estimated that the total travel and tourism contribution to GDP will rise to 12.45 trillion (11.7% of GDP) in 2028, which means a direct contribution to the GDP of about 4 billion (3.6%; WTTC 2018b). The forecasts for 10 years suggest that the T&T sector will support a total of around 414 million jobs worldwide. The sector is expected to contribute an average of nine million new jobs per year by 2028, which represents about one-quarter of the total net job creation worldwide, and with the right regulatory conditions and government support, nearly 100 million new jobs could be created over the next decade (WTTC 2018b).

The impact of tourism on a more local level has been analyzed by the World Travel and Tourism Council, and data are given below, grouped by continents.

In all of African countries, visitor growth was particularly strong (11.6%), together with the Middle East (2.2%). The total contribution of the T&T sector to the GDP was equal to 402 bn USD (5.5% of GDP) and 28 mn of employment contribution. The strongest growth of GDP has been recorded in North Africa (22.6%), with Egypt being the country in which the T&T presented the fastest growth in 2017 (72.9%).

Regarding East Asia and Oceania (so-called Asia Pacific), tourism in these regions has some of the fastest growth [contributing with 2.7 Tn USD (5.4% total contribution to GDP), and with 177 mn employment contribution]. The strongest direct growth was recorded in Northeast Asia (7.4%), with a particular increase in Mongolia (23.0%), Macao (14.2%), and China (9.8%).

Americas: According to estimates by the National Travel and Tourism Office, 2017 was a weak year for incoming tourism to the United States (currently the world's largest travel and tourism economy), where international tourist arrivals fell by around 2.3%. On the contrary, Canada recorded faster direct GDP growth from T&T than any other year since the beginning of the millennium (5.5%). Brazil's weakness in 2017 drove the Latin American countries' growth in general, while there was a much more encouraging performance in other states: Nicaragua, 21.2%; Uruguay, 11.6%; Ecuador, 7.3%; Costa Rica, 7.2%; and Paraguay, 7.2%. Despite the weak year, the total contribution to the GDP of the T&T sector in the Americas was 1.7%, corresponding to 2.2 Tn USD and 42 Mn contribution to employment.

Europe had a particularly successful year in the tourism sector in 2017, with passenger numbers rising by 8%, driven by strong visitors' growth in southern Europe. After the decline in the past few years, West Mediterranean destinations recorded particularly strong visitor export growth especially Spain (10.3%). The United Kingdom too, because of the pound weakness registered after the "Brexit" referendum (2016), experienced a strong competitive boost for price, with both an increase in visitors from overseas markets and an increase in national travel spending. From an economic point of view, the total contribution to the GDP of T&T sector in Europe was more than 2 trillion USD (3.9% of total contribution to GDP), with 37 mn USD of employment contribution.

However, the worldwide growth in tourism is expected to lead to overcrowding. Exceeding the carrying capacity of a tourism destination can generate complex situation and degradation.

Moreover, mass tourism, a large-scale tourism, which is based mainly on holiday packages offered by multinationals that focus their interests on fashionable destinations, with little attention to local communities, creates substantial disadvantages: First, the tourists' spent money is mainly paid into the country of the tour operators, contributing to a foreign economy (Trans National Corporations) more than to local communities' livelihood. For example, in the Republic of Vanuatu (Oceania), 90% of the profits go to foreign countries, and this situation—for countries with low income—may be the rule (Watson 2015). The disadvantages related to the impact of mass tourism on culture can be more or less substantial, depending on the type and volume of tourism, but they are mostly negative: Tourists can easily offend culture, traditions, and local codes of behavior in various ways (getting drunk and becoming loud and offensive, dressing inappropriately, unconsciously encouraging illegal actions, eroding the local language by relying too much on their native language, behaving in inadequate way in places of prayer) and, finally, negatively impact the environment that involves clearance of important habitats (mangroves and rain forests) to provide building sites for hotels; overuse of water (especially in small areas with limited resources); pollution of sea, lakes, and rivers by rubbish and sewage; traffic congestion; and air and noise pollution (Watson 2015).

Extremely serious is the illegal killing of animals (safari, hunting, and fishing) *by* or *for* tourists, which is even carried out in protected areas, where hunters and poachers enter and leave scot-free. For example, as early as 2011, Harrison claimed that most of the tropical nature reserves were to be considered "empty forests" due to the extinction of some animal species and/or the low density of other species. The problem is very widespread (Akani et al. 1998; Angelici et al. 1999; Brashares et al. 2004; Craige et al. 2010; Nasi et al. 2011; Petrozzi 2018) and risky not only for the harm it poses to biodiversity but also for the threat to public health due to the likelihood of zoonosis transmission (Wolf et al. 2005).

Without strong destination management, high visitor numbers can create extra pressure on local resources and an overloaded infrastructure, and this in turn can cause tension between residents and tourists and, at times, a degraded experience for the visitors (UNWTO 2018), damage to nature, and threats to culture and heritage (McKinsey and Company 2017).

In this perspective, ecotourism can be considered as a valid economic alternative or preferable land use model to other economic activities with greater environmental impact, not only because it provides the financial means for wildlife conservation and research but also because it raises awareness about different cultures of the country (Van der Duim and Caalders 2002). And when communities see direct benefits from prospering wildlife population, they have a greater stake in protecting it (Twining-Ward et al. 2018).

References

Acharya H (2014) Overview of protected areas management in Nepal, Department of National Parks and Wildlife Conservation, Nepal July 17, 2014. https://www.cbd.int/doc/meetings/ecr/cbwecr-2014-07/other/cbwecr-2014-07-presentation-day2-04-en.pdf

Acharya BP, Halpenny EA (2013) Homestays as an alternative tourism product for sustainable community development: a case study of women managed tourism product in rural Nepal. Tour Plann Dev 10:367–387. https://doi.org/10.1080/21568316.2013.779313

Adelaja AO, Hailu YG, Tekle AT (2008) The political economy of state conservation spending- the case of Michigan. Conference Paper in Journal of Environmental Monitoring and Restoration. https://doi.org/10.4029/2008jemrest5no112

African Wildlife Foundation (2019a) Rwanda. Available via https://www.awf.org/country/rwanda. Accessed 29 Jan 2019

African Wildlife Foundation (2019b) Uganda. Available via https://www.awf.org/country/uganda. Accessed 29 Jan 2019

African Wildlife Foundation (2019c) Zimbabwe. Available via https://www.awf.org/country/zimbabwe. Accessed 29 Jan 2019

African Wildlife Foundation. https://www.awf.org/country/botswana

Agence France-Presse (2015) India plans more tiger reserves on the back of jump in numbers. The Guardian, April 22, 2015. https://www.theguardian.com/environment/2015/apr/22/india-plans-more-tiger-reserves-on-the-back-of-jump-in-numbers. Accessed 09 Oct 2018

Akani GC, Luiselli L, Angelici FM, Politano E (1998) Bushmen and herpetofauna: notes on amphibians and reptiles traded in bush-meat markets of local tribes in the Niger Delta (Port Harcourt, Rivers State, Nigeria). Anthropozoologica 27:21–26

Alvarado L (2017) Costa Rica's Corcovado National Park has new infrastructure. The Costa Rica Star, February 19, 2017. https://news.co.cr/costa-ricas-corcovado-national-park-new-infrastructure/56782/. Accessed 10 Oct 2018

Amerson A, Parson ECM (2018) Evaluating the sustainability of the grey-whale-watching industry along the pacific coast of North America. J Sustain Tour. https://doi.org/10.1080/09669582.2018.1449848

Angelici FM, Luiselli L, Politano E, Akani GC (1999) Bushmen and mammal-fauna: a survey of the mammals traded in bush-meat markets of local people in the rainforests of South-Eastern Nigeria. Anthropozoologica 30:51–58

Annex 12 (2014) U.S.-China cooperative efforts for the conservation of wildlife and wildlife habitat activities for 2014–2016. https://www.fws.gov/international/pdf/activities-report-china-2014-2016.pdf

Anup KC (2017) Ecotourism in Nepal. Gaze J Tour Hosp 8:1–19

Assunção J, Gandour CC, Rocha R (2012) Deforestation slowdown in the Brazilian Amazon: price or policies? Climate Policy Initiative Rio de Janeiro. https://pdfs.semanticscholar.org/53a5/9c0ef21a0a748f02e26969df1ff9dbc249f2.pdf

Bagla P (2017) Saving 2 tigers gives more value than the cost of Mangalyaan. Press Trust of India, July 16, 2017. https://timesofindia.indiatimes.com/home/environment/green-eco-saving-2-tigers-gives-more-value-than-mangalyaan/articleshowprint/59615732.cms. Accessed 10 Oct 2018

Balmford A, Green JMH, Anderson M, Beresford J, Huang C, Naidoo R, Walpole M, Manica A (2015) Walk on the wild side: estimating the global magnitude of visits to protected areas. PLoS Biol 2015(13):1–6

Baral N, Stern MJ, Hammet AL (2012) Developing a scale for evaluating ecotourism by visitors: a study in the Annapurna Conservation Area, Nepal. J Sustain Tour 20:975–989. https://doi.org/10.1080/09669582.2012.663379

Batavia C, Nelson MP, Darimont CT, Paquet PC, Ripple W, Wallach AD (2018) The elephant (head) in the room: a critical look at trophy hunting. Conserv Lett 12:e12565. https://doi.org/10.1111/conl.12565

BBC News (2018) Nepal country profile, February 19, 2018. https://www.bbc.com/news/world-south-asia-12511455. Accessed 10 Oct 2018

Belen N (2018) Brazil received record number of international tourists in 2017. The Rio Times. https://riotimesonline.com/brazil-news/rio-business/brazil-received-record-number-of-international-tourists-in-2017/. 27 March 2018

Berghöfer A, Emerton L, Moreno Diaz A, Rode J, Schröter-Schlaack C, Wittmer H, van Zyl H (2017) Sustainable financing for biodiversity conservation – a review of experiences in German development cooperation. Study commissioned by GIZ and KfW. Full report published as: UFZ Discussion Paper 1/2017. UFZ – Helmholtz Centre for Environmental Research GmbH, Leipzig, Germany

Bergquist E (2016) Statistics Norway: tourism bypasses mining in Svalbard. High North News, December 15, 2016. http://www.highnorthnews.com/statistics-norway-tourism-bypasses-mining-in-svalbard/. Accessed 12 Oct 2018

Berry H (1997) Historical review of the Etosha region and its subsequent administration as a National Park. Modoqua 10:3–12

Bhattacharya A (2018) Nationwide tiger census kicks off with more cameras, mobile app. The Times of India, updated February 7, 2018. https://timesofindia.indiatimes.com/india/nation-wide-tiger-census-kicks-off-with-more-cameras-mobile-app/articleshow/62812334.cms. Accessed 10 Oct 2018

Biggs D, Turpie J, Fabricius C, Spenceley A (2011) The value of Avitourism for conservation and job creation – an analysis from South Africa. Conserv Soc 9:80–90

Blamey RK (2001) Principles of ecotourism. In: Weaver DB (ed) Encyclopedia of ecotourism. CABI Publishing, Wallingford

Boer B, Gruper S (2010) Legal framework for protected areas: Australia. IUCN-EPLP no. 81, May 6, 2010. https://www.iucn.org/downloads/australia_1.pdf. Accessed 12 Oct 2018

Boucher D, Elias P, Lininger K, May-Tobin C, Roquemore S, Saxon E (2011) The root of the problem: what's driving tropical deforestation today? Union of Concerned Scientists,

Cambridge, MA. Online at: www.ucsusa.org/assets/documents/global_warming/UCS_RootoftheProblem_DriversofDefore
Brashares JS, Arcese P, Sam MK, Coppolillo PB, Sinclair AR, Balmford A (2004) Bushmeat hunting, wildlife declines, and fish supply in West Africa. Science 306:1180–1183
Brazil Ministry of Environment (2015) Secretariat of Biodiversity and Forests – SBF. Fifth National report to the convention on biological diversity: Brazil. Ministry of the Environment, Brasília
BrazilGovNews (2017) Agreement will promote eco-tourism in Brazil. http://www.brazilgovnews.gov.br/news/2017/12/agreement-will-promote-eco-tourism-in-brazil. 11 December 2017
Bruha P (2015) Ecotourism Market in Brazil The Brazil Business, May 25, 2015. http://thebrazilbusiness.com/article/ecotourism-market-in-brazil
Bucaram SJ, Fernandez MA, Grijalva D (2017) Sell the oil deposits! A financial proposal to keep the oil underground in the Yasuni National Park, Ecuador. In: Arent D, Arndt C, Miller M, Tarp F, Zinaman O (eds) The political economy of clean energy transition. A study prepared by the United Nations University World Institute for Development Economics Research (UNU-WIDER). Oxford University Press, Oxford, pp 430–450
Budgetdirect.com (2018) Australian Tourism Statistics 2018. Latest available data from March, 2018. https://www.budgetdirect.com.au/travel-insurance/research/tourism-statistics.html. Accessed 12 Oct 2018
Budgetdirect.com (2019) Australian Tourism Statistics Information 2019. Available at: https://www.budgetdirect.com.au/travel-insurance/research/tourismstatistics.html. Accessed 03/03/2020
Camacho-Ruiz E, Carrillo-Reyes A, Rioja-Paradela TM, Espinoza-Medinilla EE (2016) Sustainability indicators for ecotourism in Mexico: current state. Revista Liminar Estudios Sociales y Humanisticos 14:156–168
Can ÖE, Macdonald DW (2018) Looking under the bonnet of conservation conflicts: can neuroscience help? Biodivers Conserv 27:2087–2091. https://doi.org/10.1007/s10531-018-1514-1
CBD (2012) Convention on biological diversity, COP 11. https://www.cbd.int/cop11/
CBD (2014a) Fifth national report on the implementation of the convention on biological diversity. Convention on biological diversity, Poland, Warsaw, March 2014. https://www.cbd.int/doc/world/pl/pl-nr-05-en.pdf. Accessed 12 Oct 2018
CBD (2014b) Fifth national report to the convention on biological diversity. Convention on biological diversity, the Norwegian Ministry of Climate and Environment, Norway. https://www.cbd.int/doc/world/no/no-nr-05-en.pdf. Accessed 12 Oct 2018
CBD (undated) Australia- country profile. Biodiversity Facts. Status and trends of biodiversity, including benefits from biodiversity and ecosystem services. https://www.cbd.int/countries/profile/default.shtml?country=au. Accessed 12 Oct 2018
Chapagain PS (2017) General pattern and emerging trend of tourism development in Nepal Tribhuvan University 58th Anniversary special Publication (TU Bulletin Special 2017), Tribhuvan University, pp 49–64
Ciucci P, Boitani L (2008) The Apennine brown bear: a critical review of its status and conservation problems. Ursus 19:130–145
Ciucci P, Altea T, Antonucci A, Chiaverini L, Di Croce A, Fabrizio M, Forconi P, Latini R, Maiorano L, Monaco A, Morini P, Ricci F, Sammarone L, Striglioni F, Tosoni E (2017) Distribution of the brown bear (*Ursus arctos marsicanus*) in the central Apennines, Italy, 2005-2014. Hystrix Ital J Mammal 28:86–91. https://doi.org/10.4404/hystrix-28.1-12049
Clapham PJ (2016) Managing leviathan: conservation challenges for the great whales in a post-whaling world. Oceanography 29(3). https://doi.org/10.5670/oceanog.2016.70
CleanMalaysia.com (2017) https://cleanmalaysia.com/2017/08/04/lets-make-ecotourism-work/
Climate Focus (2017) Progress on the New York declaration on forests: finance for forests – Goals 8 and 9. Assessment report. Prepared by climate focus in cooperation with the New York declaration on forest assessment partners with support from the climate and land use alliance. https://climatefocus.com/sites/default/files/NYDF%20report%202017%20FINALpdf. Retrieved 11 October 2018

Coad L, Leverington F, Knights K, Geldmann J, Eassom A, Kapos V, Kingston N, de Lima M, Zamora C, Cuardros I, Nolte C, Burgess DN, Hockings M (2015) Measuring impact of protected area management interventions: current and future use of the global database of protected area management effectiveness. Philos Trans R Soc B 370:20140281

Coast Funds (2018) The success of spirit bear lodge: how a remote, community-led business became a global model for ecotourism, March 9, 2018. https://coastfunds.ca/stories/the-success-of-spirit-bear-lodge/. Accessed 10 Oct 2018

Cota I (2012). Costa Rica passes law banning hunting as a sport. Reuters. December 11, 2012. Accessed 03/03/2020

Coltman DW, O'Donoghue P, Jorgenson JT, Hogg JT, Strobeck C, Festa-Bianchet M (2003) Undesirable evolutionary consequences of trophy hunting. Nature 426:655–658

Costello C, Gaines SD, Gerber LR (2012) A market approach to saving the whales. Nature 481:139–140

Cottar C (2015) Olderkesi wildlife conservancy, Kenia. In: Roe D (ed) Conservation, crime and communities: case studies of efforts to engage local communities in tackling illegal wildlife trade. IIED, London. http://pubs.iied.org/14648IIED.html

Coulson T, Schindler S, Traill L, Kendall BE (2017) Predicting the evolutionary consequences of trophy hunting on a quantitative trait. J Wildl Manag. https://doi.org/10.1002/jwmg.21261

Craige JD, Baillie JEM, Balmford A, Carbone C, Collen B, Green RE, Hutton JM (2010) Large mammal population declines in Africa's protected areas. Biol Conserv 143:2221–2228

Creel S, M'Soka J, Droge E, Rosenblatt E, Becker MS, Matandiko W, Simpamba T (2016) Assessing the sustainability of African lion trophy hunting, with recommendation for policy. Ecol Appl 26:2374–2357

CREST (2014) Economic impact of bear viewing and bear hunting in the great bear rainforest of British Columbia. Center for Responsible Travel, Pacific Analysis Inc., Small Planet Consulting, Washington, DC. http://pacificanalytics.ca/sites/default/files/reports/Economic_Impact_of_Bear_Viewing_and_Bear_Hunting_in_GBR.pdf

CSD-7 (undated) Sustainable development success stories, Baghmara community forest: Bufferzone plantation programme, Agenda 21, Chapter 23. https://sustainabledevelopment.un.org/content/dsd/dsd_aofw_mg/mg_success_stories/csd7/tour9.htm. Accessed 10 Oct 2018

Cunningham PA, Huijben EH, Wearing SL (2011) From whaling to whale watching: examining sustainability and cultural rhetoric. J Sustain Tour iFirst:1–19. https://doi.org/10.1080/09669582.2011.632091

Currie JJ, Stack SH, Kaufman GD (2018) Conservation and education through ecotourism: using citizen science to monitor cetaceans in the four-island region of Maui, Hawaii. Tour Mar Environ 13:65–71. https://doi.org/10.3727/154427318X15270394903273

Curtin S (2013) Lesson from Scotland: British wildlife tourism demand, product development and destination management. J Destin Mark Manag 2:196–211

Czajkowski M, Buszko-Briggs M, Hanley N (2009) Valuing changes in forest biodiversity. Ecol Econ 68:2910–2917. https://doi.org/10.1016/j.ecolecon.2009.06.016

Czajkowski M, Giergiczny M, Kronenberg J, Tryjanowski P (2014) The economic recreational value of a white stork nesting colony: a case of 'stork village' in Poland. Tour Manag 40:352–360. https://doi.org/10.1016/j.tourman.2013.07.009

D'Cruze N, Machado FC, Matthews N, Balaskas M, Carder G, Richardson V, Vieto R (2017) A review of wildlife ecotourism in Manaus, Brazil. Nat Conserv 22:1–16

Dayananda KC (2016) Tourism and its impact on Indian economy. IOSR J Hum Soc Sci (IOSR-JHSS) 21:24–28

de Witt L, van der Merwe P, Saayman M (2011) An ecotourism model for South African National Parks. In: Proceedings of the international conference on tourism and management studies – Algarve, vol II, 26–29 October. http://tmstudies.net/index.php/ectms/article/view/311/476

Desforges JP, Hall A, McConnell B, Rosing-Asvid A, Barber JL, Brownlow A, De Guise S, Eulaers I, Jepson PD, Letcher RJ, Levin M, Ross PS, Samarra F, Víkingson G, Sonne C, Dietz R (2018)

Predicting global killer whale population collapse from PCB pollution. Science 361:1373–1376. https://doi.org/10.1126/science.aat1953

Devkota A (2018) More tourists, revenue in Chitwan National Park. News NRM, June, 13 2018. http://en.newsnrn.com/2018/06/13/more-tourists-revenue-in-chitwan-national-park/. Accessed 10 Oct 2018

Di Clemente J, Christiansen F, Pirotta E, Steckler D, Wahlberg M, Pearson HC (2018) Effects of whale watching on the activity budgets of humpback whales, *Megaptera novaeangliae* (Borowski, 1781), on a feeding ground. Aquat Conserv Mar Freshwat Ecosyst. https://doi.org/10.1002/aqc.2909

Di Minin E, Leader-Williams N, Bradshaw CJA (2016) Banning trophy hunting will exacerbate biodiversity loss. Trends Ecol Evol 31:99–102

Dougill AJ, Akanyang L, Perkins JS, Eckardt FD, Stringer LC, Favretto N, Atlhopheng J, Mulale K (2016) Land use, rangeland degradation and ecological changes in the southern Kalahari, Botswana. Afr J Ecol 54:59–67

Driscoll L, Hunt C, Honey M, Durham W (2011) The importance of ecotourism as a development and conservation tool in the Osa Peninsula, Costa Rica. Report for the Centre of Responsible Tourism (CREST), April 2011. https://www.responsibletravel.org/docs/Importance%20of%20Ecotourism%20Osa%20Peninsula%20-%20English.pdf

Duffield J, Neher C, Patterson D (2006) Wolves and people in Yellowstone: impacts on the regional economy. Yellowstone Park Foundation. The University of Montana, Missoula, pp 1–67

Duffy LN, Kline C, Swanson JR, Best M, McKinnon H (2016) Community development through agro ecotourism in Cuba: an application of the community capitals framework. J Ecotour. https://doi.org/10.1080/14724049.2016.1218498

Ecuador.com (2019). Discover Ecuador. Ecuador.com. Galapagos National Park. Last updated January 14, 2020

Ecotourism Australia (2016) Nature based tourism in Australia Manifesto. https://www.ecotourism.org.au/assets/Uploads/Manifesto-v5.0.pdf. Retrieved 18 October 2018

Edake SS (2016) The protected area network of India, The Goan, 17 December 2016

Edwards S, McLaughlin WJ, Ham SH (2003). A Regional Look at Ecotourism Policy in the Americas. Chapter 15 in, D. Fennell and R. Dowling (eds.), Ecotourism: Policy and Strategy Issues 293–307. Wallingford, UK: CAB International

Emerton L, Kyin UA, Tizard R (2015) Protected areas in Myanmar-assessment and recommendations. WCS, Yangon

European Commission (2015) The EU's protected areas – Natura 2000. Available via http://ec.europa.eu/environment/basics/natural-capital/natura2000/index_en.htm. Accessed 29 Jan 2019

FAO (Food and Agriculture organization of the United Nations) (2019) Poland. Nature Conservation Act. Available via http://www.fao.org/faolex/results/details/en/c/LEX-FAOC060227. Accessed 29 Jan 2019

Ferraro PJ, Hanauer MM (2014) Quantifying causal mechanisms to determine how protected areas affect poverty through changes in ecosystem services and infrastructure. Proc Natl Acad Sci U S A 111:4332–4337. https://doi.org/10.1073/pnas.1307712111

Fletcher R, Neves K (2012) Contradiction in tourism. The promise of pitfalls of ecotourism as a manifold capitalist fix. Environ Soc Adv Res 3:60–77. https://doi.org/10.3167/ares.2012.030105

Ford MJ, Parsons KM, Ward EJ, Hempelmann JA, Emmons CK, Bradley M, Bradley Hanson M, Balcomb KC, Park LK (2018) Inbreeding in an endangered killer whale population. Anim Conserv. https://doi.org/10.1111/acv.12413

Fredman P, Haukeland JV (2016) New challenges for managing sustainable tourism in protected areas a Scandinavian perspective. In: Hammer T, Mose I, Siegrist D, Weixlbaumer N (eds) Parks of the future! Protected areas in Europe challenging regional and global change. Oekom Publisher, Munich, pp 137–147

Freeman AM III, Herriges JA, Kling CL (2014) The measurement of environmental and resource value-theory and methods, 3d edn. RFF Press/Taylor and Francis, Abingdon

Galapagos Conservation Trust (2019) Sustainable tourism 4/5: managing ecotourism in Galapagos. Available via http://www.discoveringgalapagos.org.uk/discover/sustainable-development/sustainable-tourism/managing-ecotourism-in-galapagos/. Accessed 29 Jan 2019

Gandiwa E (2011) Wildlife tourism in Gonarezhou national park, Southeast Zimbabwe: opportunities for wildlife viewing. J Sustain Dev Afr 13(1):304–315

Gandiwa E, Gandiwa P, Muboko N (2012) Living with wildlife and associated conflicts in a contested area within the northern Gonarezhou National Park, Zimbabwe. J Sustain Dev Afr 14:252–260

Garcia-Cegarra AM, Pacheco AS (2017). Whale-watching trips in Peru lead to increases in tourist knowledge, pro-conservation intentions and tourist concern for theimpact of whale-watching on humpback whales. Aquatic conservation Marine Freshwater Ecosystems. DOI 10.1002/aqc.2754

Gazzetta del Turismo (2016) Il Turismo Natura secondo Ecotur. http://gazzettadelturismo.altervista.org/il-turismo-natura-secondo-ecotur/. Accessed 11 Oct 2018

Gerber LR, Costello C, Gaines SD (2014) Conservation markets for wildlife management with case studies from whaling. Ecol Appl 24:4–14

Giergiczny M (2009) Rekreacyjna wartość Białowieskiego Parku Narodowego [Recreational value of the Białowieża National Park]. Ekonomia i Środowisko 2:117–127

Global Conservation (2018) Protecting a refuge for the last Jaguar of Costa Rica in Corcovado National Park. http://globalconservation.org/news/protecting-refuge-last-jaguar-costa-rica-corcovado-national-park/. Accessed 16 Oct 2018

Global Forest Watch (2019) Malaysia. Available via https://www.globalforestwatch.org/dashboards/country/MYS. Accessed 20 Oct 2018

Government of Canada (2017a) Canada tourism vision. https://www.ic.gc.ca/eic/site/095.nsf/eng/home.Updated 2017-07-01

Government of Canada (2017b) Protected areas. https://www.canada.ca/en/services/environment/conservation/protected-areas.html. Updated 2017-12-11

Great Barrier Reef Marine Park Authority, Australian Government (undated) Threats to the reef. http://www.gbrmpa.gov.au/managing-the-reef/threats-to-the-reef. Accessed 12 Oct 2018

Hanks J (2003) Transfrontier conservation areas (TFCAs) in Southern Africa. J Sustain For 17:127–148

Hares S (2018) Brazil protecting vast sea areas in bid to safeguard biodiversity. Reuters, March 9, 2018. https://www.reuters.com/article/us-brazil-environment-oceans/brazil-protecting-vast-sea-areas-in-bid-to-safeguard-biodiversity-idUSKCN1GL2UC. Accessed 10 Oct 2018

Hariohay KM, Jackson CR, Fyumagwa RD, Roskaft E (2018) Trophy hunting versus ecotourism as a conservation model? Assessing the impacts on ungulate behaviour and demographics in the Ruaha-Rungwa ecosystem, central Tanzania. Environ Nat Resour Res 8:33–43

Harrison RD (2011) Emptying the forest: hunting and the extirpation of wildlife from tropical nature reserves. Bioscience 61:919–924

Hawkins DE, Lamoureaux K (2001) Global growth and magnitude of ecotourism. In: Weaver DB (ed) Encyclopedia of ecotourism. CABI Publishing, Wallingford

Heinen JT, Kattel B (1992) Parks, people, and conservation: a review of management issues in Nepal's protected areas. Popul Environ 14:49–84. https://doi.org/10.1007/BF01254607

Heinen JT, Shrestha SK (2006) Evolving policies for conservation: an historical profile of the protected area system of Nepal. J Environ Plan Manag 49:41–58. https://doi.org/10.1080/09640560500373048

Heinen JT, Yonzon P (1994) A review of conservation issues and programs in Nepal: from a single species focus toward biodiversity protection. Mt Res Dev 14:61–76. https://doi.org/10.2307/3673738

Housman J (2018) Trump dumps Obama-era ocean conservation policy. Surfers. https://www.surfer.com/features/trump-dumps-obama-era-ocean-conservation-policy/. Posted June 20, 2018

Hoyt E, Hvenegaard GT (2002) A review of whale-watching and whaling with applications for the Caribbean. Coast Manag 30:381–399. https://doi.org/10.1080/089207502900273

Hoyt E, Iñígues M (2008) The state of whale watching in Latin America. WDCS/Global Ocean, Chippenham/London, 60 pp

https://www.npca.org/parks/yellowstone-national-park. Accessed 29 Oct 2019

Huang B, Abbot JK, Fenichel EP, Muneepeerakul R, Petrings C, Gerber L (2017) Testing the feasibility of a hypothetical whaling-conservation permit market in Norway. Conserv Biol 31:809–817. https://doi.org/10.1111/cobi.12916

ICT -Instituto Costarricense de Turismo (2017) Costa Rica Tourism Board Institutional site. https://www.ict.go.cr

ICT -Instituto Costarricense de Turismo (2018). http://www.visitcostarica.com

IFAW (2016) Killing for trophies. Ana analysis of global trophy hunting trade. https://www.ifaw.org/united-states/resource-centre/killing-trophies-analysis-global-trophy-hunting-trade. Retrieved 19 September 2018

IFAW (International Found for Animal Welfare) (2018) Pro-whaling countries again harpoon plans for a South Atlantic Whale Sanctuary. Available via https://www.ifaw.org/africa/news/pro-whaling-countries-again-harpoon-plans-south-atlantic-whale-sanctuary. Accessed 10 Jan 2019

Indigenous leadership initiatives (2018) RELEASE: budget 2018 proposes $1.3 billion for conservation, points to importance of Indigenous partnerships in reaching conservation goals. https://www.ilinationhood.ca/2018/02/27/budget-2018-proposes-investment-conservation/, https://www.ilinationhood.ca/2018/02/27/budget-2018-proposes-investment-conservation/. Updated February 27, 2018

Innovation Norway (2016) Key figures for Norwegian travel and tourism 2016. https://res.cloudinary.com/simpleview/image/upload/v1/clients/norway/Key_Figrues_2016_1__8402bc6c-f2e7-4851-b279-2bc639f0803d.pdf. Accessed 12 Oct 2018

International Whaling Commission (2019) Whale sanctuaries. Available via https://iwc.int/sanctuaries. Accessed 29 Jan 2019

IPCC (2013) Summary for policymakers. In: Stocker TF, Qin D, Plattner G-K, Tignor M, Allen SK, Boschung J, Nauels A, Xia Y, Bex V, Midgley PM (eds) Climate change 2013: the physical science basis. Contribution of working group I to the fifth assessment report of the intergovernmental panel on climate change. Cambridge University Press, Cambridge/New York

IUCN (2016). Informing decision on trophy hunting. A briefing decision for European Union decision-makers regarding potential plans for restriction of imports oh hunting trophies. https://d2ouvy59p0dg6k.cloudfront.net/downloads/iucn_informingdecisionsontrophyhuntingv1_1.pdf

IUCN/PACO (2009) Big game hunting in West Africa. What is its contribution to conservation? isbn: 978-2-8317-1204-8

Ivanshenko YV, Clapham PJ, Brownell RL Jr (2013) Soviet catches of whales in the North Pacific: revised total. J Cetacean Res Manag 13:59–71

Ivashchenko YV, Clapham PJ (2015) What's the catch? Validity of whaling data for Japanese catches of sperm whales in the North Pacific. R Soc Open Sci 2:150177. https://doi.org/10.1098/rsos.150177

IWC (2011) Five year strategic plan for whale watching 2011–2016. https://iwc.int/private/downloads/ZibAR4HShR6wjYdH9C8NRw/AC-002s3%20IWC%20Whale%20Booklet_HR.pdf

Jhala YV, Qureshi Q, Gopal R (2015) The status of Tigers in India 2014. National Tiger Conservation Authority, New Delhi and The Wildlife Institute of India, Dehradun. https://projecttiger.nic.in/WriteReadData/LetestNews/Document/Tiger%20Status%20booklet_XPS170115212.pdf. Accessed 10 Oct 2018

Jonas H, Kothari A, Shrumm H (2012) Legal and institutional aspects of recognizing and supporting conservation by indigenous peoples and local communities. Natural Justice Publisher, Bangalore. www.iccaconsortium.org

Jonga C, Pangeti (2015) The CAMPFIRE program in Zimbabwe. https://firstforwildlife.com/2015/08/18/the-campfire-program-in-zimbabwe/

Josephs L (2018) International tourism is booming, but not to the US. CNBC. https://uk.finance.yahoo.com/news/international-tourism-booming-not-us-144700292.html. January 17 2018

Kaplinsky R, Memedovic O, Morris M, Readman J (2003) The global wood furniture value chain: what prospects for upgrading by developing countries? The case of South Africa. UnIDo, Vienna. http://oro.open.ac.uk/8593/1/globalwoddfurniturevaluechainhttp___www.unido.pdf

Kaur CR (2007) Ecotourism in Malaysia's Marine Parks: principles, issues and the effects of 'Green Washing' practices. 5th Asia Pacific ecotourism conference, 27–28 October, Terengganu

Kenya Wildlife Service (2019) Available via http://kws.go.ke/content/overview-0. Accessed 30 Jan 2019

Kimanthi K (2018) Kenya tourism grew 20pc in 2017 despite poll jitters. Business Daily February 8, 2018. https://www.businessdailyafrica.com/corporate/industry/Kenya-tourism-grew-20pc-in-2017-despite-poll-jitters/4003110-4296860-hfgikfz/index.html. Accessed 08 Oct 2018

Kimaro M, Lendelvo S, Nakanyala J (2015) Determinants of tourists' satisfaction in Etosha National Park, Namibia. J Stud Hum Soc Sci 4(1 & 2):116–131

King J, Craig I (2015) Northern Rangelands Trust, Kenia. In: Roe D (ed) Conservation, crime and communities: case studies of efforts to engage local communities in tackling illegal wildlife trade. IIED, London. https://communitiesforwildlife.iied.org/northern-rangelands-trust. Accessed 16 Sept 2018

Kiringe JW, Okello MM (2007) Threats and their relative severity to wildlife protected areas of Kenya. Appl Ecol Environ Res 5:49–62

Knowles T, Campbell R (2011) What's a whale worth? Valuing whales for National Whale Day, a report for the International Fund for Animal Welfare (IFAW), prepared by Economists at Large, Melbourne, Australia. http://www.ecolarge.com/wp-content/uploads/2012/06/IFAW-NWD-Report-FINAL-lowres.pdf

Knowlton AR, Robbins J, Landry S, McKenna HA, Kraus SD, Werner TB (2015) Effects of fishing rope strength on the severity of large whale entanglements. Conserv Biol 30:318–328

Lane I (2018) Ecotourism 101: countries most dependent on ecotourism revenue. Greenloons, December 14, 2018 https://greenloons.com/2010/12/14/ecotourism-101-which-countries-are-most-dependent-on-ecotourism-revenue/. Accessed 11 Oct 2018

Lawson S (2010) Illegal logging and related trade: indicators of the global response. Chatham House, London. https://www.illegal-logging.info/sites/files/chlogging/uploads/0904CHAillegalloggingbriefingpaper09.0731.pdf. Accessed 10 Oct 2018

Leaper R, Miller C (2011) Management of Antarctic baleen whales amid past exploitation, current threats and complex marine ecosystems. Antarct Sci. https://doi.org/10.1017/S0954102011000708

Lengefeld K (2013) Touristic economic value and threats for emblematic species such as koalas, Pandas and Rhinos, Power Point Presentation

Lew DK (2015) Willingness to pay for threatened and endangered marine species: a review of the literature and prospects for policy use. Front Mar Sci 2(96). https://doi.org/10.3389/fmars.2015.00096

Lindsey PA, Roulet PA, Romañach SS (2007) Economic and conservation significance of the trophy hunting industry in sub-Saharan Africa. Biol Conserv 134:455–469

Lindsey PA, Balme GA, Booth VR, Midlane N (2012) The significance of African lions for the financial viability of trophy hunting and the maintenance of wild land. PLoS One 7(1):e29332. https://doi.org/10.1371/journalpone.0029332

Lindsey PA, Balme GA, Funston PJ, Henschel PH, Hunter LTB (2015) Life after Cecil: channeling global outrage into funding for conservation in Africa. Conserv Lett 9:296–301. https://doi.org/10.1111/conl.12224

Lindsontheroad (2011) Ecotourism in Ecuador: education for the future. My Natour, September 21, 2011 http://mynatour.org/destination/ecotourism-ecuador-education-future. Accessed 11 Oct 2018

Lonely Planet (2017) Best in Travel 2017: top 10 countries. https://www.lonelyplanetitalia.it/articoli/consigli-di-viaggio/best-in-travel-2017-top-10-paesi.Accessed 10 Oct 2018

Luty A (2018) Tourism is the big money maker in Costa Rica, February 22, 2018. https://www.osa-tropicalproperties.com/discover-costa-rica/tourism-costa-ricas-big-money-maker/. Accessed 11 Oct 2018

MacDonald DW (2016) Report on lion conservation with particular respect to the issue of trophy hunting. WildCRU Oxford University. https://www.wildcru.org/wpcontent/uploads/2016/12/Report_on_lion_conservation.pdf. Accessed 07 Sept 2018

Macdonald DW, Dickman A, Loveridge A, Jacobsen K, Johnson P, O'Kane C, du Preez B (2016). Report on Lion Conservation with Particular Respect to the Issue of Trophy Hunting. Report for the Secretary of State for the Environment, 28 November 2016, Wildlife Conservation Unit, Oxford University.

Main M, Warburton-Lee J (2002) Wildlife conservation | National Parks of Botswana. Travel Africa. http://www.botswana.co.za/Wildlife_Conservation-travel/botswana-national-parks.html. Accessed 07 Sept 2018, 09 Oct 2018

Malakoff D (2018) Trump's new oceans policy washes away Obama's emphasis on conservation and climate. Science. http://www.sciencemag.org/news/2018/06/trump-s-new-oceans-policy-washes-away-obama-s-emphasis-conservation-and-climate. June 19, 2018

Malik K (2013) Human development report 2013. The rise of the South: human progress in a diverse world. United Nations Development Programme. ISBN 978-92-1-126340-4. http://hdr.undp.org/sites/default/files/reports/14/hdr2013_en_complete.pdf

Markarian M (2015) Eco-tourism worth more to African economies than trophy hunting. The Blog 11/02/2015 ET | Updated November 02, 2016

Marshall (1920) quoted by Freeman III et al 2014

Martella D, Poglayen G, Gentile L, Mari E, Martini M (2003) Indagine sui coccidi presenti nel camoscio d'Abruzzo. J Mt Ecol 7(suppl):251–256

Masseti M, Salari L (2017) Late Pleistocene and Holocene chamois in Italy. Mammal Rev 47:306–313. https://doi.org/10.1111/mam.12100

Massida C, Mattana P (2013) A SVECM analysis of the relationship between international tourism arrivals, GDP and trade in Italy. J Travel Res 52:93–105. https://doi.org/10.1177/0047287512457262

Mathur S (2018) Report: India to be 3rd largest tourism economy in 10 years. The Times of India, March 23, 2018. http://timesofindia.indiatimes.com/articleshow/63421777.cms?utm_source=contentofinterestandutm_medium=textandutm_campaign=cppst. Accessed 10 Oct 2018

Mattioli L, Canu A, Passilongo D, Scandura M, Apollonio M (2018) Estimation of pack density in grey wolf (*Canis lupus*) by applying spatially explicit capture-recapture models to camera trap data supported by genetic monitoring. Front Zool 15:38. https://doi.org/10.1186/s12983-018-0281-x

Maun Self Drive 4x4 Hire For Camping, Game Viewing and off road Travel (2015) Reservations or booking of campsites in Botswana National Parks (DWNP). Available via https://www.maunselfdrive4x4.com/dwnpreservations.htm. Accessed 15 Aug 2018

Mbaiwa JE (2008) The realities of ecotourism development in Botswana. In: Spenceley A (ed) Responsible tourism: critical issues for conservation and development. Earthscan, London, pp 205–223

McKinsey and Company (2017) Coping with success. Managing overcrowding in tourist destinations. Copyright © McKinsey and Company and World Travel and Tourism Council

MET (2013) Report on the Namibia tourist exit survey 2012–2013. Ministry of Environment & Tourism. http://www.namibiatourism.com.na/uploads/file_uploads/Report_Namibia_Tourist_Exit__survey_2012_2013.pdf. Accessed 09 Oct 2018

Meynecke JO, Richards R, Sahin O (2017) Whale watch or no watch: the Australian whale watching tourism industry and climate change. Reg Environ Chang 17:477–488. https://doi.org/10.1007/s10113-016-1034-z

MiBACT (undated) Strategic plan for tourism 2017–2022 executive summary. Ministero dei Beni e delle Attività Culturali e del Turismo. http://www.cultourproject.eu/sites/cultourproject.eu/files/repository/2017/04/EN_ES_light-per-web.pdf. Accessed 10 Oct 2018

Miller C, Batibasiga A, Sharma-Gounder S, Solomona P (2015) Very low numbers of endangered Oceania Humpback whale seen in Fijian waters. S Pac J Nat Appl Sci 33:39–45

Ministerio del Ambiente (2014). Ecuador's forest reference emission level for deforestation. Submission of the Republic of Ecuador to the United Nations framework convention on climate change, Quito December, 2014. https://redd.unfccc.int/files/2014_december_frel_submission_ecuador.pdf. Accessed 11 Oct 2018

Ministerio del Ambiente (2015) National system of protected areas – Nspa. http://areasprotegidas.ambiente.gob.ec/en/info-snap. Accessed 17 Oct 2018

Ministry of Culture, Tourism and Civil Aviation (2018) Nepal tourism statistics 2017. Singha Durbar, Kathmandu May, 2018. http://tourism.gov.np/files/statistics/2.pdf. Accessed 10 Oct 2018

Ministry of Environment, Water and Climate Republic of Zimbabwe (2014) National biodiversity strategy and action plan 2014. https://www.cbd.int/doc/world/zw/zw-nbsap-v2-en.pdf

Ministry of Environment, Water and Climate Republic of Zimbabwe (2015) Zimbabwe's fifth national report to the convention on biodiversity. https://www.cbd.int/doc/world/zw/zw-nr-05-en.pdf

Mohd Rusli Y, Ahmad S, Alias R (2008) How much ecotourism development contribute to local communities? An empirical study in a Small Island. ICFAI J Environ Econ 6:54–67

Montana.gov (2018) Wolf numbers remained strong in 2017. Montana Government official state website http://fwp.mt.gov/news/newsReleases/fishAndWildlife/nr_1093.html. May 21, 2018

Moore SE (2016) Is it 'boom times' for baleen whales in the Pacific Arctic region? Biol Lett 12:20160251. https://doi.org/10.1098/rsbl.2016.0251

Morrison-Métois S, Lundgren H (2016) Forests and Sustainable Forest Management. Evaluation evidence on addressing deforestation to reduce CO2 emissions. Evaluation insights OECD Number 11, April 2016

MTPB (2018) 25.9 million international tourists visited Malaysia in 2017. Malaysia Tourism Promotion Board, Ministry of Tourism, Arts and Culture, Malaysia. March 29, 2018. https://www.tourism.gov.my/statistics

Mulvaney K, Taylor C (2013) The economics of Japanese whaling: a collapsing industry burdens taxpayers. International Fund for Animal Welfare. ISBN 978-1-939464-01-9. http://www.ifaw.org/sites/default/files/economics-of-japanese-whaling-japan-ifaw.pdf. Accessed 20 Sept 2018

Muposhi VK, Gandiwa E, Bartels P, Makuza SM (2016) Trophy hunting, conservation, and rural development in Zimbabwe: issues, options, and implications. Int J Biodivers:Article ID 8763980, 16p. https://doi.org/10.1155/2016/8763980

Mutanga CN, Muboko N, Gandiwa E, Vengesayi S (2016) Beyond a single perspective to conservation relationships: exploring factors influencing protected area staff and local community relationships in Zimbabwe. Int J Biodivers Sci Ecosyst Serv Manag 12(3):212–226

Mutanga C, Gandiwa E, Muboko N (2017) An analysis of tourist trends in northern Gonarezhou National Park, Zimbabwe, 1991–2014. Cogent Soc Sci 3:1392921

Naevdal E, Olaussen JO, Skonhoft A (2012) A bioeconomic model of trophy hunting. Ecol Econ 73:194–205

Naidoo R, Fisher B, Manica A, Balmford A (2016) Estimating economic losses to tourism in Africa from the illegal killing of elephants. Nat Commun 7:13379. https://doi.org/10.1038/ncomms13379

National Parks and Wildlife Conservation Act (1973), FAOLEX database http://www.fao.org/faolex/results/details/en/c/LEX-FAOC006220

Nasi R, Taber A, Van Vliet N (2011) Empty forest, empty stomachs? Bushmeat and livelihoods in the Congo and Amazon basin. Int For Rev 13:355–368

National Parks Conservation Association (2019) Yellowstone. Available via https://www.npca.org/parks/yellowstone-national-park. Accessed 29 Jan 2019

National Wildlife Federation (undated) The Recovering America's Wildlife Act factsheet. https://www.nwf.org/~/media/Documents/PDFs/Wildlife-Conservation/Fast-Facts-Wildlife-Crisis_10192017.ashx. Accessed 9 Sept 2018
NEA (2016) Norwegian environment agency. Protected areas. http://www.environment.no/topics/biodiversity/protected-areas/. Accessed 11 Oct 2018
Nepal SK (1997) Sustainable tourism, protected areas and livelihoods needs of local communities in developing countries. Int J Sustain Dev World Ecol 4:123–135. https://doi.org/10.1080/17451590609618095
Nepstad D, McGrath D, Stickler C, Alencar A, Azevedo A, Swette B, Bezerra T, DiGiano M, Shimada J, da Motta RS, Armijo E, Castello L, Brando P, Hansen MC, McGrath-Horn M, Carvalho O, Hess L (2014) Slowing Amazon deforestation through public policy and interventions in beef and soy supply chains. Science 344:1118–1123
Nielsen H, Spenceley A (2010) The success of tourism in Rwanda – Gorillas and more. The World Bank and The Netherlands Development Organisation) 29 p
Njugunah M (2018) Tourism sector finished strong in 2017 despite poll jitters. Capital Business, 26/04/2018. https://www.capitalfm.co.ke/business/2018/04/tourism-sector-finished-strong-in-2017-despite-poll-jitters/
Nolte C, Agrawai A, Silvius KM, Soares-Fil BS (2013) Governance regime and location influence avoided deforestation success of protected areas in the Brazilian Amazon. Proc Natl Acad Sci 110:4956–4961
Norwegian Directorate for Nature Management (undated) Protected areas in Svalbard – securing internationally valuable cultural and natural heritage. http://www.miljodirektoratet.no/Global/dokumenter/Publikasjoner/Brosjyrer/Eng_brosj_Svalbard_web_opp.pdf. Accessed 12 Oct 2018
Notarbartolo-di-Sciara G, Agardy T, Hyrenbach D, Scovazzi T, Van Klaveren P (2008) The Pelago sanctuary for Mediterranean marine mammals. Aquat Conserv Mar Freshwat Ecosyst 18:367–391. https://doi.org/10.1002/aqc.855
O'Connor S, Campbell R, Cortez H, Knowles T (2009) Whale watching worldwide: tourism numbers, expenditures and expanding economic benefits. A special report from the International Fund for Animal Welfare, Yarmouth MA, USA, prepared by Economists at Large
Okello MM, D'amour DE, Manka SG (2008a) Tourism attractions and satisfaction of Amboseli National Park, Kenya. Tour Anal 13:373–387
Okello MM, Schuttler S, D'Amour DE (2008b) The relative importance of large mammal species for tourism in Amboseli National Park, Kenya. Tour Manag 29:751–760
Oniszczuk-Jastrząbek A, Dębicka O, Gutowski T (2017) Ecotourism in Poland. https://doi.org/10.26881/sim.2017.4.20. Accessed 12 Oct 2018
Osborne L (undated) Number of native species in United States. Current results: weather and science facts. https://www.currentresults.com/Environment-Facts/Plants-Animals/number-of-native-species-in-united-states.php
Palazy L, Bonenfant C, Gaillart JM, Courchamp F (2011) Cat dilemma: too protected to escape trophy hunting? PLoS One 7:e22424. https://doi.org/10.1371/journalpone.0022424
Palazy L, Bonenfant C, Gaillard JM, Courchamp F (2012) On the use of the IUCN status for the management of trophy hunting. Wildl Res 39:711–720. https://doi.org/10.1071/WR12121
Pallin LJ, Baker CS, Steel D, Kellar NM, Robbins J, Johnston DW, Nowacek DP, Read AJ, Friedlaender AS (2018) High pregnancy rates in humpback whales (*Megaptera novaeangliae*) around the Western Antarctic Peninsula, evidence of a rapidly growing population. R Soc Open Sci 5:180017. https://doi.org/10.1098/rsos.180017
Pant K (2016) The study on the assessment of forest sector's contribution in national economy and environmental conservation in Nepal. MSFP-SSU, Kathmandu. https://asiafoundation.org/wp-content/uploads/2018/07/Conservation-and-Prosperity-in-New-Federal-Nepalpdf
Parks.it (2018) Available via www.parks.it. Accessed 8 Sept 2018
Parsons ECM (2012) The negative impacts of whale watching. J Mar Biol:Article ID 807294, 9 pages. https://doi.org/10.1155/2012/807294

Parsons ECM, Brown DM (2017) Recent advances in whale-watching research: 2015-2016. Tour Mar Environ 12:125–137
Pasape L, Mujwiga S (2017) Towards success of ecotourism networks in Tanzania: a case of Tanzania association of tour operators. J Hosp Manag Tour 8:14–24
Perez MA (2003) Compilation of marine mammal incidental rake data from the domestic and joint venture Groundfish fisheries in the U.S. EEZ of the North Pacific, 1989–2001. United States Department of Commerce, NOAA Technical Memorandum NMFS-AFSC-138, 145 p
Petrozzi F (2018) Bushmeat and fetish trade of birds in West Africa: a review. Life Environ 68:51–64
Porras I, Barton DN, Miranda M, Chacón-Cascante A (2013) Learning from 20 years of payments for ecosystem services in Costa Rica. International Institute for Environment and Development, London
Porter L, Lai HY (2017) Marine mammals in Asian societies, trends in consumption, bait, and traditional use. Front Mar Sci 4:47. https://doi.org/10.3389/fmars.2017.00047
Porter-Bolland L, Ellis EA, Guariguata MR, Ruiz-Mallén I, Negrete-Yankelevich S, Reyes-Garcia V (2012) Community managed forests and forest protected areas: an assessment of their conservation effectiveness across the tropics. For Ecol Manag 268:6–17
Possingham HP, Gerber LR (2017) The effect of conservation spending. Nature 551:309–310
Price R (2017) The contribution of wildlife to the economies of Sub Saharan Africa. https://assets.publishing.service.gov.uk/media/59ad5313ed915d78233d6474/145-Contribution-of-wildlife-to-SSA-economies.pdf. Retrieved 19 September 2018
Prokosch P (2015a) Svalbard, linking tourism and conservation, March 5, 2015. https://www.ltandc.org/svalbard/. Accessed 10 Oct 2018
Prokosch W (2015b) Namibia, linking tourism and conservation. https://www.ltandc.org/namibia/
Prokosch P (2016) Example South African National Parks (SANParks). Linking tourism and conservation, June 20, 2016. https://www.ltandc.org/south-african-national-parks-sanparks/
Prokosch P (2018) Spirit bear lodge – community-based tourism in the Great Bear Rainforest, Canada. Linking tourism and conservation, June 13, 2018. https://www.ltandc.org/spirit-bear-lodge-community-based-tourism-in-the-great-bear-rainforest-canada/
Read AJ (2008) The looming crisis: interaction between marine mammals and fisheries. J Mammal 89:541–548
Redang Island Malaysia Official (2019) Marine Park. Available via http://redangisland.co/index.php/marine-park/. Accessed 10 Jan 2019
Reeves RR, Smith BD, Crespo EA, Notarbartolo di Sciara G (2003) Dolphins, whales and porpoises. 2002–2010 conservation action plan for the World's cetaceans. IUCN/SSC Cetacean Specialist Group. IUCN, Gland/Cambridge. Ix + 139 pp
REMA (Rwanda Environment Management Authority) (2009) Rwanda state of environment and Outlook Report, Chapter VI. Forest and protected areas. http://www.rema.gov.rw/soe/chap6.php
Rickerby F, Estes O, Costea E, Cheng A (2016) Yasuní-ITT project in Ecuador. Rainforest, oil, and indigenous peoples in Ecuador: the Yasuní-ITT project. Open case studies. The University of British Columbia http://wiki.ubc.ca/Course:CONS200/2016w2/Wiki_Projects/Yasuní-ITT_Project_Ecuador. Accessed 11 Oct 2018
Robbins J (2017) The new threat to wolves in and around Yellowstone. The New York Times https://www.nytimes.com/2017/05/01/science/wolves-hunting-yellowstone-national-park.html. May 1, 2017
Rocha RC, Clapham PJ, Ivashchenko YV (2015) Empting the oceans: a summary of industrial whaling catches in the 20th century. Mar Fish Rev 76:37–48
Rodriguez J (2004) The myth of trophy hunting as conservation. Harare: Zimbabwe conservation task force. A league against cruel sports submission to environment minister, Elliott Morley MP
Sainato M (2015) Protected areas in the United States too small, disconnected to preserve biodiversity, studies find. Earth Island Journal. http://www.earthisland.org/journal/index.php/articles/

entry/protected_areas_in_usa_too_small_disconnected_to_preserve_biodiversity/. June 4, 2015
SANParks (2012) Kgalagadi Transfrontier Park. https://www.sa-venues.com/game-reserves/nc_kgalagadi-transfrontier-park.htm. Accessed 09 Oct 2018
Sawe BE (2017) Habitats and ecosystems in Poland, World Atlas, April 25, 2017. https://www.worldatlas.com/articles/habitats-and-ecosystems-in-poland.html. Accessed 12 Oct 2018
Scherffius L (2015) Ecuador supports ecotourism to stimulate local economies. Telesur, May 7, 2015. https://www.telesurtv.net/english/news/Ecuador-Supports-Ecotourism-to-Stimulate-Local-Economies-20150507-0026.html. Accessed 11 Oct 2018
Sheahan M (2018) U.S. fails to keep pace with global tourism boom. Reuters World News. https://www.reuters.com/article/us-usa-trump-travel/u-s-fails-to-keep-pace-with-global-tourism-boom-idUSKCN1GL21E. March 9, 2018
Shermans Travel (2013) Top 10 ecotourism destinations, April 18, 2013. https://www.shermanstravel.com/advice/top-10-ecotourism-destinations. Accessed 10 Oct 2018
Shuib A, Mansor WSW, Abdullah M, Zahid E (1994) The impact of tourism development on local employment. Malays J Agric Econ 11:25–41
Siyabona Africa (2017) http://www.botswana.co.za/Wildlife_Conservation-travel/botswana-wildlife-protection.html. Accessed 09 Oct 2018
Smith B (2015) Costa Rica: environmental issues, policies and clean technology. AZoNetwork, July 14 2015. https://www.azocleantech.com/article.aspx?ArticleID=554. Accessed 11 Oct 2018
South African National Parks (2019) Kruger National Park. Available via https://www.sanparks.org/parks/kruger/tourism/history.php. Accessed 29 Jan 2019
South African National Parks Annual Report (2016/17) https://www.sanparks.org/assets/docs/general/annual-report-2017.pdf
Spenceley A, Rylance A, Laiser SL (2017) Protected area entrance fees in Tanzania: the search for competitiveness and value for money. Koedoe 59(1):a1442. https://doi.org/10.4102/koedoe.v59i1.1442
Statista.com (2018a) National park tourism in the U.S. – statistics and Facts https://www.statista.com/topics/2393/national-park-tourism-in-the-us/
Statista.com (2018b) Annual number of international tourist arrivals in Norway from 2007 to 2017 (in millions). https://www.statista.com/statistics/806411/international-tourist-arrivals-in-norway/. Accessed 10 Oct 2018
Statistics Norway (2016) This is Svalbard 2016. What the figures say. ISBN 978-82-537-9487-7 https://www.ssb.no/en/befolkning/artikler-og-publikasjoner/_attachment/294354?_ts=15a12de02c0. Accessed 10 Oct 2018
Statistics Norway (2018) Protected Areas, last updated August 27, 2018. https://www.ssb.no/en/arealvern. Accessed 10 Oct 2018
Statistics Poland (2018) Tourism in 2017. Warsaw, 2018. http://stat.gov.pl/en/topics/culture-tourism-sport/tourism/tourism-in-2017,1,15.html. Accessed 12 Oct 2018
Stem CJ, Lassoie JP, Lee DR, Deshler DD, Schelhas JW (2003) Community participation in ecotourism benefits: the link to conservation practices and perspectives. Soc Nat Resour 16:387–413. https://doi.org/10.1080/08941920309177
Stolton S, Mansourian S, Dudley N (2010) Valuing protected areas. Technical report. © 2010 The International Bank for Reconstruction and Development/The World Bank, Washington, DC. https://doi.org/10.13140/RG.2.1.2296.6642
Stonich A (2016) Brazil's best kept secret: The Pantanal National Geographic June 7, 2016 https://www.nationalgeographic.com/travel/destinations/south-america/brazil/brazils-best-kept-secret-the-pantanal-wildlife-biodiversity-wetlands/. Accessed 10 Oct 2018
Tanzaniainvest (2017) Tanzania tourist arrivals increase by 12.9% in 2016 to reach 1,28 M. Available via https://www.tanzaniainvest.com/tourism/tourist-arrivals-reach-2016. Accessed 10 Sept 2018
Tanzaniainvest.com (2017) https://www.tanzaniainvest.com/tourism/tourist-arrivals-reach-2016

Tashobya A (2018) Conservation body donates land to Volcanoes Park to expand habitat for gorillas. The New Times, January 10, 2018. https://www.newtimes.co.rw/section/read/212451

Taylor A, Lindsey P, Davies-Mostert H (2016) An assessment of the economic, social and conservation value of the wildlife ranching industry and its potential to support the green economy in South Africa. Green Economy Research Report. Environmental Affairs Republic of South Africa 15 January 2016

TEEB (2009) The economics of ecosystems and biodiversity for national and international policy makers summary: responding to the value of nature. United Nations Environment Programme, Nairobi

TEEB (2017) TEEB (The Economics of Ecosystems and Biodiversity): the development of strategies to maintain and enhance the protection of ecosystem services in Namibia's State, Communal and Freehold Lands. Anchor Environmental Consultants and Namibia Nature Foundation, June 2017. https://resmob.org/wp-content/uploads/2017/11/Namibia-TEEB-Presentation-June-2017_v3-edited-for-Internet.pdf. Accessed 08 Oct 2018

Teichman K, Cristescu B, Darimont CT (2016) Hunting as a management tool? Cougar-human conflicts is positively related to trophy hunting. BMC Ecol 16:44. https://doi.org/10.1186/s12898-016-0098-4

Thakali S, Peniston B, Basnet G, Shrestha M (2018) Conservation and Prosperity in New Federal Nepal: opportunities and challenges. The Australian Government- The Asia Foundation Partnership on Subnational Governance in Nepal, July 2018. https://asiafoundation.org/wp-content/uploads/2018/07/Conservation-and-Prosperity-in-New-Federal-Nepalpdf. Accessed 10 Oct 2018

Thapa D (2010) Bio-diversity conservation in Nepal. Bibechana 6:31–36

The Daily (2017) Travel between Canada and other countries. December 2017. Statistics Canada, February 20, 2018 https://www150.statcan.gc.ca/n1/daily-quotidien/180504/dq180504a-eng.htm. Updated 2018-05-04

The Dian Fossey Gorilla Fund International (2018) Mountain gorilla population still increasing: census results released. Available via https://gorillafund.org/mountain-gorilla-population-still-increasing-census-results-released/. Accessed 3 Jan 2019

The Governor of Svalbard (2019) The Svalbard environmental protection act. Available via https://www.sysselmannen.no/en/Toppmeny/About-Svalbard/Laws-and-regulations/The-Svalbard-Environmental-Protection-Act-/. Accessed 29 Jan 2019

The International Ecotourism Society (2019) What is ecotourism? Available via http://www.ecotourism.org/what-is-ecotourism. Accessed 29 Jan 2019

The International Ecotourism Society. http://www.ecotourism.org/australia

The International Ecotourism Society. http://www.ecotourism.org/ecuador

The National Wildlife Federation (2019) Recovering America's Wildlife Act. Available via https://www.nwf.org/Our-Work/Wildlife-Conservation/Policy/Recovering-Americas-Wildlife-Act. Accessed 29 Jan 2019

The Svalbard treaty (1920) Available via http://library.arcticportalorg/1909/1/The_Svalbard_Treaty_9ssFy.pdf. Accessed 29 Jan 2019

The World Bank Group: Economical and statistical analysis of tourism in Uganda, 2013. http://documents.worldbank.org/curated/en/117851468108862384/pdf/797680WP0Touri0Box0379789B00PUBLIC0.pdf. Accessed 09 Oct 2018

Thomas PO, Reeves RR, Brownell RL Jr (2016) Status of the world's baleen whales. Mar Mamm Sci 32:682–734. https://doi.org/10.1111/mms.12281

Thome WH (2015) Tanzania's best kept tourism secret. African Travel Magazine. http://www.africa-ata.org/tz_economy.htm. July 13, 2015

Thomson G (2007) Ecuador, Worldmark Encyclopedia of Nations. https://www.encyclopedia.com/places/latin-america-and-caribbean/south-american-political-geography/ecuador. Accessed 11 Oct 2018

Tortato FR, Izzo TJ, Hoogesteijn R, Peres CA (2017) The numbers of the beast: valuation of jaguar (*Panthera onca*) tourism and cattle depredation in the Brazilian Pantanal. Glob Ecol Conserv 11:106–114. https://doi.org/10.1016/j.gecco.2017.05.0032351-9894/

Tourism Research Australia (2008a) Travel by Australians—December 2007 quarterly results of the national visitor survey. Tourism Research Australia, Canberra

Tourism Research Australia (2008b) International visitors in Australia—December 2007 quarterly results of the international visitor survey. Tourism Research Australia, Canberra

Trading Economics (2018) Zimbabwe tourist arrivals. https://tradingeconomics.com/zimbabwe/tourist-arrivals

Trading Economics (2019a) Italy Tourism Revenues. Available via https://tradingeconomics.com/italy/tourism-revenues. Accessed 29 Jan 2019

Trading Economics (2019b) Uganda Tourist Arrivals. Available via https://tradingeconomics.com/uganda/tourist-arrivals. Accessed 29 Jan 2019

Treves A, Naughton-Treves L (2005) Evaluating lethal control in the management of human-wildlife conflict. In: Woodroffe R, Thirgood S, Rabinowitz A (eds) People and wildlife: conflict or coexistence? Cambridge University Press, Cambridge

Twining-Ward L, Li W, Bhammar H, Wright E (2018) Supporting sustainable livelihoods through wildlife tourism. Tourism for Development, World Bank, Washington, DC. © World Bank. https://openknowledge.worldbank.org/handle/10986/29417. License: CC BY 3.0 IGO

U.S. Department of State Archive (2003) Paper submitted to the world ecotourism summit by U.S. Department of State. https://2001-2009.state.gov/g/oes/rls/or/19412.htm. May 2002, April 8, 2003

U.S. Fish and Wildlife Service (2019) US-China nature Conservation protocol. Available via https://www.fws.gov/international/wildlife-without-borders/east-asia/us-china-nature-conservation-protocol.html. Accessed 29 Jan 2019

Uganda Bureau of Statistics (2018) Visitor to National Parks (Citizens and Foreign), 000s. Available via https://www.ubos.org/explore-statistics/statistical-datasets/14250/. Accessed 29 Jan 2019

Uganda Eco Tours (2011) Ecotourism in Uganda. Available via http://www.ecouganda.com/ecotourism-in-uganda. Accessed 3 July 2018

UN-ECLAC (2018) Economic survey of Latin America and the Caribbean. Evolution of investment in Latin America and the Caribbean: stylized facts, determinants and policy challenges. Briefing Paper. The United Nation Economic Commission for Latin America and the Caribbean. https://www.cepalorg/en/publications/43965-economic-survey-latin-america-and-caribbean-2018-evolution-investment-latin. Accessed 11 Oct 2018

UNEP-WCMC (2018a) Protected area profile for Kenya from the world database of protected areas, September 2018. https://www.protectedplanet.net/

UNEP-WCMC (2018b) Protected area profile for Australia from the world database of protected areas, September 2018. https://www.protectedplanet.net/

UNEP-WCMC (2018c) Protected area profile for Botswana from the world database of protected areas, September 2018. https://www.protectedplanet.net/

UNEP-WCMC (2018d) Protected area profile for Brazil from the world database of protected areas, September 2018. https://www.protectedplanet.net/

UNEP-WCMC (2018e) Protected Area Profile for Canada from the World Database of Protected Areas, September 2018. https://www.protectedplanet.net/

UNEP-WCMC (2018f) Protected area profile for Costa Rica from the world database of protected areas, September 2018. https://www.protectedplanet.net/

UNEP-WCMC (2018g) Protected area profile for Ecuador from the world database of protected areas, September 2018. https://www.protectedplanet.net/

UNEP-WCMC (2018h) Protected area profile for India from the world database of protected areas, September 2018. https://www.protectedplanet.net/

UNEP-WCMC (2018i) Protected Area Profile for Italy from the World Database of Protected Areas, September 2018. https://www.protectedplanet.net/

UNEP-WCMC (2018j) Protected area profile for Malaysia from the world database of protected areas, September 2018. https://www.protectedplanet.net/
UNEP-WCMC (2018k) Protected area profile for Namibia from the world database of protected areas, September 2018. https://www.protectedplanet.net/
UNEP-WCMC (2018l) Protected area profile for Nepal from the world database of protected areas, September 2018. https://www.protectedplanet.net/
UNEP-WCMC (2018m) Protected area profile for Norway from the world database of protected areas, September 2018. https://www.protectedplanet.net/
UNEP-WCMC (2018n). Protected area profile for Poland from the world database of protected areas, September 2018. https://www.protectedplanet.net/
UNEP-WCMC (2018o) Protected area profile for Rwanda from the world database of protected areas, September 2018. https://www.protectedplanet.net/
UNEP-WCMC (2018p) Protected area profile for South Africa from the world database of protected areas, September 2018. https://www.protectedplanet.net/
UNEP-WCMC (2018q) Protected area profile for Uganda from the world database of protected areas, September 2018. https://www.protectedplanet.net/
UNEP-WCMC (2018r) Protected area profile for United States of America from the world database of protected areas, September 2018. https://www.protectedplanet.net/
UNEP-WCMC (2018s) Protected area profile for Zimbabwe from the world database of protected areas, September 2018. https://www.protectedplanet.net/
UNEP-WCMC and IUCN (2016) Protected planet report 2016. Cambridge/Gland, UNEP-WCMC and IUCN
Unicredit, Touring Club Italiano (2017) Rapporto sul Turismo 2017 realizzato da UniCredit in collaborazione con il Touring Club Italiano, July 17, 2017. https://www.unicreditgrounicreditup.eu/it/press-media/press-releases/2017/rapporto-sul-turismo-2017-realizzato-da-unicredit-in-collaborazi.html. Accessed 10 Oct 2018
UNWTO (2014a) Towards measuring the economic value of wildlife watching tourism in Africa – Briefing Paper. Published by the World Tourism Organization (UNWTO), Madrid. ISBN: 978-92-844-1675-2
UNWTO (2014b) World tourism organization: towards measuring the economic value of wildlife watching tourism in Africa – briefing paper. UNWTO, Madrid
UNWTO (2018) World tourism organization: tourism for development – volume I: key areas for action. UNWTO, Madrid. https://doi.org/10.18111/9789284419722
UNWTO (World Tourism Organization) (2019) Tourism in the 2030 Agenda. Available via http://www2.unwto.org/content/tourism-2030-agenda. Accessed 29 Jan 2019
US News and World Report. https://travel.usnews.com/Kruger_National_Park_South_Africa/. Accessed 09 Oct 2018
Uttarakhand Tourism Development Board (2019) Corbett National Park. Available via http://uttarakhandtourism.gov.in/corbett-national-park. Accessed 29 Jan 2019
UWA (Uganda Wildlife Authority) (2018) Bwindi Impenetrable National Park. http://www.ugandawildlife.org/other-parks/bwindi-impenetrable-national-park
UWA (Uganda Wildlife Authority) (2019). http://www.ugandawildlife.org/
Van der Duim R, Caalders J (2002) Biodiversity and tourism: impacts and interventions. Ann Tour Res 29:743–761
Veras O (2017) Unleashing Zimbabwe's tourism potential. https://www.howwemadeitinafrica.com/unleashing-zimbabwes-tourism-potential/57415/
Verma M, Negandhi D, Khanna C, Edgaonkar A, David A, Kadekodi G, Costanza R, Gopal R, Bonal BS, Yadav SP, Kumar S (2016) Making the hidden visible: economic valuation of tiger reserves in India. Ecosyst Serv 26:236–244. https://doi.org/10.1016/j.ecoser.2017.05.006
Vidal J (2016) Oil drilling underway beneath Ecuador's Yasuní National Park. Guardian News 2018. https://www.theguardian.com/environment/2016/oct/26/oil-drilling-underway-beneath-ecuadors-yasuni-national-park. Accessed 11 Oct 2018

Volcanoes National Park Rwanda (2019) Volcanoes National Park Rwanda/Parc National Des Volcans (PNV). Available via https://www.volcanoesnationalparkrwanda.com/. Accessed 29 Jan 2019

Walker Painemilla K, Rylands AB, Woofter A, Hughes C (eds) (2010) Indigenous peoples and conservation: from rights to resource management. Conservation International, Arlington

Watson M (2015) The impact of mass tourism. WordPress. Edexccel IGCSE geography. https://maxwatsongeography.wordpress.com/section-d/globalisation-and-migration/the-impact-of-mass-tourism/

Watts J (2018) New round of oil drilling goes deeper into Ecuador's Yasuní National Park. Guardian News 2018. https://www.theguardian.com/environment/2018/jan/10/new-round-of-oil-drilling-goes-deeper-into-ecuadors-yasuni-national-park. Accessed 11 Oct 2018

WCED (World Commission on Environment and Development) (1987) Our common future. Available at http://www.un-documents.net/our-common-future.pdf

Weller RJ (2017) Precis. In: Weller RJ, Hoch C, Huang C (ed) Atlas for the end of the world. http://atlas-for-the-end-of-the-world.com

Wilfred P (2012) Trophy hunting and trophy size in Ugalla Game Reserve, Western Tanzania. Tanzan J Sci 38:111–122

Wiltse-Ahmad A (2018) New state park to strengthen tiger protection in Malaysia. Current news, August 15, 2018. https://www.rainforesttrust.org/new-state-park-to-strengthen-tiger-protection-in-malaysia/. Accessed 10 Oct 2018

Wetlands International (2016) https://www.wetlands.org/casestudy/towards-sustainable-palm-oil/. Accessed 03/03/2020.

Wolf ND, Heneine W, Carr JK, Garcia AD, Shanmungam V, Tamoufe U, Torimiro JN, Prosser AT, LeBreton M, Mpoudi-Ngole E, McCutchan FE, Birx DL, Folks TM, Burke DS, Switzer WM (2005) Emergence of unique primate T-lymphotropic viruses among central African bushmeat hunters. PNAS 102:7994–7999

World Economic Forum (2017) The travel and tourism competitiveness report 2017. World Economic forum. Geneva, ISBN-13: 978-1-944835-08-8. http://www3.weforum.org/docs/WEF_TTCR_2017_web_0401.pdf. Accessed 09 Oct 2018

World Travel and Tourism Council: Economic Impact 2018 – March 2018. All rights reserved. Licensed under the Attribution, Non-Commercial 4.0 International Creative Commons Licence

WTTC (2018a) Travel and tourism economic impact-South Africa. March 2018. All rights reserved. Licensed under the Attribution, Non-Commercial 4.0 International Creative Commons Licence. https://www.wttc.org/economic-impact/country-analysis/country-reports/

WTTC (2018b) Travel and tourism council: tourism economic Impact 2018 Italy 2018. All rights reserved. Licensed under the Attribution, Non-Commercial 4.0 International Creative Commons Licence. https://www.wttc.org/economic-impact/country-analysis/country-reports/

WTTC (2018c) Travel and tourism council: tourism economic impact 2018 Norway 2018. All rights reserved. Licensed under the Attribution, Non-Commercial 4.0 International Creative Commons Licence. https://www.wttc.org/economic-impact/country-analysis/country-reports/

WTTC (2018d) Travel and tourism economic impact 2018 Namibia. London 2018, March 2018. All rights reserved. Licensed under the Attribution, Non-Commercial 4.0 International Creative Commons Licence. https://www.wttc.org/economic-impact/country-analysis/country-reports/. Accessed 09 Oct 2018

WTTC (2018e) Travel and tourism economic impact 2018, Kenya. World Travel and Tourism Council. March 2018. All rights reserved. Licensed under the Attribution, Non-Commercial 4.0 International Creative Commons Licence. https://www.wttc.org/economic-impact/country-analysis/country-reports/

WTTC (2018f) Travel and tourism economic impact Rwanda 2018 – March 2018. All rights reserved. Licensed under the Attribution, Non-Commercial 4.0 International Creative Commons Licence. https://www.wttc.org/economic-impact/country-analysis/country-reports/

WTTC (2018g) World travel and tourism council: tourism economic impact 2018 Tanzania. March 2018. All rights reserved. Licensed under the Attribution, Non-Commercial 4.0

International Creative Commons Licence. Travel and https://www.wttc.org/economic-impact/country-analysis/country-reports/
WTTC (2018h) World travel and tourism council: tourism economic impact 2018 Uganda 2018. All rights reserved. Licensed under the Attribution, Non-Commercial 4.0 International Creative Commons Licence. https://www.wttc.org/economic-impact/country-analysis/country-reports/
WTTC (2018i) World travel and tourism council: tourism economic impact 2018 Uganda 2018. All rights reserved. Licensed under the Attribution, Non-Commercial 4.0 International Creative Commons Licence. https://www.wttc.org/economic-impact/country-analysis/country-reports/
WTTC (2018j) World travel and tourism council: tourism economic impact 2018 Malaysia. March 2018. All rights reserved. Licensed under the Attribution, Non-Commercial 4.0 International Creative Commons Licence. https://www.wttc.org/economic-impact/country-analysis/country-reports/
WTTC (2018k) World travel and tourism council: tourism economic impact 2018 Nepal March 2018. All rights reserved. Licensed under the Attribution, Non-Commercial 4.0 International Creative Commons Licence. https://www.wttc.org/economic-impact/country-analysis/country-reports/
WTTC (2018l) World travel and tourism council: travel and tourism economic impact 2018 – March 2018. https://www.wttc.org/economic-impact/country-analysis/country-reports/
WTTC (2018m) World travel and tourism council: travel and tourism global economic impact and issues 2018. https://www.wttc.org/economic-impact/country-analysis/country-reports/
WWF (2014) Brazilian System of Protected Areas (SNUC) completes 14 years of advances and challenges. https://www.wwf.org.br/informacoes/english/?40662%252FBrazilian-System-of-Protected-Areas-SNUC-completes-14-years-of-advances-and-challenges. August 4 2014
WWF (2017) Amazon deforestation. http://wwf.panda.org/our_work/forests/deforestation_fronts/deforestation_in_the_amazon/. Accessed 27 Oct 2018
WWF (2018) Great barrier reef. http://www.wwf.org.au/what-we-do/oceans/great-barrier-reef#gs.MyF8CEw. Accessed 12 Oct 2018
WWF Australia (2018) Building a world leading protected area system for Queensland, August 2018. http://www.wwf.org.au/ArticleDocuments/353/doc-queensland-protected-area-alliance-position-paper-NPAQ-conservation-17aug18.pdf.aspx. Accessed 12 Oct 2018
Wyoming Office of Tourism (2018) Wyoming travel impacts 2004–2017. Dean Runyan Associated, Cheyenne, Wyoming, April 2018. http://www.deanrunyan.com/doc_library/WYImp.pdf
World Tourism Organization (2014), Towards Measuring the Economic Value of Wildlife Watching Tourism in Africa – Briefing Paper, UNWTO, Madrid.
Yasukawa O, Pisa K (2016) In Botswana, ecotourism and conservation draw travelers. CNN travel, 13th June 2016. https://edition.cnn.com/travel/article/botswana-ecotourism-mpa-feat/index.html
Zisadza P, Gandiwa E, van der Westhuizen H, van der Westhuizen E, Bodzo V (2010) Abundance, distribution, and population trends of hippopotamus in Gonarezhou National Park, Zimbabwe. S Afr J Wildl Res 40(2):149–157

Part V
Species Extinction

One of the most downloaded chapters of the first *Problematic Wildlife* book was an original and somehow provocative manuscript about cryptozoology (Rossi 2016). This work represents the first attempt in scientific literature to review cryptozoology, a controversial discipline considered a pseudoscience.

Results, despite unquestionably pointed on some pseudoscientific aspects, showed also how Bernard Heuvelmans, the "father" of cryptozoology, anticipated *ante litteram* current issues in the field of biological conservation, such as the "Romeo error" (Collar 1998), the "Lazarus species" (Dawson et al. 2006), and the "thylacine effect" (Macphee and Flemming 1999).

To ascertain the presence of a species is a key factor in its conservation. In fact, erroneous assumption of extinctions can be fatal for conservation and biodiversity: if a species is declared extinct, it can't be protected in any way.

There are many cases, even involving big-sized species, where declaration of extinction has proven unfounded. For example, animals like the Barbary lion (*Panthera leo leo*), the Père David's deer (*Elaphurus davidianus*), and the Schomburgk's deer (*Rucervus schomburgki*) survived for some decades after their official extinction (Black et al. 2013; Turvey et al. 2017; Schroering and Gary 2019).

In this part Black (2020) carries out an in-depth review concerning the conservation of difficult-to-observe species, listing several cases useful to improve conservation attempts, while Rossi et al. (2020) discuss case studies about species, subspecies, and populations of big cats (mainly by the genus *Panthera*) whom extinction stories are still uncertain.

Both works agree that the only possible solution to better understand the extinction process and to be more effective in species conservation is to incentivize a multidisciplinary approach involving primarily local population.

References

Black (2020) Assessing presence, decline and extinction for the conservation of difficult-to-observe species. In: Angelici FM, Rossi L (eds) Problematic wildlife II. New conservation and management challenges in the human-wildlife interactions. Springer, Cham, pp 359–392

Black SA, Fellous A, Yamaguchi N et al (2013) Examining the extinction of the Barbary Lion and its implications for felid conservation. PLOS One 8(4):e60174
Collar NJ (1998) Extinction by assumption; or, the Romeo Error on Cebu. Oryx 32:239–244
Dawson MR, Marivaux L, Li, CK et al (2006) Laonastes and the 'Lazarus effect' in recent mammals. Science 311:1456–1458
MacPhee RDE, Flemming C (1999) Requiem aeternam: the last five hundred years of mammalian species extinctions. In: MacPhee RDE (ed) Extinctions in neartime: causes, contexts, and consequences. Kluwer Academic/Plenum, New York, pp 333–371
Rossi L (2016) A review of cryptozoology: towards a scientific approach to the study of 'hidden animals'. In: Angelici FM (ed) Problematic wildlife – a cross-disciplinary approach. Springer, New York, p 573–588
Rossi L, Scuzzarella CM, Angelici FM (2020) Extinct or perhaps surviving relict populations of big cats. Their controversial stories and implications for conservation. In: Angelici FM, Rossi L eds) Problematic wildlife II. New conservation and management challenges in the human-wildlife interactions. Springer, Cham, pp 393–417
Schroering GB, Galbreath GJ (2019). Evidence of late survival of Schomburgk's Deer *Rucervus schomburgki* in Central Laos. J Bombay Nat Hist Soc. doi: https://doi.org/10.17087/jbnhs/2019/v116/142873
Turvey ST, Barnes I, Marr M, Brace S (2017) Imperial trophy or island relict? A new extinction paradigm for Père David's deer: a Chinese conservation icon. R Soc Open Sci 4:171096. https://doi.org/10.1098/rsos.171096

Chapter 11
Assessing Presence, Decline, and Extinction for the Conservation of Difficult-to-Observe Species

Simon A. Black

11.1 Introduction

Conservation management often involves working with species that are difficult to detect, for reasons of scarcity, behaviour, morphology, ecology, or habitat. Examples of cryptic species which are likely to need human intervention are many and varied, inhabiting remote or difficult-to-access ranges such as the polar icecaps, deep ocean, high mountain ranges, vast river systems, isolated island archipelagos, or inhospitable deserts, swamps, and forests. Some species have lifestyles so obscure that observation is unlikely, including nocturnal species, those with fossorial or subterranean existences, and communities around oceanic hydrothermal vents. Even relatively well-known taxa such as big cats and whales can be difficult to detect (Brassine and Parker 2015; MacLeod et al. 2005). Some species have only been described in recent years, including relatively large animals such as the antelope-like saola *Pseudoryx nghetinhensis* (Dung et al. 1993), the 2-metre-long monitor lizard *Varanus bitatawa* (Welton et al. 2010), the Burmese snub-nosed monkey *Rhinopithecus strykeri* (Geissmann et al. 2011), and the recently identified orang-utan (proposed as *Pongo tapanuliensis*) of Batang Toru, Sumatra (Nater et al. 2017). Additionally, once-considered 'extinct' species have remerged unexpectedly (Clark 1997; McDougall et al. 2009) including the black-footed ferret (*Mustela nigripes*) and night parrot (*Pezoporus occidentalis*).

The potential of relic populations as a basis for successful recoveries is an important area of learning of which the conservation community needs to be reminded. Successful cases illustrate how little knowledge is required to initiate impactful action and how cycles of investigation increase the quality of interventions which help recover species and ecosystems of concern. While past misunderstandings of species survival due to uncertain data have led to failures in required conservation

S. A. Black (✉)
Durrell Institute of Conservation & Ecology, University of Kent, Canterbury, UK
e-mail: s.black@kent.ac.uk

F. M. Angelici, L. Rossi (eds.), *Problematic Wildlife II*,
https://doi.org/10.1007/978-3-030-42335-3_11

intervention, contemporary conservation experiences provide methods which can be used to avoid these mistakes in the future. This includes development of a working knowledge of difficult-to-observe species, enhanced with new methods, technologies, and analytical tools. The experiences provide an opportunity to accelerate innovation and learning in conservation management.

To place these in scientific context, the basis of knowledge and philosophy of knowledge (i.e. epistemology) needs to be understood to provide a mental framework for considering information of varying degrees of quality and veracity (Lewis 1956). An appreciation of uncertainty in data will enable a practitioner to make reasonable consideration and decision-making in different contexts (Black and Copsey 2014), and relative knowledge possessed by conservation professionals, scientists, and local people can support an understanding of species decline, persistence, or recovery. Data-deficient cases may require alternative sources of information to support management decisions. A new type of 'boundary science' (Cook et al. 2013) is needed to bring practical conservation for vulnerable, cryptic, and less-well-understood species using emerging methods for presence detection.

An understanding of a species' status, range, population size, demographics, ecology, and threats is a vital building block in establishing conservation strategies (Jones et al. 2018). This knowledge influences decisions for deploying resources and effort to ensure a future for species and ecosystems of concern (Black et al. 2013b). A successful recent effort to prevent the extinction, based upon detection of decline and population vulnerability, was seen with recent recovery of the orange-bellied parrot in Australia (Martin et al. 2012a). While many species' presence and behaviours are patently visible and suitable for monitoring by specialists or even the lay public on behalf of conservation bodies, the converse is also common where scientists encounter difficulties associated with less-visible species.

Visibility, remoteness, density, and population size all affect detectability of a species. Some species are highly cryptic in behaviour or choose habitats that are remote or difficult for humans to access making observation difficult. An alternative situation is faced where an otherwise easily visible species becomes less likely to observe when its population density reduces due to decline, such that the simple probability of an encounter becomes unlikely (Solow 2005). Some species fit several of these criteria, naturally occurring in low densities and in difficult terrain so are less likely to be observed. For others the situation is made more difficult by the species moving to remoter areas as an adaptation to threats.

Extinctions are important indicators of biodiversity status (Keith et al. 2017). Knowing whether a species has become extinct or not is of practical importance since a surviving species may be a candidate for further conservation effort (Lee et al. 2015) and investment in its conservation will steer resources away from work on other species. Without objective judgement of presence, a species could continue to be categorised as critically endangered (possibly extinct) and continue to incur costs associated with this status (McKelvey et al. 2008). To address this, the IUCN adopted the category of 'possibly extinct' for species 'that are, on the balance of evidence, likely to be extinct, but for which there is a small chance that they may be extant' to both avoid wasted resources and also avoid underestimation of biodiversity loss (Akçakaya et al. 2017).

Knowledge of extinction, survival, decline, or recovery of species is highly variable under any circumstance. With cryptic species (whether by behaviour, habitat, or density) a shortfall in understanding means that the less-concrete end of the spectrum of knowledge becomes important, including vagaries of opinion, belief, rumour, or, at best, historical records (Black et al. 2013b). While inconvenient, this situation does not change the need for practical solutions when working with cryptic species. Conservation is interested in the boundary of proven extinction or unknown marginal persistence. The divide between scientific fact and sufficient information required to support practitioner judgements or policy decisions can become sufficiently wide to cause uneasiness, yet it is a situation where working (and workable) knowledge is of utmost concern and should be addressed.

11.2 The Value of Efforts to Recover Small Populations

Small populations may be relics of a previously abundant species cornered into a last remnant of suitable habitat over geological time, such as Nile crocodiles in the Saharan fringes of Mauritania, Egypt, and Chad (Brito et al. 2011). In recent centuries, species have been common victims of landscape changes due to human agricultural or infrastructure encroachment. Remaining individuals survive under more particular threats, such as increases in human development and industrial pollution. In any of these circumstances, it is tempting to consider each case as beyond redemption. The most important lesson in species conservation over the past 40 years is, however, that even for species which have dwindled to a surviving remnant, there is no such thing as a lost cause. In some instances, the expected principles of genetic impoverishment have been confounded by species recovery, in some cases from a last remaining breeding pair.

The case of the Mauritius kestrel (*Falco punctatus*) is probably the most pertinent. Reduced to a population of just four individuals of which there was a single breeding pair, intensive work to recover the species has today achieved a free-living wild population of more than 300 surviving on Mauritius (Jones et al. 1995; Birdlife International 2016a). The echo parakeet (*Psittacula eques*) on Mauritius (Fig. 11.1) has been recovered with similar intervention principles (Jones et al. 2018). We have also learned that wild populations can achieve similar turnarounds without help, since the Seychelles kestrel (*Falco araeus*) recovered naturally from genetic bottlenecks as severe as those seen in recent intensive human-recovered populations (Groombridge et al. 2009). In contrast, attempts to intervene and recover the Po'ouli, the rare honeycreeper on Maui, from the remaining birds were not achieved; by the 1990s individual birds' respective territories no longer overlapped (Groombridge et al. 2004). Additionally, even if successfully captured for breeding purposes, the remaining Po'oulis were likely too old to thrive in captivity and possibly beyond

Fig. 11.1 The echo parakeet (*Psittacula eques*) on Mauritius, the last surviving endemic parrot of the Mascarenes, is one species brought back from the brink of extinction (Jones et al. 2018). Around 10 birds remained when efforts were initiated by the Mauritian Wildlife Foundation, recovering to currently 500 wild birds. (Photo: S Black)

breeding age. Had this intervention been considered 10 years earlier, the outcome may have been startlingly different.

Recovering a vertebrate species from a single reproductive unit to a viable population defies many conservation paradigms. It also underlines the importance of taking isolated observations of species survival seriously, and to gain better understanding of population status where a species is poorly understood or difficult-to-observe.

Justification of conservation effort can be made on the basis or moral, scientific, social, economic grounds, simple practical convenience, or the enthusiasm of a single person or organisation. A species may at the outset appear inconsequential or even an inconvenience, in which case development of local people's social identity with the species needs to be encouraged as a lever for its recovery (as seen with the St. Lucia Parrot). Despite these nuanced reasons relating to particular cases considered worthy of conservation, the underlying criterion of importance which stands up to scientific, moral, social, and economic scrutiny is that each species has, in known or unknown ways, a relevant ecological function.

Species play a role in the success, sustainability, and balance of ecosystems. Species may function as a food source or prey, act as a control on other populations (as a predator), provide or modify habitat (e.g. plants, corals, grazing ungulates, burrowing animals), disperse seeds, or be a commensal (Jones et al. 1994). Even low numbers of a species may be important within the overall system. When a species goes into decline (or experiences a destructive population explosion), this will either destabilise the ecosystem or at the least indicate that the ecosystem itself is unstable or declining. The issue with more cryptic, hard-to-detect species generally relates to scarcity and vulnerability of surviving populations. This latter challenge requires significant innovation in conservation science for which there is now a growing body of knowledge.

11.3 The Basis for Knowledge of a Species

Conservation can involve handling decisions and action in the light of uncertain or incomplete knowledge (Black and Copsey 2014). As scientists we usually insist on dealing with facts and follow evidence-based decision-making, yet circumstances commonly arise where this is not the case, requiring leadership and decision-making in the face of uncertainty (Martin et al. 2012a). Activities with ecosystems, species, landscapes, human communities, geo-politics, climate change, and the like operate under greater uncertainty than many areas of human endeavour. This contradicts the desire for more elegant, definitive knowledge required by biological science. The situation is no truer than in the cases of poorly known or rarely encountered species, particularly where they have adapted to a suboptimal existence in unfavourable circumstances. The challenge is to avoid a Type II error, remembering that while verifying a species' presence is relatively straightforward (e.g. by sight, sound, or sign), inability to detect a species does not necessarily mean it is absent (Morrison 2002). In grappling with potential extinction or survival of a species (and the implications for conservation), the maxim that 'absence of evidence is not evidence of absence' rings true (Collen and Turvey 2009) lest we either risk neglecting a recoverable species or waste resources on a lost cause.

Knowledge in relation to the physical world of species and ecosystems runs on a continuum (Black and Copsey 2014). Scientific knowledge may range from unknown, to belief, to the perceived, to partially known, to the established fact (Lewis 1956; Godfrey-Smith 2003). Even the most concretely asserted facts are not truly 'concrete' in the dynamics of natural systems. For example, the answer to the question 'How many deer live in the wood?' is no more certainly known now than in 1 hour, 2 weeks, or 2 months. Individuals die, are born, are predated, and move in and move out of locations. Bodies decay, are eaten, and are carried away or consumed by fire, flood, or landslide. Even then, when we measure their presence, detectability to humans is affected by visibility as well as nuances of measurement technique, technology, and error.

Information relating to cryptic and rare species tends towards the less definitive end of the knowledge spectrum. Rediscoveries occur in situations and locations that are more difficult to access or survey. The review by Scheffers et al. (2011) notes a concentration of rediscoveries in tropical and subtropical forests of South America, Africa, Madagascar, India, and New Guinea, which are generally environments which are difficult to access. The probability of detection of any species will be related to the remoteness of habitat (Collen and Turvey 2009). Adding to this mix, anecdotal and inconclusive physical occurrence data, including misidentifications, can lead to serious errors in conservation decision-making and practice, with the potential to waste resources (Roberts et al. 2010).

In conservation science, it is common for people to gravitate towards ‘facts’. Yet as has been mentioned, from an epistemological perspective, namely, the Theory of Knowledge (Lewis 1956), there is no such inestimable fact. The singular point of fact only has meaning under an operational definition (including measurement metrics), but too often the operationalised definition is poorly or wrongly diagnosed. This causes problems with estimations of extinction, persistence, re-emergence, rediscovery, or small population status.

By simply siding with the ‘fact’ that ‘there is no evidence’, conservation experts are guilty of siding with opinion; the expectation of hard evidence may not fit the context being observed. Why, for example, is a folk tale or anecdote about an animal so easily rejected? Indeed recent species ‘discoveries’ by science often follow, or are specifically led by, existing traditional knowledge. For example, recent research on primates identified since 1980 revealed 40% to have been informed by local knowledge (Rossi et al. 2018). With highly cryptic species, where scientific data or knowledge is absent, one can accept the usefulness of anecdote. It has, after all, been suggested that ‘the plural of anecdote is data’ (Noll 1980).

The Theory of Knowledge expects us to become comfortable with notions of ‘reasonable belief’ as well as scientific fact (Black et al. 2013b). When we are required to use vague information, the purpose of the knowledge sets the agenda. A historical interest in the extinction of a species will be satisfied by a description of specifics around date and location. If we wish to model the decline to understand the impact of threats, disease, or some other vector, then an understanding of factors including behaviour, habitat, and circumstantial changes over time becomes important (Jones et al. 2018). The aim is to build knowledge upwards, from the vague towards the strongly defined. We should avoid scientific bias in assumptions about acceptable or unacceptable information lest it causes us to make incorrect judgements of a species status.

11.4 Key Concepts in Assessing Extinction Versus Survival

Listing a species as Extinct on the IUCN Red List requires that exhaustive surveys have been undertaken in all known or likely habitat throughout its historic range, at appropriate times (diurnal, nocturnal, seasonal, annual) and over a time frame appropriate to its life cycle and life stages. Cases have occurred where continued persistence of a species has been missed, and later reappearance occurs, such as the woolly flying squirrel *Eupetaurus cinereus* of Pakistan (Zahler 1996) and the Vietnamese warty hog, *Sus bucculentus* (Groves et al. 1997). The danger is that an extant species is neglected under the incorrect assumption that it is extinct. Four concepts need to be considered to avoid the problem.

11.4.1 The Precautionary Principle

The Precautionary Principle (Foster et al. 2000) suggests that where an action is expected to have a negative consequence for a system, then it should be rejected, but where likely to have an acceptable impact, it should be undertaken. This is an important concept in environmental management in general and particularly applicable in species conservation. Precaution aims to avoid a negative outcome for the species or ecosystem of concern. For extinction assessment, this built-in conservatism in decision-making has a point. Extinction is irreversible and recovery impossible, so considerations, including implication for initiatives aimed at retention of that species, should follow similar caution. Extinction should not be accepted if there is a reasonable level of doubt.

The reverse perspective, namely, conservatism based on economic constraints, is not reasonable. Resources are finite, but they are also interchangeable and can be redeployed, withheld, and redirected. The balance within the Precautionary Principle tips towards an over-riding priority for species' needs. Factors of proportionality, non-discrimination, consistency, action versus inaction, and scientific developments will refine our judgement, as summarised in Table 11.1.

The Precautionary Principle is not reason to accept an absence of data, but should be a prompt to seek better data to inform future action (Black et al. 2013b), as indeed is the cue provided by a 'data-deficient' categorisation under the IUCN Red List criteria (IUCN 2017). Precaution indicates, for example, when to err on the side of:

1. Accepting non-expert sightings
2. Placing weight on indirect measures of species presence
3. Potential survival when suitable habitat is known to exist
4. Persistence by long-lived species after significant absence

11.4.2 The Thylacine Effect

Working against precaution is the potential for overenthusiastic acceptance of unverified sightings on the weight of accumulated evidence. In some cases, the norms of human societies, traditions, or other incentives can cloud the motivations and behaviours of witnesses (Young et al. 2018). Tasmania's wolflike marsupial (*Thylacinus cynocephalus*), last seen reliably in the 1930s, is so closely associated with overenthusiastic misidentifications that the tendency to accept the continued presence of a species according to the frequency of unsubstantiated sightings has been termed the 'Thylacine Effect' (Macphee and Flemming 1999). This effect becomes important since acceptance of observations may be unwarranted yet still drive an unreliable analysis of probability of persistence, unless the quality of those sightings is scrutinised (Collen and Turvey 2009). Methods have since been proposed to evaluate quality and patterns of persistence to overcome this problem (Lee et al. 2015).

Table 11.1 Guidelines for applying the Precautionary Principle (Foster et al. 2000) with additional examples of conservation considerations

Principle	Conservation context
Proportionality 'Measures … must not be disproportionate to the desired level of protection… not aim at zero risk'	Assume that a species still survives using available evidence (e.g. sightings, habitat, life cycle) even if not scientifically verified
Non-discrimination 'Comparable situations should not be treated differently… different situations should not be treated in the same way, unless there are objective grounds for doing so'	Evidence requirements relating to one species need not be equivalent to evidence for others unless ecology, habitat, and behaviour relating to detectability can be reasonably compared
Consistency 'Measures…should be comparable in nature and scope with measures …in equivalent areas in which all the scientific data are available'	Reasonable assumptions about a species can be made where equivalent evidence of presence or absence was subsequently supported by scientific data
Examine the benefits and costs of action or lack of action 'This examination should include an economic cost/benefit analysis when this is appropriate and feasible. However, other analysis methods...may also be relevant'	Conservation action will be in context of likely population levels, threats, and mitigation. This will include the level of analysis of existing information or efforts to increase knowledge
Examine scientific developments 'The measures must be of a provisional nature pending the availability of more reliable scientific data "…" scientific research shall be continued with a view to obtaining more complete data'	An initial decision, based on reports of species presence, should be followed by scientific data collection or indirect and direct measures when resources or technology can inform future decisions

This is an important phenomenon with which the scientist needs to grapple honestly and authentically. The circumstances of the species will guide which type of data can best be accumulated to verify its presence. When the Po'ouli was first discovered in 1973, the initial scientific quest involved the shooting of two birds as voucher specimens, when only nine had ever been observed (Groombridge et al. 2004). This approach seems absurd when considering survival of a tiny relic population. Nowadays destructive sampling need not be used since alternative methods such as video recording and non-invasive genetic sampling is available (Groombridge et al. 2004) and both these techniques have become a staple in understanding elusive populations.

One solution, short of modern data sources, is to establish 'reliable observational evidence' (Macphee and Flemming 1999), namely, the detection by a qualified witness at recorded times and places, essentially only accepting expert sightings as reliable. For certain taxa, however, the opportunity for experts to gain enough observations can be unacceptably limited, particularly if the range of the species is large, or the habitat inaccessible other than for relatively short periods of observation, perhaps merely days or weeks at a time. This raises a question of both method and mindset. To utilise lay-witness sightings, and to avoid unwarranted scepticism, as well as to apply a more circumspect, critical evaluation of unverified reports, scientists must keep in mind two concepts: Lazarus species and Romeo's Error.

11.4.3 'Lazarus' Species

Rediscovered species have been dubbed 'Lazarus taxa' (Dawson et al. 2006), after the New Testament biblical character revived days after his death. The concept differs from the palaeontological concept of 'Elvis taxa' which concerns similar but unrelated species arising in the fossil record (Erwin and Droser 1993). There are many examples of Lazarus species where a declaration of extinction has proven unfounded. Scheffers et al. (2011) catalogued 351 rediscovered species in the previous 122 years, approximately three 're-appearing' per year (104 amphibians, 144 birds, and 103 mammals). Many pronouncements of extinction have stood for several decades or longer, including the recent notable rediscovery of the Night parrot in Australia after a 150-year absence (McDougall et al. 2009).

Examples of mammalian Lazarus species include the pygmy hog (Fig. 11.2) which was considered extinct in the 1950s but re-emerged in India in the 1970s after an opportunistic capture of four individuals in a plantation coincided with local

Fig. 11.2 The pygmy hog (*Porcula salvania*) inhabits the grasslands in Assam, India. Now part of a captive breeding and release programme, the species is never observed in the wild aside from reintroduced animals in the immediate vicinity of feeding locations days after release (G. Narayan, personal communication, 2014). The species was considered extinct until an opportunistic find in 1971. (Photo: S Black)

enquiries by a friend of Gerald Durrell (Goodall et al. 2010). Durrell's interest prompted action to find, capture, and captive-breed the species for reintroduction. In Laos, the obscure rodent *Laonastes aenigmamus* was described in 2005 as a surviving member of a rodent family known from fossil sites in Asia, a taxon rediscovered after 11 million years (Dawson et al. 2006). Other examples include the Chacoan peccary (*Catagonus wagneri*) which was known only from fossils until rediscovery in 1971 (Wetzel et al. 1975) and, more modestly after a few years of inconspicuousness, the black-footed ferret (Wisely et al. 2008).

Rediscovered reptiles and amphibians, which are often more typically cryptic and hard to detect than other terrestial taxa, include the painted frog (*Latonia nigriventer*) considered extinct since the 1950s until its rediscovery in Lake Hula Marshes in Israel in 2011 (Biton et al. 2013) and La Palma giant lizard (*Gallotia auaritae*) not seen for centuries in the Spanish Canary Islands until photographs were taken in 2007 (Nogales et al. 2001). Fish include, most spectacularly, the coelacanth (*Latimeria* spp.) only previously known in the fossil record (Scheffers et al. 2011).

For many Lazarus species, and newly discovered species, local people's existing levels of awareness remained unknown to science, so had no influence on research or policy. Science needs to consider informal sources more carefully since, clearly, species do reappear, however unlikely, and may need a conservation response sooner rather than later. Unexpected rediscovery has implications; it is not uncommon to raise the question 'What would we do if this extinct species was rediscovered?' This was addressed in practice with the black-footed ferret and pygmy hog and considered (Turvey 2008) for the Yangtze river dolphin (*Lipotes vexillifer*) and Maui's now extinct Po'ouli (VanderWerf et al. 2006). Responses to rediscovery need to be carefully managed, including how to announce the find. When new populations of Sumatran rhinoceros (*Dicerorhinus sumatrensis*) were announced in 2013, there was concern that news would attract criminal hunting, the very menace which threatened the potential exinction of the species (Meijaard and Nijman 2014).

11.4.4 Romeo's Error or Extinction by Assumption

The concept of 'Romeo's Error' (Collar 1998) is taken from the story of Romeo and Juliet by William Shakespeare. At the climax of the tragedy, Romeo mistakenly assumes Juliet as dead when she could have been revived, and in desperation he takes his own life. Wildlife conservation practitioners must not erroneously presume a species' extinction and thereafter cease to make efforts to save that same species. This places conservation biologists at a fulcrum between scepticism and hope. On one hand, while reports of new sightings should meet a high burden of proof (Roberts et al. 2010), the Precautionary Principle suggests that decisions should tend towards acceptance of presence. Scientific scrutiny demands scepticism, but scientific values nurture the hope of new discovery.

With Romeo's Error at play, the preservation or loss of certain species is open to chance. The black-footed ferret is a survivor by luck as much as judgement. Its rediscovery and recovery after declared extinction was due to a ferret being killed by a landowner's dog (Clark 1997 p23) after which the physical evidence triggered deployment of resources to attempt the species' recovery. Earlier decades saw other species less fortunate. The presumed extinction of the Carolina parakeet (*Conuropsis carolinensis*) in the early 20th century masked its likely survival, evidenced by considerable observations by local people (including details of its behaviour), into the 1950s, and possibly 1960s (Snyder 2004). Those same decades saw conservation interventions initiated for the more visible peregrine falcon (*Falco peregrinus*), for which recovery was ultimately achieved (Cade and Burnham 2003).

The presence of the Caspian tiger (*Panthera tigris virgata*) in Turkey was not recognised until a young animal was killed in Uludere in 1970, an occurrence considered as the last concrete record of the subspecies, although a local trade in hunted skins persisted in Turkey into the 1980s and tigers most likely survived in that locality into the 1990s (Can 2004). At that time tiger conservation was mobilised elsewhere in the world yet missed the opportunity for recovery of the population in northeastern Turkey, northwestern Iran, and the Caucasus. The Caspian subspecies was only recently declared officially extinct by the IUCN, in 2003. In a final opportunistic twist, genetic studies later identified that the extant Amur tiger (*P. t. altaica*) of the Russian Far East and the Caspian tiger where almost genetically indistinguishable, sharing a common ancestor less than 10,000 years ago (Driscoll et al. 2009). This precipitated current consideration of restoration of Amur tigers into Kazakhstan, the former range of the Caspian tiger (Chestin et al. 2017).

Fortunately, lessons have been learned. Efforts to catalogue species presence in key hotspots are important in defining legislation to protect habitats and landscapes. To support these approaches and better understand the presence of difficult-to-observe species, several considerations need to be addressed.

11.5 The Nature of Cryptic or Difficult-to-Observe Species

The expected rate at which a species would normally be seen affects our understanding of whether its absence (i.e. non-observation) is an indication of significant decline or extinction (MacPhee and Flemming 1999). If the normal rate of observation is high (e.g. a highly visible bird), then even a short lapse in sightings would indicate extinction, but if normal observation rate is low, then even a long lapse may not necessarily indicate disappearance (Solow 2005).

Several factors influence how difficult it is to detect, observe, and monitor a species: morphology (size, camouflage, distinctiveness), behaviour (movement, habits, calls, survival strategies), biology (life cycle, activity periods), population (size, density), habitat (cover, hazards), and human access (geographically, physically).

11.5.1 Low Population Density

The lower the density in which a species inhabits its environment, the lower chance there is of detection. Group behaviour and territories will also influence detectability. Species with solitary lifestyles will tend to be present at a lower density in a given landscape. Predators will be dependent on the density of available prey. For example, the Barbary lion lived in low densities in the low productivity landscape of the Maghreb of North Africa (Black et al. 2013a), and historical accounts rarely mention more than family groups, instead of prides familiar with lions in East African savannah. Even bachelor groups of male lions known in sub-Saharan Africa and India were rarely recorded as being observed in North Africa. Small groups of lions remained undetected for decades in remote areas in northern Algeria and southern Morocco and died out just at the time when conservation science began to emerge in the 1950s and 1960s. Felids with similar solitary or small group behaviour also prove difficult to detect.

11.5.2 Impenetrable Habitats

The pygmy hog (*Porcula salvania*) is a small cryptic mammal species which also inhabits terrain in which it is extremely difficult to make observations (Fig. 11.2). Aside from camera trap glimpses of formerly captive hogs at release sites a few days after introduction, the only photographs and film footage is of captive individuals. Its grassland habitat in Assam comprises dense cover of up to 3 m in height which engulfs even the largest species including Indian rhinoceros (*Rhinoceros unicornis*) which burrow unnoticed through the undergrowth. The pygmy hog's movements through close cover means that current methods of radio telemetry do not provide remote transmitters robust enough to remain attached for tracking individuals (Deka et al. 2009).

The olm (*Proteus anguinus*), a small aquatic salamander, occupies a very different habitat, the subterranean caves in the Dinaric Alps of Central and Southeastern Europe. The species is rarely observed above ground unless flushed out by floodwaters. Observation requires resource intensive cave-diving expeditions by highly trained specialists (Šarić and Konrad 2017). Even then, population monitoring requires indirect sensing using emerging, but effective, environmental DNA techniques which analyse water samples to detect species presence (Vörös et al. 2017).

11.5.3 Remote Locations

The Arabian leopard (*Panthera pardus nimr*) population stands at about 150 animals in the wild. Researchers in Oman who have studied the species for decades rarely observe the animals which have been captured on only a few occasions for captive breeding or radio collar fitting (Spalton and al Hikmani 2014). Home range is vast, so the choice of trails for spotting sign or placing cameras must be informed by local knowledge or sightings to avoid significant wasted time and effort. The first camera trap survey recorded leopards on average every 29 days on fixed camera traps (Spalton et al. 2006). Extended in-person field observation at such sites is unrealistic.

11.5.4 Extensive Ranges

Even with accessible ranges, vast scale makes comprehensive surveys difficult. The Yangtze river dolphin (*Lipotes vexillifer*) ranged from the Three Gorges region in Hubei, China, to the estuary at Shanghai, along the main river channel and two adjoining lake systems in Hunan and Jiangxi (Turvey 2008). Although a clearly bounded habitat, this is a vast expanse stretching nearly 1700 kilometres. When numbers of dolphins decreased rapidly, the effort required to survey the river was significant. Surface observations were also hindered by the extreme volume of boat traffic, averaging one boat every 100 metres on the last survey (Turvey 2008).

On land, the cheetah (*Acinonyx jubatus*) is ostensibly an easily recognised species but is rarely encountered over most of its range in uninhabited or in seldom-visited landscapes. The discovery of a population in the Algerian Sahara occurred only relatively recently (Busby et al. 2009).

11.5.5 Combined Problematic Factors Affecting Observation

Species are difficult to observe for a combination of factors. This has implications on the logistics of surveying and monitoring. The secretive pygmy hog lives in an impenetrable grassland (Fig. 11.3) where organisation of field-based interventions requires significant logistical arrangements including a team of mahouts with trained domestic elephants to secure researchers from encounters with tigers, Indian rhinoceros, buffalo, and elephants (G. Narayan, personal communication, 2014).

The ivory-billed woodpecker (*Campephilus principalis*) is a distinctive, visible species which is nonetheless difficult to observe (Snyder et al. 2009). Collins (2017) notes that following field experiences including discovery of a relic population of these woodpeckers in Cuba in 1948, John Dennis observed: 'It takes a couple of years to search out and find the Ivorybill in only a single swamp' and 'next to impossible to obtain photographs of an Ivorybill in a southern swamp unless a nesting site is discovered'. Collins (2017) himself claimed a few fleeting sightings from 500 visits to swamps of the Pearl River in Louisiana and Choctawhatchee River, Florida, USA.

The Nihoa finch (*Telespiza ultima*) is a small bird which lives in low numbers on the uninhabited Nihoa island, Hawaii. It frequents differing habitats, some of which

Fig. 11.3 The Terai grasslands in Assam, India. Although the edge of this habitat is accessible by vehicle and presents superficially open terrain, the dense grass obscures even the largest species. Monitoring in person is near impossible, due to risk of sudden close-proximity appearance of rhino, tiger, buffalo, and elephant. (Photo S Black)

are very difficult to make observations (Gorresen et al. 2016). This provides a challenge for how often field observations can be made and the quality of those sightings in terms of error or omission in the more difficult-to-observe habitats. Datasets on the bird's population status are relatively inconsistent as different methodologies have been applied in attempts to overcome these problems. At a practical level, a way of analysing long-term longitudinal data derived from mixed methods is needed to understand species' status (Pungaliya et al. 2018).

The Sumatran rhinoceros (*Dicerorhinus sumatrensis*) is a large mammal, 3 m long, up to 1.5 m tall at the shoulder with a body mass of 2000 kg yet is notoriously elusive. Animals in mainland Myanmar were last known from tracks in the 1990s and only indirect evidence exists for the animals in peninsular Malaysia (Kawanishi et al. 2003; Magintan et al. 2010). Tiny groups on Borneo, only discovered in recent decades, may have disappeared (Pusparini et al. 2015). All populations outside Sumatra now considered extinct (Havmøller et al. 2016). Experienced biologists spending years in the field have never seen the species, so indirect evidence is the most practical way to assess presence, using knowledge of species ecology. Patrols or camera traps set along ridges, waterways, salt licks, and mineral springs have the best chance of success (Rabinowitz et al. 1995). Detection rates remain low even with extensive, overlapping camera trap grids in areas where prints are detected (Havmøller et al. 2016).

11.6 The Importance of Relic or Rediscovered Populations

While true rediscoveries of species are outstanding, more common observations include cataclysmic decline of existing populations, the discovery of previously unknown populations or re-emergence of populations previously considered extinct. In each of these situations, four issues of importance arise: (1) the need for emergency recovery interventions on the species itself, (2) the need for habitat recovery to enable the species to thrive, (3) the designation of protection of habitats for continued survival of the species, and (4) the need for legal re-designation of the species to support its protection. Successes and failures in each of these areas of response have been important points of learning for conservation science and management. The California condor (*Gymnogyps californianus*), black-footed ferret, and Po'ouli are three cases where discovery in small numbers or being known to exist at the brink of extinction did not translate to urgent conservation priorities. Last-ditch attempts to address the four points for each of these species finally arose, but only two species survived while the Po'ouli succumbed to its fate.

11.6.1 Survival in Marginal Habitat or Preference for Degraded Habitat

A number of species cling on in habitats which remain free from threats which had historically driven the species towards its precarious state. Although a change in range and behaviour has implications for future conservation management actions, it is also a factor to consider when initially investigating presence of potentially extinct species. For example, various species of forest bird in Hawaii have moved to higher altitude habitats which are clear of avian malaria (Young et al. 2018). In some instances, a species preferentially occupies a degraded habitat, such as the Mauritius fody (*Foudia rubra*) surviving in exotic conifer *Cryptomeria* rather than native habitat fragments since *Cryptomeria* offers better protection against alien introduced mammalian predators (Safford and Jones 1998). Movement towards marginal habitats or changes in behaviour should be considered before all available options for detection are concluded. Knowledge and reports by local people can be useful and should not be rejected outright even if the area or circumstances of a sighting appear unusual for the species.

The high-altitude forest of Maui was most likely a suboptimal habitat for the Po'ouli to breed productively (Porter et al. 2006), and the few nests observed failed due to local climatic effects. Had conclusions been drawn from those observations, in addition to emergency follow-up (e.g. captive breeding in more habitable circumstances), then a recovery might have been possible. More proactive work has since taken place with successful recovery of the Maui parrotbill (*Pseudonestor xanthophrys*), including captive breeding and habitat development initiatives (Becker et al. 2010; Mounce et al. 2014). New Zealand's takahe (*Porphyrio hochstetteri*) currently exists in suboptimal grassland away from its historic mosaic of forest, grasslands, and shrub which are dominated by introduced mammalian predators (Trewick and Worthy 2001). Since climate change moves suitable vegetation to higher altitude in some regions, proactive conservation includes planting native trees in elevated locations to secure future habitat.

Mediterranean monk seals (*Monachus monachus*) were once present throughout the Mediterranean and Atlantic coast of North Africa. The Atlantic populations survive, but the Mediterranean population has been extirpated except in eastern coasts and islands of Greece and Turkey (Güçlüsoy and Savaş 2003; Karamanlidis et al. 2016). Historically seals used haul-out beaches to breed and raise pups, but these locations are now dominated by human activity. Seals have shifted to secluded beaches in underground caves inaccessible to humans. Many of these locations are much less suitable for raising pups (Gucu et al. 2004), with fewer breeding sites meaning reductions in productivity and survival rate of pups. The species suffers suboptimal habitat to avoid persecution (Karamanlidis et al. 2016). Conservation based solely on these locations is unlikely to change the fortunes of the species, so different thinking is required.

The Lord Howe woodhen (*Hypotaenidia sylvestris*) retreated to high altitude cloud forests on Lord Howe Island, quite different to its historical preferred habitat, the coastal flats (Caughley 1994). The move to marginal habitat was to enable breeding without harassment from introduced invasive pigs. By understanding woodhen behaviour and movement, including occasional forays into pig-dominated areas, the conservation actions which followed (i.e. diagnose agent of decline, neutralise agent of decline, re-establish species) have since served as models for species recovery (Caughley 1994). The population is now near the carrying capacity of available habitat, mainly palm forest, at 220 individuals (Birdlife International 2016b).

11.6.2 Presence in Unexpected Locations

Sea turtles are a common sight in waters of southwestern Indian Ocean islands of Réunion, Mauritius, and Rodrigues. Female turtles are occasionally seen on beaches (Bertrand et al. 1986; Ciccione and Bourjea 2006; Ciccione et al. 2008; Fretey et al. 2013; Reyne et al. 2017), but nesting no longer occurs due to disturbance of beaches by herdsmen, fishing, and holidaymakers. Knowing the biology of turtle species, their extended life cycle, and fidelity to breeding sites, opportunities exist to re-establish nesting beaches. From a human perspective, a turtle arriving at a location many decades after previous occurrences might be unexpected, but for a species with a reproductive cycle of many decades, it is quite normal. Re-establishment of nesting relies on knowledge of visiting female turtles, so informal surveys of local people are a sensible start point for scientific study and potential future interventions that mobilise local communities.

11.6.3 Revival of Relic Populations

There is a long history of recovery of tiny surviving populations including Pere David's deer and American bison, but until the 1980s many rare relic populations were considered quirks of nature, echoes of past natural history. Some species were considered a lost cause including the Mauritius kestrel (Myers 1979) and California condor (Phillips and Nash 1981). However, in the past 30 years, there have been significant, successful, intensively managed recoveries of plants, fish, reptiles, amphibians, birds, and mammals worldwide.

The European bison, Europe's largest land animal (Fig. 11.4), was recovered from a small zoo population based on just 12 founder animals after World War I and again after World War II. There followed several phases of breeding, release, recovery, local extinction, and reintroduction. Current projects have recovered populations in Poland, Lithuania, Belarus, Russian Federation, Ukraine, and Slovakia (Olech 2008). The California Channel Islands fox (*Urocyon littoralis*), comprising six subspecies, suffered catastrophic declines but was recovered through intensive effort in 10 years by systematic identification and elimination of various threats from disease to predation, differing on each of its native islands (Coonan et al. 2010).

Fig. 11.4 The European bison (*Bison bonasus*) was once extinct in the wild but now numbers around 1800 animals (Olech 2008) following captive breeding, release, and management. (Photo: S Black)

Some species remain on the road to recovery; the California condor and European bison are managed to support feeding and breeding. The black-footed ferret, echo parakeet, and Majorcan midwife toad (*Alytes muletensis*) have shown strong recoveries yet now face significant new challenges of emerging infectious diseases, namely, canine distemper and sylvatic plague, chytrid fungus, and beak and feather disease, respectively (Clark 1997; Jones et al. 2018; Walker et al. 2008), but are in a better state than their relic founders. In some cases, recovered populations have naturally expanded their range, with movement of Asiatic lion into new subpopulations in Gujarat, India, being one example (Singh and Gibson 2011). Despite compromises to genetic health, recoveries from tiny founder groups have proven successful after elimination of threats. With human effort, relic populations can be recovered to sustainable levels.

11.6.4 Lessons from Lost Relic Populations

A species' life cycle should be considered when assessing potential survival of relic populations, as should the persistence of similar species, before assigning extinction. For example, lions (*Panthera l. leo*) had occupied Northern Africa from the Mediterranean coast to the Atlas Mountains (Black 2016), but by the time scientists took interest in a group in Rabat zoo, lions were considered locally extinct since the 1920s (Leyhausen 1975). Since then, analysis of informal historical sightings suggests that lions actually persisted in North Africa for decades longer, into the late 1950s (Black et al. 2013a; Lee et al. 2015). While the North African situation may appear academic, similar circumstances are now faced by lions in West and Central Africa. In Gabon an individual male appeared, verified by DNA analysis as belonging to a population considered extinct for 20 years (Barnett et al. 2018; Hedwig et al. 2018). Similar, unverified sightings have arisen in Ghana (Angelici and Rossi 2017). These cases demonstrate that long-lived species are resilient to extirpation and have recovery potential when previously considered closed to further action. Consideration should be given to mammalian carnivores presumed extinct or near extinct in other regions (Black et al. 2013), with implications for habitat protection and population management.

11.6.5 Challenges with Newly Discovered and Rediscovered Species

A species' discovery or rediscovery requires situation assessment involving a priori hypotheses (Black 2018; Jones et al. 2018): What is the species' status? What interventions are required? Some species are so poorly understood that justifying a best approach is difficult. The saola, an antelope-like bovid, has been rarely seen since its discovery so very little is known about its range, behaviour, breeding biology, or ecology. Despite this it became a flagship for conservation of the Annamites Ecoregion, reflecting the excitement following the 1993 discovery of this large and unusual mammal (Hardcastle et al. 2004). Other cases face negative lobbying which questions the species' very presence, placing a heavy burden of proof on scientific confirmation of a species' presence. The possible rediscovery of the ivory-billed woodpecker (Wilcove 2005; Collins 2017) ignited significant debate concerning old growth forest in North America.

11.7 Building an Understanding of Cryptic Species Status

First and foremost, a good understanding of the species can enable planned efforts to make direct observations through targeted, cost-effective fieldwork. An example is the biologically informed but straightforward investigation of the Barbados leaf-toed gecko (*Phyllodactylus pulcher*) where its suspected extinction was dispelled by making observations in predicted rock crevice locations at night-time when similar species are active (Young et al. 2018). Essentially, the principle is to look for the species in the right place at the right time.

Since threat aggregation (e.g. poaching, invasive species) and environmental degradation (e.g. mining, deforestation, infrastructure development) is likely to overtake the pace of recovery actions, there needs to be more effective detection of early signs of critical biodiversity change so that suitable interventions can be implemented (Black 2015; Schmeller et al. 2018). Direct observation of a species may be difficult, but several methodological adaptations allow biologists enough insight into the current status of that species to initiate relevant action.

11.7.1 Using Analogue Indicators of Presence for Cryptic Species

In difficult-to-access terrain and with difficult-to-observe species, indirect measures of presence are required. In an extensive ground survey of mammals in Htamanthi Wildlife Sanctuary (Rabinowitz et al. 1995), one of the most important wildlife hotspots in Myanmar, researchers detected 21 mammal species of which only 5 were through direct visual observation (domestic water buffalo, flying squirrel, Hoolock gibbon, rhesus macaque, and pig-tailed macaque). The remainder which were not observed directly included large species such as Asian elephant, leopard, tiger, gaur, sambar deer, Asiatic black bear, and wild pigs. These were only detected by tracks, hunters' kill reports, and one species, Sumatran rhino, solely by a local eyewitness account. This illustrates the practical reliance on indirect evidence to inform a basic understanding of presence of even relatively visible species, as well as those that are more cryptic.

As previously discussed, the pygmy hog (Fig. 11.2) is an example of a species almost undetectable in the wild. To date, radio telemetry technology has not been sufficiently robust nor discrete for equipment to remain fixed to the animals in the close contact habitat of the grasslands. Camera traps or video surveillance approaches are similarly ineffective for any extended period beyond the first few days of soft release in the immediate release location. Instead, an understanding of pygmy hog presence requires identification of nests in the grasslands after controlled seasonal burning activity (Deka et al. 2009; Narayan et al. 2010). Hoof-prints have some degree of relevance, although only juvenile piglets can be definitively identified by prints (due to their small size) since adult prints resemble juvenile wild boar (*Sus scrofa*) which also inhabits the area (G. Narayan, personal communication, 2014). Despite this shortfall in precision, the birth of juveniles can be detected, allowing inferences of wild population status.

11.7.2 Opportunities from New Technologies

Technology enables data collection through direct observation, photographs, sound recordings, remote camera stills, video (including night vision), spoor, faecal samples, fur, feathers, nests, and carcass swabs. A species sighting should therefore be considered in the broadest sense, including any hard record of occurrence whether specimens, photographs, or recordings (Roberts et al. 2010). Remote detection devices including camera traps, hair traps, or drones are important for cryptic species (Young et al. 2018). Ultralight aircraft allow surveys of otherwise easily disturbed species (Jean et al. 2010) and have the advantage if an expert is housed on board to take morphological details to indicate population structure.

Historically, the potential of technologies of photography and powered flight opened new opportunities for surveying species, although not necessarily in a systematic manner. The lion unexpectedly photographed in Morocco (Fig. 11.5) is an example of an accidental find, although the evidence was not used at the time. The modern equivalent situation was addressed by Borowicz et al. (2018) using remote sensing from Landsat medium-resolution satellite imagery for the detection of previously unknown Adélie penguin colonies in Antarctica. This work combined satellite imagery with ground-based field surveys and drone (i.e. unmanned aerial vehicle or UAV) camera surveys.

Most recently, high-resolution image data now available to scientists from the WorldView-3 satellite system enables body outline and flukes of whales to be compared enabling accurate species identification (Cubaynes et al. 2018). This technology, alongside methods of visual image analysis and spectral image analysis, offers the opportunity for remote census counts of wide-ranging species not easily observed by traditional boat-based surveys. Since image analyses can be time consuming, one interesting shortcut for scientists is to engage the public in this activity as part of a citizen science approach as has been used with UAV imagery on the African savanna (Rey et al. 2017). With continued technological and analytical

Fig. 11.5 An aerial photo of a lion in the Atlas Mountains in 1925, several years after lions were last seen north of the Sahara. (Photo: M. Flandrin). This was taken during one of the first exploratory air flights across Morocco. Despite widespread use in a postcard, this evidence was not examined to inform species presence until recently (Black et al. 2013a). Note that the object jutting out immediately right of the lion's head is a feature on the terrain, not part of the animal

developments, satellite surveys could potentially apply to population counts of other species which predictably congregate in difficult-to-access locations such as seals or albatrosses.

Technologies of miniaturisation and digitisation have opened new avenues of data collection including the use of drones, radio telemetry, and analysis of satellite imagery for species presence (Pimm et al. 2015) as well as concealed in situ cameras, night vision equipment, audio recording, and associated electronic analysis.

11.7.3 Using Local Knowledge and Eyewitness Accounts

Local people can be a useful source of information on species distinctiveness, presence, status, behaviour, and ecology (Cozzuol et al. 2014; Young et al. 2018). Concern that some informants unconsciously exaggerate (e.g. to please the interviewer or for perceived reward) and an increased tendency in some socioeconomic groups for overestimation (Lunn and Dearden 2006) is one reason why, historically, science has a poor record in considering local accounts. However, serious studies of species survival beyond declared dates of extinction repeatedly identify the importance of respecting sightings by local people, including presence of Carolina parakeets (Snyder 2004) and Barbary lion (Black et al. 2013a) continuing many decades later than previously recognised by science. The Chacoan peccary (*Catagonus wagneri*) was well-known to local people but only later did scientists recognise the species as one from the fossil record (Wetzel et al. 1975). Similarly, the saola was well-enough known locally that it had a specific name, yet this spindle-horned bovid remained undiscovered in Vietnam until the 1990s (Dung et al. 1993; Schaller and Rabinowitz 1995).

Current knowledge of some isolated subpopulations of the elusive Arabian leopard (*Panthera pardus nimr*) has been prompted entirely by local sighting reports in regions not previously considered within the range of the species (Fig. 11.6). Arabian leopards have even been casually spotted by tourists in the Negev desert, Israel, in the 1990s (J. Fitchett, personal communication, 2015) and informally reported to a scientist without that eyewitness ever knowing a wild and very rarely seen relic population of these felids existed (Perez et al. 2006). Sightings of leopards in North Africa in 2007 (Jdeidi et al. 2010) are evidence of other relic felid populations.

Community studies are likely to be useful to cover large geographic areas or long time periods. Surveys of large mammals in Zimbabwe followed a systematic research design involving a large body of local volunteers (Dunham and du Toit 2013). In a more informal process, on-going assessments of cheetah presence in

Fig. 11.6 Arabian leopard caught on camera trap in the remote Dhofar Mountains. Presence of this rarely observed felid is often indicated by reports from camel herders and farmers. Knowledge grows with examination of paw prints, scrapes, faeces, kills, images, coat pattern analysis, and DNA from faecal samples (Photo: H. Al Hikmani)

remote landscapes that are rarely visited by researchers, ask local people and tourists to upload sightings using mobile phone apps (Jedersberger et al. 2018). There is a growing discussion across the scientific community about the effectiveness of these types of 'citizen science' approaches including involvement of people with limited formal education (Danielsen et al. 2014). For example, emerging evidence suggests that under some circumstances non-professional volunteers may perform as well as experts in field identification of species (Austen et al. 2016) enabling increased capacity in high volume identification activity such as camera trap analysis without excessive vulnerability to error (Gibbon et al. 2015). Now that remote sensing technologies (such as camera traps, non-invasive DNA faecal sampling techniques, drone surveys) can utilise remotely captured data, local community knowledge becomes even more pertinent. Scientists who engage closely with community members can more easily identify sensible localities to set cameras or collect samples in a timely and resource-efficient manner, or local people can be directly involved in data collection.

11.7.4 Expert Opinion

Expert judgements are helpful when resources are stretched or a fast response is required (Burgman et al. 2011). Experts routinely inform species listings under the IUCN Red List criteria (McBride et al. 2012). While the 'Thylacine Effect' suggests caution with non-expert views, similar caution should apply to expert opinion. Bias in expert opinion is present in many forms (McBride et al. 2012) both in individual and group decision-making. Methods to calibrate opinion and ensure informativeness are necessary (Martin et al. 2012b). McBride et al. (2012) used calibrating 'test' questions to assess accuracy in IUCN Red Listing panellists. Expert discussion processes allow consensus to be achieved more effectively, although there is a trade-off in time required, whether conducted face to face or not.

Several methods of consensus assessment of temporal datasets of sighting observations have been developed. For example, Lee et al. (2015) used two parallel approaches to ask experts about sighting reliability, when those experts examined reports from a variety of informants. First, they asked experts for a probability that the sighting is true. Second, they asked the experts about three distinct factors which relate to sighting reliability: distinguishability, observer competence, and verifiability. These tests allowed some verification of survival or extinction dates as defined by other linear methods.

11.7.5 Sequential, Opportunistic Evidence Gathering

As the Theory of Knowledge suggests, with cryptic species, scientific understanding may follow a pathway from rumour, belief, and supposition, through to physical evidence. In the case of the saola, initial reports by local people were followed by body parts, photographs, and later live animals to provide more concrete scientific evidence of the species. In a different context, the Arabian leopard, despite being a known, large carnivore, with consistent movements and scent marking habits, is an animal that is wary of humans and covers a vast, remote range. This requires multiple methods to understand population status (Mazzolli et al. 2017) including prints, scrapes, faeces, and direct observation. Reports of leopards in unexpected localities prompt follow-up data collection action to build knowledge.

In this sequence, scientific recognition changes when physical evidence is collected as was the case in 1981 with the black-footed ferret (Wisely et al. 2008). The first step in the rediscovery of the black-footed ferret was wholly unscientific (Goodall et al. 2010). It was the merest interest shown by the wife of a local rancher after comments by her husband who had discarded a dead animal killed by their dog. She recovered the body and showed it to a taxidermist who had enough knowledge to identify the species. This triggered a systematic search by wildlife professionals. The sequence of knowledge had not run smooth for this species. The 1981 finding was, essentially, third-time lucky, 2 years after a failed breeding programme had ended with the last animals dying in captivity (Clark 1997). Most other species will not have the same benefit of chance.

11.8 Analysis of Species Status from Available Data

The nature of hard-to-detect species makes it difficult to conduct repeatable systematic surveys within normal economic and operational constraints. Population levels often need to be estimated using occupancy or distance sampling frameworks to establish likely population counts from otherwise incomplete observations. Examples of methods involving sampling observations, line transects, and point counts of species of lemur, rodent, and owl in difficult-to-access habitats are discussed by Young et al. (2018).

The important lesson from these analytical studies is to establish longitudinal datasets. This requires conservation professionals to get into the habit of taking notes on informal sightings, dates, locations, and other context (e.g. number of individuals, behaviour), as well as who and how many people made the sighting, and the type of sighting, whether visual, aural, recorded (film, photograph, sound), physical capture, artefact (body, body part, hair, feathers, faeces, nests), tracks, or other marks.

11.8.1 Modelling Based on Habitat Suitability

In extreme cases where observation is considered unlikely, various modelling procedures which integrate biophysical principles with remote sensing technologies have been applied to identify potential areas where a species might be sustained, but remains unobserved (Porter et al. 2006). In the case of the Po'ouli, last seen as a captive specimen that died in 2004, analysis enabled insight into the bird's likely diet, distribution, and levels of environmental resources necessary for survival, growth, and reproduction, alongside other variables such as the presence of pathogens (Porter et al. 2006). The only difficulty with this approach is the relative resource-heavy and analytically intensive nature of the work. In the Po'ouli's case, the publication of research papers some years after the species had been declared most probably extinct did little to assist its conservation. Nevertheless, when used in a timely fashion, these methods offer much and are enabled by increasingly cost-effective analytical computing power.

11.8.2 Mathematical Modelling of Sightings, Survival, and Extinction

Mathematical models for assessing sightings, taking account of imprecise or uncertain records, can identify probable extinction dates. Solow (2005) used the sparse information from just five observations since 1915 of the Caribbean monk seal (*Monachus tropicalis*) to pinpoint likely time of extinction (1961, 1956, and 1964 under differing treatments). Estimations have been made for a variety of considered-extinct species, so this method could be useful for confirming extinctions in controversial cases such as the ivory-billed woodpecker (Solow et al. 2006; Elphick et al. 2010).

11.8.3 Using 'Messy' Longitudinal Datasets to Understand Species Status

While traditional understanding of species status based upon the Red List categorisation remains a useful tool, interventions even at range-wide scales are usually dependent on nuanced approaches in differing contexts relating to landscapes, human activity, threats, presence of other species, and variability in habitats.

Conservation science needs to develop a better understanding of these local situations. Existing species status assessments such as the IUCN Red List Index and the Living Planet Index are conducted on any one species at relatively long intervals, so inherently they function as retrospective late-warning indicators (Schmeller et al. 2018). Important sources of information when considering likely extinction include

(1) threat status relating to species susceptibility, (2) longitudinal (time-series) records of species presence (e.g. sightings, specimens, documented accounts), and (3) adequacy and extent of surveys for the species (Keith et al. 2017).

The extent of these sources in many instances is often fragmented, yet the need to make decisions is no less imperative. Conservation science requires better *early* detection of signs of critical biodiversity change to support proactive management intervention, and new, straightforward analytical methods are now being evaluated under the umbrella term 'Systems Behaviour Charts' or SBC (Black 2015; Black and Leslie 2018).

Several graphical treatments have been developed to support analysis of presence data to identify stability, decline, population size, and vulnerability (Black 2015). These treatments have the advantage of allowing consideration of datasets of observations of mixed quality and veracity, since exceptional changes in the data can be identified and cross-checked with circumstances of observation. SBC analyses use 'natural limits' (e.g. mean +/− 3SD, or better perhaps, mean +/- 3σ, see Black and Leslie 2018) derived from empirical metrics such as population size to provide early warning of vulnerability, decline, imminent population collapse, and risk of extinction (Fig. 11.7). SBCs also indicate biases in the data ahead of decision-making, planning, and intervention. While more research is needed to develop these techniques, the SBC approach has been applied to various taxa such as to present evidence of decline in mammal populations (Stringell et al. 2013; Black 2015; Leslie et al. 2017) and the threatened population status of endangered birds (Pungaliya and Black 2017; Pungaliya et al. 2018).

11.9 Conclusion

The doomsday narrative of the Anthropocene has become familiar: biodiversity decline, extinction, and the ravages of humanity's dominant influence on climate and the environment (Swaisgood and Sheppard 2010). Scientists and wildlife professionals need, however, to revisit this narrative, placing conservation interventions, in whatever form, at the heart of a not-so-inevitable decline. Positive experiences of rediscovery and recovery and an ambition for what is yet possible must provide hope and motivation for current and future generations. In some areas of work, this has been attained. Reversals in decline have been successfully achieved in reforestation efforts in places as diverse as Africa, the Indian Ocean, Southeast Asia, and South America. Species recovery efforts have seen remarkable turnarounds in ecosystems in Australasia, continental USA, mainland Europe, South and Southeast Asia, islands of the Indian Ocean, the Pacific, Caribbean, and Mediterranean; the list is vast.

The same degree of restoration must be achieved in contexts where people live alongside wildlife. We need expectations that enable a balance for ecosystems that

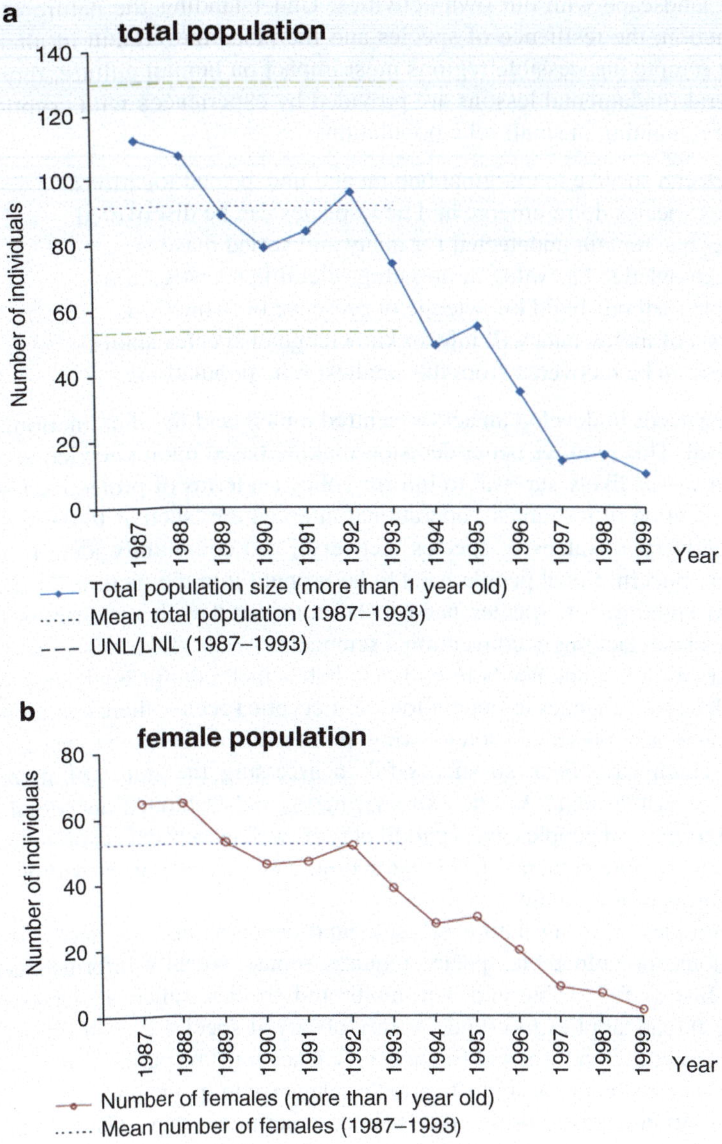

Fig. 11.7 SBCs signal a population crash in ground squirrels (*Spermophilus brunneus*) by 1994–1996 from data falling below (**a**) calculated limit (LNL) from population counts and (**b**) mean count of females (Black 2015). Without this method available, sadly, no action was taken and the population's final extinction in the winter of 1999–2000 caught scientists who were monitoring the colony 'off guard' (Sherman and Runge 2002)

share the landscape with our own activities. Understanding the nature of survival and extinction, the resilience of species and the place they retain in altered landscapes or remote inaccessible regions must impact on human culture, thought, and life. Several fundamental lessons are provided by experiences with cryptic species and those remaining in small relic populations:

- Species can survive in marginal habitat and unexpected locations.
- Lazarus species do re-emerge and new species can be discovered.
- Species can remain undetected for many and varied reasons.
- Local knowledge has value in initiating scientific investigation.
- Multiple methods build knowledge of presence or extinction.
- Analysis of messy data will inform knowledge of species status.
- Species can be recovered from the smallest relic populations.

Science needs to develop an action-centred understanding of extinction, decline, and survival. This requires better decision-making based upon knowledge of actual extinction versus likely survival to inform policy (in terms of protective legislation and prioritisation of resources), operational intervention (such as threat mitigation, habitat recovery, or intensive species recovery), and community identity with the species of concern. Local people need to be brought into the picture. Their experiences and sightings of, species need to be considered with seriousness that has sometimes been lacking in conventional science.

Conservation science needs to embrace better methods for analysing available data for detecting changes in populations. For cryptic species, there cannot be a sole reliance on typical presence surveys using 'point assessments' and transect methodologies, which have been so successful in assessing the status of more easily accessed or visible taxa. At one extreme, messy old-fashioned anecdotal reports collected from local people have a part to play, as well as new data capture technologies such as remote cameras, DNA extraction, and sequencing, providing support for new forms of longitudinal analyses.

The complex, shifting nature of ecological concerns and the need for speedy interventions for vulnerable species requires robust, soundly informed decision-making. Just as for visible and well-monitored species, timely understanding of emerging threats applies to cryptic, rarely observed species even if observational data is unstructured in form and frequency. Science must be purposeful in its analysis rather than focusing on sophistication or elegance in its method.

Conservation needs to bridge the 'science-practitioner gap' (Game et al. 2013) and utilise new methods of 'boundary science' (Cook et al. 2013) to inform action with best available knowledge. This includes taking new opportunities to engage with human communities that coexist with wildlife. This will give us a better chance to share the planet with the species and ecosystems that are so often negatively affected by our presence. If we learn to work within the functioning ecosystems of which we are part, rather than attempting to control or replace them, we have a better chance of a sustainable future. If we understand how some species cling on to survival, and how to pull them back from the brink of extinction, we will better understand our own place in avoiding biodiversity loss on this planet.

References

Akçakaya HR, Keith DA, Burgman M, Butchart SHM, Hoffmann M, Regan HM, Harrison I, Boakes E (2017) Inferring extinctions III: a cost-benefit framework for listing extinct species. Biol Conserv 214:336–342

Angelici FM, Rossi L (2017) Further lion, *Panthera leo senegalensis* Meyer, 1826, sightings in Mole National Park, Ghana, and possible first serval *Leptailurus serval* Schreber, 1776 record after 39 years (Mammalia Felidae). Biodivers J 8(2):749–752

Austen GE, Bindemann M, Griffiths RA, Roberts DL (2016) Species identification by experts and non-experts: comparing images from field guides. Sci Rep 6:33634

Barnett R, Sinding MHS, Vieira FG et al (2018) No longer locally extinct? Tracing the origins of a lion (*Panthera leo*) living in Gabon. Conserv Genet 19:611. https://doi.org/10.1007/s10592-017-1039-2

Becker CD, Mounce HL, Rassmussen TA et al (2010) Nest success and parental investment in the critically endangered Maui Parrotbill Pseudonestor xanthophrys with implications for recovery. Endanger Species Res 11(3):189–194

Bertrand J, Bonnet B, Lebrun G (1986) Nesting attempts of Chelonia Mydas at Reunion Island (S. W. Indian Ocean). Mar Turt Newsl 39:3–4. http://www.seaturtle.org/mtn/archives/mtn39/mtn39p3.shtml. Accessed 10 Feb 2018

BirdLife International (2016a) Falco punctatus. IUCN Red List of Threatened Species 2016: e.T22696373A93557909

BirdLife International (2016b) Hypotaenidia sylvestris. IUCN Red List of Threatened Species 2016: e.T22692395A93352096

Biton R, Geffen E, Vences M, Cohen O, Bailon S, Rabinovich R., Malka Y, Oron T, Boistel R, Brumfeld V, Gafny S (2013) The rediscovered Hula painted frog is a living fossil. Nature Communications 4(1):1–6

Black SA (2015) System behaviour charts inform an understanding of biodiversity recovery. Int J Ecol 2015:1. https://doi.org/10.1155/2015/787925

Black SA (2016) The challenges and relevance of exploring the genetics of North Africa's "Barbary lion" and the conservation of putative descendants in captivity. Int J Evol Biol 2016:1. https://doi.org/10.1155/2016/6901892

Black SA (2018) Planning for recovery. In: Copsey JA et al (eds) Species conservation: lessons from islands. Cambridge University Press, Cambridge, UK, pp 174–175

Black SA, Copsey JA (2014) Purpose, process, knowledge, and dignity in interdisciplinary projects. Conserv Biol 28(5):1139–1141

Black SA, Leslie SC (2018) Understanding impacts of mitigation in waterway control systems on manatee deaths in Florida. IJAWB 3(5):386–390. https://doi.org/10.15406/ijawb.2018.03.00124

Black SA, Fellous A, Yamaguchi N et al (2013a) Examining the extinction of the Barbary Lion and its implications for felid conservation. PLoS One 8(4):e60174. https://doi.org/10.1371/journal.pone.0060174

Black SA, Groombridge JJ, Jones CG (2013b) Using better management thinking to improve conservation effectiveness. ISRN Biodivers 2013:1. https://doi.org/10.1155/2013/784701

Borowicz A, McDowall P, Youngflesh C et al (2018) Multi-modal survey of Adélie penguin mega-colonies reveals the Danger Islands as a seabird hotspot. Sci Rep 8:3926. https://doi.org/10.1038/s41598-018-22313-w

Brassine E, Parker D (2015) Trapping elusive cats: using intensive camera trapping to estimate the density of a rare African felid. PLoS One 10(12):e0142508. https://doi.org/10.1371/journal.pone.014250

Brito JC, Martínez-Freiría F, Sierra P et al (2011) Crocodiles in the Sahara Desert: an update of distribution, habitats and population status for conservation planning in Mauritania. PLoS One 6(2):e14734. https://doi.org/10.1371/journal.pone.0014734

Burgman MA, McBride M, Ashton R et al (2011) Expert status and performance. PLOS One 6(7):e22998. https://doi.org/10.1371/journal.pone.0022998

Busby GB, Gottelli D, Wacher T et al (2009) Genetic analysis of scat reveals leopard Panthera pardus and cheetah Acinonyx jubatus in southern Algeria. Oryx 43(3):412–415
Cade TJ, Burnham W (2003) Return of the peregrine: a North American saga of tenacity and teamwork. The Peregrine Fund, Boise, p 394
Can OE (2004) Status, conservation and management of large carnivores in Turkey. Convention on the conservation of European wildlife and natural habitats, standing committee meeting 24, 29th November–3rd December, TPVS/Inf 8. Council of Europe, Strasbourg
Caughley G (1994) Directions in conservation biology. J Anim Ecol 63(2):215–244
Chestin IE, Paltsyn MY, Pereladova OB, Iegorova LV, Gibbs JP (2017) Tiger re-establishment potential to former Caspian tiger (*Panthera tigris virgata*) range in Central Asia. Biol Conserv 205:42–51
Ciccione S, Bourjea J (2006) Nesting of green turtles in Saint Leu, Reunion Island. Mar Turt Newsl 112:1–3. http://www.seaturtle.org/mtn/archives/mtn39/mtn39p3.shtml. Accessed 10 Feb 2018
Ciccione S, Lauret-Stepler M, Bourjea J (2008) Marine turtle nest translocation due to hurricane threat on Reunion Island. Mar Turt Newsl 119:6–8. http://www.seaturtle.org/mtn/archives/mtn39/mtn39p3.shtml. Accessed 10 Mar 2018
Clark TW (1997) Averting extinction: reconstructing endangered species recovery. Yale University Press, New Haven, p 23
Collar NJ (1998) Extinction by assumption; or, the Romeo Error on Cebu. Oryx 32:239–244
Collen B, Turvey ST (2009) Probabilistic methods for determining extinction chronologies. In: Turvey ST (ed) Holocene extinctions. Oxford University Press, Oxford, pp 181–191
Collins MD (2017) Video evidence and other information relevant to the conservation of the Ivory-billed Woodpecker (Campephilus principalis). Heliyon 3(1):e00230. https://doi.org/10.1016/j.heliyon.2017.e00230
Cook CN, Mascia MB, Schwartz MW et al (2013) Achieving conservation science that bridges the knowledge-action boundary. Conserv Biol 27(4):669–678
Coonan TJ, Schwemm CA, Garcelon DK (2010) Decline and recovery of the island fox: a case study for population recovery. Cambridge University Press, Cambridge, UK
Cozzuol MA, De Thoisy B, Fernandes-Ferreira H, Rodrigues FH, Santos FR (2014) How much evidence is enough evidence for a new species? J Mammal 95(4):899–905
Cubaynes HC, Fretwell PT, Bamford C, Gerrish L, Jackson JA (2018) Whales from space: four mysticete species described using new VHR satellite imagery. Mar Mamm Sci. https://doi.org/10.1111/mms.12544
Danielsen F, Topp-Jørgensen E, Levermann N et al (2014) Counting what counts: using local knowledge to improve Arctic resource management. Polar Geogr 37(1):69–91
Dawson MR, Marivaux L, Li CK et al (2006) Laonastes and the "Lazarus effect" in recent mammals. Science 311:1456–1458
Deka PJ, Narayan G, Oliver WL et al (2009) Reintroduced pygmy hogs (Porcula salvania) thrive a year after release of more hogs released in Sonai Rupai Wildlife Sanctuary, Assam, India. Suiform Soundings 9(1):23–28
Driscoll CA, Yamaguchi N, Bar-Gal GK, Roca AL, Luo S, Macdonald DW, O'Brien SJ (2009) Mitochondrial phylogeography illuminates the origin of the extinct Caspian tiger and its relationship to the Amur tiger. PLoS One 4(1):e4125
Dung VV, Giao PM, Chinh NN et al (1993) A new species of living bovid from Vietnam. Nature 363:443–445
Dunham KM, du Toit AJ (2013) Using citizen-based survey data to determine densities of large mammals: a case study from Mana Pools National Park, Zimbabwe. Afr J Ecol 51(3):431–440
Elphick CS, Roberts DL, Reed JM (2010) Identifying anomalous reports of putatively extinct species and why it matters. Biol Conserv 143:617–624
Erwin DH, Droser ML (1993) Elvis taxa. PALAIOS 8(6):623–624
Foster KR, Vecchia P, Repacholi MH (2000) Science and the precautionary principle. Science 288(5468):979–981

Fretey J, Frétey T, Dupré A et al (2013) Is the Hawksbill Turtle, Eretmochelys imbricata, a Regular Nester on Rodrigues Island (Indian Ocean)? Mar Turt Newsl (138)22. http://www.seaturtle.org/mtn/archives/mtn138/mtn138-8.shtml. Accessed 10 Feb 2018

Game ET, Meijaard E, Sheil D et al (2013) Conservation in a wicked complex world; challenges and solutions. Conserv Lett 7:271–277

Geissmann T, Lwin N, Aung SS, Aung ZM, Aung ZM, Hla TH, Grindley M, Momberg F (2011) A new species of snub-nosed monkey, Genus Rhinopithecus Milne-Edwards, 1872 (Primates, Colobinae), from northern Kachin State, northeastern Myanmar. Am J Primatol 73(1):96–107

Gibbon GE, Bindemann M, Roberts DL (2015) Factors affecting the identification of individual mountain bongo antelope. PeerJ 3:e1303

Godfrey-Smith P (2003) Theory and reality: an introduction to the philosophy of science. University of Chicago Press, Chicago

Goodall J, Maynard T, Hudson G (2010) Hope for animals and their world: how endangered species are being saved from the brink. Icon Books, London

Gorresen PM, Brinck KW, Camp RJ et al (2016) State-space modeling of population sizes and trends in Nihoa Finch and Miller bird. Condor 118(3):542–557

Groombridge JJ, Massey JG, Bruch JC et al (2004) An attempt to recover the Po'ouli by translocation and an appraisal of recovery strategy for bird species of extreme rarity. Biol Conserv 118(3):365–375

Groombridge JJ, Dawson DA, Burke T et al (2009) Evaluating the demographic history of the Seychelles kestrel (Falco araea): genetic evidence for recovery from a population bottleneck following minimal conservation management. Biol Conserv 142(10):2250–2257

Groves CP, Schaller GB, Amato G et al (1997) Rediscovery of the wild pig Sus bucculentus. Nature 386(6623):335

Güçlüsoy H, Savaş Y (2003) Status of the Mediterranean monk seal, Monachus monachus, in the Foça pilot monk seal conservation area, Turkey. Zool Middle East 28(1):5–16

Gucu AC, Gucu G, Orek H (2004) Habitat use and preliminary demographic evaluation of the critically endangered Mediterranean monk seal (Monachus monachus) in the Cilician Basin (Eastern Mediterranean). Biol Conserv 116(3):417–431

Hardcastle, J, Cox S, Dao NT et al (2004) Rediscovering the Saola. Proceedings of the 'rediscovering the saola – a status review and conservation planning workshop. Pu Mat National Park, Vietnam 27–28 Feb 2004. WWF Indochina Programme, Hanoi

Havmøller RG, Payne J, Ramono W et al (2016) Will current conservation responses save the critically endangered Sumatran rhinoceros Dicerorhinus sumatrensis? Oryx 50(2):355–359

Hedwig D, Kienast I, Bonnet M et al (2018) A camera trap assessment of the forest mammal community within the transitional savanna-forest mosaic of the Batéké Plateau National Park, Gabon. Afr J Ecol 56:777. https://doi.org/10.1111/aje.12497

IUCN (2017) Guidelines for using the IUCN red list categories and criteria. Version 13. Standards and Petitions Subcommittee. http://www.iucnredlist.org/documents/RedListGuidelines.pdf

Jdeidi T, Masseti M, Nader I et al (2010) Panthera pardus. The IUCN Red List of Threatened Species 2010: e.T15954A5328595

Jean C, Ciccione S, Ballorain K et al (2010) Ultralight aircraft surveys reveal marine turtle population increases along the west coast of Reunion Island. Oryx 44(2):223–229

Jedersberger J, Stern C, Wykstra M et al (2018) Supporting the monitoring of Cheetahs in Kenya by Mobile Map Apps. The Annual International Conference on Geographic Information Science. Springer, Cham, June pp 169–188. https://doi.org/10.1007/978-3-319-78208-9_9

Jones CG, Lawton JH, Shachak M (1994) Organisms as ecosystem engineers. Oikos 69(3):373–386

Jones CG, Heck W, Lewis RE et al (1995) The restoration of the Mauritius kestrel Falco punctatus population. Ibis 137(s1):S173–S180

Jones CJ, Cole N, Canessa S et al (2018) Journey to recovery. In: Copsey JA et al (eds) Species conservation: lessons from islands. Cambridge University Press, Cambridge, UK, pp 255–290

Karamanlidis AA, Dendrinos P, Larrinoa PF et al (2016) The Mediterranean monk seal Monachus monachus: status, biology, threats, and conservation priorities. Mammal Rev 46(2):92–105

Kawanishi K, Sunquist M, Sahak AM (2003) Rarity and possible new records of Sumatran rhinoceros in Taman Negara. J Wildl Parks 20:125–128

Keith DA, Butchart SH, Regan HM, Harrison I, Akçakaya HR, Solow AR, Burgman MA (2017) Inferring extinctions I: a structured method using information on threats. Biol Conserv 214:320–327

Lee TE, Black SA, Fellous A et al (2015) Assessing uncertainty in sighting records: an example of the Barbary lion. PeerJ 3:e1224. https://doi.org/10.7717/peerj.1224

Leslie SC, Blackett FC, Stalio M et al (2017) Systems behaviour charts of longitudinal data inform effective management in marine conservation. J Aquac Mar Biol 6(5):00171. https://doi.org/10.15406/jamb.2017.06.00171

Lewis CI (1956) Mind and the world order: outline of a theory of knowledge. Dover Publications, New York

Leyhausen P (1975) Preliminary report on the possibility of a breeding programme for the Atlas Lion at Temara Zoo, Morocco. Int Zoo News 21:22–23

Lunn K, Dearden P (2006) Monitoring small-scale marine fisheries: an example from Thailand's KoChang Archipelago. Fish Res 77:60–71

MacLeod CD, Perrin WF, Pitman R et al (2005) Known and inferred distributions of beaked whale species (Cetacea: Ziphiidae). J Cetacean Res Manag 7(3):271–286

MacPhee RDE, Flemming C (1999) Requiem aeternam: the last five hundred years of mammalian species extinctions. In: MacPhee RDE (ed) Extinctions in neartime: causes, contexts, and consequences. Kluwer Academic/Plenum, New York, pp 333–371

Magintan D, Rufino MBM, Cosmas N et al (2010) Some evidence of Sumatran rhinoceros presence in Temengor Forest Reserve, Perak. J Wildl Parks 26:5–10

Martin TG, Nally S, Burbidge AA et al (2012a) Acting fast helps avoid extinction. Conserv Lett 5(4):274–280

Martin TG, Burgman MA, Fidler F et al (2012b) Eliciting expert knowledge in conservation science. Conserv Biol 26(1):29–38

Mazzolli M, Haag T, Lippert BG et al (2017) Multiple methods increase detection of large and medium-sized mammals: working with volunteers in South-Eastern Oman. Oryx 51(2):290–297

McBride MF, Garnett ST, Szabo JK et al (2012) Structured elicitation of expert judgments for threatened species assessment: a case study on a continental scale using email. Methods Ecol Evol 3(5):906–920

McDougall A, Porter G, Mostert M et al (2009) Another piece in an Australian ornithological puzzle–a second Night Parrot is found dead in Queensland. Emu 109(3):198–203

McKelvey KS, Aubry KB, Schwartz MK (2008) Using anecdotal occurrence data for rare or elusive species: the illusion of reality and a call for evidentiary standards. Bioscience 58(6):549–555

Meijaard E, Nijman V (2014) Secrecy considerations for conserving Lazarus species. Biol Conserv 175:21–24

Morrison ML (2002) Wildlife restoration: techniques for habitat analysis and animal monitoring, vol 1. Island Press, Washington.

Mounce HL, Iknayan KJ, Leonard DL et al (2014) Management implications derived from long term re-sight data: annual survival of the Maui Parrotbill Pseudonestor xanthophrys. Bird Conserv Int 24(3):316–326

Myers N (1979) The sinking ark. Collins, London

Narayan G, Deka PJ, Oliver WLR, Fa JE (2010) Conservation breeding and re-introduction of the pygmy hog in N.W. Assam, India. In: Soorae PS (ed) Global re-introduction perspectives: additional case-studies from around the globe. IUCN/SSC Re-introduction Specialist Group, Abu Dhabi, pp 290–296

Nater A, Mattle-Greminger MP, Nurcahyo A et al (2017) Morphometric, behavioral, and genomic evidence for a new Orangutan species. Curr Biol 27(22):3487–3498

Nogales M, Rando JC, Valido A, Martín A (2001) Discovery of a living giant lizard, genus Gallotia (Reptilia: Lacertidae), from La Gomera, Canary Islands. Herpetologica 57(2):169–179

Noll RG (1980) The game of health care regulation. In: Gordon RS (ed) Issues in health care regulation. McGraw-Hill, New York, p 136
Olech W (2008) Bison bonasus. IUCN Red List of Threatened Species 2008. e.T2814A9484719
Perez I, Geffen E, Mokady O (2006) Critically endangered Arabian leopards Panthera pardus nimr in Israel: estimating population parameters using molecular scatology. Oryx 40(3): 295–301
Phillips D, Nash H (1981) The condor question: captive or forever free? Friends of the Earth, San Francisco
Pimm SL, Alibhai S, Bergl R et al (2015) Emerging technologies to conserve biodiversity. Trends Ecol Evol 30:685–696
Porter WP, Vakharia N, Klousie WD et al (2006) Po'ouli landscape bioinformatics models predict energetics, behavior, diets, and distribution on Maui. Integr Comp Biol 46(6):1143–1158
Pungaliya AV, Black SA (2017) Insights into the recovery of the Palila (Loxioides bailleui) on Hawaii through use of systems behaviour charts. Int J Avian & Wildlife Biol 2(1). https://doi.org/10.15406/ijawb.2017.02.00007
Pungaliya AV, Leslie SC, Black SA (2018) Why stable populations of conserved bird species may still be considered vulnerable: the Nihoa finch (Telespiza ultima) as a case study. Int J Avian & Wildlife Biol 3(2). https://doi.org/10.15406/ijawb.2018.03.00050
Pusparini W, Sievert PR, Fuller TK et al (2015) Rhinos in the parks: an island-wide survey of the last wild population of the Sumatran rhinoceros. PLoS One 10(9):e0136643. https://doi.org/10.1371/journal.pone.0136643
Rabinowitz A, Schaller GB, Uga U (1995) A survey to assess the status of Sumatran rhinoceros and other large mammal species in Tamanthi Wildlife Sanctuary, Myanmar. Oryx 29(2):123–128
Rey N, Volpi M, Joost S, Tuia D (2017) Detecting animals in African Savanna with UAVs and the crowds. Remote Sens Environ 200:341–351
Reyne M, Webster I, Huggins A (2017) A preliminary study on the sea turtle density in Mauritius. Mar Turt Newsl 152:5. http://www.seaturtle.org/mtn/archives/mtn152/mtn152-2.shtml. Accessed 10 Feb 2018
Roberts DL, Elphick CS, Reed JM (2010) Identifying anomalous reports of putatively extinct species and why it matters. Conserv Biol 24(1):189–196
Rossi L, Gippoliti S, Angelici FM (2018) The role of indirect evidence and traditional ecological knowledge in the discovery and description of new ape and monkey species since 1980. Primates 59(4): 327–337
Safford RJ, Jones CG (1998) Strategies for land-bird conservation on Mauritius. Conserv Biol 12(1):169–176
Schaller GB, Rabinowitz A (1995) The saola or spindlehorn bovid Pseudoryx nghetinhensis in Laos. Oryx 29(2):107–114
Scheffers BR, Yong DL, Harris JBC et al (2011) The world's rediscovered species: back from the brink? PLoS One 6(7):e22531. https://doi.org/10.1371/journal.pone.0022531
Schmeller DS, Weatherdon LV, Loyau A et al (2018) A suite of essential biodiversity variables for detecting critical biodiversity change. Biol Rev 93(1):55–71
Sherman PW, Runge MC (2002) Demography of a population collapse: the northern Idaho ground squirrel (Spermophilus brunneus brunneus). Ecology 83(10):2816–2831
Singh HS, Gibson L (2011) A conservation success story in the otherwise dire megafauna extinction crisis: the Asiatic lion (Panthera leo persica) of Gir forest. Biol Conserv 144(2011):1753–1757
Snyder NFR (2004) The Carolina parakeet: glimpses of a vanished bird. Princeton University Press, Princeton
Snyder NFR, Brown DE, Clark KB (2009) The travails of two woodpeckers: ivory-bills and imperials. University of New Mexico Press, Albuquerque
Solow AR (2005) Inferring extinction from a sighting record. Math Biosci 195(1):47–55. https://doi.org/10.1016/j.mbs.2005.02.001
Solow AR, Kitchener AC, Roberts DL et al (2006) Rediscovery of the Scottish polecat, Mustela putorius: survival or reintroduction? Biol Conserv 128:574–575
Spalton JA, al Hikmani HM (2014) The Arabian leopards of Oman. Stacey Publishing, London

Spalton JA, al Hikmani HM, Willis D et al (2006) Critically Endangered Arabian leopards Panthera pardus nimr persist in the Jabal Samhan Nature Reserve, Sultanate of Oman. Oryx 40(3):287–294

Stringell TB, Bamber RN, Burton M et al (2013) A tool for protected area management: multivariate control charts 'cope' with rare variable communities. Ecol Evol 3(6):1667–1676

Swaisgood RR, Sheppard JK (2010) The culture of conservation biologists: show me the hope! Bioscience 60(8):626–630

Šarić KK, Konrad PK (2017) The deepest finding of an olm (Proteus anguinus): Zagorska peć, Ogulin, Croatia. Acta Carsologica 46(2–3):347–351

Trewick SA, Worthy TH (2001) Origins and prehistoric ecology of Takahe based on morphometric, molecular, and fossil data. In: Lee WG, Jamieson IG (eds) The takahe. University of Otago Press, Dunedin, pp 31–48

Turvey S (2008) Witness to extinction: how we failed to save the Yangtze river dolphin. Oxford University Press, Oxford, UK

VanderWerf EA, Groombridge JJ, Fretz JS et al (2006) Decision analysis to guide recovery of the Po'ouli, a critically endangered Hawaiian honeycreeper. Biol Conserv 129(3):383–392

Vörös J, Márton O, Schmidt BR et al (2017) Surveying Europe's only cave-dwelling chordate species (Proteus anguinus) using environmental DNA. PLoS One 12(1):e0170945. https://doi.org/10.1371/journal.pone.0170945

Walker SF, Bosch J, James TY et al (2008) Invasive pathogens threaten species recovery programs. Curr Biol 18(18):R853–R854. https://doi.org/10.1016/j.cub.2008.07.033

Welton LJ, Siler CD, Bennett D, Diesmos A, Duya MR, Dugay R, Rico ELB, Van Weerd M, Brown RM (2010) A spectacular new Philippine monitor lizard reveals a hidden biogeographic boundary and a novel flagship species for conservation. Biol Lett 6(5):654–658

Wetzel RM, Dubos RE, Martin RL et al (1975) Catagonus, an 'extinct' peccary alive in Paraguay. Science 189(4200):379–381

Wilcove DS (2005) Rediscovery of the ivory-billed woodpecker. Science 308(5727):1422–1423

Wisely SM, Santymire RM, Livieri TM et al (2008) Genotypic and phenotypic consequences of reintroduction history in the black-footed ferret (Mustela nigripes). Conserv Genet 9:389–399

Young RP, Copsey JA, Turvey ST (2018) Measuring and diagnosing species decline. In: Copsey JA et al (eds) Species conservation: lessons from islands. Cambridge University Press, Cambridge, UK, pp 121–153

Zahler P (1996) Rediscovery of the woolly flying squirrel (Eupetaurus cinereus). J Mammal 77(1):54–57

Chapter 12
Extinct or Perhaps Surviving Relict Populations of Big Cats: Their Controversial Stories and Implications for Conservation

Lorenzo Rossi, Carmelo Maria Scuzzarella, and Francesco Maria Angelici

12.1 Introduction

The impact of modern humans (*Homo sapiens*) on wildlife has increased in historical times with the advent of the industrial era (Zalasiewicz et al. 2010), but our species has always had a negative effect on many animal populations (e.g., Braje and Erlandson 2013).

Several carnivorous species suffered strong demographic loss, such as the gray wolf *(Canis lupus*), once the most widespread mammal in Eurasia and North America (Boitani et al. 2018). All species of the genus *Ursus* also had a decline in North America, Europe, and Asia, disappearing from North Africa (McLellan et al. 2017).

Among the predators that suffered the impact of human, there are the big cats, mainly afferent to the genus *Panthera* (e.g., Dinerstein et al. 2007; Bauer et al. 2015; Jacobson et al. 2016).

These animals have always represented for human beings not only competitors but also species of great emotional impact. In fact, they appear in prehistoric art (e.g., Bar-Oz and Lev-Yadun 2012; Killin 2013), in ancient folklore (e.g., Ge 2007), religion (Benson 1998), and heraldry (Ross 2006) as symbols of strength and pride.

Some species have also been the focus of recreational (and sometimes controversial) activities in the past and present such as ancient games in arenas (Lindstrøm 2010), big game hunting (Storey 1991), ecotourism (Mossaz et al. 2015), and circus

L. Rossi (✉)
Museo dell'Ecologia di Cesena, Cesena, FC, Italy
e-mail: info@associazioneorango.com

C. M. Scuzzarella
Università di Parma, Facoltà di Scienze Politiche e delle relazioni internazionali, Parma, Italy

F. M. Angelici
Fondazione Italiana per la Zoologia dei Vertebrati (FIZV), Rome, Italy

F. M. Angelici, L. Rossi (eds.), *Problematic Wildlife II*,
https://doi.org/10.1007/978-3-030-42335-3_12

shows (Tait 2009) and as attractions for the public in the modern zoos (Bashaw and Maple 2010). Even today they are still the focus of modern urban legends (e.g., Goss 1992; Hurn 2009).

Currently the species of the genus *Panthera* have legal protection in the states where they are present in the wild, with very few exceptions such as South Africa, Namibia, Zambia, Ecuador, and French Guiana, where it is possible to hunt these felids, in a controlled manner, in some private or state areas, or even throughout the national territory (e.g., Silvius et al. 2004). Many conservation plans were initiated in the second half of the last century (Nowell and Jackson 1996), and thanks to this, in some areas, the populations are now stable or even increasing, but their future remains uncertain (Meena et al. 2014; Jhala et al. 2019).

The historical range of the tiger *(Panthera tigris)* once covered an area from Turkey to the east coast of Russia, but today it has decreased by 93% and the populations are highly fragmented (Natesh et al. 2017). Currently it is the most threatened large felid and its habitat has declined due to forestry, agriculture, and oil palm plantations (Hunter 2015). The species is also under intense threat from illegal hunting due to the direct demands of traditional Chinese medicine and increased bushmeat trafficking in Southeast Asia, which has led to a decrease in its prey.

Three subspecies are traditionally recognized to be extinct, the Javan tiger *(Panthera tigris sondaica)*, the Bali tiger *(P. t. balica)*, and the Caspian tiger *(P. t. virgata)*. The latter is now being considered the same subspecies as the Amur tiger *(P. t. altaica)* (see Driscoll et al. 2009). A fourth subspecies, the southern Chinese tiger *(P. t. amoyensis)*, has experienced a huge decline (Tilson et al. 2004) and is extinct in the wild (Hunter 2015).

Even the lion *(Panthera leo)* suffered an impressive demographic decline: over the last century, it has lost about 82% of its former distribution range (see Trinkel and Angelici 2016). Currently the largest population lives in eastern and southern Africa, while the species has declined in Central Africa and become extinct in most of West Africa (Henschel et al. 2014). It is also extinct throughout the Middle East and Asia, with the exception of the Indian population, consisting of approximately 650 individuals, although currently growing (Singh and Gibson 2011; Singh 2017; Kaushik 2017). The decline of the African lion is mainly due to the extensive loss of habitat and prey due to agriculture and cattle breeding, but also to direct persecution and poaching on potential prey (e.g., Trinkel and Angelici 2016).

In historical times, the Cape lion (*Panthera leo melanochaita*) and the Barbary lion (*Panthera leo leo*) become extinct. These populations are now considered part of the new taxonomy of southern and northern subspecies (Black 2016; Bertola et al. 2016).

The leopard *(Panthera pardus)* is the felid with the largest natural range (Jacobson et al. 2016) and the most varied diet (Hayward et al. 2006). Despite its extreme plasticity that allows it to survive in a wide variety of different environments (e.g., Nowell and Jackson 1996), and its surprising tolerance to human activities (Hunter 2015), the decline of its populations is compatible with that of the other large carnivores (Ripple et al. 2014). Currently the species occupies 25% to 35% of its historical range (Jacobson et al. 2016). The survival of the species is threatened by the loss of habitats and prey and by the persecution to which it is subjected in grazing areas. The leopard is also hunted for the skin and for the parts sold in the trafficking of traditional Asian medicine (Hunter 2015).

Despite the fact that the jaguar *(Panthera onca)* is the only species of the genus *Panthera* not considered "endangered" or "vulnerable" by the IUCN (its actual status is "NT"), since 1900 its historical range decreased from 19,000,000 to 9,000,000 km^2 (Seymour 1989), and this big cat is now considered extinct in the USA – but in recent years there has been sporadic sightings in Arizona and New Mexico (see Brown and Gonzalez 2000) and Uruguay and El Salvador (Hunter 2015). Furthermore, assessments of the conservation status of the species have indicated a decline in its current range (Sanderson et al. 2002).

Even the snow leopard *(Panthera uncia)*, despite being perceived as an animal living in remote and inhospitable areas, has frequent interactions with human and his activities (e.g., Hussain 2003; Bagchi and Mishra 2006). Also the populations of this species are in decline (McCarthy and Chapron 2003), and among the threats there is climate change (Forrest et al. 2012) and the loss of prey species (Namgail et al. 2007).

The purpose of this work is to summarize the information available on some populations and subspecies, mainly of the genus *Panthera*, extinct or considered extinct in historical times. We identify common features in these extinctions of big cat subspecies and significant populations, in terms of patterns of conservation importance and risk of biodiversity loss (i.e., unique taxa, genetics, etc.) and impact on ecosystem balance. Some historical pictures, often representing the last available visual documents of some very rare or already extinct subspecies, have also been collected.

Currently there are different viewpoints on the systematics of the genus *Panthera* with regard to the subspecies (see Kitchener et al. 2017), but in this work we have decided to adhere more to the previous approach (Wozencraft 2005), in particular to underline the importance of the concept of subpopulations and local extinction with regard to conservation.

12.2 Case Studies

12.2.1 *Cape Lion (***Panthera leo melanochaita** *Smith, 1842) EX*

The Cape lion (Fig. 12.1) inhabited the western part of the Cape Province of South Africa. It was the first *taxon* of the genus *Panthera* to be extirpated in historical time and the one with the least information available. Smith (1842) described it as a large subspecies with ears edged with black and a thick black mane that covered also the shoulders and the belly. Some lions killed near the Vaal River reached weights of

Fig. 12.1 A Cape lion at the Jardin du Plantes, Paris, 1860. (Public domain)

about 272 kg, a size comparable to that of the Barbary lion that populated North Africa (Pease 1913). Its extinction followed soon after the arrival of the European settlers, making the subsequent destruction of the habitats an irrelevant factor (Day 1981). In fact, the population was exterminated because it was considered harmful to livestock (Haagner 1920). The last known specimen was hunted down and killed by General J. Bisset in Natal (now KwaZulu-Natal), in 1865 (Day 1981). Actually only a few skeletal remains and stuffed specimen are known and stored in museums (Christiansen 2008).

12.2.2 *Barbary Lion (***Panthera leo leo** *Linnaeus, 1758) EX*

The Barbary lion (Fig. 12.2) was formerly widespread in Mediterranean Africa (Guggisberg 1963). Thousands of years ago, the Barbary lion inhabited the whole area of the present Sahara desert, once characterized by a savannah similar to that of today's East Africa. Males could reach a length of 3.5 m and a weight of 280–300 kg. The mane was black and abundantly developed around the head, neck, and shoulders and under the belly (Day 1981).

Its decline began with the desertification that started in the second millennium BC. This reduction of the range lead to three isolated subpopulations: one on the Atlas mountains between Morocco and Algeria, one in the Nile delta, and one in the Nubia mountains (Pease 1899).

Fig. 12.2 Male Barbary lion at Lincoln Park Zoo, New York, 1900. (The Field Museum Library)

The Roman conquest of North Africa further decreased the lion population: thousands of specimens were captured and imported every year for games in the arenas (Pease 1899), but the populations started to seriously dwindle in Libya by the late 1700s.

Deprived of its main habitat and prey, the Barbary lion began to feed on domestic livestock, which thus contributed to its persecution. The subsequent introduction of firearms significantly accelerated its demographic decline (Guggisberg 1963): in Algeria, lions were numerous enough for a bounty to be issued by the French colonial government (see Yamaguchi and Haddane 2002).

Between 1500 and 1700, lions were still reported in the northern Moroccan coast, and until 1830 they were still spotted on the Rif mountains and in the Mamora forest (Guggisberg 1963), but from 1880 they began to retreat on the Atlas chain and in the Saharan regions, where there was less human pressure (Cabrera 1932). Lions from Morocco survived in captivity for a certain time before the extinction in nature due to the custom of the sultans to keep specimens in the gardens of the palace in Fez. In fact, for centuries, lion cubs were offered by tribes from the Atlas mountains as tributes. In the late 1960s, to improve life for the lions, a new enclosure (which in 1973 will become the Rabat zoo) was built in Tamara (Yamaguchi and Haddane 2002).

The last lion of the Rif was killed in 1895 (Lavauden 1932), and in 1925 a male was photographed by an airplane who was flying on the Casablanca-Dakar route

(Black et al. 2013). In 1930 some footprints were found in the area of Ouiouane (Morocco), and in the summer of the same year, some lions were observed at an altitude of 3000 m on the Toubkal massif (Panouse 1957).

In 1942 a male was shot on the Tizi-N'Tichka pass, High Atlas: it was the last confirmed individual in Morocco (Guggisberg 1963). In Tunisia the last lion was shot in 1891 and rumors about the presence of the species in the Khmir mountains and near Feriana continued until the early 1900s (Guggisberg 1963). In Algeria, the last lion of the Saharan Atlas was killed around 1920 (Yamaguchi and Haddane 2002).

Another individual was shot in 1943: it was the last Barbary lion confirmed in nature (Yamaguchi and Haddane 2002). The last known sighting of a Barbary lion, however, took place in 1956 in Algeria: passengers on a bus claimed to have observed one in a forest (Haddadou 1994).

On the basis of a mathematical model, Black et al. (2013) hypothesize that small populations of lions survived until the early 1960s in Algeria (later eradicated by the Franco-Algerian war) and probably until the 1960s in remote areas of Morocco.

12.3 Tiger (*Panthera tigris*)

*12.3.1 Bali Tiger (*Panthera tigris balica *Schwarz, 1912) EX*

The Bali tiger (Fig. 12.3), the smallest subspecies of tiger, was the first to become extinct. The largest males were 220–230 cm long and weighed a maximum of 90–100 kg (Mazak 1981).

Its skull was similar in size to that of the Javan tiger *(Panthera tigris sondaica)* but with more marked zygomatic arches, and the hair was short and bright orange, with fewer black stripes than other subspecies (Schwarz 1912).

The Bali tiger lived on the Indonesian island of the same name. Because of the small size of the territory (5780 km^2), it is very likely that this tiger was never numerous, another factor that probably contributed to its rapid extinction.

In the late nineteenth century, rice and oil palm crops developed a great deal, taking advantage of the rich volcanic soil and the alluvial plain along the perimeter of the island (Seidensticker 1986). Tigers were hunted and shot for sport. The kills perpetrated by Western hunters and the disappearance of its habitat and its prey led to the extinction of the species.

On September 27, 1937, an adult female was killed in the western region of Bali: it was the last individual known in nature (Day 1981).

Based on unconfirmed reports, Seidensticker (1986) believes that some tigers could have survived until the 1940s and maybe until the 1950s. The last alleged sightings were reported in 1970 and in 1972 in the western region of the island (Maas 2010).

Fig. 12.3 Bali tigers at the Ringling Bros circus, USA, circa 1915. (Harry A. Atwell 1915)

12.3.2 *Javan Tiger (*Panthera tigris sondaica *Temminck, 1884) EX*

The Javan tiger (Fig. 12.4) was a subspecies native to the Indonesian island of Java. It was one of three island subspecies, together with the Sumatran tiger *(P. t. sumatrae)* in danger of extinction, and the Bali tiger (extinct), all originating from the Sunda Islands.

Its size was between that of the Sumatran tiger and the Bali tiger, but the color was much lighter and faded into yellow. Its footprints were larger than those of the

Fig. 12.4 Javan tiger at London Zoo, 1942. (F.W. Bond)

Fig. 12.5 Female tiger killed at Besuki, Java, 1934. (Public domain)

Bengal tiger *(P.t. tigris)*, most likely an adaptation to the mud and soft ground of the tropical forest (Seidensticker 1986).

Once widespread throughout the island, until 1850 it was still so numerous that it was considered a pest by the population (Figs. 12.5 and 12.6), but since 1940 it survived only in the most remote mountain forests (Seidensticker 1987). In 1970 the

Fig. 12.6 Tiger killed in Malingping, Banten province, western Java region, 1941. (Public domain)

only known tigers were believed to live exclusively on Mount Betiri, in the eastern end of Java: the area was declared a nature reserve in 1972, and in 1976 the last tigers were observed inside it (Seidensticker 1987).

The factors that led to the extinction of the Javan tiger were the same as the Bali tiger: a growing human population that destroyed forests by replacing them with crops and rarefying its main prey. In Java, forest cover fell from 23% in 1938 to 8% in 1975 (Seidensticker 1987).

In addition, several epidemics exterminated populations of maned sambar *(Rusa timorensis)*, the main prey of the Javan tiger, and the villagers began to attract the tigers with poisoned morsels to protect the crops (Seidensticker 1987).

After the period of civil unrest in 1965, several armed groups withdrew to the forests where they killed the last remaining tigers (Seidensticker 1987). In 1971 an old female was killed in a plantation near Mount Betiri (Seidensticker and Suyono 1980).

No offsprings have been sighted in the area since then. In 1976, on the basis of footprints, 3–5 tigers were estimated to exist in the eastern part of the reserve of Meru Betiri (Seidensticker and Suyono 1980).

Since 1979 no tiger has been sighted in Meru Betiri, but in 1987 a survey conducted by 30 students of the Agricultural University of Bogor (Java) found alleged scats and footprints (Istiadi et al. 1991).

In the western part of the island, the last confirmed shooting of a Javan tiger took place: in 1984 a tiger was killed in the Halimun nature reserve (now a national park

Fig. 12.7 Caspian tiger, Berlin Zoo, 1899. (Public domain)

of Mount Halimun Salak), and in the same area in 1989, alleged footprints were found, but a survey conducted in 1990 did not find any tiger (Istiadi et al. 1991). Further surveys were carried out between 1992 and 1994 in Meru Betiri, using camera traps, but no data were collected (Rafiastanto 1994). So, the Javan tiger was declared extinct (Kemf and Jackson 1994).

A further attempt made in 1999 again in Meru Betiri, with the help of the staff of the Sumatran Tiger project, failed in its intent (Tilson 1999).

*12.3.3 Caspian Tiger (***Panthera tigris virgata** *Illiger, 1815) Probably EX*

Once widespread from western Anatolia to the western regions of China and Mongolia, the Caspian tiger (Fig. 12.7) was among the largest: the maximum weight recorded in nature was 240 kg, and a tiger killed in the valley of the river Sumbar (mountains Kopet-Dag, Turkmenistan) in January 1954 had a skull length of 385 mm, larger than the average of Amur tigers (Heptner and Sludskii 1972).

The distribution of the Caspian tiger was associated with rivers, swamps, and lakes where the riparian forest offered cover to itself and its prey (Heptner and Sludskii 1972). The first hard blow to the population of Caspian tigers was inflicted by the Romans, who brought back thousands of individuals from Anatolia and the Caucasus for games in the arenas of the Empire: stone traps for tigers and leopards are still visible on the mountains of Taurus in Turkey (Şekercioĝlu et al. 2011).

This caused the extirpation of the Caspian tiger from a large part of the Anatolian peninsula, but the most impactful factor was the colonization of Central Asia by Tsarist Russia at the end of the nineteenth century when the tiger's hunting grounds were converted into cotton fields and an intense campaign of persecution took place. Until the First World War between the Amu-Darja and Pjandi rivers, about 50 tigers were killed every year (Prokhorov 2002). At the beginning of the twentieth century, a tiger skin could be sold for between 1500 and 2500 rubles, while, for example, a snow leopard skin was worth only 300–500 rubles (Prokhorov 2002).

The strong demand for tiger bones and parts for use in traditional Chinese medicine also encouraged hunting (Heptner and Sludskii 1972; Prokhorov 2002). The Russian army was also employed in the fight against animals considered harmful, and for each killed tiger, a large bounty was given until 1929 (Jungius et al. 2009).

Large swamps and rivers in Central Asia were infested with malaria and therefore not populated, providing a safe area for the tigers, but in the 1930s malaria was eradicated which lead to the colonization of the territories and consequent destruction of habitat for agriculture (Jungius et al. 2009).

In the meantime due to intense agricultural development, especially in the Fergana valley (eastern Uzbekistan, southern Kyrgyzstan, and northern Tajikistan), large portions of the wild territories were converted into cotton fields and farms. Several roads were built and the consequent water exploitation (with the construction of numerous canals that drew water from the Syr-Darya and the Amu-Darya, which started the inesorable decline of the Aral Lake) caused a desertification that fragmented furthermore the tiger's habitat into smaller and more distant groups (Jungius 2010).

Today only 10% of the ancient habitat of the Caspian tiger remains intact. The excessive hunting of its prey and the consequent attacks on livestock (and sometimes people) forever marked its fate: shepherds, hunters, and soldiers used traps, poison, and firearms to exterminate tigers (Jungius et al. 2009).

The only known tiger in Iraq was killed near Mosul in 1887 (Kock 1990). In western China, the tigers disappeared from the Tarim River basin in the 1920s, due to desertification (Ognev 1935). From the basin of the river Manasi, in the Tien Shan, they were eliminated during the period immediately after 1960 (Heptner and Sludskii 1972).

The last tiger in the Caucasus was killed in 1922 in Georgia, near Tbilisi, after attacking cattle (Ognev 1935). Its stuffed body is now on display at the National Museum of Georgia. In the Amu-Darya delta, tigers were common until the beginning of the twentieth century: in the early 1940s, 12–15 specimens were estimated to be in the area, and the last of them were killed in 1947. Unconfirmed sightings in the same area were reported in 1953, 1955, and 1963 (Heptner and Sludskii 1972).

In Uzbekistan, a tiger was killed near Nukus in 1938, and another was shot near the capital Tashkent in 1942. Yet near Nukus another individual was spotted twice in 1968 (Heptner and Sludskii 1972).

In 1972 a tiger was killed in Uzbekistan: it was probably the last confirmed specimen in the whole of the former Soviet Union (Heptner and Sludskii 1972). It is now exhibited at the State Museum of Karakalpakstan.

In Tajikistan, the last tiger in the Gissar valley was shot in 1938. Four more were shot in 1950 on the Tajik bank of the Pjandi River (Heptner and Sludskii 1972). In 1938 the first protected area of Tajikistan was founded: Tigrovaya Balka ("the old river channel of the tiger" in Russian). The name comes from an event that happened years earlier, when a tiger attacked two Russian army officers riding along a dry river channel.

In the early 1930s, 15–30 individuals were estimated in the reserve, but by the end of the 1940s, the number collapsed to no more than 5 (Heptner and Sludskii 1972). The last tiger in the reserve and probably the last in Tajikistan was observed in 1953. Individuals (probably roaming from neighboring Afghanistan) were reported in 1955, 1957, 1959, 1960, 1962, 1964, and 1967 (Heptner and Sludskii 1972). In the middle course of the Syr-Darya river, tigers disappeared in the mid-1930s. In the lower part of the river's valley, the last tiger was killed in 1933 (Jungius 2010). Individuals from Amu-Darya were seen in 1937 and 1945 and for the last time in the early 1950s (Jungius et al. 2009).

In Iran in the 1930s, the Caspian tiger was still present with hundreds of individuals in the northern part of the country. In 1953 a tiger was killed in the province of Golestan, and another specimen was sighted in the same area in 1958 (Firouz 2005). In 1955 a young female was captured and sent to Hamburg Zoo with a herd of onagers (Humphreys and Kahrom 1995; Firouz 2005). It died in 1960. It was most likely the last Caspian tiger kept in captivity in Europe. In a protected area of the Menkalech peninsula, on the Caspian coast, the last tiger was shot in 1957: in the 1960s, 15–20 individuals were estimated in the area and probably survived until the 1970s (Humphreys and Kahrom 1995, Firouz 2005).

In Turkey two tigers were shot in 1943 near Selcuk (western Anatolia), far from their usual range (Johnson 2002). A tiger was killed near Uludere, in the province of Sirnak, in 1970 (Can 2004). Subsequent investigations discovered that in the Far East of Turkey (Kurdistan) until the mid-1980s, one to eight tigers were killed each year and that they most likely survived until the early 1990s (Can 2004). Further reports continued in the 1990s and 2000s, with villagers and border guards reporting that they had heard roars and had observed a large striped cat with night vision equipment.

The last possible report of the Caspian tiger dates back to 1998 and came from Tajikistan, in the Babatag Mountains (Jungius et al. 2009). In Afghanistan, Soviet soldiers and Soviet border guards reported several times that they had seen tigers during the occupation (from 1979 to 1989) and the same was reported by military staff of the international coalition in 2007 (Jungius et al. 2009). A Kazakh hunter claimed to have seen a female with cubs in the Balkhash Lake region in May 2006 (Jungius et al. 2009; Jungius 2010). However, the sighting remains uncertain and unconfirmed. Very few Caspian tigers are known to have been kept in captivity in recent times. A young domesticated female was given to the Soviet ambassador to Iran in 1924 (Chikin and Tsaruk n.d.) . It died at the Moscow Zoo in 1942. The last Caspian tigers in captivity were killed in Mohammad Reza Pahlavi's personal zoo in 1979, during the Islamic Revolution (Chikin and Tsaruk n.d.).

12.4 Leopard (*Panthera pardus*)

12.4.1 Anatolian Leopard (**Panthera pardus tulliana** *Valenciennes, 1856) EX?*

The Anatolian leopard is native to western Turkey (Valenciennes 1856), and under current taxonomy, it is considered as a Persian leopard subspecies *(P. p. saxicolor)* (Stein et al. 2016).

Anatolian leopards once populated the Aegean regions, the Mediterranean coasts, and the mountain forests of central and western Anatolia. Leopards and other predators were imported in large numbers from Anatolia to Rome and to the various provinces of the Roman Empire for games in the arenas (Kullman 1967). In any case, it did not suffer particularly from human pressure until May 5, 1937, when the first hunting regulations of the newborn Republic of Turkey were passed: leopards, tigers, wolves, and many other animals were declared pests and therefore huntable at any time of the year (Ertüzün and Ergir 2017).

At the same time, intensive rural development began in Asia Minor. Large areas of forest were converted into olive groves and the main prey of the leopard began to be hunted relentlessly (Borner 1977). Deprived of their natural refuge and food source, leopards began to attack domestic livestock, thus completing the vicious circle that ended with felids being shot dead with firearms or poisoned (Borner 1977). In 1942 a cub was found by a shepherd in Urla, near Izmir. The individual, a female dubbed Zoza, was sold to a hunter who let it grow for 9 months and who then donated it to the zoo in Izmir (M. Ertüzün, pers. comm.) (see Fig. 12.8).

Fig. 12.8 Zoza, Izmir zoo, 1946 (Cafer Tayyar Türkmen)

In the 1970s, the Turkish Department for National Fund financed a 2-month survey to investigate the leopard status in western Turkey and collected several anecdotes.

In 1972 two specimens were spotted: one near the village of Catacik (Samandağ province) and the other on Mount Agri. On January 17, 1974, in the village of Bagozu (province of Beypazari, not far from Ankara), an Anatolian leopard attacked a woman who was going to the fields, inflicting several wounds. On the same day, the hunters of the village tracked down and killed the leopard. Also in 1974 a leopard was poisoned between Ikistas and Ketendere and in Samsun-Dag National Park (where a leopard had been spotted in 1972). In January and February 1975, a gamekeeper found alleged footprints, and in the same period, soldiers and road workers reported that they heard roars of leopards and that a horse was been killed in the park, presumably by a leopard. In the Spring of 1975, an individual attacked and killed a cow near the village of Kacanci: the inhabitants shot the leopard, but it managed to escape. In the same year a leopard was killed near Seferler. In the Spring of 1976, several goats were presumably killed by a leopard near the village of Viran: the area was sieved but no sign of the presence of the feline emerged. In the same period, an individual was shot near Asar, and in September of the same year, a mule was most likely killed by a leopard, according to the description of the villagers (Borner 1977; Ertüzün and Ergir 2017).

Surveys carried out recently showed that small populations of leopards still survive: in 1992 two chamois hunters saw a leopard between Yusufeli and Artvin, and in 2007, not far from this area, a leopard was sighted from a Discovery Channel operator (Spassov et al. 2016).

Photographs of a dead male specimen were taken in February 2008 in Cumhuriyet village, Bitlis province (Toyran 2018).

In 2010 a leopard was killed and skinned in the Sirnak Province (Avgan 2013a), and in 2013 another individual was killed in Diyarbakir Province (Avgan 2013b).

In September 2013, Sagdan Baskaya of Forestry Faculty of the Karadeniz Technical University, Trabzon, declared that he had obtained several camera trap pictures of leopards in the province of Trebisonda (northeast of Turkey), but the pictures clearly show feral domestic cats and not big cats (Spassov et al. 2016). Alleged leopard videos were captured by thermal cameras found in a military base in Buzulup and gendarmerie station of Karınca village (Toyran 2018).

12.4.2 *Barbary Leopard (***Panthera pardus panthera***, Schreber 1777) CR-EX?*

Very little is known about the history and the decline of these leopards. Until 1900 they were still quite common and widespread in the forests of cedar, juniper, and Aleppo pine of the Middle and High Atlas (Aulagnier et al. 2015), but already in the first decades of the last century, they disappeared from Algeria, and its strongholds remained in the mountain valleys, gorges, and dense Moroccan forests of the roughest points of the Atlas, and in the 1950s the population was estimated at 50 individuals (Panouse 1957).

Like other big cats, the Barbary leopard was intensely hunted as a trophy and for livestock losses (see Fig. 12.9). Morocco banned leopard hunting in 1952, but after a population increase (about 100 individuals), from the 1970s onwards, the population has drastically decreased: about ten individuals estimated in the 1980s (Aulagnier and Thévenot 1986) and 2–5 individuals in 1996 (Cuzin 1996).

Considered extinct due to the lack of credible data from 1994 onwards (see Cuzin 2003), almost 20 years later, there remains the possibility that a small population still persists (Purroy Iraizoz 2010).

Fig. 12.9 A leopard hunting in Morocco, no data available

12.4.3 *Zanzibar Leopard (***Panthera pardus adersi***, Pocock, 1932) EX?*

Once considered an endemic subspecies of Unguja, the largest island in the Zanzibar archipelago, the Zanzibar leopard (Fig. 12.10) was not scientifically studied until the 1920s (Walsh and Goldman 2017), when a specimen was shot and sent to the British Museum (Pocock 1932). Leopards on this island population were smaller in size and had a different coat than all other known leopards. In particular, the rosettes appeared "disintegrated" in small points very close together (Pocock 1932).

Currently there are only six specimens stored in museums, all dating back to the British colonial period (Walsh and Goldman 2008), and its taxonomic status remains uncertain, given its exclusion from a recent review of the species on genetic basis (Miththapala et al. 1996; Uphyrkina et al. 2001).

Zanzibar leopard has never been studied in the wild, and very little is known from the literature of the time (e.g., Mansfield-Aders 1920). Although it was the largest carnivore on the island, nothing is known about its diet, but it is likely that Aders's duikers (*Cephalophus adersi*), blue duikers (*Philantomba monticola*), and sunis (*Neotragus moschatus*) were among its prey.

Its decline began in the first half of the nineteenth century with the large-scale transformation of the island's landscapes (Walsh and Goldman 2007).

The main cause of its extermination was due to the complex belief system developed by the natives to explain the increasing presence of leopards near inhabited areas. In fact, there was a widespread belief that the specimens sighted

Fig. 12.10 Stuffed Zanzibar leopard specimen in Zanzibar Natural History Museum. (Peter Maas)

far from the forests were bred by witches to cause harm to people (Walsh and Goldman 2017).

The locals urged the British protectorate to control the leopard population because of attacks on people and livestock, but despite the British government's efforts to prevent leopard hunting, rural communities continued to persecute them. The peak came after the Zanzibar Revolution in 1964, when the new government supported a national campaign of shootings (Walsh and Goldman 2017).

Interest in the fate of the leopard population arose only in the 1990s (e.g., Archer et al. 1991; Archer 1994). The Zanzibar leopard is generally considered extinct (e.g., Smithers 1971; Hes 1991; Miththapala et al. 1996; Nowell and Jackson 1996). In the 1980s two sightings were reported (Swai 1983), while the last documented shooting with material evidence seems to date back to 1986 (Walsh and Goldman 2008) and is proved by two fragments of skin (now lost) in the possession of the former Secretary of the Zanzibar National Hunters.

In 1997 and 2001, some rumors circulated about the discovery of alleged excrements, but in both cases the samples were lost before being properly analyzed (Walsh and Goldman 2008). In 2018 a crew of Animal Planet, during the recordings of a television program, obtained a camera trap video of an alleged Zanzibar leopard (Li 2018). Some authorities responsible for the Zanzibar leopard do not consider this film to be reliable evidence (H. Goldman pers. comm.; Goldman and Walsh 2018) and its diffusion on the internet has been restricted to the American television newsmagazine Inside Edition and some blogs devoted to paranormal and other alleged mysteries. On the other hand, the author has defended the authenticity of the film (F. Galante, pers. comm.). The video undoubtedly shows a leopard, but the images do not allow to verify precisely the pattern of the rosettes or to determine the shooting's locality. It should also be noted that a feral African leopard released in Zanzibar is an option which can be eliminated, but only DNA evidence (such as from scats) would offer an opportunity to differentiate this animal from other leopards. Although remaining skeptical, we hope that, given the potential importance for conservation, further investigation will deepen the matter.

12.5 Discussion

This review of case studies stressed how the direct impact of human can be lethal to the conservation of animal populations. Predators at the top of the trophic chain are the first to suffer the consequences of direct persecution and habitat alteration (e.g., Henke and Bryant 1999; Rodríguez-Lozano et al. 2015), and it is for this reason that we can consider them among the most "problematic" animal species, which need special attention and targeted conservationist approaches (Sergio et al. 2005). In the case of lions, Black et al. (2013) suggest that while prey species will decline, possibly to micro population size, predators can survive by switching to livestock, causing new problematics. This trend seems related also to the cases on tigers and leopards reviewed in our work. One of the key risks relating to predators on the verge of extinction is the disappearance of species, subspecies, or populations before it is possible to investigate their taxonomic status and ecology (e.g., Spassov et al. 2016; Angelici et al. 2019).

From this point of view, an interesting case is represented by the population of leopards of the small island of Kangean (Indonesia). First reported by Hartert (1902), the only material evidence is represented by a part of the tail secured by G.C. Shortridge in 1908 (Pocock 1930) and by a skin and skull collected in January 1984 (van Helvoort et al. 1985). Currently nothing is known about the current numerical size of the population, its origin, and systematic identity although it would seem consistent with the Java leopard *Panthera pardus melas* G. Cuvier, 1809. Some authors have suggested that the Kangean leopard was smaller than the continental form (Iongh et al. 1982; Delsman 1951), but there are no studies carried out on the few samples available. Other authors hypothesize an introduction in the island of a hunting prey (van Helvoort et al. 1985; Long 2003).

A characteristic of many big cats, especially when populations are very small in number, is their high elusiveness, often helped by the habitats in which they live (see Burton et al. 2011). This can lead to prematurely declare a species as extinct, thus potentially creating serious conservation damage (see Collar 1998). The case studies of the Barbary lion, Caspian tiger, and Barbary leopard, which showed their existence for decades beyond their supposed extinction into the modern era of pro-active conservation, point out how they could have been recovered before their extinction. A current example is provided by the western lion, considered extinct in Ghana (Henschel et al. 2014), but recent reports from the Mole National Park seem to indicate the survival of a very small population that could prove strategic for conservation of the species in Ghana and neighboring countries (see Angelici et al. 2015; Angelici and Rossi 2017).

Reports of big cats years after their declared extinction are not uncommon (see Black et al. 2013), and the most recent concerns the Taiwanese clouded leopard *Neofelis nebulosa brachyura*. The last reliable data date back to 1986 (Rabinowitz 1988) and 1989 (Anonymous 1996). Then, apart from a dubious sighting in 1990 (Lue et al. 1992), subsequent surveys did not obtain results (Chiang 2007, Chiang et al. 2015). In 2013 the subspecies was officially declared extinct (Grassman et al. 2016), but in early 2019, independent sightings from Taitung County (Southeast of the island) were recorded (Hoffner 2019).

The history of the extinction and decline of these animals, in all cases, were caused by interactions with human, demonstrating the extreme fragility of ecosystems and the extreme difficulty in protecting endangered species. The fact that in many cases even legal protection has not been sufficient to help a species (see Zanzibar leopard) indicates that conservation strategies cannot only be carried on by governments and researchers and that local populations must be involved to play a role (e.g., Danielsen et al. 2007; Ocholla et al. 2016). It should be pointed out that in some cases even the application of very strict laws can be harmful. In North Africa, for example, following the killing of a leopard in the Bou Tferda region in 1983 by the locals, severe sanctions were issued. The news spread quickly in the region and locals stopped giving the researchers information about the leopard (Cuzin 2003).

12.6 Conclusion

We can summarize the main findings of our work in four principal points that can suggest actions to improve the conservation of micro populations of big cats:

– Locals are generally the last to report and to interact with relict predator populations, often due to attack on livestock. Measures to encourage locals to live with predators through virtuous practices aimed at making people understand the importance of these top predators for the entire ecosystem are thus essential. In 2010 in Kenya, through a program called "Warrior Watch", the Samburu warriors (once lion hunters) are working within their local communities to protect livestock and promote coexistence between people and lions (Gurd 2012).
– As in the case of Barbary lion in Morocco, top predators can survive in captivity after their eradication or extinction in nature. Ex situ conservation has proved useful for several big cats, such as the Chinese tiger (actually living only in captivity) and the Amur leopard *(Panthera pardus orientalis)*, which has an important population living in zoos, and this should be incentivized (e.g., Luo et al. 2008).
– Big cats can "reappear" unexpectedly even many years after their official disappearance. Research must be encouraged and the use of camera traps can be very effective, as in the case of the recent rediscovery of the lion in Gabon, formerly considered locally extinct since 2006 (Barnett et al. 2018).
– Habitat loss is the main cause of extinction processes. Protected areas must be established where there are none yet. Considering the large home range big cats need (Cushman et al. 2018; Paviolo et al. 2018), contiguous and transnational areas are important for the establishment of ecological corridors.

These hints can lead to new and feasible conservation strategies, especially when a population still exists but is so small as to be on the verge of total extinction (e.g., Tilson et al. 2004; Ahmad Zafir et al. 2011).

References

Ahmad Zafir AW, Payne J, Mohamed A et al (2011) Now or never: what will it take to save the Sumatran rhinoceros *Dicerorhinus sumatrensis* from extinction? Oryx 45:225–233. https://doi.org/10.1017/S0030605310000864

Angelici FM, Ciucani MM, Angelini S et al (2019) The sicilian wolf: genetic identity of a recently extinct insular population. Zool Sci 36(3):189–197

Angelici FM, Mahama A, Rossi L (2015) The lion in Ghana: its historical and current status. Anim Biodivers Conserv 38:151–162

Angelici FM, Rossi L (2017) Further lion, *Panthera leo senegalensis* Meyer, 1826, sightings in Mole National Park, Ghana, and possible first serval *Leptailurus serval* Schreber, 1776 record after 39 years (Mammalia Felidae). Biodivers J 8:749–752

Anonymous (1996) The mystery of the Formosan clouded leopard. Cat News 24:16

Archer AL (1994) A survey of hunting techniques and the results thereof on two species of Duiker and the Suni Antelopes in Zanzibar. Report to FINNIDA/Forestry Sector, Commission for Natural Resources, Zanzibar

Archer AL, Collins S, Brampton I (1991) Report on a visit to Jozani Forest, Zanzibar. East Afr Nat Hist Soc Bull 21:59–66

Aulagnier S, Bayed A, Cuzin F et al (2015) Mammifères du Maroc: extinctions et régressions au cours du XXème siècle. Travaux de l'Institut Scientifique, Série Générale 8:53–67

Aulagnier S, Thévenot M (1986) Catalogue des Mammifères sauvages du Maroc. Trav Inst Sci Rabat sér zool 41:1–164

Avgan B (2013a) Leopar yeniden Türkiye'de. National Geographic Magazine 144:34–43

Avgan B (2013b) Leoparın verdiği ipuçları. National Geographic Magazine 152:18–26

Bagchi S, Mishra C (2006) Living with large carnivores: predation on livestock by the snow leopard (Uncia uncia). J Zool 268:217–224. https://doi.org/10.1111/j.1469-7998.2005.00030.x

Barnett R, Sinding MHS, Vieira FG et al (2018) No longer locally extinct? Tracing the origins of a lion *(Panthera leo)* living in Gabon. Conserv Genet 19:611. https://doi.org/10.1007/s10592-017-1039-2

Bar-Oz G, Lev-Yadun S (2012) Paleolithic cave rock art, animal coloration, and specific animal habitats. Proc Natl Acad Sci 109:E1212. https://doi.org/10.1073/pnas.1200729109

Bashaw MJ, Maple TL (2010) Signs fail to increase zoo visitors' ability to see tigers. Curator Museum J 44:297–304. https://doi.org/10.1111/j.2151-6952.2001.tb01167.x

Bauer H, Chapron G, Nowell K et al (2015) Lion *(Panthera leo)* populations are declining rapidly across Africa, except in intensively managed areas. Proc Natl Acad Sci 112:14894–14899. https://doi.org/10.1073/pnas.1500664112

Benson EP (1998) The lord, the ruler: jaguar symbolism in the Americas. In: Sanders NJ (ed) Icons of power: feline symbolism in the Americas. Routledge, London

Bertola LD, Jongbloed H, Van Der Gaag KJ et al (2016) Phylogeographic patterns in Africa and high resolution delineation of genetic clades in the lion *(Panthera leo)*. Sci Rep 6:30807

Black SA (2016) The challenges and relevance of exploring the genetics of North Africa's 'barbary lion' and the conservation of putative descendants in captivity. Int J Evol Biol 2016(6901892):1. https://doi.org/10.1155/2016/6901892

Black SA, Fellous A, Yamaguchi N et al (2013) Examining the extinction of the barbary lion and its implications for felid conservation. PLoS One 8:e60174. https://doi.org/10.1371/journalpone.0060174

Boitani L, Phillips M, Jhala Y (2018) *Canis lupus*. The IUCN Red List of Threatened Species 2018: e.T3746A119623865. https://doi.org/10.2305/IUCN.UK.2018-2.RLTS.T3746A119623865.en. Accessed 12 May 2019

Borner M (1977) Leopards in Western Turkey. Oryx 14:26–30. https://doi.org/10.1017/S0030605300014769

Braje TJ, Erlandson JM (2013) Human acceleration of animal and plant extinctions: a late pleistocene, holocene, and anthropocene continuum. Anthropocene 4:14–23. https://doi.org/10.1016/j.ancene.2013.08.003

Brown DE, Gonzalez CAL (2000) Notes on the occurrences of jaguars in Arizona and New Mexico. Southwest Nat 45:537–546. https://doi.org/10.2307/3672607

Burton A, Buedi EB, Balangtaa C et al (2011) The decline of lions in Ghana's Mole National Park. Afr J Ecol 49:122–126. https://doi.org/10.1111/j.1365-2028.2010.01234.x

Cabrera A (1932) Los mamíferos de Marruecos. Trabajos del Museo Nacional de Ciencias Naturales, Serie Zoologica 57:1–362

Can OE (2004) Status, conservation and management of large carnivores in Turkey. Paper presented at the Convention on the Conservation of European Wildlife and Natural Habitats Standing Committee 24th meeting Strasbourg, 29 November–3 December 2004

Chiang PJ (2007) Ecology and conservation of Formosan clouded leopard, its prey, and other sympatric carnivores in southern Taiwan. Dissertation, Virginia Polytechnic Institute and State University

Chiang PJ, Pei KJC, Vaughan MR et al (2015) Is the clouded leopard *Neofelis nebulosa* extinct in Taiwan, and could it be reintroduced? An assessment of prey and habitat. Oryx 49:261–269
Chikin Y, Tsaruk O (n.d.) Turanian Tiger – Analysis of the modern situation. https://wwf.ru/upload/iblock/d6d/atacollectionanalyzismodernsituation_all_eng.pdf. Accessed: 8 June 2019
Christiansen P (2008) On the distinctiveness of the Cape lion (*Panthera leo melanochaita* Smith, 1842), and a possible new specimen from the Zoological Museum, Copenhagen. Mamm Biol 73:58–65. https://doi.org/10.1016/j.mambio.2007.06.003
Collar NJ (1998) Extinction by assumption, or, the Romeo Error on Cebu. Oryx 32:4
Cushman SA, Elliot NB, Bauer D, Kesch K, Bothwell H, Flyman M, Mtare G, Macdonald DW, Loveridge AJ (2018) Prioritizing core areas, corridors and conflict hotspots for lion conservation in southern Africa. PLoS One 13(7):e0196213
Cuzin F (1996) Répartition actuelle et statut des grands Mammifères sauvages du Maroc (Primates, Carnivores, Artiodactyles). Mammalia 60:101–124
Cuzin F (2003) Les grands Mammifères du Maroc méridional (Haut Atlas, Anti Atlas et Sahara): Distribution, écologie et conservation. Dissertation, Université de Montpellier
Danielsen F, Mendoza MM, Tagtag A et al (2007) Increasing conservation management action by involving local people in natural resource monitoring. AMBIO J Hum Environ 36:566–570. https://doi.org/10.1579/0044-7447(2007)36[566:icmabi]2.0.co,2
Day D (1981) The encyclopedia of vanished species. London Editions Limited, London
Delsman HC (1951) Dierenleven in Indonesie. NY Uitgeverij W. van Hoeve, The Hague
Dinerstein E, Loucks C, Wikramanayake E et al (2007) The fate of wild tigers. Bioscience 57:508–514. https://doi.org/10.1641/b570608
Driscoll CA, Yamaguchi N, Bar-Gal GK et al (2009) Mitochondrial phylogeography illuminates the origin of the extinct Caspian tiger and its relationship to the Amur tiger. PLoS One. https://doi.org/10.1371/journalpone.0004125
Ertüzün M, Ergir Y (2017) The last Anatolian Panther. http://www.ergir.com/last_anatolian_panther.htm. Accessed 6 June 2019
Firouz E (2005) Tiger. The complete fauna of Iran. I. B. Tauris, London, New York
Forrest JL, Wikramanayake E, Shrestha R et al (2012) Conservation and climate change: assessing the vulnerability of snow leopard habitat to treeline shift in the Himalaya. Biol Conserv 150:129–135. https://doi.org/10.1016/j.biocon.2012.03.001
Ge L (2007) The Tiger-Killing Hero and the Hero-Killing Tiger. Comp Lit Stud 43:39–56. https://doi.org/10.1353/cls.2006.0030
Goldman H, Walsh M (2018) "The Zanzibar leopard" on Animal Planet. https://zanzibar-leopard.blogspot.com/2018/06/the-zanzibar-leopard-on-animal-planet.html?m=1&fbclid=IwAR1H1S1z5eId4QdU4tARmE4LvWNhzhjAz7Th-pnQUveG6PGebNUwcMqmtj8. Accessed: 9 June 2019
Goss M (1992) Alien big cat sightings in Britain: a possible rumour legend? Folklore 103:184–202. https://doi.org/10.1080/0015587X.1992.9715841
Grassman L, Lynam A, Mohamad S et al (2016) *Neofelis nebulosa*. The IUCN Red List of Threatened Species. IUCN. 2016: e.T14519A97215090. https://doi.org/10.2305/IUCN.UK.2016-1.RLTS.T14519A97215090.en. Accessed 20 May 2019
Guggisberg C (1963) Simba, the life of the lion. Chilton Books, Philadelphia
Gurd H (2012) Evaluating the success of 'warrior watch': a community-based conservation initiative in Samburu, northern Kenya. Dissertation, Imperial College London
Haagner A (1920) South Africa Mammals; a short manual for the use of field naturalists, sportsmen and travellers. Whiterby, London, xx + 248 p
Haddadou MA (1994) L'Algérie mystérieuse. Alger
Hartert E (1902) The birds of the Kangean Islands. Nov Zool 9:419–442
Hayward MW, Henschel P, O'Brien J et al (2006) Prey preferences of the leopard *(Panthera pardus)*. J Zool 270:298–313. https://doi.org/10.1111/j.1469-7998.2006.00139.x
Henke S, Bryant F (1999) Effects of coyote removal on the faunal community in western Texas. J Wildl Manag 63:1066–1081

Henschel P, Coad L, Burton C et al (2014) The lion in West Africa is critically endangered. PLoS One 9:e83500. https://doi.org/10.1371/journalpone.0083500

Heptner VG, Sludskii AA (1972) Mlekopitaiuščie Sovetskogo Soiuza. Vysšaia Škola, Moskva. (In Russian). English translation: Heptner VG, Sludskii AA, Bannikov AG (1992) Mammals of the Soviet Union. Volume II, Part 2: Carnivora (Hyaenas and Cats). Smithsonian Institute and the National Science Foundation, Washington, D.C., pp 95–202

Hes L (1991) The leopards of Londolozi. Struik Winchester, Cape Town

Hoffner E (2019) Taiwan: Extinct leopard subspecies allegedly seen by rangers. https://news.mongabay.com/2019/03/taiwan-extinct-leopard-subspecies-allegedly-seen-by-rangers/. Accessed: 9 June 2019

Humphreys PN, Kahrom E (1995) The lion and the gazelle: the mammals and birds of Iran. Images Publishing, Bath

Hunter L (2015) Wild cats of the world. Bloomsbury, London, New York

Hurn S (2009) Here be dragons? No, big cats! Predator symbolism in rural West Wales. Anthropol Today 25:6–11. https://doi.org/10.1111/j.1467-8322.2009.00640.x

Hussain S (2003) The status of the snow leopard in Pakistan and its conflict with local farmers. Oryx 37:26–33. https://doi.org/10.1017/S0030605303000085

Iongh HH, Atmosoedirdjo S, Helvoort BE van et al (1982) An ecological survey of the Kangean Archipelago. PPA/VTF Report, Bogor/Amsterdam

Istiadi Y, Panekenan N, Priatna D et al (1991) Untersuchung über die Carnivoren des Gunung Halimun Naturschutzgebietes. Zoologische Gesellschaft für Arten- und Populationsschutz e.V. Mitteilungen 7:3–5

Jacobson AP, Gerngross P, Lemeris JR Jr et al (2016) Leopard (*Panthera pardus*) status, distribution, and the research efforts across its range. PeerJ 4:e1974. https://doi.org/10.7717/peerj.1974

Jhala YV, Qureshi Q, Nayak AK (eds) (2019) Status of tigers, co-predators and prey in India 2018. Summary Report. National Tiger Conservation Authority, Government of India, New Delhi and Wildlife Institute of India, Dehradun

Johnson K (2002) The status of mammalian carnivores in Turkey. Endanger Species Updat 19:232–237

Jungius H (2010) Feasibility study on the possible restoration of the Caspian Tiger in Central Asia. Survey Report. WWF Russia, 34 p

Jungius H, Chikin Y, Tsaruk O, Pereladova O (2009) Pre-feasibility study on the possible restoration of the Caspian Tiger in the Amu Darya Delta. https://wwf.ru/upload/iblock/56e/tiger_pre_feasibility_study.pdf. Accessed: 5 June 2019

Kaushik H (2017) Lion population roars to 650 in Gujarat forests. The Times of India. https://timesofindia.indiatimes.com/city/ahmedabad/Lion-population-roars-to-650-in-Gujarat-forests/articleshow/59907625.cms. Retrieved 9 Aug 2017

Kemf E, Jackson P (1994) Wanted Alive! Rhinos in the Wild. WWF International, Gland, Switzerland

Killin A (2013) The arts and human nature: evolutionary aesthetics and the evolutionary status of art behaviours. Biol Philos 28:703–718. https://doi.org/10.1007/s10539-013-9371-5

Kitchener AC, Breitenmoser-Würsten C, Eizirik E et al (2017) A revised taxonomy of the Felidae. The final report of the Cat Classification Task Force of the IUCN/ SSC Cat Specialist Group. Cat News 11:3–80

Kock D (1990) Historical record of a tiger, *Panthera tigris* (Linnaeus, 1758), in Iraq. Zool Middle East 4:11–16. https://doi.org/10.1080/09397140.1990.10637583

Kullman E (1967) Ueber Leoparden Afghanistans und ihre Parasiten. Freunde des Kolner Zoos Jahrg 10:126–135

Lavauden L (1932) Les grands fe´lins de l'Afrique du Nord et leur disparition. Le Che´ne. Soc. Forestie`re Mediterrane´enne et Coloniale 4:208–234

Li J (2018) Zanzibar leopard captured on camera, despite being declared extinct. https://www.insideedition.com/zanzibar-leopard-captured-camera-despite-being-declared-extinct-43962. Accessed: 3 July 2019

Lindstrøm TC (2010) The animals of the arena: how and why could their destruction and death be endured and enjoyed? World Archaeol 42:310–323. https://doi.org/10.1080/00438241003673045
Long JL (2003) Introduced mammals of the world: their history, distribution and influence. CSIRO Publishing, Collingwood
Lue KY, Chang WS, Hwa BZ (1992) The faunal investigation on Yu-Li Wildlife Nature Reserve. Report 79-02 Taiwan Forest Bureau Conservation and Research, Taipei
Luo SJ, Johnson WE, Martenson J et al (2008) Subspecies genetic assignments of worldwide captive tigers increase conservation value of captive populations. Curr Biol 18:592–596. https://doi.org/10.1016/j.cub.2008.03.053
Maas PHJ (2010) Bali Tiger - *Panthera tigris balica*. In: TSEW (2019). The Sixth Extinction Website. http://www.petermaas.nl/extinct. Accessed: 8 Aug 2019
Mansfield-Aders W (1920) The natural history of Zanzibar and Pemba. In: Pearce FB (ed) Zanzibar: the island metropolis of eastern Africa. Frank Cass, London, pp p326–p339
Mazak V (1981) *Panthera tigris*. Mamm Species 152:1–8. https://doi.org/10.2307/3504004
McCarthy MC, Chapron G (eds) (2003) Snow leopard survival strategy. ISLT and SLN, Seattle
McLellan BN, Proctor MF, Huber D et al (2017) *Ursus arctos* (amended version of 2017 assessment). The IUCN Red List of Threatened Species 2017: e.T41688A121229971. https://doi.org/10.2305/IUCN.UK.2017-3.RLTS.T41688A121229971.en. Accessed: 12 May 2019
Meena V, Macdonald DW, Montgomery RA (2014) Managing success: Asiatic lion conservation, interface problems and peoples' perceptions in the Gir Protected Area. Biol Conserv 174:120–126. https://doi.org/10.1016/j.biocon.2014.03.025
Miththapala S, Seidensticker J, O'Brien SJ (1996) Phylogeographic subspecies recognition in leopards (Panthera pardus): molecular genetic variation. Conserv Biol 10:1115–1132. https://doi.org/10.1046/j.1523-1739.1996.10041115.x
Mossaz A, Buckley RC, Castley JG (2015) Ecotourism contributions to conservation of African big cats. J Nat Conserv 28:112
Namgail T, Fox JL, Bhatnagar YV (2007) Carnivore-caused livestock mortality in Trans-Himalaya. Environ Manag 39(4):490–496
Natesh M, Atla G, Nigam P et al (2017) Conservation priorities for endangered Indian tigers through a genomic lens. Sci Rep 7:9614. https://doi.org/10.1038/s41598-017-09748-3
Nowell K, Jackson P (1996) Wild cats. Status survey and conservation action plan. IUCN, Gland
Ocholla GO, Mireri PC, Muoria PK (2016) Application of indigenous knowledge systems in wildlife management: a case study of the Samburu pastoral community in Kenya. Int J Appl Sci Technol 6:72–80
Ognev SI (1935) Carnivora (Fissipedia). Mammals of the U.S.S.R. and adjacent countries. National Science Foundation, Washington D.C.
Panouse JB (1957) Les Mammifères du Maroc. Primates, Carnivores, Pinnipèdes, Artiodactyles. Travaux de l'Institut Scientifique Chérifien. Série Zoologie 5:1–206
Paviolo A, Cruz P, Iezzi ME, Pardo JM, Varela D, De Angelo C, Benito S, Vanderhoeven E, Palacio L, Quiroga V, Arrabal JP, Costa S, DiBitetti MS (2018) Barriers, corridors or suitable habitat? Effect of monoculture tree plantations on the habitat use and prey availability for jaguars and pumas in the Atlantic Forest. For Ecol Manag 430:576–586
Pease AE (1899) The lion in Algeria. In: Bryden HA (ed) Great and small game of Africa. Rowland Ward Ltd., London, pp 564–568
Pease AE (1913) The book of the lion. John Murray, London
Pocock RI (1932) The leopards of Africa. Proc Zool Soc London 2:543–591
Pocock RI (1930) The panthers and ounces of Asia. J Bombay Nat Hist Soc 34:64–82
Prokhorov V (2002) Illegal international wildlife trade. http://www.yurclub.ru/docs/ecology/article13.html. Accessed: 17 Apr 2019
Purroy Iraizoz FJ (2010) El leopardo del Atlas: 'Salsero' y otras andanzas. Edilesas
Rabinowitz A (1988) The clouded leopard in Taiwan. Oryx 22:46–47. https://doi.org/10.1017/S003060530002740X

Rafiastanto A (1994) Camera trapping survey of Javan tiger and other wild animals in Meru Betiri National Park: a final report. Ministry of Forestry

Ripple WJ, Estes JA, Beschta RL et al (2014) Status and ecological effects of the world's largest carnivores. Science 343:1241484

Rodríguez-Lozano P, Verkaik I, Rieradevall M, Prat N (2015) Small but powerful: top predator local extinction affects ecosystem structure and function in an intermittent stream. PLoS One 10(2):e0117630

Ross DH (2006) The heraldic lion in Akan art: a study of motif assimilation in southern Ghana. Metrop Mus J 16:165–180. https://doi.org/10.2307/1512775

Sanderson EW, Redford KH, Chetkiewicz CLB et al (2002) Planning to save a species: the jaguar as a model. Conserv Biol 16:58–72. https://doi.org/10.1046/j.1523-1739.2002.00352.x

Schwarz E (1912) Notes on Malay tigers, with description of a new form from Bali. Ann Mag Nat Hist 10:324–326. https://doi.org/10.1080/00222931208693243

Seidensticker J (1986) Large carnivores and the consequences of habitat insularization: ecology and conservation of tigers in Indonesia and Bangladesh. In: Miller SD, Everett DD (eds) Cats of the world: biology, conservation, and management. National Wildlife Federation, Washington, D.C., pp 1–41

Seidensticker J (1987) Bearing witness: observations on the extinction of *Panthera tigris balica* and *Panthera tigris sondaica*. In: Tilson RL, Seal US (eds) Tigers of the world: the biology, biopolitics, management, and conservation of an endangered species. Noyes Publications, Park Ridge, pp 1–8

Seidensticker J, Suyono I (1980) The Javan tiger and the Meru-Beriti Reserve, a plan for management. International Union for the Conservation of Nature and Natural Resources, Gland

Şekercioĝlu ÇH, Anderson S, Akçay E et al (2011) Turkey's globally important biodiversity in crisis. Biol Conserv 144:2752–2769

Sergio F, Newton I, Marchesi L (2005) Top predators and biodiversity. Nature 436(7048):192. https://doi.org/10.1038/436192a

Seymour KL (1989) Mammalian species: *Panthera onca*. Am Soc Mammal 340:1–9

Silvius KM, Bodmer RE, Fragoso JMV (eds) (2004) People in nature: wildlife conservation in South and Central America. Columbia University Press, New York, 464 p

Singh AP (2017) The Asiatic lion (*Panthera leo persica*): 50 years journey for conservation of an endangered carnivore and its habitat in GIR protected area, Gujarat. India Indian Forester 143:993–1003

Singh HS, Gibson L (2011) A conservation success story in the otherwise dire megafauna extinction crisis: the Asiatic lion (*Panthera leo persica*) of Gir forest. Biol Conserv 144:1753–1757. https://doi.org/10.1016/j.biocon.2011.02.009

Smith CH (1842) Black maned lion *Leo melanochaita*. In: Jardine W (ed) The naturalist's library. Vol. 15 Mammalia. Chatto and Windus, London. p Plate X, 177

Smithers RHN (1971) Family Felidae. In: JAJ M, Setzer HW (eds) The mammals of Africa: an identification manual. Smithsonian Institution Press, Washington, D.C.

Spassov N, Ignatov A, Acosta-Pankov I (2016) On the status of the leopard in Turkey, again. CAT News 64:18–22

Stein AB, Athreya V, Gerngross P et al (2016) *Panthera pardus*. The IUCN Red List of Threatened Species. IUCN. 2016: 15954/102421779. https://doi.org/10.2305/IUCN.UK.2016-1.RLTS.T15954A50659089.en. Accessed: 21 June 2019

Storey WK (1991) Big cats and imperialism: lion and tiger hunting in Kenya and Northern India, 1898–1930. J World Hist 2:135–173

Swai IS (1983) Wildilife conservation status in Zanzibar. Dissertation, Univeristy of Dar es Salaam

Tait P (2009) Controversy about a human-animal big cat stunt in Fillis's circus. Early Pop Vis Cult 7:199–211

Tilson R (1999) Sumatran Tiger Project Report No. 17 and 18: July–December 1999 Grant number 1998-0093-059. Indonesian Sumatran Tiger Steering Committee, Jakarta

Tilson R, Defu H, Muntifering J, Nyhus PJ (2004) Dramatic decline of wild South China tigers Panthera tigris amoyensis: field survey of priority tiger reserves. Oryx 38:40–47

Toyran K (2018) Noteworthy record of Panthera pardus in Turkey (Carnivora: felidae). Fresenius Environ Bull 27:7348–7353

Trinkel M, Angelici FM (2016) The decline in the lion population in Africa and possible mitigation measures. In: Angelici FM (ed) Problematic wildlife: a cross-disciplinary approach. Springer, Cham, pp 45–68

Uphyrkina O, Johnson WE, Quigley H et al (2001) Phylogenetics, genome diversity and origin of modern leopard, Panthera pardus. Mol Ecol 10:2617–2633. https://doi.org/10.1046/j.0962-1083.2001.01350.x

Valenciennes A (1856) Sur une nouvelles espèce de Panthère tué par M. Tchihatcheff à Ninfi, village situé à huit lieues est de Smyrne. C R Hebd Seances Acad Sci 42:1035–1039

van Helvoort BE, de Iongh HH, van Bree PJH (1985) A leopard-skin and -skull (*Panthera pardus* L.) from Kangean Island, Indonesia. Z. Säugetierkunde 50:182–184

Walsh MT, Goldman HV (2007) Killing the king: the demonization and extermination of the Zanzibar leopard. In: Dounais E, Motte-Florac E, Dunham M (eds) Animal Symbolism. The 'Keystone' Animal in Oral Tradition and Interactions between Humans and Nature. IRD, Paris, pp 1133–1182

Walsh MT, Goldman HV (2008) Updating the inventory of Zanzibar leopard specimens. CAT News 49:4–6

Walsh MT, Goldman HV (2017) Cryptids and credulity. The Zanzibar leopard and other imaginary beings. In: Hurn S (ed) Anthropology and cryptozoology. Exploring encounters with mysterious creatures. Routledge, London, New York

Wozencraft WC (2005) Order Carnivora. In: Wilson DE, Reeder DAM (eds) Mammal species of the world. A taxonomic and geographic reference, 3rd edn. Johns Hopkins University Press, Baltimore, pp 532–628

Yamaguchi N, Haddane B (2002) The north African Barbary lion and the atlas lion project. Int Zoo News 49:465–481

Zalasiewicz JAN, Williams M, Steffen W, Crutzen P (2010) The new world of the anthropocene. Environ Sci Technol 44:2228–2231

Part VI
Zoos, Conservation, and Animal Rights

We organized the 'III Convegno Nazionale Fauna Problematica' (see introduction) with the aim of involving as many stakeholders as possible: conservation scientists, agriculture and hunting associations, and animal rights activists. The conference was a success.

Despite this, in the following weeks, some animal rights activists argued on social networks that farmers and hunters should not had been given space at a scientific conference. This episode inspired us to include in the book a section entirely dedicated to the relationship between animal rights activists and the world of conservation. The result is four chapters that deal with the subject in many ways.

The first (Perry et al. 2020) underlines the need for greater collaboration between these two worlds. In fact, say the authors, although often distant and characterized by issues difficult to reconcile, animal rights activists and conservationists share important principles.

In the second chapter, Hosey et al. (2020) address the issue of 'charismatic species' in zoos from an unusual point of view. In fact, the general attention of the public and researchers, both in terms of animal rights movements and conservation efforts, tends to focus mainly on large mammals (e.g. elephants, big apes, and cetaceans) at the expense of 'noncharismatic' species that also play an important role in ecosystems.

Finally, Robovský et al. (2020) address the age-old issue of the purpose, role, and mission of zoos in the light of the sixth mass extinction with particular regard to breeding programmes for the most endangered species.

The last chapter (Perco 2020) of this section concerns two Italian case studies on grey wolf (*Canis lupus italicus*) and brown bear (*Ursus arctos arctos* and *U.a. marsicanus*), which highlight the importance of professional communicators in supporting technicians.

References

Angelici FM, Rossi L (eds) (2016) Atti del III Congresso Nazionale Fauna Problematica (Cesena, 24–26 November, 2016). Comune di Cesena, and Associazione 'Orango', xiii + 201 p

Hosey G, Melfi V, Ward SJ (2020). Problematic animals in the Zoo: the issue of charismatic megafauna. In: Angelici FM, Rossi L (eds) Problematic wildlife II. New conservation and management challenges in the human-wildlife interactions. Springer International, Cham, pp 485–508

Perco F (2020) Communication and wildlife conservation (grey wolf and brown bear in Italy). In: Angelici FM, Rossi L (eds) Problematic wildlife II. New conservation and management challenges in the human-wildlife interactions. Springer International, Cham, pp 529–557

Perry G, Sarge MA, Perry D (2020) Alternative facts and alternative views: scientists, managers, and animal rights activists. In: Angelici FM, Rossi L (eds) Problematic wildlife II. New conservation and management challenges in the human-wildlife interactions. Springer International, Cham, pp 421–450

Robovský J, Melichar L, Gippoliti S (2019) Zoos and conservation in the Anthropocene: opportunities and problematic issues. In: Angelici FM, Rossi L (eds) Problematic wildlife II. New conservation and management challenges in the human-wildlife interactions. Springer International, Cham, pp 421–450

Chapter 13
Alternative Facts and Alternative Views: Scientists, Managers, and Animal Rights Activists

Gad Perry, Melanie A. Sarge, and Dan Perry

I thought of all
the massacres and slaughter
of persecuted insects
at the hands of cruel humans
and I cried

— Don Marquis, Archy and Mehitabel, 1927

13.1 Introduction

On 16 April 2017, a female polar bear died at SeaWorld San Diego. "Szenja" had been there for two decades. The *LA Times* (Weisberg 2017) reported that Szenja "died unexpectedly Tuesday following a brief illness. The bear had been showing signs of appetite and energy loss for about a week, but the exact nature of her illness and sudden death are still unexplained. … Citing Polar Bears International, SeaWorld said polar bears typically have a life expectancy of 15 to 18 years in the wild, although some can live as long as 30 years." "In a news release, SeaWorld said that Szenja served as an ambassador for arctic animals, raising awareness of polar bears … She also has participated in various studies related to polar bear hearing sensitivity, social habits, reproductive hormones and seasonal behavior patterns," Weisberg (2017) concluded.

G. Perry (✉)
Department of Natural Resource Management, Texas Tech University, Lubbock, TX, USA
e-mail: Gad.Perry@ttu.edu

M. A. Sarge
College of Media and Communication, Texas Tech University, Lubbock, TX, USA

The Media School, Indiana University, Bloomington, IN, USA
e-mail: msarge@iu.edu

D. Perry
Arava Institute for Environmental Studies, Kibbutz Ketura, D.N. Hevel Eilot, Israel

F. M. Angelici, L. Rossi (eds.), *Problematic Wildlife II*,
https://doi.org/10.1007/978-3-030-42335-3_13

The same event was reported rather differently in One Green Planet (Neff 2017). "Being separated from someone you care about is one of the worst feelings. You think about them constantly, wondering when you'll be able to see them again. Well, humans aren't the only ones that form such strong bonds. Polar bears do too," the story began. Two paragraphs later it concludes "let's be honest, it's pretty clear Szenja died of a broken heart." That claim originated with People for the Ethical Treatment of Animals (Peta). According to Neff (2017), "Life for wild animals in zoos is stressful, boring and most of all, miserable. These animals have been stolen from the wild, dumped behind bars in concrete enclosures, sometimes even pumping the animals full of antidepressants, and then the animals are left to live out the rest of their lives in a place they'll never be able to call home. To zoos, they are disposable, there solely to make a profit. … The best thing you can do to help these animals in captivity is to avoid visiting SeaWorld and zoos altogether."

Science Daily did not report *this* event at all. Not long before, however, it reported a different polar bear story originating in San Diego (Zoological Society of San Diego 2016). The account is based on a published scientific paper and mentions that the "study, by San Diego Zoo Global conservationists … has revealed that access to terrestrial food is not sufficient to reduce the rate of body mass loss for fasting polar bears." "The data … will provide scientists with new insights into the bears' daily behavior, movements and energy needs, and a better understanding of the effects of climate change on polar bears," it concluded.

These media reports are emblematic of the disconnect between scientists and managers, on the one hand, and animal rights advocates, on the other hand. The report in the *LA Times* emphasizes the benefits to science, society, and conservation of having the bear in captivity, and the lack of clear medical evidence about the cause of death, which happened at an age when polar bears may be expected to die in the wild. The story in One Green Planet instead delivered an emotional interpretation of the cause of death and a call to avoid zoos. Neither story touched on the perspective given in the other, however briefly. A person reading the stories without identifying information would hardly be able to tell they were reporting the death of the same bear. The Science Daily reporting might as well have originated in an alternate reality.

The lack of communication between scientists, the general public, and animal rights advocates can be both significant and problematic. Here we review progress in the decade since our previous work on this disconnect (Perry and Perry 2008a, b) and then examine two case studies and review the recent literatures of animal rights and science/management groupings. Although many examples come from the United States and, to a lesser extent, Europe, we sought case studies from around the globe, where conservation is more often addressed than issues of animal rights and welfare. We provide an analysis of communication used in both groups and in their interactions and attempt to integrate insights gained in each section into an overall picture and set of recommendations. Similar to conclusions from studies of modern communication in other contexts, an important part of our thesis is that current technology allows audiences to more thoroughly segregate themselves from one another such that they are exposed to different information, are inundated by more uniform interpretations, and are decreasingly likely to even listen to alternative viewpoints, let alone reexamine their own.

13.2 Where We Were in 2008: A Brief Recap

In the mid-1970s, two seminal publications (Regan 1976; Singer 1975) helped energize a social movement sometimes generically lumped under the titles "animal rights," "animal welfare," or "animalism." In this paper we use the term "animal rights" in its broadest sense, denoting people who are actively working for the consideration of animals in moral decision-making. While the movement is varied and diverse, ranging from those who call for "humane killing" in the industry (Vinter 1963) to those who call for the complete "liberation" of all animals (Domonick 1997), they are here all labeled under this heading. Philosophically we also include under the title of "animal rights" a variety of ethical systems that move to include animals in ethical considerations – even when those philosophers are not arguing within a moral framework of rights (Singer's utilitarian philosophy is a good example). Whereas conservation focuses on wild organisms and the ecosystems they rely on, the animal rights movement has traditionally focused on captive or domesticated animals. Since most of the suffering that animals encounter in nature is not human-caused, humans do not have a duty to intervene (Rolston III 1988). In fact, traditional philosophical views of animal rights have a lot more to say about what *not* to do than about what *should* be done (Keulartz 2016). Animal ethicists have generally also been quick to assume that the motivations of conservationists and managers are human-centered, rather than on what biologists consider good for the animals (Keulartz 2016) – not always without cause.

A decade later, after this movement had begun to affect scientific research, wildlife management, hunting, and commercial activity, Hutchins and Wemmer (1986) discussed the difference in worldviews between wildlife conservation and animal rights. Overall, they saw few areas of congruence between the utilitarian view they advocated and the animal rights/animal welfare approach, though they agreed that humane treatment of study organisms was desirable. Moreover, Hutchins and Wemmer (1986) noted, "Clearly, the animal rights/humane ethic and the environmental/ conservation ethic will lead to the same decisions in many situations. For example, both ethics would consider it wrong for humans to destroy wildlife habitat, or to pollute it with chemicals and wastes." Growth of the human population and the resulting growing extinction crisis make it "more essential that those who care about wildlife and nature, and those who care about the rights of individual animals close ranks to do battle with a common enemy," they concluded. Shortly thereafter, Soulé (1990) was one of the first to note that "Conflicts between animal rights groups and management agencies are increasing in frequency and cost … The minimization of such conflicts," he added, "will require both public education and courageous leadership."

Conservationists were slow to take up the challenge. A decade later, for example, Van Houtan (2006) discussed ethics and social input in conservation and considered virtue ethics as an approach that might allow scientific and social concerns to be simultaneously addressed. Neither the term "animal rights" not "animal welfare" appears in the article, nor are any of the works of Singer or Regan cited.

Perry and Perry (2008a) explored the intersection of animal rights and conservation/management in the context of invasive species. Conservation biologists, who define an invasive species as a non-native taxon which poses a major threat to biodiversity, human health, or economic interests, often recommend lethal control. Animal rights groups typically oppose such killing, sometimes blocking eradication attempts. Though the gaps between the views of the two groups are large, Perry and Perry (2008a) also described a small number of successful collaborations between managers and animal rights groups. They recommended that both groups attempt to better understand the other and seek areas of possible cooperation, such as joint support for alternatives to lethal control, for example, policy solutions that reduce the rate of invasive species arrival. Incidentally, such communication had just been reported on by Engeman et al. (2007). In that case study, which involved removal of Black-tailed jackrabbits from the Miami International Airport in Florida, USA, external concerns (human safety) are ultimately what led to a successful process involving live-trapping and translocation of some of the animals and killing of the rest (Engeman et al. 2007). The clearest success story recounted by Perry and Perry (2008a) likewise had a human safety component.

The call for greater understanding and cooperation, where possible, was not universally lauded. Once supporting conciliation (Hutchins and Wemmer 1986), Hutchins (2008), at the time serving as the Executive Director/CEO of The Wildlife Society, stated that "It would be wonderful if we could all get along, but it is time to recognize that some ideas are superior to others because they clearly result in the 'greatest good.' … It is time to face up to the fact that animal rights and conservation are inherently incompatible." Referencing Soulé (1990), he concluded, "Cooperation and compromise, no matter what the cost, is not courageous leadership" (Hutchins 2008). Formally, at least, the animal rights community neglected to engage altogether.

Responding, Perry and Perry (2008b) noted that in those cases where collaboration is possible, "What matters in terms of achieving conservation goals is that members of the two sides have a common objective and would both benefit if they cooperated in trying to achieve it. Whether they choose to collaborate remains to be seen, but the camps are not always as far apart as some imagine." Perry and Perry (2008b) again advocated for cooperation to address "the financial interests that drive much of the invasive species crisis (agriculture, horticulture, and the pet industry) … rather than advocating a mop-up approach for some of its consequences."

13.3 The Past Decade

Much has happened in the decade that has elapsed since publication of the work of Perry and Perry (2008a, b). In this section, we provide updates on recent perspectives on the philosophy of animal rights and in the science and management community.

13.3.1 Animal Rights

The animal rights literature has exploded over the past two decades, with hundreds of papers appearing annually and more than 500 journals including coverage of the issue from different perspectives. Many of them deal with farmed animals, and mammals consistently receive the most attention (Walker et al. 2014). Although much of the literature is oriented towards technicians, veterinarians, and others involved in the welfare of science and industry animals, some interesting philosophical ideas have also emerged or developed in the last decade. Here we discuss two that were not examined by Perry and Perry (2008a) and that have bearing on the environmental/animal debate.

The first development covered here is ecofeminism, a frame of thought that emerged in the 1970s and has lately regained popularity, especially in the field of animal studies (Graad 2011). As the name suggests, it is primarily an environmental movement. However, ecofeminism also has a well-developed animal ethics side as well, one that, judging from the number of speakers in conferences and papers published in recent years, has become much more prominent in "animal studies" circles over the past decade. Ecofeminism can be described as a collection of theories with the common claim that the historical exploitation of the environment, women, Africans, animals, and other groups all stem from the same dualist view that categorizes groups as "superior" and "inferior". Ecofeminists generally call for an approach that is based on "care" rather than "control," an attitude that usually sits well with traditional animal advocates. Many writers, such as Carol Adams and Lori Gruen, are dedicated to animal rights, whereas others are more broadly concerned with environmental issues. However, ecofeminism in general views the scientific method, which lies at the heart of traditional environmental concern, as a part of the oppressive, masculine, rational-oriented system, and treat scientific methods and scientists with suspicion (Bailey 2007).

As early as 2003, it was argued that the "ethics of care" promoted by ecofeminists is less equipped to deal with human-wild animals relationships than it is with our interactions with domesticated ones (Clement 2007). Indeed, most ecofeminist writing on animals is centered on the treatment of domesticated animals and the eating of meat. The one issue concerning wild animals that is commonly addressed is hunting, which ecofeminists are categorically against (Kheel 1996; Luke 2007; Graad 2011). Ecofeminism's relative flexibility could potentially allow for different forms of "care" and "compassion" that address the topics discussed here, though this has not happened so far. Thus, ecofeminism does not currently offer much novel insight on wildlife. Lack of a firm agenda allows for intellectual flexibility that is often missing in this area, but it does not lay an obvious path for productive engagement with scientists and managers.

Another interesting development in the field of animal studies comes from the social and political fields. In "Zoopolice," Donaldson and Kymlicka (2011) claim that we need to address our societies not as ones comprised of humans only, but rather as a mixed species society. Although the authors mostly discuss domesticated animals, "Zoopolice" holds the potential to encompass all living species and thus extend to relationships with wild animals as well. By treating animals as political entities, one can relate to both individuals and groups (think of both citizens and countries receiving attention in human affairs), something which was a problem for animal-related philosophical frameworks (see several examples in Hargrove 1992). The hope this frame of thought presents is in its potential for flexibility. While traditional animal ethics were fairly rigid, these two seem to have more room to "wiggle." After all, politics often involves a more pragmatic approach to conflict resolution, one that may involve compromise out of mutual respect, or at least shared interests.

13.3.2 Science and Wildlife Management

"[C]oncern for rights of the individual organism must be secondary to the much larger and ecologically more important concerns for species and functioning ecosystems… native species must trump exotics, even if it means individual loss and suffering by the exotic species" opined Meffe (2008), then Editor of Conservation Biology. To a large extent, this remains the core position taken by biologists and ecosystem managers (Minteer and Collins 2013). The disputes between them and animal rights advocates have been both common and profound, and current philosophical theory does not offer a workable bridge between the more extreme positions espoused on each side (Curzer et al. 2011). Yet there is often an interest in combining traditional conservation goals with improved animal rights (Fraser 2010; Minteer 2013). For example, Hampton and Forsyth (2016) evaluated the humaneness of night shooting for controlling urban kangaroos in Australia and concluded that, as death was virtually instantaneous, the method caused little suffering or stress. Focusing on the issue of subsistence hunting of wild animals for food (often known as "bushmeat"), Minteer (2013) pointed out that conservation, human health, and other interests align in attempting to address this problem and showed some cases where divergent interests align. However, although his recommendation that "Both animal rights/welfare and conservation proponents, moreover, could support greater restrictions on unnecessarily harmful and indiscriminate bushmeat hunting techniques" is very similar to that of Perry and Perry (2008a, b), Minteer (2013) does not report any actual examples of such collaboration.

Seeking a practical approach to the ongoing arrival of non-native species associated with the pet and aquarium trades (UNEP 2010), Perry and Farmer (2011) set to provide a policy approach to address invasive species prevention that could be adopted by both groups. They proposed a targeted tax on the pet trade whose proceeds would be dedicated to local and national prevention and response actions and whose level could be increased or decreased based on changes in invasive species arrival rates. Unfortunately, there is greater support and willingness-to-pay for eradication than for prevention (García-Llorente et al. 2011). Moreover, although efforts such as that of Perry and Farmer (2011) or Essl et al. (2017) can help managers better value and mitigate impacts, they do not address social conflicts about underlying ethical perceptions. When stakeholders disagree about whether invasive species require action because of different philosophical worldviews, more precise valuation of economic or ecological impact and more effective policy remedies have limited applicability.

A number of authors (e.g., Webb and Raffaelli 2008; Boudjelas 2009; Harrington et al. 2013) have stressed the importance of considering the social background for a project and taking public concerns into consideration in project design. For example, ecological considerations might suggest removal of female aoudad in the many countries where they have become established invaders, but public sentiment makes killing of large males a more acceptable management strategy (Mori et al. 2017).

Reporting of invasive species issues in the mainstream media remains biased. Stories focusing on welfare issues involving wild vertebrate, especially ones in which lethal control is a factor, are most likely to be reported by the UK media (Feber et al. 2017). Most recently, Crowley et al. (2017a, b) advocated for use of social research and a deliberative engagement model in situations of conflict between wildlife managers and the public. They argued that recognizing the inevitability of such conflicts and planning to address them early in a project reduces the severity of such conflicts and their likelihood of becoming destructively disruptive. Crowley et al. (2017a) recount in detail the initial conflict over management of hedgehogs that were harming native wading birds in Scotland. Stressing the need for public support for management recommendations, Webb and Raffaelli (2008) analyzed the language surrounding this conflict and noted that managers and animal rights advocates used incompatible arguments to talk past, instead of with one another, a disconnect exacerbated by media reporting. However, following a rocky start, the two groups developed a more collaborative model in which hedgehogs are translocated rather than killed. Managers now focus their message on bird recovery rather than invasive species control (Crowley et al. 2017a). The changes in how the managers conducted their work resulted in improved animal rights but caused some delays and additional costs. The cases of public participation detailed by Boudjelas (2009) all similarly involve local communities, not animal rights activists. Under the heading "Ethics," the only reason provided by Boudjelas (2009) for consulting the public is the need for inclusivity. "Animal welfare," which is completely missing from the book's index, only comes up at the very end of the chapter. "Every effort must be made to ensure the ethical considerations are fully addressed and the public concern with animal rights and humane behavior during control are assuaged" (Boudjelas 2009).

Crowley et al. (2017a) mention that hunters and animal-rights activists, seemingly implacable foes (Hutchins and Wemmer 1986; Hutchins 2008) at times work together to protect non-native game species. For example, it was hunter organizations such as the Texas Hog Hunters Association and the Texas State Rifle Association, supported by Environmental Defense Fund, that convinced a judge to temporarily prevent the use of warfarin baits to kill feral hogs and convinced the bait's manufacturer to withdraw registration for such use in Texas (Copeland 2017; Murphy 2017). Feral hogs "provide hunters with a food source and a significant revenue stream, as slaughterhouses will pay $30 to $180 per pig ... But it's unlikely chefs will buy poisoned pork belly, bacon and short ribs." Uncharacteristically, it was the Texas State Rifle Association which "had concerns about the poison making its way into other animals or into humans" (Copeland 2017). And just as ironically, Texas Agriculture Commissioner Sid Miller, who was the major proponent of using warfarin to kill feral hogs, was quick to blame "lawyers, environmental radicals and the misinformed … politically correct urban media hacks and naysayers," rather than the hunters or meat processors for blocking such use (Murphy 2017). And it was the scientists who conducted research on the use of warfarin for hog control who raised concerns about its humaneness (Murphy 2017). In many situations, however, hunter groups have taken an opposite view, applauding recent regulation changes, opposed by animal rights organizations and conservationists alike, which increase predator hunting opportunities in Alaska (Actamn and Bale 2017).

Animal rights concerns have at times also aligned with those of conservation advocates. For example, lethal control of native predators has long been carried out by the government on behalf of ranching and hunting interests (Cluff and Murray 1995; Van Ballenberghe 2006). However, the public has traditionally shown a preference for non-lethal control, and conservation advocates have also argued against the culling on ecological grounds (Cluff and Murray 1995; Treves and Naughton-Treves 2005). The relatively recent "compassionate conservation" movement (e.g., Bekoff 2013) is a good example of this approach. It was a combination of animal rights and conservation groups that jointly got Mendocino County in California to suspend its contract to lethally control predators with Wildlife Services, an arm of the US Department of Agriculture (Hill 2016). At times, they have led to unexpected outcomes, as in a Dutch legal case in which animal rights advocates sued the state over active management of reintroduced animals. The court sided with the plaintiffs that such animals could no longer be considered to be owned by the state, but active management, including culling of suffering individuals, has resulted from the controversy (Swart 2016).

More contentious, at times, have been manager-instituted killings and translocations of wildlife that were carried out over the last few years in southern Africa, with the partial aim of reducing animal suffering. For example, 18 elephants were to be sent from Hlane National Park in Swaziland, which was suffering from the worst drought in its history, rather than being killed by managers to alleviate starvation for multiple species (Vidal 2016). With encouragements from several conservation organizations, three US zoos have agreed to adopt them instead of having them killed and were given import permits by the US Fish and Wildlife Service. This seems like exactly the kind of case where captivity could be justified, according to Bovenkerk (2016). However, animal rights groups such as Friends of Animals and GroupElephant.com filed a lawsuit, arguing that other solutions such as export to South Africa were also available and blocking the transfer (Harris and Best 2016; Vidal 2016). At the same time, overpopulation of hippos and buffaloes in Kruger national park in South Africa, combined with the worst drought in 35 years, led the South African National Park service to call for an "offtake." Although not the only consideration, a spokesperson for the service explained that "We don't want the animals dying of hunger and rotting on the ground. We are trying to be humane in the way the animals die" (Burke 2016). The same action in Zambia was halted over concerns from animal rights advocates (Mfula 2016). In Kenya, the ongoing drought has resulted in conflicts between pastoralists, desperate to feed their starving cattle, and both ranchers and natural areas. Both ranchers and elephants are being shot by herders (Burke 2017). Cash-strapped, the authorities in Zimbabwe have been unable to provide water to the overpopulated elephants of Hwange National Park, which is suffering from the same prolonged drought. Here, however, a coalition of managers, conservation organizations, and rights advocates such as GroupElephant.com (the same group involved in an adversarial position in Swaziland) united to pay for water pumping which is helping alleviate the situation (Nkala 2017).

13.4 Case Studies

We begin this section with a more detailed look at zoos and aquariums because they form the setting for the bear example we opened this chapter with. We then return to the invasive species setting addressed by Perry and Perry (2008a), this time emphasizing a perennial friction point between conservationists and animal rights advocates, management of feral cats.

13.4.1 Zoos and Aquariums

Zoos originated as royal pastimes as early as the thirteenth century, with modern zoos emerging starting in the eighteenth century (Jamieson 1985). By the 1990s there were about 1100 such institutions around the world, and they hosted some 600,000,000 visitors a year (Baratay and Hardouin-Fugier 2003). Zoo involvement in conservation began in the nineteenth century and has greatly intensified in the late twentieth century as the scope of the extinction crisis began emerging and as public expectations of zoos changed (Frost 2011; Keulartz 2015). Interest in the welfare (and sometimes rights) of zoo animals is extensive. For example, four books focusing on this topic appeared during 2012–2013 (Walker et al. 2014).

Most animal rights and many animal welfare advocates, such as Jamieson (1985), consider zoos to be immoral because they restrict animal liberty and place wildlife in artificial and limiting environments. Beckoff and Pierce (2017, p. 3) also emphasized the importance of freedom as a basic value for animals. Although acknowledging that captive breeding can assist in wildlife restoration, Jamieson (1985) pointed out that many such efforts fail. Given the limitations of the artificial environment, Jamieson (1985) questioned whether even in "a world in which mountain gorillas can survive only in zoos … it is really better for them to live in artificial environments of our design than not to be born at all." Moreover, if housing in a zoo enabled the destruction of a habitat essential for wild existence of a species, worried Russow (2010), does not the zoo abet habitat destruction and increased endangerment? Another concern has been that it is difficult for animals to maintain their dignity in the zoo environment (Wickins-Dražilová 2006), and animal rights in (at least some) zoos needs to be greatly improved (reviewed in Maple and Perdue 2013). In contrast, proponents of zoos argue that visiting them can enhance empathy, altruism, and social success (Fraser 2009), improve scientific reasoning (Kisiel et al. 2012), play a vital role in conservation education and ex situ and in situ conservation practice (Miller et al. 2004; Jensen 2014; Keulartz 2015), and more. Much of the discussion ignores the wide range of institutions currently in existence, lumping together a broad range of facilities, education efforts, and attempts to participate in conservation activities (Fig. 13.1).

A large segment of recent literature has sought ways to enhance animal welfare in zoos, whether through empirical studies of animals and how enrichment can improve their experience, attempts to better understand the (mostly visiting) public, or philosophical exploration (e.g., Bovenkerk 2016). Following a series of papers exploring ecological ethics in natural habitats, Minteer and Collins (2013) looked at how those applied to captive animals and concluded that "ultimately a more sophisticated and candid analysis of the trade-offs and the multiple imperatives of conservation-driven research on captive populations is required." There is evidence that this attention is paying off (Shani and Pizam 2008; Maple and Perdue 2013; Kagan et al. 2015). The first item in the code of Ethics of the Association of Zoos and Aquariums (AZA: https://www.aza.org/code-of-ethics, accessed 28 April 2017) requires members to "Recognize the moral responsibilities of the individual and the institution not only to our professional associates, fellow employees and volunteers, and the public, but also to the animals under our care." Nonetheless, progress has been partial. As of April 2017, the AZA website did not have a main heading for "animal welfare," for example, and multiple news stories in the last few years reported on perceived ethical concerns about aquatic and other zoological parks holding large mammals.

On the animal rights side, progress has been even more limited. The view of Beckoff and Pierce (2017, p. 94–115) is that animals in zoos are exploited, unhappy, and unhealthy. Zoos are condemned by Bekoff and Pierce (2017) whether they do or do not let animals breed (a basic right when denied, a source of more captive animals and income when allowed); live to a ripe old age (proof of mistreatment if they do not, but the cause of old-age ailments when allowed); provided enrichment

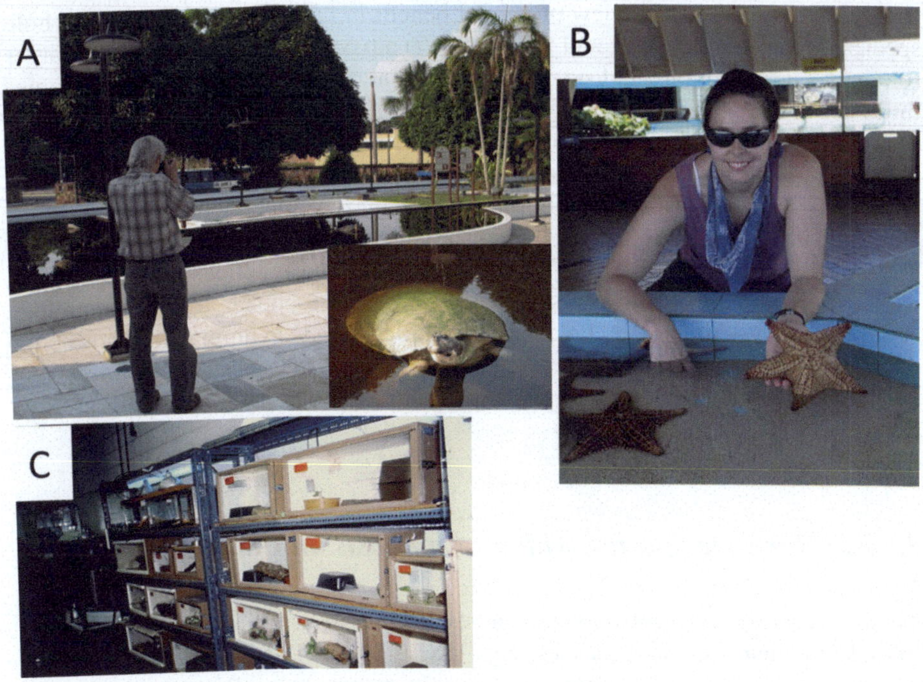

Fig. 13.1 A wide range of facilities offer members of the public an opportunity to experience wildlife. (**a**) A pool at the Tropical Manaus Ecoresort in the State of Amazonas, Brazil, displays native freshwater turtles typical of the Amazon river. Signs offer hotel guests information about each species. Inset: one of the turtles on display. (**b**) The Curaçao Sea Aquarium in the Dutch Antilles offers both traditional dolphin and seal shows and a tank in which visitors can touch harmless sea animals. (**c**) A non-display area at the Dallas Zoo in Texas, USA, is used for holding animals involved in research and captive reproduction efforts

(improving quality of life but also designed "to increase the entertainment value of the animals on display"); and so on. Prominent animal rights organization Peta maintains that "Zoos [are] An Idea Whose Time Has Come and Gone" (http://www.peta.org/issues/animals-in-entertainment/zoos/, accessed 28 April 2017). The Humane Society of the United States has an only marginally more nuanced position (http://www.humanesociety.org/issues/zoos/, accessed 29 April 2017), as do the British Colombia Society for the Prevention of Cruelty to Animals in Canada (http://www.spca.bc.ca/assets/documents/welfare/position-statements/wild-and-exotic-animals-in-permanent-captivity.pdf, accessed 27 April 2017) and the Captive Animals' Protection Society in the United Kingdom (https://www.captiveanimals.org/our-work/zoos, accessed 26 April 2017). And the bar is perhaps being raised even higher: Gjerris et al. (2016) recently advocated that zoos and aquariums should only serve visitors food that is vegetarian or vegan.

Nonetheless, Bovenkerk (2016) recently argued that there are situations where captivity could be ethically justified, most notably when doing so is to the benefit of the animal. However, she also stressed that this view might necessitate a departure from an individual-centric view of animal benefit in favor of one that takes whole species into account. Bekoff and Pierce (2017) advocated for zoos to play a larger role in rescue operations. Indeed, zoos and animal rights organizations have sometimes collaborated to translocate animals from facilities offering suboptimal conditions to more appropriate locations (Maple and Perdue 2013, pp. 172–174).

Convergence of views should also be possible at the other end of the zoo spectrum. Whereas some institutions, most notably in the developed world, have a strong education and conservation mission, there are many facilities in the world which do not. For example, Nuwer (2017) recently described the practices at a number of animal farms (calling themselves zoos) in Asia. These appear to offer appalling conditions to the animals they hold, with often illegal trade in animal products being the main goal. According to Nuwer (2017), opposition to these establishments spans the range from regulatory agencies and diplomatic missions, to animal rights groups, to conservation organizations.

13.4.2 Invasive Species, with an Emphasis on Cats

There is general agreement among scientists that invasive species are a major source of endangerment for native species (e.g., UNEP 2010). The scope and urgency of invasive species issues have only increased in recent years (Bellard et al. 2016; Mollot et al. 2017). Ironically, even as reporting on invasive species issues is increasingly common in the mainstream media, challenges to the scientific view of invasive species as a major biodiversity problem are also rising (Russell and Blackburn 2017). Some of these challenges represent academic disagreements about the nature of biotic change. Many others, however, involve the broader social context in which managers operate, not least of which is concern over animal welfare or rights (Crowley et al. 2017a), on the one hand, and economic constraints, on the other hand. For example, as Klier et al. (2017) point out in assessing control of beaver invasions in Tierra del Fuego, Argentina, "no doubt it is much less expensive to shoot a beaver than to relocate it."

One of the longest-lasting points of contention between animal rights advocates and managers involves cats (Fig. 13.2). The ongoing conflict between those focused on saving individual cats and others looking to protect the species cats prey on (Dauphiné and Cooper 2009; Marra and Santella 2016) is described in a literature too voluminous to fully cover here. Briefly, roaming and feral cats are common in both urban and rural settings in multiple countries (Medina et al. 2011). They prey on both native and problem species, and even though there is little doubt of their impact on native species (e.g., Medina et al. 2011; Loyd et al. 2013; but see Calver et al. 2011), the relative importance of impacts and benefits are sometimes contested (e.g., Wood et al. 2016; Kikillus et al. 2017). Indirect impacts, such as feral cats interbreeding with native species or serving as reservoirs for diseases that affect wildlife and humans, are also documented (reviewed in Brickner 2003). However, a distinct segment of the population cares about such cats, not just in the abstract but

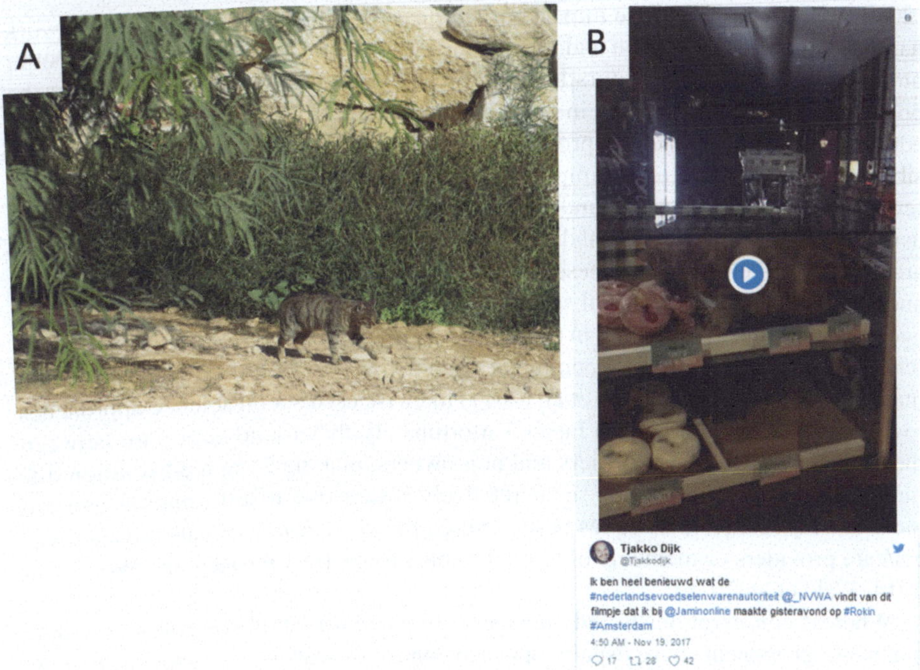

Fig. 13.2 Human activity allows cats to inhabit a wide range of habitats; from native environments they would not normally be able to survive into dense urban centers. (**a**) A feral cat in Midreshet Sde Boker, in the Negev Desert of Israel. (**b**) Cat feeding on a donut in a store in Amsterdam (Twitter post by City Councilor Tjakko Dijk)

also to the extent of spending time each day to feed as many as dozens of feral cats at multiple locations (Gunther et al. 2016). For example, when a City Councilor posted a video to Twitter showing a cat feeding on a donut inside a store closed for the night (Fig. 13.2b), responses ranged from concern about human health violations to admonitions that cats like that are necessary in order to keep at bay rats and mice (Pieters 2017).

Much of the conservation literature remains focused on the biology and ignores or minimizes the social aspects of cat and other invasive species issues. For example, the book of Perrings et al. (2010) has a section devoted to management and policy. However, it does not mention ethics, animal welfare, or animal rights in the index, nor is there a chapter devoted to the human dimensions of the invasive species problem. A recent IUCN publication focusing on urban invaders in Europe (van Ham et al. 2013) mentions "animal rights" once in its 108 pages, in the context of squirrel invasions reviewed by Perry and Perry (2008a). Even there, the only lessons seemingly learned are that animal rights groups can be a serious hindrance, and therefore that "Especially in urban areas if the invader is a 'pretty' species it is necessary to adopt a multidisciplinary approach that involves wildlife biologists and communications experts" (van Ham et al. 2013, p. 70). Similarly, Nogales et al.

(2013) recently reviewed the impacts of feral cats in island ecosystems. Their work has a section on conservation and management considerations which calls for better understanding the social factors leading to cats becoming feral on islands, but the book then abandons that issue in favor of the mechanics of eradication. "Effective education or social marketing programs are important tools for increasing the sensibility of stakeholders to the impact of feral cats on islands and to the benefits of feral cat eradication for native and endangered species," Nogales et al. (2013) stated, using wording that no doubt made some "insensible" stakeholders unhappy. There are exceptions, however, and these are becoming more common. For example, Lohr and Lepczyk (2014) used social science tools to examine the perceptions and management preferences of Hawaiian residents, identified predictors of attitudes towards feral cats, and were able to recommend approaches that would be acceptable to a majority of residents. Mameno et al. (2017) likewise used social science approaches to examine the feral cat issue in Japan. Unfortunately, they found large gaps between the attitudes shown by cat owners and non-owners, making an agreed solution difficult. Importantly, such gaps seem relatively insensitive to information and are more closely related to attitudes best captured by whether individuals are cat owners/care providers or members of animal rights groups (Dombrosky and Wolverton 2014; Wald et al. 2016).

Whereas conservationists and managers often call for lethal control of cats, animal rights proponents have usually opposed control of any invasive species, including brown treesnakes (Rodriguez 2013) and feral cats, based on the work of Singer and Regan (reviewed in Keulartz 2016). However, as Brickner (2003) pointed out, this position is numerically baffling. Woods et al. (2003) estimated that British cats alone brought home 85–100 million prey items during a several-month study. Some of these were doubtlessly city dwellers like sparrows and pigeons, though sparrows have been declining in urban settings within their native range and are therefore of some concern to conservationists (Moudrá et al. 2018). Being killed by a cat is certainly not a painless way to die, but pain is a welfare issue, not a rights issue and therefore only a concern for some. In contrast, registered hunters, who are opposed by animal rights proponents, kill approximately 20,000 mammals annually (Spedding 2000, p. 62) and researchers kill tens of millions of animals annually (Zamir 2007, p. 80). Peta does suggest that cats should not be allowed to roam (http://www.peta.org/issues/companion-animal-issues/overpopulation/feral-cats/great-outdoors-cats/, accessed 25 April 2017). However, the reason given is "the many deadly hazards that cats face outdoors," not the harm they can cause. According to Marra and Santella (2016, p. 141), some prominent UK organizations go as far as denying the impact of cats on wild bird populations.

The position of animal rights groups appears to be slowly evolving. Peta (http://www.peta.org/issues/companion-animal-issues/overpopulation/feral-cats/) elsewhere confronts predation by feral cats, which "terrorize, maim, and kill countless native birds and other small wild animals." The organization no longer supports trap-alter-and-release programs (TNR; e.g., Schmidt et al. 2009) and "managed" feral cat colonies, as it once did (http://www.peta.org/about-peta/why-peta/feral-cats/). Briefly, TNR emerged as a humane-seeming management alternative for roaming cats. Although supported by many animal rights organizations and tolerated by non-scientific conservation organizations, particularly in Europe, research has repeatedly shown that TNR is not an effective management regime (reviewed in Marra and Santella 2016, pp. 121–143). In fact, Peta now "believe[s] that these programs are not usually in cats' best interests." Nonetheless, "it can be marginally acceptable to trap, vaccinate, alter, and release feral cats when the cats are isolated from roads, people, and animals who could harm them, are regularly attended to by people who not only feed them but also provide them with veterinary care, and are kept in areas where they do not have access to wildlife and the weather is temperate" (https://www.peta.org/about-peta/why-peta/feral-cats/).

Calver et al. (2011) called for a consultative approach to cat management and specifically identified "collaborations involving conservation biologists, veterinarians, animal rights activists, concerned citizens and municipal officers" as desirable. Recent work has begun to more thoughtfully explore public perceptions related to cat predation in the hope of helping develop better strategies (Gramza et al. 2016; Kikillus et al. 2017). Such work indicates that cat owners are more open to suggestions coming from veterinarians and couched in terms of cat welfare (MacDonald et al. 2015). Nonetheless, much work remains to be done on this subject (Kikillus et al. 2017). Overall, Peta opposes the concept of pets (http://www.peta.org/about-peta/why-peta/pets/) and instead calls "for the population of dogs and cats to be reduced through spaying and neutering and for people to adopt animals." Pragmatically (but quietly), Peta shelters euthanize many animals (e.g., Markoe 2015). This is an approach that both managers and conservationists can surely support.

13.5 Effective Communication

The public, through their influence on policy makers, is an important stakeholder for both scientists/managers and animal rights advocates (Burstein 2003; Page and Shapiro 1983). The media influences the agenda of the public, who in turn has power to create and push issues for policy decision-making (McCombs 2014). Given that media play a large role in shaping public attitude and opinion, it is crucial for both groups to make efforts to increase their media literacy and become more involved in the creation and distribution of information via traditional and social media.

Current information systems often reinforce existing attitudes and promote polarization. As demonstrated in the case studies detailed above, conservationists (scientists/managers) and animal rights advocates hold different opinions on many issues related to animal treatment. Such divides exist across several other science-based issues that require collective action in order to change negative societal situations, such as alternative energy, climate change, genetically modified foods, pandemics, and stem cell research. At the core of each divide are communication barriers that, once overcome, can promote change and cooperation. In the

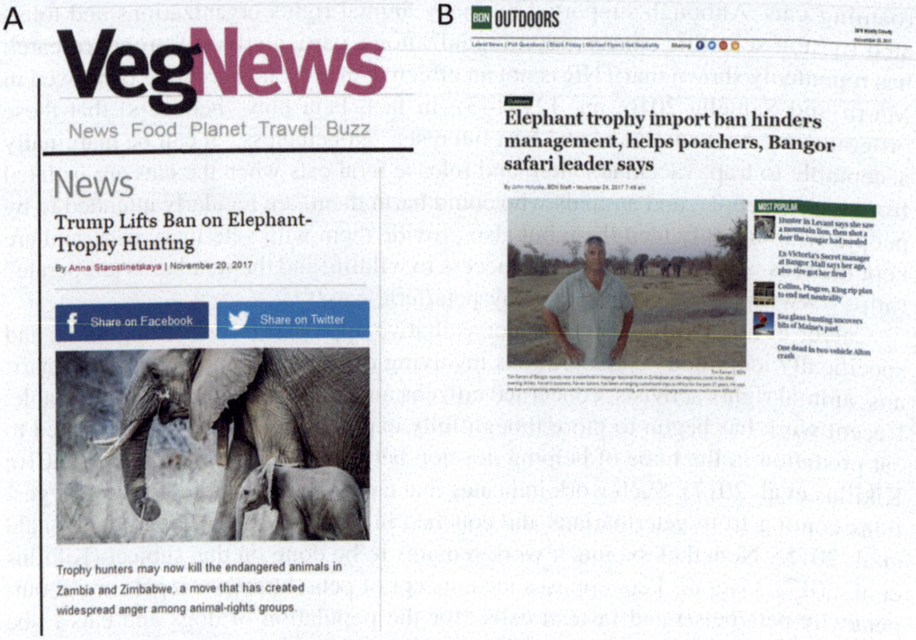

Fig. 13.3 Initial reports of the US government's recommendation to allow importation of elephant trophies from hunts conducted in Zambia and Zimbabwe, both countries where elephants are generally considered to be overpopulated. (**a**) *VegNews*, a vegan outlet, reported "widespread anger." (**b**) In contrast, the headline of the Outdoor section of the *Bangor Daily News* was highly supportive of the change

meantime, the language used by both groups during crises fuels adversarial relations rather than collaboration.

13.5.1 Information Systems: Attitude Reinforcement and Public Opinion Polarization

The polar bear example discussed in the beginning of this chapter shows that the opinions held by each group are shared in local, national, and international media markets. Similarly, when the US administration reported its initial intent to allow importation of elephant trophies from Zambia and Zimbabwe, responses diverged sharply. A spokesperson for the National Rifle Association said the decision "underscored, once again, the importance of sound scientific wildlife management and regulated hunting to the survival and enhancement of game species." The president of the Humane Society of the United States, meantime, opined that "elephants are on the list of threatened species … and now the U.S. government is giving American trophy hunters the green light to kill them" (Cama 2017). The divergent reporting of the initial announcement is demonstrated in Fig. 13.3, though many reports were more nuanced.

Unfortunately, reporting and coverage in the mainstream media can be biased (e.g., Boykoff and Boykoff 2004; see brief content bias reviews in Entman 2007; Baum and Groeling 2008), and individuals with strong preexisting attitudes are biased in their information exposure and processing (Festinger 1957; Knobloch-Westerwick 2015; Kunda 1990). This preference for attitude-congruent information can lead to avoiding media and messages from the "opposition" (e.g., Iyengar et al. 2008; Knobloch-Westerwick 2012), denying or dismissing incongruent information (e.g., Feldman et al. 2012; McCright and Dunlap 2011), or perceiving balanced information as supporting (e.g., Balcetis and Dunning 2006; Sarge et al. 2015) or opposing (e.g., Vallone et al. 1985; Gunther and Liebhart 2006) their side.

Stories that are more sensational often make up a larger portion the messages the public receives and command greater attention (Knobloch et al. 2003; Vilella-Vila and Costa-Font 2008). Scientists and managers work within institutional structures that restrict the certainty which one can have regarding their claims, weakening arguments presented. In addition, these constraints make scientific views less appealing to report and when reported, less appealing to read (Dunwoody 1999; Kitzinger 1999). On the other hand, animal rights advocates do not operate under the same constraints and thus have less to lose and more freedom to express extreme opinions, which garner attention of journalists and readers alike. For example, discussing their objection to lethal control of brown treesnakes, Rodriguez (2013) observed "A cynic might suggest that PETA's objections have less to do with reducing the amount of overall animal suffering in the world, and more to do with manufacturing controversy in an attempt to gain publicity."

Groups based on partisan opinions, like conservationists and animal rights advocates, are common across several issues, and collaboration across such groups is rare. It is extremely difficult to influence the attitudes of those with high involvement with an issue (Ajzen and Fishbein 2005). Their opinions are extreme and latitude of message acceptance is narrow, whereas their latitude of message rejection is wide (Sherif et al. 1965; Sherif et al. 1973). Thus, most messages received have a high likelihood of being rejected rather than accepted, and if debate occurs, each side's counter-arguing simply reinforces their original position through elaboration and a need for consistency (Pallak et al. 1972; Tesser and Leone 1977). However, generally the public is less informed about specific scientific issues and as such hold more ambiguous perspectives. Their attitudes are weaker, which make them subject to greater influence (Ajzen and Fishbein 2005; Fazio 1990). For instance, Jang (2014) found that participants with higher perceived science knowledge were more likely to select and spend time with attitude-congruent information, but those with low perceived science knowledge were more likely to challenge preexisting attitudes through incongruent selective exposure (Jang 2014). Yet, strong content and arguments are not likely the mechanisms by which these readers are persuaded. Audiences without high involvement in an issue or lacking knowledge regarding a topic do not have the motivation and/or ability to carefully scrutinize arguments present within messages (Petty and Cacioppo 1986; Chaiken and Trope 1999).

These audiences are more subject to persuasion through simple message cues or appeals such as use of endorsements, vivid imagery, and emotional language (e.g., Petty et al. 1983). Rights advocates are often willing to adopt strategies that are more effective for less-involved audiences because they focus less on content and arguments and more on simple cues and appeals, which scientists and managers rarely use.

Assessing the current public opinion toward a particular conservation/animal rights issue is important to determine what type of message should be crafted. It is likely public knowledge regarding the issue is low, and thus, opinions are uninformed and ambiguous. Utilizing heuristic message cues (e.g., endorsements, vivid imagery, and emotional language) would be an effective way to grab attention and increase awareness and concern (Petty and Cacioppo 1986; Petty et al. 1983). Most message recipients will not be motivated to action yet but will hopefully begin contemplating the issue, which is one step closer to behavior change (DiClemente and Prochaska 1998; Slater 1999). Animal rights organizations have figured this out. For example, Peta currently has an online campaign titled "Ink, Not Mink" (https://www.peta.org/features/ink-mink-psas/), in which athletes and musicians pose nude in an antifur statement. Another one (https://www.peta.org/features/25-celebs-wont-eating-turkey-thanksgiving/) lists "Celebrities Who Won't Be Eating Turkey on Thanksgiving." In contrast, although some celebrities have been outspoken about climate change, a Google search for "'invasive species' + celebrity" resulted in zero relevant hits. Could the star of movies with an aggressive species theme, say, David Hasselhoff who has battled sharks (Sharknado 4: The 4th Awakens, and others), piranhas (Piranha 3DD), and anacondas (Anaconda: The Offspring), be recruited for a campaign? Or how about Sigourney Weaver? Not only has she battled aliens multiple times and starred in the ecologically themed Avatar series, but she has also narrated conservation-themed documentaries.

Additionally, it should be acknowledged that rarely are people as divided on issues as we think or the media portrays. Indeed, climate change research has identified six categories that represent public opinion regarding climate change as opposed to two (Leiserowitz et al. 2011), with the two most polarized groups making up the smallest proportion of the American public. Representations in the media of conservation scientist/managers and animal rights advocates as at absolute odds with one another may not be as representative in their attitudes and beliefs either. Those that are on either extreme end might make up the minority. It would be beneficial to conduct an audience analysis to determine the broader spectrum of opinions. Then, messages tailored to their stage of involvement with the issue can be crafted, since biased information selection and processing may not be as prevailing for more moderate categories.

13.5.2 Language: Framing an Issue

Once opinions are polarized and groups are formed, the divide continues to grow, not only through biased information selection and processing but also through social identification. When "groups" are present in situations, even if these groups are meaninglessly assigned, in-group favoritism and out-group discrimination arises (Billig and Tajfel 1973). Social identity theory holds that people evaluate their in-group relative to other out-groups and participate in in-group bias and behavior in order to boost their self-concept, which is tied to their group identity (Tajfel 1978; Tajfel and Turner 1979).

Attitudes of highly involved individuals do not change overnight. To gradually change an attitude, messages must fall adjacent to someone's latitude of acceptance (Hovland et al. 1957; Sherif et al. 1965) so that the message is assimilated. Assimilation in this context means the message recipient perceives the message to be more similar than it really is to his or her own attitude, thus enhancing the persuasive impact of the argument. Efforts to land messages in the vicinity of someone's latitude of acceptance require identification of common ground, either in values or similar ends. Commonality, once found, can be used to frame your message – both the language adopted and the structure of information presented. Framing is using language and information presentation as a way to emphasize a particular consideration (Goffman 1974; Entman 1993). Common grounds used to frame messages have been shown effective in enhancing support for teaching evolution in schools (evolution science's application to medical, agriculture, industrial, and other areas: Labov and Pope 2008) and eliciting positive, hopeful reactions consistent with support for climate change policy actions (public health benefits: Maibach et al. 2010; Myers et al. 2012) among both sides of the respective issues. Nisbet (2009) reviews other successful (e.g., moral and ethical) and failed frames used to communicate climate change that could be applicable to other science-related issues. While social science has identified successful frames that cross groups with polarized views, resources including communication experts are needed to develop and test such frames in the current context.

Nisbet (2009) notes that a frame could discuss why an issue is a problem and what should be done to fix it. Similar questions seem to be the primary divide between conservationists and animal rights advocates. However, Perry and Perry (2008a) and this chapter identify ethical common ground between both groups (e.g., humane treatment of organisms, human/public safety) that, if used to frame why something is a problem (e.g., invasive species), may bring both groups together to discuss solutions. Some policy approaches, such as establishing a targeted tax with proceeds to pay for spaying and neutering, seem to be supported by both parties. Framing problems using ethical language and presenting solutions as policy need to be tested to determine if they are effective communication strategies. In addition to this formative research, efforts to identify other commonalities between both groups should be ongoing to help identify framing prospects and avoid costly mistakes. A message that is too extreme will be perceived as even more discrepant than it really is (i.e., contrasting effect; Hovland et al. 1957; Sherif et al. 1965). In such cases, the message's persuasive impact is not only reduced but also has the potential to strengthen preexisting positions (i.e., boomerang effect; Sarup et al. 1991). Effective direct communication utilizing strategically crafted messages can lead to outcomes whereby both sides may be satisfied. Cooperation as a result of successful communication requires both sides to put forth effort and commitment. People's innate drive for consistency in the face of commitment will encourage future collaborative behaviors and eventual attitude change (Bem 1972; Lokhorst et al. 2013).

13.6 Summary and Conclusions

Earlier this decade, Ricciardi et al. (2011) argued that biological invasions share important characteristics with natural disasters. Even more recently, Minteer and Collins (2013) noted that the pervasive impacts of global climate change on ecosystems and species (e.g., Glick et al. 2011) have clear ethical implications for those interested in the long-term survival of species and ecosystems. These, human population growth, the need for poverty alleviation, and other major environmental perturbations were not as apparent, and the extent of their impacts not as apparent or well understood, when both traditional animal activism and animal research and management traditions developed. Many who think about global challenges do not do so with either conservation or animal rights in mind (e.g., Hayward 2012; Macpherson 2014). Yet wild animals also face challenges that affect their welfare, and at least some of those arguably compel action from both conservationists and animalist advocates (Fraser 2010; Walker et al. 2014). That human impacts are so pervasive has led Tomasik (2015) to conclude that "humans already influence ecosystems in substantial ways, so the question is often not *whether* to intervene but *how* to intervene" (emphasis added).

Disagreements between scientists/managers and animal rights advocates result in asymmetrical conflicts. Scientists or managers are trying to achieve a goal (complete a study, protect a species, etc.) that advocates can slow or block. In contrast, animal advocates do not need the cooperation of scientists or managers to achieve most of their goals. Moreover, the conflict can serve the interests of leaders of organizations such as Peta, but offers only harm to the ability of scientists to accurately communicate with the public, influence policy, or even complete projects so they can justify current funding or ask for additional support. Thus, animal advocates have a potentially crucial impact on scientists and managers, whereas the latter do not hold any power over the former. This disparity could help explain why there has been greater apparent pressure on and movement within the science/management community towards considering and addressing the concerns of rights advocates (e.g., Harrington et al. 2013; Hampton et al. 2016), who in turn have shown relatively little change in return over the decades surveyed here.

In recent years, several authors have sought to find philosophical underpinnings that would address, at least to some degree, both types of concern. Curzer et al.

(2013a, b) set out to use the three Rs of laboratory animal ethics (Replacement, Reduction, and Refinement) as the basis for evaluating the ethical basis of studies of wildlife and ecosystems. In addition, they made explicit the need for a fourth R – Refusal – which would eliminate studies that cause excessive harm while producing insufficient gain. Their combined theory suggests that there are multiple acceptable answers for many of the relevant ethical conundrums involved and provides a way to discuss where, along a continuum of possible levels of acceptability, a particular project may lie (see also Bovenkerk and Verweij 2016). The argument is not that this side or that is wrong, but rather that there is value in both views, and the challenge is to balance them out in specific cases and in the light of individual value balancing. Indeed, the definitions of "excessive" and "insufficient" would surely vary among individuals and between animal rights advocates and scientists/managers, thus not fully resolving the kinds of disagreements we have been discussing here. Curzer et al. (2016) further extended this discussion to "conclude that the Three Rs accord animals moral standing, though not necessarily 'rights' in the philosophical sense." This first-time acknowledgement that the long-standing basis for animal research ethics contains an animal rights component should further narrow the gap between those focusing on the rights aspect and those concentrating on the research or management perspective.

Despite this progress, Ramp and Bekoff (2015) recently called conservation "ethically challenged" for killing animals and discussed ways to make conservation projects more humane. Using the catchphrase "compassionate conservation" coined by Beckoff (2010), they advocated for approaches that include individual welfare in design of conservation projects. However, they did not provide examples of this approach involving cooperation with rights advocates, and the call for change only went out to scientists, managers, and decision-makers, who should recognize the intrinsic value of individual animals, not for rights advocates to acknowledge the value of ecosystems and ecological integrity. Although the work of Curzer et al. (2013a, b, 2016) has helped narrow the philosophical divide between scientists/mangers and animal rights advocates, it is not clear to us that the majority in either side is ready to adopt a less confrontational approach. Unfortunately, it is not always clear that either approach produces the kind of results its adherents expect. For example, Martin (2012) surveyed the impacts both approaches have had on wildlife in Africa and found that neither had a particularly stellar track-record under conditions of exploding human populations, conflict, and corruption. However, an approach that allows locals to gain some benefits from the presence of wildlife – more consistent with traditional wildlife management than animal rights approaches – seems more likely to be at least partially successful in providing for long-term survival of large African wildlife (Martin 2012). Similarly, allowing locals to harvest some of the eggs of sea turtles laid on the Ostional beach in Costa Rica has benefitted not only people but also olive ridley (*Lepidochelys olivacea*) survival (Campbell et al. 2007)

There is a deep kinship between the two fields surveyed here, animal rights and conservation, in that both contain a strong component of care for animals (Fraser 2010). A number of authors (Perry and Perry 2008a, b; Fraser 2010; Beausoleil

2014; Ness 2017) have suggested the desirability of collaboration. The notion of respect and care for nature espoused by Martin et al. (2016) and Swart (2016), for example, could be one way to achieve this, and the increasingly pervasive and damaging nature of human dominance of the globe (Palmer 2016) should be a strong incentive to do so. In most cases, however, this has not occurred.

For every polar bear that dies a solitary but well-reported death in a zoo, many thousands will die as their polar habitats thaw over the coming decades. Their deaths will be caused by hunger, disease, and exhaustion caused by the disintegration of the ecosystem they rely on. These deaths will be at least as due to human actions as the captive death of "Szenja" that opened this chapter. Under those conditions, wild animals begin to lose their "wildness" (Palmer 2016) and their ability to live fully autonomous lives is reduced (Keulartz 2016). The exclusion of "natural" suffering from human ethical concern by thinkers such as Rolston III (1988) no longer applies when animal populations are cooped in parks by human actions and suffering from drought and ether ecosystem destabilization that is at least partially the result of human actions (Swart 2016). Should the urgency and overwhelming nature and impacts of climate change and other global human impacts not motivate managers, scientists, and animal rights advocates to come together and collaborate to maximize benefits and minimize harms to animals? For example, in jointly advocating for reduced consumption of red meat, reducing both animal slaughter and climate-affecting gas emissions (Conniff 2018)? Similarly, beginning conservation efforts by attempting to alter problematic human behaviors, as suggested by Dubois et al. (2017), could both be more effective in the long-term and garner support from animal rights advocates. In the absence of such support, however, it is hard to imagine managers attempting options that are more complicated and costly than the lethal approaches usually preferred.

Scientists and animal rights advocates may never agree on whether chimpanzees and other nonhuman animals should get the same legal rights as people, an issue facing the courts in the United States and elsewhere (Hajela 2017). They may continue to disagree whether a monkey deserves to own the copyright to a selfie it took (Molina 2018). Nonetheless, they could focus joint attention on stopping illegal trafficking in chimpanzees (Shukman 2017) or closing down wildlife-exploiting operations such as the animal holding facilities described by Nuwer (2017) in Asia and the United States. Closing those would enhance both animal rights and conservation efforts. What such joint efforts lack, however, is human conflict. We suspect that some individuals and organizations would prefer to maintain such conflict, perhaps with fundraising in mind, rather than achieve progress towards their stated goals.

References

Actamn J, Bale R (2017) These iconic animals could again be hunted in Alaska's refuges. National Geographic, 22 Mar 2017. Available via http://news.nationalgeographic.com/2017/03/wildlife-watch-alaska-predator-control-hunting/. Accessed 2 May 2017

Ajzen I, Fishbein M (2005) The influence of attitudes on behavior. In: Albarracin D, Johnson BT, Zanna MP (eds) The handbook of attitudes. Lawrence Erlbaum Associates, Mahwah, pp 173–221

Bailey C (2007) On the backs of animals: the valorization of reason in contemporary animal ethics. In: Donovan J, Adams C (eds) The feminist care tradition in animal ethics. Columbia University Press, New-York

Balcetis E, Dunning D (2006) See what you want to see: motivational influences on visual perception. J Pers Soc Psychol 91:612. https://doi.org/10.1037/0022-3514.91.4.612

Baratay E, Hardouin-Fugier E (2003) Zoo: a history of zoological gardens in the west. Reaktion Books, London

Baum MA, Groeling T (2008) New media and the polarization of American political discourse. Polit Commun 25:345–365. https://doi.org/10.1080/10584600802426965

Beausoleil NJ (2014) Balancing the need for conservation and the welfare of individual animals. In: Appleby MC, Sandøe P, Weary DM (eds) Dilemmas in animal welfare. CABI International, Croydon, pp 124–147

Beckoff M (2010) Conservation lacks compassion. New Sci 207:24–25

Bekoff M (ed) (2013) Ignoring nature no more: the case for compassionate conservation. University of Chicago Press

Bekoff M, Pierce J (2017) The animals' agenda: freedom, compassion, and coexistence in the Human Age. Beacon Press, Boston

Bellard C, Genovesi P, Jeschke JM (2016) Global patterns in threats to vertebrates by biological invasions. Proc R Soc B 283:20152454. https://doi.org/10.1098/rspb.2015.2454

Bem DJ (1972) Self-perception theory. Adv Exp Soc Psychol 6:1–62. https://doi.org/10.1016/S0065-2601(08)60024-6

Billig M, Tajfel H (1973) Social categorization and similarity in intergroup behaviour. Eur J Soc Psychol 3:27–52. https://doi.org/10.1002/ejsp.2420030103

Boudjelas S (2009) Public participation in invasive species management. In: Clout M, Williams PA (eds) Invasive species management. Oxford University Press, Oxford, pp 93–107

Bovenkerk B (2016) Animal captivity: justifications for animal captivity in the context of domestication. In: Bovenkerk B, Keulartz J (eds) Animal ethics in the age of humans: blurring boundaries in human-animal relationships. Springer, Dordrecht, pp 151–171

Bovenkerk B, Verweij M (2016) Between individualistic animal ethics and holistic environmental ethics blurring the boundaries. In: Bovenkerk B, Keulartz J (eds) Animal ethics in the age of humans: blurring boundaries in human-animal relationships. Springer, pp 369–385

Boykoff MT, Boykoff JM (2004) Balance as bias: global warming and the US prestige press. Glob Environ Chang 14:125–136. https://doi.org/10.1016/j.gloenvcha.2003.10.001

Brickner I (2003) The impact of domestic cat (*Felis catus*) on wildlife welfare and conservation: a literature review. With a situation summary from Israel. Available via http://132.66.16.6/lifesci/zoology/members/yom-tov/inbal/cats.pdf. Accessed 20 Mar 2017

Burke J (2016) South African national park to kill animals in response to severe drought. The Guardian, 14 Sept 2016. Available via https://www.theguardian.com/world/2016/sep/14/south-african-national-park-kill-animals-severe-drought. Accessed 7 May 2017

Burke J (2017) Inequality, drought and the deadly fight for precious grazing land in Kenya. The Guardian, 3 May 2017. Available via https://www.theguardian.com/world/2017/may/03/inequality-drought-and-the-deadly-fight-for-precious-grazing-land-in-kenya. Accessed 7 May 2017

Burstein P (2003) The impact of public opinion on public policy: a review and an agenda. Polit Res Q 56:29–40. https://doi.org/10.1177/106591290305600103

Calver MC, Grayson J, Lilith M, Dickman CR (2011) Applying the precautionary principle to the issue of impacts by pet cats on urban wildlife. Biol Conserv 144:1895–1901

Cama T (2017) Trump to allow imports of African elephant trophies. The Hill, 15 Nov 2017. Available via http://thehill.com/policy/energy-environment/360614-trump-to-allow-imports-of-african-elephant-trophies. Accessed 24 Nov 2017

Campbell LM, Haalboom BJ, Trow J (2007) Sustainability of community-based conservation: sea turtle egg harvesting in Ostional (Costa Rica) ten years later. Environ Conserv 34:122–131
Chaiken S, Trope Y (eds) (1999) Dual-process theories in social psychology. Guilford Press, New York
Clement G (2007) The etic of care and the problem of wild animals. In: Donovan J, Adams C (eds) The feminist care tradition in animal ethics. Columbia University Press, New-York
Cluff HD, Murray DL (1995) Review of wolf control in North America. In: Carbyn LN, Fritts SH, Seip R (eds) Ecology and conservation of wolves in a changing world, Occasional publication, vol 35. Canadian Circumpolar Institute, University of Alberta, Edmonton, pp 491–607
Conniff R (2018) The case for a carbon tax on beef. New York Times, 17 Mar 2018. Available at https://www.nytimes.com/2018/03/17/opinion/sunday/carbon-tax-on-beef.html. Accessed 15 Apr 2018
Copeland M (2017) Maker of feral hog poison pulling product from Texas. Waco Tribune-Herald, 25 Apr 2017. Available via http://www.wacotrib.com/news/environment/maker-of-feral-hog-poison-pulling-product-from-texas/article_a619efb1-35d5-53ed-80bd-d334b6bf3c1f.html. Accessed 30 Apr 2017
Crowley SL, Hinchliffe S, McDonald RA (2017a) Conflict in invasive species management. Front Ecol Environ 15:133–141
Crowley SL, Hinchliffe S, McDonald RA (2017b) Invasive species management will benefit from social impact assessment. J Appl Ecol 54:351–357
Curzer HJ, Wallace M, Perry G et al (2011) Teaching wildlife research ethics: a progress report. Teach Ethics, Fall 2011:1–18
Curzer HJ, Muhlberger P, Perry G et al (2013a) The ethics of wildlife research: a nine R theory. ILAR J 54:52–57
Curzer HJ, Wallace M, Perry G et al (2013b) Environmental research ethics: extensions of the three R's. Environ Ethics 35:95–114
Curzer H, Perry G, Wallace M, Perry D (2016) The three Rs of animal research: what they mean for the institutional animal care and use committee and why. Sci Eng Ethics 22:549–565
Dauphiné N, Cooper RJ (2009) Impacts of free-ranging domestic cats (*Felis catus*) on birds in the United States: a review of recent research with conservation and management recommendations. Proceedings of the Fourth International Partners in Flight Conference: Tundra to Tropics. Partners in Flight, US, pp 205–219
DiClemente CC, Prochaska JO (1998) Toward a comprehensive, transtheoretical model of change: stages of change and addictive behaviors. In: Miller WR, Heather N (eds) Treating addictive behaviors, Applied clinical psychology, pp 3–24. https://doi.org/10.1007/978-1-4899-1934-2_1
Dombrosky J, Wolverton S (2014) TNR and conservation on a university campus: a political ecological perspective. PeerJ 2:e312. https://doi.org/10.7717/peerj.312
Domonick BA (1997) Animal liberation and social revolution: a vegan perspective on anarchism or an anarchist perspective on veganism. Critical Mess Media, Syracuse
Donaldson S, Kymlicka W (2011) Zoopolice: a political theory of animal rights. Oxford University Press, Oxford
Dubois S, Fenwick N et al (2017) International consensus principles for ethical wildlife control. Conserv Biol 31:753–760
Dunwoody S (1999) Scientists, journalists, and the meaning of uncertainty. In: Friedman SM, Dunwoody S, Rodgers CL (eds) Communicating uncertainty: media coverage of new and controversial science. Lawrence Erlbaum, Mahwah, pp 59–79
Engeman RM, Constantin B, Bunting J (2007) The political, economic and management history of a successful exotic eradication: the case of black-tailed jackrabbits in one part of Florida. In: Witmer GW, Pitt WC, Fagerstone KA (eds) Managing vertebrate invasive species: proceedings of an international symposium. USDA/APHIS/ WS, National Wildlife Research Center, Fort Collins, pp 476–478
Entman RM (1993) Framing: toward clarification of a fractured paradigm. J Commun 43:51–58. https://doi.org/10.1111/j.1460-2466.1993.tb01304.x

Entman RM (2007) Framing bias: media in the distribution of power. J Commun 57:163–173. https://doi.org/10.1111/j.1460-2466.2006.00336.x

Essl F, Hulme PE, Jeschke F et al (2017) Scientific and normative foundations for the valuation of alien-species impacts: thirteen core principles. Bioscience 67:166–178

Fazio RH (1990) Multiple processes by which attitudes guide behavior: the MODE model as an integrative framework. Adv Exp Soc Psychol 23:75–109. https://doi.org/10.1016/S0065-2601(08)60318-4

Feber RE, Raebel EM, D'cruze N et al (2017) Some animals are more equal than others: wild animal welfare in the media. Bioscience 67:62–72

Feldman L, Maibach EW, Roser-Renouf C, Leiserowitz A (2012) Climate on cable: the nature and impact of global warming coverage on Fox News, CNN, and MSNBC. Int J Press/Politics 17:3–31. https://doi.org/10.1177/1940161211425410

Festinger L (1957) A theory of cognitive dissonance. Row, Peterson, Evanston

Fraser J (2009) The anticipated utility of zoos for developing moral concern in children. Curator: Mus J 52:349–361

Fraser D (2010) Toward a synthesis of conservation and animal welfare science. Anim Welf 19:121–124

Frost W (2011) Rethinking zoos and tourism. In: Frost W (ed) Zoos and tourism: conservation, education, entertainment? Channel View Publications, Bristol, pp 1–8

García-Llorente M, Martín-López B, Nunes PALD et al (2011) Analyzing the social factors that influence willingness to pay for the management of invasive alien species under two different strategies: eradication and prevention. Environ Manag 48:418–435

Gjerris M, Birkved M, Gamborg C, Brando S (2016) Eating to save wild-life: is a truly conservation minded zoo/aquarium a vegan zoo/aquarium? In: Olsson IAS, Araújo SM, Vieira MF (eds) Food futures: ethics, science and culture. Wageningen Academic Publishers, Wageningen, pp 381–386

Glick P, Stein BA, Edelson NA (eds) (2011) Scanning the conservation horizon: a guide to climate change vulnerability assessment. National Wildlife Federation, Washington, D.C.

Goffman E (1974) Frame analysis: an essay on the organization of experience. Harvard University Press, Cambridge, MA

Graad G (2011) Ecofeminism revisited: rejecting essentialism and re-placing species in a material feminist environmentalism. Fem Form 23:26–53

Gramza A, Teel T, VandeWoude S, Crooks K (2016) Understanding public perceptions of risk regarding outdoor pet cats to inform conservation action. Conserv Biol 30:276–286

Gunther AC, Liebhart JL (2006) Broad reach or biased source? Decomposing the hostile media effect. J Commun 56:449–466. https://doi.org/10.1111/j.1460-2466.2006.00295.x

Gunther I, Raz T, Even Zor Y et al (2016) Feeders of free-roaming cats: personal characteristics, feeding practices, and data on cat health and welfare in an urban setting of Israel. Front Vet Sci 3:21. https://doi.org/10.3389/fvets.2016.00021

Hajela D (2017) Appeals court says chimpanzees don't have rights of people. ABC News, 8 June 2017. Available via http://abcnews.go.com/Weird/wireStory/appeals-court-chimpanzees-rights-people-47924225. Accessed 10 June 2017

Hampton JO, Forsyth DM (2016) An assessment of animal welfare for the culling of peri-urban kangaroos. Wildl Res 43:261–266

Hampton JO, Handyman TH, Laurence M et al (2016) Animal welfare and the use of procedural documents: limitations and refinement. Wildl Res 43:599–603

Hargrove EC (ed) (1992) The animal rights/environmental ethics debate: the environmental perspective. State University of New York Press, Albany

Harrington LA, Moehrenschlager A, Gelling M et al (2013) Conflicting and complementary ethics of animal welfare considerations in reintroductions. Conserv Biol 27:486–500

Harris M, Best J (2016) Friends of Animals files lawsuit to stop import of 18 elephants to U.S. Zoos from Swaziland. News release, 9 Feb 2016. Available via https://www.friendsofanimals.org/

news/2016/february/friends-animals-files-lawsuit-stop-import-18-elephants-us-zoos-swaziland. Accessed 7 May 2017

Hayward T (2012) Climate change and ethics. Nat Clim Chang 2:843–848

Hill T (2016) Activists devise successful strategy to stop mass wildlife killings. Takepart, 29 Apr 2016. Available via http://www.takepart.com/article/2016/04/29/grassroots-plan-stop-wildlife-killing-agency. Accessed 1 May 2017

Hovland CI, Harvey OJ, Sherif M (1957) Assimilation and contrast effects in reactions to communication and attitude change. J Abnorm Soc Psychol 55:244–252. https://doi.org/10.1037/h0048480

Hutchins M (2008) Animal rights and conservation. Conserv Biol 22:815–816

Hutchins M, Wemmer C (1986) Wildlife conservation and animal rights: are they compatible? In: Fox MW, Mickley LD (eds) Advances in animal welfare science 1986/87. The Humane Society of the United States, Washington, D.C., pp 111–137

Iyengar S, Hahn KS, Krosnick JA, Walker J (2008) Selective exposure to campaign communication: the role of anticipated agreement and issue public membership. J Politics 70:186–200. https://doi.org/10.1017/S0022381607080139

Jamieson D (1985) Against zoos. In: Singer P (ed) In defence of animals. Harper and Row, New York, pp 108–117

Jang SM (2014) Seeking congruency or incongruency online? Examining selective exposure to four controversial science issues. Sci Commun 36:143–167. https://doi.org/10.1177/1075547013502733

Jensen E (2014) Evaluating children's conservation biology learning at the zoo. Conserv Biol 28:1004–1011

Kagan R, Carter S, Allard S (2015) A universal animal welfare framework for zoos. J Appl Anim Welfare Sci 18(Suppl 1):S1–S10

Keulartz J (2015) Captivity for conservation? Zoos at a crossroads. J Agric Environ Ethics 28:335–351

Keulartz J (2016) Toward an animal ethics for the Anthropocene. In: Bovenkerk B, Keulartz J (eds) Animal ethics in the age of humans: blurring boundaries in human-animal relationships. Springer, Dordrecht, pp 243–264

Kheel M (1996) The killing game: an ecofeminist critique of hunting. J Philos Sport 23:30–44

Kikillus KH, Chambers GK, Farnworth MJ, Hare KM (2017) Research challenges and conservation implications for urban cat management in New Zealand. Pac Conserv Biol 23:15–24

Kisiel J, Rowe S, Vartabedian MA, Kopczak C (2012) Evidence for family engagement in scientific reasoning at interactive animal exhibits. Sci Educ 96:1047–1070

Kitzinger J (1999) Researching risk and the media. Health Risk Soc 1:55–69. https://doi.org/10.1080/13698579908407007

Klier G, Casalderrey C, Busan TE, Pasquo F (2017) Conservación de la biodiversidad y sus vínculos utilitaristas: cercanías y distancias con Peter Singer y Gifford Pinchot. Rev Metrop Sustentabilidade 7:63–82

Knobloch S, Carpentier FD, Zillmann D (2003) Effects of salience dimensions of informational utility on selective exposure to online news. J Mass Commun Q 80:91–108. https://doi.org/10.1177/107769900308000107

Knobloch-Westerwick S (2012) Selective exposure and reinforcement of attitudes and partisanship before a presidential election. J Commun 62:628–642. https://doi.org/10.1111/j.1460-2466.2012.01651.x

Knobloch-Westerwick S (2015) Choice and preference in media use: advances in selective exposure theory and research. Routledge, New York

Kunda Z (1990) The case for motivated reasoning. Psychol Bull 108:480–498. https://doi.org/10.1037/0033-2909.108.3.480

Labov JB, Pope BK (2008) Understanding our audiences: the design and evolution of science, evolution, and creationism. CBE Life Sci Educ 7:20–24. https://doi.org/10.1187/cbe.07-12-0103

Leiserowitz A, Maibach E, Roser-Renouf C, Smith N (2011) Global warming's six Americas, May 2011. Yale University and George Mason University. Yale Project on Climate Change Communication, New Haven. Available via http://environment.yale.edu/climate/files/SixAmericasMay2011.pdf. Accessed 7 Oct 2017
Lohr CA, Lepczyk CA (2014) Desires and management preferences of stakeholders regarding feral cats in the Hawaiian Islands. Conserv Biol 28:392–403
Lokhorst A et al (2013) Commitment and behavior change: a meta-analysis and critical review of commitment-making strategies in environmental research. Environ Behav 45:3–34. https://doi.org/10.1177/0013916511411477
Loyd KAT, Hernandez SM, Carroll JP et al (2013) Quantifying free-roaming domestic cat predation using animal-borne video cameras. Biol Conserv 160:183–189
Luke B (2007) Brutal: manhood and the exploitation of animals. University of Illinois Press, Urbana and Chicago
MacDonald E, Milfont T, Gavin M (2015) What drives cat-owner behaviour? First steps towards limiting domestic-cat impacts on native wildlife. Wildl Res 42:257–265
Macpherson CC (2014) Climate change matters. J Med Ethics 40:288–290
Maibach EW, Nisbet M, Baldwin P et al (2010) Reframing climate change as a public health issue: an exploratory study of public reactions. BMC Public Health 10:299. https://doi.org/10.1186/1471-2458-10-299
Mameno K, Kubo T, Suzuki M (2017) Social challenges of spatial planning for outdoor cat management in Amami Oshima Island, Japan. Glob Ecol Conserv 10:184–193
Maple TL, Perdue BM (2013) Zoo animal welfare. Springer, Berlin
Markoe L (2015) At PETA's shelter, most animals are put down. PETA calls them mercy killings. Washington Post, 12 Mar 2015. Available via https://www.washingtonpost.com/national/religion/at-petas-shelter-most-animals-are-put-down-peta-calls-them-mercy-killings/2015/03/12/e84e9af2-c8fa-11e4-bea5-b893e7ac3fb3_story.html?utm_term=.26f4ee74643e. Accessed 29 May 2017
Martin G (2012) Game changer: animal rights and the fate of Africa's wildlife. University of California Press, Oakland
Martin J-L, Maris V, Simberloff DS (2016) The need to respect nature and its limits challenges society and conservation science. PNAS 113:6105–6112
Marra PP, Santella C (2016) Cat wars: the devastating consequences of a cuddly killer. Princeton University Press
McCombs M (2014) Setting the agenda: mass media and public opinion. Polity Press, Cambridge
McCright AM, Dunlap RE (2011) The politicization of climate change and polarization in the American public's views of global warming, 2001–2010. Sociol Q 52:155–194. https://doi.org/10.1111/j.1533-8525.2011.01198.x
Medina FM, Bonnaud E, Vidal E et al (2011) A global review of the impacts of invasive cats on island endangered vertebrates. Glob Chang Biol 17:3503–3510
Meffe GK (2008) A pragmatic ethic for the twenty-first century— Commentary on "From environmental to ecological ethics: Toward a practical ethics for ecologists and conservationists.". Sci Eng Ethics 14:503–504
Mfula C (2016) Zambia halts hippo culling after protests by animal rights activists. Reuters, 14 June 2016. Available via http://www.reuters.com/article/us-zambia-wildlife-idUSKCN0Z01JT. Accessed 7 May 2016
Miller B, Conway W, Reading RP et al (2004) Evaluating the conservation mission of zoos, aquariums, botanical gardens, and natural history museums. Conserv Biol 18:86–93
Minteer BA (2013) Conservation, animal rights, and human welfare: a pragmatic view of the "bushmeat crisis". In: Bekoff M (ed) Ignoring nature no more: the case for compassionate conservation. University of Chicago Press, Chicago, pp 77–94
Minteer BA, Collins JP (2013) Ecological ethics in captivity: balancing values and responsibilities in zoo and aquarium research under rapid global change. ILAR J 54:41–51

Molina B (2018) A monkey took a selfie, went to court — and lost the appeal. USA Today, 24 Apr 2018. Available online at https://www.usatoday.com/story/news/nation-now/2018/04/24/monkey-selfie-copyright-case-naruto-crested-macaque/545166002/. Accessed 5 May 2018

Mollot G, Pantel JH, Romanuk TN (2017) The effects of invasive species on the decline in species richness: a global meta-analysis. Adv Ecol Res 56:61–83

Mori E, Mazza G et al (2017) Strangers coming from the Sahara: an update of the worldwide distribution, potential impacts and conservation opportunities of alien aoudad. Ann Zool Fenn 54:373–386

Moudrá L, Zasadil P, Moudrý V, Šálek M (2018) What makes new housing development unsuitable for house sparrows (*Passer domesticus*)? Landsc Urban Plan 169:124–130

Murphy K (2017) A plan to poison the wild hogs of Texas. In: New York Times, 29 Apr 2017. Available via https://www.nytimes.com/2017/04/29/opinion/sunday/a-plan-to-poison-the-wild-hogs-of-texas.html. Accessed 30 Apr 2017

Myers TA, Nisbet MC, Maibach EW, Leiserowitz AA (2012) A public health frame arouses hopeful emotions about climate change. Clim Chang 113:1105–1112. https://doi.org/10.1007/s10584-012-0513-6

Neff M (2017) Polar bear dies at SeaWorld San Diego after her only companion was sent away. Available via http://www.onegreenplanet.org/http://www.onegreenplanet.org/news/polar-bear-dies-at-san-diego-zoo/. Accessed 10 Apr 2017

Ness B (2017) Anthropogenic climate change and animal welfare. In: Brotton MJ (ed) Ethoecology and nonhuman ethics in society. Lexington Books, Lanham, pp 63–80

Nisbet MC (2009) Communicating climate change: why frames matter for public engagement. Environ Sci Policy Sustain Dev 51:12–23. https://doi.org/10.3200/ENVT.51.2.12-23

Nkala O (2017) Water relief for 8,000 thirsty elephants neglected by Zimbabwe. The Guardian, 11 Jan 2017. Available via https://www.theguardian.com/environment/2017/jan/11/water-relief-for-8000-thirsty-elephants-neglected-zimbabwe. Accessed 7 May 2017

Nogales M, Vidal E, Medina FM et al (2013) Feral cats and biodiversity conservation: the urgent prioritization of island management. Bioscience 63:804–810

Nuwer R (2017) Asia's illegal wildlife trade makes tigers a farm-to-table meal. New York Times, 5 June 2017. Available via https://www.nytimes.com/2017/06/05/science/animal-farms-south-east-asia-endangered-animals.html?hpw&rref=science&action=click&pgtype=Homepage&module=well-region&region=bottom-well&WT.nav=bottom-well&_r=0. Accessed 6 June 2017

Page BI, Shapiro RY (1983) Effects of public opinion on policy. Am Polit Sci Rev 77:175–190. https://doi.org/10.2307/1956018

Pallak MS, Mueller M, Dollar K, Pallak J (1972) Effect of commitment on responsiveness to an extreme consonant communication. J Pers Soc Psychol 23:429–436. https://doi.org/10.1037/h0033363

Palmer C (2016) Climate change, ethics, and the wildness of wild animals. In: Bovenkerk B, Keulartz J (eds) Animal ethics in the age of humans: blurring boundaries in human-animal relationships. Springer, pp 131–150

Perrings C, Mooney H, Williamson M (eds) (2010) Bioinvasions & globalization: ecology, economics, management, and policy. Oxford University Press, Oxford

Perry G, Farmer M (2011) Reducing the risk of biological invasion by creating incentives for pet sellers and owners to do the right thing. J Herpetol 45:134–141

Perry D, Perry G (2008a) Improving interactions between animal rights groups and conservation biologist. Conserv Biol 22:27–35

Perry G, Perry D (2008b) Response to "animal rights.". Conserv Biol 22:816–818

Petty RE, Cacioppo JT (1986) The elaboration likelihood model of persuasion. Adv Exp Soc Psychol 19:123–205. https://doi.org/10.1016/S0065-2601(08)60214-2

Petty RE, Cacioppo JT, Schumann D (1983) Central and peripheral routes to advertising effectiveness: the moderating role of involvement. J Consum Res 10:135–146. https://doi.org/10.1086/208954

Pieters J (2017) Cat caught sampling donuts in Amsterdam store. NL Times, 20 Nov 2017. Available via https://nltimes.nl/2017/11/20/cat-caught-sampling-donuts-amsterdam-store. Accessed 24 Nov 2017
Ramp D, Bekoff M (2015) Compassion as a practical and evolved ethic for conservation. Bioscience 65:323–327
Regan T (1976) Do animals have a right to life? In: Regan T, Singer P (eds) Animal rights and human obligations. Prentice-Hall, Englewood Cliffs, pp 197–204
Ricciardi A, Palmer ME, Yan ND (2011) Should biological invasions be managed as natural disasters? Bioscience 61:312–317
Rodriguez JM (2013) Is killing an invasive species animal cruelty? Greener Ideal, 4 Mar 2013. Available via https://greenerideal.com/science/0304-is-killing-an-invasive-species-animal-cruelty/. Accessed 2 May 2017
Rolston H III (1988) Environmental ethics: duties to and values in the natural world. Temple University Press, Philadelphia
Russell JC, Blackburn TM (2017) The rise of invasive species denialism. Trends Ecol Evol 32:3–6
Russow L-M (2010) Environment. In: Comstock GL (ed) Life science ethics, 2nd edn. Springer, Dordrecht, pp 93–121
Sarge MA, VanDyke MS, King AJ, White SR (2015) Selective perceptions of hydraulic fracturing: the role of issue support in the evaluation of visual frames. Politics Life Sci 34:57–72. https://doi.org/10.1017/pls.2015.6
Sarup G, Suchner RW, Gaylor G (1991) Contrast effects and attitude change: a test of the two-stage hypothesis of social judgment theory. Soc Psychol Q 654:64–372. http://www.jstor.org/stable/2786848
Schmidt PM, Swannack TM, Lopez RR, Slater MR (2009) Evaluation of euthanasia and trap–neuter–return (TNR) programs in managing free-roaming cat populations. Wildl Res 36:117–125
Shani A, Pizam A (2008) Towards an ethical framework for animal-based attractions. Int J Contemp Hosp Manag 20:679–693
Sherif CW, Sherif M, Nebergall RE (1965) Attitude and attitude change: the social judgment-involvement approach. Saunders, Philadelphia
Sherif C, Kelly M, Rodgers HL et al (1973) Personal involvement, social judgement, and action. J Pers Soc Psychol 27:311–327
Shukman D (2017) Ivory Coast jail sentences for chimpanzee traffickers. BBC, 9 June 2017. Available via http://www.bbc.com/news/world-africa-40222022. Accessed 10 June 2017
Singer P (1975) Animal liberation: a new ethics for our treatment of animals. Random House, New York
Slater MD (1999) Integrating application of media effects, persuasion, and behavior change theories to communication campaigns: a stages-of-change framework. Health Commun 11:335–354. https://doi.org/10.1207/S15327027HC1104_2
Soulé ME (1990) The onslaught of alien species, and other challenges in the coming decades. Conserv Biol 4:233–239
Spedding C (2000) Animal welfare. Earthscan Publications, London
Swart JAA (2016) Care for the wild in the Anthropocene. In: Bovenkerk B, Keulartz J (eds) Animal ethics in the age of humans: bluring boundaries in human-animal relationships. Springer, Dordrecht, pp 173–188
Tajfel H (ed) (1978) Differentiation between social groups: studies in the social psychology of intergroup relations. Academic Press, London
Tajfel H, Turner JC (1979) An integrative theory of intergroup conflict. In: Austin WG, Worchel S (eds) The social psychology of intergroup relations. Brooks/Cole, Monterey, pp 33–47
Tesser A, Leone C (1977) Cognitive schemas and thought as determinants of attitude change. J Exp Soc Psychol 13:340–356. https://doi.org/10.1016/0022-1031(77)90004-X
Tomasik B (2015) The importance of wild-animal suffering. Relations. Beyond Anthropocentrism 3:133–152

Treves A, Naughton-Treves L (2005) Evaluating lethal control in the management of human–wildlife conflict. In: Woodroffe R, Thirgood S, Rabinowitz A (eds) People and wildlife, conflict or coexistence? Cambridge University Press, Cambridge, pp 86–106

UNEP: Secretariat of the Convention on Biological Diversity (2010) Pets, aquarium, and terrarium species: best practices for addressing risks to biodiversity. SCBD, Technical Series No. 48, Montreal

Vallone RP, Ross L, Lepper MR (1985) The hostile media phenomenon: biased perception and perceptions of media bias in coverage of the Beirut massacre. J Pers Soc Psychol 49:577. https://doi.org/10.1037/0022-3514.49.3.577

Van Ballenberghe V (2006) Predator control, politics, and wildlife conservation in Alaska. Alces 42:1–11

van Ham C, Genovesi P, Scalera R (2013) Invasive alien species: the urban dimension. Case studies on strengthening local action in Europe. IUCN European Union Representative Office Brussels

Van Houtan KS (2006) Conservation as virtue: a scientific and social process for conservation ethics. Conserv Biol 20:1367–1372

Vidal J (2016) 18 elephants to be flown to US zoos as drought puts pressure on Swaziland wildlife. The Guardian, 16 Feb 2016. Available via https://www.theguardian.com/environment/2016/feb/26/18-elephants-to-be-flown-us-zoos-drought-puts-pressure-on-swaziland-wildlife. Accessed 7 May 2017

Vilella-Vila M, Costa-Font J (2008) Press media reporting effects on risk perceptions and attitudes towards genetically modified (GM) food. J Socio-Econ 37:2095–2106. https://doi.org/10.1016/j.socec.2008.04.006

Vinter J (1963) Kind killing. Universities Federation for Animal Welfare, London

Wald DM, Lohr CA, Lepczyk CA et al (2016) A comparison of cat-related risk perceptions and tolerance for outdoor cats in Florida and Hawaii. Conserv Biol 30:1233–1244

Walker M, Diez-Leon M, Mason G (2014) Animal welfare science: recent publication trends and future research priorities. Int J Comp Psychol 27:80–100

Webb TJ, Raffaelli D (2008) Conversations in conservation: revealing and dealing with language differences in environmental conflicts. J Appl Ecol 45:1198–1204

Weisberg L (2017) Polar bear at SeaWorld San Diego dies after a brief, unexplained illness. Available via http://www.latimes.com/local/lanow/la-me-ln-seaworld-polar-bear-death-20170418-story.html. Accessed 20 Apr 2017

Wickins-Dražilová D (2006) Zoo animal welfare. J Agric Environ Ethics 19:27–36

Wood V, Seddon PJ, Beaven B, Heezik Y (2016) Movement and diet of domestic cats on Stewart Island/Rakiura, New Zealand. N Z J Ecol 40:186–190

Woods M, McDonald RA, Harris S (2003) Predation of wildlife by domestic cats *Felis catus* in Great Britain. Mammal Rev 33:174–188

Zamir T (2007) Ethics and the beast: an argument for speciesist liberationism. Princeton University Press, Princeton

Zoological Society of San Diego (2016) Land-based food not nutritionally sufficient for wild polar bears, according to new study. Available via https://www.sciencedaily.com/releases/2016/09/160913150503.htm. Accessed 10 Apr 2017

Chapter 14
Zoos and Conservation in the Anthropocene: Opportunities and Problems

Jan Robovský, Lubomír Melichar, and Spartaco Gippoliti

zoos are or should be one of the last bastions of organismal biology...
– Seidensticker and Forthman (1998)

Acronyms

AZA	Association of Zoos and Aquariums
BSC	Biological Species Concept
DFSC	Differential Fitness Species Concept
EAZA	European Association of Zoos and Aquaria
EDGE	Evolutionary Distinct and Globally Endangered (species)
EEP	European Endangered Species Programme
ESU	Evolutionarily Significant Unit
ISIS	International Species Information System
IUCN	International Union for Conservation of Nature
IZE	International Zoo Educators Association
MK	Mean Kinship
PSC	Phylogenetic Species Concept
RCP	Regional Collection Plan
SSC	Species Survival Commission

J. Robovský (✉)
Faculty of Science, University of South Bohemia, České Budějovice, Czechia
e-mail: robovsky@prf.jcu.cz

L. Melichar
Zoo Liberec, Liberec, Czechia
e-mail: melichar@zooliberec.cz

S. Gippoliti
Società Italiana per la Storia della Fauna "Giuseppe Altobello", Rome, Italy

F. M. Angelici, L. Rossi (eds.), *Problematic Wildlife II*,
https://doi.org/10.1007/978-3-030-42335-3_14

TAG	Taxon Advisory Group
WAZA	World Association of Zoos and Aquariums
ZGAP	Zoological Society for the Conservation of Species and Populations
ZIMS	Zoological Information Management System

14.1 Introduction

Homo sapiens is a relatively young species, with an extensive adaptive and invasive capacity (e.g. Cameron and Groves 2004; Glikson and Groves 2016), whose activities have enormous impacts on global ecosystems and species diversity. These impacts are currently referred to in the literature as 'the sixth mass extinction', 'defaunation' or 'biological annihilation' (e.g. Dirzo et al. 2014; Ceballos et al. 2015, 2017; see also Macdonald and Service 2007; Carroll and Fox 2008). Our relationship to nature and in particular species has varied greatly both spatially and temporally, according to the intentions and motivations of particular individuals, societies and populations of our species; in effect, we have been plundering natural resources too often in our history. Luckily, we also enjoy the beauty of nature in its various forms and undertake conservation actions to safeguard valuable species and landscapes. Edward O. Wilson (1987) recognizes in our species what he calls 'biophilia', which he defines as 'the innate tendency to focus on life and lifelike processes' (see also Kellert and Wilson 1993). Our relation to nature (natural resources) is complex, influenced by cultural and local conditions, and is sometimes full of paradoxes, which may depend on particular historical periods and the beliefs of that era. For example, we should be aware that many conservation actions in historic times were initiated or carried out by the nobility, out of their love of hunting (e.g. attempts to preserve aurochs, European wisent and Alpine ibex – e.g. Kowalski 1967) or by particular avid hunters (e.g. Theodore Roosevelt and Ernest Thompson Seton; for more on our hunting history and perceptions over time, see also Guthrie 2005).

Our species has also often exhibited ambitions for rearing and breeding nondomestic animals, alongside domesticated ones, for aesthetic, utilitarian and exhibition purposes. This has been well documented since antiquity (e.g. Belozerskaya 2006; Grigson 2016). The history of proto-collections and menageries intended to enrich the experience of an audience is long and fascinating (e.g. Kisling 2000; Rothfels 2002; Belozerskaya 2006). The associated animal trade was also relatively extensive. (For documentation on it since the Middle Ages, see, e.g. Rothfels 2002; Dittrich 2007; Grigson 2016.)

Travelling or stationary menageries, where the main ambition was to exhibit animals and captivate a paying public, evolved into zoos, which focused on breeding and exhibiting animals according to zoological and aesthetic standards, for example, by using Hagenbeck's concepts (e.g. Rothfels 2002; Kisling 2000) (Fig. 14.1). Zoos themselves have evolved into modern zoos spontaneously and gradually, thanks to their increasing body of knowledge on how to keep wild ani-

Fig. 14.1 The lions in Rome as an example of Hagenbeck's open zoo style. (Photo by Spartaco Gippoliti)

mals alive in captivity and their first-hand awareness of the deteriorating status of several species in the wild (e.g. Rabb 1994; Kisling 2000; Rabb 2004). For a detailed history and description of the species collections of some particular zoos over time, see, among others, Peel 1903; Schlawe 1969; Bridges 1974; Edwards 1996; Klös et al. 1994; Bell 2001; Kisling 2000; Gippoliti 2010; Blaszkiewitz 2005; Weigl 2005; Solski and Strehlow 2015; and Grigson 2016.

Modern zoos perform many crucial duties other than merely serving as a public educational institution. Specifically, their core missions are to conserve threatened species through coordinated ex situ breeding programmes, support in situ conservation projects and collaborate with research institutions to increase basic and applied knowledge on threatened fauna (see below) (Table 14.1, Fig. 14.2).

The literature on zoos, their purposes and conservation missions is extensive and complex. In this chapter, therefore, we merely summarize the basic roles of zoos and then focus on some conservation and management challenges associated with zoos (and other captive institutions) in the light of our experiences. We would also like to refer to readers, including young zookeepers, to important reviews, chapters and monographs, specifically Hediger 1942, 1969; Benirschke 1986; Frankham et al. 1986; Bostock 1993; Schonewald-Cox et al. 1983; Snyder 1995; Kisling 2000; Rothfels 2002; Miller et al. 2004; Kleiman et al. 2010; Frost 2011; Prichard et al. 2011; and Maple and Perdue 2013 and also to at least two inspiring, optimistic and highly readable bestsellers – Durrell (1976) and Goodall et al. (2010).

Table 14.1 Landmarks of zoo history, science and evolution as conservation actors

Year	Event
1752	The Tiergarten Schönbrunn in Vienna is established as the oldest zoo in the Western world
1793	The Menagerie at the Jardin des Plantes in Paris is established as the first public zoo
1826	Creation of the Zoological Society of London; the 'Menagerie' will be opened in 1828
1859	*Der Zoologische Garten* began publication, thanks to Frankfurt Zoo director Max Schmidt
1892	*A handbook of the management of animals in captivity in Lower Bengal* is authored by Ram Bramha Sanyal in Calcutta
1907	Tierpark Hagenbeck, the 'open zoo', is opened to the public near Hamburg
1932	The first international studbook for a wild animal established; *Bos bonasus*
1942	Heini Hediger published *Wilde Tiere in Gefangenschaft*, later translated in English in 1950
1946	Birth of the *International Union of Directors of Zoological Gardens* (IUDZG), currently WAZA
1959	First International Symposium on Zoo Medicine in Berlin (East) convened by Johannes Dobberstein Zoological Society of London began publishing *International Zoo Yearbook*, edited by Desmond Morris Gerald Durrell opened Jersey Zoo, now operated by the Durrell Wildlife Conservation Trust
1960	The *American Association of Zoo Veterinarians* is established
1964	*The management of wild mammals in captivity* is authored by Lee S. Crandall of the New York Zoological Society (now WCS)
1970	*Journal of Zoo and Wildlife Medicine*
1972	First World Conference on Breeding Endangered Species in Captivity is held in Jersey
1982	*Zoo Biology*
1993	*The World Zoo Conservation Strategy* launched by IUDZG
2005	*The World Zoo and Aquarium Conservation Strategy* launched by WAZA
2015	*Committing to Conservation: The World Zoo and Aquarium Conservation Strategy* produced by WAZA

14.2 Roles of Zoos

Zoos should realize their entire potential in order to fulfil conservation, education and other duties and roles to the highest standards, which are well expressed and comprehensively summarized in these experienced opinions, reviews, strategies and reports (in this case EAZA reports) and on the web pages of particular zoo and aquaria associations: van Bemmel 1971; Dathe 1978; Luoma 1988; Olney et al. 1994; Rabb 1994; Hutchins and Conway 1995; Kitchener 1997; Stepherdson et al. 1998; Conway 2003; Hutchins 2003; Mallinson 2003; Young 2003; Rabb 2004; WAZA 2005; Gippoliti and Kitchener 2007; Wirth 2007; Bowkett 2009; Gippoliti 2011; Gusset and Dick 2011a, b; Witzenberger and Hochkirch 2011; Gippoliti 2012; Conde 2013; Gusset and Dick 2013a, b; Tribe and Booth 2013; Heckel et al. 2013; Gippoliti 2014; Moss et al. 2014; Gusset and Lowry 2014; Mellor et al. 2015;

Fig. 14.2 The Red Panda is an example of a globally managed ex situ species. (Photo by Spartaco Gippoliti)

Keulartz 2015; Barongi et al. 2015, Gusset and Dick 2015; Wirth 2015; Annual Report 2016a EAZA; TAG Reports 2016b EAZA; and Schwartz et al. 2017.

In summary, zoos and aquariums pursue the goals listed below, in full accordance with the United Nations *Strategic Plan for Biodiversity 2011–2020.*

14.3 Ex situ Conservation

This important conservation mission is based on scientific management focused on sufficient population size, demographic stability and retention of genetic diversity over the long term (e.g. Schonewald-Cox et al. 1983; Lacy 1995; Young and Clarke 2000; Ferrière et al. 2007; Gusset and Dick 2011b; Frankham et al. 2002; Allendorf and Luikart 2007; Mills 2007; Bertorelle et al. 2009; Witzenberger and Hochkirch 2011; Lacy 2013; Leus 2018). Input pedigree and studbook data are based on identified animals in a standard way and moreover registered in the Species360 database (formerly the International Species Information System, abbreviated as ISIS).

Species360 regularly publishes and distributes the ISIS/WAZA Studbook Library DVD. The 2011 edition comprises 1540 studbooks, including 1350 regional and 190 international studbooks, plus 292 husbandry manuals and nearly 2800 other documents (WAZA web pages, downloaded in July 2018). Currently there are 130 active international studbooks, including 159 species or subspecies, which are specified regularly in the International Zoo Yearbook (WAZA web page, downloaded in July 2018; but see also Witzenberger and Hochkirch 2011).

Since the exchange of animals between regions are expensive and often difficult due to logistic, legal and veterinary issues, interregional (ideally global) and coordinated population management is highly recommended (Olney et al. 1994; see also below). This approach, for example, is the goal of Global Species Management Plans, which, however, have been applied until now to only around a dozen species (e.g. Gusset and Dick 2011b).

Ex situ conservation is often exemplified using only a few species that have been saved from extinction (e.g. Conde 2013). The full list, though, is much richer. In 2013, of the 33 animal species classified as 'extinct in the wild' on the IUCN Red List, 31 are actively bred in zoos and aquariums, and 17 are managed in studbook-based breeding programmes (Gusset and Dick 2013a). Maas (2013, 2017) also enumerated the number of animal subspecies and plant species 'extinct in the wild' according to IUCN Red List of 2013 and tried to evaluate animal and plant reintroductions. Specifically, Maas (2013) noted six successful reintroductions that achieved self-sustaining populations (e.g. *Acanthobrama telavivensis, Bos bonasus, Equus przewalskii, Mustela nigripes*); eight reintroductions that met with some success, but with wild populations not (yet) self-sustaining (e.g. *Elaphurus davidianus, Gallirallus owstoni, Nectophrynoides asperginis, Thermosphaeroma thermophilum*); five attempted reintroductions with no signs of success (yet) (only one animal species – *Corvus hawaiiensis*); and 59 species that remain solely in captivity, with no reintroduction attempted (yet) (e.g. several cichlids, 3 species of *Cyprinodon* and frogs, 14 species or subspecies of *Partula, Zenaida graysoni*). Although some updates are available, we use the original and comprehensive review made by Maas (2013).

Unfortunately, captive breeding has not been successful in saving some species and subspecies, including such unique taxa as Tasmanian tiger (*Thylacinus cynocephalus*), quagga (*Equus quagga quagga*), passenger pigeon (*Ectopistes migratorius*) and pink-headed duck (*Rhodonessa caryophyllacea*). (For a list of 22 species and subspecies where the last living member died in captivity, see Maas 2013.)

Species should be held according to high standards of animal welfare, best-practice guidelines and often species-specific husbandry (e.g. Norton et al. 1995; Shepherdson et al. 1998; Yong 2003; *WAZA Code of Ethics and Animal Welfare* adopted in 2003; Maple and Perdue 2013; also Wickins-Dražilová 2006; Melfi 2009; Hill and Broom 2009; Gippoliti 2014). The *World Zoo and Aquarium Animal Welfare Strategy* recommends that zoos and aquariums should apply a simple welfare model referring to the so-called Five Domains (Gusset and Dick 2015).

14.4 In situ Conservation and Fundraising in Support of Field Conservation

Zoos support a number of in situ conservation projects and contribute to valuable research and practical conservation efforts worldwide (e.g. AZA and EAZA Conservation Databases). As in current conservation practice, an evaluation of conservation impacts is necessary; within the WAZA community, the *Project Conservation Impact Tool* is recommended (Mace et al. 2007). Such evaluations are important for improving future procedures and strategies.

Apart from the support of zoo and aquarium staff, the financial contribution of WAZA members to wildlife conservation is estimated at over 350 million US dollars every year), making zoos and aquariums the third-highest contributor to conservation worldwide after the *Nature Conservancy* and the *World Wildlife Fund* global network (Gusset and Dick 2011a). In more detail, in 2016 AZA's institutions spent approximately $216 million on more than 3409 conservation initiatives in 127 countries, and 823 species and subspecies benefitted from conservation actions (for details, see the AZA Conservation and Research Database); as of 5 April 2017, 1340 conservation projects and the activities of 165 EAZA members (51% of EAZA's members) amounting to €58.5 million have been registered in the new EAZA Conservation Database (https://www.aza.org/field-conservation; Zimmermann 2017).

14.5 Integrated Species Conservation

Ex situ and in situ conservation missions are closely related through animal transfers and by personnel, financial and other resources (see also Lacy 2010), and so the dichotomy between these two forms of species conservation is often arbitrary, as noted, for example, by Redford et al. (2013). Such interconnections have often been realized as a spontaneous progression of the conservation work of zoos and aquariums and are currently recommended by some directives (e.g. Olney et al. 1994; Mallinson 2003; Conway 2003; Hutchins 2003; Tribe and Booth 2003; Bowkett 2009; Prichard et al. 2011; Witzenberger and Hochkirch 2011; *WAZA Vision and Corporate Strategy Towards 2020;* Conde 2013; Gusset and Dick 2013b; Keulartz 2015; McGowan et al. 2017; Schwartz et al. 2017). In effect, knowledge and experience are shared, and depending on the species, some in situ projects use ex situ principles and vice versa.

This integrated approach has been well documented based on genetic exchanges of individuals between ex situ and in situ projects. Reintroductions are also an example of integrated species conservation, and it appears that in situ conservation will probably increasingly require our active interventions (e.g. translocations of many large mammals through national borders that are now fenced (Linnell et al. 2016), to preserve gene flow and metapopulation dynamics). Concerning reintroductions, there is often a fear that animals bred in captivity are less likely to survive once released than their wild counterparts (cf. Mathews et al. 2005). On balance, reintroductions of captive animals are promising (e.g. Maas 2013), although some prerelease actions such as training could be important for the survival of reintroduced animals (Menzel and Beck 2000; Reading et al. 2013). It is also relevant to such issues to consider that many reintroductions undergo a gradual process that allows a slow readaptation to 'wild' habitats.

14.6 Research and Development of Relevant Technologies

Through their living collections, zoos and aquariums have contributed much to the documentation and understanding of morphological, chromosomal, genetic, behavioural and other life-history parameters of many animal species and subspecies (e.g. Ryder and Byrd 1984; Conde 2013; Ryder and Feistner 1995; Conde 2013; Rees 2015). Such information is important as basic theoretical or applied knowledge, which is often useful in conservation management of particular taxa. Many technologies have evolved and have improved in close cooperation with zoos and aquariums, and some are put to significant use in these institutions (e.g. Ryder and Feistner 1995; Piña-Aguilar et al. 2009).

Research for evidence-based husbandry (e.g. effectiveness of husbandry practices – diets, enrichment, housing conditions, etc.) and veterinary issues are prominent research topics in zoos (see the section "Nutrition and Veterinary Issues" in this chapter and Rees 2015).

Research associated with zoos and aquariums is also often focused on socio-economic, educational, visitor and marketing issues and results that could improve future steps in these activities (e.g. Reade and Waran 1996; Majolo et al. 2005; Ballantyne et al. 2007; Moos and Esson 2013; Roe et al. 2014; Colléony et al. 2017; Skibins et al. 2017).

Scientific contributions of authors affiliated with zoos or aquariums is significant (e.g. our survey across the Web of Science – see below; WAZA web pages, downloaded in August 2018; Loh et al. 2018). Research is often conducted in collaboration with academic institutions (Fernandez and Timberlake 2008).

Our survey across the Web of Science (downloaded in August 2018) using the address search with 'zoo' as a keyword identified approx. 12,000 records that have been published since 1972, with approx. 1000 records each year during the last 3 years. Contributions are predominantly published in this sequence of top seven journals: *Journal of Zoo and Wildlife Medicine, Zoo Biology, American Journal of Primatology, PLoS ONE, Theriogenology, Journal of Wildlife Diseases* and *Veterinary Record*. Using the Web of Science categories, contributions are predominantly associated with these fields (in this sequence of top seven): veterinary sciences (35%), zoology (22%), ecology (9%), reproduction biology (5%), biodiversity conservation (5%), multidisciplinary sciences (5%) and evolutionary biology (4%). For much thorough analysis of scientific research of AZA members, see Loh et al. (2018).

As with experience gained in zoos and aquariums that become best-practice husbandry guidelines and veterinary procedures, technologies should be shared across these institutions as much as possible.

14.7 Public Relations and (Conservation) Education

Gusset and Dick (2011a) estimated that zoos and aquariums around the world receive 700 million visitors every year. Zoos and aquariums have therefore an enormous obligation to educate these visitors about taxa held at particular institutions, the general mission of zoos and aquariums, basic topics associated with biodiversity, the conservation of our planet and environmental sustainability. In the general view, zoos and aquariums have unprecedented potential to fascinate all generations through the beauty of the animal species they exhibit, to support our 'biophilia', to educate, to inspire, to encourage visitors to engage with conservation actions, to increase their environmental awareness and to enhance a basic sense of responsibility with regard to lifestyle and consumption, often in an entertaining form. Zoos and aquariums are extremely important for urban populations that have little or no contact with nature (e.g. Gippoliti 2011), which is even more important for children and young people (WAZA 2005) (Fig. 14.3). Zoo educators are organized in the IZE, which publishes the *Journal of the International Zoo Educators' Association*. Some surveys have provided compelling evidence that zoos and aquariums contribute to increasing the number of people who understand biodiversity and to increasing actions which could help to protect it (e.g. Gusset and Lowry 2014; Moss et al. 2014 and references therein).

14.7.1 Zoos as Sanctuaries

Zoos sometimes serve as sanctuaries for injured or otherwise disabled, donated or confiscated wild animals, and such animals are sometimes included in conservation programmes or education activities carried out by zoos and aquariums (Conde 2013; Cuarón 2005). What is often not realized, however, is that a modern zoo can serve as a 'repository' for confiscated animals only occasionally. This is because zoo design exhibit has evolved from a row of cages for taxonomically similar species (monkeys, large cats and so on) into habitat exhibits intended to hold together a whole social group or even several species; and so to find adequate space for a single animal often poses more than a challenge for the zoo staff. Regrettably, in some countries modern zoos goals are not well-known, and there continues to exist a call by some sections of society such as animal rights groups to transform zoos into 'sanctuaries' in which the breeding of captive animals is not allowed. Evidently, a lack of awareness of the environmental situation at the global level hinders an understanding of how zoos and other captive-breeding facilities should today be considered true 'sanctuaries of biodiversity'.

14.7.2 Deficiencies

Naturally, the fulfilment of above-mentioned duties varies around the world's zoos and aquariums for various reasons, even among accredited members of regional zoo and aquarium associations.

As the IUCN SSC Antelope Specialist Group (2015) opposes all Intentional Genetic Manipulations of antelopes for commercial or amenity purposes, with particular reference to (i) hybridization of different species, (ii) crossing of different

Fig. 14.3 Zoos are important links with wildlife, especially for children and young people. (Photo by Spartaco Gippoliti)

subspecies and (iii) selective inbreeding of a population, zoos and aquariums should re-evaluate the breeding of colour variants of animals, which are rare to non-existent in the wild. White tigers and lions may be highly appreciated by the public, but there is a risk that visitors will get a mistaken conservation message about the relevance of these animals, which are already being exploited by private organizations as 'conservation targets'. Further these colour mutations take up zoo space that is urgently needed for conservation breeding of genetically unmodified animals. EAZA (2013) took a negative position on Intentional Breeding for the Expression of Rare Recessive Alleles directly.

Concerning database (Species360) and studbook data (e.g. Witzenberger and Hochkirch 2011), the quality of the same and the associated reasonableness of population management recommendations are closely connected with the quality of the input data. While some curators and keepers are conscientious, others are not, which greatly damages the work of the conscientious colleagues and studbook keepers. Although some animals present challenges in terms of identifying individuals and relationships, zoos and aquariums should pursue registration work to the best of their abilities, giving the required time and full institutional support to their staff.

Similarly, some zoos are very cooperative in the concept of population management proposed by coordinators and/or particular species committees (e.g. in establishing bachelor groups, other measures of birth control or providing management euthanasia), but some are less active. The system should be modified such that the cooperative and altruistic institutions receive more benefits (see also below).

14.8 Conservation and Management Challenges Associated with Zoos

14.8.1 Prioritizing of Biodiversity

Global biodiversity is not distributed equally across the Earth's surface, and different regions with greater biodiversity face different levels of threat (Myers et al. 2000). In effect, rational conservation actions should be focused predominantly on threatened regions with exceptional concentrations of endemic species, the so-called biodiversity hotspots (Myers et al. 2000), which is an approach that has resonated well in global conservation strategies and actions. Similarly, different species face different threats and do not have the same conservation status, as conventionally evaluated under *The IUCN Red List of Threatened Species* (http://www.iucnredlist.org/about/citing). In view of limited personnel and other resources, conservation priorities could be combined with evolutionary distinction based on phylogenetic diversity, under the so-called EDGE initiative (e.g. Isaac et al. 2007). As a result, the conservation of species with lower EDGE scores should not be abandoned easily, but nor should we continue to neglect species with higher EDGE scores. The first application of EDGE criteria for mammals (Isaac et al. 2007) detected that many evolutionarily distinct and globally endangered species within the 100 highest-ranking species did not benefit from existing conservation projects or protected areas, which is alarming. Currently, the EDGE approach is available for mammals, birds, reptiles, amphibians and corals (http://www.edgeofexistence.org/index.php).

14.8.2 Prioritizing of Collections

Zoos are often considered 'Noah's Arks' in that they may be able to keep animal populations safe from threats that they face in the wild. Objectively they are such 'arks', but the degree to which they are depends on how 'loaded' they are.

Current attempts to evaluate the ex situ conservation contribution of zoos have clearly shown that there is much to be improved in relation to threatened species (Conde et al. 2011; Witzenberger and Hochkirch 2011; Gippoliti 2012; Conde 2013; Conde et al. 2013; Heckel et al. 2013; Martin et al. 2014; Dawson et al. 2016; Biega et al. 2017, but see also Bowkett 2009, 2014). Specifically, Conde et al. (2013) demonstrated that only 23% (!) of terrestrial vertebrate species held in ISIS zoo are threatened, and only in Dasyuromorphia and Testudines are threatened species significantly overrepresented (i.e. the actual number of threatened species differed from the expected value, if zoo collections were taken at random). Martin et al. (2014) demonstrated that mammals and bird species held in zoos are less endemic and less threatened than their close relatives not held in zoos. On the contrary, amphibians held in zoos are equally as threatened as their close relatives not found in zoos, although ex situ institutions are not prioritizing range-restricted habitat specialists, which are species with a greater extinction risk in the future (Biega et al. 2017). Frynta et al. (2009, 2010) reported that zoos preferentially keep 'cute' species (in many vertebrate groups) and pay less attention to actual conservation needs (for species less attractive to visitors). On the other hand, reasons to why simply holding higher proportions of threatened taxa may not increase conservation impact are given in Bowkett (2014).

The major zoo and aquarium associations (e.g. EAZA, AZA) try to prioritize collections based on different criteria in the form of RCPs. Our experiences with the EAZA association, however, indicate that even some experienced colleagues are often unable to discriminate between truly threatened taxa and common taxa that have had some breeding tradition. Despite RCPs and appeals from TAG chairs, much zoo capacity often continues to be used for non-threatened taxa or stocks of unknown/mixed origin. By way of illustration, in September 2017, 1105 mouflons (*Ovis aries musimon*), a 'taxon' that was created by human's early sheep introduction in Corsica and Sardinia (cf. Gippoliti and Amori 2004), 736 Sika deer (*Cervus nippon*) of unspecified subspecies, and 987 red deer (*Cervus elaphus*) of unspecified subspecies occupied spaces in world zoos, included in Species360 Database, that could be used instead for threatened caprine or deer taxa. Some currently available, highly threatened taxa (of similar size and needs) that could occupy those enclosures include Bukhara markhor (*Capra falconeri heptneri*), West Caucasian tur (*Capra caucasica*), Transcaspian urial (*Ovis arkal*), Bukhara urial (*Ovis bochariensis*); Laristan mouflon (*Ovis laristanica*), Armenian mouflon (*Ovis gmelini*), Bactrian stag (*Cervus bactrianus*), white-lipped deer (*Cervus albirostris*), Formosan sika deer (*Cervus taioanus*) and Vietnamese sika deer (*Cervus pseudaxis*) (taxonomy follows Groves and Grubb 2011 and/or Castelló 2016).

Prioritizing is also relevant for some particular breeding lines. In case of the Przewalski horse, two main lines, A and M, are recognized, of which the A-line exhibits many morphological and genetic similarities with wild Przewalski horses (e.g. Groves 2009; Robovský 2012; Groves and Robovský in prep.). How can we explain that population management for the A-line was abandoned by the EEP (Schook et al. 2016), given that the current genomic study (Der Sarkissian et al. 2015) recognized A-line horses as virtually devoid of the admixture from domestic horses, in contrast to M-line horses, and with heterozygosity/inbreeding levels (based on genomic data) being very similar to those of M-line horses? Is it really not possible to preserve the diminishing population of A-line horses (124 animals in September 2015, though not all of these animals are housed with other A-line horses) at some target population size (approx. 100 mares, as proposed in 2008 – Yasynetska and Zharkikh 2008) alongside the M-line horses, which number over 2000? It bears keeping in mind here that A-line horses have helped to improve/standardize phenotype parameters of M-line horses throughout the captive history of the Przewalski horse (for references, see, e.g. Robovský 2012).

Making conservation actions easier by reducing or minimizing conservation options in the future is extremely risky, since concerns over the health of particular species/subspecies/line may be shaped by other motivations or targets.

Although most zoos and aquariums (at least those located in Europe and North America) should be focused on globally threatened biodiversity (Fig. 14.4 and Fig. 14.5), they should not neglect local or regional autochthonous taxa; indeed this should be desirable and inspiring, highlighting domestic environmental problems and the often overlooked local biodiversity (e.g. Gippoliti 2004; Olive and Jansen 2017). We must remember as well that some races of domesticated animals could be of conservation concern (Taberlet et al. 2008).

Some observers argue that we also need common species for 'ambassador' education and public relation purposes. This may be true, but skilful education could work well with threatened species, while the above-mentioned proportion of threatened vs. non-threatened species in zoos indicates that we already have too many non-threatened 'ambassadors'. We believe that visitors could ascribe positive values to the taxonomic diversity of a zoological collection, especially when it's presented skilfully to visitors. For excellent species-based conservation breeding and education work regarding threatened birds, see Hirschfeld et al. (2013).

Fig. 14.4 Several mammals, which need our urgent help. (**a**) Visayan warty pig (*Sus cebifrons*), photo by Roland Wirth. (**b**) Bukhara urial (*Ovis bochariensis*), photo by Lubomír Melichar. (**c**) Half-maned zebra (*Equus quagga borensis*), photo by Lubomír Melichar. (**d**) A-line Przewalski horse (*Equus przewalskii*). (Photo by Roland Wirth)

Fig. 14.5 Two examples of birds, which need our urgent help – see the EAZA Silent Forest Campaign (https://www.silentforest.eu). (**a**) Javan Pied Starling (*Gracupica jalla*), photo by Roland Wirth. (**b**) Sumatran laughingthrush (*Garrulax bicolor*). (Photo by Roland Wirth)

Another issue is the preservation of species diversity in zoos (cf. Peš and Vogeltanz 2010; Lupták 2015; Matschei 2017; Santore 2017), especially of threatened taxa. That diversity has often diminished, due to several factors, which include the loss of interest in some species, veterinary and other logistic limitations in attempts to refresh currently low stocks and homogenization of zoos potentially due to perceived visitor preferences, as usually declared, etc. (e.g. Wirth 2011; Matschei 2017), but also availability of species and prioritization of managed species. Any attempt either to keep the current diversity of threatened taxa in zoos or obtain and breed new threatened taxa should therefore be supported and commended by the zoo and conservation community and visitors. Specialized facilities, which are closed to the public, are often required for this mission. In this context, it should be noted that generally some facilities used for successful breeding of threatened species need not always meet 'optimal welfare/aesthetic criteria' from the human perspective but should at best be exclusively for the needs of particular species (cf. Heckel et al. 2013).

14.8.3 Taxonomic Instability

We are living through a revolution in the biological sciences, thanks to new evolutionary concepts, statistical and molecular phylogenetics methods and both old and new data sets. For many biologists/zoologists/taxonomists and conservationists, the traditional view on species and the parameters that define species (i.e. BSC – biological species concept) in taxonomy and management practice seems to be untenable. (For further reading, see Behie and Oxenham 2015; Groves et al. 2017;

Gippoliti et al. 2018; Gippoliti and Groves 2020, this volume and references therein).

This is at times considered negatively (e.g. Garnett and Christidis 2017; for overview, see also replies and papers citing this comment). To allay these fears of instability, however, we can now access a large body of data that often fully suffice for qualified conservation and management actions. Similarly, the current taxonomic instability (e.g. in larger mammals) is often exaggerated with regard to its impact on conservation and simplified as a conflict between BSC and PSC (phylogenetic species concept) or between 'lumpers' and 'splitters'. Under both (or any other) species concepts, taxonomic opinions are often very similar (e.g. compare Grubb 2005 and Groves and Grubb 2011 in case of ungulates). Predominantly we do have the same entity but of different evolutionary and conservation values. Although we consider PSC highly suitable, not merely because it minimizes the risk that some unique populations would be neglected, we should be open to different opinions as much as possible in order to preserve current species diversity (e.g. Hazevoet 1996).

In summary, taxonomic instability does not hamper conservation, because conservation actions can be assessed, often quite easily, under different taxonomic concepts and approaches – for example, Swayne's hartebeest is a (very) distinct and endangered hartebeest taxon that could be classified as separate subspecies of *Alcelaphus buselaphus* under BSC or separate species under PSC; irrelevant which species concept is closer to biological/evolutionary reality, it is no doubt that its existence is widely accepted by scientific and conservation communities and it deserves our full attention and adequate conservation actions. As some IUCN species specialist groups, however, insist on the BSC, a sensitive approach would require that assessments be done at the subspecies level, or at least the level of subspecies that are recognized as evolutionary distinct (ESU), or at the level of separate species under PSC when data are available. Such an approach is sometimes more difficult to be adopted with lesser charismatic creatures, such as rodents. Yet zoos should lead the way, attracting attention towards little-known overlooked lineages or taxa that are suffering decline, even in our own backyard, such as the subterranean members of genera *Spalax* and *Nannospalax* in Central Europe or the little known populations subsumed under *Castor fiber* in Europe and Central Asia (Gippoliti and Amori 2007; Csorba et al. 2015).

14.8.4 Precautionary Principle in Management

Frankham et al. (2012) advocated a DFSC, differential fitness species concept, to be applied to fragmented populations with some diagnosable differences, under which populations are considered the same species unless there are signs of outbreeding depression or fixed chromosomal differences. Other authors assert that successful (e.g. fertile) interbreeding under human control gives no indication of species status (for references, see Groves and Robovský 2011; Groves et al. 2017). Some zoos seem to consider DFSC sensible (e.g. the current mixing of different subspecies of dama gazelle (*Nanger dama*) is underway in some institutions, as proposed in Senn et al. 2014; for a critique, see Schreiber et al. 2018). The precautionary principle should be applied here, since 'Once they have been mixed up they can never be unscrambled' (Groves 1995). In practice, when the 'lumping' of some separate stocks is unavoidable, we should proceed carefully, step by step, that is, not lumping the mass, but at first only in a few institutions and insisting on a detailed documentation of somatic, reproductive and other parameters of the lumped stock.

Concerning the taxonomic status of animals in zoo and aquarium collection, zoos should also try (1) to accumulate morphological and genetic data on their captive populations and priority should be given to threatened taxa; (2) to check historical data on the origin of their stocks, and if unavailable or incomplete, genetic/morphological comparisons should be used and priority should be given to populations used for reintroductions; (3) concerning potential reintroductions and new imports, animals should be mixed only within the same evolutionary significant unit/taxon under the precautionary principle (cf. van Bemmel 1971; Dathe 1978); and (4) in general, our conservation steps should minimize potential harm/regrets and maximize potential benefits/options, as measured by reproductive fitness and adaptive evolutionary processes.

Concerning reintroductions, they should also follow species (morphological) adaptations and their historical distribution (when available), habitat and diet preferences. This is because some endangered species had/have lived or live in suboptimal-marginal habitat due to destructive human activities, e.g. European bison; these species are called 'refugee species' (e.g. Cromsigt et al. 2012; Kerley et al. 2012).

14.9 Other Management Issues

Modern zoos manage current stock based on population management (summarized in the EAZA region as *Population Management Manual* – EAZA 2015), scientific principles and data, in order to maximize genetic variability over the long-term perspective (for references, see above). Additionally, all animals are managed in order to maximize the welfare of particular animals (see above), also using various enrichment methods (Stepherdson et al. 1998; Young 2003; Mason and Rushen 2006; Shyne 2006; Hoy et al. 2010; Jonas et al. 2018). These principles are often meaningful, and some associated tools are very sophisticated, but in some cases, there may be room for common sense.

Concerning welfare and enrichment, the effectiveness of the principles should be enforced when life-history parameters (e.g. sociality, group size, diet) and associated adaptations are considered. For example, breeding within a group (of the suitable size and composition) is the best welfare enrichment for species that do live in groups; breeding should be an essential part of animals' lives in zoos, at least as a source of natural welfare/enrichment. Similarly, enrichment should be designed to meet species-specific requirements (Law and Kitchener 2002; Mason 2010) and could be highly effective when applied with common sense. In some institutions, felids and other carnivores are fed predominantly on pellets or commercially prepared diets, sometimes offered through more or less sophisticated enrichment techniques; but the most basic type of enrichment should utilize the long-lasting processing of food, by providing whole or partial carcasses that include feathers/skin, hairs and bones (e.g. cf. McPhee 2002; Bashaw et al. 2003; Skibiel et al. 2007). Similarly, the food can be scattered or hidden across the entire enclosure, and feeding schedules can be randomized (Jonas et al. 2018).

Inbreeding is another important topic in population management (e.g. Schonewald-Cox et al. 1983; Hedrick and Kalinowski 2000; Frankham et al. 2002; Holt et al. 2003; Koeninger Ryan et al. 2003; Charpentier et al. 2007). Many pages have been written about inbreeding, but our impression is that much evidence, thorough meta-analyses and a consideration of the biology and history of particular species or groups are still needed, since some results regarding the avoidance of inbreeding, inbreeding depression, etc., are often unexpected or vary according to particular species and/or life-history parameters (e.g. social and reproductive system) (e.g. Smith 1979; Schonewald-Cox et al. 1983; Hedrick 2000; Charpentier et al. 2007; Holland et al. 2007; Olson et al. 2012; Ibáñez et al. 2011, 2013; Bichet et al. 2014; Ellegren and Galtier 2016). Additionally, some captive stocks continue to prosper despite numerically limited founder stock. This may be due to previous exposure to inbreeding; bottlenecks in the history of the species caused by natural processes and/or by chance – inbreeding could fix negative but also positive allelic combinations, which could be of adaptive significance; differences between detected inbreeding values based on studbook vs. genomic data; or a relatively short time, from an evolutionary perspective, for an effect to have been felt (e.g. Kalinowski et al. 1999; Charpentier et al. 2007; Tokarska et al. 2011; Holland et al. 2007; Der Sarkissian et al. 2015; Moreno et al. 2015). In the case of European bison (*Bos bonasus*), the average inbreeding level in lowland bison is almost 50%, yet no signs of inbreeding depression have been observed. In contrast, inbreeding effects have been noticed in the lowland-Caucasian line, which has a much lower average inbreeding level (28%) (Todarska et al. 2011).

Some observers believe in 'genetic rescue' for stocks/populations or species with limited genetic diversity parameters, yet often propose to undertake outbreeding with different ESUs or species, which is more risky. All these cases require a proper consideration of historic (or prehistoric) genetic diversity; of the history of the species (some species or subspecies have been derived by isolating some segment of the paternal line, with the result that they include only some genetic variants); and of other evolutionary factors (cf. Weeks et al. 2011). For example, the observed genetic depletion in Amur tigers likely reflects a founder history via Central Asia that predates human-induced bottlenecks (Driscoll et al. 2009). In any case, sensitive conservation actions should rigorously evaluate the pros and cons of 'rescue', taking account as well of proposals published to obtain all possible feedback, evidence and arguments (e.g. Weeks et al. 2011; Hoban et al. 2013; Gippoliti et al. 2017). These actions could, moreover, deploy varying conservation approach criteria, as was done, for example, by Moodley et al. (2017) for black rhinoceroses (see also Elemental Conservation Units – Wood and Gross 2008).

The care of zoos and aquariums devoted to particular specimens is highly complex. In effect, many species prosper well under our management (e.g. Tidière et al. 2016). Concerning limited capacities of ex situ institutions, coordinators of some breeding programmes are trying to slow the growth of prospering populations. There are several basic options, which could be combined skilfully, such as separation of the sexes (ideally only for a short time; breeding every 2 years is often sufficient), having offspring remain with parents as long as possible, management euthanasia (breed and cull strategy) and contraception or castration (e.g. Gippoliti 2014; Penfold et al. 2014). Again, we tend to recommend that the natural processes and biology of particular species be considered. It is known, for example, that some species or groups (e.g. suids, rhinoceroses) are sensitive to not being allowed to breed at a young age, and the current practice, the results of which are unfortunately too seldom published (see point 'Publication of Interesting Observation and Experiences'), indicates that reproduction in a 'switch-on, switch-off' regime, as requested by coordinators using contraception, does not always work correctly (cf. Penfold et al. 2014). Stopping the breeding for the whole or majority of a population is risky, as it could reduce the reproductive success of some animals. For example, wild pig species such as the endangered Visayan warty pig *Sus cebifrons* have evolved to produce many offspring at regular intervals (most of which will not survive to adulthood in the wild), and captive management, to 'avoid this surplus' through temporarily preventing mating or temporary contraception, can render females permanently infertile within 3 years (Przybylska 2014; cf. Leus 2018).

We tend to encourage coordinators and keepers to apply 'clever' tools for population management that accord with the biology of managed species (cf. Norton et al. 1995; Leus 2018). Although the breed and cull strategy is controversial for many observers, we should be aware of its similarity and optimality from the biological perspective for the majority of species, when consideration is made of their natural mortalities owing to many factors (e.g. predation, stochastic catastrophes or harsh environmental conditions) for particular age cohorts. Considering welfare and other biological factors, animals managed under the breed and cull strategy could experience most natural behaviours, such as reproduction, care of offspring and existence within a normal social structure with all age cohorts.

It should be also mentioned that although some institutions, curators and keepers accept this strategy, they put it into practice under the assumption that the coordinator is working carefully with the population and also taking other actions to allow the population an acceptable growth rate (cf. Powell and Ardaiolo 2016). Otherwise, the repeated application of the breed and cull strategy by each cooperating institution could demotivate the institutions, curators and keepers in the future. And, as stated above, these very cooperative institutions should benefit from their altruistic (and difficult, especially from the public relation point of view) actions, e.g. by being recommended for breeding programmes in the future, obtaining animals with a higher ranking in population and a higher influence in species (e.g. EEP) commissions, etc.

14.10 Hidden Risk of the Mean Kinship Criterion?

Several approaches and criteria are used in managing captive populations to retain as much genetic variability as possible (e.g. Schonewald-Cox et al. 1983; Lacy 1995; Frankham et al. 2002; Allendorf and Luikart 2007; Witzenberger and Hochkirch 2011). The traditional approach has tried to minimize reproduction between closely related individuals based on studbook data and associated inbreeding coefficients, but currently the use of MK is considered to be a better management that should be preferred (e.g. EAZA Population Management Manual – EAZA 2015). According to the same document, breeding priority should be given to individuals with low mean kinships, which is an approach that could equalize the genetic influence of particular individuals within the population. The low mean kinship could mean two different breeding histories of the particular individual: (1) the individual has not yet had any breeding opportunity, yet could indeed have such an opportunity to reproduce, and (2) the individual has had breeding opportunities but has failed due to factors that could include poor somatic and reproduction parameters, unusual or even pathological behaviour of genetic or environmental-ontogenetic origin or previous husbandry. In effect, the support/prioritization of animals under the second variant could deteriorate a managed population over the long term (cf. Frankham et al. 1986; Massaro et al. 2013; Chargé et al. 2014). The potential negative effect of MK under this variant could be minimized by the proactive communication of the coordinator with particular keepers and between keepers before animals are exchanged. Transferring animals in suboptimal condition due to health, somatic or other parameters is considered an unethical conduct.

14.11 Nutrition and Veterinary Issues

A lot of energy, budget and personal capacities are devoted to these important issues, which exhibit a relatively good publication production (see, e.g. *Journal of Zoo and Wildlife Medicine*). As with the management issues mentioned above, the nutrition of zoo animals should reflect the biology and nutrition of particular species (cf. Clauss et al. 2009; Junge et al. 2009 Hatt et al. 2011; Clauss et al. 2013). When the seasonal availability of some parts of a diet (e.g. fruits), the daily intake and the nutritional quality of the diet in the wild are considered, some species are evidently overfed, with regard to the amount or to the nutritional quality of the daily intake (e.g. lemurs – Junge et al. 2009). This overfeeding has a significant effect on the health, longevity and reproduction of such animals (e.g. Junge et al. 2009). Considering that zoos must avoid domestication and provide relevant educational experiences to visitors, there is ample opportunity for greater collaboration between nutritionists and caretakers to increase foraging time and encourage species-typical behaviours without overfeeding the animals. To this end, particular relevance may be assigned to natural browsing of berries and older natural fruit varieties whose consumption can elicit the most appropriate feeding behaviour without compromising the general health of the concerned animals. Currently, cultivated fruits and vegetables with much higher level of sugar and lower level of fiber are often replaced by vegetables with a well-balanced content of sugar and fiber (e.g. Junge et al. 2009).

We must confess that the extensiveness of this field is beyond our scope. We therefore refer interested readers to the series *Fowler's Zoo and Wild Animal Medicine Current Therapy*, *Comparative Animal Nutrition and Metabolism* (Cheeke and Dierenfeld 2010), *Wild mammals in captivity* (Kleiman et al. 2010), publications by Prof. Marcus Clauss and materials associated with the EAZA Nutrition Group and Veterinary Committee, AZA Nutrition and Veterinary Advisory Group.

14.12 Collaboration with Other Sectors of Society

Although some zoo associations are quite restrictive in relation to working with private animal holders, when it comes to the prioritizing of collections, recommended meta-population approaches, limited space and other capacities, zoos should cooperate more with other organizations or responsible private holders that are both sensitive to conservation issues and trustworthy, in order to achieve greater population viability of particular threatened species. Organizations focused on in situ conservation, which have been increasingly supported by zoos providing funds and professional training (see above), could also recommend the establishment of some ex situ management for particular species. It should be mentioned that division between in situ and ex situ conservation is often arbitrary, as the two forms are often closely connected and use very similar tools, as could be demonstrated, for example, in the case of kakapo (*Strigops habroptilus* – Powlesland et al. 2006), the Iberian lynx (*Lynx pardinus*) (e.g. Vargas et al. 2009), Tasmanian devil (*Sarcophilus harrisii* – McCallum 2008), Western Derby's eland (*Taurotragus derbianus derbianus* – Derbianus Conservation: www.derbianus.com), the giant sable antelope (*Hippotragus niger variani* – Pinto et al. 2016) and many ZGAP projects (Wirth 2007). Synergies among various subjects could increase the effectiveness of conservation projects, and so such collaborative efforts can only be recommended.

14.13 Education and Popularization

Excellent educational and popularization work can highlight the conservation mission of a particular zoo, ideally with a regional or global perspective, and popularize each and every species held in a particular zoo, especially threatened species which are seemingly considered 'normal', 'dull' or 'less attractive' (cf. Gerald Durrell). Durrell's propagation of Rodrigues fruit bat *Pteropus rodricensis* and pink pigeon *Nesoenas mayeri* are exemplary cases of talented popularization and educational work (Durrell 1979). Excellent educational work should not shy away from an explanation of potentially controversial and ethically complicated issues under the responsible breeding management (Norton et al. 1995) such as management euthanasia, the breed and cull strategy, the feeding of carnivores with euthanized zoo animals, the significance of bachelor groups and the exchange of favourite specimens under population managements across zoos. Institutions with such agenda should win our recognition, as they allow other institution scopes for explanations, discussions and following (also Gippoliti 2014). In any case, a truthful media presentation is more valuable than populism, since it can help us to realize the best management steps in the future. We should recognize that there may be different degrees of difficulty according to prevalent cultural attitudes at regional or national levels. At present, at least in some countries, increasing attention is being paid to welfare issues and the detriment caused by other zoo activities (Maynard 2017). Zoo resources, like natural resources, are finite. Increasing the attention paid to welfare-related arguments poses the very real danger that zoos could be diverted from work on other scientific or conservation issues.

14.14 Publication of Interesting Observation and Experiences

Zoos breed a huge number of species, and the obtained knowledge and experiences are the source of a vast amount of information on all possible biological parameters (see above and Fig. 14.6). This information is often published and currently also uploaded into the ISIS – now the Species360 – ZIMS database (e.g. Schwartz et al. 2017). All zoos should be encouraged to publish interesting observations, basic biological parameters of kept animals and the observed effects of husbandry changes. Even rare observations are valuable, as when they are combined with an understanding of typical behaviour, they can give a richer picture of individual motivations and relationships (cf. Fischhoff et al. 2010). Additionally, even short and technical (descriptive) studies help provide a better understanding of the biology of particular species, which could have great importance for species-specific management and conservation actions. Publication of documented husbandry experiences minimizes duplications of activities and provides a potential for improvements to husbandry or conservation actions (e.g. Barongi et al. 2015). The category Management of Captive Animals is included in the very interesting Conservation Evidence website (conservationevidence.com) (Andrew Bowkett, pers. comm.).

The established zoo journals, *Zoo Biology*, *Journal of Zoo and Wildlife Medicine, International Zoo Yearbook* and the new *Journal of Zoo and Aquarium Research* increase the opportunities to publish results, as the reviewing process is accommodated to the specifics of zoo work and available sample sizes. Unfortunately, some established zoo journals have not been longer published – *Der Zoologische Garten, Dodo, Milu* or *Bongo*. Time limitations, a crowded agenda and 'professional blindness' often limit the prolific scientific contribution that zoos could make.

Fig. 14.6 The Nile rhinoceros (*Ceratotherium cottoni*) belongs to a taxon well-documented by zoo staff (especially at Dvůr Králové Zoo) – see references in the unique Rhino Resource Center (http://www.rhinoresourcecenter.com/). (Photo by Jan Robovský)

14.15 Collection of Material

Corpses of animals held in zoos and aquariums have been sources of valuable morphological comparisons since the early histories of zoos. Details of the anatomy of extinct or nearly extinct species are often known, thanks only to the opportunities offered by scientifically managed zoos in the 1800s (Owen 1868; Garrod 1878; Owen 1868; Beddard and Treves 1887). Even the external morphology of a multitude of mammal species is being studied in zoos, an opportunity that was exploited fully by the great zoologist Reginald Innes Pocock in London (Gippoliti et al. 2017). Kitchener (2002) and Gippoliti and Kitchener (2007) comprehensively reviewed the importance of zoo specimens stored in collections (see also Randi 2007). Unfortunately, material of great importance has often been irretrievably lost (e.g. Groves 1982).

Since some (captivity-induced) morphological changes have been described for zoo animals (e.g. O'Regan and Kitchener 2005, references therein, and also de Beaux 1923; Hilzheimer 1937; Kleinschmidt 1950; Angst 1967; Angst and Storch 1967; Rausch 1967; Velzen 1967; Klimov and Orlov 1982; Dathe 1984; Heráň 1988; Volf 1995; Duckler and Binder 1997; Spasskaya and Orlov 1999; Spasskaya 2000; Spasskaya and Kůs 2003; Wisely et al. 2005; Stuermer and Wetzel 2006; Spasskaya 2007; Clauss et al. 2007; Kaiser et al. 2009; Zordan et al. 2012; Edwards et al. 2013; Hartstone-Rose et al. 2014; Saragusty et al. 2014; Taylor et al. 2016; but see also Guay et al. 2011), zoo animals could be, after their demises, the source of much valuable data to allow a calibration of previous and current forms of husbandry (e.g. Kitchener 2002; Kitchener and MacDonald 2004; Taylor et al. 2016). In the case of captivity-induced morphological changes, it would be worth knowing which factors are responsible, whether they are inherited or obtained via ontogeny (and how often), and whether they are reversible or not, as in domestic animals (e.g. Hemmer 1990; Groves 1989, 1999; Kruska 2005; O'Regan and Kitchener 2005; Dobney and Larson 2006; Saragusty et al. 2014; see also Frankham et al. 1986). We should encourage museums, zoos and aquariums to collaborate more intensively in collecting valuable zoo animals (cf. Kitchener 1997; Gippoliti and Kitchener 2007), which should be recommended by zoo staff based on the significance of particular specimens for the global population, being representatives of some particular breeding line and/or of interest based on origin (wild, wild-born, captive) and on some specific features (known age and husbandry, veterinary procedures, standard or atypical somatic parameters). Zoos should be aware that the quality of any comparison depends strongly on representativeness and sample size; our survey of somatic parameters of the Przewalski horses (Groves and Robovský, in prep.) across worldwide collections showed that only the Scientific Museum of the Biosphere Reserve 'Askania Nova' Reserve and National Museum Praha (via Prague Zoo) have stored the representative skeletal material of this species that allowed us to compare captive lines and generations (Robovský et al. 2014).

New technologies encourage the collection of tissue cultures, gametes, embryos or tissue samples of dead or live animals, under the so-called BioBank or Frozen Zoo initiative, which could be used for many scientific and conservation goals, including artificial insemination and cloning (e.g. Holt et al. 2003; Clarke 2009; Piña-Aguilar et al. 2009). Biomaterial banks could be the source of much interesting information, such as taxonomic identity, purity, genetic diversity, paternity of particular specimens, etc. (e.g. Randi 2007; Witzenberger and Hochkirch 2011; Fienieg and Galbusera 2013). Currently, EAZA tends to concentrate the biomaterial in a network of laboratories (i.e. EAZA Biobank – Hvilsom et al. 2016). Nonetheless, all zoos and aquariums could be encouraged to collect biomaterial at their facilities or in collaboration with natural history museums, since storage is easy and inexpensive (e.g. blood/tissue samples stored in tubes with pure ethanol, hairs stored in paper envelopes, both ideally stored in cold conditions).

14.16 Zoo Design Trends

Considerable funding has been directed in recent decades towards the building of new zoo sections or even whole new zoos (e.g. Salzert 2010). The development of a true 'zoo design' industry does not facilitate a critical review of the successes and failures achieved in this field. The title chosen for the proceedings of one of the most recent meetings on zoo design *Innovation or Replication* (Plowman and Tonge 2005) identifies one of the current issues. Although zoos are often currently distancing themselves from their history, today as in the past, zoos have often copied each other's design styles, often with results that get worse over the years (Hancocks 2001). Incidentally, the relevance of (often huge) budgets devoted, for example, to very costly rock artwork design is probably one of the factors leading to an increasing commercialization of zoo operations, which is a factor that may have negative consequences for their overall conservation mission. Many 'outsiders' of the zoo world are led to think that a good zoo where animals are kept well requires millions of euros. This is simply untrue, as shown by several zoos that achieved considerable importance in the zoo world for their innovative yet low-cost design and conservation mission without having millions to spend (Jersey Zoo, La Torbiera Zoological Park in Piedmont, Due La Fontaine in the Loira, Pilsen Zoo and others). Often, these designs fully exploit the potential of the local landscape and wise use of plantings. One general concern is that some modern-style exhibits pay excessive attention to aesthetic elements, while neglecting important key factors that are functionally relevant to animal welfare, such as space, shade, soil texture, vertical climbing apparatus for arboreal animals, etc. (Gippoliti 2006).

14.17 Conclusion

In conclusion, we hope that zoos will continue to pursue their important conservation missions: to be *fully loaded* arks of *threatened biodiversity*, managed based on biologically sensible principles and in close cooperation with other (including private) individuals and organizations that are sensitive to conservation issues and trustworthy. Our management goals should focus on preserving particular threatened species or evolutionary lineages based on a (meta-) population approach (i.e. breeding should take priority over restrictions and often hypothetical fears associated, e.g. with inbreeding, disease, etc.). Critics want zoos to stop breeding as they regard zoos as prisons. This idea is gaining traction and must be countered by responsible breeding plans throughout the world. The key is improved habitats that encourage animals to thrive. In the end, the public will support zoos if the animals are perceived as being at the centre of zoo mission. Our steps should be taken as well in the spirit of Ulie S. Seal who said, 'Strategies and priorities should maximize options while minimizing regrets for species conservation', and of Gerald Durrell who said, 'In conservation, the motto should always be "never say die"'.

Acknowledgements JR is deeply indebted to many keepers, curators and zoo directors and other colleagues from zoos and universities for sharing their knowledge and experience. Additional thanks go to Luděk Čulík (Dvůr Králové Zoo), Lubomír Melichar (Liberec Zoo), the late Peter Lupták (Bojnice Zoo), Jan Vašák (Jihlava Zoo), Jan Pluháček (Ostrava Zoo), Tomáš Peš (Plzeň Zoo), the late Zdeněk Veselovský (Praha Zoo) and Evžen Kůs (Praha Zoo) for the formation of his deeper insight in the zoo world and to Colin Groves∗, Wolfgang Frey, Roland Wirth, Jens-Ove Heckel, Arnd Schreiber, Eulalia Moreno, Noam Werner, Klaus Rudloff, Christian Kern, Luboš and Luděk for their inspiring enthusiasm, conservation missions and always pleasant meetings!

SG wishes to thank those colleagues who have played a long-standing role in the development of his scientific and conservation interests: Pier Lorenzo Florio, Fabio Cassola, Ermanno Bronzini,

Jean-Marc Lernould, John Oates, Ken Kawata, the late Nicholas Gould and Lothar Schlawe. Several friends have shared their interest and concern over zoos, including Francesco Rocca, Michele Capasso, Mauro Picone, Antonio Rosati, Andreina D'Alessandro, Corradino Guacci, Harry Salomon and the late Roberto Busacchi.

Our sincere thanks belong to all previous and current zoo staff for their enthusiastic and conservation-oriented missions.

We would like to express our gratitude to Roland Wirth and two referees – Terry Maple and Andrew Bowkett – for their critical review and very valuable suggestions which greatly improved the final version of the manuscript, to Roland Wirth for the beautiful photos and to Anton Baer and Andrew Bowkett for language editing. *We dedicate this chapter to the memory of our highly regarded friend Colin Peter Groves, who passed away on 30 November 2017.

References

Allendorf FW, Luikart G (2007) Conservation and the genetics of populations. Blackwell Publishing, Oxford

Angst R (1967) Beitrag zum Formwandel des Craniums der Ponginen. Z Morph Anthrop 58(2):109–151

Angst R, Storch G (1967) Bemerkungen über den Schädel des Gorilla Abraham aus dem Frankfurter Zoologischen Garten. Nat Mus 97(10):417–420

Ballantynea R, Packera J, Hughesa K et al (2007) Conservation learning in wildlife tourism settings: lessons from research in zoos and aquariums. Environ Educ Res 13(3):367–383

Barongi R, Fisken FA, Parker M, Gusset M (eds) (2015) Committing to conservation: the world zoo and aquarium conservation strategy. WAZA Executive Office, Gland

Bashaw MJ, Bloomsmith MA, Marr MJ et al (2003) To hunt or not to hunt? A feeding enrichment experiment with captive large felids. Zoo Biol 22(2):189–198

Beddard FE, Treves F (1887) On the anatomy of the Sondaic rhinoceros. Trans Zool Soc Lond 12:183–198

Behie AM, Oxenham MF (eds) (2015) Taxonomic tapestries. The threads of evolutionary, behavioural and conservation research. ANU Press, Canberra

Bell CE (ed) (2001) Encyclopedia of the world's zoos. Fitzroy Dearborn, London/Chicago

Belozerskaya M (2006) The Medici giraffe. And other tales of exotic animals and power. Little, Brown and Company, New York/Boston/London

Benirschke K (ed) (1986) Primates. The road to self-sustaining populations. Springer-Verlag, New York/Berlin/Heidelberg/London/Paris/Tokyo

Bertorelle G, Bruford MW, Hauffe HC et al (eds) (2009) Population genetics for animal conservation. Cambridge University Press, Cambridge/New York/Melbourne/Madrid/Cape Town/Singapore/Sao Paulo

Bichet C, Penn DJ, Moodley Y et al (2014) Females tend to prefer genetically similar mates in an island population of house sparrows. BMC Evol Biol 14(1):47

Biega A, Greenberg DA, Mooers AO et al (2017) Global representation of threatened amphibians ex situ is bolstered by non-traditional institutions, but gaps remain. Anim Conserv 20(2):113–119

Blaszkiewitz B (2005) 50 Jahre Tierpark Berlin. Berliner Tierpark-Buch 44:1–348

Bostock SSC (1993) Zoos and animal rights. The ethics of keeping animals. Routledge, London/New York

Bowkett AE (2009) Recent captive-breeding proposals and the return of the Ark Concept to global species conservation. Conserv Biol 23(3):773–776

Bowkett AE (2014) *Ex situ* conservation planning is more complicated than prioritizing the keeping of threatened species in zoos. Anim Conserv 17(2):101–103

Bridges W (1974) Gathering of animals. An unconventional history of the New York Zoological Society. Harper & Row Publishers, New York/Evanston/San Francisco/London

Cameron DW, Groves CP (2004) Bones, stones and molecules. "Out of Africa" and human origins. Elservier Academic Press, Amsterdam/Tokyo
Carroll SC, Fox CW (eds) (2008) Conservation biology. Evolution in action. Oxford University Press, New York
Castelló JR (2016) Bovids of the world. Antelopes, gazelles, cattle, goats, sheep, and relatives. Princeton University Press, Princeton
Ceballos G, Ehrlich PR, Barnosky AD et al (2015) Accelerated modern human–induced species losses: entering the sixth mass extinction. Sci Adv 1(5):e1400253
Ceballos G, Ehrlich PR, Dirzo R (2017) Biological annihilation via the ongoing sixth mass extinction signaled by vertebrate population losses and declines. PNAS 114(30):e6089–E6096
Chargé R, Sorci G, Sain Jalme M et al (2014) Does recognized genetic management in supportive breeding prevent genetic changes in life-history traits? Evol Appl 7(5):521–532
Charpentier MJE, Widdig A, Alberts SC (2007) Inbreeding depression in non-human primates: a historical review of methods used and empirical data. Am J Primatol 69(12):1370–1386
Cheeke PR, Dierenfeld ES (2010) Comparative animal nutrition and metabolism. CABI, Oxfordshire
Clarke AG (2009) The Frozen Ark Project: the role of zoos and aquariums in preserving the genetic material of threatened animals. Int Zoo Yearb 43:222–230
Clauss M, Franz-Odendaal TA, Brasch J et al (2007) Tooth wear in captive giraffes (*Giraffa camelopardalis*): Mesowear analysis classifies free-ranging specimens as browsers but captive ones as grazers. J Zoo Wildlife Med 38(3):433–445
Clauss M, Wilkins T, Harley A et al (2009) Diet composition, food intake, body condition, and fecal consistency in captive tapirs (*Tapirus* spp.) in UK collections. Zoo Biol 28(4):279–291
Clauss M, Kohlschein G-M, Peemöller A et al (2013) Short-term digestible energy intake in captive moose (*Alces alces*) on different diets. Zoo Biol 32(5):484–489
Colléony A, Clayton S, Couvet D et al (2017) Human preferences for species conservation: animal charisma trumps endangered status. Biol Conserv 206:263–269
Conde DA (2013) Role of zoological gardens. In: MacLeod, Archibald JD, Levin P (eds) Grzimek's animal life encyclopedia, Gale Research Inc., Detroit, pp 207–215
Conde DA, Flesness N, Colchero F et al (2011) An emerging role of zoos to conserve biodiversity. Science 331(6023):1390–1391
Conde DA, Colchero F, Gusset M et al (2013) Zoos through the lens of the IUCN red list: a global metapopulation approach to support conservation breeding programs. PLoS ONE 8(12):e80311
Conway (2003) The role of zoos in the 21st century. Int Zoo Yearb 38(1):7–13
Cromsigt JPGM, Kerley GIH, Kowalczyk R (2012) The difficulty of using species distribution modelling for the conservation of refugee species – the example of European bison. Divers Distrib 18(12):1253–1257
Csorba G, Krivek G, Sendula T, Homonnay ZG, Hegyeli Z, Sugár S, Farkas J, Stojnić N, Németh A (2015) How scientific researches can change conservation priorities? A review of decade-long research of blind mole-rats (Rodentia: Spalacinae) in the Carpathian Basin. Therya 6(1):103–121
Cuarón AD (2005) Further role of zoos in conservation: monitoring wildlife use and the dilemma of receiving donated and confiscated animals. Zoo Biol 24(2):115–124
Dathe H (1978) The meaning and value of breeds representing pure subspecies in zoological gardens. Zool Garten N F 48(2/3):164–166
Dathe H (1984) Zur Akzeleration bei Zootieren. Zool Garten NF Jena 54(6):369–376
Dawson J, Patel F, Griffiths RA, Young RP (2016) Assessing the global zoo response to the amphibian crisis through 20-year trends in captive collections. Conserv Biol 30(1):82–91
De Beaux O (1923) A proposito di un cranio di *Theropithecus* ♀. Giornale Morfologia Uomo e Primati 4:24–29
Der Sarkissian C, Ermini L, Schubert M et al (2015) Evolutionary genomics and conservation of the endangered Przewalski's horse. Curr Biol 25(19):2577–2583
Dirzo R, You HS, Galetti M et al (2014) Defaunation in the Anthropocene. Science 345(6195):401–406

Dittrich L (2007) Der Import von Wildtieren nach Europa - Einfuhren von der fruhen Neuzeit bis zur Mitte des 20. Jahrhunderts. In: Pechlander H et al (eds) Tiere unterwegs: Historisches und Aktuelles über Tiererwerb und Tiertransporte. Braumuller, Wien, pp 1–64

Dobney K, Larson G (2006) Genetics and animal domestication: new windows on an elusive proces. J Zool 269(2):261–271

Driscoll CA, Yamaguchi N, Bar-Gal GK et al (2009) Mitochondrial phylogeography illuminates the origin of the extinct Caspian tiger and its relationship to the Amur tiger. PLoS ONE 4(1):e4125

Duckler GL, Binder WJ (1997) Previously undescribed features in the temporalis and masseteric musculature of several large felids raised in captivity. Zoo Biol 16(2):187–191

Durrell G (1976) The stationary ark. HarperCollins Publishers Ltd, London

Durrell G (1979) Golden bats and pink pigeons. Collins, Fontana

EAZA (2013) EAZA position on intentional breeding for the expression of rare recessive alleles. EAZA, Amsterdam

EAZA (2015) EAZA population manual. http://www.eaza.net/assets/Uploads/Governing-documents/201510-EAZA-Population-Management-Manual-V1.05-Council-approved-20131401.pdf. Accessed 22 Sept 2017

EAZA (2016a) Annual report 2016. EAZA, Amsterdam

EAZA (2016b) TAG reports 2016. EAZA, Amsterdam

Edwards J (1996) London zoo from old photographs 1852–1914. Edward Charles, London

Edwards CTT, Rasmussen GSA, Riordan P et al (2013) Non-adaptive phenotypic evolution of the endangered carnivore *Lycaon pictus*. PLoS ONE 8(9):e73856

Ellegren H, Galtier N (2016) Determinants of genetic diversity. Nat Rev Genet 17(7):422–433

Fernandez EJ, Timberlake W (2008) Mutual benefits of research collaborations between zoos and academic institutions. Zoo Biol 27(6):470–487

Ferriere R, Dieckmann U, Couvet D (eds) (2007) Evolutionary conservation biology. Cambridge University Press, Cambridge

Fienieg ES, Galbusera P (2013) The use and integration of molecular DNA information in conservation breeding programmes: a review. J Zoo Aquarium Res 1(2):44–51

Fischhoff IR, Sundaresan SR, Larkin HM et al (2010) A rare fight in female plains zebra. J Ethol 28:201–205

Frankham R, Hemmer H, Ryder OA et al (1986) Selection in captive populations. Zoo Biol 5:127–138

Frankham R, Ballou JD, Briscoe DA (2002) Introduction to conservation genetics. Cambridge University Press, Cambridge

Frankham R, Ballou JD, Dudash MR et al (2012) Implications of different species concepts for conserving biodiversity. Biol Conserv 152:25–31

Frost W (ed) (2011) Zoos and tourism. Conservation, education, entertainment? Channel View Publications, Bristol/Buffalo/Toronto

Frynta D, Marešová J, Landová E et al (2009) Are animals in zoos rather conspicuous than endangered? In: Columbus AM, Kuznetsov L (eds) Endangered species: new research. Nova Science Publishers, New York, pp 299–341

Frynta D, Lišková S, Bültmann S, Burda H (2010) Being attractive brings advantages: the case of parrot species in captivity. PLoS ONE 5(9):e12568

Garnett ST, Christidis L (2017) Taxonomy anarchy hampers conservation. Nature 546(7656):25–27

Garrod AH (1878) On some points in the visceral anatomy of the rhinoceros of the Sunderbunds (*Rhinoceros sondaicus*). Proc Zool Soc Lond 1877:707–711

Gippoliti S (2004) The role of captive breeding in the conservation of the European mammal diversity. Hystrix Ital J Mamm 15(1):35–53

Gippoliti S (2006) Applied primatology in zoos: history and prospects in the field of wildlife conservation, public awareness and animal welfare. Primate Rep 73:57–71

Gippoliti S (2010) La giungla di Villa Borghese. I cento anni del Giardino Zoologico di Roma, Belvedere/Latina

Gippoliti S (2011) Zoos and conservation in the XXI century: overlooked meeting points between ecology and social sciences? Museologia Scientifica nuova serie 5(1–2):168–176
Gippoliti S (2012) Ex situ conservation programmes in European zoological gardens: can we afford to lose them? Biodivers Conserv 21(6):1359–1364
Gippoliti S (2014) Bioethics in the zoo from an Italian perspective. Museologia Scientifica nuova serie 8:9–12
Gippoliti S, Amori G (2004) Mediterranean island mammals: are they a priority for biodiversity conservation? Biogeographia 25:135–144
Gippoliti S, Amori G (2007) Beyond threatened species and reintroduction: establishing priorities for conservation and breeding programmes for European rodents in zoos. Int Zoo Yearb 41:194–202
Gippoliti S, Groves CP (2020) Cryptic problematic species and troublesome taxonomists: a tale of the Apennine bear and the Nile white rhinoceros. In Angelici F.A., Rossi L. (eds) Problematic Wildlife II) – new conservation and management challenge in the human-wildlife interactions. Springer Chasm.
Gippoliti S, Kitchener A (2007) The Italian zoological gardens and their role in mammal systematic studies, conservation biology and museum collections. Hystrix Ital J Mamm 18(2):173–184
Gippoliti S, Di Francesco T, Robovský J (2017) On the shoulders of giants: Reginald Innes Pocock and integrative mammal research in museums and zoos. Mammalia 82:529–539
Gippoliti S, Cotterill FPD, Zinner D, Groves CP (2018) Impacts of taxonomic inertia for the conservation of African ungulate diversity: an overview. Biol Rev 93:115–130
Glikson AY, Groves C (2016) Climate, fire and human evolution. The deep time dimensions of the Anthropocene. Springer, Cham/Heidelberg/New York/Dordrecht/London
Goodall J, Maynard T, Hudson G (2010) Hope for animals and their world (how endangered species are being rescued from the brink). Icon Books, New York/London
Grigson C (2016) Menagerie. The history of exotic animals in England 1100–1837. Oxford University Press, Oxford
Groves C (1982) The skulls of Asian rhinoceroses: wild and captive. Zoo Biol 1(3):251–261
Groves CP (1989) Feral mammals of the Mediterranean islands: documents of early domestication. In: Clutton-Brock J (ed) The walking larder. Unwin-Hyman, London, pp 46–58
Groves CP (1995) Microtaxonomy and its implications for captive breeding. In: Gansloßer U (ed) Research and captive propagation. Filander Verlag, Fürth, pp 24–28
Groves C (1999) The advantages and disadvantages of being domesticated. Perspect Human Biol 4(1):1–12
Groves CP (2009) The A-line Przewalski horse. News Biosphere Reserve "Askania Nova" 11:97–103
Groves CP, Grubb P (2011) Ungulate taxonomy. The Johns Hopkins University Press, Baltimore
Groves C, Robovský J (2011) African rhinos and elephants: biodiversity and its preservation. Pachyderm 50:69–71
Groves CP, Cotterill FPD, Gippoliti S et al (2017) Species definitions and conservation: a review and case studies from African mammals. Conserv Genet 18:1247–1256
Grubb P (2005) Order Artiodactyla. In: Wilson DE, Reeder DAM (eds) Mammal species of the world. The Johns Hopkins University Press, Baltimore, pp 637–722
Guay PJ, Parrott M, Selwood L (2011) Captive breeding does not alter brain volume in a marsupial over a few generations. Zoo Biol 30:1–5
Gusset M, Dick G (2011a) The global reach of zoos and aquariums in visitor numbers and conservation expenditures. Zoo Biol 30(5):566–569
Gusset M, Dick G (2011b) Towards sustainable population management. WAZA Mag 12:1–44
Gusset M, Dick G (2013a) Fighting extinction. WAZA Mag 13:1–41
Gusset M, Dick G (2013b) Towards integrated species conservation. WAZA Mag 14:1–45
Gusset M, Lowry R (2014) Towards effective enviromental education. WAZA Mag 15:1–48
Gusset M, Dick G (2015) Towards positive animal welfare. WAZA Mag 16:1–44
Guthrie RD (2005) The nature of paleolithic art. The University of Chicago Press, Chicago

Hancocks D (2001) A different nature. The paradoxical world of zoos and their uncertain future. California University Press, Berkeley
Hartstone-Rose A, Selvey H, Villari JR et al (2014) The three-dimensional morphological effects of captivity. PLoS ONE 9(11):e113437
Hatt JM, Müller DWH, Bingaman L et al (2011). Life expectancy in zoo mammals: what a zoo veterinarian should know. AAZV conference, Kansas City, Missouri, USA, October 2011, p 181–183
Hazevoet CJ (1996) Conservation and species lists: taxonomic neglect promotes the extinction of endemic birds, as exemplified by taxa from eastern Atlantic islands. Bird Conserv Int 6:181–196
Heckel J-O, Holland J, Richardson DM, Wirth R (2013) Has the Ark gone off course? WAZA News 4:4–7
Hediger H (1942) Wildtiere in Gefangenschaft. Benno Schwabe & Co., Basel
Hediger H (1969) Man and animal in the zoo. Routledge & Kegan Paul, London
Hedrick PW, Kalinowski ST (2000) Inbreeding depression in conservation biology. Annu Rev Ecol Syst 31:139–162
Hemmer H (1990) Domestication: the decline of environmental appreciation. Cambridge University Press, New York
Heráň I (1988) Comments on anomalies in the frame of *foramen occipitale magnum* in the lion, *Panthera leo* (Linné, 1758). Gazella 15:97–101
Hill SP, Broom DM (2009) Measuring zoo animal welfare: theory and practice. Zoo Biol 28(6):531–544
Hilzheimer M (1937) Domestikationsmerkmale am Schädel des Gorilla Bobby. Z Säugetierkd 12:89–96
Hirschfeld E, Swash A, Still R (2013) The world's rarest birds. Princeton University Press, Princeton
Hoban S, Arntzen JW, Bertorelle G et al (2013) Conservation genetic resources for effective species survival (ConGRESS): bridging the divide between conservation research and practice. J Nat Conserv 21(6):433–437
Holand O, Askim KR, Roed KH et al (2007) No evidence of inbreeding avoidance in a polygynous ungulate: the reindeer (*Rangifer tarandus*). Biol Lett 3(1):36–39
Holt WV, Pickard AR, Rodger JC, Wildt DE (2003) Reproductive science and integrated conservation. Cambridge University Press, Cambridge/New York/Melbourne/Madrid/Cape Town/Singapore/Sao Paulo
Hoy JM, Murray PJ, Tribe A (2010) Thirty years later: enrichment practices for captive mammals. Zoo Biol 29(3):303–316
Hutchins M (2003) Zoo and aquarium animal management and conservation: current trends and future challenges. Int Zoo Yearb 38(1):14–28
Hutchins M, Conway WG (1995) Beyond Noah's Ark: the evolving role of modern zoological parks and aquariums in field conservation. Int Zoo Yearb 34(1):117–130
Hvilsom C, Pereboom Z, Schad K (2016) Banking on our future. Zooquaria 94:34
Ibánez B, Moreno E, Barbosa A (2011) No inbreeding effects on body size in two captive endangered gazelles. Mamm Biol 76(6):748–754
Ibánez B, Moreno E, Barbosa A (2013) Parity, but not inbreeding, affects juvenile mortality in two captive endangered gazelles. Anim Conserv 16(1):108–117
Isaac NJB, Turvey ST, Collen B et al (2007) Mammals on the EDGE: conservation priorities based on threat and phylogeny. PLoS ONE 2(3):e296
IUCN SSC Antelope Specialist Group (2015). IUCN SSC Antelope Specialist Group position statement on the intentional genetic manipulation of antelopes. Ver. 1.0
Jonas CS, Timbrell LL, Young F et al (2018) Management of captive animals: global evidence for the effect of selected interventions. University of Cambridge, Cambridge
Junge RE, Williams CV, Campbell J (2009) Nutrition and behavior of lemurs. Vet Clin North Am Exot Anim Pract 12(2):339–348

Kaiser TM, Brasch J, Castell JC et al (2009) Tooth wear in captive wild ruminant species differs from that of free-ranging conspecifics. Mamm Biol 74:425–437

Kalinowski ST, Hedrick PW, Miller PS (1999) No inbreeding depression observed in Mexican and Red wolf captive breeding programs. Conserv Biol 13(6):1371–1377

Kellert SR, Wilson EO (eds) (1993) The biophilia hypothesis. Sherawater Book, Shearwater

Kerley GIH, Kowalczyk R, Cromsigt JPGM (2012) Conservation implications of the refugee species concept and the European bison: king of the forest or refugee in a marginal habitat? Ecography 35(6):519–529

Keulartz J (2015) Captivity for conservation? Zoos at a crossroads. J Agric Environ Ethics 28(2):335–351

Kisling VN Jr (ed) (2000) Zoo and aquarium history. Ancient animal collections to zoological gardens. CRC Press, Boca Raton/London/New York/Washington D.C

Kitchener AC (1997) The role of museums and zoos in conservation biology. Int Zoo Yearb 35(1):325–336

Kitchener AC (2002) Using mammal collections. Biol Curator 22:24–27

Kitchener A, MacDonald A (2004, Sept) The longevity legacy. The problem of old mammals in zoos. In: Hiddinga B (ed) EAZA conference, Kolmarden, pp 132–137

Kleiman DG, Thompson KV, Baer CK (eds) (2010) Wild mammals in captivity. University of Chicago Press, Chicago/London

Kleinschmidt A (1950) Anatomische Beobachtungen an Fetthaube, Schädel und Wirbelsäule des Gorilla "Bobby". Zool Anz 145:380–405

Klimov VV, Orlov VN (1982) Sovremennoe sostoânie i problemy sochraneniâ lošadi Prževalskogo (*Equus przewalskii*) (present status and problems of conservation of *Equus przewalskii*). Zoologičeskij Žurnal 61(12):1862–1869

Klös H-G, Frädrich H, Klös U (1994) Dier Arche Noah an der Spree. 150 Jahre Zoologischer Garte Berlin. Eine tiergärtnerische Kulturgeschichte von 1844 bis 1994. FAB Verlag, Berlin

Koeninger Ryan K, Lacy RC, Margulis SW (2003) Impacts of inbreeding on components of reproductive success. In: Holt WV, Pickard AR, Rodger JC, Wildt DE (eds) Reproductive science and integrated conservation. Cambridge University Press, Cambridge, pp 82–96

Kowalski K (1967) The Pleistocene extinction of mammals in Europe. In: Martin PS, Wright HE (eds) Pleistocene extinctions. The search for a cause. Yale University Press, New Haven/London, pp 349–364

Kruska DC (2005) On the evolutionary significance of encephalization in some Eutherian mammals: effects of adaptive radiation, domestication, and feralization. Brain Behav Evol 65(2):73–108

Lacy RC (1995) Clarification of genetic terms and their use in the management of captive populations. Zoo Biol 14(6):565–578

Lacy R (2010) Re-thinking ex situ vs. in situ species conservation. Proceedings of 65th Annual WAZA Conference:25–28

Lacy RC (2013) Achieving true sustainability of zoo populations. Zoo Biol 32(1):19–26

Law G, Kitchener A (2002) Simple enrichment techniques for bears, bats and elephants – untried and untested. Int Zoo News 49(1):4–12

Leus K (2018) Ex-situ conservation of wild pigs and peccaries: roles, status, management successes and challenges. In: Melletti M, Meijaard E (eds) Ecology, conservation and management of wild pigs and peccaries. Cambridge University Press, New York, pp 420–436

Linnell JDC, Trouwborst A, Boitani L et al (2016) Border security fencing and wildlife: the end of the transboundary paradigm in Eurasia? PLoS Biol 14(6):e1002483

Loh T-L, Larson ER, David SR et al (2018) Quantifying the contribution of zoos and aquariums to peer-reviewed scientific research. Facets 3:287–299

Luoma JR (1988) A crowded ark, the role of zoos in wildlife conservation. Houghton Mifflin Co., Boston

Lupták P (2015) Zoo Bojnice: Slovenská archa biodiverzity. Zoo Bojnice, Bojnice

Maas PHJ (2013) Last stand in captivity or cultivation: successes and failures. In: TSEW (2017). The sixth extinction. Available via http://www.petermaas.nl/extinct. Accessed 22 Sept 2017
Maas PHJ (2017) Globally extinct mammals. In: TSEW (2017). The sixth extinction. Available via http://www.petermaas.nl/extinct. Accessed 22 Sept 2017
Macdonald DW, Service K (eds) (2007) Key topics in conservation biology. Blackwell Publishing, Malden/Oxford/Victoria
Mace GM, Balmford A, Leader-Williams N et al (2007) Measuring conservation success: assessing zoos' contribution. In: Zimmermann A, Hatchwell M, Dickie LA, West C (eds) Zoos in the 21st century: catalysts for conservation? Cambridge University Press, Cambridge, pp 322–342
Majolo B, Schino G, Troisi A (2005) Towards thirty years of ethological research on the Japanese Macaque (*Macaca fuscata*) colony of the Rome zoo. J Anthropol Sci 83:43–60
Mallinson JJC (2003) A sustainable future for zoos and their role in wildlife conservation. Hum Dimens Wildl 8(1):59–63
Maple TL, Perdue BM (2013) Zoo animal welfare. Springer, Heidelberg/New York/Dordrecht/London
Martin TE, Lurbiecki H, Joy JB et al (2014) Mammal and bird species held in zoos are less endemic and less threatened than their close relatives not held in zoos. Anim Conserv 17(2):89–96
Mason GJ (2010) Species differences in responses to captivity: stress, welfare and the comparative method. Trends Ecol Evol 25(12):713–721
Mason G, Rushen J (eds) (2006) Stereotypic animal behaviour. Fundamentals and applications to welfare, 2nd edn. CABI, Wallingford/Cambridge
Massaro M, Sainudiin R, Merton D et al (2013) Human-assisted spread of a maladaptive behavior in a critically endangered bird. PLoS ONE 8(12):e79066
Mathews F, Orros M, McLaren G et al (2005) Keeping fit on the ark: assessing the suitability of captive-bred animals for release. Biol Conserv 121(4):569–577
Matschei C (2017) Seltene Tiere im Zoo: Säugetiere. Schüling Verlag, Münster
Maynard L (2017) Media framing of zoos and aquaria: from conservation to animal rights, environmental communication. J Environ Comm. 12(2):177–190
McCallum H (2008) Tasmanian devil facial tumour disease: lessons for conservation biology. Trends Ecol Evol 23(11):631–637
McGowan PJK, Traylor-Holzer K, Leus K (2017) IUCN guidelines for determining when and how ex situ management should be used in species conservation. Conserv Lett 10(3):361–366
McPhee ME (2002) Intact carcasses as enrichment for large felids: effects on on- and off-exhibit. Zoo Biol 21(1):37–47
Melfi (2009) There are big gaps in our knowledge, and thus approach, to zoo animal welfare: a case for evidence-based zoo animal management. Zoo Biol 28(6):574–588
Mellor DJ, Hunt S, Gusset M (eds) (2015) Caring for wildlife: the world zoo and aquarium animal welfare strategy. WAZA Executive Office, Gland
Menzel CR, Beck BB (2000) Homing and detour behavior in Golden lion Tamarin social groups. In: Boinski S, Garber PA (eds) On the move: how and why animals travel in groups. Chicago University Press, Chicago, pp 299–326
Miller B, Conway W, Reading RP et al (2004) Evaluating the conservation mission of zoos, aquariums, botanical gardens, and natural history museums. Conserv Biol 18(1):1–8
Mills LS (2007) Conservation of wildlife populations. Demography, genetics, and management. Blackwell Publishing, Malden/Oxford/Victoria
Moodley Y, Russo IRM, Dalton DL et al (2017) Extinctions, genetic erosion and conservation options for the black rhinoceros (*Diceros bicornis*). Sci Rep 7:41417
Moreno E, Pérez-González J, Carranza J, Moya-Larano J (2015) Better fitness in captive Cuvier's gazelle despite inbreeding increase: evidence of purging? PLoS ONE 10(12):e0145111
Moss A, Esson M (2013) The educational claims of zoos: where do we go from here? Zoo Biol 32(1):13–18
Moss A, Jensen E, Gusset M (2014) A global evaluation of biodiversity literacy in zoo and aquarium visitors. WAZA executive office, Gland

Myers N, Mittermeier RA, Mittermeier CG et al (2000) Biodiversity hotspots for conservation priorities. Nature 403:853–858
Norton BG, Hutchins M, Stevens EF, Maple TL (eds) (1995) Ethics on the ark: zoos, animal welfare, and wildlife conservation. Smithsonian Institution Press, Washington DC
O'Regan HJ, Kitchener AC (2005) The effects of captivity on the morphology of captive, domesticated and feral mammals. Mammal Rev 35(3–4):215–230
Olive A, Jansen K (2017) The role of accredited zoos in the recovery process for species at risk in Canada. Can Geogr 61(3):319–333
Olney PJS, Mace GM, Feistner ATC (1994) Creative conservation. Interactive management of wild and captive animals. Springer Science+Business Media, Dordrecht
Olson LE, Blumstein DT, Pollinger JR, Wayne RK (2012) No evidence of inbreeding avoidance despite demonstrated survival costs in a polygynous rodent. Mol Ecol 21(3):562–571
Owen R (1868) Comparative anatomy and physiology of vertebrates 3. Mammals. Longmans Green, London
Peel CVA (1903) The zoological gardens of Europe. Their history and chief features. Bloomsbury, London
Penfold LM, Powell D, Traylor-Holzer K et al (2014) "Use it or lose it": characterization, implications, and mitigation of female infertility in captive wildlife. Zoo Biol 33(1):20–28
Peš T, Vogeltanz J (2010) The garden full of treasures. List of species in zoological and botanical gardens of the town Pilsen. Zoologická a botanická zahrada města Plzně and Městské knihy s. r. o, Plzeň
Pina-Aguilar RE, Lopez-Saucedo J, Sheffield R et al (2009) Revival of extinct species using nuclear transfer: hope for the mammoth, true for the Pyrenean ibex, but is it time for "conservation cloning"? Cloning Stem Cells 11(3):341–346
Pinto PV, Beja P, Ferrand N et al (2016) Hybridization following population collapse in a critically endangered antelope. Sci Rep 6:18788
Plowman AB, Tonge SJ (eds) (2005) Innovation or replication? Proceedings of the 6th international symposium on zoo design. Whitley Wildlife Conservation Trust, Paignton
Powell DM, Ardaiolo M (2016) Survey of U.S. zoo and aquarium animal care staff attitudes regarding humane euthanasia for population management. Zoo Biol 35:187–200
Powlesland RG, Merton DV, Cockrem JF (2006) A parrot apart: the natural history of the kakapo (*Strigops habroptilus*), and the context of its conservation management. Notornis 53(1):3–26
Pritchard DJ, Fa JE, Sara Oldfield S et al (2011) Bring the captive closer to the wild: redefining the role of *ex situ* conservation. Oryx 46(1):18–23
Przybylska L (2014) EEP Visayan warty pig studbook *Sus cebifrons negrinus*. Poznań Zoo, Poznań
Rabb GB (1994) The changing roles of zoological parks in conserving biological diversity. Am Zool 34(1):159–164
Rabb GB (2004) The evolution of zoos from menageries to centers of conservation and caring. Curator (Mus J) 47(3):237–246
Randi E (2007) Museum theriological collections for the study of genetic diversity. Hystrix Ital J Mamm 18(2):157–171
Rausch RL (1967) Comparative development of the canine teeth in wild and captive black bears, *Ursus americanus* Pallas. Can J Zool 45(4):465–471
Reade LS, Waran NIL (1996) The modem zoo: how do people perceive zoo animals? Appl Anim Behav Sci 47(1–2):109–118
Reading RP, Miller B, Shepherdson D (2013) The value of enrichment to reintroduction success. Zoo Biol 32(3):332–341
Redford KH, Jensen DB, Breheny JJ (2013) The long overdue death of the Ex situ and In situ dichotomy in species conservation. WAZA Mag 14:19–22
Rees PA (2015) Studying captive animals. A workbook of methods in behavior, welfare and ecology. Wiley, Chichester
Robovský J (2012) Muss die A-Linie des Przewalski-Pferdes Teil einer wissenschaftlich fundierten Erhaltungsstrategie der Art werden? ZGAP Mitt 28(2):23–28

Robovský J, Novotná A, Čulík L et al (2014) Revised catalogue of wild equids in the collection of the National Museum, Prague, and several other collections in the Czech Republic (Perissodactyla: Equidae). Lynx 45:97–138
Roe K, McConney A, Mansfield CF (2014) The role of zoos in modern society – a comparison of zoos' reported priorities and what visitors believe they should be. Anthrozoös 27(4):529–541
Rothfels N (2002) Savages and beasts. The birth of modern zoo. The Johns Hopkins University Press, Baltimore/London
Ryder OA, Byrd ML (1984) One medicine. A tribute to Kurt Benirschke from his students and colleagues. Springer-Verlag, Berlin/Heidelberg/New York/Tokyo
Ryder OA, Feistner ATC (1995) Research in zoos: a growth area in conservation. Biodivers Conserv 4(6):671–677
Salzert W (2010) Was macht Tiergarten attraktiv? Ein kleines Einmaleins des Zoogehegeplanung. Schüling Verlag, Münster
Saragusty J, Shavit-Meyrav A, Yamaguchi N et al (2014) Comparative skull analysis suggests species-specific captivity-related malformation in lions (*Panthera leo*). PLoS ONE 9:e94527
Sartore J (2017) National Geographic the photo ark: one man's quest to document the world's animals. National Geographic
Schlawe L (1969) Die für die Zeit vom 1. August 1844 bis 31. Mai 1888 nachweisbaren Thiere im zoologischen Garten zu Berlin - Amphibia - Reptilia - Aves - Mammalia. Selbstverlag, Berlin
Schonewald-Cox CM, Chambers SM, MacBryde B, Thomas WL (1983) Genetics and conservation. A reference for managing wild animals and plant populations. The Benjamin/Cummings Publishing Company Inc., London/Amsterdam/Don Mills/Ontario/Sydney/Toronto
Schook MW, Powell DM, Zimmermann W (2016) Wild equid captive breeding and management. In: Ransom JI, Kaczensky P (eds) Wild equids. Ecology, management, and conservation. The Johns Hopkins University Press, Baltimore, pp 149–163
Schreiber A, Moreno E, Groves C, Robovský J (2018) Systematics and the management units of the dama gazelle *Nanger dama*. Gnusletter 35(1):8–12
Schwartz KR, Parsons ECM, Rockwood L, Wood TC (2017) Integrating *in-situ* and *ex-situ* data management processes for biodiversity conservation. Front Ecol Evol 5(120):1–7
Seidensticker J, Forthman DL (1998) Evolution, ecology, and enrichment. Basic considerations for wild animals in zoos. In: Mellen JD, Hutchins M (eds) Second nature. Environmental enrichment for captive animals. Smithsonian Institution Press, Washington/London, pp 15–29
Senn H, Banfield L, Wacher T et al (2014) Splitting or lumping? A conservation dilemma exemplified by the critically endangered Dama Gazelle (*Nanger dama*). PLoS ONE 9(6):e98693
Shyne A (2006) Meta-analytic review of the effects of enrichment on stereotypic behavior in zoo mammals. Zoo Biol 25(4):317–337
Skibiel AL, Trevino HS, Naugher K (2007) Comparison of several types of enrichment for captive felids. Zoo Biol 26(5):371–381
Skibins JC, Dunstan E, Pahlow K (2017) Exploring the influence of charismatic characteristics on flagship outcomes in zoo visitors. Hum Dimens Wildl 22(2):157–171
Smith RH (1979) On selection for inbreeding in polygynous animals. Heredity 43(2):205–211
Snyder NFR (1995) A much improved conservation role for zoos. Zoo Biol 14(1):36–38
Solski L, Strehlow H (2015) 150 Jahre ZOO Breslau/150 lat ZOO Wrocław. Zoo Wrocław, Wrocław
Spasskaya N (2000) Revision of the osteological material of the Przewalski horses (*Equus przewalskii* Poljakov, 1881) in museums in Czech Republic. Gazella 27:71–96
Spasskaya N (2007) Investigation of the kiang (*Equus kiang*, Equidae) skull from Ladakh, India. Erforsch biol Ress Mongolei (Halle/Saale) 10:227–230
Spasskaya N, Kůs E (2003) The exterior´ peculiarities of Przewalski horse. Gazella 30:79–99
Spasskaya N, Orlov VN (1999) Patterns of cranial variability in Przewalski's horses of different lines of breeding. Vestn Zool Suppl 11:196–201
Stepherdson DJ, Mellen JD, Hutchins M (eds) (1998) Second nature. Environmental enrichment for captive animals. Smithsonian Institution Press, Washington/London

Stuermer IW, Wetzel W (2006) Early experience and domestication affect auditory discrimination learning, open field behaviour and brain size in wild Mongolian gerbils and domesticated laboratory gerbils (*Meriones unguiculatus* forma *domestica*). Behav Brain Res 173(1):11–21
Taberlet P, Valentini A, Rezaei HR et al (2008) Are cattle, sheep, and goats endangered species? Mol Ecol 17(1):275–284
Taylor LA, Müller DWH, Schwitzer C et al (2016) Comparative analyses of tooth wear in free-ranging and captive wild equids. Equine Vet J 48(2):240–245
Tidiere M, Gaillard J-M, Berger V et al (2016) Comparative analyses of longevity and senescence reveal variable survival benefits of living in zoos across mammals. Sci Rep 6:36361
Tokarska M, Pertoldi C, Kowalczyk R et al (2011) Genetic status of the European bison *Bison bonasus* after extinction in the wild and subsequent recovery. Mammal Rev 41(2):151–162
Tribe A, Booth R (2003) Assessing the role of zoos in wildlife conservation. Hum Dimens Wildl 8(1):65–74
Van Bemmel ACV (1971) Keeping and breeding of pure subspecies in zoos and National Parks. Zool Garten N F 40(3):160–162
Van Velzen SKT (1967) Does captivity influence the morphogenesis of the animal skull and teeth? Ned Tijdschr Tandheelkd 74(4):17–42
Vargas A, Breitenmoser C, Breitenmoser U (2009) Iberian Lynx Ex situ conservation: an interdisciplinary approach. Fundación Biodiversidad and IUCN Cat Specialist Group, Madrid
Volf J (1995) Variabilität, Plastizität und Selektion der Przewalskipferde, *Equus przewalskii* Poljakov, 1881. Zool Gart 65(5):284–292
WAZA (2005) Building a future for wildlife - the world zoo and aquarium conservation strategy. World Zoo and Aquarium Association, Bern
Weeks AR, Sgro CM, Young AG et al (2011) Assessing the benefits and risks of translocations in changing environments: a genetic perspective. Evol Appl 4(6):709–725
Weigl R (2005) Longevity of mammals in captivity: from the living collections of the world. A list of mammalian longevity in captivity. Kleine Senckenberg-Reihe 48:1–215
Wickins-Dražilová D (2006) Zoo animal welfare. J Agric Environ Ethics 19:27–36
Wilson EO (1987) Biophilia. Harvard University Press, Cambridge, MA, London
Wirth R (2007) 25 Jahre ZGAP. Eine ZGAP-Geburtstagsbetrachtung. ZGAP Mitt 23(1):1–8
Wirth R (2011) Biobenzin, Problemwölfe und Tigerbabys in Zoos – wie mit falschen Arumenten der Niedergang der biologischen Vielfalt beschleunigt wird. ZGAP Mitt 1:2–6
Wirth R (2015) Sind Zoos und Erhaltungszucht noch im Einklang? ZGAP Mitt 31(1):2–5
Wisely SM, Santymire RM, Livieri TM et al (2005) Environment influences morphology and development for *in situ* and *ex situ* populations of the black-footed ferret (*Mustela nigripes*). Anim Conserv 8(3):321–328
Witzenberger KA, Hochkirch A (2011) Ex situ conservation genetics: a review of molecular studies on the genetic consequences of captive breeding programmes for endangered animal species. Biodivers Conserv 20:1843–1861
Wood CC, Gross MR (2008) Elemental conservation units: communicating extinction risk without dictating targets for protection. Conserv Biol 22(1):36–47
Yasynetska NI, Zharkikh TL (2008) Hellabrunn symposiums on the a-line Przewalski horse. News Biosphere Reserve 'Askania Nova' 10:208–216
Young RJ (2003) Environmental enrichment for captive animals. Blackwell Science Ltd, Oxford
Young AG, Clarke GM (eds) (2000) Genetics, demography and viability of fragmented populations. Cambridge University Press, Cambridge
Zimmermann M (2017) EAZA conservation database report. Available via www.eaza.net. Accessed 22 Sept 2017
Zordan M, Deem SL, Sanchez CR (2012) Focal palatine erosion in captive and free-living cheetahs (*Acinonyx jubatus*) and other felid species. Zoo Biol 31(2):181–188

Chapter 15
Problematic Animals in the Zoo: The Issue of Charismatic Megafauna

Geoff Hosey, Vicky Melfi, and Samantha J. Ward

15.1 Introduction

Animals in zoos can be regarded as problematic for a number of reasons. Firstly, they indicate something to us about deep-seated problems in our relationship with the natural world, about human population growth and encroachment on natural habitats and about hunting and poaching and anthropogenic climate change. So for animals of some species which are threatened in the wild, being in a zoo might be their best prospect for survival; indeed for some, which are already extinct in the wild such as the scimitar-horned oryx *(Oryx dammah)*, Père David's deer *(Elaphurus davidianus)* and the Wyoming toad *(Anaxyrus baxteri)*, this is their only prospect of survival. This might make them problematic animals in another sense; how do we ensure the most successful and appropriate management and best welfare for these species, given that they have not evolved to live in a captive environment and in some cases we are deficient in information about their biology in the wild? Addressing these husbandry issues can be done through systematic research and an evidence-based approach (Melfi 2009), but much of this effort is directed at a small selection of large-bodied popular animals, the so-called charismatic megafauna, to the neglect of smaller, less popularly appealing species, thus rendering some species 'problematic' for this reason. Finally, for some people there are no reasons which can justify keeping any animals in zoos, or at least no

G. Hosey (✉)
University of Bolton, Bolton, UK
e-mail: gh2@bolton.ac.uk

V. Melfi
Hartpury University, Gloucester, UK

S. J. Ward
School of Animal Rural and Environmental Sciences, Nottingham Trent University, Nottingham, UK

F. M. Angelici, L. Rossi (eds.), *Problematic Wildlife II*,
https://doi.org/10.1007/978-3-030-42335-3_15

reasons to keep the 'charismatic megafauna', so animals in zoos become problematic animals simply because of the view that they shouldn't be there.

It could be argued that the charismatic megafauna receive undue attention, both in terms of public and professional concern for their welfare and in terms of the research effort devoted to them, undue in respect of the small numbers of individuals and species involved in comparison with the majority of animals kept in zoos and undue because their conservation importance, while high, is not as great as for many other relatively overlooked species. On the other hand, it could equally be argued that the cognitive capacities and self-awareness of these charismatic species make them more worthy of concern and attention, since they are more likely to experience suffering in a captive environment. In this chapter, we will examine some of these issues with respect to three notable charismatic taxa which have received a lot of attention: great apes, which include the chimpanzee (*Pan troglodytes*), the bonobo (*Pan paniscus*), two species of gorilla (western *Gorilla gorilla* and eastern *Gorilla beringei*) and two species of orangutans (Bornean *Pongo pygmaeus* and Sumatran *Pongo abelii*); elephants, comprising two of the three extant species, the African (*Loxodonta africana*) and Asian (*Elephas maximus*); and cetaceans, notably the bottlenose dolphin (*Tursiops truncatus*), white whale (*Delphinapterus leucas*) and killer whale (*Orcinus orca*). To achieve this, we will pair these taxa with morphologically and ecologically similar taxa which are not regarded as charismatic and compare how they are represented in media stories, in zoo research and in zoo populations.

15.2 Evidence-Based Animal Welfare Versus Ethics

Firstly, we must distinguish between animal welfare and animal ethics. These are two quite different things, but they are often conflated in discussions about the value of zoos and what they do.

Animal welfare can be defined as 'the state of an animal as regards its attempts to cope with its environment' (Hill and Broom 2009). Welfare can be assessed using health, physiological and behavioural measures (Dawkins 1985; Hill and Broom 2009). Interpreting what these measures tell us can be challenging and is often done from one of three different viewpoints: that good welfare means that animals are physically fit and healthy; that they have a pleasurable life, free from undue unpleasant affective states; or that they are able to perform behaviours that they would do in the wild (Fraser 2009). Integrating evidence from all three approaches is desirable (Fraser 2009), but sometimes we can reduce our assessment to answering just two questions: are the animals healthy and do they have what they want? (Dawkins 2006). Again, there are behavioural and physiological measures that can help us answer these questions (Dawkins 2004). Maintaining and enhancing high welfare for animals should thus be based upon knowledge of what the animal needs or wants, how pleasurable its life is and how healthy it is. In many instances this knowledge is based upon supposition, experience of what works, or suggestions from other people, sources which we might collectively term 'anecdotal'. Many owners of companion animals, for instance, believe they know what is best for their animals, and that belief often derives from hearsay and conversations with other people. In good zoos, however (by which we mean zoos that aspire to the standards of an accrediting body such as the Association of Zoos and Aquariums [AZA], the European Association of Zoos and Aquariums [EAZA] and others), much of that knowledge now comes from Animal Welfare Science rather than from anecdotal sources (Melfi 2009).

Ethical concerns about our behaviour towards animals largely stem from differing views about the moral status of animals relative to us (DeGrazia 2002). Such concerns may well arise from concerns over animal welfare, but they are not in themselves animal welfare concerns, nor are ethical concerns evidence of the quality of welfare, and having ethical concerns does not by itself render the holder of those concerns an animal welfare expert. There are several possible ways of viewing the moral status of animals, and perhaps the most prominent are the animal rights view and the utilitarian view. The animal rights view, forcefully enunciated in the case of zoos by Regan (1985, 1995) and Jamieson (1985, 1995) is that we have duties to animals and must therefore respect those interests which are most important to them, in which case there is a moral presumption against keeping them in captivity. The utilitarian view, advocated for zoos by Singer (1990) and Bostock (1993), argues that captivity can be morally justified if there are important benefits that can be obtained only by keeping animals in zoos. Such benefits are usually framed within the context of conservation. Most of the organisations and individuals who campaign for the closure of zoos adopt the animal rights position.

However, as with many other ethical issues, ethical choices and decisions have to be played out in the real world, where things are often more complicated, and conflicts may arise where the resolution might be contrary to the original ethical position. For example, closing all zoos would result in hundreds of thousands of animals needing to be euthanised or having their reproduction stopped, both of which are contrary to the animal rights assertion that we should respect those interests which are most important to them. Similarly, zoos are major funders of field conservation (Gusset and Dick 2011), and this funding, which ultimately comes from admission fees paid by members of the public at the gate, would dry up if zoos were closed. If it is argued that extinction is a cost that animals have to bear for not being in captivity (quoted in Bostock 1993), then equally we could argue the reverse that captivity is the cost some individuals have to bear in order to save their species from extinction.

For those of us who are involved with zoos and believe deeply that the work zoos do is helping to save species and habitats, the philosophical arguments against zoos are hard to hear, partly because they have a logic which is difficult to counter and partly because they appear to denigrate a great deal of work which is being done and seem to imply that we should just let these species go extinct. Jamieson, for example, who thinks zoos cannot be defended, advocates instead that the only hope for wild nature is 'to put large tracts of the Earth's surface off-limits to human beings and to alter radically our present life styles' (Jamieson 1995). Well, yes indeed, but while we're waiting for that to happen, and it almost certainly won't, we may well lose most of it.

What we can see, however, is that there are two quite distinct areas of knowledge and opinion here, one, animal welfare, which should be informed by scientific evidence, and the other, animal ethics, which should be informed by moral arguments. Concerns about the welfare of animals in zoos are in principle answerable from existing research or by setting up new research to answer specific welfare questions. Concerns about whether particular animals, or indeed any animals, should be in zoos can also be welfare questions but are often reflections of a particular moral position. If that moral position is a utilitarian or an animal rights one, an acceptable justification, if there were one, would most likely be on conservation imperatives that override the strict moral position. Excellent welfare in captivity should have no bearing on either moral position (i.e. if your position is that animals in zoos cannot be justified, then even if their welfare is exceptionally good, this will have no effect on your position). But many anti-zoo groups, while adopting an animal rights philosophical position, use welfare concerns as their justification, which muddies the waters considerably. This muddying is partly because of different conceptions, not just between zoos and activists but also between different activist groups as well, as to what actually constitutes animal welfare (Wuichet and Norton 1995), but also because in some respects, both zoos and their critics are 'competing for righteousness' in a battle which perhaps cannot be won (Allen 1995).

A cursory examination of the websites of prominent anti-zoo activist groups suggests that much (but not all) of their concern is directed at a small number of animal taxa. The Born Free Foundation, for example, specifically mentions 'lions, elephants, gorillas, chimpanzees, tigers, polar bears, wolves, dolphins, turtles, sharks' (Born Free 2018); People for the Ethical Treatment of Animals (PETA) have separate sections on their website for great apes, elephants, bears, big cats and dolphins (PETA 2017). The charge has been levelled at zoos that they too tend to concentrate their efforts on the same restricted range of taxa (Hancocks 1995). This concentration of attention on the so-called charismatic megavertebrates is a feature of the accusations of poor welfare from animal rights activists and of the research done on the welfare of captive animals (Kreger and Hutchins 2010).

15.3 Zoos in the Media

It is instructive to examine how stories about animal welfare in zoos are portrayed in the media. The ways that the media 'frame' stories, which in this context includes which stories to tell and the way they are told, can set agendas for the way the public view different issues (Maynard 2017). In her analysis of magazine articles covering stories about AZA zoos, Maynard (2017) found that most stories were positive about zoos (only 11% were negative) and that animal welfare was the most common frame. She also found that the majority of articles focused on 'charismatic megafauna', and notably on mammals.

What, then, are these 'charismatic megafauna'? The term 'charismatic megafauna' has acquired high usage in the conservation literature to denote large-bodied animals that have particular appeal both to practitioners and the public, though the term is rarely defined (Ducarme et al. 2013). Using information from a number of different sources, Albert et al. (2018) attempted to define what made particular taxa charismatic and found that those animals consistently being named in lists of charismatic species were characteristically described by respondents as beautiful, impressive or endangered. They were able to compile a list of the 20 most charismatic taxa, in which tigers were top of the list and all but 2 of the remaining 19 (the exceptions being crocodiles and sharks) were large mammals (Fig. 15.1). Maynard (2017) does not state exactly which species appeared most frequently in magazine articles but did find that 38 of the 44 articles surveyed were concerned with mammals.

Of course, media framing also occurs in online articles, blogs, newspapers and other reports, and while there is no systematic way of surveying these, an indication of what animal welfare themes are most picked up on, and which taxa they mostly concern, can be achieved by quick sampling using an Internet search with relevant search terms. We did such a search using the search terms 'zoo welfare concerns' and then selecting all those stories which concerned zoos accredited by AZA (mostly North America), EAZA (mostly Europe) or ZAA (mostly Australasia) but only listing each story once, regardless of how many times it was reported (Table 15.1). In our selection, most stories were negative, a probable consequence of the search terms used. While 9 of the negative stories were criticisms of zoos in general or of particular zoos, 13 were concerned with the welfare of one or more particular taxa. These are listed in Table 15.2, where it can be seen that the majority concern taxa that occupy positions on the list of the top 20 charismatic animals published by Albert et al. (2018). One way of interpreting this is that there is no need to have concerns over the welfare of non-charismatic species, either because it doesn't matter or else because their welfare is generally good. To take the point to an extreme, it is rare to hear concerns expressed over the welfare of butterflies or frogs or coral reef fish in captivity. A more plausible way of interpreting it is that the media, and presumably therefore the public, are more concerned with these animals simply because they are charismatic.

To probe this further, we have looked at the number of different welfare-related stories in the media which come up in short Internet searches (within the first 120

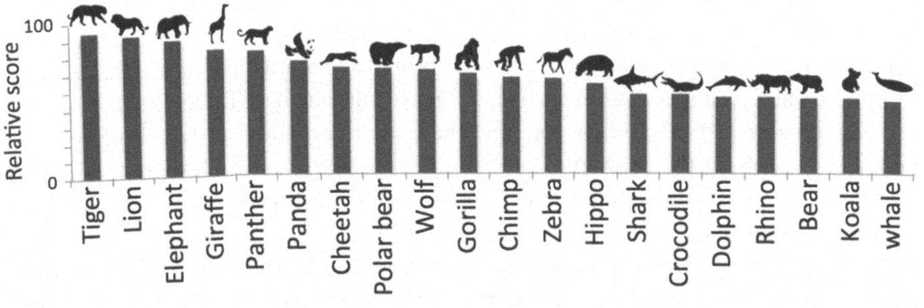

Fig. 15.1 The 20 most charismatic taxa in rank order from left to right, with their score derived from four different surveys. (From Albert et al. 2018, PLoS One 13(7): e0199149)

Table 15.1 Stories or events reported on the Internet concerning welfare of zoo animals

Number of positive stories	10
Number of negative stories	22
Negative stories which criticised zoos generally	2
Negative stories which criticised particular zoos	7
Negative stories with concerns about particular animals	13

The search terms 'zoo welfare concerns' generated 5340,00 hits; the first 300 of these generated these 32 items

Table 15.2 Taxa concerned in the negative Internet reports, with their position in the top 20 of charismatic animals defined by Albert et al. (2018)

Taxon	No. of stories	Position in charismatic top 20
Lion	4	2
Elephant	3	3
Sea lions	2	–
Penguin	2	–
Tiger	1	1
Gorilla	1	10
Sloth	1	–
Lemur	1	–
Antelope	1	–
Polar bear	1	8
Panda	1	6
Dolphin	1	16
Giraffe	1	4

The total number of stories is higher than 13 because some stories alluded to more than 1 taxon

hits for each taxon), using three typical charismatic taxa (elephants, great apes, cetaceans) in comparison with three similar non-charismatic taxa (tapirs, lesser apes or gibbons and manatees; see Fig. 15.2). The results of this (Table 15.3) show that this relationship holds true here too, that there are many more stories, both positive and negative, about the charismatics than the non-charismatics. It would seem that these are the animals we want to hear about, as well as those we are most likely to be concerned about.

Elephants are undeniably charismatic. They appear to show high levels of intelligence and self-awareness (Irie and Hasegawa 2009; Irie-Sugimoto et al. 2008; Plotnik et al. 2006), and they live in structured social groups and seem to display empathy (Bates et al. 2008). For these and other reasons, many animal activist

Fig. 15.2 Our examples of three charismatic (left) and three non-charismatic taxa (right), shown here by (**a**) Asian elephant, *Elephas maximus*; (**b**) Malayan tapir, *Tapirus indicus*; (**c**) western lowland gorilla, *Gorilla gorilla gorilla*; (**d**) yellow-cheeked gibbon, *Nomascus gabriellae*; (**e**) bottlenose dolphin, *Tursiops truncatus*; and (**f**) West Indian manatee, *Trichechus manatus*. (Photos: Geoff Hosey)

groups say they should not be kept in zoos (e.g. PETA 2018; One Green Planet 2018a). Of course, zoos are not the only organisations who keep captive elephants. Much concern has also been expressed over elephants used in the tourist trade in South-East Asia (Guardian 2017; BBC 2017) and in logging (One Green Planet 2018b). Unfortunately, life in the wild may also be poor for many elephants, where they may come into conflict with humans through crop-raiding or causing human

Table 15.3 Comparison in the number of different welfare-related stories in the first 120 hits of a literature search for selected charismatic taxa (elephants, great apes, cetaceans), compared with similar non-charismatic taxa (rhinoceros, gibbons, manatees)

	Charismatic taxa		Non-charismatic taxa	
Taxa	Positive stories	Negative stories	Positive stories	Negative stories
Elephants/tapirs	1	12	0	2
Great apes/gibbons	3	7	2	1
Cetaceans/manatees	3	14	3	1

deaths through attack (Sitati et al. 2003) or be killed by trophy hunters (Leader-Williams et al. 2001) or poachers (Douglas-Hamilton 2009). These also generate negative media stories, but compared to the stories about zoos, the stories of wild elephants have mostly been general demands for elephants to be moved to sanctuaries, or have revolved around accusations of cruel handling, deaths of young elephants, or movement of elephants between zoos.

By contrast, tapirs, which are not on the list of top 20 charismatic animals, generated only two welfare related stories in the online media. Both of these were negative, but one of them, where a Brazilian tapir *Tapirus terrestris* attacked and injured a mother and her child during a supervised encounter, was not explicitly to do with the welfare of the animal. Tapirs may not seem an obvious choice for a matched comparison with elephants, but the other large pachyderms which would appear more suitable, hippopotamus and rhinoceros, are both in the charismatic list. Clearly something about large pachyderms makes them charismatic. Nevertheless, hippos and rhinos only generated two negative welfare-related stories each in our search, and one of those was about a zoo rhino being killed for its horn by intruders, something which happens all too frequently in the wild too. It appears that elephants are the ones people worry most about.

Of the great apes, two species, western lowland gorilla and chimpanzee, appear in the charismatic animals list. There should be no surprise in this. These are our closest living relatives; they share with us many behaviours and signals and have a body form similar to ours. Together these two taxa generated seven negative welfare-related zoo stories and three positive stories. There was no particular theme to the negative stories, with a variety of events eliciting criticism, but there was a flurry of Internet comment following the shooting of Harambe the gorilla in 2016 after a child entered his enclosure (BBC 2016). Of the many comments which followed, there was both criticism and support of the zoo. For a matched comparison, we looked at the gibbons, the so-called lesser apes, which do not appear in the charismatic list. These generated only one negative story, where a zoo was cited by the US Department of Agriculture for two violations of the Animal Welfare Act in relation to its housing for gibbons.

Finally we looked at media stories concerning the welfare of cetaceans. These appear in the charismatic list under the two general terms 'dolphin' and 'whale', and in captivity the stories relate to three different species, the bottlenose dolphin, killer whale or orca and white whale or beluga. These generated 14 negative and only three positive welfare-related stories. Most of these stories relate to perceptions that captive conditions do not, and never could be adequate for these animals, especially in relation to pool size, and negative views regarding the purpose and role of shows for the public. Again, one prominent event was followed by a great deal of online comment, in this case the release of the film *Blackfish* in 2013, which was concerned with the captive conditions of killer whales, and in particular the events surrounding the killing of several people by the male Tilikum, who was described as having a 'negative history with trainers in the water' (Seaworld 2010). The nearest match we could find among non-charismatic animals were manatees, which generated only one negative and three positive stories. It's not obvious why cetaceans should be seen as charismatic and manatees not, although cetaceans are usually portrayed as highly intelligent and very social, and a kind of mythology has grown up around them as a sort of human-loving, benevolent super-intellect (Montagu 2003; Gregg 2013).

So it would seem that society's concerns are mostly about those kinds of animals which are perceived as big, beautiful, rare and clever. Concern can also occur when individual animals are named and stories constructed around them. Thus we have a great deal of comment, both positive and negative, about Tilikum the killer whale and Harambe the gorilla, as mentioned above. We could add examples such as Marius the giraffe, who was euthanised and then publicly dissected by his zoo in 2014 (The Independent 2014), or Fritz, the polar bear cub who died suddenly and unexpectedly (The Telegraph 2017) in 2017. However, animals which do not tick all of those boxes of big, beautiful, rare and clever, which after all are the majority, also deserve and need the best welfare, and although public concern for them may be lower, they should still command evidence-based welfare programmes, so we can now turn to examining how animal welfare research is distributed across these taxa.

15.4 Zoo Welfare Research

Zoo welfare science began in the 1980s to try and alleviate animals performing abnormal behaviours, attributed to poor housing and husbandry (Morris 1964; Meyer-Holzapfel 1968; Boorer 1972). Since then, zoo welfare research has fuelled the desire to follow evidence-based practice. For example, studies on environmental enrichment, which aims to enhance the captive environment of animals, started in the late 1970s (Markowitz et al. 1978; Markowitz 1982) and still continues today (Grunauer and Walguarnery 2018; Shapiro et al. 2018), emphasising its importance for so many animal lives. This is likely to have contributed to enrichment provision changing from an 'if we have time' husbandry practice to part of standard daily husbandry routines in most zoos around the world.

Welfare audits and welfare assessment tools are also now starting to become common practice within zoos due to the advances in research. For many zoos and legal zoo inspection teams, these were focussed on resource-based provision, i.e. how much space do the animals have, what the temperatures and humidity levels are, what climbing structures do they have, etc. These measures, while being important, do not show how the animals respond to these components or if the animals actually utilise them to their benefit. Thanks to zoo welfare research, assessment tools are able to additionally utilise animal-based measures such as behavioural and physiological indicators and clinical/pathological signs. These animal-based measures would simply not be possible if it wasn't for zoo researchers testing and validating them. For example, Clegg et al. (2015) developed an animal welfare assessment tool for bottlenose dolphins *(Tursiops truncatus)* utilising animal-based indices, as did Asher et al. (2015) for captive elephants. Both of these species assessment tools are now utilised in the daily management of elephants (in the UK) and dolphins (in Europe) at institutions where these animals are housed. In fact, in the UK, the elephant welfare assessment has helped shape the updated legislation and elephant management guidelines (Defra 2017) which are enforced by the Zoo Licencing Act 1981. Currently the only other welfare assessment tool that has been scientifically published is for Dorcas gazelle *(Gazella dorcas;* Salas et al. 2018), which is the only one of these three taxa not listed as a charismatic animal according to Albert et al. (2018). This accentuates the need for more research across a variety of species.

Melfi (2009) identified a variety of gaps in zoo science, and species bias was listed as one of these. Data showed that over a period of 10 years (1998–2009), of 774 projects from British and Irish Association of Zoos and Aquariums (BIAZA) zoos, 690 were undertaken on mammals and of these 490 focused on primates. However, these projects were only from BIAZA member zoos, so does not include non-BIAZA zoos in the UK and other zoos around the world. In addition, this was a general search and did not focus specifically on welfare. Azevedo et al. (2007) however reviewed environmental enrichment publications between 1985 and 2004 as a measure of zoo welfare science and found 744 publications during this time frame. Of these, 90.2% of them focused on mammals. Of the mammal publications, 54.14% were enriching rodents, 21.72% on primates, 6.29% on artiodactyls and 5.86% on carnivores. However, it is important to understand here that these studies were not restricted solely to zoos and included laboratory and farm research. This explains the high number of studies conducted on rodents (laboratory mice and rats) and artiodactyls (farm pigs), with only the carnivores being predominantly zoo animals.

We conducted a recent survey of published literature focussing exclusively on zoo animal welfare from 1 January 1992 to 31 December 2016 using Google Scholar and Web of Science and search terms 'zoo animal welfare'. The search was limited to the journals *Zoo Biology, Animal Welfare* and *Journal of Applied Animal Welfare Science* to ensure the results most likely represented the welfare of animals in zoos. Papers were then tabulated, checked to ensure they were not repeated and evaluated according to the species involved in the research. There was a total of 241 publications, with 2 publications in 1992 increasing to 32 by the end of 2016. Figure 15.3 shows the increasing trend of publications in the field and suggests that research in zoo animal welfare is increasing at an encouraging rate. This could be due to the increased recognition of animal welfare as a scientific discipline over the time period. But also there is increased recognition of the validity of zoo animal

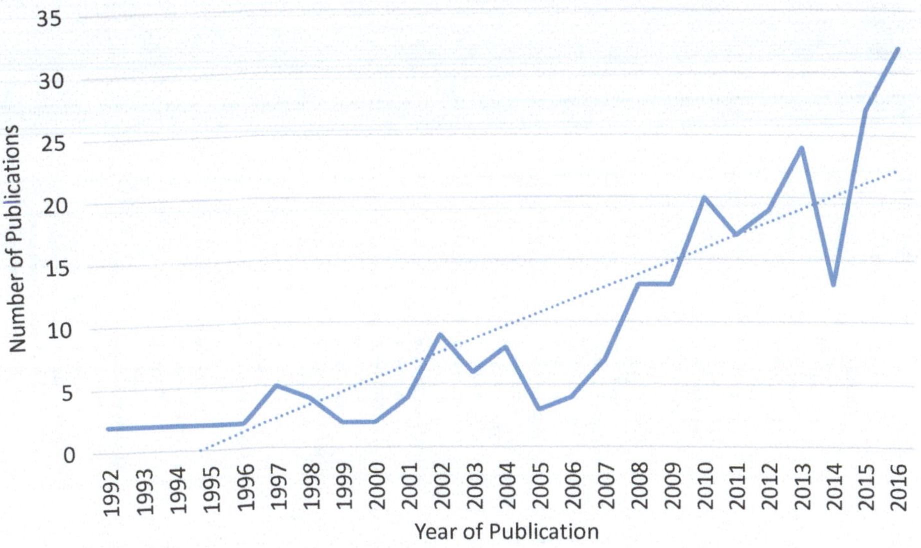

Fig. 15.3 Number of zoo animal welfare publications during 1992 until 2016. The dotted line represents the trend in publications across this time period

science, which in earlier years was not necessarily recognised to be robust science due to small sample sizes and the fear that variables such as individual zoo and enclosure effects were confounding results. Additionally, the increase could be linked to an increased need; with an estimated 10,000 zoos worldwide (Fravel 2003), there are thousands of animals held in captivity and it is up to 'us' to ensure they are kept in suitable conditions. We are in need of science to work out what these may be.

As with previous research (Azevedo et al. 2007; Melfi 2009), there is still a strong bias towards the welfare science of mammals (Fig. 15.4). Of the 241 publications in our search, 52 were not linked specifically to a species/taxon, and so were removed from these calculations. Of the remaining 189 publications, 87.30% (165) were on mammals, 9.52% on birds, 2.65% on reptiles, 1.06% on fish and none researched amphibian or invertebrate welfare. This shows that even though we are aware that there are taxonomic gaps in the literature, it is still the case that researchers focus predominantly on mammals. If we remove other taxa and concentrate this search on our selected charismatic mammals, great apes received 16.97% of the research attention, elephants 16.36% and dolphins 3.03%. There were no publications on the welfare of killer whale or beluga. Nevertheless, in these three major zoo science journals, our three selected charismatic taxa accounted for more than one third of welfare-related studies. In contrast, research focussing on our selected non-charismatic comparison species: gibbons 2.42%, tapirs 0.61% and none on manatees. There is a highly significant association in these data between the number of papers published and whether or not the taxon is charismatic ($\chi^2 = 46.54$, df = 1, $p < 0.001$). Data here suggest that two of our three groups of charismatic species

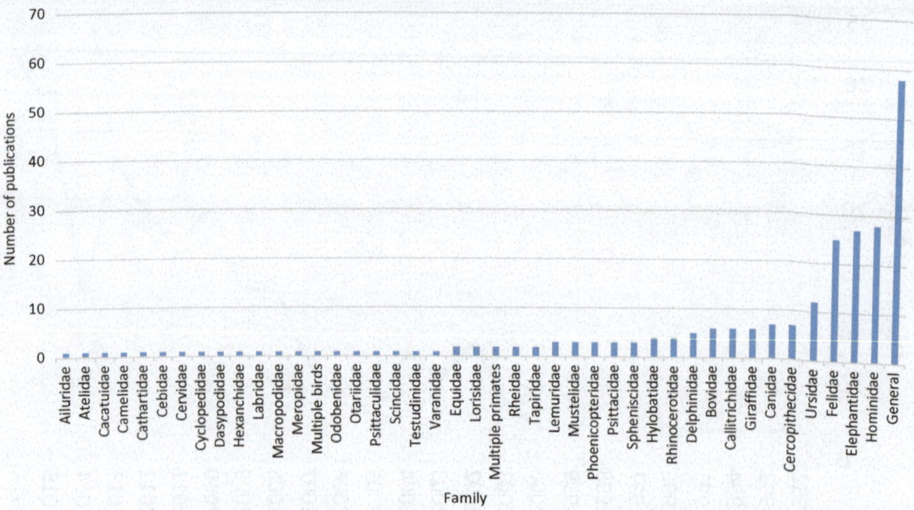

Fig. 15.4 Taxa (families) involved in zoo animal welfare research from publications during 1992 to 2016. Note that two of our selected charismatic taxa (Elephantidae and Hominidae) occupy the highest positions (extreme right of the chart) in terms of number of published studies

have much higher numbers of publications dedicated to providing information about animal welfare needs especially in comparison to our alternative species. From delving deeper into these charismatic species papers, they do not outline any specific welfare needs or issues of being housed in captivity, so why do we select these species to study? Why is there such a skew towards welfare science of these charismatic species? One potential argument as to why is linked to our knowledge of their cognitive ability. Research has shown that animals that have an increased brain-body size ratio and an encephalization quotient (EQ) of above 1 are more cognitively aware (Anderson et al. 2016). Humans are said to have an EQ of 7.4–7.8, dolphin 4.14, chimpanzee 2.2–2.5 and elephant 1.3 (Roth and Dickie 2005). Additionally, it has been documented that all of these species are able to pass mirror-self recognition tests (Plotnik et al. 2006), a measure to determine self-awareness and an indication of high cognitive ability (Gallup 1970). With high intelligence and additionally complex social systems, it might be suggested that these species have a greater propensity to suffer and therefore their welfare is of greater concern in zoos.

Two species that do not follow the trend in high research outputs and yet are recognised as cognitively and socially intelligent are the killer whale, which has an EQ of 2.57–2.59 (Marino 2007), and the European magpie (*Pica pica*), known to pass the mirror recognition test (Prior et al. 2008). As outlined above, our search showed that welfare research on birds and killer whales was low (9.52% and 0%, respectively), and so although we recognise that some species are highly intelligent, this is not necessarily reflected in the research outputs. It is therefore difficult to fully understand the justification behind why certain species are selected for welfare research and others left out. What we can gather, however, is that with such a strong bias towards testing these megafauna over other taxa so we know and can understand more about them and their needs in captivity, we do not necessarily know or understand if the lesser studied taxa can suffer in captivity and how we can alleviate this.

15.5 Animals in Zoos

Back in the early days of zoos, around the time of black-and-white televisions and before mobile phones and the Internet (yes, there really was a time like this), animals were collected from the wild and brought back to zoos to allow members of the public to be amazed at such wonderful specimens. Famous naturalists such as the late Gerald Durrell and Sir David Attenborough were among the teams of people visiting 'strange' lands and bartering with locals to bring back animals that people may only have heard about, let alone have the opportunity to see. Nowadays, removing animals from the wild to stock zoos is mostly unheard of as we are now aware of the plummeting population numbers and increase in conservation issues facing a high number of species and habitats. Zoo animal populations are now comprised predominately of captive bred stock (Fig. 15.5) with some wild caught individuals that may have been captured years ago and are still living, e.g. Aldabra tortoise *(Aldabrachelys gigantea)*, blue-and-yellow macaw *(Ara ararauna)* or Borneo orangutan *(Pongo pygmaeus)*, to name a few long-lived (>50 years) species.

As previously mentioned, there are an estimated 10,000 zoos worldwide (Fravel 2003), and this number is seemingly increasing year on year. With this in mind, the number of animals that are housed within zoos is large. Figure 15.5 shows the total number of individual animals currently housed in member zoos of Species360, who manage an online database known in the zoo industry as ZIMS (Zoological Information Management System). Membership to Species360 is not compulsory for every registered zoo; however it is a requirement for those that belong to zoo associations such as AZA and EAZA to increase knowledge transfer, share good husbandry practices and maintain long-term animal records. Species360 has over 1100 member zoos and so it is possible to estimate that these numbers may represent 10% of the total zoo animal population. However it may represent a high proportion of, if not all, zoos that are linked to an association of some kind.

Figure 15.5 also shows the number of individuals held worldwide by zoos per taxonomic group, with the highest number of captive individuals being the invertebrate taxa (299,816), followed by mammals (237,377), which interestingly is not too dissimilar to the number of birds (230,859), a taxon which, as already discussed, is low on welfare research outputs. When we then focus on our selected charismatic animals and non-charismatic animal alternatives (Table 15.4), the number according to the ZIMS database suggests a large skew towards housing charismatic species. With 11 charismatic species selected, there are currently 4351 individuals compared to our 10 non-charismatic species, whereby only 2013 individuals are currently housed. Additionally, there are a slightly higher number of zoos that house

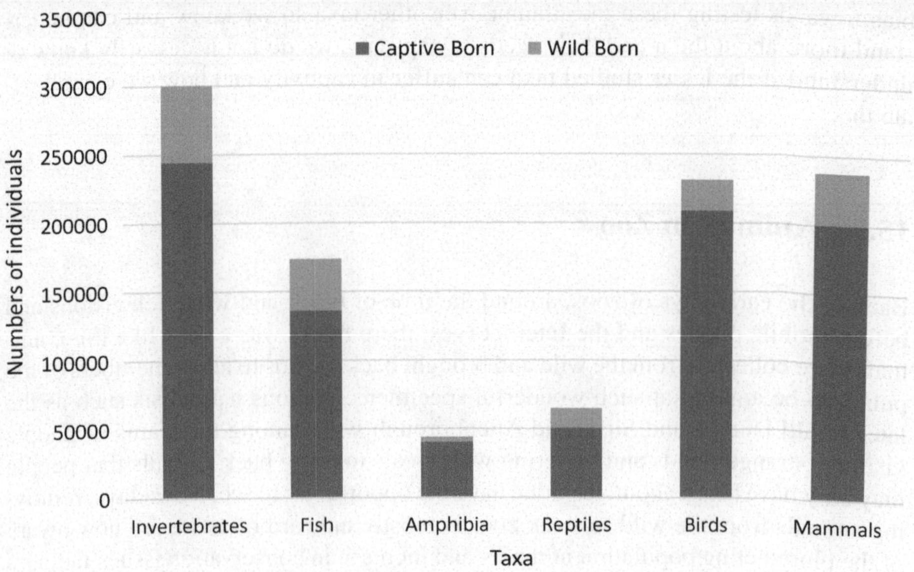

Fig. 15.5 Number of individual animals that are captive or wild born that make up the current captive population of zoo animals. (Data source: Species360, ZIMS as of 18 September 2018)

our charismatic species compared to our non-charismatic species (771 compared to 669, respectively); however, this is not as large a difference as you may expect. If we correlate these numbers with the number of papers published on these taxa, the association fails to reach significance ($r = 0.763$. 4df, not significant), largely because of the lack of publications on tapirs and manatees.

One potential reason behind these results could be linked to the collection planning and what the visitors want to see, but alternatively, it may be that the availability of these species, as part of breeding programs, for example, is similar. A study by Whitworth (2012) investigated which factors impact on visitor numbers; he found that the types and numbers of the species housed at the zoo were positively correlated with visitor number, whereas demographic variables of visitors, such as the age structure and locality of the local population, were not. Additionally, Moss and Esson (2010) found that taxonomic grouping was the most significant predictor of visitor interest (i.e. mammals were more interesting to visitors than any other taxa). Additional predictors in this study were body size (i.e. larger animals were more interesting), animal activity levels (i.e. the more active, the more interesting) and whether the animal was a flagship species (defined as a species that can excite public attention and can therefore also raise public support and perhaps generate funds; Hosey et al. 2013) or not (i.e. more interest if the animal was). Therefore, it might be that zoos plan their animal collections around what the visitors want to see. One step beyond this is a recent study (Mooney 2018) which investigated the conservation impact of zoos according to the visitor numbers and species held. Mooney found that the number of animals, mean species body mass and the 10 km radius

Table 15.4 Numbers of charismatic and non-charismatic animals housed at Species360 member zoos worldwide and the species IUCN conservation status where LC is least concern, V is vulnerable, EN is endangered and CR is critically endangered. Data derived from Species360: Zoological Information Management System (ZIMS) on 17 September 2018

Species	Male	Female	UNK	Total	No. holders	IUCN status
Charismatic species						
Asian elephant (*Elephas maximus*)	231	555	0	**786**	162	E
African elephant (*Loxodonta*	104	296	0	**400**	103	V
Eastern gorilla (*Gorilla beringei*)	0	1	0	**1**	1	CR
Western gorilla (*Gorilla gorilla*)	366	436	3	**805**	123	CR
Chimpanzee (*Pan troglodytes*)	525	849	3	**1377**	188	E
Bonobo (*Pan paniscus*)	84	117	2	**203**	16	E
Bornean orangutan (*Pongo pygmaeus*)	203	260	3	**466**	108	CR
Sumatran orangutan (*Pongo abelii*)	94	158	0	**252**	60	CR
Bottlenose dolphin (*Tursiops truncatus*)	11	27	0	**38**	5	LC
Beluga whale (*Delphinapterus leucas*)	5	5	0	**10**	2	LC
Killer whale (*Orcinus orca*)	6	7	0	**13**	3	Data Deficient
Total				**4351**	**771**	
Non-charismatic species						
Malayan tapir (*Tapirus indicus*)	88	90	0	**178**	60	EN
South American tapir (*Tapirus terrestris*)	231	229	3	**463**	193	V
Agile gibbon (*Hylobates agilis*)	18	19	2	**39**	18	EN
Lar gibbon (*Hylobates lar*)	298	287	63	**648**	177	EN
Grey gibbon (*Hylobates muelleri*)	13	18	0	**31**	16	EN
Hoolock gibbon (*Bunopithecus hoolock*)	16	8	5	**29**	5	EN
Black crested gibbon (*Nomascus concolor*)	4	3	0	**7**	4	CR
Buff-cheeked gibbon (*Nomascus gabriellae*)	94	72	16	**182**	47	EN
Siamang (*Symphalangus syndactylus*)	186	157	17	**360**	130	EN
Manatee (*Trichechus*)	41	35	0	**76**	19	V
Total				**2013**	**669**	

human population all significantly impacted on the zoo attendance. This is yet to be understood how or why, although it could be related to a larger body mass, larger number of animals available to be housed and larger local human populations. This then had a direct impact on the in situ conservation contributions made by that zoo, i.e. the higher these components were for a zoo, the more money that particular zoo was able to donate to conservation. Therefore, it could be suggested that more and larger-sized animals, such as our charismatic species, could provide an increased contribution towards conservation.

Table 15.4 also outlines the IUCN red list conservation status of our selected animals. IUCN-threatened species are those which are classed as endangered or critically endangered. Both our charismatic and non-charismatic species contain seven threatened species. This could just be due to chance regarding the species that were selected, but it does suggest that there are still non-charismatic species housed in zoos that are under threat and therefore are in need to be housed in zoos for captive breeding and maintaining genetic diversity.

15.6 Discussion

It is clear that society pay much more attention to charismatic than non-charismatic animals. Media stories about the welfare of animals in the Top 20 list of charismatic animals greatly outnumber those of taxa that are not on that list. Among these, the majority of stories are negative, reflecting greater concern by the public over the welfare of those taxa. Scientists also publish more peer-reviewed papers about the welfare of these top 20 charismatics than we do about those who are not on the list. That is not to say that these animals do not receive attention in the zoos themselves. For example, a great deal of animal welfare good practice, including provision of enrichments, takes place for captive cetaceans (Fig. 15.6), but these interventions are either not systematically evaluated or else are only published in the grey

Fig. 15.6 Artificial kelp used as part of an enrichment plan for captive bottlenose dolphins. Although the zoo has monitored the effectiveness of this, the results have not been published in the peer-reviewed literature. (Photo: Geoff Hosey)

literature, but not in peer-reviewed journals (Brando and Hosey, *in preparation*). The same is likely true for other taxa, both charismatic and non-charismatic. Many of these charismatics are of conservation concern, but many of the non-charismatics are as well, and the numbers of those kept in captivity do not necessarily reflect their conservation importance. For example, of the 19 currently recognised species of gibbons (Mittermeier et al. 2013), 1 is vulnerable, 13 are endangered, 4 are critically endangered and 1 has not yet been assessed. Of these, only seven species are held in zoos contributing to the ZIMS database, and the total of all of those gibbons held is less than the total number of just chimpanzees. This leads us to ask whether this emphasis on the charismatic animals is warranted for welfare and conservation reasons, and if so, do they add sufficiently to the goals of captive collections.

Balmford et al. (1996) discussed the need for zoos to refocus priorities for ex situ conservation from larger, less likely to breed species to smaller-bodied taxa. They suggested that zoos should select species that reflect biological and economic realities of what they are trying to achieve. Our data have shown that even 20 years later, there is a stronger focus towards megafauna compared to smaller-bodied individuals and this has now expanded into media coverage, research and visitor popularity. Interestingly, Balmford et al. (1996) suggested that by housing larger individuals, conservation priorities would be negatively impacted; however, research has shown that the opposite is true and that by zoos housing larger, more charismatic animals, they are able to contribute more towards in situ conservation.

One component of housing such large and complex species is of course the cost associated. Not only do these larger, more charismatic species need more space, complex enclosures, social groupings and enrichment regimes but their lighting, heating and food bills are also pretty high, depending on the species. For example, in the UK, Chester Zoo is possibly the largest and one of the most popular zoos with many large charismatic species including chimpanzees and elephants. In 2017, they welcomed over 1.8 million visitors, which generated an income of £18.4 million. The animal and botanical collection expenditure during this year was £21.8 million (Chester Zoo 2017). From the information available, it is impossible to allocate which species received more or less money; however it can be assumed that due to dietary and housing requirements for many charismatic species, a large proportion would have been allocated to them. One thing to remember here is that this does not necessarily mean that non-charismatic species suffer or receive less in terms of their needs; it might just be that the non-charismatic species do not need as much while still receiving what they need. Figure 15.7 shows enclosures designed for two of our comparative primate species, and it is clear that both species indeed have enclosures that suit their behavioural and cognitive needs, however they may differ. This suggests that modern zoos invest just as much effort into the housing and husbandry for non-charismatics as they do for charismatic species, and therefore zoos are not neglecting non-charismatics in their operation, it is just the research and public concern that show the bias.

So is this bias justified? A successful case study, which we have already referred to earlier, is that of the captive elephants in the UK. With a high mortality rate and issues raised by public concern, driven by the media, the UK government and

Fig. 15.7 Photos showing elaborate and complex enclosure designs for both (**a**) non-charismatic species, northern white-cheeked gibbon (*Nomascus leucogenys*), and (**b**) charismatic species, western lowland gorilla *(Gorilla gorilla gorilla)*. (Photos: Samantha J. Ward)

BIAZA invested significantly in understanding if the provision for our captive elephants was sufficient to allow animals to thrive. This led to researchers investigating various components of captivity, which therefore led to an increase in the number of research outputs. This has subsequently led to changes in legislative guidelines which impact on how elephants in the UK are housed. This is a great example of how the public, media and researchers can work together to improve the lives of animals in captivity. But the question still remains, why are charismatic species the main focus as it seems as though it's not that non-charismatics are always less important in terms of conservation or educational value, is it just that we like bigger animals more (Ward et al. 1998)?

Of course, the conservation role of zoos is much more than the maintenance of captive populations of threatened species. Although the concept of the zoo as a kind of 'ark', a safe haven for populations of rare species until such time as they could be safely reintroduced into the wild, was promoted in the 1980s and 1990s (Tudge 1992), zoos have moved on considerably from this and now recognise that an important part of their conservation role is with members of the public as much as with the animals themselves. Enormous numbers of people visit zoos, estimated to be more than 700 million per year worldwide, and as a consequence, zoos spend about $350 million per year on wildlife conservation (Gusset and Dick 2011). This has a significant impact on the success of in situ conservation, that is, conservation of wild-living populations of threatened species (Gusset & Dick 2010), and is only possible by attracting visitors into the zoo.

Equally important is the educational role zoos can play in informing the public about conservation issues, raising public awareness of these issues and what can be done about them, and hopefully as a consequence changing public behaviour to support more conservation initiatives (Zimmermann et al. 2007; Fa et al. 2011). There are many aspects to this role, and their success is not easy to demonstrate, so early attempts to show an educational benefit for zoo visitors (Falk et al. 2007) were subsequently criticised for methodological weaknesses (Marino et al. 2010). There is now, however, convincing evidence that visiting the zoo can lead to enhanced knowledge and understanding of environmental and conservation issues and increased commitment from visitors to engage in environmentally responsible behaviours (Esson and Moss 2014; Pearson et al. 2014; Moss et al. 2014). Indeed, the mere act of visiting a zoo and as a result getting close-up experiences of animals leads to a positive affective response in visitors (Luebke et al. 2016), and this in turn can strengthen biodiversity and conservation awareness (Powell and Bullock 2014). It is, however, not clear whether charismatic and non-charismatic taxa have differential contributions to this. Zoo visitors have clear preferences and like to see animals which are endangered, active and intelligent (Carr 2016), particularly large mammals, in other words, charismatic taxa. Visitors at a great ape exhibit showed greater knowledge of the animals on exit than on entry to the exhibit (Lukas and Ross 2005). There is also evidence that demonstrations and interactive events enhance knowledge and caring attitudes and behaviour in visitors in, for example, great apes (Price et al. 2015) and dolphins (Miller et al. 2013). There is, then, support for the notion that many zoo visitors go to the zoo particularly to see the large charismatic mammals and that as a result, they are likely not only to get a positive emotional response when encountering the animal but will also come away knowing and caring more about conservation and environmental issues. Whether these outcomes are as likely or as powerful in response to non-charismatic animals isn't clear. Zoo demonstrations with these can also be popular (Fig. 15.8), but little analysis of their impact has been done.

Fig. 15.8 Demonstrations and interactive events in the zoo can have far-reaching positive influences on conservation awareness and knowledge, but research on this has neglected demonstrations with non-charismatic taxa, even though these can be popular. (Photo: Geoff Hosey)

So, where does this leave us? Is there a problem with problematic animals in the zoo, and if so, what should we do about it? As we saw in Sect. 15.2, anti-zoo and animal rights groups tend to focus their attention on the large charismatic species, and many advocate that they should not be held in zoos. These views are generally not evidence-based, but stem from philosophical arguments. But the animal rights view that we should respect those interests that matter to animals (in this case being captive) is of no help in deciding what to do, as potential solutions (euthanasia or preventing breeding) do not respect the interests of the animals either. Similarly the assertion that these animals should be moved from zoos to sanctuaries is of no help, as the animals would still be captive, and there is no evidence that welfare is better in sanctuaries than in zoos or that they are actually any different. Indeed the evidence presented here in Sect. 15.4 shows that a disproportionate amount of effort goes into researching the welfare needs of the charismatics in zoos, which implies that if they are to be kept captive at all, then zoos are the place to keep them.

Given that large charismatic mammals are the animals that most zoo visitors most want to see and that this draws in large numbers of visitors, many of whom leave the zoo more knowledgeable and aware of conservation issues, and with a positive emotional experience from encountering animals close up, perhaps we should conclude that these animals should be getting a lot of attention. But the non-charismatics are important too, and if we do the research that's currently largely lacking, we might find that these are of great importance to the zoo experience as well. They also are deserving of more research attention on their welfare needs in captivity. Hopefully we can then move to a situation where none of the species housed in zoos can be regarded as 'problematic animals'.

References

Albert C, Luque GM, Courchamp F (2018) The twenty most charismatic species. PLoS One 13(7):e0199149
Allen K (1995) Ethical parameters for marketing and public relations. In: Norton BG, Hutchins M, Stevens EF, Maple TL (eds) Ethics on the ark. Smithsonian Institution Press, Washington DC, pp 289–296
Anderson R, Waayers R, Knight A (2016) Orca behavior and subsequent aggression associated with oceanarium confinement. Animals 6(8):49
Asher L, Williams E, Yon L (2015) Developing behavioural indicators, as part of a wider set of indicators, to assess the welfare of elephants in UK zoos. Defra Project WC1081: final report. https://irep.ntu.ac.uk/id/eprint/30334/
Azevedo CS, Cipreste CF, Young RJ (2007) Environmental enrichment: a GAP analyses. Appl Anim Behav Sci 102:329–343
Balmford A, Mace GM, Leader-Williams N (1996) Designing the ark: setting priorities for captive breeding. Conserv Biol 10(3):719–727
Bates LA, Lee PC, Njiraini N, Poole JH, Sayialel K, Sayialel S, Moss CM, Byrne RW (2008) Do elephants show empathy? J Conscious Stud 15:204–225
BBC (2016) Harambe gorilla killing: zoo defends action. Web site: https://www.bbc.co.uk/news/world-us-canada-36414813. Accessed 18 Sept 2018
BBC (2017) Elephant tourism is 'fuelling cruelty'. Web site: https://www.bbc.co.uk/news/science-environment-40501667. Accessed 13 Sept 2018
Boorer M (1972) Some aspects of stereotyped patterns of movement exhibited by zoo animals. Int Zoo Yearb 12:164–168
Born Free (2018) Web site: http://www.bornfree.org.uk/about-us/. Accessed 13 Aug 2017
Bostock SSC (1993) Zoos and animal rights: the ethics of keeping animals. Routledge, London/New York
Carr N (2016) An analysis of zoo visitors' favourite and least favourite animals. Tour Manag Perspect 20:70–76
Chester Zoo (2017) Trustees' report and financial statements, year ended 31st December 2017. https://www.chesterzoo.org/corporate-information/annual-reports/
Clegg ILK, Borger-Turner JL, Eskelinen HC (2015) C-well: the development of a welfare assessment index for captive bottlenose dolphins *(Tursiops truncatus)*. Anim Welf 24:267–282
Dawkins MS (1985) The scientific basis for assessing suffering in animals. In: Singer P (ed) In defence of animals. Basil Blackwell, Oxford, pp 27–40
Dawkins MS (2004) Using behaviour to assess animal welfare. Anim Welf 13:53–57
Dawkins MS (2006) A user's guide to animal welfare science. Trends Ecol Evol 21:77–82
Defra (2017) Secretary state standards of modern zoo practice. Appendix 8, Specialist exhibits, elephants
DeGrazia D (2002) Animal rights: a very short introduction. Oxford University Press, Oxford
Douglas-Hamilton I (2009) The current elephant poaching trend. Pachyderm 45:154–157
Ducarme F, Luque DM, Courchamp F 2013 What are "charismatic species" for conservation biologists? Biosci Master Rev 10:1–8
Esson M, Moss A (2014) Zoos as a context for reinforcing environmentally responsible behaviour: the dual challenges that zoo educators have set themselves. J Zoo Aquarium Res 2:8–13
Fa JE, Funk SM, O'Connell D (2011) Zoo conservation biology. Cambridge University Press, Cambridge
Falk JH, Reinhard EM, Vernon CL, Bronnenkant K, Heimlich JE, Deans NL (2007) Why zoos and aquariums matter: assessing the impact of a visit to a zoo or aquarium. Association of Zoos and Aquariums, Silver Springs
Fraser D (2009) Assessing animal welfare: different philosophies, different scientific approaches. Zoo Biol 28:507–518

Fravel L (2003) Critics question zoos' commitment to conservation. [Online] Accessed from: https://www.nationalgeographic.com/animals/2003/11/news-zoo-commitment-conservation-critic/ Accessed on: 17.09.2018
Gallup G (1970) Chimpanzees: self recognition. Science 167:86–87
Gregg J (2013) Are dolphins really smart? Oxford University Press, Oxford
Grunauer PP, Walguarnery JW (2018) Relative response to digital tablet devices and painting as sensory enrichment in captive chimpanzees. Zoo Biol 37(4):269–273
Guardian (2017) The dark side of wildlife tourism: thousands of Asian elephants held in cruel conditions. Web site: https://www.theguardian.com/environment/2017/jul/06/thousands-elephants-exploited-tourism-held-cruel-conditions. Accessed 13 Sept 2018
Gusset M, Dick G (2010) 'Building a future for wildlife'? Evaluating the contribution of the world zoo and aquarium community to *in situ* conservation. Int Zoo Yearb 44:183–191
Gusset M, Dick G (2011) The global reach of zoos and aquariums in visitor numbers and conservation expenditures. Zoo Biol 30:566–569
Hancocks D (1995) Lions and tigers and bears, Oh No! In: Norton BG, Hutchins M, Stevens EF, Maple TL (eds) Ethics on the ark. Smithsonian Institution Press, Washington DC, pp 31–37
Hill SP, Broom DM (2009) Measuring zoo animal welfare: theory and practice. Zoo Biol 28:531–544
Hosey G, Pankhurst S, Melfi V (2013) Zoo animals behaviour, management and welfare, 2nd edn. Oxford University Press, Oxford
Irie N, Hasegawa T (2009) Elephant psychology: what we know and what we would like to know. Jpn Psychol Res 51:177–181
Irie-Sugimoto N, Kobayashi T, Sato T, Hasegawa T (2008) Evidence of means-end behavior in Asian elephants. Anim Cogn 11:359–365
Jamieson D (1985) Against zoos. In: Singer P (ed) In defence of animals. Basil Blackwell, Oxford, pp 108–117
Jamieson D (1995) Zoos revisited. In: Norton BG, Hutchins M, Stevens EF, Maple TL (eds) Ethics on the ark. Smithsonian Institution Press, Washington DC, pp 52–66
Kreger MD, Hutchins M (2010) Ethics of keeping mammals in zoos and aquariums. In: Kleiman DG, Thompson KV, Baer CK (eds) Wild mammals in captivity: principles and techniques for zoo management, 2nd edn. University of Chicago Press, Chicago, pp 3–10
Leader-Williams N, Smith RJ, Walpole MJ (2001) Elephant hunting and conservation. Science 293(5538):2203–2204
Luebke JF, Watters JV, Packer J, Miller LJ, Powell DM (2016) Zoo visitors' affective responses to observing animal behaviors. Visitor Stud 19:60–76
Lukas KE, Ross SR (2005) Zoo visitor knowledge and attitudes toward gorillas and chimpanzees. J Environ Educ 36:33–48
Marino L (2007) Cetacean brains: how aquatic are they? Anat Rec 290:694–700
Marino L, Lilienfeld SO, Malamud R, Nobis N, Broglio R (2010) 'Do zoos and aquariums promote attitude change in visitors?' A critical evaluation of the American zoo and aquarium study. Soc Anim 18:126–138
Markowitz H (1982) Behavioral enrichment in the zoo. Van Nostrand Reinhold, New York
Markowitz H, Schmidt M, Moody A (1978) Behavioural engineering and animal health in the zoo. Int Zoo Yearb 18:190–195
Maynard L (2017) Media framing of zoos and aquaria: from conservation to animal rights. Environ Commun. https://doi.org/10.1080/17524032.2017.1310741
Melfi VA (2009) There are big gaps in our knowledge, and thus approach, to zoo animal welfare: a case for evidence-based zoo animal management. Zoo Biol 28:574–588
Meyer-Holzapfel M (1968) Abnormal behaviour in zoo animals. In: Fox MW (ed) Abnormal behavior in animals. WB Saunders, Philadelphia, pp 476–503
Miller LJ, Zeigler-Hill V, Mellen J, Koeppel J, Greer T, Kuczaj S (2013) Dolphin shows and interaction programs: benefits for conservation education? Zoo Biol 32:45–53

Mittermeier RA, Rylands AB, Wilson DE (2013) Handbook of the mammals of the world, vol 3 Primates. Lynx Edicions, Barcelona

Montagu A (2003) The history of the dolphin. In: Frohoff T, Peterson B (eds) Between species: celebrating the dolphin-human bond. Sierra Club Books, San Francisco, pp 27–40

Mooney A (2018) Zoos: does size matter? In: Proceedings of the 20th anniversary research conference, 9–11th July 2018. Living Coasts, Paignton Zoo and South Devon College

Morris D (1964) The response of animals to a restricted environment. Symp Zool Soc Lond 13:99–118

Moss A, Essen M (2010) Visitor interest in zoo animals and the implications for collection planning and zoo education programmes. Zoo Biol 29:715–731

Moss A, Jensen E, Gusset M (2014) Evaluating the contribution of zoos and aquariums to Aichi biodiversity target 1. Conserv Biol 29:537–544

One Green Planet (2018a) Why elephants don't belong in zoos! Web site: http://www.onegreenplanet.org/animalsandnature/why-elephants-dont-belong-in-zoos/. Accessed 13 Sept 2018

One Green Planet (2018b) New elephant sanctuary will provide safe home for elephants retired from brutal logging industry. Web site: https://www.onegreenplanet.org/news/elephant-sanctuary-retired-brutal-logging-industry/. Accessed 13 Sept 2018

Pearson EL, Lowry R, Dorrian J, Litchfield CA (2014) Evaluating the conservation impact of an innovative zoo-based educational campaign: 'Don't palm us off' for orang-utan conservation. Zoo Biol 33:184–196

PETA (2017) Web site: https://www.peta.org/issues/animals-in-entertainment/zoos/. Accessed 13 Aug 2017

PETA (2018) Get elephants out of zoos. Web site: https://www.peta.org/issues/animals-in-entertainment/zoos/get-elephants-zoos/. Accessed 13 Sept 2018

Plotnik JM, de Waal FBM, Reiss D (2006) Self-recognition in an Asian elephant. Proc Natl Acad Sci 103:17053–17057

Powell DM, Bullock EVW (2014) Evaluation of factors affecting emotional responses in zoo visitors and the impact of emotion on conservation mindedness. Anthrozoös 27:389–405

Price A, Boeving ER, Shender MA, Ross SR (2015) Understanding the effectiveness of demonstration programs. J Mus Educ 40:46–54

Prior H, Schwarz A, Güntürkün O (2008) Mirror-induced behavior in the Magpie (Pica pica): evidence of self-recognition. PLoS Biol 6(8):e202. https://doi.org/10.1371/journal.pbio.0060202

Regan T (1985) The case for animal rights. In: Singer P (ed) In defence of animals. Basil Blackwell, Oxford, pp 13–26

Regan T (1995) Are zoos morally defensible? In: Norton BG, Hutchins M, Stevens EF, Maple TL (eds) Ethics on the ark. Smithsonian Institution Press, Washington DC, pp 38–51

Roth G, Dicke U (2005) Evolution of the brain and intelligence. Trends Cogn Sci 9:250–257

Salas M, Manteca X, Abaigar T, Delclaux M, Ensenat C, Martinez-Nevado E, Quevedo MA, Fernandez-Bellon H (2018) Using farm animal welfare protocols as a base to assess the welfare of wild animals in captivity-case study: Dorcas Gazelles (*Gazella dorcas*). Animals 8(7). https://doi.org/10.3390/ani8070111

SeaWorld (2010) Tilikum. Seaworld animal profiles [online]. Available at: http://www.orcahome.de/swprofiles1.pdf

Shapiro ME, Shapiro HG, Ehmke EE (2018) Behavioral responses of three lemur species to different food enrichment devices. Zoo Biol 37(3):146–155

Singer P (1990) Animal liberation, 2nd edn. Random House, New York

Sitati NW, Walpole MJ, Smith RJ, Leader-Williams N (2003) Predicting spatial aspects of human-elephant conflict. J Appl Ecol 40:667–677

The Independent (2014) The killing of Marius the giraffe opens an important debate about genetics, animal rights and zoo inbreeding. Web site: https://www.independent.co.uk/news/world/europe/the-killing-of-marius-the-giraffe-opens-an-important-debate-about-genetics-animal-rights-and-zoo-9120219.html. Accessed 18 Sept 2018

The Telegraph (2017) Berlin's beloved polar bear cub Fritz diesin Tierpark Zoo. Web site: https://www.telegraph.co.uk/news/2017/03/07/berlins-beloved-polar-bear-cub-fritz-dies-tierpark-zoo/. Accessed 14 Sept 2018

Tudge C (1992) Last animals at the zoo: how mass extinctions can be stopped. Oxford University Press, Oxford

Ward PI, Mosberger N, Kistler C, Fischer O (1998) The relationship between popularity and body size in zoo animals. Conserv Biol 12(6):1408–1411

Whitworth AW (2012) An investigation into the determining factors of zoo visitor attendances in UK zoos. PLoS One 7(1):e29839. https://doi.org/10.1371/journal.pone.0029839

Wuichet J, Norton G (1995) Differing conceptions of animal welfare. In: Norton BG, Hutchins M, Stevens EF, Maple TL (eds) Ethics on the ark. Smithsonian Institution Press, Washington DC, pp 235–250

Zimmerman A, Hatchwell M, Dickie LA, West C (eds) (2007) Zoos in the 21st century: catalysts for conservation? Cambridge University Press, Cambridge

Chapter 16
Cryptic Problematic Species and Troublesome Taxonomists: A Tale of the Apennine Bear and the Nile White Rhinoceros

Spartaco Gippoliti and Colin P. Groves

16.1 Introduction

Among a broader definition of 'problematic wildlife' (Angelici 2016), we think that a special section should be reserved for those 'taxa' whose conservation strategies depend upon the taxonomic status accorded to them by the scientific community. In recent decades, it has often been argued that a taxonomy of large mammals based on the BSC (biological species concept) is favourable to conservation policies as it allows artificially mediated gene flow among distant populations belonging to the same species (Frankham et al. 2012; Heller et al. 2013; Zachos et al. 2013; Ralls et al. 2018). Linked to this view seems to be the hypothesis that conservation problems mainly concern the genetic viability of small populations that remain isolated owing to human-caused destructive behaviours such as hunting and habitat alteration (small population paradigm, cf. Caughley 1994). However, a finer definition of 'evolutionary species units' can inform a conservation strategy that might make sense if past unique evolutionary trajectories in populations are being conserved (Weeks et al. 2011). Weeks and co-authors are also concerned that if these populations are not really 'unique' at all but that they are managed as distinct units, we may increase their risk of extinction due to reductions in genetic diversity and loss of population fitness. Actually we share this concern and, additionally, we do not wish to allocate precious resources to 'phantom taxa' (cf. Gippoliti and Amori 2002a) when there is so much neglected biodiversity nor is it our wish to let species vanish because of depauperate genetic pools.

S. Gippoliti (✉)
Società Italiana per la Storia della Fauna "Giuseppe Altobello", Rome, Italy

C. P. Groves (Deceased)
School of Archaeology & Anthropology, Australian National University, Canberra, ACT, Australia

F. M. Angelici, L. Rossi (eds.), *Problematic Wildlife II*,
https://doi.org/10.1007/978-3-030-42335-3_16

Previously we advocated integrative taxonomy as the way to identify evolutionary species beyond the sole use of neutral molecular markers to identify evolutionary lineages in ungulate mammals (Groves and Grubb 2011; Cotterill et al. 2014; Groves et al. 2017; Gippoliti et al. 2018b). Users of taxonomic data are all too often ignorant of how imprecise or wrong current taxonomies can be and consequently how dangerous it is to utilize current taxonomy as a yardstick to generalize about issues such as inbreeding and outbreeding depression, especially in the absence of taxonomic revisions in the last half a century (Goldstein et al. 2000).

Supporters of 'genetic rescue' (e.g. Vander Wal et al. 2013) credit this technique with the restoration of populations of cougars in Florida (*Puma concolor*), bighorn sheep (*Ovis canadensis*) and adders (*Vipera berus*). We note that the last two cases refer to translocation between the same putative subspecies. In the case of the Florida cougar, generally considered a distinct subspecies, *Puma concolor coryi*, it has been re-stoked with individuals belonging to the Texas subspecies *P. concolor stanleyana*. It seems that all North American forms of cougars have a recent origin, and genetic diversity among subspecies is low (Culver et al. 2000) so the risks of outbreeding depression (Templeton 1986) in this case was probably low (Hedrick and Fredrickson 2010). But this is a much greater risk in the case of evolutionary species that are artificially lumped into 'biological' species (see below), or little-known subspecies that have never been thoroughly taxonomically revised in recent decades (cf. Groves et al. 2017).

Outbreeding depression indicates a fitness reduction following hybridization (Templeton 1986). According to the so-called Dobzhansky–Muller model, isolated populations gradually accumulate neutral or advantageous mutations over time, and selection for positive epistasis may result in the development of unique coadapted gene complexes within each isolated population (Edmonds 2007). Thus, when mating occurs between populations, segregation and recombination can break up these coadapted gene complexes and bring together mutations that have not been 'tested' together and potentially have harmful effects (Turelli et al. 2001). This is not to deny the potential evolutionary importance of hybridization that has influenced even our own species (Harris and Nielsen 2016) but rather to highlight our scepticism towards a single cure – genetic rescue – which has been studied in the laboratory but has not been tested in the field.

As a case in point, we draw attention to one successful conservation story: the Arabian oryx *Oryx leucoryx*. After the species was saved from extinction by captive breeding and reintroduced into protected areas located in its original homeland, signs of outbreeding depression were discovered in the surviving population (Marshall and Spalton 2000; Ochoa et al. 2016). Why should this be? It is of some interest that in 1934 Pocock described a subspecies, *Oryx leucoryx latipes* from Wadi Ghudun, Southern Arabian Peninsula (now in Oman), remarking how the skin he received showed 'markedly wider and differently shaped hoofs' compared to all other available specimens then in the British Museum (Pocock 1935: 464). He compared the hoofs to those of *Addax nasomaculatus* and suggested that this would be a character of specific relevance resulting from adaptation to loose sand. Considering that he had only one specimen at hand, he preferred to name a new subspecies merely 'to draw attention to its interest' (Pocock 1935: 466). Regrettably, the age of 'taxonomic inertia' was beginning at that time and Ellerman and Morrison-Scott (1951: 386) simply commented, 'We regard this form as of doubtful validity'. No surprise then that most present-day researchers (Ochoa et al. 2016) dismissed the hypothesis that there was any ancient isolation in *Oryx leucoryx* driven by geographic and ecological factors. Today we have the technology to test if the Arabian oryx shows cryptic historical patterns of genomic and taxonomic diversity. Specifically, a genomic study should analyse not only available museum specimens from the historic range but the extant captive population – including the founders – to ascertain if two or more cryptic evolutionary units have effectively been swamped together.

16.2 Taxonomic Oversimplification

For most of the twentieth century, under the BSC with its acceptance of a plethora of polytypic species, most described taxon names were synonymized or, at best, accepted to be of subspecific validity. But, again, this was often accomplished without the help of a true taxonomic revision (e.g. critical examination of most available museum specimens and type specimens at the end to assess taxonomic variability in a taxonomic group (genera, species)). Natural history museums and traditional taxonomists had in the meantime fallen under the spell of the Modern Synthesis, or rather of what was interpreted as its taxonomic corollary, especially in Western Europe (Gippoliti and Groves 2018). This process of simplification is described below by one of the fathers of the modern synthesis, Ernst Mayr (1982): 'While the morphological criterion of intergradation had previously been the exclusive subspecies criterion, "geographical representation" now became the yardstick. As Stresemann (1975) has described so well, every isolated 'species' was now scrutinized for the possibility that it was simply a 'geographical representative' of some other species, in which case it was reduced to the rank of subspecies. The subspecies was subsequently defined as a member of a polytypic species, not simply as a 'slightly different' local population. The new way of looking at geographical isolates, particularly the downgrading of every isolate to subspecies rank, even if these were not clearly connected by intermediates, resulted in an extraordinary simplification of taxonomy at the species level. Among the 607 species of North American birds alone, 315 taxa that had originally been described as full species were subsequently reduced to subspecies status. The newly recognized polytypic species were much more distinct, real entities of nature compared to the purely morphologically defined 'species' of the 1880s. Morphological difference was replaced as species criterion by reproductive isolation' (Mayr 1982: 593–594).

Another consequence of this process is that under the same category of 'subspecies', there can now be found a heterogeneous array of evolutionary stages (Gippoliti and Amori, 2007). True species has been artificially lumped into a polytypic species, while other subspecies are essentially points on a cline. 'In retrospect, it has become clear not only that many of the so-called subspecies described from the 1920s to the 1940s did not differ in the slightest, but also that the recognition of minutely differing populations served, in most cases, no good purpose' (Mayr 1982: 594). This has important consequences because many researchers assign a spurious objectivity even to the subspecies category, lamented by Lorenzo Camerano as long as a century ago (Camerano 1916).

Interestingly, it seems that the oversimplification of taxonomy through the polytypic species concept has found a use in at least one other field of environmental conservation: ecological connectivity. Human-induced habitat degradation and fragmentation is certainly one of the major causes of a population's loss of viability (Lindenmayer and Fisher 2007), so the maintenance or creation of a network of ecological corridors has often become one of the main goals of wildlife conservation, although the validity of this approach has been rarely tested in a scientific way (Gippoliti and Battisti 2017). What is more interesting from our perspective is that isolated populations of large mammals have almost invariably been assumed to be the result of historically human-mediated extirpation of connection between populations by either direct persecution or by habitat degradation. This belief is repeated in almost all papers dealing with molecular aspects of large mammals (i.e. Angelone-Alasaad et al. 2017), even in those cases where genetic or morphological data could in fact support (or at least be tested) a different scenario, namely, that a given population was already distinctive (thus isolated) before the peak of anthropogenic destructive activities (see also Sexton et al. 2014).

16.3 Troublesome Taxonomists

Rojas (1992) was among the first to recognize the existence of a problem in the application of different species concepts for biodiversity conservation policy. Much earlier, the pioneer conservationist and mammal taxonomist Oscar de Beaux (Fig. 16.1; Gippoliti 2006) was probably the first to recognize how the 'new taxonomy' posed more than a challenge for biodiversity conservation. He dedicated the last years of his life to a monograph on the wild goats of the genus *Capra* (de Beaux, 1956), in which he refuted the taxonomic vision proposed by Ellerman and Morrisson-Scott (1951) which, in that specific case, was even less simplified than those already advanced by Schwarz (1935). As early as 1930, when de Beaux published his *Etica Biologica*, later translated into English (de Beaux 1932) and cited by Aldo Leopold (Leopold 1933), he stressed the importance of preserving the results of evolutionary history. De Beaux included among the conservation priorities the distinctive populations of Maremman roe deer, those of the Gargano and Calabria (Southern Italian Peninsula), the Maremman wild boar and those living in Southern Italy even if then unstudied scientifically (de Beaux 1930).

It is perhaps no coincidence that one of the most known (and unsuccessful) cases of *ante-litteram* genetic rescue/translocation involved members of the genus *Capra*. An assemblage of *Capra* taxa, *Capra ibex*, *Capra ibex* x *Capra nubiana* and *Capra*

Fig. 16.1 Oscar de Beaux

aegagrus, which at that time were regarded as members of one polytypic species, were released on the Tatra Mountains, but owing to the lack of the right birth seasonality and lack of human care during the Second World War, the experiment failed (Turček 1951). This remarkable early unsuccessful example, fuelled by the excessive synthetic approach of Schwarz (1935), is a powerful reminder that species cannot be discriminated on the basis of breeding compatibility and, if this is done, there would be some grave problems in conservation biology. Interestingly, de Beaux believed humans had a moral duty to not interfere with the existing diversity pattern well below species level, an issue that is critically reviewed by Rohwer and Marris (2016).

We have observed that after decades of 'taxonomic inertia', it is not easy to propose 'revolutionary' changes in the way people see the world. The revisionary work on the world's ungulates by Groves and Grubb (2011) was often received with hostility and the criticisms often failed to focus on true scientific issues. But even in more parochial (but yet important) issues, we see how difficult it is to open a frank discussion among different researchers and disciplines. Regarding the conservation issues of the so-called woodland caribou *Rangifer tarandus caribou*, only the single authoritative voice of Valerius Geist (Geist 2007) called for an urgent revision of the taxonomy, an issue that has so far remained unheard. How neglected taxa can become critical for biodiversity conservation is demonstrated by the case of the Bale monkey *Chlorocebus djamdjamensis* in Ethiopia. Ignored by most mammalogists and primatologists – with the exception of Pierre Dandelot – during the twentieth century, this species was first 'rediscovered' in the Harenna Forest by one of us (Carpaneto and Gippoliti 1994) who reviewed its taxonomic history and highlighted that this was a uniquely mountain bamboo-forest subspecies of the otherwise savannah-living *aethiops* complex. Subsequently, both Kingdon (1997) and Groves (2001) raised *djamdjamensis* to full species status. Further genetic data established that under the name *djamdjamensis* two genetically distinct populations exist (Haus et al. 2013), one of which shows traces of extensive introgression with other *Chlorocebus* species. Further research has confirmed the distinctiveness of the two ESUs (Mekonnen et al. 2018). What is of relevance is that the presence of an endemic primate species made international funding available for research and conservation on this highly restricted species, making it an effective umbrella and flagship species for montane forest conservation in Bale and Sidamo regions of Ethiopia (Mekonnen et al. 2012, 2017), a habitat so far neglected by conservationists.

16.4 History of Translocations in Conservation Biology

From a conservation biologist's perspective, the era of taxonomic inertia opened the door to a number of appealing and apparently successful conservation operations, such as the translocation of wildlife species in regions from where they had been long extirpated. In 1976 WWF Italy organized a meeting on *Reintroductions: techniques and ethics* (Boitani 1976), from which, however, taxonomic considerations were almost completely absent. The attached 'manifesto on animal re-introductions' specified that, in cases of restocking, '…the animal must be of the same race as those in the population into which they are released'. In addition, the manifesto specified that '…the animals reintroduced must be of the closest available race to the original stock' (Boitani 1976: 300). Given that we may consider the term race – in this instance anyway – a synonym of subspecies, the document is in agreement with the attitude of past times, which assumed that all we need to know about subspecific variation was already known and available to wildlife conservationists. We note that many researchers have always emphasized the importance of maintaining unaltered existing patterns of geographic variation inside biological species. In the case of Caprinae, it has been stressed that a subspecies extinction by human-mediated introgression is a really extinction, and proposed conservation recommendations are at the subspecies level (cf. Shackleton and Lovari 1997). In 1987 an IUCN Statement on the Translocation of Animals was produced (IUCN 1987) and in 1988 an IUCN/SSC Re-introduction Specialist Group was established that published guidelines for re-introductions in 1998 (IUCN 1998). In 1999 Seddon and Soorae (1999) proposed guidelines for subspecific substitutions in translocations, when the original subspecies is extinct.

According to the IUCN/SSC (2013), founders should show characteristics based on genetic provenance and on morphology, physiology and behaviour that are assessed as appropriate through comparison with the original or any remaining wild populations. In some cases the original species or subspecies may have become extinct both in the wild and in captivity. A similar, related species or subspecies can be substituted as an ecological replacement, provided that the substitution is based on objective criteria such as phylogenetic closeness, similarity in appearance, ecology and behaviour to the extinct form. Regrettably, extinction of a 'subspecies' is not a rare phenomenon and a reasonable replacement, as we will see, is not always possible as we deal, in fact, with unique lineages adapted to very specific ecological and geographical niches. The whole issue is much more complicated (Gippoliti et al. 2018a, b), but we would highlight here that the abandonment of the subspecies category as observed in some IUCN Specialist Groups will certainly have consequences for translocation/restocking policies.

16.5 The Nile or Northern White Rhinoceros

Groves et al. (2010) revised the taxonomy of living *Ceratotherium* (white rhinoceros) and concluded that the two living taxa are best treated as separate species, the southern white rhino (SWR) *Ceratotherium simum* in Southern Africa and the northern (or Nile) white rhino (NWR) *Ceratotherium cottoni* in Central Africa.

IUCN's African Rhino Specialist Group (AfRSG) reacted negatively to this proposition:

> This conclusion is being contested by, amongst others, African rhino genetics expert Colleen O'Ryan who has informed the AfRSG that she and her colleagues are working on a detailed rebuttal of Groves et al.'s paper based on findings derived from larger sample sizes, and using what she feels are more appropriate genes (Brooks 2010: 14).

This represent an obvious misunderstanding, as no scientist would work on an already predetermined rebuttal, and no geneticist would refer in such a context to any gene as 'more appropriate' than any other. Nonetheless, O'Ryan's research group has more recently conducted such a test, sequencing the entire mtDNA genomes of four NWR and three SWR (Harley et al. 2016), and compared them to Rhinocerotidae from GenBank as well as selected outgroup taxa, including mtDNA genomes of modern *Homo sapiens* and what they referred to as *Homo sapiens neanderthalensis* and *Homo sapiens denisova* (that is to say, the Late Pleistocene Neanderthal people of Europe and western Asia and the mysterious Denisovans, known only by a tooth and a manual phalanx). The results corroborated the main findings of Groves et al. (2010), namely, that NWR and SWR are reciprocally monophyletic (with 100% support). Despite this, Harley and colleagues disputed their status as separate species, the main arguments being that, first, accepting them as different species would be 'a problem for conservation' (Harley et al. 2016: 1286) because they maintained that if some of the genes of NWR were to be saved, the only hope seemed to be that the last survivors should be interbred with SWR. Their second argument was that the genetic distances (p-distances) between SWR and NWR were less than those between modern humans and either Neanderthals or Denisovans. The first point comes up against the probability of outbreeding depression in any hybrids; as for the second, Harley et al. (2016) were under the impression that Neanderthals and Denisovans are considered subspecies of *Homo sapiens,* which is not the case generally. They likewise claimed that the PSC 'would also lead to the requirement for *H. sapiens* to be divided into a large number of separate species' (Harley et al. 2016: 1290); this frumpy statement has been dealt with by Groves (2012, 2014) and Groves & Robovský (2011).

Fig. 16.2 Part of the Nile white rhinoceros group at Dvur Kralove Zoo in 1991. (Photo S. Gippoliti)

In summary, the combined evidence for unambiguous diagnosibility of *Ceratotherium cottoni* versus *C. simum* is beyond doubt. The two are (or were, alas) clearly individuated evolutionary lineages. Yet the fact that it was almost universally recognized only as a subspecies of *C. simum* and that the latter taxon flourished under strict protection (Rookmaaker 2000) makes awareness of the importance of conserving the Nile white rhinoceros (*C. cottoni*) less urgent. Although first imported to zoos in 1950 (much earlier than its southern congeneric), the captive population of the NWR was fragmented among a number of zoos that kept pairs that never breed in such a deprived social condition.

Only Dvur Kralove Zoo (in then Czechoslovakia) (Fig. 16.2) operated its own innovative acquisition programme that in retrospect offered at least a chance to create a self-sustaining ex situ population. In 1975 a small herd was captured in South Sudan, before the poaching rise after 1980 (Hilmann-Smith et al. 1986). The history of the status of the species in the Garamba National Park (Democratic Republic of Congo) (Fig. 16.3) was summarized by Hilmann-Smith et al. (1986). Interestingly, differences between NWR and SWR were already evidenced at the time: 'The skull shape is distinct in the field, and the head is held higher than in southern whites. Body proportions are also different, with the northern white rhinos tending to be shorter, and there are almost certainly *ecological* and *behavioural* differences between the two subspecies' (Hilmann-Smith et al. 1986: 20). We evidenced some words in italics because too many laboratory researchers and zoologists, at most familiar with one studied population, are unaware of how subtle ecological conditions coupled with geographic isolation may produce phenotypically different populations, which is quite unappreciated to the unexperienced eye.

Groves et al. (2017) reviewed the issue and another important aspect is that while in Southern Africa the SWR was sympatric with the black rhinoceros *Diceros bicornis*, this was not the case in Central Africa. This suggests that the NWR could be less strict grazers than the SWR. Interestingly, differences between the two taxa were also found when studying the social behaviour of captive rhinoceros and their

Fig. 16.3 Nile white rhinoceros in the Garamba National Park. (Photo Francesco Germi)

vocalizations (Kuneš and Bičík 2002; Cinková and Policht 2014. It seems that either the wish of the Democratic Republic of Congo (formerly Zaire) to maintain in situ a 'national treasure', and the fall of the ark paradigm in zoos' conservation work (Gippoliti 2011) and the adopted taxonomy conspired to halt the preparation of a 'plan B' for conservation of NWR. In some cases, political turmoil prevented the conservation of the last wild stronghold of the taxon. In a sense, the northern white rhino was first a victim of colonialism, 'scientific inquiry' and trophy hunting, with several hunters (some very famous such as Theodore Roosevelt, Winston Churchill, Frederick C. Selous, Vittorio Emanuele of Savoy-Aosta, Herbert Lang and Powell Cotton, obviously) bringing back not one or two but several specimens of these – then – little-known species; 14 were taken by the Roosevelt expedition alone (Heller 1913). Later, it fell victim to post-colonialism and the decline of the new independent states, the heritage of colonialism. If we reflect on a history of now 60 years (Curry-Lindahl 1972) of political turmoil in Central Africa (and no bright future insight), we should ask ourselves if a strategy for this extraordinary species should have been developed when we had more options available. In 1976 the population in Garamba had risen to some hundreds of individuals and the capture of a small herd for a zoo should have had no effect on the wild population. Furthermore, it seems that no interest was shown in conducting surveys and eventually establishing protect areas in Chad and Central African Republic where the species was reported by local hunters (Owen-Smith 2013). Inevitably, geopolitics needs to be considered in conservation biology.

16.6 Apennine Bear

A similar case occurred in Italy, where the morphologically distinctive Apennine brown bear population has been generally treated just as a southern population of *Ursus arctos*. The history of the conservation of the Apennine bear (and of the Apennine chamois *Rupicapra ornata* Neumann, 1899) offers a powerful demonstration of how taxonomy can have a positive influence on wildlife conservation, as previously indicated by Cotterill et al. (2014) and Gippoliti et al. (2018a, b and references therein). When the Parco Nazionale d'Abruzzo was finally established in Central Italy, Altobello (1921) had just described the local bear population as a new subspecies, *Ursus arctos marsicanus*. Although some proposals for a National Park had been made before, safeguarding the stronghold of two unique mammal taxa – the Apennine bear and the Apennine chamois, with their high touristic potential value – was a crucial element in the establishment of the park (Sipari 1926). The founder of the park, Erminio Sipari, discussed at length the taxonomic status of the Apennine bear, consulting not only with Altobello but with several other mammalogists such as Giuseppe Lepri, Enrico Festa, Paul Matschie and Theodor Knottnerus-Meyer (Sipari 1926). Altobello had only an adult female skull at hand (and two skulls of juveniles), so his original description (Altobello 1921) appeared so weak to Pocock (1932) that he relegated *marsicanus* to the synonymy of *arctos*. Already in Sipari's work new evidence was collected and Enrico Festa communicated to Sipari on the basis of material in the Turin University Museum that the superior profile of the skull in the frontal region was more convex than in *Ursus arctos* (Sipari 1926: 29). Ellerman and Morrison-Scott (1951) had no choice other than to follow Pocock's arrangement.

An original and (at the time) 'shocking' taxonomic view was offered by the paleontologist Sergio Conti. While studying and describing a particular 'variety' of cave bear from Liguria (North-West Italy) which he named *Ursus spelaeus* var. *ligustica* (Conti 1954), he made some comparisons with modern Alpine and the Apennine bears. After suggesting that his *ligustica* seems an intergrade between *spelaeus* and *arctos*, he also concluded that (having at hand an adult male skull of *marsicanus*) the Apennine bear was more related to *ligustica* than the Alpine bear, to the degree that he recognized *marsicanus* as a full species. Although Conti's sample was very limited, it is clear that the unique skull he had is fully representative of the Apennine population as a whole (Vigna Taglianti 2003; Colangelo et al. 2012) (Figs. 16.4 and 16.5) and his proposal should be taken seriously – evidently he had utilized the diagnostic version of the phylogenetic species concept (Wheeler and Platnick 2000) to identify evolutionary species. Differently from North America, in Europe the concept of 'subspecies' among mammalogists has received so far scanty attention. Although sometimes listed in the more accurate checklist (i.e. Amori et al. 1999), there has generally been no attempt to revise mammal taxonomies at the subspecies level, even utilizing the increasingly available data derived from phylogeographic studies. These data often indicated a more complex situation than showed by

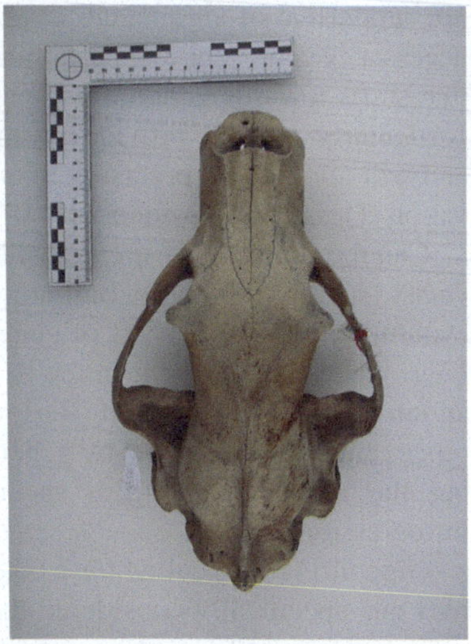

Fig. 16.4 Dorsal view of the skull of the Alpine brown bear. (Photo Jacopo Conti)

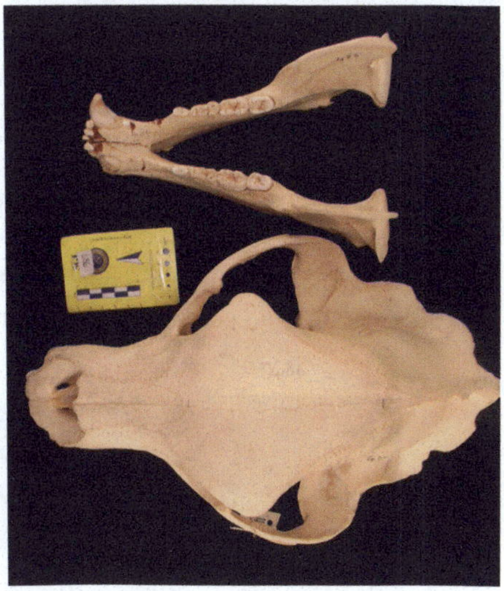

Fig. 16.5 Dorsal view of the skull and mandible of the Apennine bear *Ursus arctos marsicanus* (photo Jacopo Conti)

traditional taxonomy and this is of potentially great importance as a basis for conservation strategies (Gippoliti and Amori 2002b).

The low level of genetic distinctiveness of Apennine bears from Balkan bears led Colangelo et al. (2012) to explain that the particular skull of the Apennine bear was mainly the result of genetic drift in a small population isolated for '240–720 years' from the Alpine population. We have already referred above to the 'belief' of humans, especially utilized in Europe, to be the only agent of range fragmentation and isolation. We also know that the southern European peninsula hosts a number of unique lineages that found refuge there during glacial periods (Bilton et al. 1998). In the case of the Apennine bear, early genetic data (mtDNA, microsatellites) are not supportive of an ancient separation from other bears, yet a gene tree does not necessarily represent a species tree (Ferguson 2002). In addition, it is well known that shifts towards genome-wide single-nucleotide polymorphisms (SNPs) will be particularly useful to provide a comprehensive assessment of genetic distinctiveness and will allow more precision in targeting what deserves special efforts for conservation (Desalle and Amato 2017).

Can the presence of *spelaeus*-like characters in the skull and dentition of *marsicanus* (see also Capasso Barbato et al. 1993) really be explained by convergent evolution during 700 years of genetic drift? Or might it be possible that some *spelaeus* genes were present in the bear population south of the Alps, possibly due to an introgression episode (cf. Barlow et al. 2018)? These genes could be masked by other genes which got lost by genetic drift. So it should not be convergent evolution but the reappearance of a special set of genes due to genetic drift and/or introgression.

Gippoliti (2016) called for a taxonomic revision of the whole *Ursus arctos* complex and specifically considered the available evidence sufficient to rank *marsicanus* as a distinct ESU and, provisionally, a valid subspecies. Following a first call for action for the taxon, the need for a more vigorous approach to its conservation has been reiterated (Guacci et al. 2013; Gippoliti and Guacci 2017) given that the population size does not exceed 65 individuals including some 13 breeding females (Fig. 16.6) (Ciucci et al. 2015).

Although never openly debated, it is obvious that most bear experts in Europe believe that the first measure to ensure the long-term conservation of bears on the Apennine is restocking with a few individuals from the Balkans to allow 'genetic rescue'. This is in fact a continent-wide strategy, already utilized in the Alps and Pyrenees. Partly explaining the overt neglect of taxonomic issues – in a recent revision of 'Genetics and conservation of European brown bear *Ursus arctos*', the term

Fig. 16.6 An adult female Apennine bear. (Photo Antonio Macioce)

'taxonomy' is never used (Swenson et al. 2011). But if the unique characters of the skull of *marsicanus* are due to local adaptations as a result of directional selection instead of genetic drift (Colangelo et al. 2012), and evidence of inbreeding depression has never been observed, the picture that emerges is one of a lineage well adapted to a Mediterranean mountain habitat. A conservation strategy should, therefore, prioritize population expansion and not take the risk that the potentially unique gene pool is spoiled by introgression of genes from possibly 'less adapted' populations (outbreeding depression). The social organization of brown bears is characterized by female philopatry and male dispersal (Støen et al. 2005). Since male Apennine brown bears wander over several mountain systems of Central Italy, an important alternative conservation strategy would be to create new female breeding nuclei in other protected areas of the Apennine Mountains.

Given the taxonomic status of *marsicanus*, it has also been suggested that a bank of reproductive samples (semen, eggs) should be created – as has been planned in Spain for the Cantabrian brown bears (Anel et al. 2011; see also Saragusty et al. 2016) – and that if some bears have to be taken out of the wild population because they are 'problematic' (i.e. visiting villages to eat fruits and chicken), or are found orphaned in the wild, they should be included in an ex situ breeding program (Guacci et al. 2013). Benazzo et al. (2017) performed whole-genome sequencing of six Apennine bears comparing *marsicanus* with Iberian and Balkan *Ursus arctos* and divergence time was estimated at 3000–4000 years. They found evidence of two evolutionary processes with opposite outcomes: active maintenance of variation of specific families of genes and fixation by drift of several deleterious alleles. Their results thus support the view that, even in small populations, the random loss of variation does not affect all sites in the same way. Their work further contributes to the general debate about the relative role of drift and selection when the effective population size is very small. Interestingly, Benazzo et al. (2017)s conclusion is that 'On the other hand, the recognition of the Apennine bear as an Italian iconic endangered taxon, the possible risk of introducing aggressiveness genes and deteriorating the relatively peaceful human–bear coexistence in central Italy, and the current levels of variation at relevant immune and olfactory genes suggest avoiding genetic rescue' (Benazzo et al. 2017: 9595). This fully overlaps with Gippoliti's suggestion (2016) and strongly departs from the orthodoxy of bear management in Europe.

Although there is widespread concern for inbreeding depression, in the case of *marsicanus* there has been no national effort to conserve vouchers (skulls, skeletons, skins, etc.) of the many bears found dead due to human persecution. These materials should be critical to assess the health of the endemic Apennine bear. Hopefully, the slowly increasing evidence of the distinctiveness of *marsicanus* (see also Meloro et al. 2017) will lead to the development of a conservation plan that will effectively guide the recovery of the most endangered Italian mammals.

Incidentally, at least in the historic stronghold of the species, bears are considered by local inhabitants as a regional heritage, and in these regions, bears have never been aggressive towards humans and have a limited impact on human activities. In the end, we hope that social considerations too will help direct conservation strategy away from so-called genetic rescue.

16.7 Implications for Management and Conservation

There is no doubt that among the consequences of an overly simplified view of biodiversity, there is an oversimplified – and assuring – view of the environmental situation of the planet. The recognition of a given polytypic species with an enormous geographical distribution as a lower risk species completely overlooks the conservation status of several local populations/subspecies (Morrison III et al. 2009; Thakur et al. 2018) which, it is assumed, in case of continuing decline, could be 'restocked' with animals from other more healthy local populations (Frankham et al. 2012).

Promoting human-mediated genetic introgression, as supported by some biologists and philosophers (Sgrò et al. 2011; Rohwer and Marris 2016), is a step towards integrating and homogenizing conservation biology in an 'Anthropocene science' where we accommodate biological evolutionary history to the will of one species – *Homo sapiens*. Concern over the risks of homogenization of biological diversity of African bovids for commercial reasons has already been expressed by the IUCN/SSC Antelope Specialist Group (IUCN SSC Antelope Specialist Group 2015). Translocation and genetic rescue is potentially an important technique that has been successfully utilized in several conservation projects, for example, with desert bighorn sheep (Hedrick and Wehausen 2014; Buchalski et al. 2016). But it can also be a waste of meagre conservation resources, not to mention of individual animals, in some instances such as the unsuccessful translocations of the 'woodland caribou' in Canada (Leech et al. 2017). In other words, before applying the 'magic bullet' of genetic rescue, look carefully at what is a species and what is not. What are the special adaptations of a declining population? Has 'genetic purging' occurred? And why has that population (or species) declined? Conservation is a very complex field. The decline of a population (or species) may have little or nothing to do with genetic impoverishment (Peer and Taborsky, 2005) and outbreeding depression may follow the 'top-up' of an endangered population by the introduction of fresh genes (Tallmon et al. 2004). An outdated and oversimplified taxonomy is one of the fundamental causes of a wrong-headed conservation strategy (Gippoliti et al. 2018a). Conservationists' motivations against adoption of an evolutionary species concept (Wiley 1978) seem to be due to a scarce knowledge of the 'species problem' outside the community of systematic biologists (Padial et al. 2010; Gutierrez and Helgen 2013; Raposo et al. 2017). We also reaffirm our opinion that the goal of conservation biology must be the maintenance of biodiversity biogeographical patterns that are as much as possible similar to those developed in the planet's history aside from human interventions.

Evolutionary thinking certainly justifies an increase in the number of mammal species that are recognized, including those that are threatened and those that are 'problematic' in their relationship with humans. But a clearer picture of existing biodiversity will, we hope, motivate the conservation community to direct resources through a transparent prioritization system. Once discovered, mammal lineages are particularly effective to serve as flagship and umbrella species for conservation of otherwise neglected habitats. Further, it is urgent that – as was already evidenced by Caughley (1994) – a real integration between disciplines will be achieved. The belief that conservation problems stem automatically from small population sizes is a reductionist approach that may divert attention from true problems and the real solutions to species conservation. In the end, we think that such a finer approach to mammal taxonomy offers more opportunities for local communities' involvement in conservation and sustainable utilization is the only way we can hope to maintain a diverse and healthy planet.

Acknowledgements We wish first to thank Francesco Maria Angelici and Lorenzo Rossi for inviting us to deal with this hot issue in conservation biology. Our sincere thanks to a number of colleagues who, in the last years, shared with us the desire to innovate mammalian taxonomy as a critical tool for conservation strategies: Fenton (Woody) Cotterill, Boris Krystufek, Peter Taylor, Jan Robovský and Mostafa Saleh. SG wishes to thank Giorgio Boscagli, Mauro Ferri, Corradino Guacci and Sandro Lovari for the long discussions about the future of Apennine bears. Kees Rookmaaker and Jan Robovský commented on an early draft, while Jacopo Conti, Francesco Germi and Antonio Macioce provided some of the pictures. SG wishes to warmly thank Phyll Groves for helping in revising the English style and Eva Bärmann and Dietmar Zinner for their thorough suggestions.

References

Amori G, Angelici FM, Boitani L (1999) Mammals of Italy: a revised checklist of species and subspecies (Mammalia). Senckenberg Biol 79:271–286

Anel L, Alvarez M, Anel E, Martinez-Pastor F, Martinez F, Chamorro C, de Paz P (2011) Evaluation of three different extenders for use in emergency salvaging of epididymal spermatozoa from a Cantabric brown bear. Reprod Domest Anim 46(1):e85–e90. https://doi.org/10.1111/j.1439-0531.2010.01646.x

Angelici FM (ed) (2016) Problematic wildlife. A cross-disciplinary approach. Springer, Cham

Angelone-Alasaad S, Biebach I, Peârez JM, Soriguer RC, Granados JE (2017) Molecular analyses reveal unexpected genetic structure in Iberian ibex populations. PLoS One 12(e0170827). https://doi.org/10.1371/journal.pone.0170827

Alexander Ochoa, Stuart A. Wells, Gary West, Ma'en Al-Smadi, Sergio A. Redondo, Sydnee R. Sexton, Melanie Culver, (2016) Can captive populations function as sources of genetic variation for reintroductions into the wild? A case study of the Arabian oryx from the Phoenix Zoo and the Shaumari Wildlife Reserve, Jordan. Conservation Genetics 17(5):1145–1155

Barlow A, James A, Cahill JA, Hartmann S et al (2018) Partial genomic survival of cave bears in living brown bears. Nat Ecol Evol. https://doi.org/10.1038/s41559-018-0654-8

Benazzo A, Trucchi E, Cahill JA et al (2017) Survival and divergence in a small group: the extraordinary genomic history of the endangered Apennine brown bear stragglers. Proc Natl Acad Sci 114(45):E9589–E9597

Bilton T, Mirol PM, Mascheretti S, Fredga K, Zima J, Searle JB (1998) Mediterranean Europe as an area of endemism for small mammals rather than a source for northwards postglacial colonization. Proc R Soc Lond B 265:1219–1226

Boitani L (ed) (1976) Reintroductions: techniques and ethics. Serie Atti e Studi n. 2. WWF Fondo Mondiale per la Natura, Rome

Brooks M (2010) African rhino specialist group report. Pachyderm 48:8–15

Buchalski MR, Sacks BN, Gille DA, Penedo MCT, Ernest HB, Morrison SA, Boyce WM (2016) Phylogeographic and population genetic structure of bighorn sheep (*Ovis canadensis*) in North American deserts. J Mammal 97:823–838

Camerano L (1916) I caratteri del cranio, della colorazione e delle corna nella distinzione dei Camosci in specie e sottospecie. Riv Antropol 20:3–14

Capasso Barbato L, Cerilli E, Petronio C (1993) Differenze morfologiche e morfometriche nei crani di *U. spelaeus* e *U. arctos*. Quaternario 6:67–76

Carpaneto GM, Gippoliti S (1994) Primates of the Harenna Forest, Ethiopia. Primate Conserv 11:12–15

Caughley G (1994) Directions in conservation biology. J Anim Ecol 63:215–244

Cinková I, Policht R (2014) Contact calls of the Northern and Southern white rhinoceros allow for individual and species identification. PLoS One 9:e98475

Ciucci P, Gervasi V, Boitani L, Boulanger J, Paetkau D, Prive R, Tosoni E (2015) Estimating abundance of the remnant Apennine brown bear population using multiple noninvasive genetic data sources. J Mammal 96:206–220

Colangelo P, Loy A, Huber D, Gomerčić T, Vigna Taglianti A, Ciucci P (2012) Cranial distinctiveness in the Apennine brown bear: genetic drift or ecophenotypic adaptation? Biol J Linn Soc 107:15–26

Conti S (1954) Morfologia comparata craniale ed encefalica degli orsi pleistocenici della Liguria. Correlazioni con alcune forme attuali (*U. arctos, U. marsicanus, U. horribilis*). Mem Mus Civ St Nat "G. Doria" Genova 1:1–66

Cotterill FPD, Taylor PJ, Gippoliti S, Bishop JM, Groves CP (2014) Why one century of phenetics is enough: response to "are there really twice as many bovid species as we thought?". Syst Biol 63:819–832

Cotterill FPD, Groves CP, Taylor PJ (2017) Taxonomy: refine rather than stabilize. Nature 547:162

Culver M, Johnson WE, Pecon-Slattery J, O'Brien SJ (2000) Genomic ancestry of the American puma (*Puma concolor*). J Hered 91:186–197

Curry-Lindahl K (1972) War and the white rhinos. Oryx 11:263–267

de Beaux O (1930) Etica biologica. Tentativo di risveglio di una coscienza naturalistica. Temi, Trento

de Beaux O (1932) Biological ethics. An attempt to arouse a naturalistic conscience. The Italian Mail and Tribune, Florence

de Beaux O (1956) Posizione sistematica degli stambecchi e capre selvatiche viventi (*Capra* Linneo 1758) e loro distribuzione geografica. Atti Accad Sci Lett Genova 12:123–228

Desalle R, Amato G (2017) Conservation genetics, precision conservation, and de-extinction. Hast Cent Rep 47:S18–S23

Edmands S (2007) Between a rock and a hard place: evaluating the relative risks of inbreeding and outbreeding for conservation and management. Mol Ecol 16:463–475

Ellerman JR, Morrison-Scott TCS (1951) Checklist of Palaearctic and Indian mammals 1758 to 1946. Trust of the British Museum, London

Ferguson JWH (2002) On the use of genetic divergence for identifying species. Biol J Linn Soc 75:509–516

Frankham R, Ballou JD, Dudash MR, Eldridge MDB, Fenster CB, Lacy RC, Mendelson JR III, Porton IJ, Ralls K, Ryder OA (2012) Implications of different species concepts for conservation biodiversity. Biol Conserv 153:25–31

Garnett ST, Christidis L (2017) Taxonomy anarchy hampers conservation. Nature 546:25–27

Geist V (2007) Defining subspecies, invalid taxonomic tools, and the fate of the woodland Caribou. Rangifer 17:25–28

Gippoliti S (2006) Oscar de Beaux: a noteworthy Italian mammalogist and conservationist. Ital J Zool 73:285–289

Gippoliti S (2011) Ex situ conservation programmes in European zoological gardens: can we afford to lose them? Biodivers Conserv 21:1359–1364

Gippoliti S (2016) Questioning current practice in brown bear *Ursus arctos* conservation in Europe that undervalues taxonomy. Anim Biodivers Conserv 39:199–205

Gippoliti S, Amori G (2002a) Mammal diversity and taxonomy in Italy: implications for conservation. J Nat Conserv 10:33–44
Gippoliti S, Amori G (2002b) Anthropochorous mammal taxa and conservation lists. Conserv Biol 16:1162–1164
Gippoliti S, Amori G (2007) Beyond 'threatened species' and 'reintroduction': establishing priorities for European rodents conservation and breeding programs in zoos. Int Zoo Yearb 41:194–202
Gippoliti S, Battisti C (2017) More cool than tool: equivoques, conceptual traps and weaknesses of ecological networks in environmental planning and conservation. Land Use Policy 68:686–691
Gippoliti S, Groves CP (2012) "Taxonomic inflation" in the historical context of mammalogy and conservation. Hystrix, It J Mamm 23:8–11
Gippoliti S, Groves CP (2018) Overlooked mammal diversity and conservation priorities in Italy: impacts of taxonomic neglect on a biodiversity hotspot in Europe. Zootaxa 4434:511–528
Gippoliti S, Guacci C (2017) Il più minacciato Mammifero italiano: l'orso marsicano. Un approccio interdisciplinare per la sua conservazione. Natura e Montagna 64:29–35
Gippoliti S, Cotterill FPD, Zinner D, Groves CP (2018a) Impacts of taxonomic inertia for the conservation of African ungulate diversity: an overview. Biol Rev 93:115–130
Gippoliti S, Cotterill FPD, Groves CP, Zinner D (2018b) Poor taxonomy and genetic rescue are possible co-agents of silent extinction and biogeographic homogenization among ungulate mammals. Biogeographia J Integr Biogeogr 33:41–54
Goldstein PZ, DeSalle R, Amato G, Vogler AP (2000) Conservation genetics at the species boundary. Conserv Biol 14:120–131
Groves CP (2001) Primate taxonomy. Smithsonian University Press, Washington
Groves CP (2012) The nature of species: a rejoinder to Zachos et al. Mamm Biol 78:7–9
Groves CP (2014) Primate taxonomy: inflation or real? Annu Rev Anthropol 43:27–36
Groves CP, Grubb P (2011) Ungulate taxonomy. Johns Hopkins University Press, Baltimore
Groves CP, Robovský J (2011) African rhinos and elephants: biodiversity and its preservation. Pachyderm 50:69–71
Groves CP, Fernando P, Robovský J (2010) The sixth rhino: a taxonomic re-assessment of the critically endangered northern white rhinoceros. PLoS One 5:e9703. https://doi.org/10.1371/journal.pone.0009703
Groves CP, Cotterill FPD, Gippoliti S, Robovský J, Roos C, Taylor PJ, Zinner D (2017) Species definitions and conservation: a review and case studies from African mammals. Conserv Genet 18:1247–1256
Guacci C, Ferri M, Gippoliti S (2013) Un manifesto pro conservazione 'ex situ' dell'orso bruno marsicano *Ursus arctos marsicanus* Altobello, 1921. Biologia Ambientale 27(2):55–58
Gutierrez EE, Helgen KM (2013) Mammalogy: outdated taxonomy blocks conservation. Nature 495:314
Harley EH, de Waal M, Murray S, O'Ryan C (2016) Comparison of whole mitochondrial genome sequences of northern and southern white rhinoceroses (*Ceratotherium simum*): the conservation consequences of species definition. Conserv Genet 17:1285–1291
Harris K, Nielsen R (2016) The genetic cost of Neanderthal introgression. Genetics 203:881–891
Haus T, Akom E, Agwanda B, Hofreiter M, Roos C, Zinner D (2013) Mitochondrial diversity and distribution of African green monkeys (*Chlorocebus* Gray, 1870). Am J Primatol 75:350–360
Hedrick PW, Fredrickson R (2010) Genetic rescue guidelines with examples from Mexican wolves and Florida panthers. Conserv Genet 11:615–626
Hedrick PW, Wehausen JD (2014) Desert bighorn sheep: changes in genetic variation over time and the impact of merging populations. J Fish Wildl Manag 5:3–13
Heller E (1913) The white rhinoceros. Smithsonian Miscell Collect 61(1):1–56
Heller R, Frandsen P, Lorenzen ED, Siegismund HR (2013) Are there really twice as many bovid species as we thought? Syst Biol 62:490–493
Hilmann-Smith K, Oyisenzoo ma M, Smith E (1986) A last chance to save the northern white rhino? Oryx 20:20–26

IUCN (1987) The IUCN position statement on translocation of living organisms. IUCN, Gland
IUCN (1998) Guidelines for re-introductions. Prepared by the IUCN/SSC Re-introduction Specialist Group. IUCN, Gland
IUCN SSC Antelope Specialist Group (2015) IUCN SSC Antelope Specialist Group position statement on the intentional genetic manipulation of antelopes. Ver. 1.0
IUCN/SSC (2013) Guidelines for reintroductions and other conservation translocations. Version 1.0. Gland, Switzerland: IUCN Species Survival Commission, viiii + 57 pp
Kingdon J (1997) The kingdon field guide to African mammals. Collins, London
Kuneš M, Bičík V (2002) Social and sexual behaviour in captive breeding groups of white rhinoceros. Acta Univ Palacki Olumuc Biol 39–40:81–99
Katherine Ralls, Jonathan D. Ballou, Michele R. Dudash, Mark D. B. Eldridge, Charles B. Fenster, Robert C. Lacy, Paul Sunnucks, Richard Frankham, (2018) Call for a Paradigm Shift in the Genetic Management of Fragmented Populations. Conservation Letters 11 (2):e12412
Leech H, Jelinski DE, DeGroot L, Kuzyk G (2017) The temporal niche and seasonal differences in predation risk to translocated and resident caribou (*Rangifer tarandus caribou*). Can J Zool 95:809–820
Leopold A (1933) The conservation ethic. J Forestry 31:634–641
Lindenmayer DB, Fisher J (2007) Tackling the habitat fragmentation panchreston. Trends Ecol Evol 22:127–132
Marshall TC, Spalton JA (2000) Simultaneous inbreeding and outbreeding depression in reintroduced Arabian oryx. Anim Conserv 3:241–248
Mayr E (1982) Of what use are subspecies? Auk 99:593–595
Mekonnen A, Bekele A, Fashing PJ, Lernould JM, Atickem A, Stenseth NC (2012) Newly discovered Bale monkey populations in forest fragments in Southern Ethiopia: evidence of crop raiding, hybridization with grivets, and other conservation threats. Am J Primatol 74:423–432
Mekonnen A, Fashing PJ, Bekele A et al (2017) Impacts of habitat loss and fragmentation on the activity budget, ranging ecology and habitat use of Bale monkeys (*Chlorocebus djamdjamensis*) in the southern Ethiopian Highlands. Am J Primatol 9999:e22644. https://doi.org/10.1002/ajp.22644
Mekonnen A, Rueness EK, Stenseth NC, Fashing PJ, Bekele A, Hernandez-Aguilar RA, Missbach R, Tanja Haus T, Zinner D, Roos C (2018) Population genetic structure and evolutionary history of Bale monkeys (*Chlorocebus djamdjamensis*) in the southern Ethiopian Highlands. BMC Evol Biol 18:106
Meloro C, Guidarelli G, Colangelo P, Ciucci P, Loy A (2017) Mandible size and shape in extant Ursidae (Carnivora, Mammalia): a tool for taxonomy and ecogeography. J Zool Syst Evol Res 55:269–287
Morrison WR III, Lohr JL, Duchen P, Wilches R, Trujillo D, Mair M, Renner SS (2009) The impact of taxonomic change on conservation: does it kill, can it save, or is it just irrelevant? Biol Conserv 142:3201–3206
Owen-Smith RN (2013) Ceratotherium simum - white rhinoceros (grass rhinoceros, square-lipped rhinoceros). In J Kingdon and M Hoffmann (Eds), Mammals of Africa - Volume V (pp. 446–454). London, New Delhi, New York, Sydney: Bloomsbury
Padial JM, Miralles A, De la Riva I, Vences M (2010) The integrative future of taxonomy. Front Zool 7:16. https://doi.org/10.1186/1742-9994-7-16
Peer K, Taborsky M (2005) Outbreeding depression, but no inbreeding depression in haplodiploid ambrosia beetles with regular sibling mating. Evolution 59:317–323
Pocock RI (1932) The black and brown bears of Europe and Asia. J Bombay Nat Hist Soc 35:771–823
Pocock RI (1935) The mammals collected in S.E. Arabia by Mr. Bertram Thomas and Mr. H. St. J. Philby. Ann Mag Nat Hist (10)15:441–467
Raposo MA, Stopiglia R, Brito GRR, Bockmann FA, Kirwan GM, Gayon J, Dubois A (2017) What really hampers taxonomy and conservation? A riposte to Garnett and Christidis (2017). Zootaxa 4317(1):179–184

Rohwer Y, Marris E (2016) Is there a prima facie duty to preserve genetic integrity in conservation biology? Ethics. Policy Environ 18(3):233–247
Rojas M, (1992) The Species Problem and Conservation: What are We Protecting? Conservation Biology 6(2):170–178
Rookmaaker K (2000) The alleged population reduction of the southern white rhinoceros (*Ceratotherium simum simum*) and the successful recovery. Säuget Mitt 45:55–70
Saragusty J, Diecke S, Drukker M, Durrant B, Friedrich Ben-Nun I, Galli C, Göritz F, Hayashi K, Hermes R, Holtze S, Johnson S, Lazzari G, Loi P, Loring JF, Okita K, Renfree MB, Seet S, Voracek T, Stejskal J, Ryder OA, Hildebrandt TB (2016) Rewinding the process of mammalian extinction. Zoo Biol 35:280–292
Schwarz E (1935) On ibex and wild goat. Ann Mag Nat Hist 16:433–437
Seddon PJ, Soorae P (1999) Guidelines for subspecific substitutions in wildlife restoration projects. Conserv Biol 13:177–184
Sexton JP, Hangartner SB, Hoffmann AA (2014) Genetic isolation by environment or distance: which pattern of gene flow is most common? Evolution 68:1–15
Sgrò CM, Lowe AJ, Hoffmann AA (2011) Building evolutionary resilience for conserving biodiversity under climate change. Evol Appl 4:326–337
Shackleton DM, Lovari S (1997) Classification adopted for the Caprinae survey pp. In: Shackleton DM (ed) Wild sheep and goats. Status survey and conservation action plan. IUCN/SSC Caprinae Specialist Group, Gland, pp 9–14
Sipari E (1926) Relazione. Ente Autonomo del Parco Nazionale d'Abruzzo, Pescasseroli, 309 pp
Støen O-G, Bellemain E, Sæbø S, Swenson J (2005) Kin-related spatial structure in brown bears *Ursus arctos*. Behav Ecol Sociobiol 59:191–197
Stresemann E (1975) Ornithology. Harvard University Press, Cambridge, MA
Swenson JE, Taberlet P, Bellemain E (2011) Genetics and conservation of European brown bear *Ursus arctos*. Mammal Rev 41:87–98
Tallman DA, Luikart G, Waples RS (2004) The alluring simplicity and complex reality of genetic rescue. Trends Ecol Evol 19:489–496
Templeton A (1986) Coadaptation and outbreeding depression. In: Soulé (ed) Conservation biology. Sinauer Associates, Sunderland, pp 105–116
Thakur M, Wullschleger Schättin E, McShea WJ (2018) Globally common, locally rare: revisiting disregarded genetic diversity for conservation planning of widespread species. Biodivers Conserv. https://doi.org/10.1007/s10531-018-1579-x
Turcek FJ (1951) Effects of introductions on two game populations in Czechoslovakia. J Wildl Manag 15:113–114
Turelli M, Barton NH, Coyne JA (2001) Theory and speciation. Trends Ecol Evol 19:490–496
Vander Wal E, Garant D, Festa-Bianchet M, Pelletier F (2013) Evolutionary rescue in vertebrates: evidence, applications and uncertainty. Philos Trans R Soc B 368:20120090
Vigna Taglianti A (2003) *Ursus arctos* – note di sistematica. In: Boitani L, Lovari S, Vigna Taglianti A (eds) Fauna d'Italia. Mammalia III. Carnivora – Artiodactyla. Edizioni Calderini de il Sole 24 ORE, Bologna, pp 87–92
Weeks AR, Sgrò CM, Young AG, Frankham R, Mitchell NJ, Miller KA, Byrne M, Coates DJ, Eldridge MDB, Sunnucks P, Breed MF, James EA, Hoffmann AA (2011) Assessing the benefits and risks of translocations in changing environments: a genetic perspective. Evol Appl 4:709–725
Wheeler QD, Platnick NI (2000) The phylogenetic species concept (sensu Wheeler and Platnick). In: Wheeler QD, Meier R (eds) Species concepts and phylogenetic theory: a debate. Columbia University Press, New York, pp 55–69
Wiley EO (1978) The evolutionary species concept reconsidered. Syst Zool 27:17–26
Zachos FE, Clutton-Brock TH, Festa-Bianchet M, Lovari S, Macdonald DW, Schaller GB (2013) Taxonomy: species splitting puts conservation at risk. Nature 494:35

Rosa et al. [illegible] (2016) [illegible] to [illegible] Biol [illegible]

[illegible] (2004) The Slaughter [illegible] and [illegible] [illegible]

Rookmaaker [illegible] (2011) The [illegible] and population reduction of the [illegible] [illegible]

Savolainen [illegible] [illegible]

Snow [illegible] (2013) Cambridge [illegible] [illegible]

Sutherland [illegible] (2006) Guidelines for [illegible] in wildlife [illegible]

Somers [illegible] [illegible] (2012) [illegible]

[illegible] (2013) Building [illegible] [illegible] Appl [illegible]

Shackleton DM [illegible] (1997) [illegible] In: Shackleton DM (ed) Wild sheep and goats and their relatives. Status survey and conservation action plan for Caprinae. IUCN/SSC Caprinae Specialist Group, Gland, pp [illegible]

[illegible] Biol Conserv [illegible]

[illegible] (2011) [illegible] Europe [illegible]

[illegible]

[illegible] (1986) [illegible] Sweden [illegible]

[illegible] In: [illegible] [illegible]

[illegible] (2003) [illegible] [illegible]

Watts [illegible] [illegible]

[illegible] [illegible]

Whittaker [illegible] [illegible]

Wilson [illegible] [illegible]

[illegible] (2008) [illegible] [illegible]

Chapter 17
Communication and Wildlife Conservation (Grey Wolf and Brown Bear in Italy)

Franco Perco

17.1 Introduction

Communicating well is a problem in itself.

And the word 'well' needs to be defined. We could understand it as communicating effectively.

This said, advertising, demagogy and show business also communicate in appropriate ways, to obtain precise results, but they do so regardless of their content and for purposes that cannot (always) be shared.

Communicating correctly is certainly better.

Here, by 'correctly', however, it must be understood in a non-formal way. What would you think of a student who, asked to 'Tell me about Kant…' were to answer, 'Kant is a four-letter word and starts with the letter "k"…'? We would think he's wanting to be silly and make fun of people. The correct communication must therefore also be pertinent to the case in point, to the case under consideration, 'in tone' one might say (Baroncelli 1996; Watzlawick 2007).

Communicating in regard to wildlife involves another set of difficulties. The relationship of *Homo sapiens* (HS) with both wild and other animals has always been a very problematic one. Ideologies, cultures, habits and traditions influence human thought, and this can provoke continual recourse to prejudices or summary judgments, without taking account of the accuracy of the news, the truthfulness of a fact or its likely outcome. It is therefore essential to research the sources, their reliability and their impartiality, regardless of each one's reference values.

But even this is insufficient. Firstly, because it is necessary to emerge from the reductive interpretation that communicating is done exclusively via the word. Watzlawick et al. (1978) have argued, on the subject of humankind, that we cannot not communicate and that every behaviour represents communication. Promoting a

F. Perco (✉)
Montelupone (MC), Italy

F. M. Angelici, L. Rossi (eds.), *Problematic Wildlife II*,
https://doi.org/10.1007/978-3-030-42335-3_17

sporting event in an area sensitive for brown bear means communicating, and it is bad communication in the sense that it is harmful to the species and/or to the concerned environment and/or human sensitivity towards the same. Cutting wood in a grey wolf rendezvous area is (also) bad communication, regardless of its conservationist consequences.

We can therefore almost copy the evangelical precept: one communicates with words, deeds and omissions, that is, with any kind of behaviour towards a given problem.

Secondly, it should be pointed out that, as stated by Watzlawick et al. (1978), 'every communication has an aspect of content and an aspect of relationship so that the second classifies the first.'

In fact, Bateson and Ruesch (1951) distinguish communication in the news (report) and 'command' that it can also be defined as a prescription or an invitation, in the sense that the latter informs the way in which a subject should take on the information.

Simplifying greatly, therefore, information is objective (but not identical between the communicant and the recipient) and the subjective prescription. And one could examine the ethics of responsibility (Weber 2000): we are still responsible for what others do as a result of our actions. It is not true people's actions are their concern solely and that we have acted correctly. Their actions are our concern as well.

Since every human communication, in addition to transmitting information, implies a commitment among the communicator and the recipient and defines the nature of their relationship, we should consider whether this also applies to the brown bear, *Ursus arctos* (UA), and the grey wolf, *Canis lupus* (CL), which are certainly able to empathize with humans. Among other things, in the case of CL, given the extraordinary ability of its descendant, the domestic dog, *Canis lupus familiaris* (CLF), to have extremely complex relationships with HS, it is not useless to reflect that these could be much more complex than how it is generally believed.

As a first conclusion, we can state that it is necessary not to forget that wildlife communication involving UA and CL must be addressed in two different ways:

- Between human beings (with regard to these two species)
- Between human beings and the two species in question

And it should be pointed out that the communication between HS also concerns how UA and CL respond to the communication of HS, often accompanied by CLF (inter-HS-UA and inter-HS-CL and vice versa).

In this work, we will mainly deal with verbal communication between HS, without forgetting other communication types. For reasons of economy, the technical modalities of the communication will not be addressed, and mainly direct face-to-face communication will be considered: communicator(s) *versus public*.

17.2 Communication Involving *Ursus arctos* and *Canis lupus*

Communication regarding these two species has particular aspects. Strictly speaking, this communication is always tendentious in the sense that it aims to mitigate conflicts.

Now, mitigation cannot consist of consoling through good words but aims to solve problems, promising something, or by changing the point of view, providing a glimpse of another perspective, stimulating doubt and reflection.

It therefore requires an action on the part of the sufferer, albeit merely psychological and internal.

But it is necessary to have something to show. We need good examples, acceptable situations. If these are not existing, it is very difficult to mitigate a conflict. And if there are, it is easy and appropriate to say ‘OK, let’s do as they do’.

The first requisite, therefore, is some pre-existent quality: having virtuous examples to illustrate.

Secondly, if one promises (if A occurs, then B will be done; before C takes place, D will be carried out), we must consider (by communicating) that the promise must be realistic and verifiable (e.g. on time). Making a promise without keeping it, is an obvious self-goal. But this is not sufficient. The communicator must have ensured the full readiness of others to keep their promise, in the event that it is not him, or her or the organization that they represent, that must or can fulfil the promise made. Therefore, there is a second requirement: the seriousness of the promises in the communication. Good communication is not advertising, passing off an action as valid when it is not, when its qualities are lower than described or, worse still, when we are unsure of its value.

Thirdly, it is necessary to reflect on the fact that communication is a representation, according to Goffman’s sense (1959). On the basis of what Goffman claims, the communicators are real actors, and the public is not inert but tries to obtain information about the actors as well as about itself, so that, in turn, the public is an actor as well. All this involves precise rules on the quality of the ‘performance’ of the team (the organization), their degree of sincerity, the level of self-awareness of the actors, the dangerousness of any misleading representations and so forth. In essence, communication is a spectacle.

Finally, because we communicate to change something and not for the sake of it, the communication process is also a negotiation process, even when bargaining is not directly and immediately perceptible. Not infrequently, communication is subliminal bargaining, and the communicator does not aim (perhaps) as much to persuade but to instil thoughts, reflections and doubts. A good communicator dismantles mental closures and tries to deconstruct prejudices with emotions (Watzlawick 1976). Communication is negotiation!

Good examples to illustrate seriousness, honest spectacle and negotiation: communication is this as well.

17.3 How Are UA and CL Doing in Italy?

An assessment of their population is difficult, a priori, and in the Italian case made even more problematic by the absence of obligations and initiatives by public authorities. The data below are the result of independent research and reports of bodies that include the National Park of Abruzzo, Lazio and Molise (Central Italy), the Autonomous Province of Trento (North Italy), etc. as well as several EU-funded Life projects.

To the references (various authors 2010, 2011; Boitani 1984, 2000; Boitani and Salvatori 2019; Boitani et al. 2003; Boscagli 1985, 1999; Ciucci et al. 2013, 2015; Galaverni et al. 2015; Groff et al. 2013, 2018; Latini et al. 2017; Marucco 2014, 2016; Mattioli et al. 2014; Perco 2014; Perco and Forconi 2016; Zimen and Boitani 1975), we have added some personal and original assessments.

The grey wolf increased consistently after World War II, especially following the ban on its hunting (since 1974) and the subsequent European Union regulations (see Fig. 17.1).

It should be noted that some specialists or experts in the sector have denied until recently the need to have numerical data on this species, or they have insisted on very cautious evaluations, probably not to create social alarm. But this is created by the increase in damage and the visibility of the species, hence the perceptibility of the phenomenon.

At least 85% of the population of CL is found in the Apennine chain. About a quarter of Italy's wolves (500) are located within/around national parks (5.5% of the national surface) (Fr Perco, unpublished, 2014). It is also probable that the figure of 2000 subjects should be considered a prudent one when Piemonte alone has an estimated 150 wolves in 27 packs and 2 pairs and the province of Trentino 6 packs with

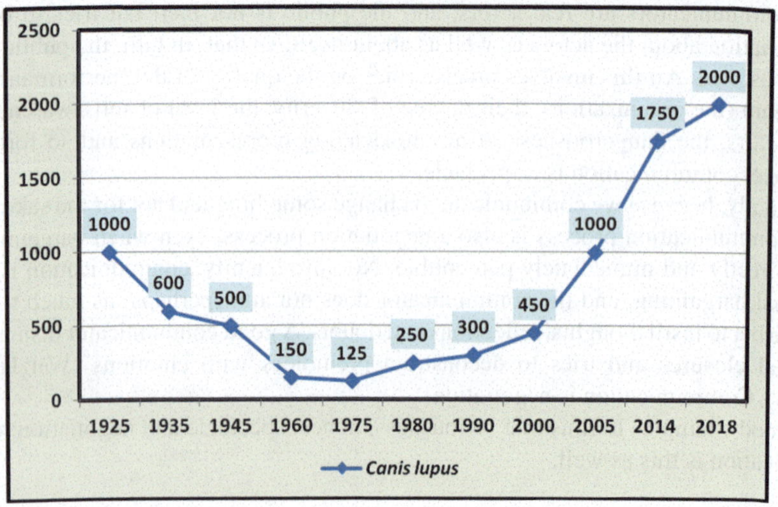

Fig. 17.1 Estimates of the numbers of *Canis lupus* in Italy (1925–2018)

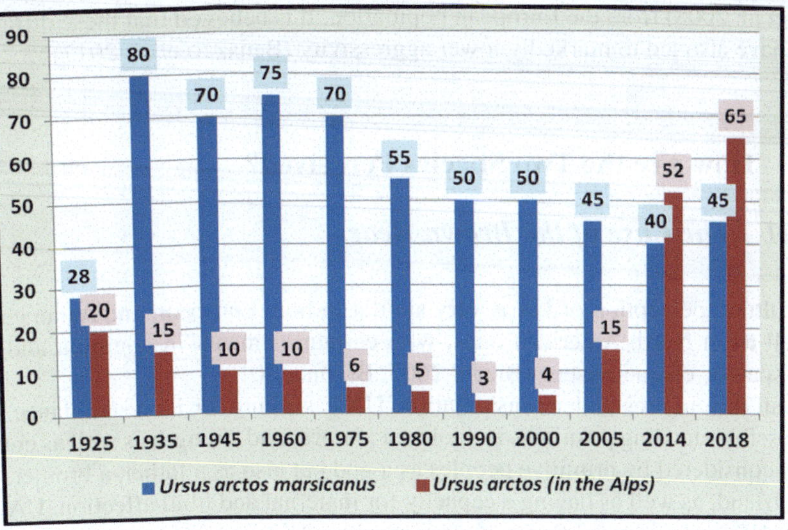

Fig. 17.2 Estimates of the numbers of *Ursus arctos* in Italy (1925–2018)

just as many in Lessinia (Veneto). An estimate for the Alpine area as a whole (Marucco 2016) is over 40 packs and 6 pairs (Zimen and Boitani 1975; Boitani 1984; Boscagli 1985, 2014, pers comm; Ciucci and Boitani 1998; Kaczensky et al. 2013; Marucco 2014; Mattioli et al. 2014; Galaverni et al. 2015; Perco and Forconi 2016; Boitani and Salvatori 2019).

The estimated 2000 wolves represent about 15% of Europe's CL (excluding Russia) and 18% of the EU total (a number only surpassed by Romania) (Kaczensky et al. 2013; Marucco 2014).

Regarding UA, this species owes its current situation mainly to the increase in the Trentino population, following a reintroduction using Slovenian bears (ten individuals), carried out by the Autonomous Province of Trento between 1999 and 2002.

Currently and in the Alps more generally, UA numbers about 50–55 animals in the province of Trento (Groff et al. 2018). For the Autonomous Province of Bolzano, there are about 4–8 bears, 1–2 in the Veneto and 5–10 straddling the border with Slovenia and Austria in Friuli Venezia Giulia (personal communication) (see Fig. 17.2).

The population of Central Italy, also estimated before the yearly births (Tosoni et al. 2016), has as its heart the National Park of Abruzzo, Lazio and Molise (ALMNP). It seems stationary but decreasing in the medium term, even after the establishment of this park in 1922 (Fig. 17.2). About 4–5 individuals are found in the Majella National Park (MNP) and perhaps 5 others in neighbouring areas bordering the park itself. This UA belongs to a distinct subspecies (*Ursus arctos marsicanus*, Altobello 1921) and, during a prolonged period of isolation, has undergone genetic (Randi et al. 1994; Lorenzini et al. 2004) and morphological differentiation

(Loy et al. 2008) from the European population. It is believed that these differentiations have also led to markedly lower aggressivity (Benazzo et al. 2017).

17.4 How Are the Two Species Perceived?

17.4.1 The Case of the Brown Bear

The cultural perception of UA is very similar in both European and Asian cultures as well as in North American ones, with certain elements in common and other more specific characteristics (Bieder 2007; Brunner 2010).

In all the cultures under consideration, UA is striking for its resemblance to HS (its possible standing position), its violence (furore) and strength as well as courage. It was considered by primitive peoples as a god but also as a father, a brother, a son and a friend, as well as having a capacity for maternal and filial affection. UA often recalls HS, and this may bring it close to monkeys (Kipling 2011), but it differs from the latter in its virile independence or, better, as a subject that puts itself to the test alone, and this makes it stand out from Kipling's always well-illustrated 'monkey gossip', almost a caricature of our species.

UA is therefore always heroic with the mother bear being likewise: loving with children, fierce in defending them and protecting them but always alone and therefore all the more admirable.

In the course of history, UA has therefore maintained a substantially strong (Pastoureau 2002, 2008) but not a bad image (strength, daring, pride but also good-naturedness and ability to forgive). More recently, Theodore Roosevelt's 'Teddy Bear' has even become a good-natured character and a stuffed animal, but in this, he did nothing but recover what had already been expressed in ancient fairy tales, which, in any case, depend upon basic symbols (Cardini 2000).

The pseudo-anthropological connotations of UA are manifold: from the dancer at funfairs to a possible rapist of young girls and represented in a human position.

Further proof of UA's cultural success comes from heraldry, from the cities named after it (Bern and Berlin) and in many surnames and first names: Orso, Ursula, Ours, Urs, Björn, Bjarne, Bernard, etc., and even Arthur, which is derived from the Greek word 'ἄρκτος' (*Arctos).*

17.4.2 The Case of the Grey Wolf

Unlike UA, CL has undergone a profound modification of its primitive image.

That CL is seen as a valiant competitor, worthy of honour as well as respect – especially after a victory – in native populations who support themselves mainly through hunting, which makes it quite logical. While, as it is obvious, shepherds consider it an evil, treacherous marauder that should be eliminated where possible.

In ancient Rome (Ortalli 1997), despite a solid propensity of the Latins (in general) and the Romans (in particular) towards pastoralism, the grey wolf was a symbol of virility and strength by associating the species with a positive vision (the Capitoline symbol, i.e. she-wolf-suckling *Romulus* and *Remus*, the mythological founders of the city).

Moreover, CL seems to have enjoyed a sort of religious protection in Central Italy and therefore in ancient Rome. CL never appeared in games and circuses, while bears, lions, leopards and other species did. The consideration of CL was not ideological and did not lead to an effective veneration (Rissanen 2014) as much as a kind of good-natured protection. A different attitude can be encountered in the Roman areas influenced by Greek culture (Sicily, Calabria, etc.) where CL was, instead, hunted and subject to sacrifice (Rissanen 2014). The fact remains that CL was considered, together with the golden eagle, one of the two totem animals of Rome and every year, in March, the *Lupercalia* (Pastoureau 2008), in which both the god *Luperco* (a divinity of the harvests and herds) and the goddess *Luperca* (the nurse wolf) were celebrated.

With the abandonment of pagan cults and the advent of Christianity, the situation changed until the early Middle Ages, which closed 'inversely to how it opened, with a defensive start, ever more threatening towards the species' (Ortalli 1997). This thus created an accentuated hostility towards nature precisely because people realized that they had (once again) a low level of environmental control, a situation not only Italian but pan-European.

After the millennium AD, in Europe, HS resumes its demographic expansion, but the relationship with CL had now completely broken down, due once again to the influence of Christianism in which CL is perceived as an evil lord, a prototype of unrestrained bestiality as well as a sinner and a heretic (Ortalli 1997). The image is a negative one, but the danger (for human activities) is also real. Around 1100, lycanthropy (werewolf mania) broke out, and this served to strengthen the animal's negative image (wolves = informers, in Latin *lupare*). According to Dante Alighieri, who lived in an era of merchants, avarice was the greatest mortal sin, and the wolf was its symbol. The negative idea of CL was strengthened, and our current view of CL, at a European level, is what was invented in the Middle Ages (Ortalli 1997). One has only to think of the title of the Martin Scorsese's film, with one of Leonardo DiCaprio's finest performances, *The Wolf of Wall Street*.

This also occurred in the Nordic cultures without forgetting that CL was not good by definition (Fenris, the primordial wolf, ate the world; Sighinolfi 2004). Today, except for environmentalists and ecologists, CL represents absolute damage without geographical or global exemption. Even the symbology and fairy tales have assisted in this process, although the names and surnames that recall the animal itself are very numerous. For example, in Europe, Lupo, Raul, Lupe, Ulf and Rudolph; Volk, Vuk and Vukašin (Slavic languages); Bleddyn (Welsh); Farkas (Hungarian surname); Wolfgang, Wolfram and Adolf (Edel-Wolf); and even Kurt (Turkish) as well as Boris (perhaps from the Turkish Bogoris).

The negative image of CL is strengthened by the fact that it has an excellent brother, the domestic dog. CL is all the more harmful and bad compared to its brother, the useful dog, and therefore good. To this, there is the added problem of hybrids, seen as bad and like all 'half-breeds' thought to possess only the negative characteristics of their parents, that is to say aggressiveness, eternal hunger, opportunism and the ability to deceive.

As a final connotation, it should not be forgotten that CL, even if in ethological terms might be considered good, if severe a father, hunts in an often numerous pack, the effect of the pack does not do it any favours (today). The concept of the pack, in fact (rather than that of the family), makes its attacks less courageous, at the limit of cowardly, especially when compared to the solitary courage of UA, an authentic hero, while CL on its own is seen as ignoble.

UA is the solitary hero, sometimes even bad, like all things pure (the Nordic furore). It elicits emulation and virility when it doesn't elicit compassion and tenderness. It is noble.

On the contrary, CL represents the infernal mob. Brave only when in a group. He steals and sneaks by stealth. He cannot be tackled on his own and does not raise noble sentiments or even a desire for emulation, only disgust and hatred. It is cowardly.

These considerations, if considered valid, should always be born in mind when addressing the problem of communication with regard to these animals.

17.5 Assumptions

When communicating, it is necessary to know why something is being done and what the intentions are. These are completely obvious elements, but they should put us on our guard about impromptu communication. Here a dichotomy presents itself: either communication is a project shared with other specialists and the opportunities are therefore planned or promoted (and never endured) or the event is a random one but one has a certain number of scripts to which to appeal to. No improvisation, it's all already in place.

In any case, it is essential to speak honestly without, as they say, sweetening the pill or providing false information. In this specific case, communicating in regard to UA and CL and softening the truth is exactly the same as lying about them. This is true both in urban and/or rural/mountain contexts as well as, I would suggest, in convivial ones.

Saying things that are true and, when in doubt, always being sceptical, that is to say, not insistent or evasive, is therefore essential. But the problem is not only being truthful but also to appear as such and not to provide as Goffman (1959) says: 'a misleading representation, which is perceived as such' (i.e. false).

In addition to demonstrating a willingness to listen, an essential requirement is to demonstrate that one has no preconceived position. It is therefore necessary that any communication is timely but also endowed with the requirement of transparency, i.e. clarity together with an evident absence of any desire for concealment or secrecy. In other words, it is very bad to be hermetic and to give the impression that something must be kept hidden (Carotenuto and Zibordi 2016). This said, it is, however, not necessary to communicate everything!

It is clear that the expressive skills and knowledge of the communicator come into play here. But these are the prime factors of major importance, although attempting to get away with an 'I don't know' is not a good move. The capacities mentioned above do not refer so much to the usual standards as to the adequacy of the audiences' understanding, or rather to the latter's somewhat modest level, but without lowering the intervention's quality excessively.

It is evident, therefore, that the communicator must be up to the job professionally and that communicators should not improvise.

17.6 Communication as a Project

As we said, improvisation is dangerous. Communicating on UA and CL means facing, albeit differently, very sensitive topics full of tension who are never indifferent to people.

This said, there can be a modest, or, better still, a generic emotional participation when one goes on to talk about subjects that are far removed from the possibility of verification and whose effects on the public are irrelevant.

(Citizen) communication is therefore informative at the limit of entertainment, as, for example, in a case in which one goes to speak in Rome about the problems of CL management in Alaska. It is not that talking about wolves in contexts, in which one is not directly involved, is completely useless as it might be considered a sort of training of citizens' palates for the good future digestion of the need to conserve CL.

Scientific communication follows other rules, and it is not appropriate to deal with it in this work.

Neither citizen nor science communication is even particularly useful in a discussion on the conservation of the two species in question for a very simple reason: I mean that the conflict to be mitigated not only isn't there but is not even conceivable. A truly useful communication must rather take place on the battlefield, that is to say, where the problem of UA and/or CL actually exists and is an example, therefore, of where good communication is a priority.

It should be made clear that it is not true that public communication, as it has been defined, is always wrong or without results. It is happily used by both environmental and animal rights associations to convince public opinion of the importance of the two species from an iconic and sacral perspective. That said, however, the message that is conveyed subliminally, and often not very subliminally, is that in general, any management intervention of the two species, of a limiting nature, is either useless or harmful, up to and including extreme cases such as the animals' capture and subsequent radiotracking. Since the urban public (please forgive the simplification, in many countries 'far from the front line' amounts to the same) matters greatly at an electoral level, the public authorities feel entitled to discount measures that might prove unpopular, over and above their necessity. This is a typical Italian case, but we do not believe it to represent an exception in a global context.

As a conclusion on this aspect, it would also be good to communicate well, 'far from the front'. This does take place, but its effects on overall public opinion can be contradictory.

In all other situations, communication is part, and an essential one, of a project with specific objectives. Like all projects, it must obey to certain rules with clarity on its goals, analysis, timescales, costs, checks and balances, without going into more details of the entire 'to-do list'.

In addition to the aforementioned assumptions, it is suggested that communication as a project ought not have a final date. In another sense, the need to communicate is eternal and does not have a deadline because it is supposed to, and this is the same goal for the preservation of UA and CL, that it does not have an end date but should continue forever. This reflection is difficult to digest by those with the task of managing the entire project because:

- Resources are/will be limited
- After what is considered a success, the tendency is to stop pedalling and trust one's fate to destiny
- Staying alert eternally is difficult and culturally demanding as well as burdensome
- It is much more reassuring to reach a goal and face the problem once again only if forced to do so
- In this last case, the need to face the problem once again is willingly avoided when possible or in any case it does not tend to elicit the same intensity as before
- An 'eternal' project also requires new technical contributions (changes to it, insertions and deletions of policy) and the reconstitution of an efficient team, which remains a problem within the problem
- General changes in local and national society, as well as in the economy and culture, represent further difficulties to add to those listed above

A remedy for this type of problem is possible if different phases are established, with periodic revisions and an obligation to relaunch the project.

It should be reiterated that giving a final deadline to a wildlife project (and not only with regard to UA and CL) and especially in the case of specific problems such as those described represents a sound basis for failure.

'Failing to plan means you're planning to fail' (attributed to both Benjamin Franklin and Sir Winston Churchill).

In other fields, e.g. the reintroduction of 'non-problematic' ungulates such as chamois (*Rupicapra rupicapra*) (Perco 1997), the problems are the same, and, generally, the 'heroic' phase (reintroduction and success) is passed through without it being at all ready to face to any emerging issues.

Another example is the reintroduction of the red deer (*Cervus elaphus*) in the Abruzzo, Lazio and Molise National Park (ALMNP), done during 1974, which led, precisely as a result of its enormous success, to contradictory, positive and negative effects, ranging from the very high visibility of the species which represented an excellent tourist image for the park to opposition from a section of the residents (too many deer in my garden, Nimby syndrome) as well as to possible competition with the Apennine chamois (*Rupicapra pyrenaica ornata*), a specially protected species (Ferretti et al. 2015). Currently, this problem is not addressed and is left to its natural evolution, the direction of which is ignored.

It therefore needs to be reiterated that communication:

1. Should not be improvised
2. Must be planned
3. Is an essential part of the project
4. Best takes place 'on the front line'
5. Should not have a fixed term
6. Must be structured in phases

17.7 Internal, External and Organizational Communication

It is commonly believed that communication is primarily aimed at users or stakeholders and overseen by the organization or staff, which act as a project group.

However, the partners of the organization always need to define the lines of action, exchange information with each other and verify the results and review the timescales and modalities of the aforementioned actions as necessary. All this involves internal communication (Strati 2013) which is, however, only one aspect of organizational communication that will later (as well as during the project) be addressed to interested parties who are outside the self-same organization (external communication).

The ways in which the organization communicates within itself have relational and structural aspects that cannot be random but must be part of a scheme that governs the way in which they are carried out. In conclusion, internal communication also needs a plan to be able to communicate equally well with the outside world.

In the management project for the two species, two levels need to be distinguished.

The first level concerns its analysis, the second its application, which from the point of view of communication concerns making it known and enabling a certain sharing at various geographical, social, productive and cultural levels, or at least a reasonable level of knowledge, from the perspective of transparency.

It is clear that the first level can only exist at a very broad, possibly national level, albeit with diversified application throughout the country. It is therefore necessary that a national or at least regional management plan of the two species be adopted by the geographical area affected (or possibly affected) with cogent value for all the institutions and/or subjects in whom power of intervention can be recognized.

17.8 Management Plans

The management plans (and therefore communication plans), at least according to the Italian experience, can be divided into three types:

(a) ORG: the plan is drawn up by a technical organization, in a certain sense autonomous, which will apply it (EU Life project, etc.) through the planned actions. For example: Life Wolfalps, LIFE12 NAT/IT/000807 (transnational project, 2013–2018), Life EX-TRA on CL and ungulates (Protected Areas of the Central Apennines, 2009–2012, etc.). Wolfsalps is a typical example of a possible structure of an ORG plan. It consists of 12 partners: 7 protected areas (6 Italian and 1 Slovenian), 2 regions, 1 university (Ljubljana, Slovenia), 1 museum (MUSE, Trento, Italy) and Italy's State Forestry Police.
(b) AUT: the plan is drawn up directly by the institution that will apply it through its technical offices, perhaps with some external consultants. For example: the 'Piano Orso' of the Autonomous Province of Trento.
(c) DEF: The plan is drawn up by subjects other than the institution that will have to guarantee its realization (State, regional authorities, etc.). The institution is, in this sense, more or less removed from its application and does not adopt legislation with binding force to apply the plan, leaving its execution to other parties. The level of presence and initiative may vary, from a formal support (plan of conservation and management of CL in Italy, by Boitani and Salvatori (2019), commissioned by the Italian Zoological Union 2015–2017 and then blocked due to disagreement between the stakeholders and the Ministry) to a more incisive one, as, for example, that of Life Arctos (2010–2014) which had as partners four regions (one of which autonomous) and an autonomous province, two national parks, one university ('Sapienza', Rome), the Worldwide Fund for Nature (WWF Italy) and Italy's State Forestry Police. In the case of the DEF plan as well, the Authority (here the State) can instruct its own formal reference to draw it up, as in the case of the National Plan for the Conservation of the grey wolf (*Canis lupus*), formulated by Italy's Higher Institute for Protection and Environmental Research (ISPRA) in 2002. In conclusion, in a DEF hypothesis, the Authority may also promote more than an outlined plan consisting of recommendations, without any obligations towards the parties that have to apply it or without supervising its application.

With the hypothesis of an ORG plan, the advantages are, as a rule:

- Scientific accreditation
- Elasticity and immediate approaches to problems
- Competence in resolving them
- Responsibility, being able to deliver what has been promised, to the extent that the organization is entrusted with this task
- Proximity to local problems
- Reliability

On the other hand, there are some disadvantages:

- Duration (a Life project, e.g. has temporal deadlines).
- An approach too focused on research and too few resources devoted to management.
- Structural skills deficit in some fields (see above indemnities, but not only).
- Bureaucratic requirements.
- Financial coverage is sometimes inadequate.

An ORG plan must however move within the laws of the State. In a sense, a pure ORG plan is impossible because there is always the public authority to constrain it in some way.

An AUT plan, managed as it is directly, by the institution (public authority, State, region, etc.) has clear advantages. The institution puts itself forward to execute all the actions and is responsible for them. Possible/probable defects include a certain, desired, lack of transparency and self-referencing, that is to say, annoyance in the face of criticisms which the institution directly interprets as being directed at itself and for itself. Moreover, an AUT plan is, however, always a political risk for the administration and works only when the institution has adequate technical skills and available structures.

A DEF plan may have problems, characteristic of any relationship, of promotion and advice, as indeed is the case. In many cases, the plan may be stranded in foreseeable difficulties, due to the opinions of the essential stakeholders and its dependence on their influence. The theoretical advantage of the DEF plan is that once it has been prepared, the Authority (the State) applies it. In this sense as well, a DEF plan is a case that can only be recommended for decisive and well-functioning public administrations.

In the three hypotheses, there is an internal organizational communication that is substantially similar in all cases.

Only in the DEF hypothesis does the external communication have two phases, a preliminary and a final one.

The first is preliminary and technical, with insiders. These are the specialists (researchers, wildlife managers, sociologists, anthropologists, etc.). In the ORG and AUT hypotheses as well, it may be that the project staff need to acquire external opinions, but this is reduced external communication, if any, of a particular type.

The final phase is rather social and concerns all three hypotheses. It refers to stakeholders at the level of the scope of application, i.e. administrators, associations (environmentalists, hunter associations, farmers and animal breeders), together with scientific institutes not involved in the first phase, in order to proceed with the approval of the plan. This phase is the responsibility of the organization in the ORG hypothesis but also in the AUT hypothesis as well, because the organization is represented by the technical staff of the institution, even if from the perspective of its political aspects there may be certain differences.

In the DEF hypothesis, on the other hand, the organization is stripped of all power except its advisory one, which is, however, facultative.

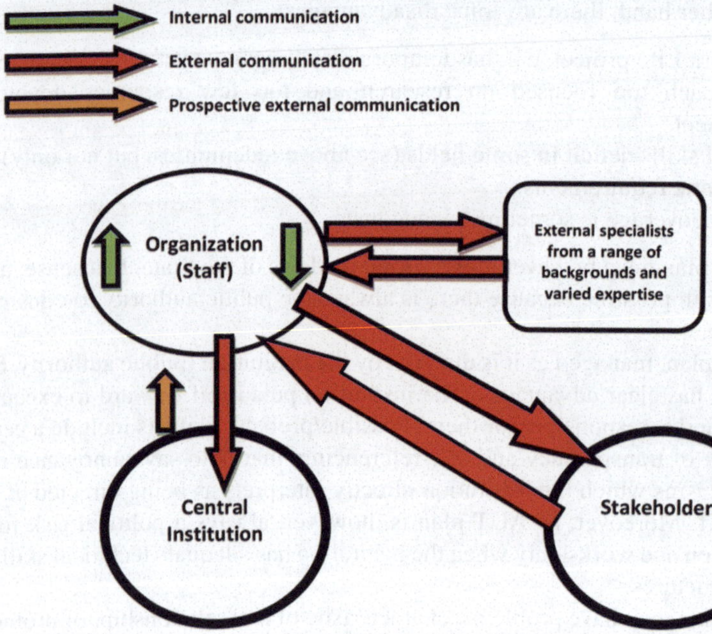

Fig. 17.3 ORG hypothesis. An organization implements the project (e.g. a Life) with central institution overseeing it

In Figs. 17.3, 17.4, and 17.5, the mechanisms and the relationship between the parts of the three possibilities inherent to the proposal and institution of projects are schematized.

17.9 Communication Problems in the DEF Hypothesis

17.9.1 The Institution

Being a piece of national planning, it is the State (or the competent Authority) that must take the initiative. But not them alone.

It should, as they say, 'stick its neck out'. In other words, the Authority, while delegating the task of the technical formulation of the plan to a group of specialists, should pre-accredit it, for example, by asserting its role, at least in the convening of the parties, and ensure its formal and organizational aspects.

If it does not, it communicates badly, and it is perceived, in this case, that the institution is cold or neutral and might even declare itself, or however to appear, in dissension with the plan.

If, however, the institution is present, making its authority felt from the beginning, it assures the parties in four fundamental ways:

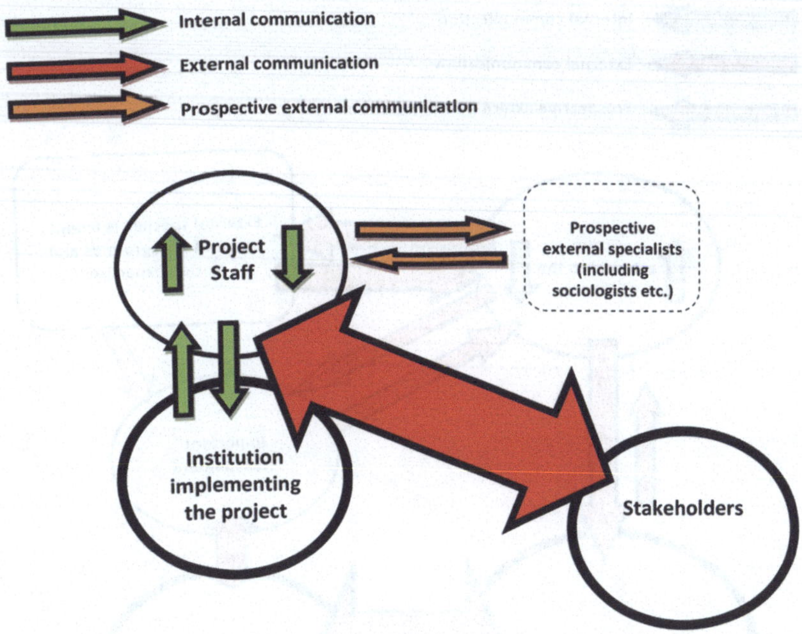

Fig. 17.4 AUT hypothesis. The institution implements the project using its own offices

1. Their determination to approve and execute the plan
2. The preconsideration of social and cultural as well as economic difficulties, which will then represent the basis for the final negotiation
3. The analysis of critical issues in the light of scientific and technical contributions of unquestionable value
4. A willingness to immediately discuss any (they are always encountered!) critical points

In doing so, the final negotiation, this time external or true, is facilitated, and in some cases, it might in some way simply become a formal moment of ratification of an already prearranged agreement.

It should be reiterated that the extraneousness of the empowered and/or perceived institution (the Ministry responsible or its offices) may allow for its contracted parties to be exposed to, scrutiny (via external negotiation) often only secondarily, excessive autonomy, which exposes them to pressures of all kinds and not only emotional (e.g. lobbying). It may cause, not only, that the plan gets buried but that it is even turned into a real 'dog's breakfast', meaning that during the real negotiations, variants can be included that may render the plan not only useless but even counterproductive.

On the contrary, good communication, in which the public authority (State or Ministry) is fully involved, may stimulate the final contractors (administrators and

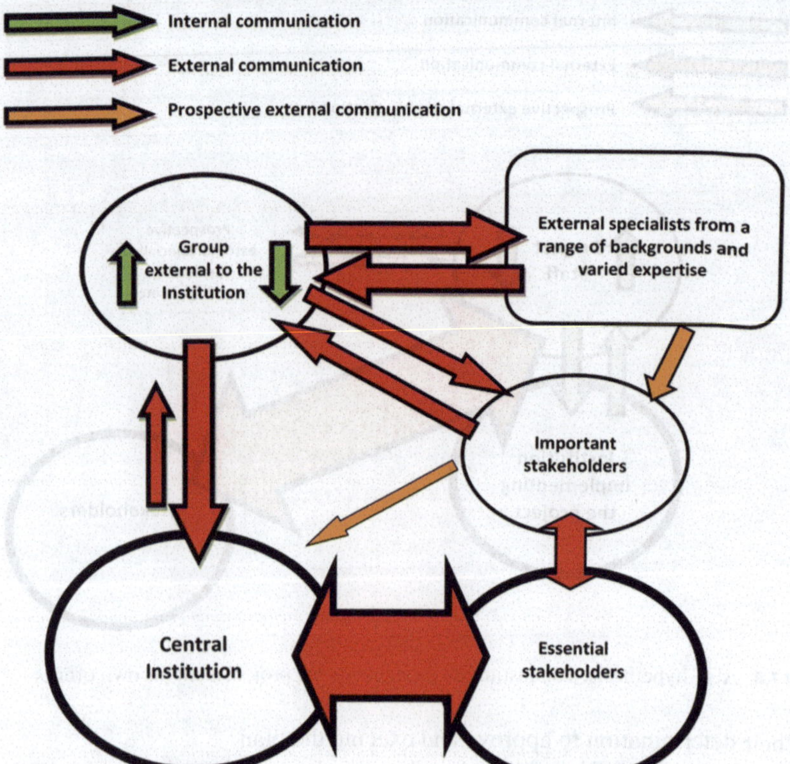

Fig. 17.5 DEF hypothesis. The central Institution entrusts the project an external group

associations) not to reject the plan a priori, as they have been involved in its development.

17.10 Technical and Social Communication

It is necessary to clearly separate the technical or preliminary communication from any social or final communication.

Best practice is to hold a certain number of extended meetings asking participants for a proposal document, based on a first draft drawn up by the project group.

Generally, it is preferable to carry out a series of extended meetings and then examine the proposals in plenary sessions several times. Another strategy might be to divide the meetings into working groups according to skills. In the latter case, the associations are also contacted and not necessarily only the technicians. Obviously, there are also mixed solutions.

The former has the advantage of being able to be worked through quickly. The second, a much longer and more elaborate process, that of measuring the availability of future contractors at arriving at something approaching a pre-agreement.

The approach chosen (plenary, working groups or mixed), which also possess paradoxical elements if there is ministerial participation as observers, has however some advantages if there are representatives of associations who are able to influence the final phase. However, this is rare, as elements of political opportunity will tend to predominate.

At the end of the meetings, the document will be presented to the competent Authority with the adhesions and non-dissonant observations of the participants, when they feel able to commit themselves. But the game is merely postponed.

This approach was followed in Italy for the formulation of the Piano di conservazione e gestione del lupo in Italia (2015–2017) for CL, currently modified and reformulated (Boitani and Salvatori 2019).

The final, or social phase in the strict sense, is very complex because of the fabric of the relationships that are a prelude to the final agreement, and therefore we are unable to examine all the permutations. The need for separate discussion tables might still apply while the technical aspects should be incorporated into the plan proposal. In this case, however, the figure of a professional communicator/facilitator, well supported by technicians, becomes important, but it is more likely that the institution proceeds directly to negotiations, with the drawbacks that this entails, more precisely, a lack of professionalism.

The delicacy of the final phase is also a result of the fact that technical and social needs are often mixed, confused and conditioned. Only an institution that is absolutely determined to adopt the plan will manage it, but if support has been half-hearted from the start, a favourable outcome is unlikely. This also derives from the fact that the decision-maker chose not to get involved at the start and therefore ignores, both rationally and emotionally, how things arrived to that point. The decision-maker (perhaps a minister, either national or regional or a top-ranking official) is then brought to convince everyone, with obvious consequences.

Conclusion: to decide well, you have to be involved in the problem, to 'have skin in the game' and not let yourself be led by others.

17.11 Communication as Negotiation

We are of the opinion that it is impossible not to have to negotiate in the case of UA and CL. But it is good to understand each other. Negotiation is not just about activities, such as, for example, with associations of livestock farmers and/or hunters, but it also means negotiating an approval or a non-refusal.

Communication acts as a prerequisite to discovery, exercising the critical sense, to emerge with fewer certainties and to enrich the desire for knowledge. It is therefore a process that has its own rules.

The rules of communication (on nature) have been well illustrated by Tosoni et al. (2016) and can be summarized as follows:

UA, CL and the Eurasian lynx (*Lynx lynx*) have ecosystemic, evolutionary, cultural, economic and existence values that may be lost. From this, it follows that, especially in this case, communication is the project and represents an integral part of it.

0. Inform in a worldly way.
1. Communicate transparently both towards the outside and within, building a relationship of trust.
2. Provide for ongoing internal training.
3. Base any communication on the 'Venice Model', that is to say, promoting communication actions as bridges to unite 'islands' (public bodies, stakeholders, political decision makers, researchers, etc.).
4. In addition, adopt new forms of communication that act on emotiveness and empathy.
5. Yes to social media as well as to a multiplicity of tools.
6. It is essential that every communication action is subject to an overall strategic plan, the latter defined in detail before the start of the project. The plan must be shared among all project partners, must include roles, skills, budgets and tools and must be periodically updated.
7. Communication must include effective indicators (e.g. SMART: specific, measurable, achievable, relevant and time-bound).
8. The communication must be set up as a continuous process.
9. Involve professionals in the field as much as possible.

Alongside this decalogue, Tosoni et al. (2016) also recommend the following:

1. Avoid the rhetoric of a nature that is good and unreal (soft toys, 'Disneyism').
2. Do not hide conflicts, damage and possible attacks.
3. Avoid being subject to the media without reacting.
4. Reject improvisation.
5. Be proactive and in any case take the initiative.
6. Do not expect to persuade the public.
7. Create a climate of trust between operators and stakeholders.
8. Create a critical sense.
9. Do not waste too much energy and resources, a priori, on people opposed to you.
10. Do not stand on one side and be impartial.

To these rules, we can add those most typical of negotiations involving wildlife (an adaptation from Fisher and Ury 1981), that is to say:

0. Each act of management should foresee a formative relapse.
1. Ensure that empathy with a category (animal rights activists, animal breeders, hunters, environmentalists, etc.) does not endanger the values of those who negotiate (communicate).
2. Do not negotiate from positions.

3. Separate people from the problem under discussion.
4. The values of others and their interests represent a major part of the problem.
5. It is good to concentrate on interests and not on positions or, worse still, on ethics.
6. Be soft with people but tough with the problem in hand.
7. Insist on objective criteria.
8. Create alternatives (for yourself) and invent solutions that are advantageous for both parties.
9. Not actively worsening the relationship with one's opponent is good for that relationship.
10. Changing things, not people or their ideas, is the goal.

There now follows some considerations, for each of the 11 points:

Undoubtedly, communication (which is the conservation plan) is an act of management (point 0), and empathy, which in the sense of Hofmann (2008) is the understanding and/or sharing of the sufferings of the different categories of stakeholder (which is not the same as saying that the suffering is real), is not dangerous (p. 1) and should not be considered as such (as if it were almost giving up what one believes in). Empathy is different from sharing.

Negotiating from positions produces badly drawn-up agreements (p. 2) because the position ends up being translated into the image that one has of oneself and renouncing it means 'losing face', above and beyond the content of the agreement. Moreover, for this reason, a separation of the person with whom we negotiate from the content of the negotiation itself (p. 3) is important, and, for the person, the adversary on this occasion, we can reserve empathy. We are dealing with a fellow human being, while it is the problem that must be solved, and conservation is the objective, not the personality (the figure) of the adversary.

If the values and interests do not coincide, it is not possible to ignore these in counterpart (p. 4). These are the problem, at least a part, and must be taken into account in order to reach an agreement.

To obtain an agreement, it is essential to understand the interests of the parties (p. 5). For example, the idea of safety and the prompt payment for damages. Understanding them and providing them without a positive response without harming our own are much better than basing oneself on certain positions (CL will not be touched, CL has to disappear) or on ethics (one doesn't negotiate with, the environmentalists are the best).

Separating the people from the problem also means having a light relationship with humans but not giving up and being strict about the problem under discussion (p. 6). Often, however, this leads to the opposite being done and being hard with people (taking offence) and soft on the problem (seeking agreement at any cost).

Objective facts and criteria apply to everyone (p. 7). We must identify them, explain them and see that they are shared. Such impartiality always ends up having extra momentum in discussions.

The strategy should not be 'I win, you lose!' (Dixit and Nalebuff 1991) but 'I win, you win!' (win-win, p. 8). Otherwise, we are vulnerable to revenge because the image that a subject has of himself has been damaged. Inventing advantageous solutions for everyone is the key, but we must also see that they are appreciated.

In negotiations, the best opponent is the one who has renounced certain negative behaviours. Improving them is therefore important, and former adversaries are often the best of allies (p. 9). This is also in accordance with points 6 and 8.

The idea of redeeming and/or converting opponents, leading them to accept our position, is very misleading. They will continue thinking that CL 'would be better if it were not' and that UA 'is stuff for the greens', but we can provide advantages for them, which, to be obtained, require virtuous behaviour (p. 10). If we honestly take care of their interests, it may even be that their values change in the future. But this is not our goal.

The objective is the conservation of UA and CL.

17.12 The Communicator

An important point concerns the figure of the communicator. When asked who should interpret this role, it is easy to answer: the person who best knows how to do it.

To identify some characteristics, however, it is useful to distinguish between the various types of communicators according to their skills (Carotenuto and Zibordi 2016):

1. Scientists and researchers
2. Wildlife managers (WM, wildlife management technicians)
3. Communication professionals
4. Amateurs, nature enthusiasts
5. Other interested parties (associations, groups of various kinds)

This list does not include officials who either fall into the second category or into the fifth which is however generic and says nothing about their level of specific knowledge and skills.

Generally, the first category is rarely very suitable, unless in possession of special skills. Scientists and researchers often have arid (scientific) expository methods and not only find themselves in difficulty in public discussions but they themselves may even feel a sense of annoyance towards the listeners, their questions and their needs. Communication is not their job.

The same could be said of the second category, which however presents many exceptions since WMs are, by profession, close to the concrete problems of management. Very often, they work and collaborate with contacts in a range of categories, a circumstance that makes them more able and more adaptable. Above all, in this case, personal attitudes and practices apply, even if it is, in any case, another type of professionalism.

The professional, the third category, is certainly more suitable and in a certain sense also the ideal figure for this purpose. The problem, however, is that a professional is often totally ignorant of the subject or has only superficial understanding of it. In this sense, it is like a driver who knows everything about the operation and performance of his vehicle but does not know where to go, how and when. A professional of this kind who works specifically within the organization is still an indispensable figure and ought to be flanked by support technicians who can supplement his or her cognitive deficiencies, obviously with the necessary tact and diplomacy, depending on the meeting in question.

Category four is the most problematic. An amateur who believes him or herself to be an expert and for various reasons has access to the media is often tempted to impose his or her particular vision of the problem or even an ideology. The purity of intent (which does not always exist) is often interpreted by others as aggression and invasiveness. The passionate amateur is in short an uncomfortable and clumsy communicator, which normally creates problems for everyone even, or above all, when they are on the right side. In this last case, it is also difficult for the communicator to blame them as the amateur is usually a figure of an interlocutor who never plays in the team. In short, they represent a subject to 'mitigate', often present in the debates, where their closeness to the team is in many cases as harmful as it is sincere. A different but similar case is the amateur in opposition, but they are easier to deal with as one is obviously not dealing with an inconvenient ally but an eminent adversary, especially if intelligent, with the exclusion of the characters clearly acting in bad faith. An intelligent and sincere opponent brings out the real problems, allowing them to be debated. Identifying them and even making use of them are appropriate.

The fifth category is essentially the public, with its strengths and weaknesses. In this case, we cannot speak of communication in the project because the public represents the counterpart, those for whom we are speaking.

A diagram on the figure of the communicator and related problems is set out in the following table:

Type of public	Main communicator	Support technician (experienced in the sector)	*Main* risks to avoid	*Preferential* thread to follow
Friends and good acquaintances	Individual and voluntary	No	Idyllic visions, jokes, hasty solutions	Noting the complexity of the problem, block extreme positions
Informal meetings, casual, with few known individuals	Individual (also sought after), may be a technician	No (it's random). Seize the opportunity!	Do-goodism, technicality, contempt for the 'non-believers'	Speaking only of what you know well, to express doubts, to sow curiosity. Showing yourself as balanced, ready to see all aspects

Type of public	Main communicator	Support technician (experienced in the sector)	*Main* risks to avoid	*Preferential* thread to follow
Administrators	Professional	One or two. Good if it is a former administrator and one who can guarantee promises	Passing over real problems. Getting involved in partisan interests	Developing the positive sides of the problem with great tact. Placing yourself as a guarantor
Stakeholders with specific interests	Professional	At most two, in the sector. No environmentalists. Good whoever can guarantee promises	Not having the necessary skills to reply effectively. Being arrogant. Giving the impression of not taking on 'their' problems	Creating personal relationships and maintaining them. Guaranteeing promised interventions.
Local communities (interested in the phenomenon)	Professional	More than one, if possible not from the are and/or too well known locally. No environmentalists.		Respecting and sharing problems emotionally even if these are 'exaggerated'. Governing the debate and emotions well
Public citizen close to the phenomenon	Professional	One or two, well-known and well-respected technicians. Environmentalists are always advised against	Taking sides. Clearly taking the side of certain groups (for or against)	Demonstrating impartiality and elegantly allowing minorities to express themselves, especially if they are unpleasant. Using humour and irony
Public citizen far from the phenomenon	Professional	One: well-known and respected technician, possibly also an environmentalist but in a subordinate position	Insulting local communities labelling them as insensitive. Speaking only for 'believers'	Stating that their problems are our problems. Being practical/realistic in the solutions proposed. Promoting social and cultural involvement

It is important to remember that communication also takes place in casual meetings as well as at a family level or with friends.

It is almost always inadvisable for environmentalists to communicate on UA and CL.

An environmentalist is understood to be the person who takes care of the environment professionally within an association of this type. Apart from personal inclinations, an environmentalist is obliged to defend the association's positions (e.g. the untouchable status of CL in Italy), and this position affects his or her ability to be convincing in a range of contexts that are not purely public gatherings and physically distant from the problems that CL management involves. If, therefore, the figure is known as such, he/she is not listened to, and the quality of any debate is worsened. It is a different situation if the environmentalist described above is taking part in public meetings as an interlocutor and not in its organization. However, it is good to bear in mind that, in various contexts, their interventions are not useful in achieving the objective. The main problem of the credibility of the affirmations of an environmentalist is that the declarations are perceived (to be underlined: 'perceived') as lacking empathy, if not with an outright ignorance and underestimation of the real problems.

By way of further clarification, it seems preferable never to resort, except in special cases, to technicians from the sector. There are professionals for this, and it is about getting them better trained in the subject.

17.13 The Italian Model

Communication must follow different orientations depending upon the national context in addition to the local one. This is not only because culture, traditions, values and therefore expectations are not homogeneous within a country (a State) but also because it is the system itself that conditions them. In a certain sense, it is a framework with which it is necessary to take account of things.

In the case of Italy, the system has particular characteristics that are not unique however and in some way are similar to certain other European nations (at least) for which these considerations might perhaps hold true. Some of these, at least those that are considered directly in this case, have been stated.

The wildlife situation is quite good overall, but with quarry species being the non-transferable property of the State, with the possibility of hunters to hunt them without the consent of landowners, the latter are excluded from their management. The environmental associations (71 recognized ones) have about half a million members and are generally quite active, with a significant contribution from animal rights sympathizers whom are both within and outside these environmental associations with their own associations.

The general picture, already summarized in Chap. 3, has generated a considerable distrust of public power. Now, since it is claimed that communication is, in this case above all, negotiation or that, in any case, the negotiating aspects of it are important, it is evident that there are obvious difficulties.

A hypothetical control of the population of CL is generally seen with total hostility, precisely because it is not believed that this control can be conducted according to appropriate rules. On the other hand, the eradication of the species is invoked at the local level, and the animal activist world is opposed to any restrictive intervention even of single problematic individuals. Poaching is rampant at least locally.

The case of Daniza is exemplary. Daniza was a female UA, reintroduced from Slovenia within the framework of the project managed by the Autonomous Province of Trento. She died as a result of anaesthesia during a capture operation carried out to transfer her into captivity due to a protocol (three attacks on people when she had cubs). Despite the fact that the operation was carried out according to the rules, the media and public opinion, not just animal rights activists, reacted with extreme outrage 'Let's sabotage Trentino, the story of a mother, a mother killed, Daniza victims of human idiocy, Who is the beast: we want justice' were just a few of the expressions bounced around in the press and on the Internet (blogs, social media and the web more generally), with thousands insults and likes for the comments in question. There were even invocations asking for a death sentence or life imprisonment (for those responsible) and all this in a situation, the one in Trentino, which can be defined as the best in Italy from the management and communication perspective (Sustersic 2016). This means that Italy 'is still unprepared to face a public discussion on complex social issues' (Sustersic 2016) regarding the subject of wildlife management.

Faced with this emotional wave, the fake news on the danger posed by the two species, on the total number of wolves, on their origin (which is said to be fraudulent, e.g. due to reintroductions rather than to natural recolonization) and on the abandonment of pastoralism and livestock breeding as a response to these 'citizen luxuries' are abound. 'On the skin of the mountain people,' they say!

To this is added the political and electoral exploitation and the paralysis of management plans at a national level, as evidenced by the blocking of the plan for conservation and management of the CL in Italy (2017).

In response to this latter case, the Autonomous Provinces of Trento and Bolzano have taken (July 2018) the initiative with special measures, to:

Provisions for the fulfilment of the obligations of the Autonomous Province of Trento deriving from Italy's membership of the European Union - European Law 2018. Implementation of the art. 16 of the directive n. 92/43 / EEC on the protection of the Alpine farming system

ART. 1. Measures of prevention and intervention involving large carnivores for the protection of the provincial Alpine farming system.

1. *In order to preserve the farming system of the province's mountain territory and in order to protect its characteristic wild fauna and flora and conserve its natural habitats, to prevent serious damage, specifically to crops, livestock farming, forests, fish stocks, water and other forms of property, to ensure the interests of public health and safety or for other imperative reasons of overriding public interest, including social or economic reasons, and for reasons that lead to positive consequences of primary importance for the environment, the President of the Province may, having obtained the opinion of the Higher Institute for Environmental Protection and Research (= ISPRA), limited to the species* Ursus arctos *and* Canis lupus, *authorize their taking, capture or the killing, provided that there is no other valid solution and that their taking does not prejudice the maintenance of the population of the species concerned in its natural area of distribution in a satisfactory state of conservation. The Autonomous Province of Trento assures the necessary information for the fulfilment of the obligations of communication of the State to the European Commission.*

17.14 Conclusions

The possibility of good communication depends not only on the capacity of the organization that communicates but on the reliability of the institution for which the organization works, in order to implement the conservation plan.

If the institution is perceived as not very reliable or is perceived in such a way during the course of the process, it is almost useless to communicate.

A positive image of the two species, an image obtained after a long struggle in some sectors that do not strictly involve the public, is usually ineffective or even harmful, at a strictly local level. And it is at the local level where acts that hamper conservation take place (e.g. above all poaching). This can then take on the characteristics of a struggle against a central power and its supporters who consider themselves vexatious and hostile (the former) as well as incompetent and snobbish (the latter).

Organizing even excellent city events in favour of UA and CL but far from the 'battlefield' has no effect on this because they also reinforce the idea, within the local institutions, that it is better to leave things more or less as they are, so as not to lose consensus. The votes of the public are numerically very significant, and communication should therefore address the institution itself, that is, its officials, but this is altogether another matter.

The result of this undesirable reluctance to engage, on the part of the institution, also communicates this. It suggests, above all to the stakeholders, that it is better to make arrangements on their own, that is, to place all their energies into obtaining some advantages for themselves, against the other stakeholders ('I win, you lose').

It is worth stressing again that both UA and CL are problematic for their symbolic value, particularly the latter (see above). This characteristic should not be underestimated, and communication must, in a certain sense, be curative but starting from the emotions. Merely taking care of the economic interests and not the fears risks breeding a generation of more astute locals who are both aware and ready to change sides if they get less than 99% (or even less than 110%) of what they have been promised.

Although in this work we have spoken almost exclusively about verbal communication in the traditional sense, that is, what happens in different types of meetings, it should be remembered that they are acts and actions as well as inaction that communicate. And in many cases in a much more effective way.

It is good communication if facts follow words, or precede them, and are, in any case, 'in time'. For example, in the event of damage, timely inspection and settlement are worth much more than the level of indemnity.

It is bad communication that makes us believe that there is nothing we can do, resigning ourselves to ineluctability. And the same is true if a species is absolutely untouchable, even if there are cases of problematic individuals, in respect of which it would be necessary to intervene in some way. In this way, communication satisfies animal rights activists, rendering the political opponents of the institution complacent, while being an anti-conservation act.

In these cases, responsibility lies with the institution, which reveals not only its unwillingness to conserve the resource but also its own weakness.

Finally, a synthesis of some priority principles are set out below, with some repetitions. It is important to:

0. *Treat the fears, not only the wallets*
1. *Communicate 'on the battlefield'*
2. *Always communicate, even at table*
3. *Not to be 'teachers' (do not be pedantic, those who teach you how to live)*
4. *Not to be goody-goodies*
5. *Predict crises before they arrive*
6. *Communicate through actions*
7. *Act immediately*
8. *Be physically present as much as possible*
9. *Always strive to negotiate*
10. *Show empathy with opponents, especially if they are intelligent*

It is likely the tenth commandment will not convince everyone.

But what's more satisfying than finding yourself in agreement with a cunning and charismatic opponent?

Acknowledgements I am particularly grateful to Giorgio Boscagli (formerly of the Sirente-Velino Regional Park and Foreste Casentinesi National Park, Italy), Mauro Fattor (GLAMM of ATit), Luca Pedrotti (the Stelvio National Park and the Autonomous Province of Trento, Italy), Stefano Filacorda (Udine University, Italy), Paolo Molinari (KORA, SCALP) Dario Febbo (formerly of the Abruzzo, Lazio and Molise National Park and Gran Sasso Laga National Park) as well as Paolo Forconi (Studio Chiros) and Corradino Guacci (S.I.Sto.Fa.'Giuseppe Altobello'), for a range of essential information.

Francesca Marucco (Life Wolfsalps), Luigi Boitani ('Sapienza' University, Rome, Italy), Piero Genovesi (ISPRA, Italy), Duccio Berzi (*Canis lupus Italia*) and Silvano Toso (formerly ISPRA, Italy) all provided important comments.

I must warmly thank Filippo Zibordi (NatCom) and Luciana Carotenuto (Tuscia University, Viterbo, Italy, and Regione Lazio) for their comments and suggestions on the paper.

A further thank you should go to Massimo Baldacci (AFV Miemo-Italy) and Guenther Mittenzwei for their reflections.

Special thanks need to go to Paul Tout who magisterially reviewed the manuscript, improving the English no end.

References

Altobello G (1921) Fauna dell'Abruzzo e del Molise. Mammiferi. IV. I Carnivori, Carnivora. Campobasso, Edizioni Colitti e Figlio. 61 p

Baroncelli F (1996) Il razzismo è una gaffe. Eccessi e virtù del 'politically correct'. Donzelli, Roma. 105 p

Bateson G, Ruesch J (1951) Communication. The social Matrix of Psychistry. WW Norton, New York, 314 p

Benazzo A, Trucchi E, Cahill JA, Delser PM, Mona S, Fumagalli M, Bunnefeld L, Cornetti L, Ghirotto S, Girardi M, Ometto L, Panziera A, Rota-Stabelli O, Zanetti E, Karamanlidis A, Groff C, Paule L, Gentile L, Vilà C, Vicario S, Boitani L, Orlando L, Fuselli S, Vernesi C, Shapiro B, Ciucci P, Bertorelle G (2017) Survival and divergence in a small group: the extraordinary genomic history of the endangered Apennine brown bear stragglers. Proc Natl Acad Sci 114:E9589–E9597. https://doi.org/10.1073/pnas.1707279114

Bieder RE (2007) Orso. Urra, Milano. XVI + 186 p

Boitani L (1984) Genetic considerations on wolf conservation in Italy. Bollettino di Zoologia 51:37–41

Boitani L (2000) Action plan for the conservation of wolves in Europe (Canis lupus); Nature and environment, No. 113

Boitani L, Salvatori V (eds) (2019) Piano di conservazione e gestione del lupo in Italia. Ministero dell'Ambiente, e della Tutela del Territorio e del Mare, ISPRA, Unione Zoologica Italiana

Boitani L, Lovari S, Vigna Taglianti A (eds) (2003) Fauna d'Italia. Mammalia III-Carnivora, Artiodactyla. Calderini, Bologna

Boscagli G (1985) Il Lupo. Lorenzini, Udine, 263 p

Boscagli G (1999) Status and management of the brown bear in Central Italy (Abruzzo). In: Servheen BC, Herrero S, Peyton B (eds) Status survey and conservation action plan IUCN/SSC Bear and Polar Bear Specialist Groups. IUCN, Gland, pp 81–84, 309 p

Brunner B (2010) Uomini e orsi. Una breve storia. Bollati Boringhieri, Torino. 224 p

Cardini F (2000) Il simbolismo dell'orso. http://www.centrostudilaruna.it/simbolismodellorso.html

Carotenuto L, Zibordi F (eds) (2016) Linee guida per la comunicazione faunistica. Edizioni Belvedere (Latina), le scienze (27) - A.T.It. Onlus, 80 p

Ciucci P, Boitani L (1998) Il Lupo. Elementi di biologia, gestione, ricerca. Documenti tecnici 23. Istituto Nazionale per la Fauna Selvatica, Ozzano dell'Emilia, 111 p

Ciucci P, Gervasi V, Boulanger J, Altea T, Boitani L, Paetkau D, Privee R, Sammarone L, Sulli C, Tosoni E (2013) Non invasive, integrated datasource survey of the core Apennine bear population (*Ursus arctos marsicanus*) in 2011. Project LifeNAT/IT/000160 'ARCTOS' - Action E3. Comincini M, La bestia feroce. Quando i lupi mangiavano i bambini nell'Italia padana. Diakronica srl, Abbiategrasso. 190 p

Ciucci P, Gervasi V, Boitani L, Boulanger J, Paetkau D, Prive R, Tosoni E (2015) Estimating abundance of the remnant Apennine brown bear population using multiple noninvasive genetic data source. J Mammal 96:206–220

Dixit A, Nalebuff B (1991) Thinking strategically. The competitive edge in business, politics and everyday life. WW Norton & Company, New York

Ferretti F, Corazza M, Campana I, Pietrocini V, Brunetti C, Scornavacca D, Lovari S (2015) Competition between wild herbivores: reintroduced red deer and Apennine chamois. Behav Ecol 26:550–559

Fisher R, Ury W (1981) Getting to Yes. Negotiating Agreement without Giving in. Penguin Group, London

Galaverni M, Caniglia R, Fabbri E, Milanesi P, Randi E (2015) One, no one, or one hundred thousand: how many wolves are there currently in Italy? One, no one, or one hundred thousand: how many wolves are there currently in Italy? Mamm Res 61:13–24

Goffman E (1959) The presentation of self in every day life. University of Edinburgh, Social Sciences Research Centre, Edinburgh, Monograph No. 2

Groff C, Bragalanti N, Rizzoli R, Zanghellini P (ed) 2013 - Rapporto Orso 2012. Servizio Foreste e Fauna, Provincia Autonoma di Trento. 56 p

Groff C, Angeli F, Asson D, Bragalanti N, Pedrotti L, Rizzoli R, Zanghellini P (eds) (2018) Rapporto Grandi carnivori 2017 del Servizio Foreste e fauna della Provincia Autonoma di Trento. 56 p

Hoffman M (2008) Empatia e sviluppo morale. Il Mulino, Milano. 370 p

Kaczensky P, Chapron G, Von Arx M, Huber D, Anfrén H, Linnel J (2013) Status management and distribution of large carnivores – bear, lynx, wolf and wolverine – in Europe. IUCN/SSC Large Carnivores Initiative for Europe (LCIE), www.lcie.org

Kipling R (2011) Il libro della giungla. Mondadori, Milano. 240 p

Latini R, Tosoni E, Gentile G, Scillitani L (2017) Conta cumulativa delle unità familiari di orso bruno marsicano per la stima della produttività della popolazione nel Parco Nazionale d'Abruzzo Lazio e Molise e Zona di Protezione Esterna Anno 2017. Rapporto 1 Servizio Scientifico, Ente Parco Nazionale d'Abruzzo, Lazio e Molise . 32 p

Lorenzini R, Posillico M, Lovari S, Petrella A (2004) Non-invasive genotyping of the endangered Apennine brown bear: a case study not to let one's hair down. Anim Conserv 7:199–209. https://doi.org/10.1017/S1367943004001301

Loy A, Genovesi P, Galfo M, Jacobone MG, VignaTaglianti A (2008) Cranial morphometrics of the Apennine brown bear (*Ursus arctos marsicanus*) and preliminary notes on the relationships with other southern European populations. Ital J Zool 75:67–75. https://doi.org/10.1080/11250000701689857

Marucco F (2014) Il Lupo. Biologia e gestione sulle Alpi e in Europa. Piviere, Gavi, 177 p

Marucco F (2016) Proceedings II Conference LIFE WolfAlps – The wolf population in the Alps: status and management, Cuneo 22nd January 2016, Project LIFE 12 NAT/IT/00080 WOLFALPS. 27 p

Mattioli L, Forconi P, Berzi F, Perco Fr (2014) Wolf (*Canis Lupus*) Population Estimate in Italy and monitoring perspectives. Poster convegno ATiT 2014, Civitella Alfedena (AQ)

Ortalli G (1997) Lupi genti culture. Uomo e ambiente nel medioevo. Einaudi, Torino, 209 p

Pastoureau M (2002) Les animaux célebres. Bonneton, Paris, 255 p

Pastoureau M (2008) L'orso. Storia di un re decaduto. Einausi, Milano. xxi + 348 p

Perco Fr (1997) Problemi gestionali delle popolazioni immesse. In: Spagnesi M, Toso S, Genovesi P (eds) Atti del III Convegno Nazionale dei Biologi della Selvaggina. Suppl. Ric. Biol. Selvaggina 27, pp 123–153

Perco Fr (2014) Le nuove relazioni tra Fauna e Uomo. Relazione presentata alla Conferenza di informazione sul progetto Life 'Ursus' e la sua evoluzione. Consiglio della Provincia Autonoma di Trento, 3/10/2014. Idem. Pubblicata per estratto (seconda parte) su Cacciare a Palla, n 9, 2015

Perco Fr, Forconi P (2016) Seasonal pattern of the wolf (*Canis lupus*) population in Italy. In: Angelici FM, Rossi L (eds) Atti del III convegno nazionale Fauna Problematica. Cesena (Italy), 24-26 novembre 2016

Randi E, Gentile L, Boscagli G, Huber D, Roth HU (1994) Mitochondrial DNA sequence divergence among some west European brown bear, (*Ursus arctos* L.) populations. Heredity 73:480–489

Rissanen M (2014) Was There a Taboo on Killing Wolves in Rome? Quaderni urbinati di cultura classica 107 (136 serie continua) 2: 125-150

Sighinolfi C (2004) I guerrieri-lupo nell'Europa arcaica. Aspetti della funzione guerriera e metamorfosi rituali presso gli indoeuropei. Il Cerchio, Homo absconditus, Milano, 104 p

Strati A (2013) La comunicazione organizzativa. Carocci, Milano, 117 p

Sustersic A (2016) (64-65) In: Carotenuto L, Zibordi F (eds), 2016. Linee guida per la comunicazione faunistica. Edizioni Belvedere, Latina, le scienze (27) - A.T.It. Onlus, 80 p

Tosoni E, Zibordi F, Salvatori V, Carotenuto L (coll) (2016) La conservazione dei grandi carnivori attraverso la comunicazione: decalogo per una comunicazione efficace sui grandi carnivori. (p 59-63). In Carotenuto L, Zibordi F (eds) (2016) Linee guida per la comunicazione faunistica. Edizioni Belvedere, Latina, le scienze (27)) A.T.It. Onlus, 80 p

Various Authors (2010) Piano d'Azione interregionale per la Conservazione dell'Orso bruno nelle Alpi centro- orientali - PACOBACE. Quad. Cons. Natura, 33, Min. Ambiente - ISPRA

Various Authors (2011) Piano d'azione nazionale per la tutela dell'Orso bruno marsicano (PATOM). Quad. Cons. Natura, 37, Min. Ambiente - ISPRA

Watzlawick P (1976) La realtà della realtà. Comunicazione Disinformazione Confusione. Astrolabio – Ubaldini, Roma
Watzlawick P (2007) Il linguaggio del cambiamento. Elementi di comunicazione terapeutica. Feltrinelli, Milano, 168 p
Watzlawick P, Beavin HJ, Jackson DD (1978) Pragmatica della comunicazione umana. Studio dei modelli interattivi, delle patologie e dei paradossi. Astrolabio, Ubaldini, Roma, 288 p
Weber M (2000) L'etica della responsabilità (with notes by Volonté P). La Nuova Italia, Firenze, 200 p
Zimen E, Boitani L (1975) Number and distribution of wolves in Italy. Z Säugetierkunde 40:102–112

Part VII
Humans and Herpetofauna

In the scenario of human-animal conflicts, the herpetofauna plays a rather delicate role.

In fact, unlike mammals and birds, few species (maybe with the exception of some sea turtles such as *Caretta caretta*) are considered "charismatic."

As a result, managing and resolving problems at various levels can be difficult (Sullivan et al. 2015; Teixeira et al. 2015).

Among reptilians and amphibians, there are also highly venomous and poisonous species potentially lethal to humans, which generates atavistic fear (LoBue and DeLoache 2008).

Perry et al. (2020) present a review of the perceived and real danger of the venomous species around the world and their role in popular traditions.

The relationships between humans and large constricting snakes are also manifold and complex. Murphy (2020) covers many fields: cases of predation on humans, hunting, breeding, and the release into the wild of alien species and their impact on habitats.

The problem of invasive herpetofauna is subsequently investigated in two case studies.

The first (Kahl et al. 2020) concerns the expansion of the brown tree snake (*Boiga irregularis*) in the island of Guam (USA) and risk assessment modeling as a tool to minimize the impact in the island and to prevent the invasion of the continental USA.

The second (Licata et al. 2020) is dedicated to the recent report of an invasive population of Asian common toad (*Duttaphrynus melanostictus*) in Toamasina, Madagascar, and stresses the importance of the synergy between research, public awareness, and policy to limit the risks related to endemic fauna unique in the world of the island.

References

Kahl SS, Henke SE, Britton D et al (2020) Risk assessment model for brown treesnake introduction into the continental United States. In: Angelici FM, Rossi L (eds) Problematic wildlife II. New conservation and management challenges in the human-wildlife interactions. Springer International, Cham, pp 603–616

Licata F, Andreone F, Freeman K et al (2020) The asian toad (Duttaphrynus melanostictus) in Madagascar: a report of an ongoing invasion. In: Angelici FM, Rossi L (eds) Problematic wildlife II. New conservation and management challenges in the human-wildlife interactions. Springer International, Cham, pp 617–638

LoBue V, DeLoache JS (2008) Detecting the snake in the grass: attention to fear-relevant stimuli by adults and young children: research article. Psychol Sci. https://doi.org/10.1111/j.1467-9280.2008.02081.x

Murphy JC (2020) Giant snake-human relationships. In: Angelici FM, Rossi L (eds) Problematic wildlife II. New conservation and management challenges in the human-wildlife interactions. Springer International, Cham, pp 581–602

Perry G, Lacy M, Das I (2020) Snakes, snakebites, and humans. In: Angelici FM, Rossi L (eds) Problematic wildlife II. New conservation and management challenges in the human-wildlife interactions. Springer International, Cham, pp 561–580

Sullivan BK, Nowak EM, Kwiatkowski MA (2015) Problems with mitigation translocation of herpetofauna. Conserv Biol. https://doi.org/10.1111/cobi.12336

Teixeira CP, Passos L, Goulart VDLR et al (2015) Evaluating patterns of human-reptile conflicts in an urban environment. Wildl Res. https://doi.org/10.1071/WR15143

Chapter 18
Snakes, Snakebites, and Humans

Gad Perry, Mark Lacy, and Indraneil Das

18.1 Introduction

Indiana Jones is a fictional intrepid archeologist who, in a series of Hollywood movies starting in the early 1980s, faced a variety of perils. He dodged bullets, faced evildoers, and escaped cunning traps set by ancient civilizations to protect assorted treasures. But in *Raiders of the Lost Ark*, he seems to meet his match: "Snakes! Why did it have to be snakes?" he rants, after dropping a torch into a chamber full of nonvenomous snakes, legless lizards, and animatronic ophidians (rest assured, he escapes intact, having achieved his mission and shown us yet again how scary snakes are). Thirty-five years later and reporting the recent scientific discovery (Dinets 2017) that Cuban boas (*Chilabothrus angulifer*) positioning themselves to hunt cave bats take into account where other snakes are located, the mass media report (McKirdy 2017) began with a similar sentiment: "Get ready to update your nightmares." Snakes consistently get a bad rap in the Western world and elsewhere, but this is not a universal viewpoint (Morris and Morris 1965; Pandey et al. 2016). How snakes are perceived is one of three main topics we cover in this chapter. We begin by updating data on snakebites around the world, treating developed countries separately from the developing world because of differences in reliability

G. Perry (✉)
Department of Natural Resource Management, Texas Tech University, Lubbock, TX, USA
e-mail: Gad.Perry@ttu.edu

M. Lacy
Department of Internal Medicine and Pediatrics, Texas Tech University Health Sciences Center, Lubbock, TX, USA
e-mail: mark.lacy@ttuhsc.edu

I. Das
Institute of Biodiversity and Environmental Conservation, Universiti Malaysia Sarawak, Kota Samarahan, Sarawak, Malaysia
e-mail: idas@unimas.my

F. M. Angelici, L. Rossi (eds.), *Problematic Wildlife II*,
https://doi.org/10.1007/978-3-030-42335-3_18

of statistics, prevalence of bites, and efficacy of treatment. We use the same separation in the next section, where we discuss the current knowledge about treatment of snakebite. Finally, we return to public perceptions and folkloristic depictions of snakes around the world.

18.1.1 Current Snake Taxonomy

Taxonomy and phylogeny are rapidly changing, with new taxa being proposed or sunk at a hard-to-follow rate. Not only are species added or removed, but the proposed relationships among them, what families they are put into, and how those are related to one another can change disturbingly frequently. To make matters even more confusing for the noninitiate, there are sometimes strong disagreements among scientists in the field, with some groups using certain names and relationships, while others adhere, just as strongly, to others. Disagreements can last for years, and snake taxonomy is among the less well understood, compared to many other taxa.

Traditionally, most venomous snakes have been placed into two families, Viperidae (Old and New World vipers and pit vipers; approximately 350 species) and Elapidae (cobras, kraits, coral snakes, and sea snakes and their allies – approximately 370 species). These two groupings are consistently upheld in various analyses, although some (e.g., Pincheira-Donoso et al. 2013) retain them as families, whereas others (e.g., Pyron et al. 2011) reduce some of these to subfamilies or provide additional divisions. Atractaspididae (mole vipers, approximately 25 species), recognized by older taxonomies such as Vidal and Hedges (2002), is reduced to subfamily status in Pyron et al. (2011) and absent in Pincheira-Donoso et al. (2013). Consequently, the exact numbers reported are of relatively little value. However, the rough numbers above, based on the frequently updated Reptile Database website (http://www.reptile-database.org/) maintained by Uetz and Hošek, are likely to be qualitatively correct. Similarly, although the names and exact rank of species groupings may change, the broad division into three main groups of venomous snakes has remained stable for quite some time. Finally, there is a rich literature on the toxins of snakes and their effects, and readers interested in additional details should examine a recent contribution, such as Mackessy (2016).

18.2 Snakebites Around the World

There is a common perception that snakebites are a frequent and dangerous phenomenon. For example, an item on WRAL, a news station in North Carolina, USA, was entitled "Snake bites common in N.C." and opened with the statement "North Carolina leads the nation in the number of people annually bitten by snakes, both venomous and nonvenomous" (Mask 2010). Similarly, a recent CBS news item was entitled "Snakebites are on the rise, and these states are the riskiest" (Rauf 2016).

"More than 1300 U.S. children suffer snakebites each year on average, with one in four attacks occurring in Florida and Texas, a new study reveals," it began. Yet such statements need to be put in perspective. In the USA, where those stories were reported, 71 individuals died from snakebite between 1950 and 1954, or 0.009 per 100,000 people (Parrish 1957). In the late 1970s, the annual number of snakebite fatalities was 9–14 (Russell 1980), and similar numbers have been reported by Chippaux (1998), some two decades later. Current numbers are about five mortalities per year in the nation of roughly 300 million people (National Institute for Occupational Safety and Health 2016). By comparison, over 600 per 100,000 died in the USA in 2014 from heart disease, 40/100,000 from car accidents, 14/100,000 from the flu and pneumonia, and 10/100,000 from firearms (National Center for Health Statistics 2015). Over the decade between 2006 and 2015, over 30 people died in the USA annually, on average, from lightning strike, according to the National Oceanic and Atmospheric Administration (2016). In that context, snakebites appear much less alarming.

Although estimated numbers for some developing countries are much higher (see below), the top 235 causes of death compiled by Lozano et al. (2012) include no animal sources. According to World Health Organization (2017a) statistics, ten causes, ranging from heart disease to road mortality, each were responsible for over one million fatalities in 2015. Over 3000 adolescents died across the world every day in 2015, with road injuries leading the fatalities at almost 10/100,000 (World Health Organization 2017b). Clearly, the obsession of popular culture with ophidians means we are disproportionally worried about snakebites. But that does not mean there is no reason to be concerned. In the mid-1900s, Swaroop and Grab (1954) estimated "the total number of snakebite deaths in the world (excluding China, the USSR, and central European countries)… to be between 30,000 and 40,000 annually." Roughly five million people are now bitten by venomous snakes each year (Kasturiratne et al. 2008; World Health Organization 2015). About half of them are envenomated, and some 100,000–125,000 die (World Health Organization 2015; Groneberg et al. 2016). The modern number is higher, despite improved availability of better treatments, for two likely reasons. First, the world population has grown tremendously, from an estimated 2.5 billion people in 1950 to over 7.6 billion today, with most of the growth occurring in the developing world. Second, current estimates, though still very approximate for some developing nations (e.g., Chippaux 1998), are based on much better data than earlier statistics. Though snakebite is not as big a health concern as heart disease or even road mortality of adolescents, it does cause considerable suffering, at least in some geographical areas and among certain population groups.

18.2.1 Developed Nations

The state of affairs in the USA is emblematic of the rest of the developed world as well, perhaps partially because such a large percentage of the population lives in urban settings and rarely spends time in snake-inhabited locations. In fact, of 1610 animal-caused fatalities recorded in the USA in recent years, some 57% were caused by nonvenomous species, with mammals leading the list and dogs standing out at 17% – much greater than the 3% reported for reptiles (Forrester et al. 2018). In Europe as a whole, Chippaux (1998, 2012) considered snakebites "relatively rare." Incidence of snakebite across the continent ranged from close to zero to about eight per 100,000, with an overall average of about 1/100,000. Mortality averaged 0.0006/100,000 (Chippaux 2012).

In Australia, Isbister et al. (2013), likewise, characterized snakebites as "rare." This evaluation is supported by data documenting 35 fatalities between 2000 and 2016, for an approximate rate of 0.0015/100,000 (Welton and Liew 2017). Deaths between 1992 and 1994 were similar, with 12 deaths reported over the 3-year period (Sutherland and Leonard 1995). In the 1940s, Swaroop and Grab (1954) reported a considerably higher but still globally low rate of 0.07 per 100,000.

In Asia, an in-hospital mortality rate of 0.2% of 1670 in-patients was reported from Japan by Yasunaga et al. (2011) – about 0.005/100,000 in the current decade, compared to 0.57/100,000 in the mid-twentieth century (Sawai 1980) and 0.13 per 100,000 in the 1940s (Swaroop and Grab 1954). Bites were often associated with agricultural or yard-related activities but, unlike in developing countries in the region, usually involved older individuals (Sawai 1998). Low snakebite frequencies were also reported from South Korea and Hong Kong, which had a mortality rate of 0.09/100,000 (Sawai 1980). The only death reported by Moore (1977) from South Korea occurred when a patient left a military facility against medical advice.

In Israel, recent numbers reported by Werner (2016) suggest few bites occur and mortalities are few: four between 1998 and 2009, under 0.05/100,000.

The only African nation identified as "developed" by the CIA (2017) is the Republic of South Africa (which is not included under the "advanced economies" list of the International Monetary Fund (CIA 2017) and is less urbanized than the countries surveyed above). The 1930s data reported by Swaroop and Grab (1954) only include the "European population" and stand at 0.57/100,000. One would assume much higher rates for other racial groups at that time, given much larger population size, greater exposure, and relatively poor medical care. One recent study from one of the more snakebite-prone regions of the country (Darryl et al. 2016) suggests snakebite incidents of roughly 6–90/100,000 and a mortality rate of about 0.28/100,000.

The developed world has a novel snakebite problem that is associated with the growing trade in non-native reptiles (Perry and Farmer 2011). In particular, European and North American hobbyists keep venomous snakes and are at times careless in their husbandry. As a consequence, Warrick et al. (2014) reported 258 envenomations by non-native snakes in North America between 2005 and 2011. Most of those bitten were young, male, and envenomated at their residence. There were few fatalities, however (Warrick et al. 2014). Schaper et al. (2009) and Valenta et al. (2014) report similar incidents from Europe.

18.2.2 Developing World

Information about snakebite frequency and outcome in the developing world is notoriously fragmentary and unreliable (Swaroop and Grab 1954; Chippaux 1998; Kasturiratne et al. 2008). Moreover, some countries, such as Brazil, are much better represented in the publication record than others (Groneberg et al. 2016). Several surveys have attempted to collate such information (Swaroop and Grab 1954; Sawai 1980; Chippaux 1998; Kasturiratne et al. 2008), and we do not intend to duplicate their efforts. Instead, we provide below updated information for each region, as well as recently reported country-specific statistics (Table 18.1) that may not have been included in older work. Although the list represents an extensive literature search, it is not likely to be comprehensive. Most developing nations do not seem to be represented in the recent (2013 or newer) peer-reviewed literature reported in Google Scholar, and we did not seek to locate data that governments may have released in other media.

Table 18.1 Recent estimates of annual snakebite incidence and resulting mortality in developing countries

Country	Bites	Bites/ 100,000	Deaths	Deaths/ 100,000	Source
Africa					
Benin			117		Habib et al. (2015)
Burkina Faso			299		Habib et al. (2015)
Cameroon			263		Habib et al. (2015)
Chad			198		Habib et al. (2015)
Cote d'Ivoire			264		Habib et al. (2015)
Gambia			31		Habib et al. (2015)
Ghana			310		Habib et al. (2015)
Guinea Bissau			24		Habib et al. (2015)
Guinea			159		Habib et al. (2015)
Liberia			37		Habib et al. (2015)
Mali			238		Habib et al. (2015)
Morocco	218	2.65	5.4	0.0–0.2	Chafiq et al. (2016)
Niger			264		Habib et al. (2015)
Nigeria			1927		Habib et al. (2015)
Senegal			192		Habib et al. (2015)
Sierra Leone			92		Habib et al. (2015)
Togo			78		Habib et al. (2015)
Asia					
Bangladesh	15,372	9.8	1709	1.22	Hossain et al. (2016)
India				4.1	Mohapatra et al. (2011)
Iran	5379	4.5–9.1	6.7		Dehghani et al. (2014a)
Iran	2586	6.9	3.5	0.36	Dehghani et al. (2014b)
Malaysia	3658				Ismail (2015)

(continued)

Table 18.1 (continued)

Country	Bites	Bites/ 100,000	Deaths	Deaths/ 100,000	Source
Sri Lanka	>80,000	398	>400	2.3	Ediriweera et al. (2016)
Taiwan	965.5	430*	0.75		Mao and Hung (2015)
Taiwan	929.4	404.9	0.4		Chen et al. (2015)
Americas					
Argentina	808	1.8	0.3	0.24	Dolab et al. (2014)
Belize	50	15.2			Gutiérrez (2014)
Bolivia		8		<4	Chippaux and Postigo 2014
Brazil	27,183				Cupo 2015
Brazil				0.05	Gutiérrez (2013)
Colombia	~4000	6–8.5		0.06–0.26	Otero-Patiño (2014)
Costa Rica				0.02–0.15	Gutiérrez (2013)
Costa Rica	600	12.9			Gutiérrez (2014)
Ecuador				0.05	Gutiérrez (2013)
El Salvador	50	0.8			Gutiérrez (2014)
Guatemala	600	4.2			Gutiérrez (2014)
Honduras	600	7.2			Gutiérrez (2014)
Nicaragua	600	10.5			Gutiérrez (2014)
Panama	2800	79.8			Gutiérrez (2014)
Panama				0.5	Gutiérrez (2013)
Venezuela				0.1-0.2	Gutiérrez (2013)

*Paper reports "4.3 cases per 100,000 person-years"

The countries and regions covered in this section generally suffer higher rates of bites and offer less effective medical treatment. Although antivenom is an effective treatment for snakebites and is readily available in most developed countries, it is often too expensive or simply unavailable in developing nations (World Health Organization 2015). Because many developing countries are located in warm zones, where snakes are more common, people are at greater risk. Moreover, developing countries tend to be poorer, with a greater percentage of the population engaged in activities such as agricultural work that bring them in closer contact with snakes (e.g., Chaves et al. 2015), perhaps with inadequate footwear (Swaroop and Grab 1954). Finally, the typically more fragile economies of developing countries often cannot support effective medical care for many conditions, including snakebite (Williams 2015). Overall, the estimates of snakebite numbers and deaths in developing countries have been fairly stable over time (Kasturiratne et al. 2008). However, the effects of climate change are likely to include increased risk of snakebite in parts of Latin America (Yañez-Arenas et al. 2016) and possibly also elsewhere.

18.2.2.1 Africa

Summarizing early 1900s data, Swaroop and Grab (1954) found it "difficult to make even an approximate estimate" for the continent but provided an overall estimate of "annual total of snakebite deaths is around 400–1,000." By the time of the review of Kasturiratne et al. (2008), African populations have grown considerably. The authors provided separate estimates for northern and sub-Saharan Africa. Some 3000–80,000 snakebite cases occur in northern Africa and the Middle East, with up to 100 fatalities (Chippaux 1998; Kasturiratne et al. 2008). This compares with about 91,000–420,000 bites and 3500–32,000 fatalities in the more populous and wetter sub-Saharan region, with West Africa and East Africa having the second-highest mortality rates in the world (Kasturiratne et al. 2008). Chippaux (1998) provided somewhat higher estimates for bites (roughly one million) and similar numbers for fatalities (about 20,000). Most bites are caused by Old World viperids and elapids.

18.2.2.2 Americas

Of the ~300,000 snakebites that occur in the developing countries of the Americas annually, almost all happen in tropical Latin America. Some 80,000–100,000 envenomations and up to 3500 deaths occur annually (Kasturiratne et al. 2008; Yañez-Arenas et al. 2016). Most bites are caused by crotalids (New World viperids), with fatalities estimated at 540–2300 (Kasturiratne et al. 2008), possibly in excess of 5000 (Chippaux 1998).

18.2.2.3 Asia

Swaroop and Grab (1954) considered Asia to be the region with the highest number of snakebite fatalities. Mortality rates in the mid-twentieth century were approximately 0.9/100,000 for Thailand, 2.7 for Burma (now Myanmar), 0.2 for Malaysia, 0.3 for Taiwan, 4.0 for Sri Lanka, 0.8 for the Philippines, and 2.1 for India (Sawai 1980). Bites were often associated with agricultural activities and typically affect males in their teens or twenties (Sawai 1998). Overall, between nearly 240,000 and 4 million bites occur in Asia each year (Chippaux 1998; Kasturiratne et al. 2008). Bites are caused by both Old World viperids and elapids, with fatalities estimated at 2800–7600 (Kasturiratne et al. 2008) or even as high as 100,000 (Chippaux 1998). South Asia, including India and nearby countries, is consistently reported to have the largest number of fatalities worldwide (Swaroop and Grab 1954; Kasturiratne et al. 2008).

In India, Shukla et al. (2017) recently reported a growing phenomenon we have yet to see described from other countries: intentional, recreational envenomation intended to elicit "euphoria, relaxation and other psychotropic effects." The snakes involved are believed to be elapids, many of which are often lethal. Such use can apparently sometimes rise to the level of addiction (Das et al. 2017).

18.2.2.4 Australasia

In developing parts of the Pacific (as defined by Kasturiratne et al. 2008), such as Papua New Guinea, about 360–4600 bites occur each year, with fatalities estimated at roughly 200–500 (Kasturiratne et al. 2008). As in Australia, bites are predominantly caused by elapids.

18.2.3 Other Dangerous Snakes

Although venomous snakes are responsible for nearly all snake-caused fatalities, attacks by large constrictors can also, albeit rarely, be lethal. Most constrictors are too small to pose any risk, but a few reach large enough size to be dangerous. In the West, most such fatalities are associated with escaped pets. For example, an escaped African rock python (*Python sebae*) "more than 10 feet (approximately 3 meters) long" killed two Canadian children (Smith and Mullen 2013). In the developing world, fatalities are more likely to result from predation in rural circumstances, as was the case for a villager found inside "a 7-meter-long python" (*Malayopython reticulatus*) in Indonesia (Anonymous 2017). In Bangkok, Thailand, the Fire and Rescue Department, which is tasked with removing snakes from homes, has taken away nearly 32,000 individual snakes in 2017 (Paddock and Jirenuwat 2017). Many of these snakes are nonvenomous. "There are only a few cases where snakes come into people's houses and hurt them," an official is quoted as saying, and urban snakes perform important pest control functions (Paddock and Jirenuwat 2017). Thus, the reaction of the public is more often a function of fear and prejudice (see Sect. 18.4), rather than real risk. Unfortunately, the ongoing growth of the urban environment (see Chap. 5) makes such encounters more frequent every year (Paddock and Jirenuwat 2017).

18.3 Clinical Facets of Snakebites

Of the 3700 species of snakes currently named, only about 10% are considered dangerously toxic to humans. Even venomous snakes sometimes inflict so-called dry bites, when no venom is injected. Such an event occurs in at least 20% of pit viper bites and an even greater number in elapids and sea snake bites (Grenvik et al. 2000). Bites without envenomation can sometimes lead to palpitations, dizziness, chest discomfort due to anxiety, and heightened sympathetic response. Two puncture wounds are generally indicative of poisonous snakebites, and nonvenomous snakes may leave an arc of small puncture wounds, but the number of puncture wounds should not be used as a definitive indication of the source of a bite.

All snake venoms are proteins, some with the capacity to cause injury. Hemotoxic venoms, most common in Old and New World viperids, may lead to rupture of red cells, damage to capillaries and blood vessel cells resulting in extravasation of plasma and cells into adjacent tissues with necrosis, and sometimes derangement of blood clotting systems. This pathology manifests within several minutes, initially as pain and swelling. Neurotoxic envenomation, usually associated with cobras and their relatives, is caused by polypeptide nonenzymatic proteins. Symptoms from these bites may begin with minimal pain and then progress to numbness, palsies, respiratory muscle paralysis, and death. Some snake venoms cause lowering of blood pressure, via several mechanisms. Some pharmaceuticals, such as the angiotensin-converting enzyme inhibitors used for the treatment of hypertension, were inspired by these venoms. The potency of toxin is not always associated with the degree of pain. For example, krait and sea snake bites may be painless, but their venoms are among the most potent in the animal kingdom. Children are more vulnerable to poor outcomes due to their smaller size and volume relative to envenomation.

There are many traditions of treating snakebite, some of which are now known to worsen rather than assist the victim. The application of electric shocks has been a popular intervention in developing countries. This is potentially hazardous, has never been shown to improve outcomes, and is not advised. What can be helpful is having the patient avoid excessive activity, immobilize the bitten extremity, and remove constrictive clothing. If the wound is on the foot, remove shoes; if on the hand, remove jewelry. Administration of alcohol or pain medications in the field is to be avoided. A low-level tourniquet to slow superficial venous and lymphatic flow may be of value, especially if applied by a medical professional. Identifying the type of snake should be attempted if this can be safely done, as it may help select an appropriate antivenin. However, snake capture is ill advised as the snake may bite again, and delivery of care may be delayed. Toxicologists do not advise using ice packs, constrictive tourniquets, or venom extractors (http://azpoison.com/venom/rattlesnakes. Accessed 6/19/17). While mechanical suction has been recommended for decades or longer, this is ineffective in removing significant amounts of venom from the wound site and probably causes focal tissue injury (Hardy 1992).

18.3.1 Developed Nations

Most snakebites in North America occur during the warmer times of the year and at night. Almost all are caused by Crotalidae (New World pit vipers), and 95% of those are caused by diamondback rattlesnakes. Copperhead bites, moderately common in the Eastern USA, have relatively weak venom and rarely lead to complications (Juckett and Hancox 2002). Less than 1% of snakebites in North America result in fatalities (Arizona Poison Control 2017). Over 1000 bites and several deaths per year are reported in Australia. Exotic venomous snakes are a trendy venture in Western countries; the pet owners are among the snakebite victims in these regions. After basic supportive care, the priority for bite management in countries with abundant resources is transfer to the nearest medical facility for definitive therapy and, depending on severity, antivenom.

18.3.2 Developing World

Most bites occur to the lower extremities in resource-limited areas of the world where agricultural workers are active in environments inhabited by snakes. Some cobras and land kraits are known to enter housing at night and bite people sleeping on the floor or ground. There is some seasonality to snakebites with increased incidence during heavy rainfall and floods. Most bites in Africa are due to vipers, puff adders, and spitting cobras (Warrell and Arnett 1976). The morbidity in Africa is difficult to estimate but is likely at least 5% of envenomations, with the puff adder causing the most deaths on that continent (Mallow et al. 2004). Some hunter-gatherer tribes in South America are also at high risk for snakebite.

The Indian subcontinent reports more snakebites than any region in the world, with thousands of annual deaths occurring, mostly in rural areas. Venom of Asian pit vipers is vasculotoxic and may have severe necrolyzing tissue effects. Krait bites often do not cause a local reaction but may cause cramping, abdominal pain, diarrhea, and collapse. Spitting cobra venom may lead to eye injury as well as systemic effects from the toxin. Following an elapid bite, the most common initial signs include vomiting, blurred vision, perioral numbness, headache, and dizziness with cranial nerve palsy resulting in ptosis. Descending paralysis has been observed in Asian patients following cobra bites. The onset of these problems may be within 15 min or delayed for more than 10 h. Depending on the venom load, paralysis of palatal, laryngeal, deglutition, and neck muscles may occur. Airway obstruction and paralysis of the respiratory muscles may ensue and result in respiratory failure (Mehta and Sahdindran 2003).

18.4 Folklore

The association between humans and snakes is an ancient one. In fact, there is evidence of fear of snakes in nonhuman primates (Lee et al. 1963; Kawai and Koda 2016). A powerful representation in many cultural artifacts, from the Mayans and Incas to contemporary art and culture, its presence has been explained by evoking a multitude of authorities, from Genesis to Freud. The views of snakes in human cultures have received extensive attention (e.g., Wake 1888; Morris and Morris 1965; Mundkur 1983), and below we provide a brief summary.

18.4.1 West/Developed World

One of the earliest depictions of snakes is the fertility god, Ningizzida from Sumeria, the first urban civilization centered in southern Mesopotamia, who was associated with vegetation and the underworld. Ningizzida later became the god of healing and is represented by a pair of snakes entwined around a rod. Greek mythology too is well stocked with snake-inspired gods, starting with the Minoan Snake Goddess (circa 1600 BCE) and including Medusa, a winged female with venomous snakes serving as hair. Of more relevance to us is Asclepius (Aesculapius in Latin), reputed as a healer and, according to some legends, a son of Apollo and therefore a demi-god, especially after the retreat of plague in 420 BCE (Nayernouri 2010). A parallel can be found in the caduceus (a herald, in the shape of two snakes entwined on a short rod, set under a pair of wings and carried by Hermes). The Rod of Asclepius or staff of knowledge (and/or the caduceus of Hermes, the messenger of the gods) serves as the alternate symbol of the medical profession. Associated at once with sin and death as well as rejuvenation and resurrection, the healing symbol can be traced from ancient Middle Eastern cultures (including Hebraic, passing on to Christianity and Islam), whose arrival subsumed many pagan beliefs. It has even been argued that the rod has been replaced by the cross and the serpent by Christ himself (Antoniou et al. 2011).

The source of much of mediaeval natural history is the work of the Roman encyclopedist, Pliny the Elder (23–79 A.D.), entitled *Natural History* (in 37 volumes, the first 10 of which were published in 77 A.D. and the remainder after his death). Information on snakes therein can be imaginative. For example, snakes of India were reported to be large enough to swallow elephants, and some grew to 80 and 140 cubits (1 cubit = 46–56 cm [18–22 inches]) in length (Rackham et al. 1967–1971). In Pliny's defense, early Greek natural history was probably primarily written to entertain a Greek audience. Huge snakes, nonetheless, abound in multiple mythologies (Fig. 18.1).

Fig. 18.1 The sea serpent was perhaps the European sailor's ultimate terror in the Middle Ages. The Swedish writer, Olaus Magnus (1490–1557), in his *Historia de Gentibus Septentrionalibus* (published in 1555, from Rome), depicted what an encounter may have looked like in the minds of seafarers

Fig. 18.2 (**a**) The biblical creation story of Adam, Eve, and the snake in the Garden of Eden is depicted on the wall of Abreha Atsbeha (also known as Abreha we Atsbeha), a rock-hewn church in the Tigray Region of northern Ethiopia. (**b**) Surprisingly, the story of Saint George slaying the dragon is depicted throughout Ethiopia, in this case at the Betre Maryam Church on Lake Tana in the Rift Valley

At the core of much of the animus toward snakes in the west, however, is the role of the snake in the creation mythology of the Judeo-Christian-Muslim tradition (Fig. 18.2). With minor variations related to language and sectarian tradition is the story of how the snake helped tempt the first woman into the first sin, resulting in the casting of humanity from paradise, the loss of the snake's legs, and an eternal enmity placed between snakes and humans by God himself. Even the Bahá'í religion, a relatively modern derivation which generally preaches kindness to animals, teaches that "harmful animals, such as the bloodthirsty wolf, the poisonous snake and other injurious animals are excepted, because mercy towards these is cruelty to man, and other animals" (Bahá'í Reference Library 1976).

18.4.2 *East/Developing World*

Snakes have been object of worship ("ophiolatry") from the earliest time in the Indian subcontinent, and the same reverence can be witnessed in many parts of India to this day. The cobra, in particular, is often associated with Shiva, one of the major gods in the Hindu pantheon (Smith 1996). Other parts of Asia, especially where Buddhism is practiced, hold the multiheaded cobra in high esteem. After all, wasn't the Buddha sheltered by one such snake while he was meditating? Such behavior among the faithful can be seen in Sri Lanka and Burma and even Japan (Kelsey 1981). Archaeological evidence from the Jomon period (ca. 2000 BC) supports the worship of snakes in ancient Japan, where the ability of snakes to shed their skin was construed as proof of rebirth (Bérczi et al. 2001). Ophiolatric practices can be encountered in many non-Western populations worldwide, on all continents. They are presumably driven by a mixture of fear and wonderment in humans who share landscapes with these animals (Figs. 18.2 and 18.3).

Also important in Asia is trade supplying the culinary business of medicine and nutrition. The "She Gong" (snake soup, Cantonese style), in particular, remains a popular offering in specialty restaurants in Hong Kong, as are snake-based drugs of Chinese traditional medicine (Read 1934). Snakes preserved in "wine" are common in Eastern Asia's many roadside pharmacies (Fig. 18.4; Somaweera and Somaweera 2010).

Apep, the great serpent of the ancient Egyptians, was a powerful adversary of the sun god and was also hostile to humans. In fact, Egyptians may have been the earliest to have employed charms and incantations to keep their dead safe from snakes (Budge 1904). Further south, the San people of southern Africa associate snakes with rainfall and classify them as rain animals (Lewis-Williams and Pierce 2004). The puff adder (*Bitis arietans*) is one such species, associated not only with rainfall and water but also with certain altered states of consciousness.

Overall, although snakes are often feared for their lethal potential in the East, and in many places in the developing world are killed on a regular basis, there is nonetheless a more nuanced view of serpents outside the West.

Fig. 18.3 Image from Hambley (1931. Field Mus. Nat. Hist. Anthropol. Ser. 21), showing temple for python worship in the Kingdom of Dahomey (corresponding to modern-day state of Benin, West Africa)

Fig. 18.4 Bottles of "snake wine" confiscated by customs agents on the island of Guam in the Pacific. The contents included a variety of venomous and nonvenomous snakes, as well as lizard, plants, and other organisms

18.4.3 Eating Snakes

Snakes are eaten in many cultures (Klemens and Thorbjarnarson 1995). However, studies in China (Wang et al. 2014) and elsewhere indicate that snakes can harbor parasites that can sicken people. For example, eating snakes was officially prohibited by the Korean army in the 1980s because of resulting intestinal parasite infections (Huh et al. 1994).

A poorly covered cultural phenomenon is consumption of snakes and/or snake blood by soldiers (Fig. 18.5). Military forces worldwide have used ophiological examples for combat training, fitness, and nutrition for survival (You et al. 2013; Deuster et al. 2013). Since snakes are often perceived as fearsome, associating with them, and especially controlling or killing them, continues to be used as a sign of being fearsome. Although the connection between snakes and violence in popular culture is well established (Morris and Morris 1965, pp. 104–114), only one short note (Werner 1988) appears to have ever focused on public snake eating by military forces, presumably as an intimidation tool.

Fig. 18.5 Soldiers using mastery of snakes to prove their fierceness in screen captures of stories from around the globe. (**a**) A video (available at https://www.youtube.com/watch?v=nBJj1hJf5VI) uploaded by the *Times of Oman* shows Indonesian soldiers handling and drinking the blood of cobras in front of US Defense Secretary Jim Mattis in 2018. (**b**) US marines drinking cobra blood in an exercise, part of a video (https://www.youtube.com/watch?v=rvx9scwRQUM) uploaded by the AFP News Agency in 2018. (**c**) A Lebanese commando shown in the Daily Mail (Wyke 2015). (**d**) A Pakistani commando preparing a snake for consumption on a video (https://www.dailymotion.com/video/x2y61cp) originating from *Express News* in Pakistan

18.5 Summary and Conclusions

A brief overview of global perception of snakes shows no fixed pattern of good versus evil in European-based cultures (European, modern Australasia, and North America) against Eastern or those of the rest of the world (South America, Africa, Asia, and Australasia). Ancient cultures in all regions have often treated snakes essentially as of positive influence. However, religious narratives have sometimes resulted in alternative perceptions in their followers, leading to disparate reactions (ranging from fear and loathing to reverence and worship). For example, Pope Francis recently "denounced fake news as evil, comparing it to the snake in the Garden of Eden, and urged journalists to make it their mission to search for the truth" (Anonymous 2018). Traditional lifestyles, typically steeped in religious belief, continue to decline in the face of globalization, and so do many habits and practices that concern conservation, often under the generic phrase "respect of nature." Erosion of such traditions has been associated with failing conservation action in numerous cases worldwide, and we are aware of one specific to snakes in Japan (see Sasaki et al. 2010). Thus, the study of human-snake associations remains relevant for effective conservation plans for endangered species.

Popular perceptions of snakes as a common source of mortality are not accurate for today's developed world. In Europe, North America, and a few other regions, venomous snakes rarely come in contact with increasingly urbanized people. When bites occur, rapid access to antivenin tremendously reduces mortality rates. In the developing world, however, and particularly in parts of Africa and Asia, rural populations regularly come into contact with a wide variety of venomous snakes, some of them potentially lethal. Regular contact is compounded by less access to medical care in general and to antivenin in particular, resulting in much greater mortality rates. Even in those areas, however, snakebite is a much less frequent cause of health concern compared to many other problems, some traditional and some recent, which do not receive quite as much popular attention.

Acknowledgments We thank H.B. Lillywhite and M. Zuffi for helpful comments on earlier versions of this chapter. ID thanks Universiti Malaysia Sarawak for support.

References

Anonymous (2017) Indonesian man found dead in belly of 7m-long python. The Jakarta Post. Available on line at http://www.thejakartapost.com/news/2017/03/29/indonesian-man-found-dead-in-belly-of-7m-long-python-.html. Accessed 8 Oct 2017

Anonymous (2018) Pope Francis compares fake news to snake in Garden of Eden. The Guardian, 24 January 2018. Accessed at https://www.theguardian.com/world/2018/jan/24/pope-francis-fake-news-snake-garden-of-eden on 5 Dec 2018

Antoniou SA, Antoniou GA, Learney R, Granderath FA, Antoniou A (2011) The rod and the serpent: history's ultimate healing symbol. World J Surg 35:217–221

Arizona Poison Control Center (2017). http://azpoison.com/venom/rattlesnakes. Accessed 19 June 2017

Bahá'í Reference Library (1976) Available via http://reference.bahai.org/en/t/c/BWF/bwf-72.html. Accessed 26 Sept 2017

Bérczi S, Sano O, Takaki R (2001) Snake patterns in Eurasia/Japan and their implications. Forma 16:279–287

Budge EAW (1904) The gods of the Egyptians or studies in Egyptian mythology, vol Vol. I. Methuen & Co., London. xxiii + 525 pp

Chafiq F, El Hattimy F, Rhalem N et al (2016) Snakebites notified to the poison control center of Morocco between 2009 and 2013. J Venomous Anim Toxins Incl Tropl Diseases 22:8. https://doi.org/10.1186/s40409-016-0065-8

Chaves LF, Chuang T-W, Sasa M, Gutiérrez JM (2015) Snakebites are associated with poverty, weather fluctuations, and El Niño. Sci Adv 1:e1500249

Chen CK, Lin CC, Shih FY et al (2015) Population-based study of venomous snakebite in Taiwan. J Acute Med 5:38–42

Chippaux J-P (1998) Snake-bites: appraisal of the global situation. Bull World Health Org 76:515–524

Chippaux J-P (2012) Epidemiology of snakebites in Europe: a systematic review of the literature. Toxicon 59:86–99

Chippaux J-P, Postigo JR (2014) Appraisal of snakebite incidence and mortality in Bolivia. Toxicon 84:28–35

CIA (Central Intelligence Agency) (2017) The world factbook: developed countries (DCs). Available via https://www.cia.gov/library/publications/the-world-factbook/appendix/appendix-b.html. Accessed 9 May 2017

Cupo P (2015) Bites and stings from venomous animals: a neglected Brazilian tropical disease. Rev Soc Bras Med Trop 48:639–641

Darryl W, Sartorius B, Hift R (2016) Estimating the burden of snakebite on public hospitals in KwaZulu Natal, South Africa. Wilderness Environ Med 27:53–61

Das S, Barnwal P, Maiti T et al (2017) Addiction to snake venom. Subst Use Misuse 52:1104–1109

Dehghani R, Fathi B, Shahi MP, Jazayeri M (2014a) Ten years of snakebites in Iran. Toxicon 90:291–298

Dehghani R, Dadpour B, Mehrpour O (2014b) Epidemiological profile of snakebite in Iran, 2009–2010 based on information of Ministry of Health and Medical Education. Int J Med Toxicol Forensic Med 4:33–41

Deuster PA, Kemmer T, Tubbs L, Zeno S, Minnick C (2013) The Special Operations Forces nutrition guide. US States Special Operations Command, Homestead Air Reserve Base, Florida

Dinets V (2017) Coordinated hunting by Cuban boas. Anim Behav Cognition 4:24–29

Dolab JA, de Roodt AR, de Titto EH et al (2014) Epidemiology of snakebite and use of antivenom in Argentina. Trans R Soc Trop Med Hyg 108:269–276. https://doi.org/10.1093/trstmh/tru038

Ediriweera DS, Kasturiratne A, Pathmeswaran A et al (2016) Mapping the risk of snakebite in Sri Lanka – a national survey with geospatial analysis. PLoS Negl Trop Dis 10:e0004813. https://doi.org/10.1371/journal.pntd.0004813

Forrester JA, Weiser TG, Forrester JD (2018) An update on fatalities due to venomous and nonvenomous animals in the United States (2008–2015). Wild Environ Med 29:36–44

Grenvik AKE, Ayers SM, Holbrook PR, Shoemaker WC (eds) (2000) Injuries by venomous and poisonous animals. In: Textbook of critical care, Philadelphia: WB Saunders, 4th edn. pp. 224–233

Groneberg DA, Geier V, Klingelhöfer D et al (2016) Snakebite envenoming – a combined density equalizing mapping and scientometric analysis of the publication history. PLoS Negl Trop Dis 10:e0005046. https://doi.org/10.1371/journal.pntd.0005046

Gutiérrez JM (2013) Snakebite envenoming in Latin America and the Caribbean. In: Gopalakrishnakone P et al (eds) Clinical toxinology. Springer, Dordrecht, pp 1–20

Gutiérrez JM (2014) Current challenges for confronting the public health problem of snakebite envenoming in Central America. J Venomous Anim Toxins Incl Trop Dis 20:7. https://doi.org/10.1186/1678-9199-20-7

Habib AG, Kuznik A, Hamza M et al (2015) Snakebite is under appreciated: appraisal of burden from West Africa. PLoS Negl Trop Dis 9:e0004088. https://doi.org/10.1371/journal.pntd.0004088

Hardy DL (1992) A review of first aid measure for ptiviper bite in North America with an appraisal of Extractor™ suction and stun gun electroshock. In: Campbell JA, Brodie ED (eds) Biology of pit vipers. Selva, Tyler, pp 405–414

Hossain J, Biswas A, Rahman F et al (2016) Snakebite epidemiology in Bangladesh – a national community based health and injury survey. Health 8:479–486

Huh S, Lee SU, Huh SC (1994) A follow-up examination of intestinal parasitic infections of the Army soldiers in Whachon-gun, Korea. Korean J Parasitol 32:61–63

Isbister GK, Brown SGA, Page CB (2013) Snakebite in Australia: a practical approach to diagnosis and treatment. Med J Aust 199:763–768

Ismail AK (2015) Snakebite and Envenomation Management in Malaysia. In: Gopalakrishnakone P et al (eds) Clinical toxinology in Asia pacific and Africa. Springer, Dordrecht, pp 71–102

Juckett G, Hancox J (2002) Venomous Snakebites in the United States. Am Fam Physician 65:1367–1375

Kasturiratne A, Wickremasinghe AR, de Silva N et al (2008) Estimating the global burden of snakebite: a literature analysis and modelling based on regional estimates of envenoming and deaths. PLoS Med 5(11):e218. https://doi.org/10.1371/journal.pmed.0050218
Kawai N, Koda H (2016) Japanese monkeys (*Macaca fuscata*) quickly detect snakes but not spiders: evolutionary origins of fear-relevant animals. J Comp Psychol 130:299–303
Kelsey WM (1981) Salvation of the snake, the snake of salvation: Buddhist-Shinto conflict and resolution. Japanese J Relig Stud 8:83–113
Klemens MW, Thorbjarnarson JB (1995) Reptiles as a food resource. Biodivers Conserv 4:281–298
Lee RW, Ordy JM, Dillman A (1963) Monkeys' fear of snakes: a study of its basis and generality. J Genetic Psych 103:207–226
Lewis-Williams JD, Pierce DG (2004) Southern African San rock paintings as social intervention. A study of rain-control images. Afr Arch Rev 21:199–228
Lozano R, Naghavi M, Foreman K et al (2012) Global and regional mortality from 235 causes of death for 20 age groups in 1990 and 2010: a systematic analysis for the Global Burden of Disease Study 2010. Lancet 380:2095–2128
Mackessy SP (ed) (2016) Handbook of venoms and toxins of reptiles. CRC Press, Boca Raton
Mallow D, Ludwig D, Nilson G (2004) True vipers: natural history and toxinology of old world vipers. Krieger Publishing Company, Malabar
Mao YC, Hung DZ (2015) Epidemiology of snake envenomation in Taiwan. In: Gopalakrishnakone P et al (eds) Clinical toxinology in Asia pacific and Africa. Springer, Dordrecht, pp 3–22
Mask A (2010) Snake bites common in N.C. WRAL.com, 7 July 2010. Available via http://www.wral.com/lifestyles/healthteam/story/7918182/. Accessed 15 May 2017
McKirdy E (2017) Wait, what? Scientist discovers snakes that hunt in packs. CNN, 26 May 2017. Available via http://www.cnn.com/2017/05/26/americas/cave-snakes-hunt-in-packs-trnd/index.html. Accessed 26 May 2017
Mehta SR, Sahdindran VK (2003) Clinical features and management of snake bite. Med J Armed Forces India 58:247–249
Mohapatra B, Warrell DA, Suraweera W et al (2011) Snakebite mortality in India: a nationally representative mortality survey. PLoS Negl Trop Dis 5:e1018. https://doi.org/10.1371/journal.pntd.0001018
Moore TC (1977) Snakebite from the Korean pit viper. Military Med 142:546–549
Morris R, Morris D (1965) Men and snakes. McGraw-Hill, New York
Mundkur B (1983) The cult of the serpent: an interdisciplinary survey of its manifestations and origins. State University of New York Press, Albany
National Center for Health Statistics (2015) Health, United States, 2015: with special feature on racial and ethnic health disparities. National Center for Health Statistics, Hyattsville
National Institute for Occupational Safety and Health (2016) Venomous snakes. Available via https://www.cdc.gov/niosh/topics/snakes/. Accessed 9 May 2017
National Oceanic and Atmospheric Administration (2016) Natural hazard statistics. Available via http://www.nws.noaa.gov/om/hazstats.shtml. Accessed 9 May 2017
Nayernouri T (2010) Asclepius, caduceus, and simurgh as medical symbols part I. Arch Iranian Med 13:61–68
Otero-Patiño R (2014) Snake bites in Colombia. In: Gopalakrishnakone P, Faiz SMA, Gnanathasan CA (eds) Clinical toxinology. Springer, Dordrecht, pp 1–43
Paddock RC, Jirenuwat R (2017) Lurking in toilets, swimming the streets: snakes of Bangkok Move In. New York Times, 28 November, 2017. Available at https://www.nytimes.com/2017/11/28/world/asia/bangkok-snakes.html. Accessed 1 Dec 2017
Pandey DP, Pandey GS, Devkota K, Goode M (2016) Public perceptions of snakes and snakebite management: implications for conservation and human health in southern Nepal. J Ethnobiol Ethnomed 12:22. https://doi.org/10.1186/s13002-016-0092-0
Parrish HM (1957) Mortality from snakebites, United States, 1950–54. Public Health Rep 72:1027–1030

Perry G, Farmer M (2011) Reducing the risk of biological invasion by creating incentives for pet sellers and owners to do the right thing. J Herpetol 45:134–141

Pincheira-Donoso D, Bauer AM, Meiri S, Uetz P (2013) Global taxonomic diversity of living reptiles. PLoS One 8(3):e59741

Pyron RA, Burbrink FT, Colli GR et al (2011) The phylogeny of advanced snakes (Colubroidea), with discovery of a new subfamily and comparison of support methods for likelihood trees. Molec Phylog Evol 58:329–342

Rackham W, Jones HS, Eicholz DE (English translation 1967–1971): Pliny, natural history. Cambridge University Press, Cambridge, MA/London. 10 vols

Rauf D (2016) Snakebites are on the rise, and these states are the riskiest. CBS News, October 20, 2016, Available via http://www.cbsnews.com/news/snakebites-kids-riskiest-states-florida-texas/

Read B (1934) Chinese Materia Medica. Dragon and snake drugs. Peking Nat Hist Bull 4:297–357

Russell FE (1980) Snake venom poisoning in the United States. Ann Rev Med 31:247–259

Sasaki K, Sasaki Y, Fox S (2010) Endangered traditional beliefs in Japan: influences on snake conservation. Herpetol Conserv Biol 5:474–485

Sawai Y (1980) Studies on snakebites in the Asian areas. In: Eaker D, Wadsröm T (eds) Natural toxins: proceedings of the 6th international symposium on animal, plant, and microbial toxins, Uppsala, August 1979. Pergamon, Oxford, pp 25–32

Sawai Y (1998) Venomous snakes and snakebite treatment in Asia. Russ J Herpetol 5:103–112

Schaper A, Desel H, Ebbecke M et al (2009) Bites and stings by exotic pets in Europe: an 11 year analysis of 404 cases from Northeastern Germany and Southeastern France. Clin Toxicol 47:39–43

Shukla L, Reddy SS, Kandasamy A, Benegal V (2017) What kills everyone, gives a high for some – recreational snake envenomation. Asian J Psychiatr 25:106–108

Smith D (1996) The dance of Siva. Religion, art and poetry in South India. Cambridge University Press, Cambridge, 191 pp

Smith M, Mullen J (2013) Snake kills two boys during sleepover, Canadian police say. CNN. Available on line at http://www.cnn.com/2013/08/06/world/americas/canada-snake-deaths/index.html. Accessed 8 Oct 2017

Somaweera R, Somaweera N (2010) Serpents in jars: the snake wine industry in Vietnam. J Threatened Taxa 2:1251–1260

Sutherland SK, Leonard RI (1995) Snakebite deaths in Australia 1992–1994 and a management update. Med J Austr 163:616–618

Swaroop S, Grab B (1954) Snakebite mortality in the world. Bull Wld Hlth Org 10:35–76

Valenta J, Stach Z, Michalek P (2014) Exotic snake bites in the Czech Republic – epidemiological and clinical aspects during 15-year period (1999 – 2013). Clin Toxicol 52:258–264

Vidal N, Hedges SB (2002) Higher-level relationships of caenophidian snakes inferred from four nuclear and mitochondrial genes. C R Biol 325:987–995

Wake CS (1888) Serpent-worship, and other essays. George Redway, London

Wang F, Li W, Hua L et al (2014) Spirometra (*Pseudophyllidea*, Diphyllobothriidae) severely infecting wild-caught snakes from food markets in Guangzhou and Shenzhen, Guangdong, China: implications for public health. Sci World J 2014:1–5

Warrell DA, Arnett C (1976) The importance of bites by the saw-scaled or carpet viper (*Echis carinatus*). Epidemiological studies in Nigeria and a review of the world literature. Acta Trop (Basel) 33:307–341

Warrick BJ, Boyer LV, Seifert SA (2014) Non-native (exotic) snake envenomations in the US, 2005–2011. Toxins 6:2899–2911

Welton RE, Liew D (2017) Incidence of fatal snake bite in Australia: a coronial based retrospective study (2000–2016). Toxicon 131:11–15

Werner YL (1988) Snakes die to prove manhood of commando girls. British Herp Soc Bull 24:34–35

Werner YL (2016) Reptile life in the land of Israel. Edition Chimaira, Frankfurt am Main

Williams DJ (2015) Snake bite: a global failure to act costs thousands of lives each year. BMJ 351:h5378. https://doi.org/10.1136/bmj.h5378
World Health Organization (2015) Snake antivenoms. Available via http://www.who.int/mediacentre/factsheets/fs337/en/. Accessed 14 May 2017
World Health Organization (2017a) The top 10 causes of death. Available via http://www.who.int/mediacentre/factsheets/fs310/en/. Accessed 9 May 2017
World Health Organization (2017b) Adolescents: health risks and solutions. Available via http://www.who.int/mediacentre/factsheets/fs345/en/. Accessed 16 May 2017
Wyke T (2015) Lebanese commandos grit their teeth and eat LIVE SNAKES as they display their strength and daring during a security convention in Beirut – but don't look like they're enjoying it. Daily Mail, 9 September 2015. Accessed at https://www.dailymail.co.uk/news/article-3228067/Lebanese-commandos-grit-teeth-eat-LIVE-SNAKES-display-strength-daring-security-convention-Beirut-don-t-look-like-enjoying-it.html on 5 Dec 2018
Yañez-Arenas C, Peterson AT, Rodríguez-Medina K, Barve N (2016) Mapping current and future potential snakebite risk in the new world. Clim Chang 134:697–711
Yasunaga H, Horiguchi H, Kuwabara K et al (2011) Venomous snake bites in Japan. Am J Trop Med Hyg 84:135–136
You CW, Lin YP, Lai YW, Chen YL, Tang Y, Chou SP, Chang HY, Zappalorti RT, Lin SM (2013) Return of the pythons: first formal records, with a special note on recovery of the Burmese python in the demilitarized Kinmen Islands. Zool Stud 52:8

Chapter 19
Giant Snake-Human Relationships

John C. Murphy

19.1 Introduction

On a warm October day in suburban Chicago, crowds gathered at a convention center to see and buy captive-bred reptiles. Snakes, lizards, and turtles were for sale by more than 100 vendors, and thousands of people walked through the trade show during the weekend. One booth attracted more people than any other, Gaspar Reptiles. Under the tabletop that displayed a variety of captive-bred hatchling pythons was Twiggy. As people discovered the 6.2 m, 100 kg reticulated python, facial expressions in the crowd shifted from curiosity to fascination. However, there was the rare facial expression that revealed a different emotion, terror. It came when the person realized they were looking at an animal that had the ability to kill and swallow a human. Hominins shared much, if not all, of their evolutionary history with the African pythons *Python natalensis* and *Python sebae*, two of the largest extant snakes. A Pleistocene site in the Buia area of Eritrea contained remains of both an early hominin and a python (Delfino et al. 2004). The python vertebrae were about 17 mm long, and *Python sebae* can have as many as 293 vertebrae in their body, and another 82 smaller tail vertebrae. Thus, it seems likely that the Buia python exceeded 5 m. The hominin present was *Homo erectus*, a species that attained 185 cm in maximum height and a maximum weight of 68 kg (Anton 2003). While this size and weight may be reached by some individuals, they are maximums, and most *H. erectus* would have been smaller as children, adolescents, or even adults. Weights reported for the African python are in the 55–65 kg range and may be greater (Pitman 1974). Thus, hominins and giant snakes were inhabiting the same biotic communities since the appearance of the hominins in the fossil record.

As hominins moved out of Africa and into Asia, they encountered other large serpents. *Homo ergaster/erectus* fossils from temperate China suggest it had

---

J. C. Murphy (✉)
Science & Education, Field Museum, Chicago, IL, USA

F. M. Angelici, L. Rossi (eds.), *Problematic Wildlife II*,
https://doi.org/10.1007/978-3-030-42335-3_19

colonized the area by 1.63 million years ago (MYA) (Zhu et al. 2015). The emigration would likely have taken *H. erectus*, and later *Homo sapiens*, into the geographic ranges of the Burmese python, *Python bivittatus*, and the reticulated python, *Malayopython reticulatus*. While the maximum size of these two snakes remains controversial, both species have been reported to prey on humans (Headland and Greene 2011; Wall 1921).

The body proportions of early hominins were significantly different from those of modern humans. Humans evolved from small arboreal species to larger terrestrial species. By 1.5 MYA, body proportions, together with body size, in *H. ergaster/erectus* were well within the range of modern human variation. From this date on, a terrestrial lifestyle can be inferred, with changes in body shape viewed as adaptations to environmental variables, particularly climate. Hominins display a considerable variation in body sizes. This may vary within and between populations. Individuals with smaller body sizes may be more likely to be considered prey by large snakes (Ruff 2002).

Boas and pythons are powerful constrictors, using coils of their body to restrict blood flow and breathing in their prey. The exact cause of death to an animal being killed by constriction has been slightly controversial. Recently, Penning et al. (2015) found the giant constrictors can exert pressures significantly higher than their prey's blood pressure, suggesting that constriction stops circulatory function and perhaps kills the prey quickly by over-pressurizing the brain and disrupting neural function. They summarized four possible mechanisms for rapid prey death due to constriction.

Boas and pythons are capable of swallowing prey that is equal to, or slightly greater than, their own body weight. Shine et al. (1999) report three reticulated pythons that contained prey that were between 1.1 and 1.2× the mass of the snake. Given that snakes can weigh more than 150 kg, they can be expected to take humans as prey.

Relationships between modern humans and the largest snakes are more complicated than predator and prey. There is evidence that humans use some of the giant species to control rodents – protecting stored food and reducing rodent-borne diseases (Cantor 1847; St. John 1863; Worcester 1898; Whitaker and Bhaskar 1978; Branch 1988b; You et al. 2013). Conversely, giant constrictors are significant predators on many human commensals such as dogs, cats, and livestock, making giant snakes competitors of humans (de Lang 2010; Headland and Greene 2011).

Snakes, unusually large species and dangerously venomous species, are common cultural symbols involved in myths and storytelling. In recent centuries, humans use giant snake's skins in the fashion leather industry, and in the last 50 years, humans began culturing giant snakes for unusual color patterns as living works of art. A large number of captive giant snakes have led to an increase in escaped animals or released pets as well as an increased number of human deaths from captive snakes. In South Florida, at least one of the giant species the Burmese python, *Python bivittatus*, has become invasive with a subsequent negative impact on the local ecosystem. A second species, the North African python, *Python sebae*, has been found in southern Florida, but its population status is unclear (Reed et al. 2010). About 42

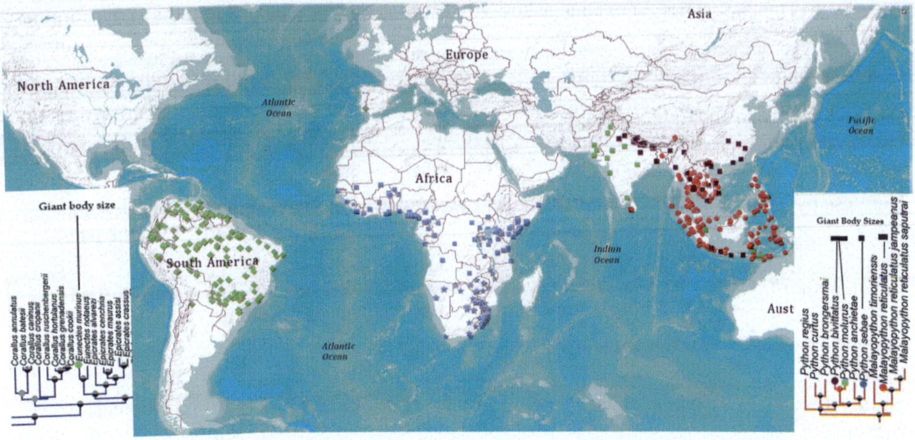

Fig. 19.1 The map illustrates the distribution of five of the largest snake species. Localities based upon the literature and museum data. The trees on each side are from Reynolds et al. (2014). The nodes on the tree are color coded with the map symbols for species identification

specimens have been found to date, the most recent found in August of 2017. The *boa constrictor* has also been introduced into Florida, but it is not a giant snake and is not further discussed here.

Of the ten largest extant snake species, there is evidence that five or six are potential predators on humans. The geographical distribution of the largest snake species (Fig. 19.1) is more or less restricted to the latitudes between 30^0 N and 30^0 S with some variation. As hominins emigrated out of Africa, they encountered giant snakes in tropical Asia and Australia.

The characteristic needed to be recognized as a giant species was defined by Murphy and Henderson (1997) as the ability to reach 6.1 m or 20 feet in total length. While this may seem a readily determined benchmark, it is not. Accounts of measuring or estimating the size of the largest snakes have frequently been showing to be dramatically inaccurate (Murphy and Henderson 1997; Barker et al. 2012). Table 19.1 lists the 12 longest extant constricting snakes, their estimated lengths, weights, and geographical distributions.

Extant giant snakes have evolved multiple times in several different clades. Of the four species of boids in the genus *Eunectes*, only the green anaconda, *Eunectes murinus*, is known to exceed 6.1 m. Other members of the genus tend to be large, but *E. murinus* is the largest (but not the longest) extant snake.

Extant giants within the family Pythonidae appear to have evolved six times if the 6.1 m definition is applied, and we include species that have not yet been documented to reach that size but probably do so. The reticulated python, *Malayopython reticulatus*, is the longest extant snake and in a different lineage than the large African and Asian *Python* species. There are four large members in the genus *Python* that likely meet the 6.1 m length. In Africa, the North African python, *Python sebae*, and the Southern African python, *Python natalensis*, were long

Table 19.1 The twelve largest extant snakes with estimates of weight and lengths and a summary distribution

Common Name	Scientific Name	Family	Weights	Lengths	Distribution
Boa constrictor[a]	*Boa constrictor*	Boidae	~30 kg	~3.6 m	South America
Green anaconda[b]	*Eunectes murinus*	Boidae	~200 kg	6.7 m	South America
Yellow anaconda[c]	*Eunectes notaeus*	Boidae	50 kg	4.5 m	South America
Burmese python[d]	*Python bivittatus*	Pythonidae	182 kg	?8.22 m 5.75 m	South Asia Southeastern Asia
Indian python[e]	*Python molurus*	Pythonidae	52.2 kg	?6.4 m 4.6 m	South Asia
South African python[f]	*Python natalensis*	Pythonidae		5.6 m	South Africa
North African python[g]	*Python sebae*	Pythonidae	113 kg	~ 5.72 m	Tropical Africa
Reticulated python[h]	*Malayopython reticulatus*	Pythonidae	158 kg	?10 m 7.9 m	Southeast Asia
Scrub python[i]	*Simalia kinghorni*	Pythonidae	91 kg	~8 m 5.6 m	NE Australia
Western olive python[j]	*Liasis olivaceus barroni*	Pythonidae	9.3 kg	5.5 m	Western Australia
Papuan python[k]	*Apodora papuana*	Pythonidae	22.5 kg	5.13 m	New Guinea
Oenpelli python[l]	*Morellia oenpelliensis*	Pythonidae	6 kg	4.57 m	Northern Australia

[a]I have included this species because it is so often thought to be a giant snake by the public. The size is based upon Mole (1924) and Glaw and Franzen (2016). Glaw and Franzen had a skin that was 455 cm long; snake skins stretch at least 20%. Thus, their skin probably belonged to a 364 cm snake
[b]The 6.7 m is based on a 22 ft. female measured by Mole (1924). The maximum size of this snake has been highly exaggerated, but a maximum length greater than 6.7 m can be expected
[c]This is based on Dirksen (2002)
[d]This size is based on Barker et al. (2012). It should be noted that the excessive weight of this species is based upon a captive specimen that was dramatically overfed
[e]This record is based upon Minton (1966)
[f]Based upon Branch (1988a)
[g]Based on Lanza and Nistri (2005)
[h] Based on a captive held at the New York Zoological Society
[i]Based on Fearn and Sambono (2000)
[j]Based upon Whitlock (1923)
[k]Based upon Tryon and Whitehead (1988)
[l]Based upon Gow (1989)

considered subspecies of *P. sebae*. They form a parapatric species pair, with *P. sebae* living in more mesic conditions and *P. natalensis* living in more xeric habitats,

A similar situation occurs in Asia. The Indian python, *Python molurus molurus*, and the Burmese python, *Python bivittatus bivittatus*, were long considered subspecies of *P. molurus* along with the Ceylonese python, *Python m. pimbura*. The Indian python, *Python molurus*, lives in the more xeric environments of the Indian peninsula, and it is smaller, lighter in color, and lacks the subocular scale of its sister species, *P. bivittatus*. The Burmese python, *P. bivittatus*, uses the more mesic habitats north of Indian peninsula, the Indochinese Peninsula, and southern China. It is larger, darker in color, and has a subocular scale. There is no evidence that the two species hybridize in nature. *Python molurus molurus* is not likely to reach 6.1 m. It is also unclear if *Python natalensis* reaches the size needed to be a giant species. *Python bivittatus* likely exceeds the 6.1 meters, although Barker et al. (2012) declared the largest known specimen to be 5.74 m, although both likely achieve this size given the literature reports of large specimens of both species.

There are several other very long pythons in Australasia, the largest of which is the scrub python, *Simalia kinghorni*. However, it may not attain the 6.1 meters (see Table 19.12).

19.2 Giant Snakes as Human Commensals

Humans create environments that increase the number of mice and rats. Additionally, they keep dogs, cats, goats, sheep, and other livestock, as well as ducks, geese, and chickens. They grow crops that improve the food supply for the rodents. Snakes, in turn, take advantage of the microhabitats created and the increased food supply.

The reticulated python has been long recognized as a human commensal. Flower (1899) described this species as common in Bangkok and present in private houses and businesses. He also recognized that it is much more common in cities than in the forests and attributes this to numerous hiding places in and around buildings combined with the readily available food supply. Even the Royal Palace had pythons preying on the royal cats.

Today, snakes, including the reticulated python, are still prevalent in urban Bangkok. Paddock and Jirenuwat (2017) report the Bangkok Fire and Rescue Department responded to 29,900 calls in 2017 to remove snakes from houses and office buildings. They do not break down the species involved in the calls, but pythons are discussed at length in the article. Cox (1997) and Cota (2010) have added to the discussion urban reticulated pythons in Thailand. The North African Python, *P. sebae*, has also been associated with urbanized areas (Luiselli et al. 2001; Akani et al. 2002), as has the Burmese python, *P. bivittatus* (Cota 2010), and the Indian python, *P. molurus* (Purkayastha et al. 2011).

19.2.1 Giant Snakes as Predators on Humans and Their Commensals

Naturalist and writer Roger Caras (1964) was skeptical about constricting snakes being able to kill and consume a human, and he was unable to locate anyone who had witnessed the death of a person in the coils of a constrictor.

Table 19.2 Documented attacks and deaths of humans from six species of giant snakes. This table is not meant to be all inclusive of every incident

The green anaconda, *Eunectes murinus*
Bates (1863:236) reported an attack by an anaconda on a 10-year-old boy playing in shallow water. The boy was rescued by his father
Blomberg (1956) reported two attacks by anacondas. A man on the Napo River in Ecuador was killed while swimming; his body was found downstream. Apparently, the snake was not large enough to swallow him. A second attack occurred at the mouth of the Yasuni River; a 13-year-old boy was attacked, swallowed and later regurgitated. The boy's father killed the snake
February 9, 2007. Cosmorama, Brazil. 8-year-old Joaquim Pereira was attacked by a wild 16 ft. anaconda. His grandfather saved him by beating and stabbing the snake (AP story)
Reticulated python, *Malayopython reticulatus*
Hagenbeck (1910) described four of these giant pythons simultaneously attacking one of his sons as he entered their cage. He survived the attack, but the pythons were not easily subdued
Kopstein (1927) reported a 14-year-old was swallowed by a 5.2 m reticulated python on Selebaboe Island (Talaud Islands Group), Indonesia. Kopstein also says an Indonesian woman was eaten by a 9.2 m python
April 1960. An 8-year-old boy was swallowed by a 6.1 m python while working in a rice paddy in Cox Bazaar, Bangladesh. Both the Burmese python and reticulated python occur at this location, so either species could have been involved
September 1965. Reuters reported a 10-year-old girl stepped on a 20 ft. reticulated python. The snake attacked. A gardener saved the girl
September 1965. Reuters reported an 8-year-old boy swallowed by a reticulated python in Ye Village in lower Burma. The boy's body was retrieved when villagers killed the snake
1972. In lower Burma (Myanmar), an 8-year-old boy was swallowed, but there was no data on the size of the snake or the species involved
September 28, 1979. *Asia Week* magazine reported a farmer named Ajobuka from a village in Central Sulawesi was attacked by a 6 m reticulated python while walking home. The snake was found after it had swallowed the man [Hamadryad, 1984 6(2);10–11]
August 1984. A captive 10 ft. reticulated python killed an 11-month-old boy in Ottumwa, Iowa (Chicago Tribune, August 28, 1984)
1992. Kuala Lumpur, Malaysia. A boy was swallowed by a reticulated python on a rubber plantation. The story was reported in a Thai magazine with photographic documentation, showing the snake to be *M. reticulatus*
September 6, 1995. Segamat, Johor, West Malaysia. A 29-year-old man was found in the coils of a reticulated python. The snake had swallowed his head and shoulders. The snake was 6.1 m long and weighed 140 kg
May 19, 1996. A Thai newspaper carried a story of a giant reticulated python that had eaten a large meal. The snake was being kept by monks. A set of child's underwear was found nearby, and it was suspected that the snake had eaten a child, but no missing children had been reported
de Lang and Vogel (2005:210) document five cases of humans being eaten by reticulated python and suggest that in Sulawesi, one person per year is consumed by this snake
October 25, 2008. The Virginia-Pilot (Day, 2008) reported the death of a Virginia Beach pet shop owner constricted by her 13 ft. reticulated python. The 25-year-old woman was apparently trying to administer medication to the snake. Virginia Beach Police believe the 13 ft. reticulated python strangled its owner when she tried to give it medication

(continued)

Table 19.2 (continued)

Muthusamy and Gopalakrishnakone (1990) report a 46-year-old farmer who dropped his flashlight while riding a bike. He stopped to pick it up and was attacked by a 14–15 ft. reticulated python. The snake severely lacerated his hand
Muthusamy and Gopalakrishnakone (1990) report a 45-year-old farmer who was thrown from his motorcycle when he hit a 15 ft. reticulated python. The snake attacked him, biting his right forearm, and when he pulled away, the snake grabbed the left forearm
March 30, 2017. *USA Today* reports the death of 25-year-old Akbar Salubiro in Sulawesi from constriction by a 23 ft. snake
Burmese python, *Python bivittatus*
July 22, 1993. Derek James Romero, 15 years old, was killed by an 11 ft., 80 pound Burmese python in his Denver, Colorado, home (Chicago Tribune, 1993)
May 19, 1993. William Bassett was found dead from a struggle with a 16 ft., 200 pound python (presumably this species) in Harahan (suburban New Orleans), LA (Reuters, Chicago tribune, 5/19/1993)
August 27, 2001. Amber Mountaian, an 8-year-old living in Irwin Pennsylvania, was killed by a 10 ft. Burmese python. The snake had apparently escaped from its cage and had wrapped its self around the girl's neck
South African python, *Python natalensis*
January 18, 1961. Hurly was a miner working at the alpine mine in the eastern Transvaal. He caught a python by the tail. The snake coiled around him. He was rescued and returned to work. The following day he went to the hospital not feeling well, and he died the next day from a ruptured spleen and kidneys
November 1979. In the northern Transvaal, a 13-year-old, 45 kg, and 130 cm tall boy was herding cattle when he was attacked by a 4.5 m python [Hamadryad, 1984 6(2):10–11]
September 18, 1982. Maritzburg, South Africa. A farm worker was attacked by an 11 ft. python (herald news AP story 9/18/1982)
January 10, 2001. Lucas Sibanda, a 57-year-old, was attacked by a python near his home in Pretoria. No size was given. Sibanda bit the snake until it released him (Reuters)
November 24, 2002. The *Sunday Telegraph* reported a 20 ft. African rock python killed and swallowed a 10-year-old boy near Durbin. The attack and ingestion were witnessed by other children, including 11-year-old Khaye Buthelezi who climbed a tree for fear the snake would attack. The children watched the snake eat their friend for 3 h
North African python, *Python sebae*
Arthur Loveridge (1947) recounts a story of a woman killed by a python on Ukerawe Island in Lake Victoria, Tanganyika
August 29, 1999. Jesse Altom, 3-year-old boy living in Carlyle, Illinois, was killed by a 7.5 ft. African rock python after it had escaped from its cage while the parents slept. The parents were tried for child endangerment but were cleared at a trial
August 5, 2013. Campbellton, New Brunswick, Canada. Two boys ages 4 and 6 were killed while sleeping by a 4.3 m, 45 kg python. The father of the boys was keeping the snake in an enclosure in the house when it escaped and accessed the boy's room. The boys had been with farm animals earlier in the day, and the odor's left on the boys may have triggered feeding behavior in the snake
August 25, 2017. Stubbs and Clarke-Billings (2018) document the death of Dan Brandon who died of asphyxiation at his home in Church Crookham, Hampshire, UK, on August 25, 2017. His 2.43 m *Python sebae* constricted his neck
Scrub python, *Simalia kinghorni*

(continued)

Table 19.2 (continued)

May 1, 2005. There is at least one human death attributed to this species. Erik Attmarrsson, a 28-year-old professional snake handler, was apparently was killed by a pet scrub python that was 5 m long in his home in Tanunda, Australia. However, an autopsy was inconclusive about the cause of death (Riches and Littely 2005; Anonymous 2005)
December 26, 2011. Shears and Parsons (2011) describe an unsuccessful predatory attack by a 4 m wild scrub python on a 4-year-old occurred when the child went to retrieve a ball in his suburban yard in northern Queensland. The attack happened at 8 PM

Skeptics may deny snake predation on humans and interpret the stories as tall tales or folklore, often stating that snakes could never expand their jaws around a human's pectoral girdle. However, the evidence for giant snake predation on humans is substantial. Additionally, variation in human body size and shape is quite variable today as it was in the past (Ruff 2002). People with large body sizes may be less likely to be attacked than smaller individuals; nonetheless, even today large snakes can and do take humans for food, albeit rarely.

Giant snakes will take humans as food, but they will also defend themselves against humans if they perceive the person as a threat. The following story is undoubtedly a defensive attack.

In early February 1997, Lou Daddono, 36-year-old owner of a tourist attraction Serpent Safari, was attacked by a reticulated python that was about 22 feet long. Daddono was cleaning a cage when the wild-caught python reacted to Daddono pinning its head with a snake hook. The snake escaped and bit Daddono in the hand and knee and then knocked the snake keeper off his feet before coiling and starting constriction. Daddono had stopped breathing, and his heart had stopped beating. At this point, the snake lost interest in Daddono and attacked Daddono's partner, Paul Keeler. Keeler had been stabbing the snake trying to get it to release Daddono. He managed to pull the snake out of the cage and ran a short distance. At that point, the snake let go and slid under another exhibit. Keeler performed CPR on Daddono. An ambulance came, and the paramedics got Daddono's heart started again. Daddono also suffered two broken ribs, some stretched cartilage around his rib cage, and a fractured right hand. The snake required 250 stitches. Daddono was quoted as saying that he learned a lesson. "The snake was acting the way it normally does," he said. "I pushed it too far." While this account seems to be more defensive than an act of predation, it is a reminder that the commonly kept captive reticulated python can be a threat to their human captors (Mills 1997; Barker et al. 2012).

People living in the developed western world consider giant snakes a novelty; those living at subsistence level in the Neotropics and the tropical Eastern Hemisphere have a different perspective (Fig. 19.2). Perhaps the earliest literary reference to a human being swallowed by a snake was in Bosman (1705:310) where he describes finding a person in an African python. Documentation of humans being eaten by snakes is not numerous but continues to the present. Table 19.2 lists some documented attacks of large constrictors on humans, and in some cases, the human prey was swallowed.

Fig. 19.2 This figure is from *The Bestiarium of Aloys Zötl* (1831–1887). It was initially titled The Boa Constrictor, (1867). It is likely based on a story from the *Bombay Courier* of August 31, 1799, and recounted in Gosse (1850). A ship headed for Ambon could not reach the island by nightfall, and they anchored near Celebes (now Sulawesi). One of the crew went ashore in search of betel nuts. When he returned to his skiff, he laid down and fell asleep. He was attacked by a reticulated python, and his screams brought some of his shipmates who killed the snake. However, it was too late, and the betel nut collector was killed by the snake (public domain)

The aquatic specializations of the Neotropical green anaconda, *Eunectes murinus*, distinguish it from the Eastern Hemisphere pythons. While all of the large pythons are known to use bodies of water for concealment to ambush prey and thermoregulate, they are not nearly as aquatic as the green anaconda and other members of the genus *Eunectes*. Rivas (2000) radio tracked 48 green anacondas in the Llanos of Venezuela and found them in the water 86% of the time and at the water's edge 14% of the time. The highly aquatic habits and its reluctance to leave the water likely lessen anaconda-human interactions. Yet, it too has been implicated in attacks on humans, but documented human deaths are absent or very few in number. More often unsuccessful attacks by the green anaconda are reported. Naturalist and explorer Henry Bates (1863) described an anaconda attacking a boy who was rescued by his father. Kingsley (1890) described an unsuccessful attack by an anaconda on Trinidad. Herpetologist Rivas (1998) describes two attacks on researchers while they were wading in the water. Blomberg (1956) reports that an *E. murinus* had successfully swallowed a person.

A BBC headline on 29 March 2017 stated, "Indonesian police say a farmer in Indonesia has been eaten by a python, which was later cut open to retrieve the man's body." Combining information from various press stories, the snake was 7 m long and weighed 158 kg. The snake killed and swallowed Akbar Salubiro, a 25-year-old male, on his way to work to harvest palm oil in West Sulawesi. After Salubiro had disappeared, people searching for him found the python. The snake was dissected, and the man's body was removed from the snake's gut. This was documented with video.

De Lang (2010) summarized the attacks on humans in Indonesia and Sarawak and found only 20 well-documented cases in the previous 150 years but suggests this is an underestimate because knowledge of many incidents remained in local communities.

A remarkable glimpse into the lives of hunter-gathers living alongside giant constrictors is provided by Headland and Green (2011). They describe a society of 120 Philippine Agta Negritos and their relationship to the reticulated pythons that inhabit the same forests. The Agta inhabit the Sierra Madre of the Aurora Province, on the island of Luzon. Until the 1970s, the Agta hunted deer, warty pigs, long-tailed macaques, and pythons, but not other snakes. Using interviews, Headland documented 26% of adult males had survived predation attempts by reticulated pythons; six fatal attacks were known from 1934 to 1973. An adult male Agta weighs about 60% of an adult female reticulated python.

On March 23, 1973, a reticulated python entered a hut where three sibling children (a 4-year-old female, a 3-year-old male, and a 6-month-old female) were sleeping. The snake killed the two older children and coiled around and swallowing one when the father entered and killed the snake.

Because the Agta and pythons preyed upon deer, wild pigs, and monkeys, the two species were reciprocally prey, predators, and competitors. This same complex of relationships was suggested by de Lang (2010).

Giant body size appears to have evolved in the Australasian clade at least five times, four times in extant species and at least once in a fossil lineage. While these are huge snakes, they rarely prey on humans. However, they will prey on livestock and other human commensals. Yet there are three stories this author is aware of the Australasian pythons attempting to prey on humans, and one human death from the scrub python, *Simalia kinghorni* (see Table 19.2). Australasia had its share of enormous birds and mammals in the late Tertiary and Pleistocene (Stuart 2015). It seems likely some of these snakes evolved large body size to exploit these food resources.

19.3 Human Predation on Giant Snakes

While local people may kill and eat the giant pythons on occasion, skin hunters are likely responsible for the removal of many more giant snakes from the wild each year. However, not all giant snakes experience extreme hunting pressure.

19.3.1 The Green Anaconda

There is little evidence that the green anaconda (*Eunectes murinus*) is being hunted for skins on a large scale although they are killed and occasionally marketed for meat, hides, and traditional medicines. Rivas (2007) noted 2138 skins were confiscated in Holland between 1988 and 1990. The CITES database reports 50 anacondas were exported from countries that have wild populations in the time span of 2013–2017, and 45 (90%) were live specimens headed for the pet trade or zoos.

Conservationists often assume that people kill predators to reduce the risk of being attacked or killed and to reduce predation on livestock. Miranda et al. (2016) tested hypotheses about the feeding habits of two species of anacondas in the genus *Eunectes*, human perception of risks, and human attitudes toward the snakes in light of the person's educational level. They used 330 Internet videos of human-anaconda encounters from ten different South American countries. They found visual evidence of a recent meal (a snake with a distended abdomen) and predation on domestic animals did not affect the probability of the anaconda being killed. However, the likelihood of human retaliation against the snake increased as the educational level decreased. Although retaliatory killing is described as one of the leading causes of animal mortality following human-wildlife encounters, the results suggest that killing of anacondas is not retaliatory or related to economic losses but preventive, because these snakes are seen as life-threatening animals.

19.3.2 The African Pythons

Trade in South African python, *Python natalensis*, from 2013 to 2017 was tabulated from the CITES website for all countries having wild populations. Only 29 snakes were reported as exported, and most were live specimens. In Namibia, *Python natalensis* is occasionally killed for its skin, food value, traditional medicinal products, and its threat to livestock; there is a negligible illicit trade in the species (Branch and Griffin 1996). In Maputo, Mozambique, Williams et al. (2016) reported that South African pythons (*Python natalensis*) are involved in traditional medicines and that they are the most frequently sought-after reptiles in markets. The authors were unable to ascertain how the pythons were used, but several traders told them they sold python fat.

Trade in North African python, *Python sebae*, from 2013 to 2017 was tabulated from the CITES website for all countries having wild populations. The total number of snakes exported over 5 years was 19,614. Of these, about 2000 (10.2%) were live animals, and the remainder were skins, skulls, and trophies.

19.3.3 *Asian Pythons*

Unlike the African pythons, Southeast Asian pythons are heavily exploited for skins, traditional Chinese medicines, and food. Kasterine et al. (2012) combined all of the skins of Southeast Asian species and found the annual exports are almost a half a million skins. This included two of the giant pythons, *Python bivittatus* and *Malayopython reticulatus*, as well as several smaller species.

The reticulated python (*Malayopython reticulatus*) is the most exploited species. The CITES website reports 5,727,592 reticulated pythons exported between 2013 and 2017. Only about 17,000 of these were live specimens. The others were used for leather products, garments, and carvings. Natusch et al. (2016) evaluated the harvest in Sumatra for sustainability by examining the size range of more than 4200 snakes taken for the commercial leather industry in northern and southern Sumatra. Surprisingly, they found the numbers, mean body sizes, clutch sizes, sizes at maturity, and proportion of large specimens have not decreased between a survey done in 1995 and repeat in 2015. They concluded that the harvest appears to be sustainable.

The number of Burmese python (*Python bivittatus*) exported between 2013 and 2017 was tabulated from the CITES website and found to be 1,693,508. About 14,600 of these were alive. The remainder were skins, garments, leather products, and carvings.

19.4 Farming Pythons

Captive breeding pythons for the skin trade has been reported in Cambodia, China, Indonesia, Lao PDR, Malaysia, Thailand, and Vietnam (Natusch et al. 2016). The economic and biological practicality of breeding pythons for the commercial trade has been questioned but Kasterine et al. (2012). They found python farms may have a system in place that makes captive breeding economically and biologically feasible. Large python farms sell hatchling snakes to smaller satellite farms that raise the snakes to market size and then sell the snakes back to the large farms for slaughter. This supposedly keeps management costs down. The breeding farms suggest that Burmese pythons can be fed for 1 year on chicken necks and heads for US$ 24. One breeder claimed he fed his stock piglets, obtained from a local pig farm for US$ 0.50/kilogram.

According to these breeders, the Burmese pythons that are being bred in Vietnam for the skin trade can reach slaughter size (>2.5 m) within 10 months. Kasterine et al. (2012) note this figure has been contested by breeders in Indonesia maintaining the growth rate is not possible. However, an investigation of python breeders in Europe and the USA suggested that it is indeed possible to feed Burmese and reticulated pythons so that they reach 3 m in less than 12 month – as long as they were males (Natusch et al. 2016).

Fig. 19.3 A reticulated python, *Malayopython reticulatus*, from the Tirta Gangga Water Palace in eastern Bali. The snake was estimated to be 4.5 to 5.0 m by the photographer, Michael J. Jowers

19.4.1 Pythons in the Bushmeat Trade in Africa and Asia

There seems to be little evidence that the giant Asian pythons are heavily involved in the bushmeat trade. A survey of wild meat markets in northeast India reported no Burmese pythons were present, even though local people were known to consume the species (Bhupathy et al. 2013) (Fig. 19.3).

Python sebae and *P. natalensis* are likely involved in the bushmeat trade across its range. In Equatorial Guinea (Fa et al. 2000) and in Ghana (Ockerman and Basu 2009), it composes only a fraction of the total harvest. Wright and Priston (2010) also report pythons are infrequently taken in the bushmeat trade and attribute the lack of interest to the belief that some humans can transform themselves into pythons.

19.5 The Australasian Pythons

Kinghorn's scrub python, *Simalia kinghorni*, is Australia's largest extant snake, and Lumholtz (1889) reported it in the diet of the aboriginal people. He described how the indigenous people hunted the giant pythons in winter in the tops of trees. The scrub python has a distribution restricted to the wet tropics of northeastern Queensland where it uses closed canopy forests as well as grassland habitats (Harvey et al. 2000)

19.5.1 Giant Snakes as Competitors

Philippine Agta Negritos likely realized they were competing with the reticulated python for food when they killed the snakes and observed pigs, deer, and monkeys in the snake's digestive systems. Thus, by killing and eating the pythons, they were removing a potential predator and competitor from their immediate environment. Farmers likely came to the same realization when pythons consume their livestock and pets (Headland and Greene 2011).

Throughout the distribution of all giant snakes, predation on domesticated animals has been reported. However, in relatively few places has the predation been quantified. Goursi et al. (2012) studied the diet of *Python molurus molurus* in Pakistan's Deva Vatala National Park during a 6-month period. A total of 91 python attacks on livestock were reported resulting in 74 livestock deaths and 17 injured. The authors found the local cattle herding practices damaged habitat and resulted in the declined the python's natural prey. The increased predation of the pythons resulted in negative perceptions of pythons in local community. Habitat destruction due to the forest cutting, overgrazing, fodder and fuel wood collection, and illegal python trade are found to be the major threats to this species (Fig. 19.4).

19.6 Invasive Populations

Colonization by invasive species takes time; invasions may start and fail before they actually succeed in establishing a population. Almost any day of the week or any time of the year, it's possible to find recent news stories about escaped large constrictors. They are often stories about people looking for escaped pets or, more likely, someone finding a giant constrictor at a location where it does not belong. The *Tribune Media Wire* carried a story on December 23, 2016, about a large (about 5 m) Burmese python found in Ohio's Chagrin River (Anonymous 2016). The snake was almost frozen. Stories like this one suggest large constrictors are continually being released, and if they can survive at the location, they may eventually establish a population.

Fig. 19.4 Upper photo. A hatchling Burmese python, *Python bivittatus*, from Central Thailand. This snake was found along a road in an area with dry tropical forest and adjacent rice paddies. Lower photo. A Burmese python caught in a fish trap in Central Thailand. The fisherman was trying to sell the snake to the author. JCM

Burmese pythons were showing up in South Florida in the 1970s according to news stories (Anonymous 2017). However, the oldest Florida specimens of *Python bivittatus* in the Florida Museum of Natural History date to 1991. Escaped or released pets were undoubtedly responsible for the early Florida specimens but maybe not the current population. The invasive populations of giant snakes in the USA have been summarized by Reed and Rodda (2009).

One event that authors often point to is Hurricane Andrew in 1992. This storm has been hypothesized as the point in time when many escaped *P. bivittatus* were simultaneously released as homes and businesses were destroyed. However, Willson et al. (2011) suggested the spatial and demographic patterns of python captures are not consistent with such scenarios. Their examination of python encounters in the Everglades and the location where wild reproduction was first confirmed is the mangrove forests and saline glades of the southern portion of Everglades National Park, located at least 30 km from the nearest reptile breeder or importer.

Twice, the Florida Fish and Wildlife Conservation Commissions (FFWCC) invited people to search for Burmese pythons in the Everglades during the "Python Challenge." A month-long event in midwinter (January to February). In 2013 more than 1600 volunteer snake hunters located 95 snakes (=0.059 snakes per hunter).

The 2016 challenge had about 1035 participants who managed to find 106 snakes (0.102 snakes per hunter). This was not FFWCC's attempt at reducing the python population. It was, however, a way to raise public awareness of the problem (FWC website). Knowledge of invasive species may reduce the number of people who would intentionally release invasive species in the future.

The invasive population will have implications for the Florida ecosystem and has implications for disease transmission to native species. Reeves et al. (2018) determined that three species of *Culex* mosquitoes took blood meals from *Python bivittatus* and that *Culex erraticus* took more blood meals from *P. bivittatus* than from any other available host. The study used captive animals. The authors suggest that if the mosquitoes behave similarly in southern Florida, *P. bivittatus* may be involved in the transmission networks of mosquito-vectored pathogens. *Culex erraticus* and *C. quinquefasciatus* are medically important vectors of arboviruses and parasites in the southeastern USA. *Culex erraticus* is suspected to be an important vector for the eastern equine encephalitis virus. Another mosquito in the study, *Culex quinquefasciatus*, may be a vector for the Zika virus or Chikungunya virus.

19.7 Designer Snakes

While giant snakes and humans consider each other predator and prey, there is a relatively new relationship between them. Humans keep and breed giant snakes not only for pets but as living works of art. The production of snakes for aesthetically pleasing colors and patterns is a contemporary trend. Designer morphs, as they are called, are sold in virtually all pet stores and at national and international trade shows and are hugely popular in the USA and Europe.

The herpetoculture of designer morphs has roots that may be a century old. The city of Iwakuni, Japan, supports an albino population of the Japanese rat snake (*Elaphe climacophora*). Tonkunaga and Akagishi (1991) estimated the number of albino snakes to be 1000 in 1925. The snakes occupied houses, storehouses, river banks, and stone walls, and they could be seen in streets, gardens, and fields. In 1924, the snake's habitat was designated a national monument, but in 1972, the snake itself was given national monument status. The number of albinos declined from 1970 to 1990, and the city of Iwakuni appealed to volunteers to conserve the snake and construct a breeding facility.

In the early 1950s, Swiss animal collector Peter Ryhiner obtained an adult leucistic Burmese python. The python, named Serata, gained notoriety as the guest of honor at a New York cocktail party when it was discussed in *The New Yorker* magazine. Eventually, the snake ended up at the Staten Island Zoo in the care and custody of herpetologist Carl Kauffeld. The publicity this snake received seems to be the genesis for much of the interest in designer morph snakes and the industry that has grown up around designer morph boas and pythons today.

Interest in color and pattern morphs increased in the following decades, and, by the 1980s, a number of different species of colubrids, as well as pythons and boas,

Fig. 19.5 A designer morph of the reticulated python commonly called the tiger morph. Clark (2002) reported the first tiger morph reticulated pythons appeared in 1992. A female owned by Karl Hermann of St. Paul, Minnesota, produced a litter with half the offspring demonstrating the female's unusual pattern. Hermann named this morph the "tiger" retic. The morph shows varying degrees of striping and some had lateral duplication of the pattern. Some of the black line on the top of the head is missing, and the white spots on the sides were enlarged and elongated. This was the apparent start of interest by hobbyists in the reticulated python, and apparently, this was the first reticulated python morph to be regularly bred in the USA. JCM

were being bred for color and pattern variations. Berry (2006) attempted to summarize the morphs of medium-sized boas and pythons. He lists more than 200 color morphs and hybrids. But it did not include the morphs of the Burmese python nor the reticulated python, which may total another 50 or more. Note that the hybrids are usually not hybrids between species but hybrids between morphs (Fig. 19.5).

The North American Reptile Breeding Conference (NARBC) has three or four trade shows each year. Thousands of people line up to see and buy the latest designer snakes, some of which sell for $10,000 or more. Ball pythons (*Python regius*) are by far the most common snakes at the shows, but morphs of many other species can also be seen. Why these snakes should be so popular is something of a puzzle, and many of the people who buy them seem to know little about snakes as organisms. Their interest lies in collecting aesthetically pleasing objects, which, in this case, happen to be live snakes. There is also a certain amount of obsession for some, to have an unusual animal that others don't have. Possessing rare objects is a desire of many. David Barringer (2008), wrote about designer snakes for AIGA, the professional association of design noted, "The duality of the hate-love relationship between humans and snakes is shifting. People's attitudes about snakes are changing for the better. The attitudinal adjustment is an excellent sign for the environment. Tolerance of snakes indicates not only a better understanding of nature but its value."

19.8 The Value of Giant Snakes

Giant snakes have been used for rodent control, leather products, and a source of proteins by humans. The ecosystem services they provide to humans are likely substantial but understudied. The number of rodents they consume in agricultural and urbanized habitats within their natural distributions is expected to be significant. Giant snakes have been and are valued as pets by many and are used in exotic animal shows, zoo displays, and other forms of entertainment.

Pythons have been used in traditional ethnic medicines for millennia. Only recently, however, have pythons become the subjects of research for an understanding of their unique physiological adaptations to feeding on large prey. The Burmese python is now providing scientists with insights into processes that are potentially very useful in modern medicine. For a recent overview of this, see Engber (2017).

Feeding on huge prey followed by long periods of fasting has resulted in a myriad of adaptations in snakes that involves downregulating its metabolism during fasting and upregulating its metabolism immediately after consuming a meal that may equal its own body weight. Giant snakes may consume 50,000 or more calories in a single meal.

After swallowing a huge meal, the production of insulin rises rapidly, the intestines double their thickness, the stomach produces hydrochloric in copious amounts, the liver and kidneys almost double their mass, snake body temperature rises about 6 °F, the weight of the heart increases by 40%, the pulse triples, many organs enhance their performance by 10 to 40 times, and metabolites and hormones in the blood may increase 100 times their normal levels. Of particular interest, the snake's pancreas doubles in mass, and the amounts of insulin and glucagon in the blood dramatically increase. All of this occurs within 24 h. Once the food has been digested, the organs return to their normal sizes.

In 2014, Secor et al. asked the question, is the Burmese python's extreme physiology relevant to diabetes? Working with Secor are Amit Choudhary and Bridget Wagner. They hope to find the molecule or molecules in the Burmese python that allows it to replace the beta cells of the pancreas. The loss of beta cells in humans results in diabetes, and a molecule from *Python bivittatus* may eventually allow humans to regenerate those cells.

Hominins and giant snakes were in an evolutionary arms race for much of their evolutionary history. They were predators, prey, and competitors for some of the same resources. Today, humans have turned giant snakes into pets, exploit them for pharmaceutical products and fashionable leather, and expanded their distribution. The reasonable conclusion for this is that giant snakes are well on their way to becoming human commensals.

19.9 Conservation

The African and Asian pythons are hunted and involved in trade. While the IUCN Red List has considered *Python bivittatus* in decline and vulnerable within its native range (Stuart et al. 2012), the other large species are waiting for an assessment, but all pythons are considered CITES Appendix II because of the difficulty in distinguishing between species. Many of the other giant snakes are in protected areas, where hunting is supposedly regulated. As the human population grows toward eight billion people, it seems likely that all large animals will come under continuing pressure as a source of protein and raises the question, will giant snakes be able to survive humans?

References

Akani GC, Eyo E, Odegbune E, Eniang EA, Luiselli L (2002) Ecological patterns of anthropogenic mortality of suburban snakes in an African tropical region. Isr J Zool 48(1):1–1

Anonymous (2005) Autopsy fails to confirm if Swedish snake handler was crushed by pet python. Yahoo Canada & Associated Press, May 5, 2005

Anonymous (2016) Fishermen find 16-foot snake in frozen river. http://wgntv.com/2016/12/23/fishermen-find-16-foot-snake-in-frozen-river

Anonymous (2017) Python fever': hunting for snakes in the Florida Everglades CNN Wire Service April 5, 2017

Antón SC (2003) Natural history of Homo erectus. Am J Phys Anthropol 1;122(S37):126–170

Barker DG, Barten SL, Ehrsam JP, Daddono L (2012) The corrected lengths of two well-known giant pythons and the establishment of a new maximum length record for Burmese pythons, *Python bivittatus*. Bull Chicago Herp Soc 47(1):1–6

Barringer D (2008) Designer snakes. AIGA, the professional association for design. http://www.aiga.org/designer-snakes

Bates HW (1863) The Naturalist on the River Amazons. J. Murray, London

Berry J (2006) Designer morphs: the complete guide to medium sized Python and Boa morphs. Privately Published, Palm Beach

Bhupathy S, Kumar SR, Thirumalainathan P, Paramanandham J, Lemba C (2013) Wildlife exploitation: a market survey in Nagaland, North-Eastern India. Trop Conserv Sci 6(2):241–253

Blomberg R (1956) Giant snake hunt. Nat Hist 65:92–97

Bosman W (1705) A new and accurate description of the coast of Guinea, divided into the gold, the slave, and the ivory coasts. Sir Alfred Jones, London

Branch WR (1988a) Field guide to the snakes and other reptiles of Southern Africa. Ralph Curtus Books, Sanibel Island, 328 pp

Branch WR (1988b) South African red data book-reptiles and Amphibians Council for Scientific and Industrial Research, South African National Scientific Programmes Report No. 151. PRETORIA

Branch WR, Griffin M (1996) Pythons in Namibia: distribution, conservation, and captive breeding programs. Advances in herpetoculture. International Herpetological Symposium Special Publication 1996 (No. 1, pp. 93–102)

Cantor T (1847) Catalogue of reptiles inhabiting the Malayan peninsula and islands. J Asiatic Soc Bengal (16) 607–656, 897–952, 1026–1078

Caras RA (1964) Dangerous to man: wild animals: a definitive study of their reputed dangers to man. Chilton Books

Clark B (2002) Reticulated pythons. Reptiles Magazine, July, pp 48–63

Cota M (2010) Geographical distribution and natural history notes on *Python bivittatus* in Thailand. Thailand Nat History Museum J 4(1):21–30

Cox MJ (1997) Reticulate pythons (*Python reticulatus*) in Bangkok. Reptile & Amphibian Magazine 1997(Mar/Apr):20–24

de Lang R (2010) The reticulated python (B*roghammerus reticulatus*) and man (*Homo sapiens*) eat each other: animals, enjoy your meal! Litteratura Serpentium 30(4):254–269

de Lang R, Vogel G (2005) The snakes of Sulawesi. A field guide to the land snakes of Sulawesi with identification keys. Edition Chimaira, Frankfurt am Main, 312 pp

Delfino M, Segid A, Yosief D, Shoshani J, Rook L, Libsekal Y (2004) Fossil reptiles from the Pleistocene Homo-bearing locality of Buia (Eritrea, northern Danakil depression). Rivista Italiana di Paleontologia e Stratigrafia (Research In Paleontology and Stratigraphy) 110:51–60

Dirksen L (2002) Anakondas. Natur und Tier Verlag, Münster, 190 pp

Engber D (2017) When the lab rat is a snake. New York Times Magazine, May 21:65–71

Fa JE, Yuste JE, Castelo R (2000) Bushmeat markets on Bioko Island as a measure of hunting pressure. Conserv Biol 14(6):1602–1613

Fearn S, Sambono J (2000) A reliable size record for the scrub python *Morelia amethistina* (Serpentes: Pythonidae) in north east Queensland. Herpetofauna 30:2–6

Flower SS (1899) Notes on a second collection of reptiles made in the Malay Peninsula and Siam, from November 1896 to September 1898, with a list of the species recorded from those countries. Proc Zool Soc London 1899:600–697

Glaw F, Franzen M (2016) On the maximum length of *Boa constrictor*. Spixiana 39:264

Gosse PH (1850) Natural history: reptiles. Society for Promoting Christian Knowledge, London. 269 pp

Goursi UH, Awan MS, Minhas RA, Ali U, Kabir M, Dar NI (2012) Status and conservation of Indian Rock Python (*Python molurus molurus*) in Deva Vatala National Park, Azad Jammu and Kashmir, Pakistan. Pak J Zool 44(6):1507–1514

Gow GF (1989) Graeme gow's complete guide to Australian snakes. North Ryde: Angus & Robertson Publishers, 171 pp

Hagenbeck C (1910) Beasts and Men. (An abridged translation by H. S. R. Elliot and A. G. Thacker). London: Longmans, Green, and Co.

Harvey MB, Barker DG, Ammerman LK, Chippindale PT (2000) Systematics of pythons of the *Morelia amethistina* complex (Serpentes: Boidae) with the description of three new species. Herpetol Monogr 1:139–185

Headland TN, Greene HW (2011) Hunter–gatherers and other primates as prey, predators, and competitors of snakes. Proc Natl Acad Sci 108(52):E1470–E1474

John SS (1863) Life in the forests of the Far East: or, travels in Northern Borneo. Smith, Elder, and Company, London

Kasterine A, Arbeid R, Caillabet O, Natusch D (2012) The trade in South-East Asian Python skins. International Trade Centre (ITC), Geneva

Kingsley C (1890) At last: a Christmas in the West Indies. Macmillan, New York, 334 pages

Kopstein F (1927) Over het verslinden van menschen door Python reticulatus. Overdruk uit De Tropische Natruur 1927:65–67

Lanza B, Nistri A (2005) Somali Boidae (genus Eryx Daudin 1803) and Pythonidae (genus Python Daudin 1803)(Reptilia Serpentes). Tropical Zoology 18(1):67–136

Loveridge A (1947) Tomorrow's a holiday. Harper & Bros, New York

Luiselli L, Angelici FM, Akani GC (2001) Food habits of Python sebae in suburban and natural habitats. Afr J Ecol 39(1):116–118

Lumholtz CS (1889) Among cannibals: account of four years travels in Australia, and of camp life with the aborigines of Queensland. London: J. Murray. [Translation by R. B. Anderson]

Mills S (1997) A breathless encounter, Lou Daddono stills loves his snakes, even after a python bit him and almost choked him to death. Chicago Tribune February 7, 1997, http://articles.chicagotribune.com/1997-02-07/news/9702070377_1_giant-snake-new-snake-snakes-and-turtles

Minton SA Jr (1966) A contribution to the herpetology of West Pakistan. Bull Am Mus Nat Hist 134:29–184
Miranda EB, Ribeiro RP Jr, Strüssmann C (2016) The ecology of human-anaconda conflict: a study using internet videos. Trop Conserv Sci 9(1):43–77
Mole RR (1924) The Trinidad snakes. Proceedings of the General Meetings for Scientific Business of the Zoological Society of London 1924, pp 234–278
Murphy JC, Henderson RW (1997) Tales of giant snakes. Krieger Publishing, Malabar
Muthusamy E, Gopalakrishnakone P (1990) Python bites–case reports. The Snake 22:113–119
Natusch DJD, Lyons JA, Mumpuni, Riyanto A, Shine R (2016) Jungle giants: assessing sustainable harvesting in a difficult-to-survey species (Python reticulatus). PLoS ONE 11(7):e0158397. https://doi.org/10.1371/journal.pone.0158397
Ockerman HW, Basu L (2009) Undomesticated food animals hunted and used for food. Agric Sci 2009:1
Paddock RC, Jirenuwat R (2017) A scourge slithers across a city (and up its toliets). The New York Times International, November 29, 2017
Penning DA, Dartez SF, Moon BR (2015) The big squeeze: scaling of constriction pressure in two of the world's largest snakes, *Python reticulatus,* and *Python molurus bivittatus*. J Exp Biol 218(21):3364–3367
Pitman CR (1974) A. guide to the snakes of Uganda. Wheldon & Wesley, Ltd Codicote
Reed RN, Rodda GH (2009) Giant constrictors: biological and management profiles and an establishment risk assessment for nine large species of pythons, anacondas, and the boa constrictor: U.S. Geological Survey Open-File Report 2009–1202, 302 p
Reed RN, Krysko KL, Snow RW, Rodda GH (2010) Is the Northern African python (*Python sebae*) established in southern Florida? IRCF Rept Amphib 17(1):52–54
Reeves LE, Krysko KL, Avery ML, Gillett-Kaufman JL, Kawahara AY, Connelly CR et al (2018) Interactions between the invasive Burmese python, *Python bivittatus* Kuhl, and the local mosquito community in Florida, USA. PLoS One 13(1):e0190633. https://doi.org/10.1371/journal.pone.0190633
Reynolds RG, Niemiller ML, Revell LJ (2014) Toward a tree-of-life for the boas and pythons: multilocus species-level phylogeny with unprecedented taxon sampling. Mol Phylogenet Evol 71:201–213
Riches S, Littely B (2005) Crushed by Pet Python. Brisbane Courier Mail May 02, 2005
Rivas JA (1998) Predatory attacks of green anacondas. *Eunectes murinus*. Herpetol Nat Hist 2:157–159
Rivas JA (2000) Life history of the green anaconda (Eunectes murinus) with emphasis on its reproductive biology. Ph.D. thesis, University of Tennessee, Knoxville.
Rivas JA (2007) Conservation of green anacondas: how Tylenol conservation and macroeconomics threaten the survival of the world's largest snake. Iguana 2007:75
Ruff C (2002) Variation in human body size and shape. Annu Rev Anthropol 31:211–232
Shears R, Parsons C (2011) Boy, 2, narrowly cheats death after mother finds him in back garden being squeezed by large python. DailyMail on Facebook. 29 December 2011
Shine R, Ambariyanto, Harlow PS, Mumpuni (1999) Ecological attributes of two commercially-harvested *Python* species in northern Sumatra. J Herpetol 33:249–257
Stuart AJ (2015) Late quaternary megafaunal extinctions on the continents: a short review. Geol J 50(3):338–363
Stuart B, Nguyen TQ, Thy N, Grismer L, Chan-Ard T, Iskandar D, Golynsky E, Lau MWN (2012) *Python bivittatus*. The IUCN Red List of Threatened Species 2012: e.T193451A2237271. https://doi.org/10.2305/IUCN.UK.2012-1.RLTS.T193451A2237271.en. Downloaded on 19 February 2018
Stubbs G, Clarke-Billings L (2018) Snake owner was suffocated by 8ft pet python making him first person in Britain to be killed by breed. Mirror, January 24, 2018
Tokunaga S, Ono Y, Akagishi S (1991) The Iwakuni shirohebis, a group of albino Elaphe climacophora. Herpetological Review 22:120–121

Tryon BW, Whitehead J (1988) Reproduction in a little-known New Guinea Python, *Liasis papuanus* (Peters and Doria). Zoo Biol 7:371–379

Wall F (1921) Ophidia taprobanica or the snakes of ceylon. Colombo Mus. (H. R. Cottle, govt. printer), Colombo, 581 p

Whitaker R, Bhaskar S (1978) Rodent control in the Lakshadweep islands. Rodent Newsletter 2:1

Whitlock FL (1923) A trip to the Fortescue River and Hamersley ranges, north-West Australia. Emu-Austral Ornithol 22(4):259–273

Williams VL, Moshoeu TJ, Alexander GJ (2016) Reptiles sold as traditional medicine in Xipamanine and Xiquelene Markets (Maputo, Mozambique). S Afr J Sci 112:1–9

Willson JD, Dorcas ME, Snow RW (2011) Identifying plausible scenarios for the establishment of invasive Burmese pythons (*Python molurus*) in southern Florida. Biol Invasions 13(7):1493–1504

Worcester DC (1898) The phillipine islands and their people. The McMillian Co., London

Wright JH, Priston NE (2010) Hunting and trapping in Lebialem Division, Cameroon: bushmeat harvesting practices and human reliance. Endanger Species Res 11(1):1–2

You CW, Lin YP, Lai YH, Chen YL, Tang Y, Chou SP, Chang HY, Zappalorti RT, Lin SM (2013) Return of the pythons: first formal records, with a special note on recovery of the Burmese python in the demilitarized Kinmen islands. Zool Stud 52(1):8

Zhu ZY, Dennell R, Huang WW, Wu Y, Rao ZG, Qiu SF, Xie JB, Liu W, Fu SQ, Han JW, Zhou HY (2015) New dating of the *Homo erectus* cranium from Lantian (Gongwangling), China. J Hum Evol 78:144–157

Chapter 20
Risk Assessment Model for Brown Treesnake Introduction into the Continental United States

Samantha S. Kahl, Scott E. Henke, David Britton, and Gad Perry

Invasive species are a growing problem worldwide, currently costing the United States approximately $120 billion per year (Marbuah et al. 2014). A primary challenge for wildlife managers is determining when and where an invasive species will spread (Holcombe et al. 2010), and research has shown propagule pressure and climate match as important predictors in the establishment of non-native reptiles and amphibians (Bomford et al. 2009; van Wilgen and Richardson 2012). The increase in transportation pathways due to the high worldwide demand for commodities, in addition to human travel and the pet trade, have made global species transport a major concern for ecosystem management (Hulme 2009). Invasibility of a species may, especially regarding reptiles and amphibians, be related to the species' detectability within their introduced environment. Species with lower detection probabilities may have more time following introduction to successfully become naturalized and then to become invasive and ecologically problematic within that environment (Mehta et al. 2007).

The brown treesnake (*Boiga irregularis*; BTS) is a venomous, nocturnal, invasive snake on the island of Guam and considered a potential ecological threat to other ecosystems (Perry and Vice 2009). BTS are relatively long and thin snakes, though they have been known to reach up over 3.1 m in length, with longer snakes maintaining a larger body circumference. They exhibit a variation of patterns, including muted blotched pattern to solid (patternless) design, and several different color morphs, including a light golden tan, green, brown, grey, or even yellow

S. S. Kahl (✉) · G. Perry
Department of Natural Resource Management, Texas Tech University, Lubbock, TX, USA
e-mail: samantha.kahl@ttu.edu

S. E. Henke
Caesar Kleberg Wildlife Research Institute, Texas A&M University-Kingsville, Kingsville, TX, USA

D. Britton
San Marcos Aquatic Resources Center, U.S. Fish & Wildlife Service, San Marcos, TX, USA

F. M. Angelici, L. Rossi (eds.), *Problematic Wildlife II*,
https://doi.org/10.1007/978-3-030-42335-3_20

(Rodda et al. 1992), with some more vibrant colorations on individuals in Australia. Many of the color and pattern combinations are relatively well-camouflaged in its jungle-like semi-arboreal habitat, as well as in brown cardboard boxes and shipping materials, which an individual may easily enter and remain for several months sans food or water (Rodda et al. 1999a). BTS were accidentally introduced to Guam following World War II on cargo from the Admiralty Islands, north of Papua New Guinea (Rodda et al. 1992), and since that time their ecological and economic impacts in the invaded range have been extensive (Savidge 1987). BTS have already been found as hitchhikers on shipments coming from Guam to multiple different US locations, including Anchorage, AK; Corpus Christi, TX; and most recently in a shipment arriving in McAlester, OK, all of which further exhibit the propagule pressure (or introduction and invasion potential) for this species to arrive in other areas of the world from Guam.

A US military relocation from Japan to Guam (a US territory) began in 2010 with an original expectation of increasing the overall human population of Guam from approximately 178,000 to over 210,000 (Department of Defense 2010). More current estimates predict that between 2021 and 2023, the population of Guam will increase 5.6% more than what it would have naturally in a regular 2-year period, without the military relocation (Department of Defense 2015). High densities and low detectability of BTS due to the general ecology of the species, along with expected changes in Guam's transportation network over time, may greatly affect and increase the potential for BTS dispersal. Prevention efforts should be the first priority and should be intense within identified high-risk vectors (Bax et al. 2001), such as military cargo shipments and flights and commercial flights. The illegal pet trade may be another vector of concern for BTS, since reptiles tend to be subject to illegal trade and species are mainly those that are rare or considered dangerous (Bush et al. 2014), and although BTS aren't rare in Guam, they are venomous and "rare" in other areas of the world. In Guam, cargo and cargo carriers are already searched at least once, and sometimes multiple times prior to leaving the island, but an earlier model of BTS dispersal indicated that inspection and dispersal prevention efforts in Guam may be insufficient to prevent future spread of the species to off-island locations (Perry and Vice 2009), and locations at high risk of BTS arrival have been identified (Kahl et al. 2012). However, the risk of establishment at such locations has yet to be assessed with an estimate of propagule pressure. Although there have not been any successful introductions in the continental United States to date, the information on arrival locations, quantifying the potential for introduction in those locations and identifying the potential for establishment, offers an estimate of propagule pressure.

In-depth studies of modeling programs and comparisons between methods have developed risk assessment and niche modeling into an expanding field of study within ecology. Peterson et al. (2003) found an ecological niche modeling using current distribution data for an invasive species that can accurately predict possible locations for future invasions. Divergent conclusions based on climatic suitability modeling of invasive pythons (*Python molurus*) in Florida, which showed suitable areas throughout the southern United States (Rodda et al. 2009; Pyron et al. 2008), exemplified the potential for both under-fit and over-fit ecological niche models. Program MaxEnt (Phillips et al. 2006) is a freely available technology that is well suited for modeling species distributions using presence-only data, when absence data for the species are unavailable. The program uses a maximum entropy modeling approach to determine similar, or suitable, locations for species according to the climatic or habitat variables used in the model (Phillips et al. 2006). MaxEnt has been used successfully as a tool for mapping species distributions or climatic or habitat suitability for species even with small sample sizes and widely scattered data (Kumar and Stohlgren 2009; Holcombe et al. 2010; Gregr 2011) and is generally equal or in some cases superior to alternatives such as the Genetic Algorithm for Rule Set Prediction (GARP; Phillips et al. 2006, Peterson et al. 2007, Papeş and Gaubert 2007, Hernandez et al. 2006) for models with small sample sizes. MaxEnt tends to be superior to GARP especially with a priori identification of biologically relevant variables (Rodda et al. 2011), at low input thresholds, and when dealing with potential omission error, though different uses can present different challenges and levels of transferability (Townsend Petersen et al. 2011).

Previous examinations of the BTS climate envelope (Rodda et al. 2007; Rödder and Lötters 2010a) have shown climatic suitability for the snake in the Mariana, Hawaiian and Caroline Islands, as well as central Africa, Central America, and South America. Within the continental United States, BTS-suitable climate has been predicted throughout the southeast region, particularly along the coastline (Rodda et al. 2007, Rödder and Lötters 2010a). Our objective was to examine the potential climatic suitability for BTS worldwide under current climatic conditions and identify the most likely locations for brown treesnake introduction and survival within the continental United States (CONUS). In order to improve upon previous approaches for BTS climate matching, such as minimal convex polygon analysis (Rodda et al. 2007) and previous MaxEnt models (Rödder and Lötters 2010a), we have combined different aspects of these two studies to improve niche prediction. We hypothesized that climatically suitable locations for BTS are in the southern coastal United States, where higher temperatures and comparable precipitation occur. The focus of previous studies of BTS risk assessment has been on defining transport vectors and general areas of potential dispersal, using only climate to define biologically relevant potential introduction locations. Our examination of specific final destinations for shipments from Guam describes a measure of survivability for BTS, based on the climate of the arrival port, to identify locations with the highest potential propagule pressure.

20.1 Methods

We obtained BTS collection localities acquired from ten museum collections: California Academy of Science, Los Angeles County Museum of Natural History, Bernice Pauahi Bishop Museum, Arctos Museum of Vertebrate Zoology, Harvard University Museum of Comparative Zoology, the Smithsonian Institute National

Museum of Natural History, Staatliches Museum für Naturkunde Stuttgart, Yale University Peabody Museum, University of Kansas Biodiversity Institute, and the Queensland Museum. Examining long-term collections from several locations allowed us to obtain a broad climatic suitability envelope by representing confirmed presence over a long period of time and wide geographic scope. We chose to use museum specimen data and eliminate data based on unconfirmed sightings to avoid modeling climatic suitability for misidentified individuals (Lozier et al. 2009) or descriptions that were simply too obscure and broad to be localized to one geographic location. Geographic locality data were absent for some of the older records used, but the description of the location was sufficient to enable us to determine locations in decimal degrees using Google Earth (Google Inc. 2009). To reduce the bias of spatial autocorrelation, we removed repeated values and pooled values within 1 degree of latitude and longitude (Rodda et al. 2007). We examined the distribution of presence data for individuals in the final dataset and assumed no misidentification of individuals (Newbold 2010) based on BTS distribution as described in previous literature (Rodda et al. 1992). In the following analyses, we obtained a total of 83 specimen localities from the native range (Australia, Papua New Guinea, and the Solomon Islands) and 44 specimen localities from the invaded range (Guam).

To predict climatic suitability for BTS, we used the WORLDCLIM (www.worldclim.org, Hijmans et al. 2005) database, selecting 6 of the 19 possible bioclimatic variables at a 10 arc-minute resolution to provide a reduced model. The six variables we selected included variables likely to be biologically relevant to a mostly tropical snake species (Rödder and Lötters 2010a, b): maximum temperature during the warmest month, minimum temperature during the coldest month, annual precipitation, total precipitation during the wettest month, total precipitation during the driest month, and total precipitation during the warmest quarter. We avoided including variables that combined temperature and precipitation measures due to difficulty of interpretation (Elith et al. 2013), and we kept the total number of variables low to avoid overfitting and underpredicting (over-conservative models) our model (Rotenberry et al. 2006; Dormann et al. 2013).

We used the program MaxEnt (Phillips et al. 2006) to predict the potential suitability of habitats to BTS. We modeled suitability based on records from the native range and the invaded range (Guam), including all presence data. MaxEnt models can differ in complexity simply by changing or adding model parameters, or feature types, and complexity tends to be better supported in models with larger sample sizes (Elith et al. 2013). To decrease the potential for overfitting the model, leading to a highly conservative and potentially restricted distribution, we used the hinge feature class transformation with a tuning regularization value of 0.5 and random background treatments, as recommended by Phillips and Dudik (2008) and Elith et al. (2011). The hinge feature class offers relatively smooth fit to samples >15, in comparison to the more abrupt cutoffs of the threshold features, and functions more like a generalized additive model (GAM), making linear and threshold features redundant (Elith et al. 2011). Prior to model production, we randomly separated the presence data into two subsets. Training data consisted of 75% of the presence records, and test data consisted of the remaining records. We also applied a threshold rule of a fixed cumulative value of 1. We ran 25 replicates of each dataset, allowing us to examine the average, maximum, and minimum predictions and provide a measure of result dispersal. We present one standard deviation from the mean.

We used ArcGIS 10 (Environmental Systems Research Institute, Inc., California, USA), applying a stretch with an equalized histogram smoothing option and 15 breaks in coloration to the MaxEnt output. We overlaid these distributions with data on final destinations for shipments leaving Guam (Wisniewski 2010; Kahl et al. 2012), georeferenced using Google Earth (Google Inc. 2009) and split into seven categories based on number of shipments and on natural breaks in distribution of the arrival data. Our models depict the most at-risk areas within the continental United States for receiving shipments from Guam (the human-aided transport factor of potential introduction), as well as the climatic suitability (the biological relevance factor for the species).

20.1.1 *Statistical Analysis*

We used the jackknife procedure in program MaxEnt to determine relative importance of each climactic variable in the native, invaded, and combined range of BTS. For model evaluation we used the area under the curve (AUC) values of a receiver-operating characteristics (ROC) plot, which represents the probability that a random presence locality would be ranked higher than a randomly chosen background locality for BTS (Phillips et al. 2006). An AUC value of 0.5 represents a random prediction, and an AUC value of 1.0 represents a perfect prediction (Swets 1988).

20.2 Results

Total precipitation during the warmest quarter was the variable with the highest contribution (59.8%) to the climate model (Table 20.1). The southeastern United States, from the Gulf Coast of Texas eastward through Florida and South Carolina, had the highest climatic suitability in our MaxEnt models. The results for predicted climatic suitability for BTS worldwide (avg. AUC = 0.962, $S^2 = 0.004$), using current climatic conditions and all presence records, show that areas of highest climatic suitability are located in every continent except Antarctica (Fig. 20.1). Climatically suitable habitat for BTS is more widespread below the equator, with some additional (though very limited) suitable areas exhibited in Europe, Southeast Asia, and the southernmost parts of North America (Fig. 20.1). Climatically suitable areas for BTS in North America are mostly located in the southeastern and coastal eastern United States (Fig. 20.2), as well as much of Mexico.

Table 20.1 Variable contribution (%) to calculating predicted brown treesnake (BTS) climatic suitability in Maxent models, from the native and invaded range data included, based on six chosen bioclimatic variables

Variable	Contribution (%) to average native range model
Precipitation of the warmest quarter	59.8
Minimum temperature of the coldest month	23.6
Annual precipitation	6.8
Precipitation of the driest month	6.7
Precipitation of the wettest month	1.7
Maximum temperature of the warmest month	1.4

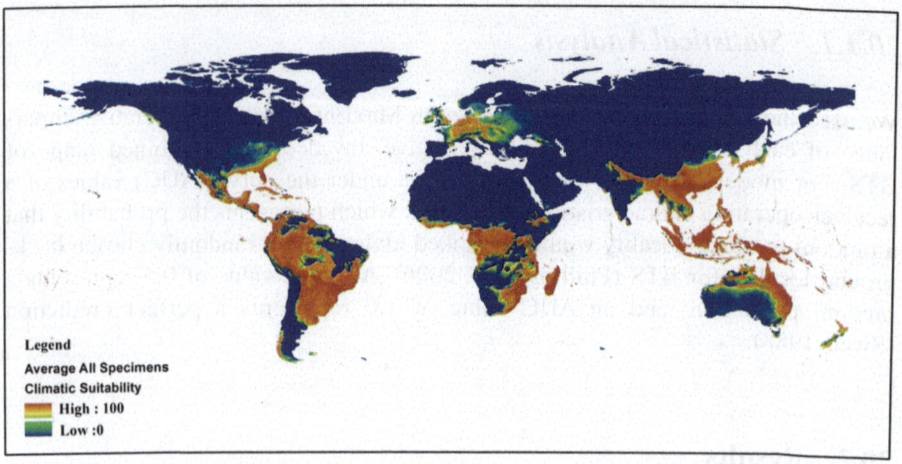

Fig. 20.1 Average worldwide climatic suitability for brown treesnakes (BTS) calculated by MaxEnt based on all presence locality data from the native and invaded ranges. Areas of high climatic suitability are shown in red and areas of low climatic suitability are shown in blue, with moderate climatic suitability shown in green, yellow, and orange

Many sites in North America receive shipments from Guam, but most of them arrive in areas of relatively low climatic compatibility (Fig. 20.3). San Diego receives the most shipments from Guam, but is characterized by low predicted climatic suitability for BTS (Fig. 20.3). Thus, no single location in North America has both high arrival rates and high climactic suitability. However, many sites in Texas, Georgia, Florida, North Carolina, South Carolina, and Virginia are moderate-to-high risk shipping locations within the predicted areas of high climatic suitability (Figs. 20.2 and 20.3).

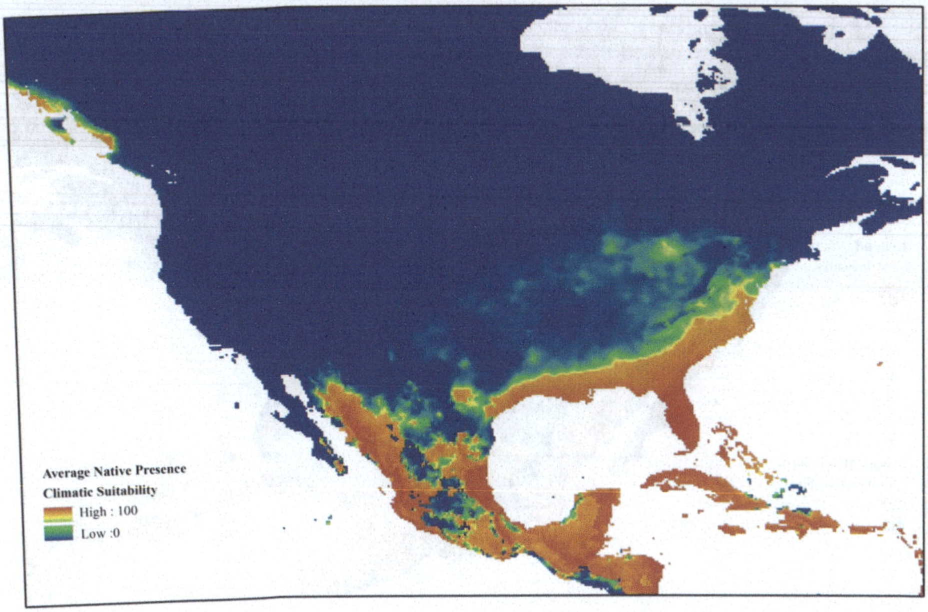

Fig. 20.2 Average predicted climatic suitability of brown treesnakes (BTS) within North America based on presence data from the native range. Areas of high climatic suitability are shown in red and areas of low climatic suitability are shown in blue, with moderate climatic suitability shown in green, yellow, and orange

20.3 Discussion

Our study is unique in quantifying both climate compatibility and propagule pressure for BTS. Our findings suggest that the southeastern United States (specifically San Antonio, Texas; Pensacola and Jacksonville, Florida; and Hampton, Arlington, and Virginia Beach, Virginia) have high climactic suitability for BTS and therefore are at relatively high risk for BTS introduction and survival. Moreover, San Diego, California, should be considered a high-risk area for BTS introduction, despite relatively low climatic suitability, because it greatly surpasses other destinations in receiving the most shipments from Guam each year (Kahl et al. 2012). Although the climate of San Diego is typically too dry for BTS long-term survival, the high propagule pressure in that area could raise concerns about the potential for BTS to enter CONUS in San Diego and then unintentionally hitchhike to a new location while in search of refugia from the dry climate. Although the likelihood of this type of travel may be low or seemingly far-fetched, keeping in mind that only a single individual may be required to become invasive in a new environment, San Diego could still be a location for raised awareness of BTS. Establishment of BTS in any US location, and especially a major export port, could serve as a stepping-stone to further spread. Perhaps of most immediate concern would be BTS introduction into Mexico. If BTS became established in Mexico, then CONUS would be at high risk of BTS

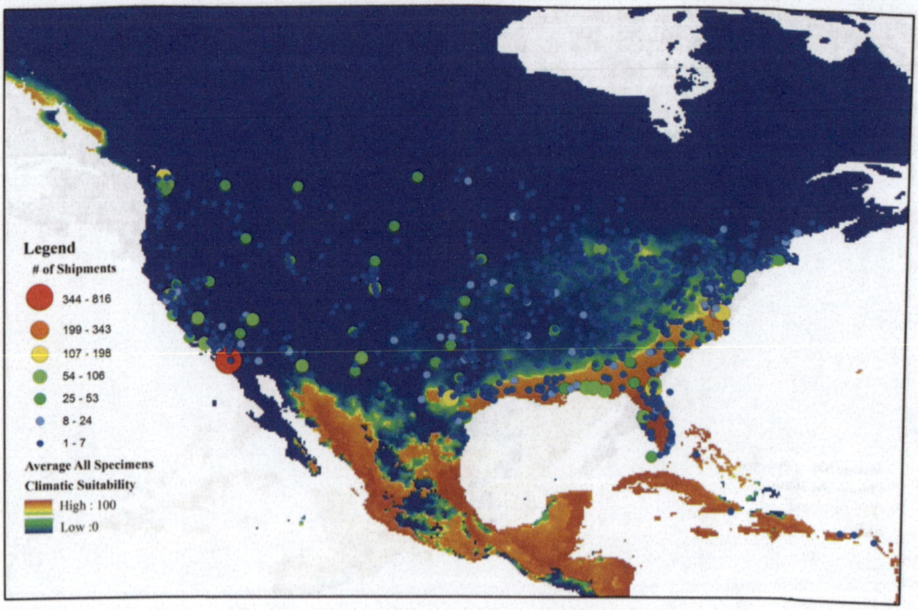

Fig. 20.3 Average climatic suitability for brown treesnakes (BTS) in North America based on presence data from the native and invaded range, overlain with information on possible propagule pressure from Kahl et al. (2012). Areas of high climatic suitability are shown in red and areas of low climatic suitability are shown in blue, with moderate climatic suitability shown in green, yellow, and orange. The size and color of each spot indicate the number of shipments received from Guam on an annual basis

invasion from the trucking industry. Trucks transport large quantities of materials into the United States via the newly established Interstate Highway 69 that begins in southern coastal Texas and travels through the Midwestern United States into Canada. Although this example is just one possible transportation vector for BTS invasion into the United States, the large area identified by the climate model (Figs. 20.1 and 20.2) as suitable for BTS highlights a number of alternative introduction routes that could result in establishment.

The use of both the invaded and native range presence data is recommended in order to provide the most comprehensive predictions (Jiménez-Valverde et al. 2011). Predicting which habitat or climate variable/s will be the primary driver or limitation of BTS spread in the future is inherently difficult; therefore we find validity in examining predictions using presence data from the entire known range. Consistent with the precautionary principle, we therefore suggest that the maximum predicted suitable climate models be used as the basis for prevention efforts. Phillips et al. (2006) note that area predictions from MaxEnt may be larger than the realized spread of the species; however, Rodda et al. (2011) found that a more specific fitting of the realized climatic suitability can produce a more limited and conservative prediction of the species' tolerance. Low climatic variability in our presence data from Guam may have limited prediction of fundamental climate space (Rodda et al.

2011), and we agree that some questions regarding the transferability and extrapolation of data to distant geographic space remain (Townsend Petersen et al. 2011). Nevertheless, we do not aver that BTS will eventually exist in all areas that are defined in our model as high risk, nor that BTS will never exist in areas predicted to have low or no climatic suitability. Distribution modeling for invasive species works best when the predicted area of distribution is very similar in habitat and climate characteristics to that of the native range (Elith et al. 2013); however, when considering long-range human-aided dispersal events, it is not prudent to wholly limit consideration to like-climates and habitats. Our models show BTS climatic suitability in some areas that might typically be considered too dry for the species to survive; however we suggest the liberal model produced here, including the propagule pressure data, is preferable to to more conservative models when examining potential locations for introductions of known invasives. Modeling approaches that use climate matching assume species retain their climatic niche in the newly invaded range; however, species may undergo rapid evolution and shift niches, thus occupying novel climates and habitats (Broennimann et al. 2007; Prentis et al. 2008; Rödder and Lötters 2010b; Chapman et al. 2017; González-Moreno et al. 2015). For example, Rodda et al. (1999b) hypothesized that BTS have potential to exist in suburban environments where they can seek refuge in garages, sheds, and homes during inclement weather and persist on a diet of house geckos (*Hemidactylus frenatus*), house mice (*Mus musculus*), sparrows (*Passer montanus*), and pigeons (*Columba livia*). Therefore, for the purpose of preventing invasive species introduction, a broader consideration of the possible areas of spread may be preferable to an underestimation of inhabitable area, which is common in niche models based on climatic data alone (González-Moreno et al. 2015; Rodda et al. 2009).

Our models have not captured the full scope of variables that may be important to BTS establishment. Propagule pressure assessments may have low predictive power in assessing whether or not a species can invade a particular area (Sagata and Lester 2009), unless competition is lacking or absent in the new habitat (Chadwell and Engelhardt 2008; Hulme 2009). We have not modeled all variables that may be important to BTS establishment. For example, the BTS is a generalist predator (Savidge 1988), and we have not assessed prey availability. Because the invaded range of BTS is lacking other snake species, excluding the insectivorous blind snake (*Ramphotyphlops braminus*) (Savidge 1988), competition for resources in the invaded range may be reduced in comparison to the native range. Fairly common species such as the black rat snake (*Pantherophis obsoleta*), which can be found throughout areas we identified as climatically suitable for BTS in the continental United States, may potentially exclude establishment of introduced snakes through competition, though this is the closest niche overlap BTS may have with continental US snakes and prior success of BTS as an invasive even in urban areas may imply added potential for success, with or without black rat snakes. If we can broadly focus on specific areas revealed here as high risk, future mechanistic species niche models may concentrate focus on smaller coverage areas with more detail. Further details may include additional drivers, such as competition and prey availability variables, and could incorporate additional detail about niche shifts during invasion

(Chapman et al. 2017) and human disturbance (Zhu et al. 2017; Beans et al. 2012). Some species may be restricted by human presence through additional management efforts in heavily populated areas (Zhu et al. 2017), but BTS are known stowaways and hitchhikers; heavily populated urban areas could enhance potential for human-aided transport of this species.

Gallardo and Aldridge (2013) also indicate that inclusion of more sociological factors (when data are available) in niche models could expand suitability predictions by up to sixfold, implying that results from liberal climatic suitability models may be more realistic than conservative estimates. Modeling presence-only data gives us some insight into the potential climatic suitability of the species, but the addition of absence data and abundance data (Kery 2002; González-Moreno et al. 2015) would contribute to the study of BTS within the native range by offering more information on preferred climate and habitat. We also recommend that future work examine whether using climatic databases such as Saga may provide improved assessment of rainfall patterns (Soria-Auza et al. 2010).

Guam provides an invasion pathway to North America. Previous invasions and human alteration of habitat have made Guam a disturbed ecosystem (and therefore less stable and more prone to invasion under the biotic acceptance hypothesis (Possel et al. 2013)). The BTS is only one of many species that have the potential to spread from Guam to North America via military shipments and other methods of human-aided dispersal. Other non-native species already in Guam also could invade North America and other locations through human-aided transport or transport via a natural disaster, such as a hurricane in which high winds and heavy rains can relocate species. One species example is the curious skink (*Carlia fusca*), a crucial food source to the BTS, or several insect species such as the coconut rhinoceros beetle (*Oryctes rhinoceros*), which is considered an invasive pest species and has been a major subject of biological control issues (Marshall et al. 2017). Other species not yet in Guam, such as the habu (*Trimeresurus flavoviridis*), which is a dangerous invasive venomous snake species in Japan (Mishima et al. 1999), could arrive in Guam through military movements and shipments from Okinawa and subsequently threaten invasion to North America. The habu is of high concern as a potential hitchhiker because of its impacts on human health in Okinawa due to snakebites, which may be vastly underestimated (Yasunaga et al. 2011). Under the "guilty until proven innocent" train of thought regarding known predatory or invasive species (Ruesink et al. 1995), prevention efforts for such species also should be prioritized. Efforts to protect Guam also will help control an important transport pathway to North America. Preventing further invasions of Guam should be a high priority.

Finally, climate change brings about new issues with species invasions and may allow individuals on the "edge" of a specific climate to have an increased ability to adapt and expand their acceptable range (Holcombe et al. 2010; Moreno-Rueda et al. 2012). Therefore, changes in future climate are expected to alter climatic suitability for the BTS throughout the southern United States and elsewhere and even to potentially alter the transport vectors for this and other species (Marshall et al. 2008). We recommend incorporating such "edge" species into future models and preemptively design and implement altered inspection protocols. However, these potential avenues for future improvement should not prevent current action. As Beissinger and Westphal (1998) stated, "Uncertainty is inherent in decision-making but is not an excuse for not making decisions."

20.3.1 Management Implications

Public awareness and rapid response program setup are essential within defined areas of interest for species introduction (Bax et al. 2001; Perry and Farmer 2011). High-risk areas should therefore be targeted for public education and awareness and should have rapid response members in place (e.g., the North America Brown Tree Snake Control Team (NABTSCT)). These actions don't have to be costly, especially when dollars are limited. Additionally, because the vast majority of cargo is transported by sea (IMO 2008; Hulme 2009), we suggest that high-risk shipping ports in climatically suitable habitat become a higher priority for investment in BTS prevention. BTS educational materials provided by the NABTSCT (available at www.nabtsct.net) should be distributed at ports and airports, as well as in wildlife parks and reserves, zoos, and science classrooms, in areas identified as high risk. The BTS is already a high-profile invasive species; therefore, general information regarding the species is relatively easy to find and may be less difficult to incorporate into a typical classroom or other educational setting. Consistent with the recommendation of Perry and Farmer (2011), we suggest that NABTSCT actively seek support and recruit members for rapid response teams within local herpetological communities and pet trade. Difficulties persist in promoting BTS awareness in North America; therefore, education and public information are just as important at the point of origin (i.e., Guam and Australasia) for BTS.

Changes in future climatic suitability of BTS throughout the southern United States may increase this snake's ability to persist in CONUS in the future. The need is dire for inspection protocols to be implemented *now* by CONUS port authorities and other governmental organizations to prevent species introductions. Additional educational information in the form of flyers, cards, website advertisement, or even a seminar at CONUS port authorities may also help to increase inspectors' BTS knowledge base and understand the issue. We understand that some educational materials already exist (though general public knowledge of them may be limited), that inspections in Guam are intensive and regular (though some inspections are missed (Kahl et al. 2012)), and that inspection protocol already exist with several US port authorities for BTS (though the risks aren't always understood), and we understand that dollars are precious. What we are suggesting is that an increase or revamp in some of those things, when affordable, may aid in preventing further ecosystem damage due to invasives. The military buildup in Guam is currently well underway, and overall shipments to CONUS are expected to increase. Immediate preventative measures are warranted.

Acknowledgments We thank Wildlife Services in Guam for providing shipment inspection data. We also thank E. Redeker for the advice and guidance on the use of ArcGIS technology. Funding was provided by the North America Brown Tree Snake Control Team through the US Fish and Wildlife Service; Caesar Kleberg Wildlife Research Institute at Texas A&M University, Kingsville; and the Hispanic Leaders in Agriculture and the Environment fellowship. This is manuscript number 12-119 of the Caesar Kleberg Wildlife Research Institute.

Literature Cited

Bax N, Carlton JT, Mathews-Amos A, Haedrich RL, Howarth FG, Purcell JE, Rieser A, Gray A (2001) The control of biological invasions in the world's oceans. Conserv Biol 15:1234–1246

Beans CM, Kilkenny FF, Galloway LF (2012) Climate suitability and human influences combined explain the range expansion of an invasive horticultural plant. Biol Invasions 14:2067–2078

Beissinger SR, Westphal MI (1998) On the use of demographic models of population viability in endangered species management. J Wildl Manag 62:821–841

Bomford M, Kraus F, Barry SC, Lawrence E (2009) Predicting establishment success for alien reptiles and amphibians: a role for climate matching. Biol Invasions 11:713–724

Broennimann O, Treier UA, Müller-Schärer H, Thuiller W, Peterson AT, Guisan A (2007) Evidence of climatic niche shift during biological invasion. Ecol Lett 10:701–709

Bush ER, Baker SE, Macdonald DW (2014) Global trade in exotic pets 2006–2012. Conser Biol 28(3):663–676

Chadwell TB, Engelhardt KAM (2008) Effects of pre-existing submersed vegetation and propagule pressure on the invasion success of *Hydrilla verticillata*. J Appl Ecol 45:515–523

Chapman DS, Scalone R, Stefanic E, Bullock JM (2017) Mechanistic species distribution modeling reveals a niche shift during invasion. Ecology 98:1671–1680

Department of Defense (2010) Record of decision for Guam and CNMI military relocation including relocating marines from Okinawa, transient nuclear aircraft carrier berth, and air and missile defense task force. Department of Defense: Department of the Navy and Department of the Army, Washington, DC

Department of Defense (2015) Record of decision for the final supplemental environmental impact statement for Guam and Commonwealth of the Northern Mariana Islands military relocation. Department of Defense: Department of the Army, Washington, DC

Dormann CF, Elith J, Bacher S, Buchmann C, Carl G, Carré G, García Marquéz JR, Gruber B, Lafourcade B, Leitão PJ, Münkemüller T, McClean C, Osborne PE, Reineking B, Schröder B, Skidmore AK, Zurell D, Lautenbach S (2013) Collinearity: a review of methods to deal with it and a simulation study evaluating their performance. Ecography 36:27–46

Elith J, Phillips SJ, Hastie T, Dudik M, Chee YE, Yates CJ (2011) A statistical explanation of MaxEnt for ecologists. Divers Distrib 17:43–57

Elith J, Simpson J, Hirsch M, Burgman MA (2013) Taxonomic uncertainty and decision making for biosecurity: spatial models for myrtle/guava rust. Australian Plant Pathol 42:43–51

Gallardo B, Aldridge DC (2013) The 'dirty dozen': socio-economic factors amplify the invasion potential of 12 high-risk aquatic invasive species in Great Britain and Ireland. J Appl Ecol 50:757–766

González-Moreno JM, Diez D, Richardson M, Vilà M (2015) Beyond climate: disturbance niche shifts in invasive species. Glob Ecol Biogeogr 24:360–370

Google Inc. (2009) Google Earth Version 5.1.3535.3218 [Software]. Available from http://earth.google.com/

Gregr EJ (2011) Insights into North Pacific right whale *Eubalaena japonica* habitat from historic whaling records. Endanger Species Res 15:223–239

Hernandez PA, Graham CH, Master LL, Albert DL (2006) The effect of sample size and species characteristics on performance of different species distribution modeling methods. Ecography 29:773–785

Hijmans RJ, Cameron SE, Parra JL, Jones PG, Jarvis A (2005) Very high resolution interpolated climate surfaces for global land areas. Int J Climatol 25:1965–1978

Holcombe TR, Stohlgren TJ, Jarnevich CS (2010) From points to forecasts: predicting invasive species habitat suitability in the near term. Diversity 2:738–767

Hulme PE (2009) Trade, transport and trouble: managing invasive species pathways in an era of globalization. J Appl Ecol 46:10–18

IMO (2008) International shipping and world trade: facts and figures. International Maritime Organization, Maritime Knowledge Centre, London

Jiménez-Valverde A, Peterson AT, Soberón J, Overton JM, Aragón P, Lobo JM (2011) Use of niche models in invasive species risk assessments. Biol Invasions 13:2785–2797

Kahl SS, Henke SE, Hall MA, Litt AR, Perry G, Britton D (2012) Examining a potential brown treesnake transport pathway: shipments from Guam. Hum-Wildl Interact 6:204–211

Kery M (2002) Inferring the absence of a species: a case study of snakes. J Wildl Manag 66:330–338

Kumar S, Stohlgren TJ (2009) MaxEnt modeling for predicting suitable habitat for threatened and endangered tree *Canacomyrica monticola* in New Caledonia. J Ecol Nat Environ 1:094–098

Lozier JD, Aniello P, Hickerson MJ (2009) Predicting the distribution of sasquatch in western North America: anything goes with ecological niche modelling. J Biogeogr 36:1623–1627

Marbuah G, Gren I, McKie B (2014) Economics of harmful invasive species: a review. Diversity 6:500–523

Marshall JD, Blair JM, Peters DPC, Okin G, Rango A, Williams M (2008) Predicting and understanding ecosystem responses to climate change at continental scales. Front Ecol Environ 6:273–280

Marshall SDG, Moore A, Vaqalo M, Noble A, Jackson TA (2017) A new haplotype of the coconut rhinoceros beetle, *Oryctes rhinoceros*, has escaped biological control by *Oryctes rhinoceros* nudivirus and is invading Pacific Islands. J Invertebr Pathol 149:127–134

Mehta SV, Haight RG, Homans FR, Polasky S, Venette RC (2007) Optimal detection and control strategies for invasive species management. Ecol Econ 61:237–245

Mishima S, Tanaka H, Sawai H (1999) The biology of the habu (*Trimeresurus flavoviridis*). In: Rodda GH, Sawai Y, Chiszar D, Tanaka H (eds) Problem snake management: the habu and the brown treesnake. Cornell University Press, Ithaca, pp 29–43

Moreno-Rueda G, Pleguezuelos JM, Pizarro M, Montori A (2012) Northward shifts of the distribution of Spanish reptiles in association with climate change. Conserv Biol 26:278–283

Newbold T (2010) Applications and limitations of museum data for conservation ecology, with particular attention to species distribution models. Prog Phys Geogr 34:3–22

Papeş M, Gaubert P (2007) Modeling ecological niches from low numbers of occurrences: assessment of the conservation status of poorly known viverrids (Mammalia, Carnivora) across two continents. Divers Distrib 13:890–902

Perry G, Farmer M (2011) Reducing the risk of biological invasion by creating incentives for pet sellers and owners to do the right thing. J Herpetol 45:134–141

Perry G, Vice D (2009) Forecasting the risk of brown tree snake dispersal from Guam: a mixed transport-establishment model. Conserv Biol 23:992–1000

Peterson AT, Papeş M, Kluza DA (2003) Predicting the potential invasive distributions of four alien plant species in North America. Weed Sci 51:863–868

Peterson AT, Papeş M, Eaton M (2007) Transferability and model evaluation in ecological niche modeling: a comparison of GARP and MaxEnt. Ecography 30:550–560

Phillips SJ, Dudik M (2008) Modeling of species distributions with MaxEnt: new extensions and a comprehensive evaluation. Ecography 31:161–175

Phillips SJ, Anderson RP, Schapire RE (2006) Maximum entropy modeling of species geographic distributions. Ecol Model 190:231–259

Possel SA, Beard KH, Callahn CM, Ferreira RB, Stevenson ET (2013) Biotic acceptance in introduced amphibians and reptiles in Europe and North America. Glob Ecol Biogeogr 22:192–201
Prentis PJ, Wilson JRU, Dormontt EE, Richardson DM, Lowe AJ (2008) Adaptive evolution in invasive species. Trends Plant Sci 13:288–294
Pyron RA, Burbrink FT, Guiher TJ (2008) Claims of potential expansion throughout the U.S. by invasive python species are contradicted by ecological niche models. PLoS One 3(8):e2931
Rodda GH, Fritts TH, Conry PJ (1992) Origin and population growth of the brown tree snake *Boiga irregularis*, on Guam. Pac Sci 1:46–57
Rodda GH, Fritts TH, McCoid MJ, Campbell EW (1999a) An overview of the biology of the brown treesnake (*Boiga irregularis*), a costly introduced pest on Pacific islands. Problem snake management: the habu and the brown treesnake. Cornell University Press, Ithaca, pp 44–80
Rodda GH, McCoid MJ, Fritts TH, Campbell EW (1999b) Population trends and limiting factors in *Boiga irregularis*. Problem snake management: the habu and the brown treesnake. Cornell University Press, Ithaca, pp 236–253
Rodda GH, Reed RN, Jarnevich CS (2007) Climate matching as a tool for predicting potential North American spread of brown treesnakes. USDA National Wildlife Research Center Symposia: Managing Vertebrate Invasive Species: Proceedings of an International Symposium
Rodda GH, Jarnevich CS, Reed RN (2009) What parts of the U.S. mainland are climatically suitable for invasive alien pythons spreading from Everglades National Park? Biol Invasions 11:241–252
Rodda GH, Jarnevich CS, Reed RN (2011) Challenges in identifying sites climatically matched to the native ranges of animal invaders. PLoS One 6:e14760
Rödder D, Lötters S (2010a) Potential distribution of the alien invasive brown tree snake, *Boiga irregularis* (Reptilia: Colubridae). Pac Sci 64:11–22
Rödder D, Lötters S (2010b) Explanative power of variables used in species distribution modelling: an issue of general model transferability or niche shift in the invasive greenhouse frog (*Eleutherodactylus planirostris*). Naturwissenschaften 97:781–796
Rotenberry JT, Preston KL, Knick ST (2006) GIS-based niche modeling for mapping species' habitat. Ecology 87:1458–1464
Ruesink JL, Parker IM, Groom MJ, Kareiva PM (1995) Reducing the risks of nonindigenous species introductions, guilty until proven innocent. Bioscience 45:465–477
Sagata K, Lester PJ (2009) Behavioural plasticity associated with propagule size, resources, and the invasion success of the Argentine ant *Linepithema humile*. J Appl Ecol 46:19–27
Savidge JA (1987) The ecological and economic impacts of an introduced snake on Guam and its threat to other Pacific islands. PIes 3:29–34
Savidge JA (1988) Food habits of *Boiga irregularis*, an introduced predator on Guam. J Herpetol 22:275–282
Soria-Auza RW, Kessler M, Bach K, Barajas-Barbosa PM, Lehnert M, Herzog SK, Böhner J (2010) Impact of the quality of climate models for modeling species occurrences in countries with poor climatic documentation: a case study from Bolivia. Ecol Model 221:1221–1229
Swets K (1988) Measuring the accuracy of diagnostic systems. Science 240:1285–1293
Townsend Petersen A, Papes M, Eaton M (2011) Transferability and model evaluation in ecological niche modeling: a comparison of GARP and Maxent. Ecography 30:550–560
van Wilgen NJ, Richardson DM (2012) The roles of climate, phylogenetic relatedness, introduction effort, and reproductive traits in the establishment of non-native reptiles and amphibians. Conserv Biol 26:267–277
Wisniewski SS (2010) Preventing brown treesnake introduction to the continental United States through education, awareness, and risk assessment modeling. Dissertation. Texas A&M University-Kingsville, Kingsville
Yasunaga H, Horiguchi H, Kuwabara K, Hashimoto H, Matsuda S (2011) Venomous snake bites in Japan. Am J Trop Med Hyg 84:135–136
Zhu G, Li H, Zhao L (2017) Incorporating anthropogenic variables into ecological niche modeling to predict areas of invasion of *Popillia japonica*. J Pest Sci 90:151–160

Chapter 21
The Asian Toad (*Duttaphrynus melanostictus*) in Madagascar: A Report of an Ongoing Invasion

Fulvio Licata, Franco Andreone, Karen Freeman, Sahondra Rabesihanaka, Eric Robsomanitrandrasana, James T. Reardon, and Angelica Crottini

21.1 Introduction

The need to identify biodiversity hotspots emerged as a direct consequence of increasing negative anthropogenic effects on ecosystems, driving scientists and experts to pinpoint priorities for conservation actions (Mittermeier et al. 1998; Myers et al. 2000). The 36 world regions currently identified as biodiversity hotspots represent just 2.4% of Earth's land surface although they support more than half of the world's endemic plant species and nearly 43% of all endemic bird, mammal, reptile and amphibian species (Conservation International 2019). Madagascar was classified as one of the richest and most threatened biodiversity hotspots worldwide, with only about 10% of its ecosystems persisting in a pristine or semi-pristine state (Myers et al. 2000; Goodman and Benstead 2005). Madagascar's uniqueness is characterized not only by the total number of species but also by the higher level of phylogenetic distinctiveness of its endemic families and genera (Ganzhorn et al. 2014). For example, among the extant vertebrates, the majority of the lineages succeeded in colonizing Madagascar through sea dispersal, during the Cretaceous – Cenozoic times, when ocean currents were particularly favourable for dispersion

F. Licata · A. Crottini (✉)
CIBIO, Research Centre in Biodiversity and Genetic Resources, InBIO, Universidade do Porto, Porto, Portugal

F. Andreone
Museo Regionale di Scienze Naturali, Torino, Italy

K. Freeman
Madagascar Fauna and Flora Group, Toamasina, Madagascar

S. Rabesihanaka · E. Robsomanitrandrasana
Ministère de l'Environnement et du Développement Durable, Antananarivo, Madagascar

J. T. Reardon
Department of Conservation, Fiordland District Office, Te Anau, New Zealand

F. M. Angelici, L. Rossi (eds.), *Problematic Wildlife II*,
https://doi.org/10.1007/978-3-030-42335-3_21

(Samonds et al. 2012) and when Madagascar underwent major adaptive radiations, which determined the composition of the present-day vertebrate assemblage (Vences et al. 2003; Goodman and Benstead 2003; Samonds et al. 2012, 2013; Crottini et al. 2012; Ganzhorn et al. 2014).

Anthropogenic habitat destruction and fragmentation have been particularly severe during the last decades (Harper et al. 2007; Vieilledent et al. 2018) and continue to be the biggest threat to the survival of Madagascar's biota. Among other threats, the impact of invasive species is emerging as an important factor impacting Madagascar's unique biota (Kull et al. 2014). Biological invasions take place when alien species are introduced to new regions, establish themselves in new environments and cause economic and environmental harm (Invasive Species Advisory Committee 2006). This human-mediated phenomenon has severe impacts on human society and is one of the major direct drivers of ecosystem changes, having an ever-increasing impact on biodiversity and ecosystem services (Millennium Ecosystem Assessment 2005). More than 50 animal and 1200 plant alien species are now documented in Madagascar (e.g. Binggeli 2003; Irwin et al. 2010; Kull et al. 2012, 2014; Rakotomanana et al. 2013; Ghulam et al. 2014; Goodman et al. 2017), and only a small fraction has been classified as invasive (Kull et al. 2014; Goodman et al. 2017). Complicating the issue further is the severe lack of border biosecurity practices and infrastructure that could help to prevent alien species introduction at the boarder (Reardon et al. 2018).

The Asian common toad (from here onwards referred to as 'Asian toad') *Duttaphrynus melanostictus* (Schneider 1799) (Fig. 21.1) was first discovered by the scientific community in Madagascar in March 2014 (Kolby et al. 2014), but its initial introduction is believed to have begun towards the end of the first decade of the twenty-first century (McClelland et al. 2015; Moore et al. 2015). This first report prompted some researchers to speculate on the possible risks associated with the management of this invasion (Mecke et al. 2014), who were calmed down by a following contribution by Andreone et al. (2014). The Asian toad invasion is likely the result of the accidental introduction of a few individuals via commercial shipping

Fig. 21.1 (**a**) *Duttaphrynus melanostictus* in Toamasina; (**b**) male *D. melanostictus* showing secondary characters, namely, the yellow-orange subgular vocal sacs and the nuptial pads on the first and second fingers. (Photo A by F. Andreone; B by A. Crottini)

containers from Southeast Asia (McClelland et al. 2015; Vences et al. 2017). After a lag phase of a few years, the Asian toad population expanded its range exponentially, and the apparent rate of spread continues unabated (Moore et al. 2015; Licata et al. 2019). Conservation biologists fear that this invasion will parallel the infamous invasion history of the cane toad *Rhinella marina* (Linnaeus 1758) in Australia, which has impacted native species communities (Shine 2010).

21.2 Natural History of the Asian Toad

The Asian toad belongs to the large and widely distributed family Bufonidae (Van Bocxlaer et al. 2010), also known as true toads. True toads have a natural worldwide distribution that includes the Americas, Eurasia and Africa, and their evolution has been mostly explained by a gradual adaptation towards a phenotype that is particularity suitable for colonizing new habitats (Duellman and Trueb 1994; Pennisi 2010; Van Bocxlaer et al. 2010).

The Asian toad is terrestrial, has crepuscular/nocturnal habits, is an ecological generalist, possesses parotoid glands that secrete bufadienolide toxin and has a strong invasive potential due to its tolerance for disturbed environments, as already documented in many other regions worldwide, both in its native and invasive range (Csurhes 2016; Reilly et al. 2017). This species is native to Southeast Asia, but after accidental introductions, it had invaded many other areas, which now include Bali, New Guinea, Sulawesi, Timor-Leste, Maldives, several other small islands of the Sundaic region and Madagascar (Gardiner 1906; Church 1960; van Dijk et al. 2004; Trainor 2009; Kolby et al. 2014; Reilly et al. 2017, Fig. 21.2). Recent phylogeographic analyses have revealed this species to be a complex of at least three distinct evolutionary lineages that are genetically and ecologically different and have discrete distributions (Asian mainland, coastal Myanmar and the Sundaic islands; Wogan et al. 2016). The Asian toad is most commonly found in disturbed lowland habitats (Hendrix et al. 2008), although they have been observed in deep forest in Thailand, Borneo and Sri Lanka (JR, pers. obs.). It is commonly found in rural and urban areas, often associated with human habitation (van Dijk et al. 2004), and it can be considered a habitat generalist in terms of breeding requirements (Daniel and Sekar 1989; Saidapur and Girish 2001).

The Asian toad is a typical large and stout toad, with rather short and strong legs and thick, dry skin covered with warts, featured by characteristic black bony ridges on the head. Individuals can grow up to 200 mm snout-vent length, with an average length of 78 mm (FL, pers. obs.). Females are generally larger than males, as is standard for most anuran species. The colouration of *D. melanostictus* varies from light cream to almost black, with brick-red tinges, but the most common pattern is pale yellow-brown streaked with black or reddish brown spots. Males show a yellowish orange subgular vocal sac, and during the breeding season, they also exhibit blackish nuptial pads on the first and second fingers (Fig. 21.1).

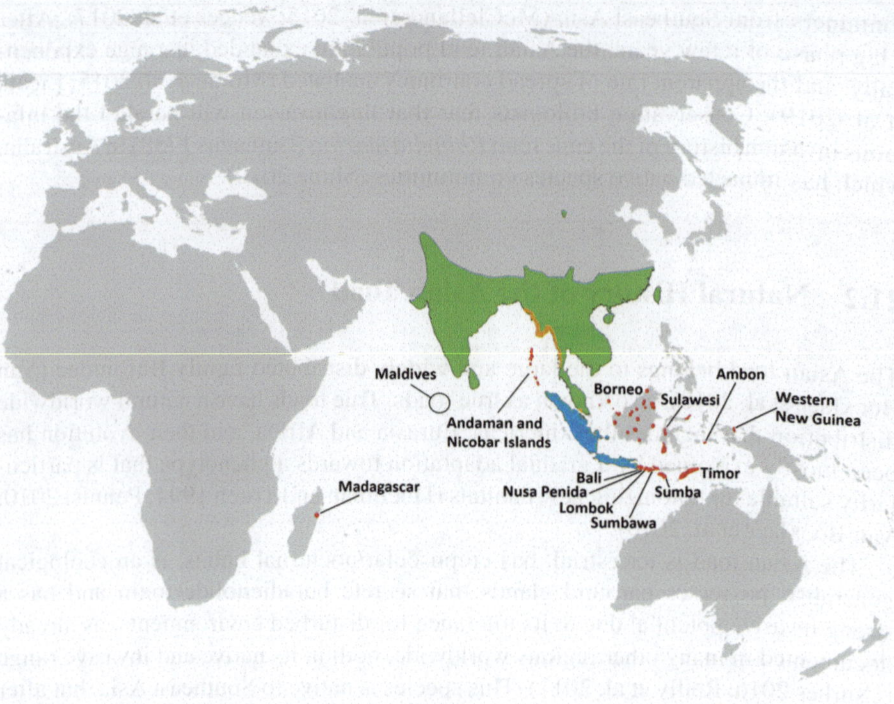

Fig. 21.2 Map showing the invasive (red) and the native distribution of the three principal clades of the *Duttaphrynus melanostictus* complex. In green, mainland lineage; in orange, coastal lineage; in blue, insular lineage. (Modified from van Dijk et al. 2004 and Wogan et al. 2016)

As for many other bufonid species, females usually reach sexual maturity at a later age than males (Ngo and Ngo 2013). A skeletochronological study on a native population of Asian toad in India reported that this species can live up to 12 years in the wild (Nayak et al. 2007). Interestingly, preliminary skeletochronological analysis of the invasive population in Madagascar failed to identify *lines of arrested growth* (*LAGs*) in either phalanges or femurs, suggesting that in Madagascar they don't undergo seasonal changes in physiological growth (F.M. Guarino, pers. comm. 2015). In Vietnam, Asian toads breed once per year, following heavy rainfall (Ngo and Ngo 2013). In Madagascar it is still not certain if the invasive toad population breeds only once or it continues to breed throughout the year. Breeding and oviposition usually occur in still water, but also in slow-flowing rivers and ephemeral water bodies, where females can lay up to 10,000 eggs (FL, pers. obs.) arranged in long double-row strings. In their native range, tadpoles require 18–20 days to reach metamorphosis (Mahapatra et al. 2017) and they are known to metamorphose faster when they are in kin groups (Saidapur and Girish 2000, 2001). Tadpoles have an opportunistic diet, and cannibalism has been reported on conspecific eggs, tadpoles (of different developmental stages) and adult carrion, as well as on tadpoles of other species (Mahapatra et al. 2017). Adult toads are considered to be generalist feeders, and their diet is mainly composed of terrestrial invertebrates, with a strong prevalence of ants and termites (Berry and Bullock 1962; Mercy 1999; Hui 2015; Döring et al. 2017).

One of the characteristics that make this species a successful invader is its toxicity (Marshall et al. 2018). Most bufonids secrete potent toxins (bufadienolides) to defend themselves from predators (Chen and Kovaříková 1967; Ujvari et al. 2015). These toxins are secreted by the skin, in particular by the post-orbital parotoid glands (Vitt and Caldwell 2013). The toxin inhibits the sodium-potassium pump of cells, thus impeding the ion-transport mechanism, causing potentially lethal cardiotoxic effects (Lingrel 2010). Despite numerous instances of resistance to bufadienolides across the animal kingdom, a recent study revealed the likely widespread vulnerability of potential predators to the bufotoxins of the introduced Asian toad in Madagascar (Marshall et al. 2018).

21.3 Ecological Impacts of Toad Invasions

Most examples of ecological impacts of toad invasions come from the invasion history of cane toads (*Rhinella marina*) in Australia. Imported in 1935 to control sugar cane pests in Queensland, cane toads have spread throughout much of the wet-dry tropical habitats of North Australia, recently crossing the border into Western Australia (Urban et al. 2008). Due to its generalist and opportunistic diet, large size, high fecundity, rapid development and tolerance of broad climatic and environmental conditions, cane toads succeeded in reaching very high densities (Freeland 1986) and have successfully colonized a large proportion of Australia (Urban et al. 2007). The large array of interactions this invasive species has with Australian native ecosystems gave rise to one of the most extensive scientific investigations of a biological invasion (e.g. Lever 2001; Shine and Brown 2007; Phillips et al. 2010; Shine 2010; Doody et al. 2015) and provides us with an insight into the problems that could emerge in Madagascar with the invasion of *D. melanostictus*. The first and most serious negative effect invasive cane toads have on Australian native communities is the lethal poisoning of native predators following the ingestion of the toxic toad (Shine 2010; Ujvari et al. 2015). Australian predators have no evolutionary defence against toad toxins, likely because bufonids have not naturally colonized Australia, and although some species showed some degree of tolerance to bufotoxins (Phillips et al. 2003), the large majority of Australian native species are susceptible (Shine 2010). Cane toads are toxic across all life stages, and their toxicity is effective both in aquatic and terrestrial environments (Lever 2001; Shine 2010). In the aquatic environment, mass mortality events or declines have been documented among freshwater fishes (Grace and Sawyer 2008), tadpoles of native frog species (Crossland et al. 2008; Crossland and Shine 2010), and aquatic invertebrates such as snails and leeches (Crossland and Alford 1998). These lethal effects are repeated in terrestrial environments, where dramatic declines of predator communities have been observed, especially among crocodiles, lizards, snakes, birds and mammals (Lever 2001; Shine 2010). Other documented negative effects of this invasion include the alteration of native invertebrate communities. Due to high cane toad densities, they have a strong impact on the abundance and species richness of native arthropods, and because they are almost 'invulnerable' to predators in their invasive ranges, they reduce the rates of nutrient recycling in the ecosystems, by acting as a major nutrient sink (Greenlees et al. 2006). Cane toads are also known to directly interfere with the behaviour of native species. For instance, Australian native frogs tend to avoid refuge sites previously used by cane toads because of the chemical cues produced by their skin that persist in the ecosystem (Pizzatto and Shine 2009). Other negative effects are related to both cane toads' high fecundity and abundance (Shine 2010). In particular, high densities of cane-toad tadpoles, especially in temporary water bodies, can promote the expression of the cannibalistic phenotype (Polis 1981; Crump 1983), increase competition for natural resources and affect the growth of native tadpoles that occur syntopically (Williamson 1999; Crossland et al. 2009).

Last but not least, alien amphibians can bring pathogens and parasites from their native ranges and transfer them to the invaded ecosystems (Delvinquier and Freeland 1988; Dubey and Shine 2008), potentially causing epidemic infections among native amphibians (e.g. Daszak et al. 2004; Garner et al. 2006; Yap et al. 2018).

21.4 A Critical Analysis of the Management of the Toad Invasion in Madagascar: What Went Wrong?

The unexpected discovery in March 2014 of a naturalized population of the Asian toad, locally called Radaka boka, around Toamasina (on the east coast of Madagascar) created great concern among the scientific community. There was a call for rapid efforts to eradicate the Asian toad (Andreone et al. 2014; Kolby et al. 2014) and a rapid mobilization of the conservation community working with the amphibians of Madagascar to identify resources to develop a strategic plan for the eradication of the invasive species (Crottini et al. 2014b). The concerted effort of amphibian and invasive species biologists led to the realization of a feasibility plan for eradication, released the following year (McClelland et al. 2015), which was followed by field tests of potential eradication methods and to provide further information for evaluating eradication feasibility (Reardon et al. 2018). In the meantime, the multiple calls for action coming from the scientific community (e.g. Kolby et al. 2014; Kull et al. 2014; Pearson 2015) and the media led some authors to publish a note which 'caution[ed] against disproportionate countermeasures' (Mecke et al. 2014). This was immediately followed by Andreone et al. (2014), advocating for the implementation of swift, although carefully weighed actions (a standard procedure for invasive-species management experts) to tackle this toad invasion, as a rapid response is often the only opportunity to obtain effective results (Clout and Williams 2009; Kraus 2009). A key point raised by Reardon et al. (2018) was that so long as the invasive species remained contained within habitat with low ecological value and no rare or locally endemic species, any nontarget impacts of eradication tools may be ethically justified. However, with ongoing delays in making operational decisions, the risk that the toad would spread to occupy more ecologically intact habitats where eradication tools would become limited due to unacceptable nontarget risks would increase and therefore seriously diminish any probability of eradication success.

After the first hectic months of planning and fund-raising, in June 2014 official working groups were set up in an effort to streamline communications and decision-making. The groups covered topics such as communication, fundraising, distribution surveys, education, the eradication feasibility study, disease screening and genetic studies. These working groups later became the backbone of a national committee tasked with managing the response for the invasive toad phenomena. The committee consisted of representatives of major stakeholders from the Malagasy government, private enterprises and the conservation NGO sector and from August 2015 was coordinated by Madagascar Fauna and Flora Group (MFG).

Visual-encounter surveys to identify the extent of occurrence of the invasive toad started in May 2014, as part of the collection of preliminary information needed for development of the eradication feasibility study (Crottini et al. 2014b; Moore et al. 2015; McClelland et al. 2015). These delimitation surveys showed that the Asian toad already occupied an area of at least 108 km^2 in the south and southwest areas of Toamasina (Moore et al. 2015). The centre of distribution of the toad—and the earliest reports of the toads' occurrence—was identified in very close proximity to a nickel and cobalt refinery based in the south of Toamasina. Based on the reports of local residents, it was suggested that the toads arrived with construction equipment sometime between 2007 and 2011, when an enormous amount of material was imported to this rural site. The consistency of these observations, along with the outcomes of the field-based research, helped to narrow down the probable point of introduction of the species (Moore et al. 2015).

With the release of the eradication feasibility study (McClelland et al. 2015), it became clear that the alien toad population in Madagascar was already extremely large, with a density of up to 51 toads per 100 m^2 reported for the urban area and around 580/ha across all sites surveyed (McClelland et al. 2015). The toad was reported in four different ecotypes: urban, rural/agricultural, palm-oil plantation and non-production forest habitat, locally called savoka (consisting of degraded forest and mixed scrubland, McClelland et al. 2015; Moore et al. 2015; see Fig. 21.3). These preliminary results led to the conclusion that an eradication would likely no longer be feasible, due to the lack of easily implemented, effective and scalable methods proven to efficiently eradicate large amphibian populations, and the lack of available techniques to reliably detect individual toads at low densities and the infeasibility of introducing effective biosecurity measures within Madagascar to prevent transport of the toad to areas outside the present invaded area (McClelland et al. 2015). In light of these considerations, it was suggested that the window of opportunity for succeeding in a full eradication was closing fast (McClelland et al. 2015). This was later confirmed by Reardon et al. (2018), who also highlighted social, political and logistical challenges (including lack of suffi-

Fig. 21.3 Range of habitats occupied by the Asian toad in Toamasina. (**a**) urban; (**b**) degraded urban habitat with garbage pile on outskirts of town; (**c**) rural area mixed with degraded forest and mixed shrubland, or 'savoka'; (**d**) oil-palm plantation. (Photos by F. Licata)

cient funds) as reasons to regard the eradication of Asian toads from Madagascar as infeasible. Reardon et al. (2018) also identified the urgent need for improved border biosecurity policy and infrastructure to prevent further incursions of this and other invasive species. With the eradication considered unfeasible, the national committee decided that the next steps should be to look at identifying site-specific programs, focusing on mitigation efforts to protect key areas of particular conservation concern and native species at risk, using these efforts to gain more experience on potential management options for this invasive species, integrating the tools tested by Reardon et al. (2018). Initiatives were directed to also promote the development of border and internal biosecurity programs aiming to prevent new invasions or reinvasions.

There are valuable lessons to be learned from the difficulties encountered during the initial years of the Asian toad invasion response that should be heeded to allow a faster and more effective response in the event of a new invasive species being introduced to Madagascar. For example, the national committee faced a number of challenges that ultimately led to its efficacy being greatly hampered: despite initial good intentions to be inclusive, the committee's size meant that it was difficult to reach consensual decisions, and the national committee was never officially ratified by the Malagasy government, so there was no official authority to execute and enforce agreed-upon urgently needed plans; all people working on the toad issue were doing so in their spare time in addition to their full-time jobs, which led to a frustratingly slow pace of progress, in 2014, and at the time of the first report of the toad in Madagascar, there was no designated government agency specifically responsible for dealing with invasive-species issues, and it remains unclear whether the toad issue falls under the jurisdiction of the regional or national government authorities. However, most of the encountered limitations are common to the ones encountered during other attempted herpetological eradications elsewhere (Kraus 2009). In the specific case of Madagascar, the limited governmental function and the partial administrative decentralization of political decision-making powers has resulted in a lack of clear jurisdictional authority between the regional and national government bodies in some instances. This exacerbates the debate between the different (local and central) institutions, which sometimes results in delays in the execution of local policies (Tahina 2015; World Bank 2018), such as has been the case with the current toad invasion response. Also, there remains no clear mechanism whereby international help and support from entities such as the IUCN can easily be accessed and operationalized to effectively tackle urgent invasive-species issues and which are the possibilities for the organizations operating in the invaded areas.

Although we are aware of the difficulties in developing something of this kind, we believe that the creation of international rapid-response capacity for invasive species could be extremely useful, at least to provide advice and initial support to countries lacking their own protocols or biosecurity networks. There are a number of international list servers that function in this way but all rely on the initiative of in-country actors. To enable timely responses within a timeframe that allows eradication to be a viable option, it is necessary to have a designated and dedicated ministerial team whose sole responsibility is to deal with broader biosecurity and invasive-species issues. To this end, the authors welcome the recent appointment of a professional figure within the Ministry of Environment and Sustainable Development (*Ministère de l'Environnement et du Développement Durable*, MEDD) to be responsible for invasive-species issues. Capacity-building for invasive species management and biosecurity is required to support this post and to develop a network of specialist experts within Madagascar itself to deal with arising threats from invasive species and to work towards strengthened border biosecurity and supporting policy.

21.5 The Risk-Assessment Framework and a Plea for Applied Research

21.5.1 Latest News from the Invasion Front

Faced with a very difficult (if not impossible) eradication, it became imperative to identify and evaluate suitable eradication tools to finalize the assessment of feasibility and to provide data from which mitigation efforts could be informed. Many different methods have been tested on cane-toad populations in Australia, among which the use of parotoid-gland secretions to attract tadpoles (Crossland and Shine

2011) and acoustic lures for adult toads (Schwarzkopf and Alford 2007), which stood out as potentially effective methods to employ, especially in contrast to the efficacy of traditional control tools, such as drift fences and pitfall traps. In early 2016, a series of techniques was tested on Asian toads in Madagascar to assess the efficacy of potential eradication tools (Reardon et al. 2018). Tested methods included hand capture, pitfall traps and drift fences, the use of baited tadpole traps and citric-acid spraying (Reardon et al. 2018). The results of this comparative study showed that chemical luring of tadpoles was ineffective, but Reardon et al. (2018) recommended further tests at different stages of larval development. The study showed that citric-acid spraying was surprisingly effective in killing adult and subadult toads, similar to its successful use for controlling other invasive anurans (Beachy et al. 2011). However, citric-acid spray, as well as being socially and logistically complex to employ, was regarded as a viable choice only in areas of extreme ecological degradation due to nontarget impacts on other fauna (Reardon et al. 2018). Drift fencing and pit-fall trapping proved to be moderately effective and simple to employ and also providing the chance to collect important demographic information (Reardon et al. 2018). These data were used to infer toad densities, which were found to be the highest (1266 toads/ha) in degraded urban habitats, whereas in forested habitats, the number estimated on the basis of counted individuals dropped down to zero (Reardon et al. 2018). Total abundance in 2016 was estimated to be 7.2 million toads, in comparison to the 3.77 million toads estimated with data collected in 2014 (see McClelland et al. 2015; Reardon et al. 2018). Sadly, this number is expected to increase more quickly each year, confirming again the tremendous efforts and rapid intervention needed for eradication. Hand capture was also tested within non-contained, replicated 9ha areas. This effort was largely a response to the frequent but doubtful suggestion that a public bounty for toads could facilitate eradication. Putting aside the issue of bounty subversion (the possibility that toads would be sustainably harvested or farmed for pay in such an economically poor environment), the method failed to demonstrate a significant decline in capture rates (Reardon et al. 2018).

The most recent field data (collected in September 2017) reports alarming rates of dispersal, with toads found more than 20 km away from the presumed point of introduction (Fig. 21.4b), meaning that the invasion front has advanced around 3.3 km/year, if we assume that the invasion started in 2010 (Licata et al. 2019). This highlights the marked increase of invasion rate for this species in comparison to the first surveys from 2014 (Moore et al. 2015), for which the dispersal rate was inferred to be 2 km/year (McClelland et al. 2015). However, this value is still consistently lower than the spreading rates of cane toads in Australia, where the invasion front advances up to 60 km per year (Phillips et al. 2007; Urban et al. 2008; Estoup et al. 2010; Brown et al. 2011).

Toad populations are currently expanding in all landward directions probably facilitated by the widespread presence of human settlements in the area, with more than 2000 villages or small groups of huts counted in the invaded area (FL, pers. obs.), which are known to be favourable habitats for the species (Licata et al. 2019). The southward invasion appears to have been facilitated by the presence of multiple

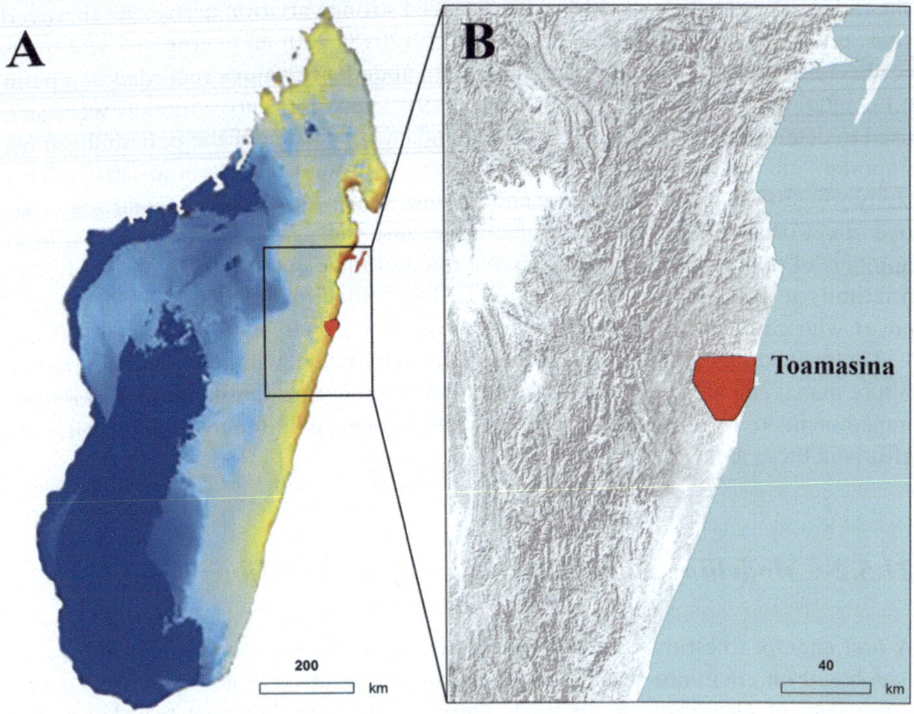

Fig. 21.4 (**a**) Species-distribution model based on occurrence data of the distribution range of the Southeast Asian lineage of the *Duttaphrynus melanostictus* complex. Warmer colours indicate climatically suitable areas. (Modified from Vences et al. 2017); (**b**) Close-up of the eastern coast of Madagascar showing the currently estimated invaded range of the Asian toad (based on occurrence data collected up to late September 2017). (Modified from Licata et al. 2019)

man-made canals and waterways of lentic character (Rahel 2007), which allowed the Asian toad to cross the Ivondro River by 2014 (Moore et al. 2015). Conversely, it seems that the northward invasion has been slowed down by the Ivoloina River, raising the possibility that this river might function as an ecological barrier, as already reported in the literature for other river systems (Leblois et al. 2000; Li et al. 2009). However, considering the relatively slow flow of this river, it seems that if it is acting as a barrier, it is likely to hold back the dispersal of toads for only a brief time. The river is crossed by several bridges and the quantity of freight travelling north from the invaded area guarantees the spread of the toads north even if the river acts as a temporary barrier to dispersal. The toad's invasion also seems to be advancing rapidly towards the interior of Madagascar, getting closer to some of the best-known biodiversity strongholds of the eastern coast of Madagascar, such as the Betampona Strict Nature Reserve (now less than 30 km northwest of the invasion front). Between the end of 2016 and the beginning of 2017, N-mixture models (Royle 2004; Ficetola et al. 2018) were used to estimate more accurately the size of the Asian toad population (Licata et al. 2019). Abundances were computed at six sites across the invaded area and results showed strong variation across the surveyed sites, reflecting the findings of Reardon et al. (2018) with an average of 184 toads per hectare (Licata et al. 2019) and maximum abundance values recorded in a palm oil plantation (559 toads/ha, Licata et al. 2019). These correlative models were also used to determine the influence on toad abundances of some of the potentially most important environmental factors observed on the ground (Licata et al. 2019). This study suggests that rubbish dumps and organic waste presence have a strong positive effect on toad abundances, probably because they provide safe shelters, high numbers of potential prey and favourable microclimate conditions. Conversely, the proximity to roads seems to be related to lower toad abundances, as does forest cover, whose influences, however, need to be further investigated (Licata et al. 2019).

Understanding the spatial and demographic parameters will help to strategize the future management of the species, since it will enable the identification of possible impediments to its dispersion and it will help inform containment and delimitation efforts in these areas.

21.5.2 *Modelling Future Scenarios of the Invasion*

A first attempt to estimate the invasive potential of the Asian toad in Madagascar was based on environmental niche modelling, which used bioclimatic and occurrence data of *D. melanostictus* in its native range, revealing that the climate of Madagascar was particularly suitable for the Asian toad to thrive (see Fig. 21.1 in Pearson 2015). Subsequently, Wogan et al. (2016) revealed that *D. melanostictus* consists of at least three distinct evolutionary lineages, each having a narrow geographical range and ecological niche. The data that became available from this study help to study the invasion history of the invasive population of *D. melanostictus* in Madagascar, which was suggested to result from a single introduction that had the natural populations from Vietnam or Cambodia as its likely source (probably the Ho-Chi-Minh-City area; see Fig. 21.1 in Vences et al. 2017). Species distribution models derived from the use of only the bioclimatic and occurrence data of the Southeast Asian lineage, restricted the suitable climate space for the Asian toad in Madagascar to the lowlands of the eastern and north-western coasts of Madagascar, an area which is almost entirely included into the predicted climate space of the Southeast Asian lineage (see Fig. 21.4a).

21.5.3 *Investigating Potential Impacts*

Ecosystem changes determined by this invasion seem likely to have a major impact on the native fauna of Madagascar. The poisoning of native predators is certainly one of the most feared consequences of this invasion (Kolby et al. 2014; Crottini et al. 2014b), especially now that Marshall et al. (2018) predicted the vulnerability of most Madagascan predatory birds, mammals, amphibians and reptiles to the toads' toxins (see Fig. 21.1 in Marshall et al. 2018).

At present it is difficult to predict the extent of impacts on native species, depending, as it does, on the natural histories of the native predators and on the toad's ability to adapt to different habitats, such as rainforest (Marshall et al. 2018). Using a multivariate niche modelling approach, Brown et al. (2016) forecast that the niche overlap between the invasive Asian toad and six potential carnivore predators in eastern Madagascar to be sufficient that they were at high risk of rapid decline if the Asian toad becomes part of their diet (Brown et al. 2016). The authors suggested that a strategic plan to monitor the endemic and most threatened carnivores should be planned to prevent their extinction and proposed that monitoring the brown-tailed mongoose, *Salanoia concolor* (I. Geoffroy Saint-Hilaire 1837), should be a priority due to its limited distribution range (Goodman 2012). To strengthen the efficacy of this proposed conservation measure, the development and implementation of a plan for the early detection and rapid response of recently established populations of toads should also be considered. We think this can possibly be achieved with the establishment of citizen-science campaigns, aimed at informing the community on the importance of the unique biodiversity of Madagascar to maintain the ecosystem services and on the importance of reporting the presence of a dangerous invasive species that can dramatically modify these services, although there are significant barriers to such a strategy due to the limited communications infrastructure and poverty of the rural population.

Further, we believe it is imperative to design containment methods to defend ecologically sensitive areas that are, or will soon be, encroached upon by the invasion front. This may consist of the establishment of fences or barriers against toads, which will, however, need constant monitoring, as a single breach would compromise their functionality (Tingley et al. 2017; Reardon et al. 2018). Other possibilities to evaluate include the use of male advertisement calls to attract females, targeted female removal and traps with acoustic lures during the breeding season (Tingley et al. 2017).

Besides the aforementioned threats, the Asian toads may have also introduced new parasites and pathogens, and it will be important to determine whether, for example, they are carrying *Batrachochytrium dendrobatidis* and *Ranavirus,* among other potential pathogens. Recent results indicate striking differences between the bacterial communities of *Duttaphrynus melanostictus* and a native co-occurring species (Santos et al. 2018), with the toad microbiome being enriched for xenobiotic biodegradation and metabolism processes (Santos et al. 2018). The microbiome on amphibian skin plays an essential role in disease resistance, host health and adaptation to biotic and abiotic stresses (Jiménez and Sommer 2017; Wang et al. 2017; Campbell et al. 2019), and the observed differences could possibly be related to the higher capacity of the toad to cope with environmental alteration and anthropogenic stress (Claus et al. 2016).

Other studies are currently underway: radiotelemetry is currently applied to ascertain whether the toad is likely to use forested habitats and to gather information on its dispersal capabilities and the toads' interactions with native predators. These data, combined with fine-scale habitat mapping through remote sensing techniques (Comiso 2010) and toad presence-absence data, will be used to inform specific individual-based models of dispersal processes and will enable the prediction of likely future spread scenarios and identify the areas of higher risk of invasion.

21.6 Future Steps

The Asian toad is emerging as a very successful invader (Reilly et al. 2017). Being a human commensal, it is often found close to human settlements, further increasing its chances of being accidentally translocated to new areas. Stowing away in freight and shipping containers is one of the most frequent paths of introduction for many alien species and is likely to become an ever-increasing risk during the forthcoming years (Hulme 2015; Tingley et al. 2018). Asian toads are frequently reported in shipping containers, raising the fear this species will invade other areas (Tingley et al. 2018; Mo 2018). This poses a threat both for countries with active biosecurity programmes that are able to modify or strengthen their protection tools to prevent new invasions and much more so for countries lacking biosecurity infrastructure and capacity, such as Madagascar. On the national scale, the seaport city of Toamasina has more than 170,000 containers per year passing through its port facilities (World Bank 2018), and tons of goods are moved daily from Toamasina across the entire island, highlighting fears that this will result in the translocation of this (and likely other) invasive species to other parts of the country or to other countries. Many products (such as lychees and bananas) are regularly exported from Toamasina and could easily reach other jurisdictions. Lychees represent an important fruit export from Madagascar, and many countries (including several EU countries) receive fruits collected and stocked from the Toamasina area (Jahiel et al. 2014). The potential presence of living Asian toads within these will become increasingly likely and should be the focus of internal biosecurity practices. On a more local scale, agricultural practices are thought to play a critical role in promoting Asian toad colonization. For example, in a palm-oil plantation close to Toamasina, where toads reach high densities (see Sect. 21.5.1), tons of plant remains are used as mulch for young palm groves (Hamdan et al. 1998). Asian toads profit from the shelter this substrate provides (FL, pers. obs.), and during the relocation of the mulch across plantations, toads are very likely to be moved into new palm oil parcels.

Non-governmental organizations are important civil-society actors and could represent key players in mitigating biodiversity crises. The local NGO 'Madagascar Fauna and Flora Group' (MFG), with its headquarters in Toamasina, has played a strategic role in responding to this invasion since its first report, setting up the basis for the risk-assessment framework that was described here and carrying out initial distribution surveys in collaboration with another local NGO, Association Mitsinjo. MFG promptly launched a campaign using radio broadcasts, posters (Fig. 21.5) and

Fig. 21.5 Poster prepared by Madagascar Fauna and Flora Group during the 'Radaka Boka' awareness campaign

village visits to raise local awareness about the presence of this noxious species; they also created a toll-free hotline number that local people could use to report the presence of the toad during the initial stages of the delimitation effort for the toad, which is now suspended because it was found to be of limited value.

Very close to the incursion area are two highly biodiverse sites: Parc Ivoloina and Betampona Strict Nature Reserve. The former is a 282-hectare forestry station located 15 km north of Toamasina in an area originally covered by lowland rainforest and now dominated by exotic tree plantations, marshlands, rice paddies, lakes and a small patch of native lowland forest (Ramasindrazana 2009) that however hosts a moderately rich amphibian community (Crottini et al. 2014a). The Asian toad has apparently not yet (May 2019) arrived there but it is currently reported to be just 300 m from the southern border of the park (R. Mahasoa pers. comm.), so its arrival is likely to be imminent. We believe that Ivoloina is a suitable site for a pilot study to test containment or protective measures that can be later applied to other sensitive areas. Here, MFG is currently working to create a pilot exclusion zone to protect Parc Ivoloina and to learn about the impacts of the toad on the native wildlife in the event of an invasion. Betampona Strict Nature Reserve, located 40 km northwest of Toamasina, is a 2228-hectare forest patch of lowland rainforest surrounded by agriculture and degraded land. It is extremely biodiverse (e.g. Rosa et al. 2012) and is one of the conservation priorities of the area and, indeed, of the country as a whole. Assuming a spread rate of the Asian toad of 3.3 km/year, according to the most-recent estimates (Licata et al. 2019), Betampona will most likely be reached by this invasion within the next decade if nothing is done to arrest its spread, and the Ankeniheny-Zahamena corridor, one of the largest blocks of rainforest in Madagascar, located at a distance less than 50 km from the current front of the invasion, is likely to undergo a similar fate.

Almost 6 years have passed since the first detection of this invasive population of *D. melanostictus* in Madagascar. Many things have been done, but much more needs to be done to try to mitigate this invasion. At present there is the primary need to (1) develop and implement a national biosecurity system, (2) develop an integrated national plan for the long-term management of this biological invasion and (3) establish a small but effective 'National Working Group' whose leadership would be ratified at the national level and whose role should be overseeing the implementation of the national management plan. Efforts should also be invested to facilitate and simplify the administrative procedures to obtain research and collection permits pertaining to urgent time-bounded research, such as is needed when actively trying to engage with an invasive-species incursion. And in the event that the National Working Group becomes a reality, we suggest that the approval of such requests should receive a fast favourable or unfavourable response from the National Working Group to ensure prompt actions can be developed in the field.

The national management plan must be implemented quickly and with an adaptive management approach that ensures that it learns from the successes and failures of its implementation and undergoes regular revision and review to incorporate those learnings into the active strategy.

Acknowledgements We are particularly indebted to I. Porton for her useful comments on the manuscript and continuous support for actions to help save Malagasy biodiversity; to the Malagasy authorities (in particular to the former Ministère de l'Environnement, Ecologie et des Forêts, now Ministère de l'Environnement et du Développement Durable) for multiple research and export permits; to J. Ratovondrahona, J.M. Razafimanantsoa, J.C. Rafanomezanjanahary and G. Ramarolahy for their help with toad surveys and sample collection in Madagascar; to R. Mahasoa, A.L. Rafilipo, C.C. Razafimbelo, P. Razanajaza, F.R. Harison, Z.M. Raharisoa, J.F. Randrianasolo, R.F. Bredison, B. Andriantianan and L. Rabenantenaina; and to all the local people who helped us in the field and through reporting of the toad's presence via the dedicated hotline or other forms. We thank the IUCN SSC Amphibian Specialist Group Madagascar and Amphibian Surveillance Alliance for their help in the first hectic months when the toad was first reported. The Portuguese National Funds through FCT (Foundation for Science and Technology) support the Investigador FCT grant to AC (IF/00209/2014) and the doctoral fellowships of FL (SFRH/BD/131722/2017). This work was partially funded by National Funds through FCT under the IF/00209/2014/CP1256/CT0011 Exploratory Research Project. Financial support in the early stages of the toad emergency was provided by Parc Zoologique de Paris, Saint Louis Zoo, Taronga Zoo and Biodiversity Conservation Madagascar.

References

Andreone F, Rabibisoa N, Randrianantoandro C, Crottini A, Edmonds D, Kraus F, Lewis JP, Moore M, Rabemananjara FCE, Rabemanantsoa JC, Vences M (2014) Madagascar: risk review is under way for invasive toad. Nature 512:253

Beachy JR, Neville R, Arnott C (2011) Successful control of an incipient invasive amphibian: *Eleutherodactylus coqui* on O'ahu, Hawai'i. In: Veitch CR, Clout MN, Towns DR (eds) Island invasives: eradication and management. IUCN, Gland

Berry PY, Bullock JA (1962) The food of the common Malayan toad, *Bufo melanostictus* Schneider. Copeia 1962:736–741

Binggeli P (2003) Introduced and invasive plants. In: Goodman SM, Benstead JP (eds) The natural history of Madagascar. University of Chicago Press, Chicago, pp 257–268

Brown GP, Kelehear C, Shine R (2011) Effects of seasonal aridity on the ecology and behaviour of invasive cane toads in the Australian wet–dry tropics. Funct Ecol 25:1339–1347

Brown KA, Farris ZJ, Yesuf G, Gerber BD, Rasambainarivo F, Karpanty S, Kelly MJ, Razafimahaimodison JC, Larney E, Wright PC, Johnson SE (2016) Modeling co-occurrence between toxic prey and native predators in an incipient invasion. Biodivers Conserv 25:2723–2741

Campbell LJ, Garner TWJ, Hopkins K, Griffiths AGF, Harrison XA (2019) Outbreaks of an emerging viral disease covary with differences in the composition of the skin microbiome of a wild UK amphibian. Front Microbiol 10:1245

Chen KK, Kovaříková A (1967) Pharmacology and toxicology of toad venom. J Pharm Sci 56:1535–1541

Church G (1960) The invasion of Bali by *Bufo melanostictus*. Herpetologica 16:15–21

Claus SP, Guillou H, Ellero-Simatos S (2016) The gut microbiota: a major player in the toxicity of environmental pollutants? Npj Biofilms Microbiomes 2:16003

Clout MN, Williams PA (eds) (2009) Invasive species management: a handbook of principles and techniques. Oxford University Press, Oxford

Comiso J (2010) Satellite remote sensing techniques. In: Comiso J (ed) Polar oceans from space. Springer, New York, pp 73–111

Conservation International (2019) Hotspots. http://www.conservation.org. Accessed 02 Jul 2019

Crossland MR, Alford RA (1998) Evaluation of the toxicity of eggs, hatchlings and tadpoles of the introduced toad *Bufo marinus* (Anura: Bufonidae) to native Australian aquatic predators. Aust J Ecol 23:129–137

Crossland MR, Shine R (2010) Vulnerability of an Australian anuran tadpole assemblage to the toxic eggs of the invasive cane toad (*Bufo marinus*). Austral Ecol 35:197–203

Crossland MR, Shine R (2011) Cues for cannibalism: cane toad tadpoles use chemical signals to locate and consume conspecific eggs. Oikos 120:327–332

Crossland MR, Brown GP, Anstis M, Shilton CM, Shine R (2008) Mass mortality of native anuran tadpoles in tropical Australia due to the invasive cane toad (*Bufo marinus*). Biol Conserv 141:2387–2394

Crossland MR, Alford RA, Shine R (2009) Impact of the invasive cane toad (*Bufo marinus*) on an Australian frog (*Opisthodon ornatus*) depends on reproductive timing. Oecologia 158:625–632

Crottini A, Madsen O, Poux C, Strauß A, Vieites DR, Vences M (2012) Vertebrate time-tree elucidates the biogeographic pattern of a major biotic change around the K–T boundary in Madagascar. Proc Natl Acad Sci U S A 109:5358–5363

Crottini A, Bollen A, Weldon C, Dalton DL, Kotzé A, Noël J, Iambana B, Andreone F (2014a) Amphibian survey and current absence of *Batrachochytrium dendrobatidis* (Bd) in Ivoloina Park, Toamasina (eastern Madagascar). Afr J Herpetol 63:70–78

Crottini A, Andreone F, Edmonds D, Hansen CM, Lewis JP, Rabemanantsoa JC, Moore M, Kraus F, Vences M, Rabemananjara F, Randrianantoandro C (2014b) A new challenge for amphibian conservation in Madagascar: the invasion of *Duttaphrynus melanostictus* in Toamasina province. FrogLog 111:46–47

Crump ML (1983) Opportunistic cannibalism by amphibian larvae in temporary aquatic environments. Am Nat 121:281–289

Csurhes S (2016) Pest risk assessment: Asian spined toad (*Bufo melanostictus*). The State of Queensland, Department of Employment, Economic Development and Innovation. https://www.daf.qld.gov.au/__data/assets/pdf_file/0003/56658/IPA-Asian-Toad-Risk-Assessment.pdf. Accessed 02 Jul 2019

Daniel JC, Sekar AG (1989) Field guide to the amphibians of western India, part IV. J Bombay Nat Hist Soc 86:194–202

Daszak P, Strieby A, Cunningham AA, Longcore JE, Brown CC, Porter D (2004) Experimental evidence that the bullfrog (*Rana catesbeiana*) is a potential carrier of chytridiomycosis, an emerging fungal disease of amphibians. Herpetol J 14:201–208

Delvinquier BLJ, Freeland WJ (1988) Protozoan parasites of the cane toad, *Bufo marinus*, in Australia. Aust J Zool 36:301–316

Doody JS, Soanes R, Castellano CM, Rhind D, Green B, McHenry CR, Clulow S (2015) Invasive toads shift predator–prey densities in animal communities by removing top predators. Ecology 96:2544–2554

Döring B, Mecke S, Kieckbusch M, O'Shea M, Kaiser H (2017) Food spectrum analysis of the Asian toad, *Duttaphrynus melanostictus* (Schneider, 1799) (Anura: Bufonidae), from Timor Island, Wallacea. J Nat Hist 51:607–623

Dubey S, Shine R (2008) Origin of the parasites of an invading species, the Australian cane toad (*Bufo marinus*): are the lungworms Australian or American? Mol Ecol 17:4418–4424

Duellman WE, Trueb L (1994) Biology of amphibians. JHU press, Baltimora, 696 p

Estoup A, Baird SJ, Ray N, Currat M, Cournet JM, Santos F, Beaumont MA, Excoffier L (2010) Combining genetic, historical and geographical data to reconstruct the dynamics of bioinvasions: application to the cane toad *Bufo marinus*. Mol Ecol Resour 10:886–901

Ficetola GF, Barzaghi B, Melotto A, Muraro M, Lunghi E, Canedoli C, Lo Parrino E, Nanni V, Silva-Rocha I, Urso A, Carretero MA, Salvi D, Scali S, Pennati R, Andreone F, Manenti R (2018) N-mixture models reliably estimate the abundance of small vertebrates. Sci Rep 8:10357

Freeland WJ (1986) Populations of cane toad, *Bufo marinus*, in relation to time since colonization. Wildl Res 13:321–329

Ganzhorn JU, Wilmé L, Mercier JL (2014) Explaining Madagascar's biodiversity. In: Scales IR (ed) Conservation and environmental management in Madagascar. Routledge, New York, pp 41–67
Gardiner JS (1906) Notes on the distribution of the land and marine animals, with a list of the land plants and some remarks on the coral reefs. In: Gardiner JS (ed) The fauna and geography of the Maldive and Laccadive archipelagos. Cambridge University Press, Cambridge, pp 1046–1057
Garner TW, Perkins MW, Govindarajulu P, Seglie D, Walker S, Cunningham AA, Fisher MC (2006) The emerging amphibian pathogen *Batrachochytrium dendrobatidis* globally infects introduced populations of the North American bullfrog, *Rana catesbeiana*. Biol Lett 2:455–459
Ghulam A, Porton I, Freeman K (2014) Detecting subcanopy invasive plant species in tropical rainforest by integrating optical and microwave (InSAR/PolInSAR) remote sensing data, and a decision tree algorithm. ISPRS J Photogramm Remote Sens 88:174–192
Goodman SM (2012) Les carnivora de Madagascar. Association Vahatra, Antananarivo, 158 p
Goodman SM, Benstead JP (eds) (2003) The natural history of Madagascar. University of Chicago Press, Chicago
Goodman SM, Benstead JP (2005) Updated estimates of biotic diversity and endemism for Madagascar. Oryx 39:73–77
Goodman SM, Raselimanana AP, Andriniaina HA, Gauthier NE, Ravaojanahary FF, Sylvestre MH, Raherilalao MJ (2017) The distribution and ecology of invasive alien vertebrate species in the greater Toamasina region, central eastern Madagascar. Malagasy Nat 12:95–109
Grace B, Sawyer G (2008) Dead fish and frogs associated with cane toads near Darwin. N T Naturalist 20:222–225
Greenlees MJ, Brown GP, Webb JK, Phillips BL, Shine R (2006) Effects of an invasive anuran (the cane toad, *Bufo marinus*) on the invertebrate fauna of a tropical Australian floodplain. Anim Conserv 9:431–438
Hamdan ABT, Tarmizi AM, Tayeb MD (1998) Empty fruit bunch mulching and nitrogen fertilizer amendment the resultant effect on oil palm performance and soil properties. PORIM Bull 37:1–14
Harper GJ, Steininger MK, Tucker CJ, Juhn D, Hawkins F (2007) Fifty years of deforestation and forest fragmentation in Madagascar. Environ Conserv 34:325–333
Hendrix R, Nguyen QT, Böhme W, Ziegler T (2008) New anuran records from Phong Nha-Ke Bang National Park, Truong Son, central Vietnam. Herpetol Notes 1:23–31
Hui YC (2015) Diet of five common anurans found in disturbed areas in northern peninsular Malaysia. Dissertation, Universiti Sains Malaysia, Penang
Hulme PE (2015) Invasion pathways at a crossroad: policy and research challenges for managing alien species introductions. J Appl Ecol 52:1418–1424
Invasive Species Advisory Committee (2006) Invasive species definition clarification and guidance white paper. http://www.invasivespeciesinfo.gov/docs/council/isacdef.pdf. Accessed 25 Aug 2018
Irwin MT, Wright PC, Birkinshaw C, Fisher BL, Gardner CJ, Glos J, Goodman SM, Loiselle P, Rabeson P, Raharison J-L, Raherilalao MJ, Rakotondravony D, Raselimanana A, Ratsimbazafy J, Sparks JS, Wilmé L, Ganzhorn JU (2010) Patterns of species change in anthropogenically disturbed forests of Madagascar. Biol Conserv 143:2351–2362
Jahiel M, Andreas C, Penot E (2014) Experience from fifteen years of Malagasy lychee export campaigns. Fruits 69:1–18
Jiménez RR, Sommer S (2017) The amphibian microbiome: natural range of variation, pathogenic dysbiosis, and role in conservation. Biodivers Conserv 26:763–786
Kolby JE, Kraus F, Rabemananjara FCE, Rabesihanaka S, Rabibisoa N, Rafanomezantsoa J, Randrianantoandro C, Randrianiana RD, Raxworthy CJ, Razafimahatratra B, Robsomanitrandrasana E (2014) Stop Madagascar's toad invasion now. Nature 509:563
Kraus F (2009) Alien reptiles and amphibians: a scientific compendium and analysis, vol 4. Springer, Dordrecht, 571 p

Kull CA, Tassin J, Moreau S, Ramiarantsoa RH, Blanc-Pamard C, Carrière SM (2012) The introduced flora of Madagascar. Biol Invasions 14:875–888

Kull CA, Tassin J, Carrière SM (2014) Approaching invasive species in Madagascar. Madag Conserv Dev 9:60–70

Leblois R, Rousset F, Tikel D, Moritz C, Estoup A (2000) Absence of evidence for isolation by distance in an expanding cane toad (*Bufo marinus*) population: an individual-based analysis of microsatellite genotypes. Mol Ecol 9:1905–1909

Lever C (2001) The cane toad: the history and ecology of a successful colonist. Westbury Academic & Scientific Publishing, Otley

Li R, Chen W, Tu L, Fu J (2009) Rivers as barriers for high elevation amphibians: a phylogeographic analysis of the alpine stream frog of the Hengduan Mountains. J Zool 277:309–316

Licata F, Ficetola GF, Freeman K, Mahasoa RH, Ravololonarivo V, Solofo Fidy Niaina JF, Koto-Jean AB, Nahavitatsara ER, Andreone F, Crottini A (2019) Abundance, distribution and spread of the invasive Asian toad *Duttaphrynus melanostictus* in eastern Madagascar. Biol Invasions 21:1615–1626

Lingrel JB (2010) The physiological significance of the cardiotonic steroid/ouabain-binding site of the Na, K-ATPase. Annu Rev Physiol 72:395–412

Mahapatra S, Dutta SK, Sahoo G (2017) Opportunistic predatory behaviour in *Duttaphrynus melanostictus* (Schneider, 1799) tadpoles. Curr Sci 112:1755–1759

Marshall BM, Casewell NR, Vences M, Glaw F, Andreone F, Rakotoarison A, Zancolli G, Woog F, Wurster W (2018) Widespread vulnerability of Malagasy predators to the toxins of an introduced toad. Curr Biol 28:R654–R655

McClelland P, Reardon JT, Kraus F, Raxworthy CJ, Randrianantoandro C (2015) Asian toad eradication feasibility report for Madagascar. Te Anau, New Zealand, 75 p

Mecke S, Kieckbusch M, Schuster K, Ernst R, O'Shea M, Rödel M-O, Köhler G, Grillitsch H, Köhler J, Haas A, Hallermann J, Kaiser H (2014) Invasive species: review risks before eradicating toads. Nature 511:534

Mercy M (1999) Studies on some aspects of the biology and ecology of the common Indian toad *Bufo melanostictus* Schneider (Class Amphibia, Order Anura). Dissertation, Mahatma Gandhi University, Kerala

Millennium Ecosystem Assessment (2005) Ecosystems and human Well-being. Volume 1. Current state and trends. Island Press, Washington, DC, x + 137 p

Mittermeier RA, Myers N, Thomsen JB, Da Fonseca GA, Olivieri S (1998) Biodiversity hotspots and major tropical wilderness areas: approaches to setting conservation priorities. Conserv Biol 12:516–520

Mo M (2018) Keeping Asian black-spined toads *Duttaphrynus melanostictus* out of Australia. FrogLog 120:14–16

Moore M, Solofo Niaina Fidy JF, Edmonds D (2015) The new toad in town: distribution of the Asian toad, *Duttaphrynus melanostictus*, in the Toamasina area of eastern Madagascar. Trop Conserv Sci 8:440–455

Myers N, Mittermeier RA, Mittermeier CG, Da Fonseca GA, Kent J (2000) Biodiversity hotspots for conservation priorities. Nature 403:853–858

Nayak S, Mahapatra PK, Mishra S, Dutta SK (2007) Age determination by skeletochronology in the common Indian toad *Bufo melanostictus* Schneider, 1799 (Anura: Bufonidae). Herpetozoa 19:111–119

Ngo BV, Ngo CD (2013) Reproductive activity and advertisement calls of the Asian common toad *Duttaphrynus melanostictus* (Amphibia, Anura, Bufonidae) from Bach Ma National Park, Vietnam. Zool Stud 52:12

Pearson RG (2015) Asian common toads in Madagascar: an urgent effort to inform surveys and eradication efforts. Glob Chang Biol 21:9

Pennisi E (2010) 'Toadness' a key feature for global spread of these amphibians. Science 327:633

Phillips BL, Brown GP, Shine R (2003) Assessing the potential impact of cane toads on Australian snakes. Conserv Biol 17:1738–1747

Phillips BL, Brown GP, Greenlees M, Webb JK, Shine R (2007) Rapid expansion of the cane toad (*Bufo marinus*) invasion front in tropical Australia. Austral Ecol 32:169–176

Phillips BL, Greenlees MJ, Brown GP, Shine R (2010) Predator behaviour and morphology mediates the impact of an invasive species: cane toads and death adders in Australia. Anim Conserv 13:53–59

Pizzatto L, Shine R (2009) Native Australian frogs avoid the scent of invasive cane toads. Austral Ecol 34:77–82

Polis GA (1981) The evolution and dynamics of intraspecific predation. Annu Rev Ecol Evol Syst 12:225–251

Rahel FJ (2007) Biogeographic barriers, connectivity and homogenization of freshwater faunas: it's a small world after all. Freshw Biol 52:696–710

Rakotomanana H, Jenkins RKB, Ratsimbazafy J (2013) Conservation challenges for Madagascar in the next decade. In: Sodhi NS, Gibson L, Raven PH (eds) Conservation biology: voices from the tropics. Wiley, London, pp 33–39

Ramasindrazana B (2009) Bat inventory of the Ivoloina Forestry Station, Atsinanana region, Madagascar. Afr Bat Conserv News 21:7–10

Reardon JT, Kraus F, Moore M, Rabenantenaina L, Rabinivo A, Rakotoarisoa NH, Randrianasolo HH (2018) Testing tools for eradicating the invasive toad *Duttaphrynus melanostictus* in Madagascar. Conserv Evid 15:12–19

Reilly SB, Wogan GO, Stubbs AL, Arida E, Iskandar DT, McGuire JA (2017) Toxic toad invasion of Wallacea: a biodiversity hotspot characterized by extraordinary endemism. Glob Chang Biol 23:5029–5031

Rosa GM, Andreone F, Crottini A, Hauswaldt JS, Noël J, Rabibisoa NH, Randriambahiniarime MO, Rebelo R, Raxworthy CJ (2012) The amphibians of the relict Betampona low-elevation rainforest, eastern Madagascar: an application of the integrative taxonomy approach to biodiversity assessments. Biodivers Conserv 21:1531–1559

Royle JA (2004) N-mixture models for estimating population size from spatially replicated counts. Biometrics 60:108–115

Saidapur SK, Girish S (2000) The ontogeny of kin recognition in tadpoles of the toad *Bufo melanostictus* (Anura, Bufonidae). J Biosci 25:267–273

Saidapur SK, Girish S (2001) Growth and metamorphosis of *Bufo melanostictus* tadpoles: effects of kinship and density. J Herpetol 35:249–254

Samonds KE, Godfrey LR, Ali JR, Goodman SM, Vences M, Sutherland MR, Irwin MT, Krause DW (2012) Spatial and temporal arrival patterns of Madagascar's vertebrate fauna explained by distance, ocean currents, and ancestor type. Proc Natl Acad Sci U S A 109:5352–5357

Samonds KE, Godfrey LR, Ali JR, Goodman SM, Vences M, Sutherland MR, Irwin MT, Krause DW (2013) Imperfect Isolation: Factors and Filters Shaping Madagascar's Extant Vertebrate Fauna. PLoS ONE 8(4): e62086

Santos B, Vences M, Cocca W, Rakotonanahary T, Ndriantsoa S, Licata F, Noël J, Solofo Niaina Fidy JF, Crottini A (2018) Characterization of the bacteria community of the invasive Asian Toad (*Duttaphrynus melanostictus*) in Madagascar and comparison with a native species. In: Abstract of the 10th international conference on biological invasions, Dún Laoghaire, Dublin, 3–7 September 2018

Schwarzkopf L, Alford RA (2007) Acoustic attractants enhance trapping success for cane toads. Wildl Res 34:366–370

Shine R (2010) The ecological impact of invasive cane toads (*Bufo marinus*) in Australia. Q Rev Biol 85:253–291

Shine R, Brown GP (2007) Adapting to the unpredictable: reproductive biology of vertebrates in the Australian wet–dry tropics. Philos Trans R Soc B 363:363–373

Tahina RN (2015) Decentralization: problems and solutions–Madagascar evidence. J Econ Financ 6:1–9

Tingley R, Ward-Fear G, Schwarzkopf L, Greenlees MJ, Phillips BL, Brown G, Clulow S, Webb J, Capon R, Sheppard A, Strive T, Tizard M, Shine R (2017) New weapons in the Toad Toolkit: a review of methods to control and mitigate the biodiversity impacts of invasive cane toads (*Rhinella marina*). Q Rev Biol 92:123–149

Tingley R, García-Díaz P, Arantes CRR, Cassey P (2018) Integrating transport pressure data and species distribution models to estimate invasion risk for alien stowaways. Ecography 41:635–646

Trainor CR (2009) Survey of a population of black-spined toad *Bufo melanostictus* in Timor-Leste: confirming identity, distribution, abundance and impacts of an invasive and toxic toad. Charles Darwin University, Darwin, 48 pp

Ujvari B, Casewell NR, Sunagar K, Arbuckle K, Wurster W, Lo N, Denis OM, Beckmann C, King GF, Deplazes E, Madsen T (2015) Widespread convergence in toxin resistance by predictable molecular evolution. Proc Natl Acad Sci U S A 112:11911–11916

Urban M, Phillips BL, Skelly DK, Shine R (2007) The cane toad's (*Chaunus marinus*) increasing ability to invade Australia is revealed by a dynamically updated range model. Proc R Soc Lond 274:1413–1419

Urban MC, Phillips BL, Skelly DK, Shine R (2008) A toad more traveled: the heterogeneous invasion dynamics of cane toads in Australia. Am Nat 171:134–148

Van Bocxlaer I, Loader SP, Roelants K, Biju SD, Menegon M, Bossuyt F (2010) Gradual adaptation toward a range-expansion phenotype initiated the global radiation of toads. Science 327:679–682

Van Dijk PP, Iskandar D, Lau MWN, Huiqing G, Baorong G, Kuangyang L, Wenhao C, Zhigang Y, Chan B, Dutta S, Inger RF, Manamendra-Arachchi K, Khan MS (2004) *Duttaphrynus melanostictus*. The IUCN Red List of Threatened Species. www.iucnredlist.org. Accessed 24 Jul 2018

Vences M, Vieites DR, Glaw F, Brinkmann H, Kosuch J, Veith M, Meyer A (2003) Multiple overseas dispersal in amphibians. Proc R Soc Lond 270(1532):2435–2442

Vences M, Brown JL, Lathrop A, Rosa GM, Cameron A, Crottini A, Dolch R, Edmonds D, Freeman K, Glaw F, Grismer LL, Litvinchuk S, Milne M, Moore M, Solofo JF, Noel J, Nguyen TQ, Ohler AM, Randrianantoandro C, Raselimanana AP, Van Leeuwen P, Wogan GO, Ziegler T, Andreone F, Murphy RW (2017) Tracing a toad invasion: lack of mitochondrial DNA variation, haplotype origins, and potential distribution of introduced *Duttaphrynus melanostictus* in Madagascar. Amphibia-Reptilia 38:197–207

Vieilledent G, Grinand C, Rakotomalala FA, Ranaivosoa R, Rakotoarijaona JR, Allnutt TF, Achard F (2018) Combining global tree cover loss data with historical national forest cover maps to look at six decades of deforestation and forest fragmentation in Madagascar. Biol Conserv 222:189–197

Vitt LJ, Caldwell JP (2013) Herpetology: an introductory biology of amphibians and reptiles. Academic, Cambridge, 776 p

Wang B, Yao M, Lv L, Ling Z, Li L (2017) The human microbiota in health and disease. Engineering 3:71–82

Williamson I (1999) Competition between the larvae of the introduced cane toad *Bufo marinus* (Anura: Bufonidae) and native anurans from the Darling Downs area of southern Queensland. Aust J Ecol 24:636–643

Wogan GO, Stuart BL, Iskandar DT, McGuire JA (2016) Deep genetic structure and ecological divergence in a widespread human commensal toad. Biol Lett 12:20150807

World Bank (2018) Container port traffic (TEU: 20 foot equivalent units). https://www.data.worldbank.org. Accessed 31 Jul 2018

Yap TA, Koo MS, Ambrose RF, Vredenburg VT (2018) Introduced bullfrog facilitates pathogen invasion in the western United States. PLoS One 13(4):e0188384

Index

F. M. Angelici, L. Rossi (eds.), *Problematic Wildlife II*,
https://doi.org/10.1007/978-3-030-42335-3

The manufacturer's authorised representative in the EU is Springer Nature Customer Service Centre GmbH, Europaplatz 3, 69115 Heidelberg, Germany. If you have any concerns regarding our products, please contact ProductSafety@springernature.com

Printed and bound by CPI Group (UK) Ltd, Croydon, CR0 4YY
05/10/2026
02228120-0019